W9-CHY-847

# GEOMORPHOLOGY

## NOTE OF APPRECIATION

If the many kind expressions of approval which have come to the author and publisher of this book since its first appearance are at all justified, they are in large measure due to the inspiration of the author's former teacher and present colleague, Professor Douglas Johnson. His influence, undiminished after twenty-five years, may be discerned on many pages, and it is a pleasure to record the indebtedness that is felt toward him for whatever degree of success this work has attained. The author has benefited not only from the opportunity of studying the principles of geomorphology under the tutelage of Professor Johnson, but also through permission to use for illustrative purposes many photographs from his collections. Although numerous references to important contributions by Professor Johnson have been made in the text, the author wishes to acknowledge a larger debt to his writings than can be indicated by specific citations.

Thanks are also extended to those interested readers whose courteous letters have brought occasional helpful suggestions. In this reprinting it has been possible to make only certain small alterations, but it is hoped that some of these valuable suggestions may be followed in a possible future edition.

A. K. Lobeck.

Lalanne

THE CHALK CLIFFS OF NORMANDY

(Frontispiece.)

# GEOMORPHOLOGY

*An Introduction to the Study of Landscapes*

BY

A. K. LOBECK

*Professor of Geology, Columbia University*

McGRAW-HILL BOOK COMPANY, INC.

NEW YORK AND LONDON

1939

xi

THE MAPLE PRESS COMPANY, YORK, PA.

# PREFACE

*Geomorphology* to most people is a closed book. And yet, with the abundant opportunities for travel now open to all, it would seem that an understanding of landscapes should have an unusual appeal, if not a practical value to many people. With this thought in mind I have undertaken the writing of this book. It is intended to serve as an introductory text for use in colleges and schools, and to convey in an interesting way to the serious reader the main outlines of the subject.

A radical departure in the plan of this book is immediately apparent. The text and explanatory illustrations are placed on the same page or opposite each other, so that each two or occasionally four pages constitute a unit. This arrangement keeps the illustration in view while the text relating to it is being read. As there are many variations in the application of this device, monotony is avoided. The text actually constitutes only one-third the bulk of the book. The illustrations, especially the diagrams, are as important as the text in providing information, and as much time should be devoted to studying them as to reading the text.

An unusually large number of photographs has been introduced in order to put before the reader an actual image of the features described. These pictures have been assembled usually at the beginning of each chapter instead of being distributed through it. This has been done because in many cases a photograph illustrates several aspects of the same general topic. It is left to the reader to correlate the photographs with the appropriate parts of the text, and it is expected that students will give the photographs serious study.

The synopsis at the beginning of each chapter is designed not so much as a summary of what the chapter contains as a linking together of the various ideas of the chapter into one all-embracing point of view.

Following the text of each chapter is a page or more of maps. These maps have been selected from foreign sources because they are likely to be less accessible to users of this book than American maps are. They include representative examples from most of the important countries of Europe. Attention is called in the text to many topographic sheets of the U. S. Geological Survey, to the charts of the U. S. Coast and Geodetic Survey, and to those of the U. S. Hydrographic Office. These may be secured so readily that reproduction of them seemed superfluous. The study of topographic maps and coast charts in their original form should constitute an inherent part of any course in which this text is used.

The questions and topics for investigation are intended to suggest additional ideas and lines of thought not sufficiently touched upon in the text. They will probably be found exasperating and occasionally unanswerable. In some cases they can be reasoned out, but in other cases wider reading or study is necessary. In any event they should not be ignored. The various topics for investigation will yield interesting results from outside reading and are designed to provide suitable matter for term papers.

The references have been selected from a vast array of titles. The list includes only those which are likely to be available to the average college class. Foreign books and journals have therefore not generally been cited. Preference has been given to such American journals as *The Bulletin of the Geological Society of America*, *The Journal of Geology*, *The Geographical Review*, *The American Journal of Science*, the publications of the U. S. Geological Survey, reports of the various state geological surveys, and the journals of the more important scientific societies. The titles have been grouped under various topic heads, and an occasional comment has been added.

It is problematical whether a reference list in a book of this type will receive the attention it deserves. It should call attention to the work of the many writers on geomorphology and invite the student to further reading. Incidentally it indicates my indebtedness to their industry. The most important contributions should be made familiar to all who are seriously interested in geomorphology. In many instances they are examples of literary expression noteworthy for the clarity and logic of their exposition and for the beauty and charm of their descriptions.

During the inception of this book twelve years ago I received many welcome suggestions from my colleagues in the Department of Geology and Geography at the University of Wisconsin, and especially from Dr. Guy-Harold Smith, now Professor of Geography at The Ohio State University. In its final stages I had the able assistance of Dr. W. G. Valentine of Brooklyn College, who checked over the long reference lists and read the entire proof, as well as similar help from Messrs. M. Hall Taylor, Richard H. Mahard, and Charles W. Carlston, graduate students and assistants in Columbia University.

The selection of illustrations has been a formidable task. I believe every photograph is properly accredited, but I regret that I failed to keep track of the sources from which some of my sketches were made; it is also impossible for me to mention the numerous people from whose studies I have derived so many of the ideas presented in this book.

A. K. LOBECK.

COLUMBIA UNIVERSITY,
NEW YORK, N.Y.,
*May*, 1939.

# CONTENTS

## CHAPTER XIX



## CHAPTER XX

# I
# INTRODUCTION

*Jackson, U. S. Geological Survey*

THE HAYDEN SURVEY PARTY OF 1870

F. V. Hayden, the leader with full beard and no hat, is seated at the table in the center of the picture. This party, which was active for several seasons laid the foundation for the unsurpassed studies in geomorphology in the western United States. It was due directly to their glowing reports that the Yellowstone National Park was established.

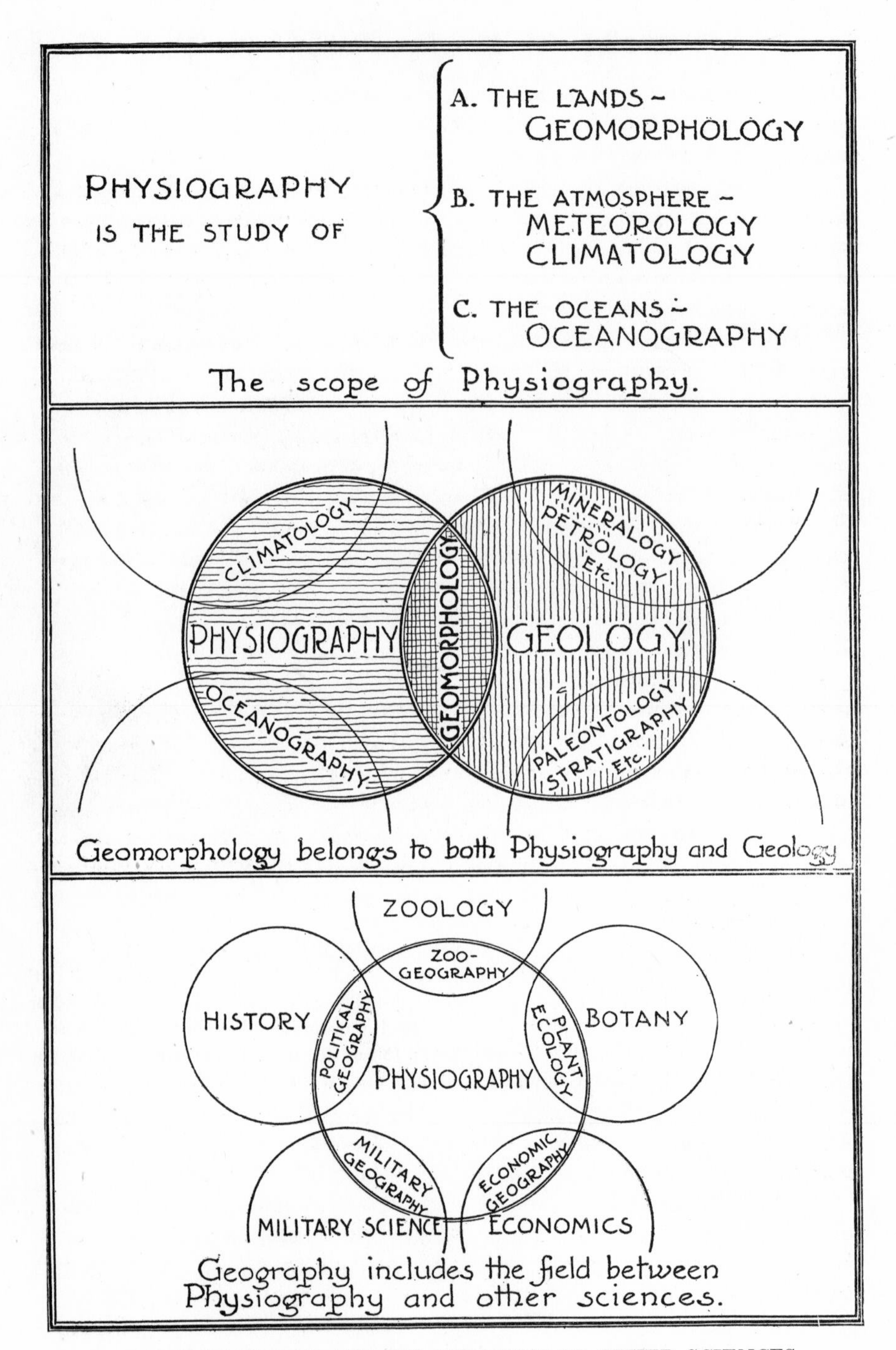


GEOMORPHOLOGY AND ITS RELATION TO OTHER SCIENCES

## GEOMORPHOLOGY AND OTHER SCIENCES

PHYSIOGRAPHY. It is usual in developing a subject to outline its scope and to indicate the relation which exists between it and other cognate fields of knowledge.

Through custom, physiography (Figure) has come to be applied to the three major physical divisions of the globe: the lands, the atmosphere, and the oceans. The study of the lands constitutes *geomorphology;* the study of the atmosphere constitutes *meteorology;* and the study of the oceans is *oceanography.*

Most textbooks in physiography deal with all three phases of the subject, but in common parlance the term physiography suggests especially the study of the lands. This is better termed geomorphology.

RELATION TO GEOLOGY. Geomorphology, or the study of land forms, is a branch of geology (Figure) sometimes considered coordinate with mineralogy and petrology, and with paleontology and stratigraphy. Structural and dynamical geology contribute toward an understanding of geomorphology by explaining the evolution of the earth's features. Geomorphology, then, like the other subjects just mentioned, is also a branch of geology—that branch which deals with the surface features of the earth's crust.

RELATION TO GEOGRAPHY. The subject of geography, considered in its broadest scope (Figure), may be defined as the study of the relationships existing between life and the physical environment. The study of the physical environment alone constitutes physiography. This study should not be called physical geography, as is often done, because the idea of this relationship of life to physical environment is not within the scope of physiography. Geography represents those fields where physiography overlaps the biological and social sciences. Physiography then, whether it be geomorphology, climatology, or oceanography, explains the environment which to some extent determines the distribution and behavior of the animal and human world.

Finally it is well to note that no science keeps strictly within its bounds, as above suggested, but frequently, as it moves along, grasps those portions of related subjects which seem helpful at the moment. So it will be here. Geographical illustrations will be introduced at each step to give that piquancy to the study of geomorphology which comes from noting the connection it has with other fields of thought.

RELIEF FEATURES OF THE EARTH. To the casual observer, going about the earth's surface, the number of land forms seems so great and the variety of types so infinite as to be bewildering. Rarely does it occur to the beginner that this apparent chaos will give way to order and understanding if the different elements of the scenery are classified and considered with some regard for their mode of origin. To accomplish this, the relief features of the earth may be thought of as having three orders of magnitude.

## RELIEF FEATURES OF THE FIRST ORDER

The largest features of the earth's surface are those of the first order: the continents and the ocean basins. Although the geologist cannot yet say just how the continents came to assume their present form and position on the earth's crust, all agree that the great continental masses were blocked out very early in the history of the earth. Geological changes have not destroyed old continents nor introduced entirely new ones. They have only altered the form of those great land masses which have always maintained a certain permanency of character. Each continent consists not only of the land areas now rising above sea level but also of certain fringing belts, known as *continental shelves*, which are only slightly submerged and whose margins drop off abruptly to the great depths of the ocean. There are also extensive shallow estuaries and gulfs which are rightfully parts of the continents.

Several theories have been proposed to account for the actual arrangement of the continents on the earth's surface. Lowthian Green's tetrahedral theory assumed that the contraction of the globular earth produced a tetrahedral form with the main continents occupying the four corners. This theory is generally discredited, since the tetrahedron is not a figure of equilibrium for a rotating earth and even a slight approximation to it cannot be retained. Suess showed that certain substantial areas of the earth have always been rigid and unyielding, as, for example, the Canadian and Baltic Shields of America and Europe, the eastern Siberia Shield of Asia, India, most of Africa, and the central portion of South America. Around these unyielding areas the continents have been built. Recently Wegener introduced the conception of "drifting continents," in which he sees the Western Hemisphere torn away from the Eastern, but this view has not been substantiated.

The fishing banks off Newfoundland and New England, the broad submerged platform above which the British Isles project, the North Sea, the Mediterranean, and the Gulf of Mexico are parts of the various continental masses with which they are contiguous. During past geological time the rising and heaving of the earth's crust has elevated and lowered parts of continents with respect to the sea, and in this manner the shapes of the continents have suffered some change. Vast portions of continents, now dry land, have at times been submerged, but not so deeply as to entitle them to be considered parts of the ocean basins. Mountain ranges have come into being and have been destroyed; extensive plains have appeared, only to be again submerged and reelevated; volcanic disturbances have left their impress; but the bolder outlines of the continents and ocean basins have stood for untold ages of time and constitute the relief features of the first order of magnitude.

# GEOLOGICAL ASPECTS OF CONTINENTS

The great antiquity of continents renders almost impossible any well-authenticated theory as to their origin. It is firmly established, however, that the present distribution of the continents on the earth's surface was accomplished very early in geological time. Nevertheless, the continents have continually been subject to change in outline because of slight and irregular elevation or depression over extensive areas. These changes have not been fortuitous. They appear to have followed a system and the result has been a definite and orderly growth in the forms of the respective continents. We can recognize, for example, in most of the continents a very ancient portion consisting of the oldest rocks; this is bordered by a region of younger rocks, which in turn is bordered by a still younger belt.

About two-thirds of the earth's surface consists of the abysmal basins of the ocean, some two to three miles deep. Above these depths rise more or less abruptly the continental platforms, whose upper faces slope gently upward to the actual coast line and thence ascend into the various elevations of the land. The water line does not represent, nor has it ever represented, the actual margin between continent and ocean.

The diastrophic or mountain-making movements of the earth's crust seem to have been periodic; that is to say, the great movements which have resulted in mountain building have occurred at more or less regular intervals in geological time. They have not been continuous. Moreover, they have been widespread, affecting not only one continent but the whole world. The result is that, in the main events, the various continents have experienced somewhat similar histories.

These movements, too, have been localized around the margins of continents and hence it is that we find the great mountain belts of the world more or less rimming the continental areas, thus giving them a basinlike form.

It is true that minor movements have occurred between the major periods of activity, and it is also true that movements have occurred outside the zones of greatest crustal deformation. These departures from the general rule, however, only serve to emphasize the systematic development which has resulted in the present configuration of the continents.

In the light of these facts the pattern of physiographic provinces in the different continents takes on a more simple aspect than if it is explained entirely as the result of chance. In Europe, for example, the fifty-two or more different physiographic units can be relegated to four categories, thus bringing order out of an apparent chaos of details.

The student of physiography rarely deciphers the entire geological history of any particular region he studies, but he realizes, nevertheless, that each unit or physiographic province which goes to make up a continent is part of a systematic scheme which embraces the whole world.

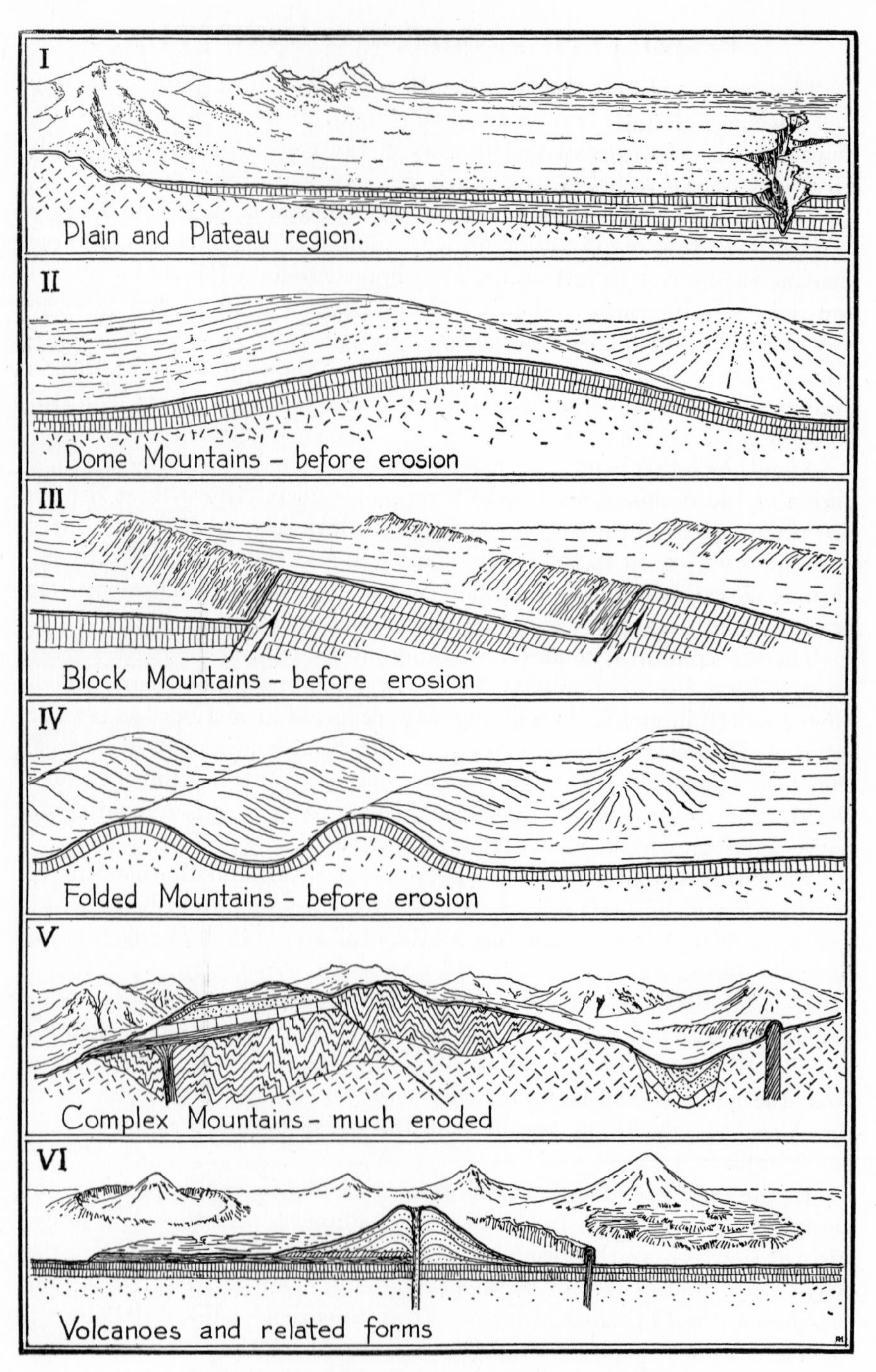


RELIEF FEATURES OF THE SECOND ORDER:
THE CONSTRUCTIONAL LAND FORMS

## RELIEF FEATURES OF THE SECOND ORDER

Constructional Forms. The continental areas may be conceived as made up of units of mountains and plains, large subdivisions sometimes termed *physiographic* or *geologic provinces*. They are the relief features of the second order. They vary from vast plains, a thousand miles or more in extent and more or less uniform in character throughout, to small mountain masses, plains, or basins, only a few miles long but markedly in contrast with surrounding areas.

The relief features of the second order have been brought into being by the action of internal forces beneath the earth's crust. The general term *diastrophism* comprehends all such movements of deformation. If such movements affect all or a large part of a continent, they are termed *epeirogenic* from the Greek *epeiros*, a continent. The term thus means "giving birth to a continent." Smaller disturbances related to mountain building are designated as *orogenic* from the Greek *oros*, a mountain.

Such earth features are often termed Constructional Land Forms, because they have been brought into being by the great constructive forces within the earth's bosom. The features thus developed by the constructive forces may be grouped very simply in accordance with the prevailing structure of the region in question. Thus those regions which have underlying them stratified formations in horizontal beds make up the class of Plains and Plateaus (Fig. I). The original, undeformed layers of rock of such regions have simply been raised bodily above the sea floor by an earth movement, often of vast extent.

A second large group comprises the class of Mountains. In these, the rock structures have been deformed, producing several types of mountains. If sedimentary layers of rock have been bowed up locally to produce a dome considerably higher than the surrounding plain, the resulting feature constitutes a Dome Mountain (Fig. II). If the earth's crust is broken into blocks, followed by their dislocation, elevation, and tilting, Block Mountains result (Fig. III). When sedimentary rocks undergo lateral pressure so that they become more or less regularly corrugated, Folded Mountains are produced (Fig. IV). Finally, the combination of many kinds of disturbances produces Complex Mountains (Fig. V).

A third group is represented by those features due to volcanic activity —Volcanic Forms (Fig. VI).

In general, the Constructional Forms are of so great size as not to be comprehended in one glance. They constitute the mountain ranges and the plains areas of the earth's surface. These forms, in their initial stages, are devoid of detail. Only after they have lain exposed to the destructive agents on the earth's surface do they acquire those elements of sculpture which give charm to the scenery of the world.

RELIEF FEATURES OF THE THIRD ORDER:
THE DESTRUCTIONAL LAND FORMS

## RELIEF FEATURES OF THE THIRD ORDER

Destructional Forms. The multitudinous details of the earth's scenery constitute the relief features of the third order. They are the minor forms, present in infinite number, which embellish and ornament the larger blocklike masses of the constructional forms. The relief features of the third order are due to the destructive agencies operating on the earth's surface. During the course of their work they leave erosional features in the form of valleys and canyons; they leave residual features in the form of peaks and summits which have not yet been attacked; and, third, they leave depositional features, temporary deposits of the debris they have removed, in the form of deltas and moraines.

The various destructional forms can readily be classified according to the agent which has produced them. Four major agents act in this capacity, namely, streams, glaciers, waves, and wind. Accompanying all of them is the universal activity of weathering, which may be best considered as an agent preparing the rocks for the actual removal to come later.

The forms produced by streams are of three types: (*a*) the erosional forms such as gullies, valleys, gorges, and canyons; (*b*) the residual forms such as peaks, monadnocks, and summit areas; and (*c*) the depositional forms such as alluvial fans, flood plains, and deltas.

The forms produced by glaciers are (*a*) the erosional forms, such as cirques and glacial troughs; (*b*) the residual forms such as the Matterhorn peaks, arêtes, and roches moutonnées; and (*c*) the depositional forms such as moraines, drumlins, kames, and eskers.

Waves likewise (*a*) erode sea caves; (*b*) leave wave-cut cliffs, benches, stacks, and arches as residual forms; and (*c*) with the help of currents deposit beaches and bars.

The wind too (*a*) erodes and thus forms blowholes on the plains or in sandy districts; (*b*) leaves residual masses having characteristic forms such as pedestal and mushroom rocks; and, finally, (*c*) deposits vast quantities of sand and silt in the form of dunes and loess.

This threefold conception of the various destructional forms provides a system of classification which accords closely with the actual facts of nature. This classification, too, is a genetic one and is based upon the mode of origin of the forms and not upon superficial points of resemblance.

Lastly, provision should be made for activities of certain organic forms. Animal life is essentially a destructive force, tending to break up rock masses on the earth's surface. This destructive effect, however, is the least conspicuous phase of organic behavior. The depositional activities, on the other hand, result in forms, such as coral reefs and ant hills, which may be of real size and topographic importance.

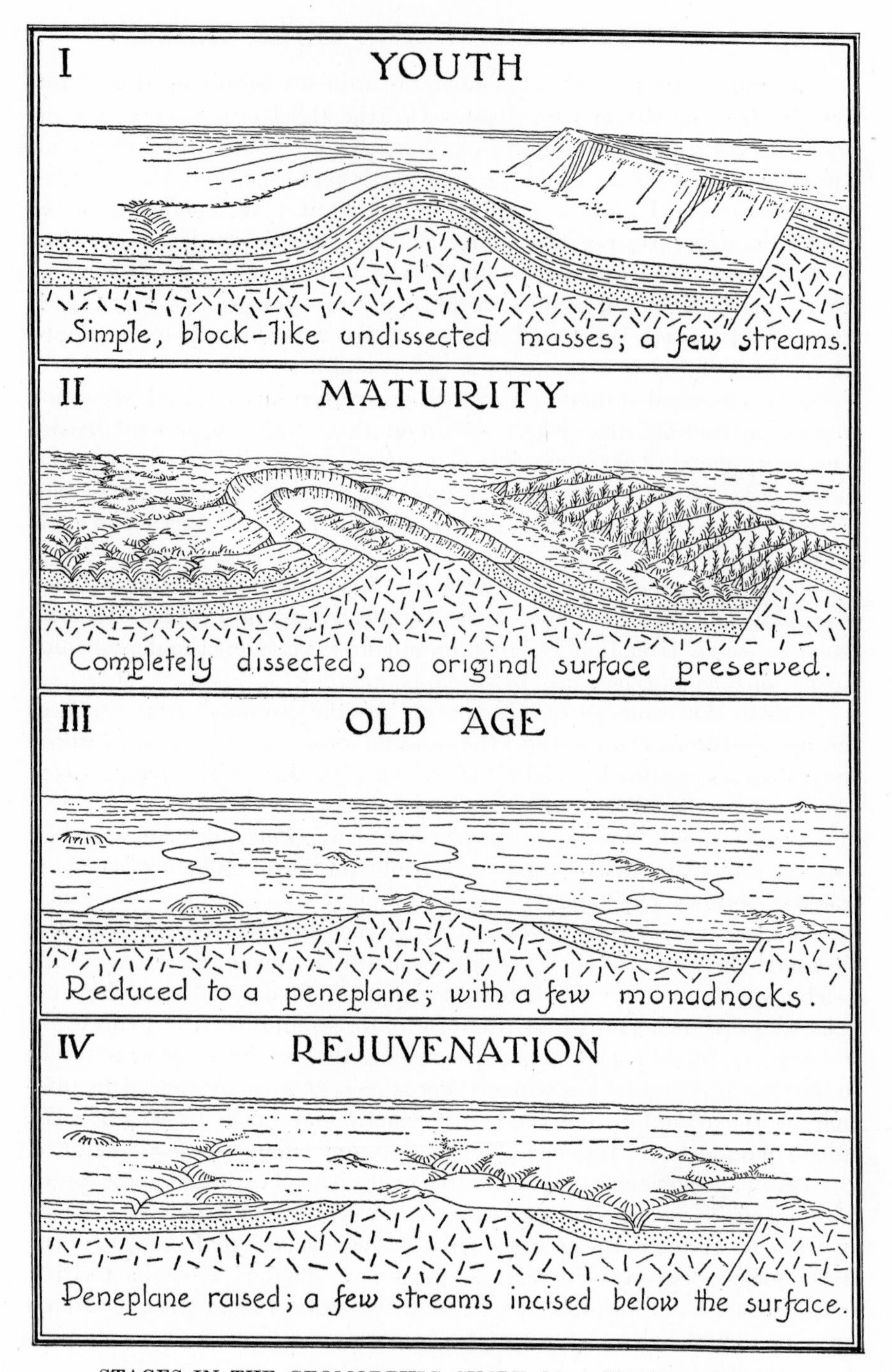


STAGES IN THE GEOMORPHIC CYCLE OF A HUMID REGION

## THE GEOMORPHIC CYCLE IN A HUMID CLIMATE*

All land forms pass through an orderly procedure of development. That is, they go through a life history. The various stages are termed *youth*, *maturity*, and *old age*. For each of these stages certain characteristics are recognized.

In *youth* (Fig. I), land forms are undiversified in their appearance. A young plain recently elevated from the sea is flat and featureless. Few streams flow over its surface. This is true of the other constructional forms such as block mountains, dome mountains, folded mountains, and volcanoes. In youth their outlines are unrelieved by the diversity which later dissection brings.

As development proceeds, and as drainage systems grow and streams increase in number and length, the original smooth slopes and planes become intricately cut up by valleys of every conceivable size and shape. Eventually a stage is reached when little of the original surface remains. The region, then, is one of ridges and valleys at the most rugged and diversified period in its history. In this condition it is said to be *mature* (Fig. II). Thus we may speak of mature plains, mature plateaus, mature block mountains, mature dome mountains, maturely dissected volcanoes. In each case we refer to a constructional land form having much variety in its topographic aspect.

Beyond maturity, the topography becomes more subdued. The destructive forces denude the land until a time when the area is worn down virtually to sea level. This is *old age* (Fig. III). The level toward which the land has been reduced is known as the *base-level* of the region. The resulting monotonous surface is called a *peneplane*, which means "almost a plane surface." Occasional remnants of the original mass may persist, owing to their superior resistance or to their remoteness from the main drainage lines. Such scattered hills are termed *monadnocks* after Mount Monadnock in New Hampshire, the type example.

REJUVENATION. Interruptions may occur at any stage in the life cycle of a region. If the cycle is already completed, a new cycle may be inaugurated by an uplift of the entire region. An elevated land tract of this type becomes an *upland*. Below its surface, the streams immediately incise themselves. Thus, a new cycle is begun. Such a region is said to be young in its second cycle of development. Youth then passes into maturity as the dissection of the area increases. With the erosion of a later peneplane near the new base-level, old age is again attained.

In any of the cycles of erosion the peneplane may not embrace the entire region. Before rejuvenation occurs, it may be developed only locally where the rocks have the least resistance, or near the sea where erosion has first reduced the area to base-level.

* The term *geomorphic cycle* is used here instead of the old term *geographical cycle* which is less accurate.

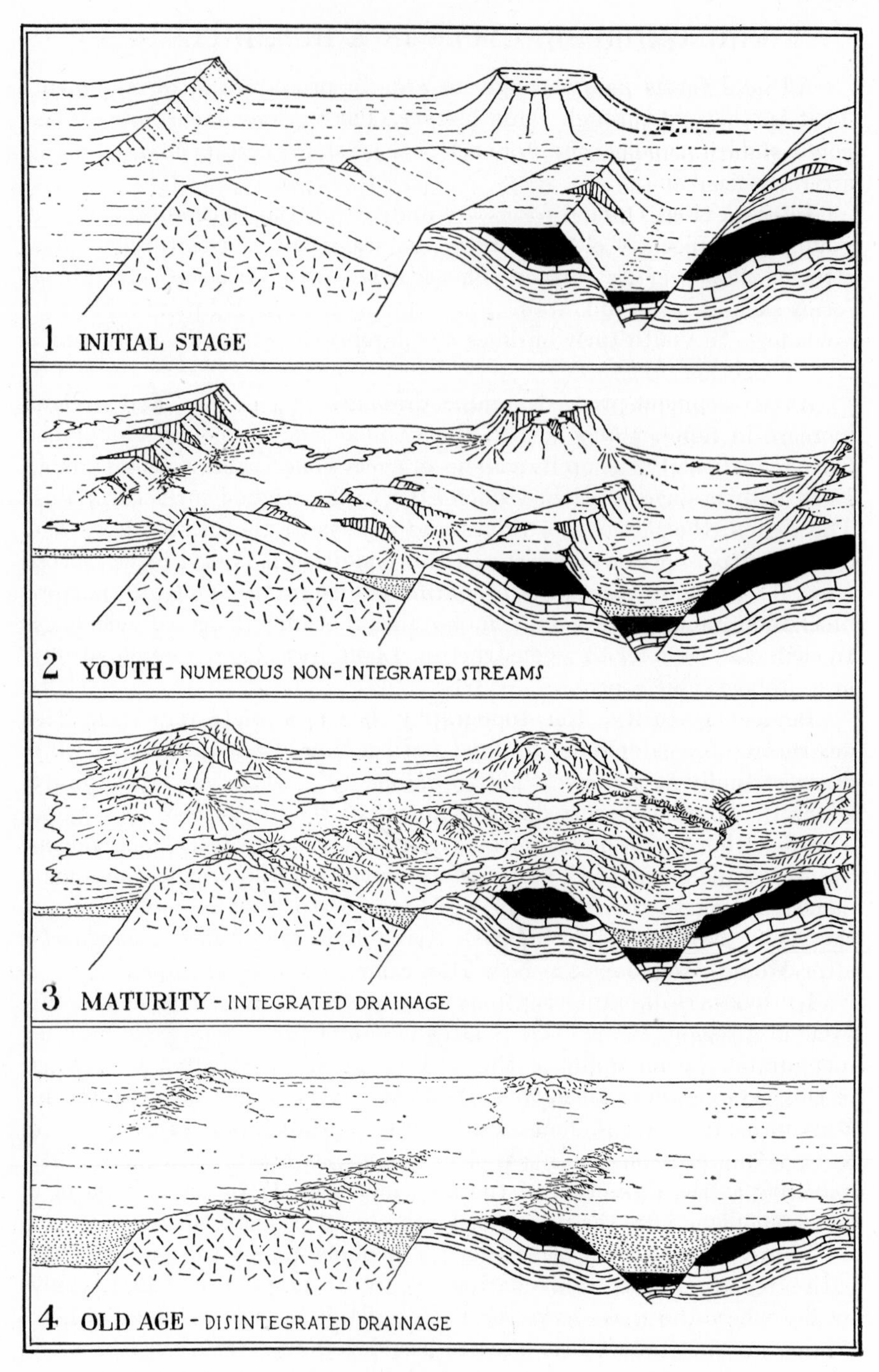


STAGES IN THE GEOMORPHIC CYCLE OF AN ARID REGION

## THE GEOMORPHIC CYCLE IN AN ARID CLIMATE

*Under humid conditions* the stream systems are integrated. Erosion prevails over deposition. Regions normally become more and more dissected until peneplanation results. The eroded material is carried beyond the confines of the region in question. In youth the initial form of the land mass is still largely preserved; in maturity the initial surfaces have largely disappeared and a high degree of relief exists; in old age peneplanation has been brought about. These stages do not concern the so-called *ages of the streams*. They are solely the characteristics of the region as a whole.

*Under arid conditions* a similar sequence of forms may be noted but with some modification. The modification is due to the inability of the streams to carry the material away.

In any region of diverse structure (Fig. 1), preferably not already flat because under arid conditions such a region would not be changed very much, there will be many independent streams flowing into the various basins. The basins will occasionally contain shallow playa lakes but they will not overflow from one basin to another. The initial forms of the different blocks or structural units are still preserved. The intermittent streams erode the higher areas and deposit in the lower ones.

With the advance of youth (Fig. 2) the various structural forms are dissected by canyons but their initial outlines and much of their initial surfaces remain. The various basins, however, become filled, and the local base-levels all rise. Wind action may be important in transporting material beyond the arid region from any of the slopes. It may deposit material even at high levels within the district, as determined by local and obscure conditions. The wind, however, is not the chief agent of erosion.

As maturity approaches (Fig. 3), streams from one basin find their way into an adjacent basin. This joining of basins may result from headward erosion from the lower level or by spilling over from the upper level due to aggradation. Thus the drainage systems of two basins coalesce. When integration of drainage lines from several basins is fully developed, the mature stage is reached. Perfect integration would mean that all the basins drained into the lowest basin. The various higher local base-levels would thus be replaced by a single lower base-level. At the same time the mountain forms are so completely dissected as to have lost their initial blocklike form. They are mature.

Under these conditions the relief of the district is constantly diminished as the lowest basin is filled up and the higher levels are worn down. Finally, the rivers become less and less effective and wind erosion and deposition assume a larger role. Drainage lines are thus disturbed and disintegration sets in. This is the beginning of old age (Fig. 4), which eventually results in peneplanation.

# CLASSIFICATION OF LANDSCAPES

## CONSTRUCTIONAL

| STRUCTURE | | | PROCESS | | | | STAGE | | |
|---|---|---|---|---|---|---|---|---|---|
| GEOLOGY | | LAND FORM | DESTRUCTIVE FORCE | | | | YOUNG | MATURE | OLD |
| SIMPLE | HORIZONTAL | PLAIN Low Relief | STREAMS | GLACIERS | WAVES | WIND | Young plain | Mature plain | Old plain |
| SIMPLE | HORIZONTAL | PLATEAU High Relief | | | | | Young plateau | Mature plateau | Old plateau |
| DISTURBED | DOME | DOME MOUNTAIN | | | | | Young dome mountain | Mature dome mountain | Old dome mountain |
| DISTURBED | FAULTED | BLOCK MOUNTAINS | | | | | Young block mountains | Mature block mountains | Old block mountains |
| DISTURBED | FOLDED | FOLDED MOUNTAINS | | | | | Young folded mountains | Mature folded mountains | Old folded mountains |
| DISTURBED | COMPLEX | COMPLEX MOUNTAINS | | | | | Young complex mountains | Mature complex mountains | Old complex mountains |
| DISTURBED | VOLCANIC | VOLCANOES | | | | | Young volcano | Mature volcano | Old volcano |

## DESTRUCTIONAL

| FORCE | EROSIONAL | RESIDUAL | DEPOSITIONAL |
|---|---|---|---|
| WEATHERING | Holes, pits | Exfoliation Domes | Talus cones<br>Landslides |
| STREAMS | Valleys<br>Canyons | Mountains<br>Divides | Deltas<br>Alluvial fans<br>Flood plains |
| GLACIERS | Cirques<br>Glacial troughs | Matterhorn peaks<br>Arêtes | Moraines<br>Drumlins<br>Eskers |
| WAVES | Sea caves<br>Clefts | Platforms<br>Cliffed Headlands | Bars<br>Beaches |
| WIND | Blow holes | Rock pedestals | Dunes<br>Loess |
| ORGANISMS | Burrows | | Coral reefs<br>Ant hills |

A GENETIC CLASSIFICATION OF LAND FORMS

## STRUCTURE, PROCESS, AND STAGE

The present-day aspect of all landscapes has been determined largely by the factors *structure*, *process*, and *stage*, discussed separately in the previous pages.

First, the *structure* of the region must be understood. If the region is a unit with one prevailing type of structure, one or two words will suffice to define it. Such terms as *plain* or *plateau*, *block mountain*, *dome mountain*, *folded mountain region*, *complex mountains*, *volcano* have already been mentioned. These terms are not only *descriptive* ones but also *explanatory*, inasmuch as they suggest the method of origin of the land mass in question. They deal with the fundamental and original qualities of the region rather than its present appearance.

Second, it is necessary to indicate the *process* employed to modify the original form. What destructive forces have been at work there? If one process has prevailed, a simple statement such as *dissected by streams*, *glaciated*, *wind-eroded*, or *attacked by waves* will suffice.

Finally, it is necessary to state how far the work of destruction has gone, that is, the *stage* of development reached. For this, such terms as *young* or *youthful*, *submature*, *mature*, *postmature*, and *old* are employed.

A very simple description, then, might read, "a mature (stage), stream-dissected (process), plain (structure)." For a more involved region it might be necessary to say, "a submaturely (stage) dissected (process) lava plain (structure) surmounted by cinder cones (structure) and later subjected to continental glaciation (process)." For a still more involved region the description might be, "a complex mountain area (structure) peneplaned (stage) in its first cycle of development, later uplifted and maturely (stage) dissected (process) in the second cycle, followed by incipient (stage) glaciation (process) in the headward portions of the valleys." Thus by a judicious use of the various terms pertaining to structure, process, and stage it is possible to present a fairly clear picture of even those regions which may have passed through a rather involved history. A single word connotes much in the way of ideas. The result is brevity, as well as a clear explanation of the genesis of the region.

In practice, certain looseness of presentation may creep in. It is not always possible, for example, to state definitely just what the succession of events may have been. The words "apparently" and "as if" may then be conveniently used. Thus, "a complex mountain region maturely dissected by youthful streams and then *apparently* slightly rejuvenated," or "surmounted by a row of monadnocks, *as if* controlled by rock structure."

The fact that streams have been the chief eroding agent is usually omitted, the simple term *dissected* being deemed sufficient. Stream erosion is thus considered the *normal* process.

## EMPIRICAL AND EXPLANATORY DESCRIPTIONS OF LAND FORMS

EMPIRICAL DESCRIPTION. An empirical description of a region is one which uses simple commonplace expressions, with no attempt at explanation. Hills, valleys, mountains, and other features of the landscape are described in terms of shape, size, position, and color. The picture thus presented to the reader is a difficult one to carry in mind because it involves a mass of detail with little apparent relationship between the different facts. Great detail is indeed necessary if a complete picture is to be presented. Each fact, moreover, stands by itself unsupported by its neighbors.

AN EXAMPLE OF AN EMPIRICAL DESCRIPTION. Following such a plan, Davis has described the coastal plain of northeastern Italy thus:

> In northern Italy south of Ancona there is a strip of hilly country, made up of sand and clay, and extending between the Apennines and the Adriatic. It is about 20 or 30 km. broad and lies 200 or 250 m. above the sea in its inner portion, where 10 or 20 km. further landward the mountains rise to about the same height, but not entirely so, those near the coast being smaller . . . . etc.

A description of this kind, to be thoroughly understood, must be interpreted by the reader who at every step constantly raises the question, "What is the reason for this and that?"

EXPLANATORY DESCRIPTIONS. Explanatory descriptions employ terms which are much more precise in character than are empirical terms and which convey some idea concerning the genesis of the form in question. Whereas the term *hill* is empirical, the term *dune* is explanatory. A dune is a hill of specific type whose origin is dependent upon wind action. In a similar manner such terms as *drumlin, kame, esker, cinder cone, moraine* are all explanatory terms descriptive of hills of various methods of origin. Such terms, too, convey more than the simple fact of origin, for they connote also much regarding the matter of shape, size, composition, as well as location with relation to other forms.

AN EXAMPLE OF EXPLANATORY DESCRIPTION. The region described above has also been described by Davis in explanatory terms as follows:

> In northeastern Italy, south of Ancona, there lies a well dissected coastal plain 20 to 30 km. broad, made up of unconsolidated beds of sand and clay, through which flow the master consequent streams, now in a late mature stage of development, with many short insequent tributaries. . . .

DESCRIPTIONS OF PRESENT FORMS RATHER THAN ACCOUNTS OF PAST PROCESSES. Finally, the best description of a region is one which gives statements framed in the present tense thus: "The region is an uplifted and dissected peneplane" or "a folded mountain region now in its second cycle of development" or "a warped and maturely dissected coastal plain." This, for the purpose of pure geographical description, is preferable to such statements as "the region was peneplaned, then uplifted and dissected" or similar expressions couched in terms of past history.

## THE ORIGIN OF PHYSIOGRAPHIC TERMS

While it is true we distinguish between empirical and explanatory terms, the mere words in themselves are not necessarily one type or the other. Explanatory terms have acquired a precise meaning because either custom or a purposeful restraint in their use has limited their application to some special kind of feature. For instance, the term *plateau*, coming from the French, means a "flat" and usually elevated tract of country. In geomorphology we purposely limit the application of this word to regions of flat-lying or horizontally bedded strata. Level-topped uplands not underlain by flat-lying rocks are not plateaus, even though they are so called in popular usage, as in the case of the Piedmont "Plateau," the Laurentian "Plateau," and the Central "Plateau" of France.

Physiographic terms have come about in a variety of ways. Some terms originally were the proper names of individuals. From Geysir in Iceland comes the name geyser; Vulcano in the Lipari Islands is the original volcano; *meander* comes from the Meander River in Asia Minor. Recently physiography has adopted such terms as monadnock (a residual on a peneplane) from Mount Monadnock; morvan (intersecting peneplanes) from the Morvan region of France; mendip (islandlike hills of the oldland projecting through a coastal plain) from the Mendip Hills of England; unakas (groups of monadnocks) from the Unaka Mountains of the southern Appalachians.

Some physiographic terms have been coined purposely and given very precise meanings, such as peneplane (almost a plane); conoplain; consequent, subsequent, obsequent, resequent, and insequent streams; antecedent streams, superimposed streams; hanging valleys; fault-line scarps.

Many terms have been adopted from other languages, richer than our own in words distinguishing different land forms. From the Spanish come such words as *cuesta*, *sierra*, *arroyo*, *canyon*, *playa*, *cuchilla*, and *bajada*. In Spanish *cuesta* means a hill or plateau steep on one side but sloping gently on the other. In geomorphology it means a hill of that shape formed by the erosion of dipping formations on a coastal plain. Similar hills, like block mountains, are not physiographic cuestas.

Our language is particularly poor in terms descriptive of desert forms, because few English-speaking people live in desert regions. We use Arab words such as *barkhan* (crescent-shaped dune), *wadi* (dry river course), and *hammada* (a rock-floored desert). For glacial forms we get from the Scandinavians such words as *osar*, *esker*, and *fiord*. From the Slavic we get *dolina* and *polje*, valley types in karst regions. From the French we have words for Alpine forms, such as *col*, *arête*, *cirque*, and *névé*. The Italian gives us *tombolo*, from *tumulus*, a little hill, referring to the little sand dunes which occur on the bar of tombolos. Terms like these, when used with precision, enrich our scientific vocabulary.

## CATASTROPHISM, UNIFORMITARIANISM, AND EVOLUTION

CATASTROPHISM. Since time immemorial it has been but natural for man to account by sudden and abrupt events for the things he sees in nature. Canyons and narrow gorges appear to be gashes in the rocks formed by a sudden splitting apart of the earth's crust. Volcanoes with their terrifying explosions have always seemed capable of greatly changing the aspect of the earth's surface, and mighty earthquakes have ever been a source of terror. Cataclysms such as these have always appealed to the popular fancy, and even at the present day there are many who resort to such events to explain all that they see. Such people are termed *cataclysmists* or *catastrophists*. Even Cuvier, the naturalist, fell under the sway of such beliefs and gave vent to the conclusion that the former mountain-making revolutions were so stupendous that "the thread of nature's operations was broken by them, that her progress was altered, and that none of the agents which she employs today could have sufficed for the accomplishment of her ancient works." Not only in geology but also in biology did these ideas prevail. The different groups of animals were thought to have arisen *de novo* and independently of other groups, only later to become suddenly extinct. Cuvier believed that some sudden catastrophe befell the surface of the earth some five or six thousand years ago, whereby the countries inhabited by man were devastated and their inhabitants were destroyed. At this time, he thought, there were a great inundation of the sea and then a retreat of the waters. Concerning this he says:

> It left behind also in northern countries, the carcasses of the great quadrupeds which are found embedded in the ice and preserved down to the present day intact with their hair, hides and flesh. It was at one and the same instant that these animals perished and that the glacial conditions came into existence.
>
> This change was sudden, instantaneous not gradual, and that which is so clearly the case in this last catastrophe is not less true of those which preceded it. The dislocation and overturning of the older strata show without any doubt that the causes which brought them into the position which they now occupy, were sudden and violent; and in like manner testimony to the violence of the movements which influenced the waters is seen in the great masses of debris and rounded pebbles which in many localities are found intercalated between beds of solid rock.
>
> Life upon the earth in those times was often overtaken by these frightful occurrences. Living things without number were swept out of existence by catastrophes. Those inhabiting the dry lands were engulfed by deluges; others whose home was in the waters perished when the sea bottom suddenly became dry land; whole races were extinguished leaving mere traces of their existence, which are now difficult of recognition, even by the naturalist. The evidences of those great and terrible events are everywhere to be clearly seen by anyone who knows how to read the record of the rocks.

UNIFORMITARIANISM. Long before Cuvier, however, there were those who felt that the little almost unobservable changes now occurring on the earth's surface were quite sufficient, if continued long enough, to produce results of great magnitude. In Germany, for example, during the eighteenth century, there were those who thought that the changes in the past had been of no abnormal kind but resembled those which might quite possibly occur now, and that this planet had always presented phenomena similar to those of the present time. The finest development of these ideas, however, came from Hutton, a Scotchman. They were recorded by his associate Playfair in 1802 under the title of *Illustrations of the Huttonian Theory of the Earth,* now cherished as one of the classics of geological literature. For here, in the most beautiful, precise, and clear English is presented the grand conception that the past history of our globe must be explained by what can be seen to be happening now, or to have happened only recently. The dominant idea in this philosophy is that *the present is the key to the past.* The notion of cataclysms was clearly refuted in such passages as follows:

> If a river consisted of a single stream without branches, running in a straight valley, it might be supposed that some great concussion, or some powerful torrent, had opened at once the channel by which its waters are conducted to the ocean; but, when the usual form of a river is considered, the trunk divided into many branches, which rise at a great distance from one another, and these again subdivided into an infinity of smaller ramifications, it becomes strongly impressed upon the mind that all these channels have been cut by the waters themselves; that they have been slowly dug out by the washing and erosion of the land.

This idea of uniformitarianism was carried to its extreme development by Charles Lyell. With unwearied industry he marshalled in admirable order all the observations that he could collect in support of the doctrine that the present is the key to the past. He refused to allow the introduction of any process which could not be shown to be a part of the present system of nature; he would not even admit that there was any reason to suppose that the degree of activity of the geological agents had ever seriously differed from what it has been within human experience. Few geologists, however, hold to this extreme form of the idea of uniformitarianism.

Evolution. Bound up with the idea of a uniform set of conditions is also the belief that great changes will result through the operation of such activities for long periods of time. Sediments are deposited in the sea, slowly buried to great depths, become compressed, are gradually elevated and sheared, changing by infinitesimal stages to schist through the action of heat and pressure. Thus even with rocks, as in the animate world, evolution plays a part in adjusting them to their environment. Landscapes, as well, evolve gradually, passing through stages in their development. It is recognized that the present aspects of the earth's surface are only transitory. The everlasting hills are slowly disappearing. The minute changes, undetected in the life span of a single individual, are cumulative in their effects. It is difficult at the outset for the human mind to conceive of the vast periods of time with which geology must deal. Only when we throw aside the conception of the years which make up our own short lives and think in terms of larger magnitude, are we prepared to deal with ideas so noble in their proportions.

## SCIENTIFIC METHOD

The ultimate aim in geomorphology as in most scientific study is to explain phenomena which have been observed. Such explanations take the form of hypotheses or theories, or they may be formulated as laws. Four distinct steps are recognized in developing a theory, although in actual practice these steps may not be kept separate from each other.

First Step: Observing the Facts. The first step always consists in the securing of a certain amount of data. These data may be classified for convenient reference. If geological in nature, this organization of the data may consist of making a geological map showing rock outcrops together with collections of specimens and notebook records and descriptions. Facts of dip and strike, thickness of beds, lithological characteristics, topographic associations are all elements of this stage of the work.

Second Step: Formulation of Hypotheses to Explain the Observed Facts. The separate disconnected facts are then assembled with a view to discovering some general principle which will tie them together. Thus a number of different outcrops exhibiting certain dips and strikes might be explained as parts of an anticlinal structure. One explanation or surmise is not entirely safe, however. If the facts are meager, it is quite probable that they can be explained in a number of ways; hence, the importance of developing several hypotheses to explain them. In short, the scientific worker should evolve as many plausible hypotheses as possible in the hope that thus he is more likely to hit upon the correct one. This formulation of hypotheses constitutes what is known in logic as the *inductive process*. It consists in working from the particular to the general, in recognizing some quality, characteristic, or method of behavior which all the facts have in common. This sifting out the significant detail from that which is irrelevant demands a keen mind, sometimes indeed a high order of genius.

Third Step: The Deduction of Further Facts. Each hypothesis is next examined and studied as an abstract idea, with a view to determining what additional consequences must result if it be true. This procedure, the working from the general to the particular, is a *deductive process* of logic. The deductions are secured only by deliberate effort. Rarely do they flash across the mind. To make them, the geologist often sits down and writes out in detail his theories with their respective consequences. In actual practice, hypotheses are evolved and deductions made simultaneously with the recording of observations. The importance of developing more than one hypothesis is readily comprehended when it is realized that with only a single hypothesis to defend, the investigator is prone to believe that it be the true one. He is prejudiced then in his search for further facts and is in danger of noting only those which support his hypothesis, to the exclusion of others.

Fourth Step: Testing the Hypotheses and Eliminating the Invalid Ones. The final step consists in finding out which of the deduced facts are actually true. If one of the hypotheses demands the existence of certain facts and upon investigation these facts are found to be either missing or true in the opposite sense, then the hypothesis demanding those conclusions is eliminated. In this manner the false hypotheses are one by one discarded until there is but a single remaining hypothesis which stands up under the test. If the proof is sufficient, this hypothesis may then rank as a theory, or even as an established fact.

It will be noted that throughout this whole procedure the investigator clearly distinguishes between actual fact and inference. An inference is merely a very weak and unsupported hypothesis, thoroughly commendable if used with discretion. And, again, it is obvious that this procedure involving deduction and the search for new facts immensely sharpens the wits of the observer and sets him on the track of additional data.

Some scientific writers, presenting the results of their study of a problem, acquaint the reader first with all the facts of the situation, following this with the theory which explains them. To keep the reader in suspense while recounting facts not obviously connected with each other involves an unnecessary burden and, moreover, is likely not to leave so clearly in the reader's mind the elements of the situation. Many writers prefer to present their conclusions deductively, first giving the broader outlines of the theory and following it up with the facts upon which it depends. In one of his essays Davis says:

> It should be understood that the deductive character of the succeeding paragraphs is more apparent than real. Many features here presented as deductions were discovered by observation. It is true that an expectation of certain occurrences had been aroused by the deductive consideration of certain processes, but there has been so continual an interweaving of observation and theory during the growth of these pages that it is now rather difficult to determine the order in which the various items here recorded came to mind.

So, taking our cue from the splendid method so frequently used by Davis and other writers, we shall in the succeeding pages of this book present the principles of geomorphology in a deductive order. In bold strokes the broader outlines of the subject will be announced, followed by the details and examples which support them.

MANY SCIENTIFIC METHODS. In common parlance, scientific method is often regarded as the one technique employed by those engaged in research in the sciences. As a matter of fact, there are many methods which scientists employ and some of them are applicable to the humanities as well as to the physical sciences. The experimental method so common in chemistry, physics, and the biological sciences finds only limited application in geology and physiography. Geological changes are so slow, and the factors so beyond control, that the techniques employed in the chemistry laboratory cannot be applied in geology or physiography. The method outlined on these pages is classical in geological investigations and may be regarded as the most fruitful method to be employed in geomorphology.

## EXAMPLES OF METHOD

SCIENTIFIC INVESTIGATION AND SCIENTIFIC PRESENTATION. It will now be apparent to the student that scientific investigation—the discovery of hypotheses and explanations—is carried out largely by induction but that the presentation of the results can best be made deductively. The careful author, mindful of the reader's point of view, will tell his story in a deductive way. This enables him to achieve a continuous presentation of his theories and explanations, postponing the evidence until later. At the end of such a deductive presentation Davis says:

> Largely deductive as the preceding portion of this essay is in its present form, the reader should not suppose that it was prepared independently of observation. The actual progress through the problem has involved repeated alternations of external and internal work; the collection of observations and the induction of generalizations on the one hand, and on the other hand the deduction of their consequences, the confrontation of deductions with generalizations, the evaluation of agreements, and the repeated revision of the whole process

He then proceeds to give the detailed facts which led to the generalizations first presented. A deductive presentation followed by an enumeration of the original observations is the method most pleasing to the reader.

Little continuity of thought can be achieved by those who recount their observations in a chronological order after the manner of travelers telling what they saw, or of investigators who dwell upon the ramifications into which their studies led them. Let us now note a few examples of both deductive and inductive presentation.

DEDUCTIVE. A leading sentence at the beginning of a paragraph is the sign of deductive presentation. In Semple's *Influences of Geographic Environment*, many of the paragraphs begin with a sentence which epitomizes the whole paragraph. For example, we read:

> A coast region is a peculiar habitat, inasmuch as it is more or less dominated by the sea.

Then follows an elaboration of this idea with specific statements as to how the coast is dominated by the sea, the citation of actual places in the world where this is true. But the main idea is in the first sentence. Or again we read:

> Food is the urgent and recurrent need of individuals and of society.

The rest of the paragraph is simply a restatement of this in various ways. Or read the first sentence from one of the paragraphs in Huxley's essay, *On a Piece of Chalk*. It is brief and to the point, as follows:

> A great chapter of the history of the world is written in the chalk.

Huxley then proceeds to indicate what this history has been.

INDUCTIVE. In contrast with the deductive method is the way of the writer who begins his paragraph with a succession of facts apparently unrelated, leaving the reader "holding the bag," as it were, until he gets to the point of sorting out the several ideas and unifying them with some kind of summary. This is especially true when the author is working up to a climax or to some important generalization which he wishes you to see grow under his hand. In the following lengthy quotation the gist of the matter is given in the final sentence after a long recounting of the facts upon which it rests:

The total volume of water flowing into the sea in one year, for all the land areas of the earth, is about 7,000 cubic miles. The total amount of various kinds of salts carried in solution in one cubic mile of river water is about 750,000 tons. This makes a total of about 5,000,000,000 tons of mineral matter carried into the sea annually. Average river water contains about 326,000 tons of calcium carbonate per cubic mile; and about 112,000 tons of magnesium carbonate; 34,000 tons of calcium sulphate; 31,000 tons of sodium sulphate; and 20,000 tons of potassium sulphate: a total of about 85,000 tons of sulphates. Of the chlorides, sodium, lithium and ammonium, there are about 20,000 tons. Sea water contains 117,000,000 tons of sodium chloride per cubic mile and 16,000,000 tons of magnesium chloride; 7,000,000 tons of magnesium sulphate; 5,000,000 tons of calcium sulphate; and 4,000,000 tons of potassium sulphate: a total of 16,000,000 tons of sulphates. Carbonates about 500,000 tons. The total volume of the sea water in the world has been estimated by Krümmel at 1,300 million cubic kilometers (area 361,000,000 square kilometers; average depth of 3,600 meters, or 2,000 fathoms).

From these figures it may be shown therefore that at the present rate it would take about 50,000 years to provide the sea with its present amount of carbonates; 10,000,000 years to provide the sulphates; and about 300,000,000 years to bring it the chlorides. It would seem, therefore, that the chlorides, being highly soluble, were introduced very rapidly at the beginning of the earth's history and that the carbonates, being least soluble, are being removed now about as rapidly as they are brought in.

(NOTE: It is suggested that the student synthesize the preceding figures and determine the correctness of the conclusions drawn from them).

EXAMPLES OF METHOD USED IN THIS BOOK. For convenience now, attention may be called to some of the pages of this book to illustrate examples of deductive and inductive presentation, and to suggest that throughout his reading of this text the student analyze each page. While doing so, he should try to decide whether the author purposely used the method he did and, if so, why he did; or whether he paid no attention to the method of presenting his facts and of developing his ideas.

The first page of the next chapter on "Rocks and Their Structures," the Synopsis, is a good example of deductive presentation. Each paragraph on this page starts with a leading sentence or two. You can get a fair idea of what the page is about by reading only these leading sentences. The second page on "Igneous Rocks" is developed in the same way. The two pages on "Sedimentary Rocks" are equally satisfactory in giving the gist of the subject at the start of each paragraph.

Ordinarily an author passes alternately from one method to the other. As soon as a student finds himself lost, he can be reasonably certain that the author is giving him some information for which he was not prepared; that is, he is presenting his material inductively. In this book the present author has attempted to use the deductive method as much as possible in the belief that it would greatly expedite the reading of the text and result in a quick comprehension by the student of the author's ideas.

## QUESTIONS

1. Is physiography the same as physical geography? Why should a distinction be made?
2. Does physiography include the study of rocks and rock structures?
3. What theories can you think of to explain the present arrangement and form of the continents? Do you think that any theories regarding the origin of the continents should also take into consideration the origin of the earth?
4. If a geologist knows that active mountain making occurred in North America at the end of Paleozoic time, is he surprised to learn therefore that similar disturbances occurred in Europe about the same time?
5. What is a physiographic province?
6. Can an ideal physiographic province include both plateaus and mountains?
7. What features usually give most of the character to a landscape, those of the second order or those of the third order?
8. In the Adirondack Mountains of New York there are many bald, rounded summits. Should these be called dome mountains?
9. What kinds of plains are features of the second order, and what kinds are features of the third order? What is a flood plain, for example? A lake plain? An outwash plain? A coastal plain?
10. In what ways do you suppose the geomorphic cycle may be interrupted?
11. Suppose a region in youth is elevated, does this change its age? If it is mature and then elevated, does this alter its age? If it is old and then elevated, how does this change its age?
12. Do you suppose that a region after reaching old age could be buried? What would cause that? When a region is in old age, is the general level of the country near sea level or high above sea level?
13. Consult the Classification of Landscapes in this chapter. Where would you classify a railroad cut made by man? a hole made by a meteor? a pile of rocks formed by a landslide?
14. Write a brief empirical description of an imaginary region, being careful not to suggest in any way how the various land forms were produced.
15. Why do you suppose the English language is relatively poor in words pertaining to desert landscapes and to high mountains but is rich in terms describing subdued topography, such as dale, vale, valley, gorge, gulch, gully, ravine, swale, hollow, dip, depression, all of which have reference to more or less the same sort of feature?
16. Did the belief that the earth was created about 4000 B.C. make it difficult to accept the doctrine of uniformitarianism? Do any people still believe in catastrophism? Just what do we mean when we say that "the present is the key to the past"?
17. Can a scientific method be applied to ordinary affairs of life, outside scientific studies? For example, suppose you have lost your gloves. Draw up a number of theories, and for each one make the necessary deductions, and show how you would test out each theory.
18. In Chapter III on "Weathering," what method of presentation is used on the two pages describing Soils?
19. When does a hypothesis become a theory? Why do we speak of the planetesimal hypothesis, but of the glacial theory?
20. Name some region that you have seen in your travels, or in photographs, or have read about that you would consider in the stage of mature dissection.

## TOPICS FOR INVESTIGATION

1. Stratigraphy. Its importance in geomorphology.
2. Physical geology. Difference between this and geomorphology.
3. The subject matter of geology. A comparison between the treatment in different texts.
4. The popular presentation of geology. How should this differ from the textbook method?
5. Scientific presentation. Examples of various methods, and their relative merits.

# REFERENCES

## General Geomorphology

First of all, every student of geomorphology should be familiar with W. M. Davis's *Geographical essays* (1909) Boston, 711 p., in which there are articles on *The geographical cycle, The arid cycle, Base-level,* and various aspects of river and glacier erosion. They are easy to read and to understand and are written in a robust and provocative manner calculated to inspire thought and action. In German, Davis's *Erklärende Beschreibung der Landformen* (1912) Leipzig, 565 p., is easily grasped, though the student have only a moderate knowledge of that language. This is his most comprehensive work on geomorphology. Davis's *Elementary physical geography* (1902) Boston, 401 p., is a simpler treatment of the subject. Among the many standard texts on land forms are W. H. Hobbs's *Earth features and their meaning* (1931) New York, 517 p., illustrated with many pleasing sketches and excellent for supplementary reading; R. D. Salisbury's *Physiography* (1919) New York, 676 p., which presents the subject in a manner quite different from that of Davis; R. S. Tarr and L. Martin's *College physiography* (1914–28) New York, 837 p., a valuable detailed treatment with long and useful reference lists; R. S. Tarr and O. D. von Engeln's *New physical geography* (1929) New York, 688 p., an attractive book, well illustrated, with reference lists and various exercises. C. A. Cotton's *Geomorphology of New Zealand,* part I, Systematic (1926) Wellington, 462 p., is an outstanding treatment of the principles of geomorphology, following the Davis technique. A more recent book is by S. W. Wooldridge and R. S. Morgan on *The physical basis of geography. Geomorphology* (1937) London, 445 p. These are British authors who have given much attention to the Davis school of geomorphology. In addition, the book takes account of much recent work by later writers, such as Johnson on shore lines and sediments, and Davis and Daly on glacial control of coral reefs, as well as certain recently developed ideas by W. S. Glock on the cycle of development in drainage systems. Penck's idea of the erosion cycle is clearly set forth in several pages. E. de Martonne's *Shorter physical geography* (1927) London, 338 p. (in English), touches on all aspects of physiography and presents something of the French school of thought.

Students of geography wishing only a brief treatment of land forms will find V. C. Finch and G. T. Trewartha's *Elements of geography* (1936) New York, 782 p., satisfactory. It includes climatology as well as much that is purely geography. A book covering all aspects of physiography is *New physiography* (1927) by A. L. Arey, F. L. Bryant, W. W. Clendenin, and W. T. Morrey, Boston, 613 p., designed for high-school use.

## General Geology

Among the numerous excellent textbooks on geology are the following:


Agar, W. M., Flint, R. F., and Longwell, C. R. (1929) *Geology from original sources.* New York, 527 p.

Branson, E. B., and Tarr, W. A. (1935) *Introduction to geology.* New York, 470 p.

Chamberlin, R. T., and MacClintock, P. (1930) *College geology.* New York, 916 p.

Cleland, H. F. (1916) *Geology, physical and historical.* New York, 718 p.

Emmons, W. H., Theil, G. A., Stauffer, C. R., and Allison, I. S. (1932) *Geology.* New York, 514 p.

Grabau, A. W. (1920) *Comprehensive geology,* part I, Physical geology. New York, 864 p. A masterly piece of work, well suited as a reference work.

Grabau, A. W. (1913–24) *Principles of stratigraphy.* New York, 1185 p. Many pages are strictly geomorphological in character.

Longwell, C. R., Knopf, A., and Flint, R. F. (1932) *Textbook of geology,* part I, Physical geology. New York, 514 p.

Longwell, C. R., Knopf, A., and Flint, R. F. (1937) *Outlines of physical geology.* New York, 356 p.

Lyell, C. (1875) *Principles of geology.* London, vol. 1, 655 p.; vol. 2, 652 p. A classic which may still be read with profit.

Scott, W. B. (1932) *Introduction to geology,* vol. 1, Physical geology. New York, 604 p.

Popularized Geology

Bradley, J. H. (1935) *Autobiography of earth.* New York, 347 p.

Croneis, C., and Krumbein, W. C. (1936) *Down to earth.* Chicago, 501 p.

Fenton, C. L. (1938) *Our amazing earth.* New York, 346 p.

Mather, K. F. (1932) *Old mother earth.* Cambridge, Mass., 177 p.

Shand, S. J. (1938) *Earth-lore.* New York, 144 p.

Scientific Method

L. E. Saidla and W. E. Gibbs's *Science and the scientific mind* (1930) is a compilation of essays by many men of science. In this book, Huxley's essay on "The Method of Scientific Investigation" is especially to be commended. There are several other essays of almost equal value. A charming analysis of thinking processes, certain to delight those interested in scientific thought, is to be found in William James's *Principles of psychology* (1890), chapter on "Reasoning," vol. 2, p. 325. Refer also to A. L. Jones's *Logic, inductive and deductive* (1909) for an analysis of the scientific method. Several suggestive essays by D. W. Johnson on *Studies in scientific method* appeared in 1938 in the Journal of Geomorphology, vol. 1; especially p. 64–66, 147–152.

Cooley, W. F. (1912) *Principles of science.* New York.

Downing, E. R. (1927) *The elements and safeguards of scientific thinking.* Sci. Monthly, vol. 25, p. 19–24.

Minot, C. S. (1911) *The method of science.* Science, vol. 33, p. 119–131.

Ritchie, A. D. (1923) *Scientific method.* New York.

Uniformitarianism

A. Geikie's famous book on *The founders of geology* (1905) London, 486 p., tells about the gradual rise of uniformitarianism as a scientific doctrine, replacing the ideas of the cataclysmists. Finally, every student of geomorphology should know of J. Playfair's *Illustrations of the Huttonian theory* (1802) Edinburgh, 528 p., and read some of the wonderfully lucid paragraphs setting forth the simple idea of uniformitarianism in a manner which has caused it to rank as one of the masterpieces of scientific literature.

*The birth and development of the geological sciences* (1938) Baltimore, 506 p., by Frank D. Adams is the most recent book on the history of geology. It reviews many of the quaint ideas from classical times down to the present. See also N. M. Fenneman (1939) *The rise of physiography.* Geol. Soc. Am., Bull. 50, p. 349–360.

Note: Many of the books mentioned in the foregoing list of references have been freely consulted in the preparation of this text. Davis's *Erklärende Beschreibung der Landformen* and Davis's *Essays* have been studied with particular care, for in these two works may be found the substance of Davis's philosophy, which it has been the aim of this book to present.

Grabau's *Comprehensive Geology* provided numerous examples of geomorphic interest, for that magnificent work is almost as much a treatise on geomorphology as it is on geology.

Beyond this it is impossible to make further mention because, as a matter of fact, virtually every one of the numerous references listed at the ends of the various chapters was read at one time or another and their essence has so filtered into the text of this book as to make all workers in this science to some extent responsible for what is set forth in these pages.

# II
# ROCKS AND STRUCTURES

*U. S. Forest Service*

"FLYING BUTTRESS" GALLATIN NATIONAL FOREST, MONTANA
Detail of weathering in horizontal beds of conglomerate.

*Publishers' Photo Service*

OLEAN ROCK CITY, OLEAN, NEW YORK

Rectangular jointing in coarse, massive conglomerate. Widening of joints caused by movement of blocks down-valley.

CANYON OF LITTLE COLORADO, ARIZONA

Strong rectangular jointing in massive limestone. Canyon system determined by arrangement of joints.

*American Museum of Natural History*

*Jackson, U. S. Geological Survey*

THE DEVILS SLIDE ON THE YELLOWSTONE

Resistant beds of sandstone dipping almost vertically. Intervening weak beds of shale removed by erosion.

QUEENS CANYON, COLORADO SPRINGS, COLO.
Dipping limestone beds resting unconformably upon underlying strongly jointed granite complex.

*Chamber of Commerce, Colorado Springs*

*Fox Photos*

DARTMOOR, SOUTHWESTERN ENGLAND

Rolling upland underlain by large granite boss. Granite outcrops in foreground with jointing that simulates bedding.

"THE GHOSTS," WHEELER NATIONAL MONUMENT, RIO GRANDE NATIONAL FOREST

Unusual weathering features in tuff, producing "craggy" topography.

*U. S. Forest Service*

*Darton, U. S. Geological Survey*

OVERTURNED ANTICLINE, PANTHER GAP, VIRGINIA

The thickening of the beds on the axis of the fold is distinctly shown.

BEVELED LIMESTONE BEDS, KNOX COUNTY, TENNESSEE

This is part of an eroded pitching anticline.

*Keith, U. S. Geological Survey*

*British Geological Survey*

MONOCLINAL FOLD, CENTRAL ENGLAND
Massive sandstone beds with weaker shale partings all underlying the coal measures.

ANTICLINE IN GNEISS, MITCHELL COUNTY, NORTH CAROLINA
Showing sedimentary beds strongly contorted and metamorphosed into a banded gneiss.

*Keith, U.S. Geological Survey*

*Walcott, U. S. Geological Survey*

SYMMETRICAL ANTICLINAL FOLD NEAR HANCOCK, MD.

Massive sandstone beds separated by weaker shales which have been weathered out.



SYMMETRICAL SYNCLINAL FOLD NEAR HANCOCK, MD.

Massive shale beds showing a strong slaty cleavage.

*Walcott, U. S. Geological Survey*

*U. S. Forest Service*

BADLANDS NEAR ASHLAND, CUSTER NATIONAL FOREST, MONTANA
Weathering along vertical joints in massive horizontal beds of clay. Thinner indurated beds standing out in relief.



WHITE CROSS-BEDDED SANDSTONES, CENTRAL UTAH
Showing ancient dune deposits of unusually pure quartz sand.

*Jackson, U. S. Geological Survey*

*U. S. Forest Service*

CHIMNEY ROCK ALONG THE HIGHWAY BETWEEN CODY AND YELLOWSTONE PARK

Vertical joints in horizontal beds of volcanic tuff and agglomerate.

## ROCKS AND THEIR STRUCTURES

Synopsis. To the geomorphologist rocks are important because they constitute the materials out of which land forms are carved. A rock should be conceived as a product of its environment. When the environment is changed, the rock changes.

In general, rocks are formed under two environments, either below the crust of the earth or at its surface. The former are usually the igneous rocks; the latter the sedimentary. When deep-seated rocks are exposed at the surface, or surface-formed rocks subjected to deep-seated conditions, changes are bound to occur. Each type may thus be changed into the other. The change is usually only partially accomplished and the result is a metamorphic rock.

It is the behavior of surface rocks under the action of weathering and erosion which concerns the student of land forms. Inasmuch as these changes are both chemical and mechanical, it is important to consider both the chemical and the structural character of rocks.

Granites and other coarse-grained igneous rocks disintegrate rapidly where temperature changes are great. Hence granites are weak rocks in deserts and on mountain tops but in humid regions are more resistant than most other rocks.

Metamorphic rocks, like schist and gneiss, are highly resistant to chemical change. Their weakness lies in their tendency to split apart or to cleave, but this usually does not result in such small fragments as are formed by disintegrating granite.

Among sedimentary rocks quartzite tops the list because it is subject to neither chemical nor mechanical change. It is even more resistant than the crystalline rocks already mentioned. Sandstone stands high for the same reason but, if it is not tightly bound together in a quartz matrix, it may yield rapidly to disintegration.

Shale, like clay, is strictly a surface product and suffers virtually no chemical change whatever upon exposure. It is, however, a weak rock because its fine grains are not tightly held together and are rapidly carried away by erosive forces.

Limestone is perhaps the weakest of all rocks, because of its solubility. In dry regions it is a resistant rock because it is less liable to changes of temperature than most other rocks are.

As for rock structures, these may be classed as major: the various types of folds and faults, occasionally joint systems, and unconformities; and minor: those produced at the time the rock was first formed, such as cross-bedding, ripple marks, mud cracks, and possibly concretions. They are rarely of topographic significance, whereas the major structures determine the pattern of the topographic features which result from erosion.

| | ROCK NAME | RESISTANCE | PHYSIOGRAPHIC FORMS |
|---|---|---|---|
| **IGNEOUS**<br>FINE TEXTURED<br>DARK (Basic) | BASALT | Usually resistant except when bearing olivine. Exfoliates readily | Columnar jointing, dikes and escarpments |
| MEDIUM | ANDESITE | Usually resistant | Not wide-spread enough to form typical landscapes |
| LIGHT (Siliceous) | RHYOLITE | Usually resistant, but sometimes decomposes badly | Bluffs and cliffs |
| COARSE TEXTURED<br>DARK (Basic) | GABBRO DIABASE | Usually very resistant except when much jointed and when containing olivine | Dikes and escarpments |
| MEDIUM | SYENITE | Similar to granite but lack of quartz renders less resistant | Uplands |
| LIGHT (Siliceous) | GRANITE | Usually resistant. Disintegrates readily in arid regions | Exfoliation domes. Bosses and uplands |
| **SEDIMENTARY**<br>FINE GRAINED: ARGILLACEOUS<br>LOOSE | CLAY | Weak but usually coherent enough to form vertical walls | Bluffs and badlands |
| CONSOLIDATED | SHALE | Usually weak | Gentle slopes, valleys and lowlands |
| FINE GRAINED: LIMY<br>LOOSE | MARL | Very weak | Low valleys |
| CONSOLIDATED | LIMESTONE | Weak in humid regions. Resistant in arid | Karstland, sink holes, high escarpments |
| COARSE GRAINED<br>LOOSE | SAND | Usually weak | Lowlands but sometimes caps uplands |
| CONSOLIDATED | SANDSTONE | Resistant if well consolidated | Cliffs and plateaus |
| VERY COARSE<br>LOOSE | GRAVEL | Moderately resistant because of porosity | Often caps uplands |
| CONSOLIDATED | CONGLOMERATE | Very resistant | Ridges and mountains |
| **METAMORPHIC**<br>CHANGED FROM SEDIMENTARY ROCKS<br>from<br>SHALE<br>by moderate pressure | SLATE | Weak but more resistant than limestone | Lowlands |
| LIMESTONE<br>by pressure | MARBLE | Weak | Lowlands unless associated with more resistant metamorphics |
| SANDSTONE<br>by cementation | QUARTZITE | Very resistant. Perhaps the most resistant rock | Residual ridges, knobs and monadnocks |
| CHANGED FROM EITHER IGNEOUS OR SEDIMENTARY ROCKS<br>BANDED | GNEISS | Usually very resistant | Uplands |
| SCHISTOSE | SCHIST | Usually resistant | Uplands and ridges |

CLASSIFICATION OF ROCKS AND THEIR REACTION TO WEATHERING AND EROSION

## KINDS OF ROCKS

The crust of the earth is built of rocks. Some of these rocks are hard masses; others loose and unconsolidated. Some rock bodies are large, others small. Some are uniform, others varied. Obviously the nature of the rocks, their arrangement, their character and structure, must be important in affecting the behavior of streams and other external forces during the process of wearing down a land mass. The surface features—the destructional forms, both of erosion and of deposition—depend largely upon the nature of the rocks which underlie the region.

Three kinds of rocks make up the crust of the earth: *igneous, sedimentary,* and *metamorphic.* The igneous and the metamorphic together are sometimes called the *crystalline* rocks. This threefold classification is, for most purposes, a satisfactory one, but a thorough understanding of rocks demands a much more elaborate scheme than is here presented.

From *the igneous rocks* the other groups eventually are derived. These rocks result from the cooling of a molten mass. If molten lava is poured out on the surface of the earth, it constitutes an *extrusive* flow. If the molten mass or magma penetrates other rocks and cools within the crust, it becomes an *intrusive* body. Igneous rocks vary greatly in mineral and chemical composition, texture, and mode of occurrence. Those formed by intrusion may be seen only after erosion has removed the rocks above them. The composition, texture, structure, and shape of the deep-seated rock mass affect the erosional forms produced upon them. Hence the importance of the characteristics of igneous rocks from the physiographic standpoint.

*Sedimentary rocks* usually occur in layers:

*a. Mechanical sediments* are most commonly deposited in water or air. They are formed of small particles of sand or clay carried by streams or swept by the wind to those places where it accumulates to form sandstone, shale, conglomerate, and certain limestones.

*b. Chemical sediments* are less extensive. They are deposited by evaporation or for other causes from solution in water, as, for example, rock salt, gypsum, cave deposits, and calcite.

*c. Organic sediments:* such as coral, shell-limestones, chalk, and coal, formed by organisms, either plant or animal, and deposited in either water or air.

Sedimentary rocks at present constitute much of the earth's surface. In many instances, the formations do not lie in the position they had at the time of deposition. As a result of folding, breaking, and other disturbances, mountains have been produced and there the sedimentary rocks usually are inclined in attitude.

*The metamorphic rocks* constitute the third division. These rocks, originally either igneous or sedimentary, have become altered by pressure and heat or by the infiltration of other material, so that the original characteristics have been lost or replaced by different ones. Most metamorphic rocks show a pronounced crystalline structure and some are highly resistant to the agencies of weathering and erosion. Gneiss, schist, and marble are the most common rocks of this type.

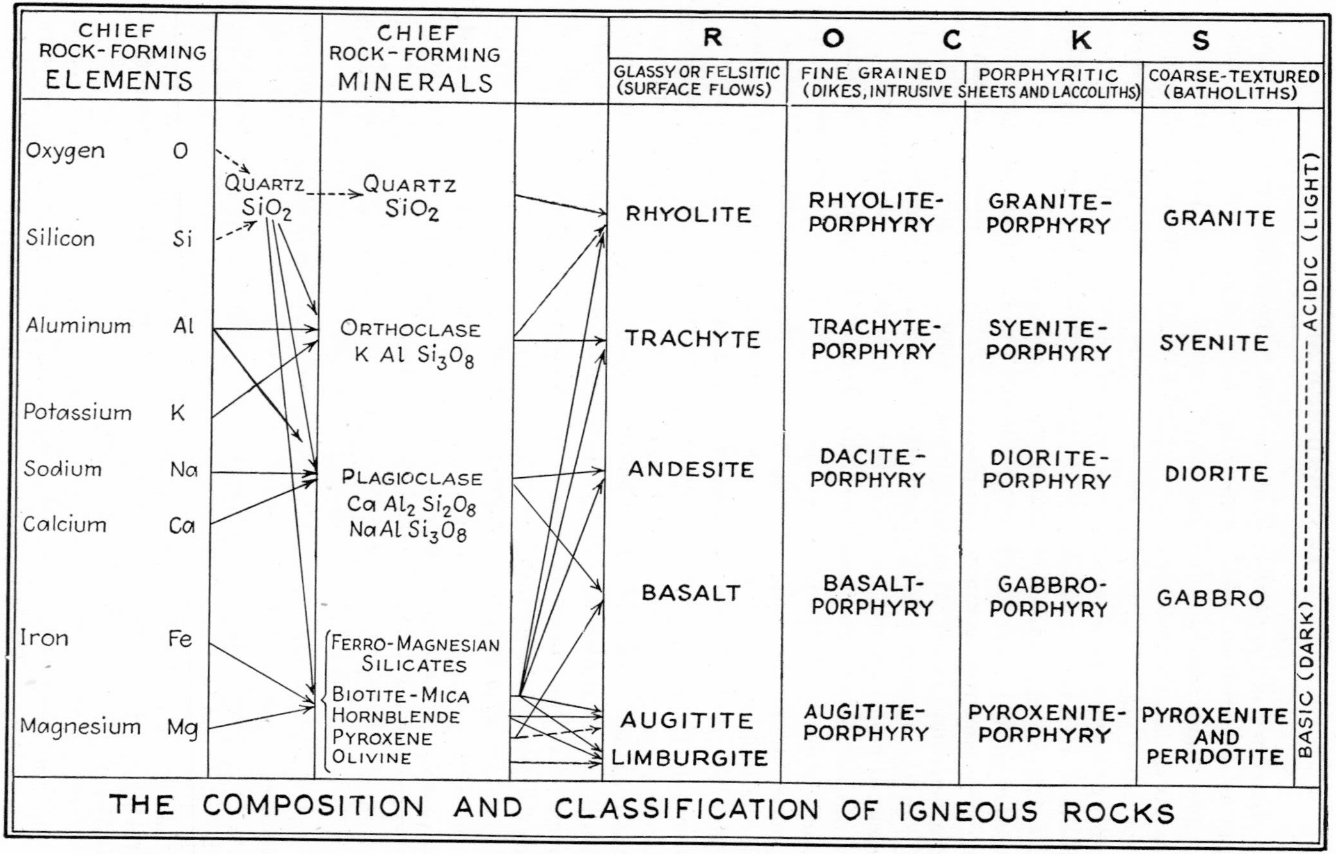

| CHIEF ROCK-FORMING ELEMENTS | |
|---|---|
| Oxygen | O |
| Silicon | Si |
| Aluminum | Al |
| Potassium | K |
| Sodium | Na |
| Calcium | Ca |
| Iron | Fe |
| Magnesium | Mg |

Quartz $SiO_2$

| CHIEF ROCK-FORMING MINERALS |
|---|
| Quartz $SiO_2$ |
| Orthoclase $KAlSi_3O_8$ |
| Plagioclase $CaAl_2Si_2O_8$ $NaAlSi_3O_8$ |
| Ferro-Magnesian Silicates: Biotite-Mica, Hornblende, Pyroxene, Olivine |

| ROCKS: Glassy or Felsitic (Surface Flows) | Fine Grained (Dikes, Intrusive | Porphyritic Sheets and Laccoliths) | Coarse-Textured (Batholiths) |
|---|---|---|---|
| RHYOLITE | RHYOLITE-PORPHYRY | GRANITE-PORPHYRY | GRANITE |
| TRACHYTE | TRACHYTE-PORPHYRY | SYENITE-PORPHYRY | SYENITE |
| ANDESITE | DACITE-PORPHYRY | DIORITE-PORPHYRY | DIORITE |
| BASALT | BASALT-PORPHYRY | GABBRO-PORPHYRY | GABBRO |
| AUGITITE LIMBURGITE | AUGITITE-PORPHYRY | PYROXENITE-PORPHYRY | PYROXENITE AND PERIDOTITE |

ACIDIC (LIGHT) ---------- BASIC (DARK)

THE COMPOSITION AND CLASSIFICATION OF IGNEOUS ROCKS

DIAGRAMMATIC SCHEME TO SHOW RELATION BETWEEN CHEMICAL ELEMENTS, MINERALS, AND ROCKS

## IGNEOUS ROCKS

Igneous rocks are classified on the basis of their mineral composition and of their texture. The texture may range from glassy or very fine grained to extremely coarse grained. The glasses and fine-grained rocks are those which have formed by the cooling of molten lavas on the surface of the ground. If a rock magma cools very quickly, it may solidify without any crystallization, forming a glass. Obsidian and pumice are rocks of this type. If cooling is not quite so rapid, incipient crystallization may take place, forming a stony or dense rock. These rocks, termed *felsitic*, make up the great lava fields and plateaus of the world.

A porphyritic rock is one having a rather fine or dense groundmass but with some larger crystals scattered through it.

In the coarsest grained rocks, there is no groundmass, the mineral crystals are all nearly uniform, and the different minerals are clearly distinguishable by the naked eye. Coarse-grained rocks have almost invariably been formed by the cooling of molten masses or magmas deep beneath the surface of the earth. The slow cooling gives time for the several kinds of minerals to crystallize out independently of each other. These deep-seated rocks are visible now only because the material once overlying them has been removed by erosion.

In the matter of composition, igneous rocks may vary from acidic to basic. Acidic rocks are those which contain large amounts of silica, alumina, and potassium.* Basic rocks contain less of these elements and more of sodium, calcium, magnesium, and iron. Acidic rocks generally have light-colored and light-weight minerals predominating; basic rocks have dark-colored and heavy minerals. Acid lavas are stiff and viscous, and their contained gases escape with explosive violence as the magma approaches the surface. Basic lavas flow readily at much lower temperatures. Most of the extensive lava flows are basic.

Because of the great range of textures and mineral types, there are hundreds of kinds of igneous rocks, grading imperceptibly into each other.

The scheme on the opposite page should make clear the main groups and the commoner types. *Granite*, for example, is a coarse-grained igneous rock containing quartz (silica), orthoclase (potassium aluminum silicate), and only small quantities of mica (potassium aluminum silicate with some iron and magnesium), and other basic minerals. *Gabbro*, like granite, is coarse grained but contains basic plagioclase (calcium aluminum silicate), pyroxene (calcium iron magnesium silicate), and other basic minerals, possibly even magnetite (iron oxide).

In a similar manner *rhyolite*, *trachyte*, *dacite*, and *andesite* are the terms applied to the series of fine-grained rocks, corresponding in mineral composition to the granite-gabbro series just mentioned.

* Silica is silicon dioxide; alumina is aluminum oxide.

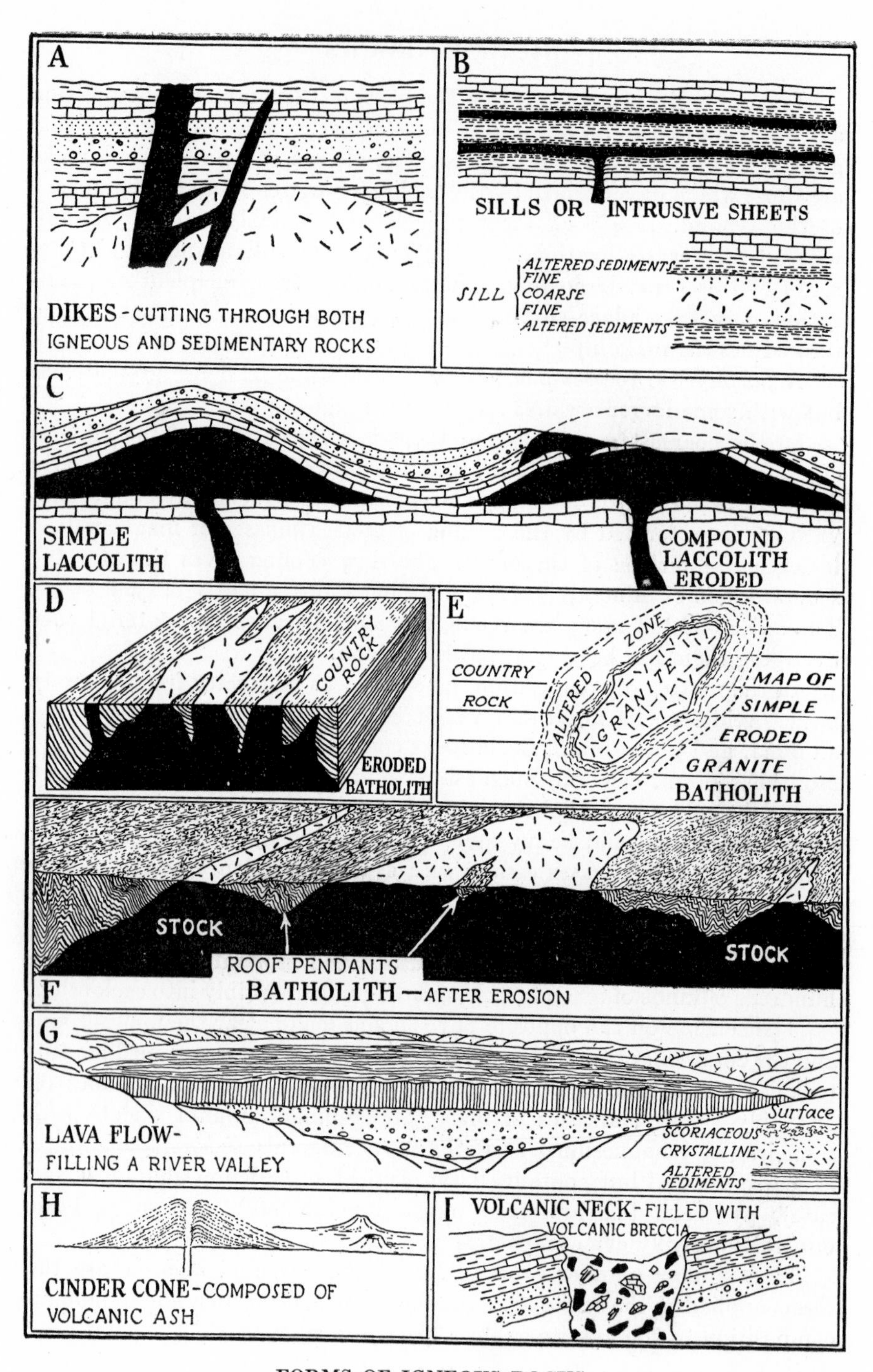


FORMS OF IGNEOUS ROCKS

## THE FORMS OF IGNEOUS ROCKS

Igneous rocks may be either intrusive or extrusive. Intrusive rocks have in a molten condition worked their way upward into overlying and surrounding formations. The nature of the rocks into which an intrusion is made often determines the shape of the intrusion: A *dike* is simply a wall-like or "tabular" mass which has entered a crack in the rocks and there cooled. Many dikes are more resistant than the rocks into which they have been intruded. When erosion occurs, they may persist as walls or ridges while the surrounding rocks weather away. Or they may decay more rapidly and produce a trench or ditch in the topography.

Dikes vary in thickness from an inch or two up to many feet and usually consist of the darker colored, more basic rocks. There is every gradation, too, from simple dikes to irregular intrusions of all shapes and sizes running off to small veins which ramify through the country rock of the region.

When the magma forces its way between the layers of stratified rocks, it forms an intrusive *sheet* or *sill.* These may vary from a few inches up to several hundred feet in thickness. If the intrusion is more or less lens shaped and bows up or raises the overlying strata, it is called a *laccolith*, meaning a lake of rock. The term *batholith*, a deep rock, has reference to the great intrusive masses, usually granites, now visible because of erosion. Some of the large mountain areas of the world reveal granite cores of this type. The terms *stocks* and *bosses* are applied to masses somewhat smaller than batholiths. Such forms produce rounded hills and knobs in the topography, if the granite of which they are composed proves to be more resistant than the rock into which it was intruded.

Extrusive igneous rocks are those deposited on the surface of the ground. When molten rock or lava flows over the surface of the ground, it constitutes a *lava flow* or an *extrusive sheet.* Such sheets may accumulate to a thickness of thousands of feet, sometimes alternating with sands and clays deposited during intermediate periods. Flows of this type result from quiet eruptions. The rocks thus formed are usually dark and basic in composition, and in the molten form much more fluid in character than the explosive type.

Upon reaching the surface in the crater of a volcano, the gas contained in the magma escapes with great violence and the rock is blown to fragments. Thus deposits of so-called *unconsolidated volcanic ash* are formed. If very coarse and consolidated, it is called *volcanic breccia;* or if extremely fine and consolidated, it is known as *tuff*. Such material, deposited in water, may form layers which are better classed with sedimentary than with igneous rocks. Some rocks of this type contain not only the fragments of the torn magma but also pieces of the rocks through which the eruption has come.

VARIOUS TYPES OF JOINTING IN IGNEOUS ROCKS

## STRUCTURES OF IGNEOUS ROCKS

In addition to the larger forms assumed by masses of igneous rock, there are also the smaller structures which characterize them and may affect the features produced by erosion.

Inasmuch as igneous rocks lack the regular arrangement which sedimentary formations possess, the effect of pressure and disturbance is revealed, not by folds, but almost entirely by faults and joints.

Joints are more abundant near the surface than deep below, since near the surface rocks have opportunity to expand under the forces of weathering. Here joints tend to be widened and become conspicuous, even though they continue below as tight and less apparent structures.

The forces which produce jointing in igneous rocks include contraction due to cooling of an igneous mass; expansion and contraction due to weathering; pressure as a result of other intrusions; the relief of pressure when overlying material is removed; as well as the larger movements of the earth's crust due to isostatic adjustment, whereby mountains are formed and continents raised and lowered.

Joints usually occur in parallel groups or sets. Very commonly different sets of joints intersect at right angles, or nearly so. Joints are usually flat or almost plane surfaces but in great homogeneous igneous masses they may be curved or warped. Most joints are vertical in attitude, especially those due to earth movements, but many are inclined and horizontal.

Joints in igneous rocks which are due to contraction on cooling are perpendicular to the cooling surface. In lava flows and sheets, such joints are vertical and account for the columnar or palisade structure characteristic of basaltic formations such as the Palisades and the Giant's Causeway. However, the jointing in dikes, which is also at right angles to the cooling surface. may constitute a horizontal set of cracks when the dikes have a vertical dip.

Sometimes joints become filled with mineral matter which may later resist weathering better than the rock mass itself. Such veins or "dikelets" will appear then as ribs or ridges on the weathered rock surfaces. Most dikes are more resistant than the rocks into which they are intruded and consequently form walls, ridges, or irregular hills when subjected to erosion. Some dikes, especially those containing much olivine, weather easily and this produces furrows or depressions. Along the coast, the etching out of the weaker dikes may commonly be observed.

Faults, like joints, are planes of weakness in rocks. Along such planes erosion is likely to be concentrated. Valleys which follow fault lines are called *fault-line valleys*. They are usually straight for considerable distances and receive their tributaries often at right angles because of the intersecting joint pattern. In fact, a strongly developed right-angled drainage pattern in igneous rocks is usually ascribed to a well-developed joint or fault system.

## SEDIMENTARY ROCKS

It is probable that, taking the land surface of the world as a whole, a greater area is made up of sedimentary rocks than of igneous and metamorphic rocks combined. This indicates that very large parts of the continental areas have at one time been under the sea. However, there are extensive sedimentary formations which were not formed by deposition under water. For example, broad alluvial plains, stretching sometimes for a hundred or more miles, have been laid down at the foot of mountains in some places to thicknesses of thousands of feet.

Sedimentary rocks may be consolidated or loose, depending upon whether the different grains are held together by any binding material, and they may grade from very coarse to very fine. Four different types will be described, but they are not sharply delimited from each other and many formations are known which stand on the borderline between the different groups.

GRAVEL AND CONGLOMERATE. A gravel formation is made up of pebbles and boulders with more or less sand, and perhaps small quantities of clay. The pebbles and boulders, as well as the sand, may consist largely of quartz, or of resistant rock which has withstood much wear and rounding by stream action. Weaker rocks under such conditions would have broken up and disappeared. Although gravel occurs in beds or layers, such deposits are apt to lack uniformity, both in thickness and in the character of the material. That is, gravel formations may vary greatly in thickness or be lens shaped. They also may contain beds of sand, which replace the gravel layers at different levels. When gravel is firmly cemented by quartz which binds the pebbles and sand strongly together, it becomes a highly resistant rock, known as *conglomerate.* Indeed, there are few rocks which are more resistant than conglomerate if the matrix of sand, the included pebbles and boulders, as well as the cementing material, are all composed of highly resistant silica. Even loose gravel may withstand erosion in an unexpected manner owing to its porous nature, which permits water to run through it rather than to flow over its surface and thereby wear it down.

SAND AND SANDSTONE. A formation of sand consists of separate grains about the size of those in granulated sugar. Usually these grains are composed of quartz, though in some regions, as on coral beaches, the sand may be calcareous, that is, of lime derived from fragments of corals and shells. Sandstones are produced when sands are cemented by some binding material. If the binding cement is oxide of iron, that is, limonite, the resulting sandstone is red or brown, this being very common. The white, buff, and gray sandstones have a binding matrix of calcium carbonate, and sometimes of silica. This last sandstone is highly resistant to weathering and erosion.

On the other hand, some sandstones are very weakly cemented so that upon weathering they crumble readily into their component grains.

Formations of this character often account for lowlands. Even firmly cemented sandstones are quite porous, so it is these formations which constitute the greatest reservoirs of artesian water.

Silt, Clay, and Shale. Silt and clay represent very fine materials not yet consolidated into rock. Moist clay becomes very compact indeed and may form steep vertical cliffs which maintain their position almost like a solid rock. The impervious nature of silt and clay usually means, however, that rills of water running over such formations are concentrated in surface flow, thus obtaining a high degree of erosive ability. For that reason clay formations frequently display badland topography.

Another result of the imperviousness of both shale and clay is that ground water, seeping down through overlying rocks, is arrested when it reaches a clay or shale horizon and moves outward at that level until it can escape in springs along valley walls and on the slopes of cliffs.

Shales result from the induration or consolidation of clay and silt. Shale usually breaks up into thin laminae which are due to the slight original differences in character of the accumulating material at different periods of deposition. Because of this, shale is apt to succumb rapidly to weathering and erosion.

Marl and Limestone. Unconsolidated calcareous deposits constitute marl or limy mud. When indurated, these become limestone or dolomite. Limestone is composed of calcium carbonate, and because of its solubility it gives way readily to weathering. Indeed, a pure limestone is one of the rocks least able to withstand weathering in regions of relatively abundant moisture. As a result of surface weathering, only the calcium carbonate is leached out. This leaves a reddish clay soil, often with nodules of chert, a form of silica, concentrated from the small amounts which were originally present in the limestone. Where the climate is prevailingly arid, limestone may show a marked resistance to destruction by weathering. It may be superior even to igneous rocks under such conditions. This is due to the fact that limestone is more or less homogeneous in its make-up, and changes of temperature do not cause differential expansion and contraction of the various crystals such as occurs with granite when its surface is alternately exposed to the rays of the sun and to freezing temperatures. Even a much jointed limestone in arid districts does not show the tendency to break up into blocks along the joints to the extent that it does in humid regions where penetrating water may turn to ice and pry apart the rock masses. Dolomite is magnesium calcium carbonate and, being much less soluble than limestone, occasionally makes great cliffs and mountains.

Chalk is a soft, porous variety of limestone. Much of the water which falls upon its surface immediately penetrates into the rock mass and there is consequently little erosion of the chalk surface. Some chalk uplands, like the Downs of England, are notoriously devoid of water and are used as grazing lands for sheep.

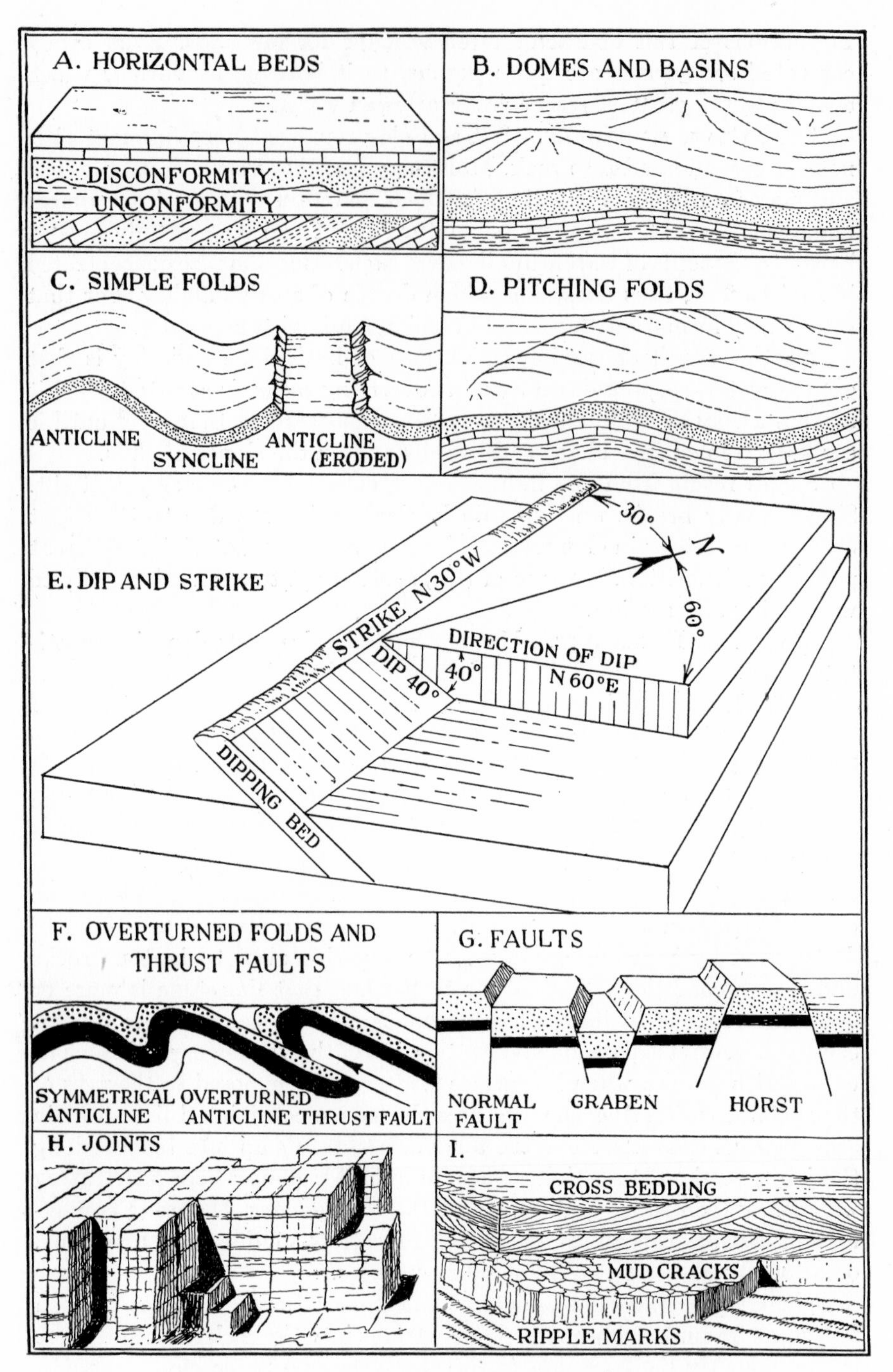


THE STRUCTURES OF SEDIMENTARY ROCKS

## THE STRUCTURES OF SEDIMENTARY ROCKS

FOLDS. Sedimentary rocks rarely occur in a strictly horizontal position. Slight warping of the region causes the development of domes or basins which may vary from a few miles to a length along the major *axis* of one hundred or more miles. Occasionally, however, pressure may cause regular folding, producing parallel undulations of alternating arches and troughs. The arches are termed *anticlines*, and the depressions *synclines*. Such folds, however, rarely continue indefinitely but taper off at the ends (Fig. *D*). The anticlines resemble cigars. They pitch away at the two ends. The synclines resemble canoes and at the ends pitch in toward the center of the syncline. Eroded rock layers presenting their edges at the surface of the ground are said to *strike* across the country, the strike being the direction or trend which a rock layer takes across a horizontal stretch of country (Fig. *E*). Such a layer also has a *dip*. The dip is the number of degrees the bed slopes down, away from the horizontal. This angle must always be measured at right angles to the strike, to give the maximum angle of slope. Occasionally the folding shows the strong effect of pressure, making the folds overturned or asymmetrical (Fig. *F*). This profoundly affects the forms later produced by erosion.

FAULTS. Extreme pressure may completely break the fold, producing a *thrust fault*. The plane of such a fault is usually not far from horizontal, but the plane of a *normal fault* is apt to be more nearly vertical. A normal fault occasionally passes gradually into a *monoclinal fold*. If a block of the earth's crust is dropped down between two normal faults, such a feature is termed a *graben*. If a block is raised between two faults, such an elevated mass is called a *horst*.

JOINTS. All sedimentary rocks are characterized by jointing. The joints or fractures are usually at right angles to the bedding planes and, in the case of more massive rocks like sandstones and shales, are important in determining the angular form of cliffs produced in such formations.

BEDDING STRUCTURES. Frequently tilted sedimentary formations are overlain by horizontal ones, as if the first series after tilting had been worn down to a level plane and covered by later sediments. A relationship of this kind between two series of beds constitutes an *unconformity* (Fig. *A*). Some unconformities reveal a great angular difference in dip between the rocks above and those below. In other cases, the two beds may be almost parallel and then, from the physiographic viewpoint, the unconformity becomes inconsequential. This is usually termed a *disconformity*.

Among the minor structures of sedimentary rocks are cross-bedding, ripple marks, mud cracks, and concretions. These are not of direct physiographic significance but, since the larger physiographic principles often turn back to the smaller geological details for their understanding, these phenomena may be of great, although indirect, physiographic value.

*Keith, U. S. Geological Survey*

STRONGLY FOLDED DOLOMITIC LIMESTONE WITH ALTERNATING SCHISTOSE AND SLATY BANDS; RUTLAND, VT.

An example of beds only slightly metamorphosed.



WALL OF MARBLE QUARRY, RUTLAND, VT.

Showing strongly folded and fractured bed of dolomite in white and banded marble.

*Keith, U. S. Geological Survey*

## METAMORPHIC ROCKS

Metamorphic rocks are those that have resulted from a change in the form and character of preexisting rocks that originally were quite different. Igneous and sedimentary rocks may, by various means or geological processes, be changed into metamorphic rocks. If the metamorphism has been very great, it may be difficult if not impossible to determine what the original rocks were like, or even whether they were igneous or sedimentary. The various types of metamorphic rocks may therefore be considered in regard to their special characteristics without much concern for the precise manner of origin.

*Schist* is one of the most common types of metamorphic rock. It usually contains large quantities of flat platy minerals, like mica, which have their cleavage planes all lying in the same direction. It is due to this fact that schist is so extraordinarily fissile. That is to say, schist cleaves readily along the planes of schistosity. Schists also contain considerable amounts of quartz, which makes them highly resistant to weathering. Some schists contain talc and chlorite, and occasionally hornblende is the most important mineral in the rock. These varieties of schist are apt to be much weaker when exposed to weathering than are the quartz schists. Most schists show such a distinct bedded character as to suggest very strongly that they originated from sedimentary rocks. It seems probable that when shales are subjected to great pressure, with perhaps heating and possible mashing and folding, the particles of clay and other mineral matter present become transformed into mica and other flat minerals, thus producing a schist.

Topographically, a region underlain by schist shows a rough tendency for the ridges and valleys to run more or less parallel. The inequalities in the schist weather out so as to produce a "grain" in the country. Often these irregularities are only a few feet across, but in some places the ridges are a good fraction of a mile apart.

*Gneiss* is a rock which resembles schist in that it is foliated; that is, it contains mica and other flat minerals lying more or less parallel. But it is usually a much coarser rock and may be described as having a banded appearance. This is especially striking when wide layers of white granular siliceous rock alternate with dark schistose bands. Gneiss, in general, is a very resistant rock, surpassing schist in its ability to withstand weathering. Topographically, the forms produced upon gneiss are more rounded in outline and do not have the corrugated aspect which is characteristic of a region underlain by schist.

*Quartzite* is a rock composed of quartz grains firmly cemented together with quartz. This condition may be produced in various ways, but the result in every case is a rock highly resistant to weathering and erosion. Quartzite is probably the most resistant of all rocks. Under weathering it usually breaks up along the joint planes into huge rectangular blocks which form great talus masses below almost vertical cliffs. The hill

*Keith, U. S. Geological Survey*

SLATY CLEAVAGE IN STRONGLY FOLDED SYNCLINE, BLOUNT COUNTY, TENN.

The cleavage planes are parallel with the axial plane of the fold and diagonal to the bedding.

summits are usually well rounded, presenting very little in the way of small details in the topography.

*Slate* is one of the most peculiar of the metamorphic rocks in that the slaty layers break apart, not along the bedding planes of the rock, as shales do, but along planes of cleavage which have been produced in the rock by pressure. These planes of cleavage may lie at any angle to the bedding planes. They lie, however, at right angles to the direction of pressure which produced them. Thus, in the illustration on the opposite page, the beds of slate exhibit clearly the slaty cleavage, crossing the bedding planes at different angles in different parts of the fold, but everywhere parallel to themselves and transverse to the direction of the pressure which produced the fold.

Slates, in general, are not so resistant as schist and gneiss, but they usually are somewhat more resistant than limestone or even marble. In fact, the topography produced upon slate rarely has any striking characteristics, being usually subdued and rolling and quite like that developed upon shales. When slates are associated with limestone, that is, if limestone beds have been folded into the series of slates, erosion will etch out the limestone to produce linear valleys bordered by hills of slate. If sandstones are included in the series, these will stand up as ridges or knobs above the more subdued slate topography.

Some slates have undergone so much pressure as to have developed a certain amount of mica and thus grade into schists. These are called *phyllites*.

Practically all sedimentary formations which have undergone folding exhibit slaty cleavage in the shaly beds. The more resistant sandstone layers, however, fracture more nearly at right angles to the bedding planes, regardless of the direction of pressure. In other words, the shaly beds act like cushions to take up the movement between the more massive layers, as the rocks are folded. The more pronounced and closer the folding, the more nearly does the slaty cleavage come to lie parallel to the bedding planes.

*Marble* results from the metamorphism of limestone by pressure and heat. This causes the calcium carbonate of the limestone to crystallize into grains of calcite. In this way, coarsely crystalline limestones may be produced. These disintegrate readily into a sugary mass which is easily removed by erosion. True marble, however, is more compact, the grains being better knitted together. Marble is usually massive and shows no cleavage such as slate does, but it is sometimes broken into large blocks by joints. Owing to its solubility, marble is a rock which usually succumbs readily to weathering in humid regions. But in arid and semiarid regions it may endure the agencies of destruction even better than the usually more resistant igneous and sedimentary rocks.

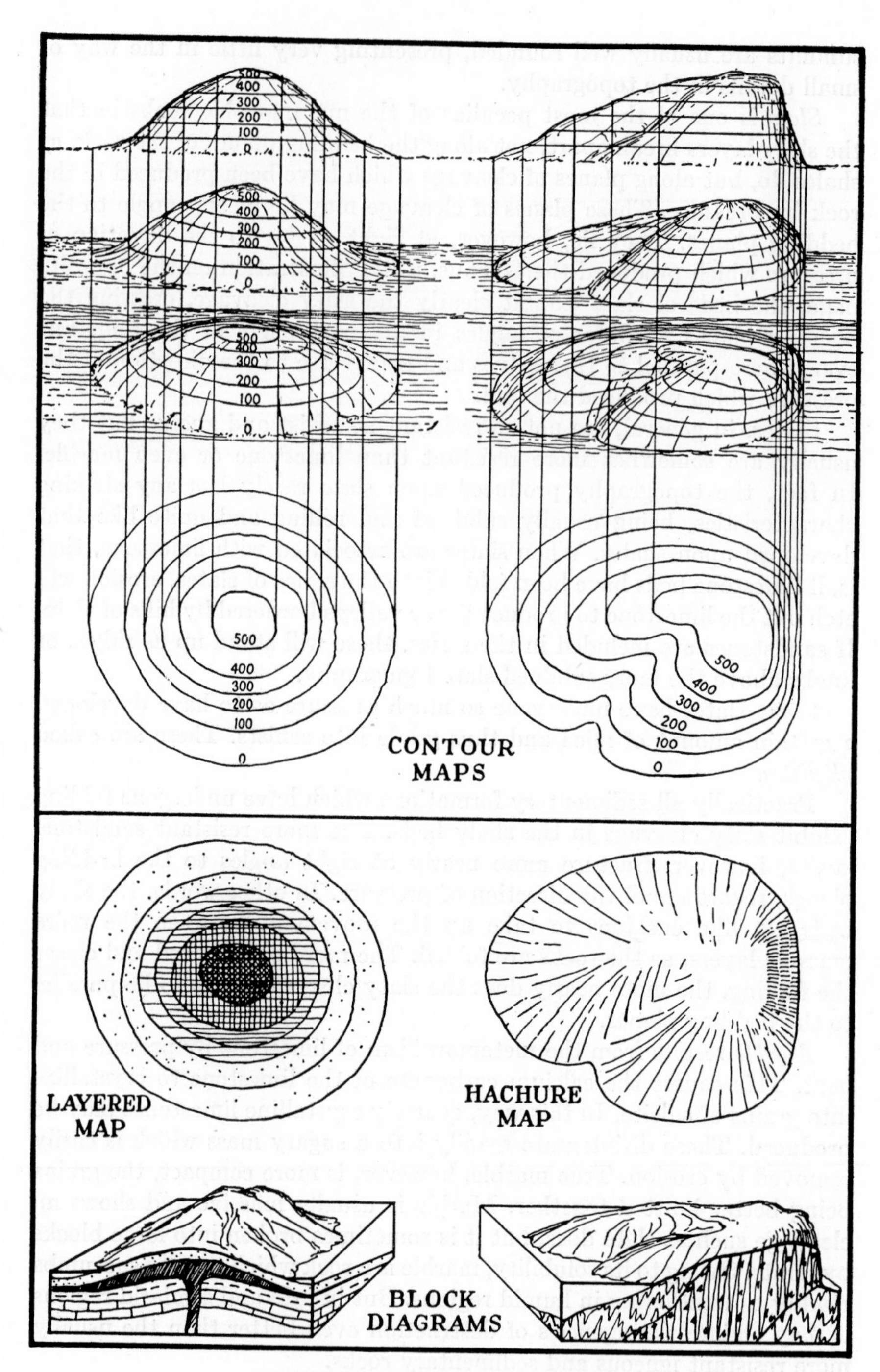


METHODS OF SHOWING RELIEF

Relief Maps; Contour Maps. The relief, that is to say, the form of the surface of the earth, or the topography, is shown on maps in several ways. The most widely used and the most satisfactory method is by contours. A *contour* is a line which everywhere represents the same distance above sea level. The zero contour line is the seashore. The 10-foot contour line is a line which would be the seashore if the sea were to rise 10 feet. The 100-foot contour is the line which would be the seashore were the sea to rise 100 feet. It is customary to space the contours so that the successive lines represent equal differences in elevation. Figures opposite show two forms of islands, each one as seen from four different points of view. The first three views in each case are simply pictures as the island might appear to a person approaching it in an airplane. The fourth view in each case is directly from above. This constitutes a map.

A few of the characteristics of contour lines should be observed:

*a.* Where a slope is uniform, the contours representing that slope are equidistant from each other.

*b.* Where a slope is irregular, that is, in some places steep and in some places gentle, the contour lines are unevenly spaced, in some places close together, and in other places far apart.

*c.* Where a slope is gentle, the contours are far apart.

*d.* Where a slope is steep, the contours are close together.

*e.* A truly vertical slope is represented by contours crowded so close together as to appear as a single line.

*f.* Contours swing out around the ends of spurs and promontories.

*g.* Contours swing up valleys and ravines.

*h.* If the relief of a region is very gentle, a small contour interval is used. For example, a 5-foot interval is used on the Mississippi Delta maps. If great detail is desired, the interval may be as small as 1 foot.

*i.* If the relief of a region is very bold, a large contour interval is used. On the Grand Canyon maps, for example, the interval used is 100 feet or in some cases even 250 feet.

*j.* Likewise, if the scale of the map is small, the contour interval is large; and if the scale of the map is large, the contour interval is small.

Hachure Maps. Hachure maps represent the relief by means of hachures, or small straight lines which run directly down the slopes and therefore transverse to the direction of the contour lines. Where the slopes are gentle, the hachures are made thin and far apart. Where the slopes are steep, the hachures are made thick and near together. The use of hachures alone does not permit portraying the relief as accurately as does the use of contours, but hachures perhaps convey to the average person a little better idea of what the country looks like. Sometimes both hachures and contours are used on the same map.

Layering. A third way of showing relief is by the so-called *layered* map. This is simply a contour map in which the spaces between the contours are shaded or colored to show the different degrees of elevation. For wall maps and small-scale maps this serves a useful purpose but nothing is gained over the use of contours alone, except greater legibility.

Block Diagrams. Block diagrams are pictures of the topography of a region drawn as if a block were cut out of the earth's crust.

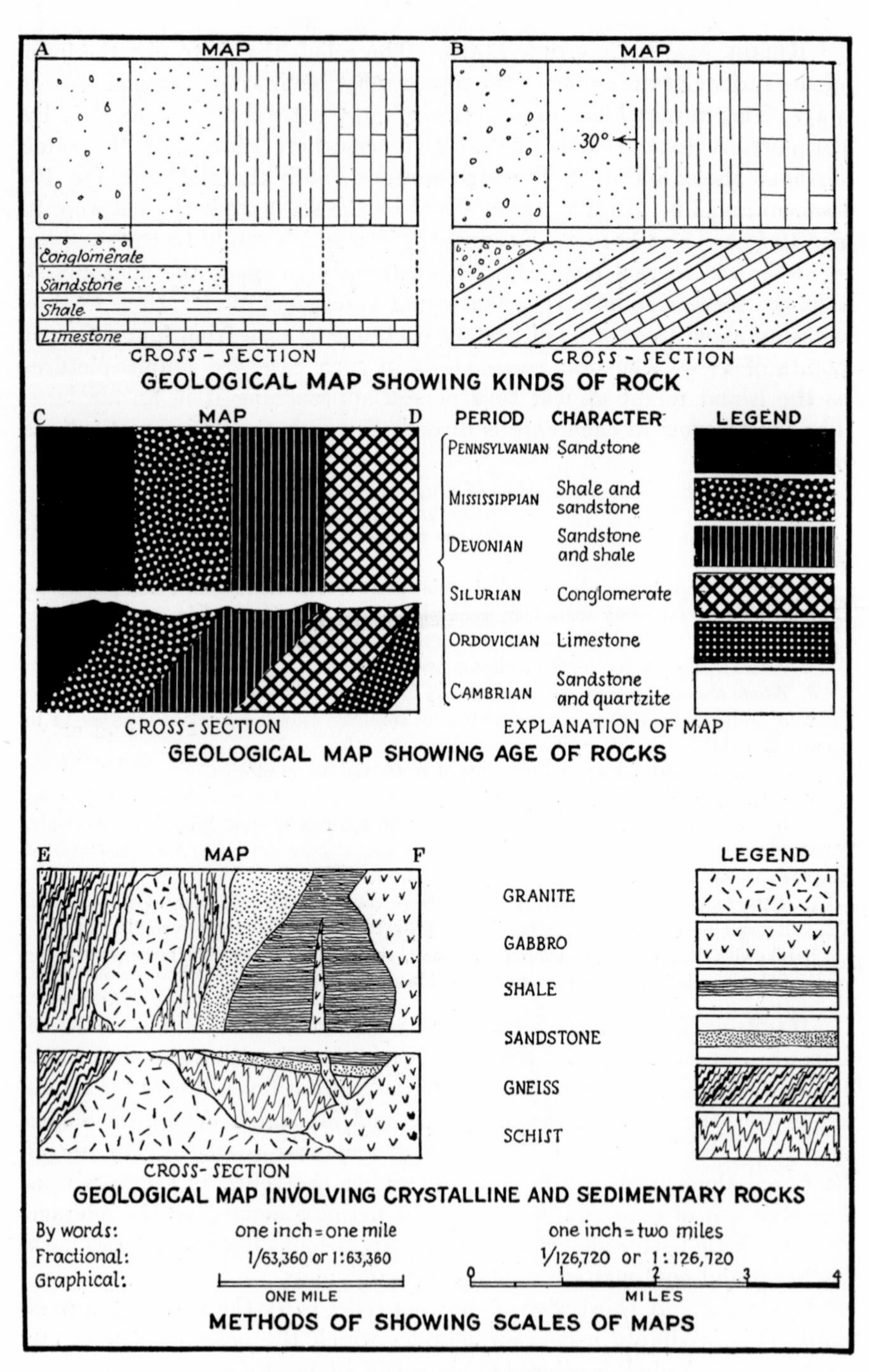


SYMBOLS USED ON GEOLOGICAL MAPS

Geological Maps. Geological maps show geological formations on the surface of the ground. Geological cross sections show the structure beneath the ground. The geology may be shown in combination with the relief, and in color or in black and white. The symbols used on geological maps may represent the kind of rocks, as, for example, sandstone, shale, or limestone (Fig. *A*); or the symbols may represent only the age of the formations, such as Triassic, Jurassic, Cretaceous, without any suggestion as to what the rocks actually are (Fig. *C*).

Even without geological cross sections a person familiar with geological maps can often interpret the structure, especially if there are symbols which indicate which way the rocks dip and strike as in *B*. But the size and shape of outcrops on maps *A* and *B* are exactly the same though the structure is different, and the correct interpretation of the structure is possible only by knowing what the topography or what the dip of the rock is.

Geological Legends. Attached to geological maps is a legend which shows the correct sequence of formations. The oldest formation is shown at the bottom, the youngest at the top. This constitutes a geological column. Any symbols and any colors may be used to represent rocks of different geological periods. Nevertheless, certain standard colors have been adopted by the U. S. Geological Survey and other national surveys.

Geological sections may be introduced anywhere on a map and may cross it in any direction so as best to show important facts concerning the structure. When specific types of rocks are indicated, it is customary to use certain approved symbols for the different types, as follows:

Sandstone is shown by dots (Fig. *A*); conglomerate by dots and circles; shale by horizontal dashes; and limestone by a brick-wall pattern. Igneous rocks are shown by a pattern of dashes, crosses, or V-shaped marks; and schist is represented by a characteristic wavy design. In short, the pattern used is supposed to suggest the character of the rock.

Other symbols used on maps are in general self-explanatory. When colors are used, the relief (*i.e.*, contours or hachures) is usually shown in brown; rivers, lakes and other water bodies in blue; and the culture (*i.e.*, roads, railroads, boundary lines, and all names) in black. Some of the foreign maps represent everything in black and when there is much cultural detail, such a map is very difficult to read.

Scale of Map. The scale of a map may be indicated in several ways: first, simply by words, as, for instance, "1 inch to a mile," which means 1 inch on the map is equivalent to 1 mile on the ground; "½ inch to a mile" would be a smaller scale and "2 inches to a mile" would be a larger scale. Second, the scale may be in fractional or proportional form, as 1:63,360. This means that one unit on the map is equivalent to 63,360 units on the ground. This is the same as 1 inch to a mile. A scale of ½ inch to a mile would be 1:126,720. Likewise a scale of 2 inches to a mile would be represented 1:31,680.

Third, the scale may be shown in a graphic manner by means of a little scale or ruler upon which are marked the correct distances. Detailed maps may be drawn on a scale as large as 1:1,000; and general maps may use 1:1,000,000 or even smaller scales.

## QUESTIONS

1. What is the difference between a mineral and a rock?
2. If a volcano ejects great quantities of fine dust into the air and this settles down over the landscape or falls into the sea and later becomes a solid rock by the cementation of the particles, is the result an igneous or a sedimentary rock?
3. Should coal be considered a rock, and, if so, what kind of a rock is it?
4. Is ice a rock? And, if so, is water a rock? Is cement a rock, and, if so, what type?
5. If a series of shales with interbedded coal seams is greatly deformed and altered by pressure, what does the shale become? What does the coal become?
6. Draw a single geological section, using appropriate symbols which will contain the following: layers of sandstone, limestone, shale, conglomerate; an unconformity, disconformity, anticline, syncline; batholith of granite, dike, sheet, laccolith, normal fault, thrust fault.
7. Why do some dikes during the erosion of a region become trenches or troughs instead of ridges or walls?
8. Draw a contour map of a hill 500 feet high above its base. Use any scale, but use a contour interval of 25 feet. Show two valleys on sides of a hill.
9. Draw a contour map of a perfect pyramid (like one of the Egyptian pyramids) 1 mile square at its base. It is 400 feet high. Use a contour interval of 100 feet and a scale 1 inch = 1 mile.
10. Draw a contour map of a conical hill 400 feet high with a contour interval of 100 feet. The hill is 1 mile in diameter at its base and the scale of the map is 1:63,360.
11. A map has a scale of 1:1,000,000. How many miles to an inch is this, approximately?
12. The scale of a certain map is 1:63,360. The contour interval is 20 feet. If successive contours on this map are ¼ inch apart, what is the slope represented, in feet per mile?
13. Exercises on actual topographic maps. Acquaintance with the topographic sheets of the U. S. Geological Survey is of paramount importance to the student of geomorphology. The following types of exercises are suggested:

    *a.* Draw profiles to natural scale from one given point to another.

    *b.* Draw profile with vertical exaggeration of 5 ×; of 10 ×.

    *c.* Draw projected profiles. This widely used method for revealing remnants of peneplanes is illustrated by the following paper: Lobeck, A. K. *Position of New England peneplane in the White Mtns.*, Geog. Review, vol. 3, 1917.

    See also Rich, J. L. (1938) *Recognition and significance of multiple erosion surfaces.* Geol. Soc. Am., Bull. 49, p. 1695–1722, for a criticism of this method.

    *d.* Select a map, preferably on a scale of 1:62,500, with a contour interval of 20 feet. Then, using the 100-foot contours only, tint in the spaces between the contours so as to make a layered map. For this purpose various tints of green (for the low levels) and brown (for the high levels) are better than many different colors.

    *e.* Select a part of the area with fairly rugged relief and change the map into a hachure map, either by tracing or by drawing the hachures in black over the contours.

    *f.* Determine the visibility of places with relation to each other. A map of mountainous country is best for this.

    *g.* A vast number of exercises of a geological nature is possible. Some of these are suggested in *Exercises in topographical and structural geology,* The Geographical Press, New York.

    *h. The interpretation of topographic maps* is to the geomorphologist one of the most important fields of study. This differs from the mere "reading" of a topographic map. Boy scouts and novices read maps by noting position of places, distances, gradients, etc. But the geologist and geographer reads between the lines, as it were, and sees more than even the man who made the map intended. He interprets the structure from the topography. He determines the dip and character of the rocks in different places. From the surface conditions he sees beneath the surface. Many maps will be cited in the following pages for use in this way.

14. Which of the pictures at the beginning of this and of other chapters portray igneous rocks? Sedimentary rocks? Metamorphic rocks?
15. Prepare a graphic scale for all the maps in this book.

## TOPICS FOR INVESTIGATION

1. The recognition of rocks by sight. Criteria used.
2. The recognition of minerals by sight. Criteria used.
3. Igneous rocks. Principles of classification.
4. Sedimentary rocks. Principles of classification.
5. Metamorphic rocks. Principles of classification.
6. Topic for debate: Resolved that sedimentary rocks are more important than igneous rocks to human welfare.
7. Geologic structures. Major structures; minor structures.
8. Geologic maps. Kinds, symbols used, colors, scale, cross sections.
9. Topographic maps. Scales, relief, color, projection. The topographic maps of various countries.

## REFERENCES

### Rocks

DALY, R. A. (1933) *Igneous rocks and the depths of the earth.* New York, 598 p. An advanced and thorough study of the theoretical aspects of the origin and relationships of igneous rocks, written in a solid scientific manner.

GROUT, F. F. (1932) *Petrography and petrology.* New York, 522 p.; 24 p. of exceedingly well-classified bibliography. Drawings and photographs are superior to those found in most texts. Useful discussions of problems in petrography. Excellent suggestions to the student.

HARKER, A. (1909) *The natural history of igneous rocks.* New York, 384 p., 112 illustrations, and 2 plates. Once an authoritative study of the subject but has since (1933) been superseded by Daly's work on the same topic.

HARKER, A. (1932) *Metamorphism.* London, 360 p. A profusely illustrated and well-organized up-to-date study of metamorphism of rocks as revealed by extensive microscopic investigations. Differs from Leith and Mead's book in utilizing a newer classification of metamorphic processes.

HARKER, A. (1935) *Petrology for students.* Cambridge, England, 300 p. Written for use with the microscope, contains much valuable information for the lithologist.

HOLMES, A. (1928) *The nomenclature of petrology.* London, 284 p. A book of definitions of rock terms, with brief French and German glossaries.

IDDINGS, J. P. (1903) *Chemical composition of igneous rocks expressed by means of diagrams, with reference to rock classification on a quantitative chemico-mineralogic basis.* U. S. Geol. Surv., Prof. Paper 18, 98 p.

IDDINGS, J. P. (1909–13) *Igneous rocks.* New York, vol. 1, 454 p., takes up the theory of formation, composition, and classification; vol. 2, 685 p., is devoted to descriptions and localities.

KEMP, J. F. (1929) *A handbook of rocks.* New York, 6th ed., 300 p. Very complete glossary of petrographic terms, 114 p. Very condensed classification and descriptions of more common rocks, with stress on chemical variations.

LEITH, C. K., and MEAD, W. J. (1915) *Metamorphic geology.* New York, 337 p. A good book, written principally for the graduate student. Discusses katamorphism and anamorphism as each affects the different rock species.

LOOMIS, F. B. (1931) *Field book of common rocks and minerals.* New York, 6th ed., 278 p. An elementary book with numerous beautiful illustrations, designed for use primarily by the layman.

MERRILL, G. P. (1906) *Rocks, rock-weathering, and soils.* New York, 2d ed., 400 p. A brief review of the nature of the various rock species, followed by a lengthy discussion of the many agents of destruction and their effects upon rocks.
MILNER, H. B. (1929) *Sedimentary petrography.* London, 2d ed., 514 p. A highly practical and useful book on the sedimentary rocks, profusely illustrated with excellent photomicrographs and drawings. Also several handy tables for recognition of minerals under the microscope.
SHAND, S. J. (1927) *Eruptive rocks.* London, 360 p. Many references. An excellent survey of the igneous rocks, largely from a quantitative compositional standpoint, in order to bring petrology to the fore as a practical science.
SHAND, S. J. (1931) *The study of rocks.* New York, 224 p. An orderly brief review of the broader classification of rocks, with helpful hints on collection, study, and chemical analysis. Also a comparison of the leading classifications of rocks.
TWENHOFEL, W. H. (1932) *Treatise on sedimentation.* Baltimore, 2d ed., 926 p. The most comprehensive work of its kind in the world, citing approximately 1,500 papers.
TYRRELL, G. W. (1929) *Principles of petrology.* New York, 2d ed., 349 p. Acknowledged by graduate students as the most comprehensive and useful of all available books on the subject.
VAN HISE, C. R. (1904) *A treatise on metamorphism.* U. S. Geol. Surv., Mon. 47, 1286 p.
WASHINGTON, H. S. (1917) *Chemical analyses of igneous rocks published from* 1884 *to* 1913, *inclusive, with a critical discussion of the character and use of analyses.* U. S. Geol. Surv., Prof. Paper 99, 1201 p. A description of the quantitative classification of igneous rocks, with tables for the calculation of the norm.

STRUCTURES

BALK, R. (1937) *Structural behavior of igneous rocks.* Geol. Soc. Am., Mem. 5, 177 p.
BUCHER, W. H. (1933) *The deformation of the earth's crust; an inductive approach to the problems of diastrophism.* Princeton, 518 p.
LEITH, C. K. (1923) *Structural geology.* New York, 390 p.
NEVIN, C. M. (1936) *Principles of structural geology.* New York, 348 p.
STOČES, B., and WHITE, C. H. (1935) *Structural geology.* London, 460 p. Splendidly illustrated.
WILLIS, B., and WILLIS, R. (1934) *Geologic structures.* New York, 544 p.

MAPS

ATWOOD, W. W., and SALISBURY, R. D. (1908) *Interpretation of topographic maps.* U. S. Geol. Surv., Prof. Paper 60, 84 p.
DAKE, C. L., and BROWN, J. S. (1925) *Interpretation of topographic and geologic maps, . . .* New York, 355 p.
LOBECK, A. K. (1924) *Block diagrams.* New York, 206 p.
LOBECK, A. K. (1932) *Atlas of American geology.* New York, 100 p.
RAISZ, E. J. (1938) *General cartography.* New York, 370 p.

# III
# WEATHERING

*Chamber of Commerce, Colorado Springs, Colorado*

"BALANCED ROCK" IN THE GARDEN OF THE GODS, COLORADO

Remnant of an extensive sandstone formation which was removed largely by weathering and wind action.

*Edgar Orr, Atlanta*

STONE MOUNTAIN, GA.

A granite monadnock standing on the Piedmont upland. An exfoliation dome upon which the Confederate War Memorial is being carved in the center of the view.

EXFOLIATION DOME SURMOUNTING BLUE RIDGE UPLAND NEAR ASHEVILLE, N. C.
A strongly resistant mass of granite resembling the domes in Yosemite Park.

*Fisher, Asheville*

*Courtesy of William Allen, Hasbrouck Heights, N. J.*

INDIAN HEAD, FORT LEE, N. J.

Basaltic columns of the Palisades, photographed in 1886.

*Courtesy of William Allen, Hasbrouck Heights, N. J.*

INDIAN HEAD, FORT LEE, N. J., 1926

Showing the insignificant amount of weathering during forty years, except for the toppling off of two blocks.

*Turner, U. S. Geological Survey*

EXFOLIATION OF GRANITE, ALPINE COUNTY, CALIF.

The lack of vegetation and the high altitude are both conducive to rapid rock disintegration. Granite is relatively weak under these conditions.



GLACIAL ERRATIC ON PEDESTAL, SIERRA NEVADA, CALIF.

This large boulder has protected the underlying granite from disintegration while the surrounding area has been weathered down two or three feet.

*Matthes, U. S. Geological Survey*

*Cross, U. S. Geological Survey*

LANDSLIDE TOPOGRAPHY NEAR SILVERTON, SAN JUAN MOUNTAINS, COLO.

The slide contains large numbers of dead trees. Many young trees have developed since the slide ceased.



SNOWSLIDE TRACKS THROUGH HEAVY TIMBER, ROCKY MOUNTAINS, COLO.

A form of weathering closely resembling rock slides in its effects.

*U. S. Forest Service*

*Russell, U. S. Geological Survey*

DECOMPOSED TRAP ROCK IN THE TRIASSIC LOWLAND, N. J.

Showing gradation from soil to less weathered rock below. The simulation of boulders is due to weathering along the joints in the igneous rock. Its rich olivine content favors chemical weathering of this character.

## WEATHERING

SYNOPSIS. The surface of the earth is everywhere crumbling to pieces under the influence of rock decay. Hills are by no means everlasting, and even the most resistant rock hardly deserves to be called the rock of ages. Weathering prepares the rocks for removal by other agents. It operates over the entire earth's surface. The effect of streams, glaciers, and waves, the so-called *destructive forces*, however, is localized. The efficacy of the wind, another of these forces, is also restricted and may be negligible where heavy vegetation prevails.

Weathering includes a large variety of destructive processes: heating, cooling, freezing, thawing; the prying action of ice, plants, and animals in minute cracks; and other forms of mechanical disintegration. It includes also a multitude of chemical changes, such as oxidation which affects many metals, notably iron, to form various iron oxides, familiar as iron rust; hydration which affects feldspars to form kaolin or clay, and which also affects iron to form limonite; carbonation by which carbon dioxide affects potassium feldspars to form potassium carbonate or potash, an important soluble plant food. It includes the solution and leaching of rocks and soil and removes much lime and other carbonates from the rocks, as well as large amounts of silica. These chemical changes are aided by high temperature and organic acids, both plant and animal.

Mechanical disintegration, induced largely by temperature changes, is most pronounced in arid regions and on exposed mountain summits; chemical changes are greatest in regions of warm temperature and high humidity.

There is no sharp line between the work of weathering and the work of streams. The stream systems, resembling the veins of a leaf, are only those places where the work of removal is more concentrated than elsewhere, but it is one great process which covers the whole surface of the earth like the entire blade of a leaf.

Under the influence of weathering, slopes are said to be young when there are scarps, steep declivities, and many rock outcrops, and the removal of waste is rapid. Landslides and rock falls are common under such conditions. Mature slopes are said to be in equilibrium, which means that the slopes have been established by weathering rather than by erosive processes. Such slopes are subdued. The forces of decay and of removal are so nicely balanced that no sudden changes occur and at any given moment of time it would appear that no changes were taking place.

Young slopes are characteristic of young land forms; mature slopes of mature land forms. This does not mean, however, that mature regions may not have young slopes with cliffs and rock exposures. The stage of development of slopes, while interrelated with the land form as a whole, need not necessarily be identical with the stage of development of the land form.

Weathering, an Adjustment to the Environment. The term *weathering* is applied to the process of rock disintegration and decomposition. This results not only in the breakdown and destruction of rock masses, but also in the development of certain topographic forms peculiar to these processes. For instance, the great rounded granitic domes of the Yosemite region are the result of weathering rather than of erosion by any of the destructive erosional forces.

Weathering, as a matter of fact, is merely the readjustment of rocks to new environmental conditions. Rocks formed by the cooling of molten magmas deep within the earth are later exposed by erosion to conditions utterly different from those to which they have been accustomed. Granite, for example, can exist almost forever if left in the environment where it was formed, but, if brought to the surface of the earth, it starts to go to pieces at once. The reverse is also true: products resulting from weathering at the surface of the earth remain stable under the environment in which they were formed but, if subjected to the conditions existing in the interior of the earth where granite is at home, they change their character in response to the new environment, becoming schist or some other type of metamorphic rock better adapted to the conditions of great heat and pressure found at depth.

Every mass of rock pushed up by the faulting and folding of the earth's crust, exposed by denudation, or erupted as molten matter from the earth's interior, finds almost at once that its various elements, in their existing combinations, are not in harmony with the new environment. The summer's heat and winter's cold, the chemical action of the atmosphere, and acidulated rains combine their forces; a breaking up ensues, to be succeeded by new combinations and perhaps reconsolidations more in keeping with the then existing circumstances. An intermediate product in all this endless cycle of change, of disintegration and recombination, is the soil. It is a comparatively thin, superficial mantle of loose debris, which, mixed with more or less organic matter, nearly everywhere covers the land.

Factors Influencing Weathering. Weathering of rocks depends upon several factors. First, it depends upon the kind of rock, that is, its mineral composition and its structure. Second, it depends very much upon the climatic conditions, whether it is dry or humid, cold or hot, uniform or changing. Third, it depends upon the presence or absence of vegetation. Finally, there are such fortuitous conditions as slope of the land and exposure to sun and rain.

It is customary to think of the forces of weathering operating in either of two ways: *mechanically* or *chemically*. Mechanical changes are wrought by changes in the temperature of the air or of the water which fill the cracks and pores of the rock, and also by the action of plants and other organisms. These agents break up the rock mass into smaller blocks, pieces, or particles.

Chemical changes, induced by these same agencies, result in actual chemical reactions and usually involve the addition of such substances as oxygen or carbon dioxide to the elements of the original rock.

Weathering, therefore, is the work of the air, of the water, and of organisms covering the surface or occupying spaces within the rock mass, all tending to break down the rock by mechanical and chemical means. The term weathering does not embrace all changes brought about by atmosphere and organisms. If these forces produce changes by virtue of their movement, they become agents of erosion and are then classed with the destructive forces which change the face of the earth in a large way. The atmosphere then becomes the wind, the water becomes a stream, and the ice or snow becomes a glacier. In some instances it is difficult to draw a clear line of distinction between weathering and erosion. The constant effect of gravity always pulls the products of weathering down the slopes and, if the quantity thus set in motion at a given place is great, it may constitute an actual flow of debris, which, except for the relative amount of water and load, is not unlike a flowing stream. This movement is called *rock flowage*, *soil creep*, or *solifluction*.

RELATIVE SUSCEPTIBILITY OF IGNEOUS AND SEDIMENTARY ROCKS TO WEATHERING. In line with the facts just given, the following generalization is of especial physiographic significance:

Igneous rocks are much more susceptible to chemical changes than are sedimentary rocks. At the surface of the earth igneous rocks are out of harmony with their surroundings and are subjected to constant physical and chemical changes which result in products more in harmony with existing conditions, and hence more stable. Sedimentary rocks are themselves the actual products of these adjustments. The conglomerates, sandstones, shales, and argillites are but the detrital remains of eruptive rocks, which under the various weathering influences have become disintegrated and decomposed, their more soluble constituents quite removed, and the residues, namely, sand and clay, laid down and consolidated under conditions such as today exist upon or near the surface of the earth or under shallow bodies of water. The usual weakness of sedimentary rocks, under erosion, therefore, as compared with igneous rocks, is due not to greater liability to chemical change but to the natural tendency of the sedimentary bed to return to the particles of sand or clay of which it was originally made. This is due usually to the weakness or scarcity of the material which cements the grains together. Firmly indurated sediments, like conglomerates, sandstones, or quartzites, in which the separate grains are firmly held together by a matrix of silica, are notably resistant and always dominate the topography, in preference to igneous rocks. Limestones constitute the only sedimentary rocks which decompose readily by chemical change.

## MECHANICAL CHANGES

THE EFFECT OF TEMPERATURE CHANGES. Mechanical changes result from either the compression or the tearing apart of the rock. This may be brought about by changes in the temperature of the rock body itself, due to its exposed position, or it may be due to the formation of ice within the interstices of the rock. Very firm, compact, crystalline rocks, like fresh granite, have virtually no pore space and are not so ready a prey to the ice which forms between the grains as are sedimentary rocks. Sedimentary rocks, such as porous sandstone, are readily disintegrated by the formation of ice between the grains.

On the other hand, clastic rocks, those formed by the breaking up of igneous rocks, like sandstone and shale, do not succumb readily to changes of temperature in the rock itself. In such rocks the grains are frequently separated from each other by a thin layer of calcareous, ferruginous, or siliceous matter which acts as a cushion and permits a certain amount of movement between the grains. In crystalline rocks, however, the constituents are practically in contact and, as temperatures rise, each and every grain expands and crowds with almost irresistible force against its neighbor. As temperatures fall, a corresponding contraction takes place. A rise in temperature of 150°F. produces in granite a lateral expansion of 1 inch for every 100 feet of distance. This expansion lessens the cohesion of the crystals and tears the upper portion from that lying more deeply. In massive, close-grained rocks disintegration from this cause is most pronounced. Andesitic and basaltic rocks and glassy obsidian flake off in small chips, having beautiful concave and convex surfaces. The surface left by the springing off of these flakes is fluted as though done with a carpenter's gouge. Granite, however, being of coarse texture, does not spall off in this fashion but disintegrates grain by grain. The different crystals of feldspar, quartz, and other minerals expand and contract at different rates and so are torn apart. The resulting soil made up of the fresh crystal grains is frequently called a *natural gravel*. The road up Pikes Peak and many other Colorado highways are surfaced with this material. Sedimentary rocks are far less liable to disintegration in this manner because of their uniform mineral make-up.

THE EXPANSIVE FORCE OF ICE. The effect of dry heat and cold upon rocks, however, is slight in comparison with the destructive action of freezing temperatures upon rocks saturated with water. The expansive force of water—no matter what the amount, as the pressure of water is constant—passing from the liquid to the solid state has been graphically described as equal to the weight of a column of ice a mile high, or equal to about 150 tons to the square foot. Water, upon freezing, expands about 10 per cent of its volume. Inasmuch as nearly all rocks contain moisture and this is especially true in those parts of the world where the temperature hovers around the freezing point, it is easy to see that this is one of the most potent forces tending to disrupt rocks. Sedimentary rocks, of course, are most open to these influences because of the greater proportion of water which they carry. But even granite contains an appreciable amount, some of it occupying innumerable minute cavities within the quartz grains.

Not only in the interstices between the grains does the water occur, but in even larger quantities in veins and joints and other lines of weakness, such as along bedding planes of sedimentary rocks.

Mechanical weathering due to the freezing of absorbed water is naturally confined to frigid and temperate latitudes. It is most pronounced where freezing and thawing alternate at frequent intervals. On mountain tops drenched with rain and snow, this is the rule throughout most of the temperate zone.

Mechanical weathering due to expansion and contraction of the rock mass itself under changing temperatures is most active in desert regions and on mountain summits where vegetation is lacking. In situations where the soil is borne away as rapidly as it is formed, the fresh rock surface is constantly exposed anew to the forces of disintegration. The wind is an effective aid in this way. Steep slopes also, because they support but little vegetation and because the dislodged particles are immediately removed by gravity, are prone to rapid weathering through temperature changes alone.

The slow prying action of plants and animals constitutes another important mechanical force which tends to disrupt rocks. Lichens and mosses send their minute rootlets into every crack and crevice; and the higher plants, the trees and shrubs, project their roots deep into the joints and cracks of larger rocks and, merely by their gain in bulk, serve to enlarge the rifts and furnish more ready access for water. The work of animals, too, especially ants and earthworms, is in some regions very effective in bringing about the reduction of rock particles to smaller and smaller sizes.

There is still another factor which causes rocks to split, namely, the relief of pressure due to the removal by erosion of mountain masses. It is usually impossible, however, to ascribe any particular joints to this cause. In quarries, granite blocks split apart and spall off with startling suddenness, presumably because the pressure from one side has been released by quarry operations. In deep tunnels the same phenomenon may be observed.

In view of all of these considerations it is less surprising to learn that mountains in time wither away and that even granite, the symbol of eternity, finds itself doomed when it meets the forces of weathering at the surface of the earth.

Above all, however, we have noted that different rocks react differently to different conditions. This results in the marked contrasts in topographic form produced by similar rocks, but under unlike climatic environments.

## CHEMICAL CHANGES

The chemical changes which take place during weathering involve four processes, namely, *hydration*, *carbonation*, *oxidation*, and *desilication*.

HYDRATION. Hydration means the taking up of water as a chemical constituent. Most minerals which result from rock decay contain large percentages of water. The commonest of these minerals is clay, which results from the hydration of feldspar. Other minerals, too, such as mica, take up water to form hydrated minerals. The taking up of water changes the rock from its original crystalline form, giving forth a clear ringing sound when struck with a hammer, to a dull lusterless mass which goes to pieces by a kind of slaking process when exposed to the air. Hydration involves also a great increase in the volume of the rock, almost doubling it in bulk. The transition from granite into soil, if no material is lost, results in an increase of 88 per cent. This change tends strongly to break up rock masses because of the strains set up between the crystal grains. Igneous rocks are most readily altered in this manner. Sedimentary rocks are themselves derived from igneous rocks by weathering changes, and further action of this kind proceeds far less effectively than on the original igneous rocks.

Some sandstones contain abundant unaltered mica. When this weathers by hydration, the sandstone may rapidly break down into its constituent grains. Muscovite mica, however, may resist weathering nearly as long as or longer than quartz.

OXIDATION. Oxidation is most effective in rocks carrying compounds of iron. Iron occurs in igneous rocks in the form of iron sulphide or pyrite; as ferrous oxide or magnetite; or as ferromagnesian silicates, such as mica and hornblende. Oxidation of these minerals produces ferric compounds, those which contain more oxygen than the ferrous compounds do. A slight oxidation produces hematite which is reddish in color; but if hydration goes on at the same time, then limonite results. Limonite is yellow or ocher colored and is the common constitutent of iron rust. The yellow or orange color of soils is due largely to the hydrated iron oxides which are present. Reducing agents or chemical compounds which take up oxygen, as, for example, acids like those which are produced by plants growing in the soil, may change ferric compounds back to the ferrous form, thus removing the red or yellow colors and imparting a black appearance to the soil.

Complete desiccation of yellow clay by heating, by burning as in the case of brick making, or by the removal of water by plants through transpiration, dehydrates it and causes it to become red. Most bricks are red; most desert soils are red.

CARBONATION. Carbonation involves the formation of carbonates which in general are very soluble. For instance, iron sulphide or pyrite is attacked by water containing carbon dioxide, and it is changed into iron carbonate and sulphuric acid. The iron carbonate is a soluble mineral and is readily removed in solution. Limestone or calcium carbonate is also attacked by water containing carbon dioxide and changes to calcium bicarbonate which is many times more soluble than limestone. One of the most important results of carbonation is the formation of potash or potassium carbonate, a valuable plant food, because it provides potassium in soluble form. This comes from the destruction of orthoclase which is a potassium silicate. As a constituent of orthoclase the potassium is unavailable for plants because it is insoluble.

DESILICATION. Desilication has to do with the removal of silica. Igneous rocks, like granite, contain much silica. Some of this occurs as free quartz. Most of the other minerals in granite are silicates. They are more readily broken down by chemical action than is the quartz, and in this process some silica is liberated to be carried off in solution or as a colloidal suspension. That explains the fact that ground water and the water of streams draining regions of igneous rocks carry in solution more silica than does the water of streams draining regions of sedimentary rocks; for in the latter areas the silica is not so readily removed, being largely in the form of almost insoluble quartz. Even the so-called *basic* igneous rocks, like basalt, syenite, and gabbro, contain much silica even though free quartz is absent. It is easy to see, therefore, that in crystalline areas much silica is liberated to find its way into the crevices of the rocks where it may be redeposited as veins of quartz.

In the comparison between the so-called *acid* igneous rocks, like granite, which contain much quartz, and the basic igneous rocks, like basalt, which contain no free quartz but have a larger proportion of bases, such as calcium, iron, magnesium, aluminum, sodium, and potassium, it is worthy of note that the basic rocks are more susceptible to chemical changes than are the acidic types. The absence of free quartz in basic rocks, the smaller proportion of combined silica, and the corresponding larger proportion of the bases make the basic rocks more readily attacked by the agents of weathering, which are mainly acid in their nature. More energy is liberated, too, by weathering of such rocks than of acid rocks. The weathering of basic rocks, therefore, proceeds more rapidly than that of acid rocks. When the two are in association, the weathered basic rocks often are in topographic depressions, while the more resistant acid rocks form the ridges. Basic dikes cutting through granite areas, instead of standing out in relief, are often etched away to form trench-like depressions. In some lava flows or trap sheets olivine occurs in great abundance, sometimes in restricted zones. This mineral, which contains much iron and magnesium, decays very quickly and rapidly causes a breakdown of the rock mass to form gentle slopes which are in contrast with the vertical walls of the less weathered trap.

THE EVOLUTION OF SOILS. When first formed, soils partake of the character of the rocks from which they were derived. But, like other things in nature, they pass through a cycle of development, involving youth, maturity, and old age. External influences modify their original character. So strong are these outside influences that the characteristics which they impress upon a soil are more dominant than are the characteristics which the soil inherits from the original rock mass. The most important influence, by far, is climate. Second to this is vegetation, itself determined largely by climate. The most potent single climatic factor is precipitation. The potency of precipitation may be seen from the fact that all the soils in the eastern humid section of the United States belong to one class and practically all those in the western arid section make up a second class, in spite of the diversity of rocks from which the different groups were derived. In other words, it makes little difference where the soils originated; if left to adjust themselves to the environment, those existing under similar conditions, in the long run, come to be very much the same. Irrespective of original derivation or past geologic history, soils which have reached maturity in one environment will be more alike than those derived from the same source but maturing under contrasting conditions.

PEDALFER SOILS. Soils developing under humid conditions are known as *pedalfers*, which means soils containing aluminum and iron. These soils develop where the rainfall is 25 to 30 inches or more. This means that the soil as well as the parent rock beneath is continually moist, down to the permanent water table. The prevailing natural vegetation under such conditions is forest. Under these circumstances the soluble materials upon which the fertility of soils depends are, to a greater or lesser degree, carried out of the soil and lost in the drainage waters. This removal takes place not only during the process of evolution of the soils but also as soluble materials are formed through further decomposition. As a result, the pedalfers are relatively impoverished soils. Their content of soluble mineral matter, especially the significant potassium, calcium, and phosphorus, is low. As a whole, the pedalfers are deficient in all the chief nutrient elements, both organic and inorganic. They possess, in short, the less desirable chemical features. Although the parent materials were rich in those elements making for high fertility the pedalfers have been deprived of those materials and reduced to soils of relatively low fertility. The extreme example of the pedalfers are the *laterite* soils of the tropics where the heavy rainfall causes intensive leaching. The soils are poor and can be cultivated continuously only with the aid of fertilizers. That is one reason why, in the equatorial rain forest belt, small clearings are cultivated for only a year or two and then allowed to revert to their natural condition.

PEDOCAL SOILS. The soils which develop under arid conditions are called *pedocals*, which means soils containing calcium. They occur throughout the western United States, except along the more humid Pacific Coast, but nowhere in the east. These soils develop where the precipitation is less than 25 inches, where the moisture supply is insufficient to maintain a continuous downward movement of moisture to an indefinite depth. The natural vegetation is nonforest, consisting prevailingly of grasses, brush, and shrub. Chemically the pedocalic soils have retained practically all the soluble substances upon which fertility depends, although there has been some transfer of materials from one horizon to another. The physical features also present no significant or unusually difficult problem in cultivation or management. In short, they have attained all those features which good soils possess.

In the eastern United States the great problems in agriculture have to do with the proper treatment of the soil. Artificial fertilizers are used in tremendous quantities in the east, but little in the west. The success of agriculture in the eastern half of the country is due, not to good soils, but to the abundant rainfall.

In the western United States the conditions are just the opposite. Lack of water is the great handicap. It costs more, in fact, to supply water to the soils of the western United States than to supply fertilizers to the pedalferic soils of the east. Only about 5 per cent of the western soils under cultivation is irrigated. The extensive cultivation of these dry soils is due largely to their high quality as plant foods. If the characteristic eastern soils (pedalfers) existed everywhere in this western country with dry conditions as at present, it is certain that almost none of the soils would have been brought under cultivation. The cost of fertilizers, combined with the rainfall risk, would make farming in general prohibitive. Under present conditions, however, the western soils are essentially unleached and they can be cultivated on a large scale at low costs. The great problem in the western areas has to do not with soils, but rather with crop improvement. Drouth-resisting grains, hay crops, and tree crops are sought, or new varieties are bred. The virgin soils are moderate to high in all essential soluble organic and inorganic substances except phosphorus; and the proportion of this mineral substance is higher than that which prevails in typical pedalferic soils.

MATURING OF SOILS OF THE PEDOCAL TYPE. The pedocal soils vary from the black earths or *chernozems*, which contain abundant organic matter, to the gray-desert soils, which have little or no organic matter present. The black earths develop in regions of fair rainfall, 25 inches or so, and constitute the great grain areas of the world. There are also chocolate-brown and brown soils intermediate between the black earths and the desert soils, the depth of color depending upon the amount of humus present.

VARIOUS FORMS PRODUCED BY WEATHERING

## FORMS PRODUCED BY WEATHERING

DIFFERENTIAL WEATHERING. This term is applied to the process of weathering which etches out the weaker parts of a rock mass. It may result in a pitted or fretted surface or bring out in strong relief the bands or layers of sedimentary rock best fitted to withstand disintegration and decomposition. Pedestal rocks and mushroom rocks are thus produced.

The weathering of volcanic breccia and even glacial till sometimes results in pillars or columns capped by the larger boulders or fragments which serve to protect the loose, poorly consolidated mass beneath. These pillars are fancifully termed *demoiselles*.

TALUS. The debris dislodged by weathering of steep slopes, which accumulates at their base, is known as *talus*, or *scree*. If very coarse and steeply conical in form, as if radiating out from a center of accumulation, it is called a *talus cone*. Coarse material produces much steeper talus slopes than does fine. The upper portion of a talus slope is usually steeper than the lower portion because the lower slopes are made up of the finest material. Some talus slopes, however, have coarse boulders and blocks near the outer margin, piled up in a ridge, the finer material lying between it and the cliff. This often occurs at the foot of north-facing rock walls where snow banks accumulate. Rocks, falling on the smooth even surface of the snow, roll much farther from the base of the cliffs before coming to rest, than if they fall directly upon a similar slope of rock debris. During the summer, ordinary simple talus slopes are formed. These, combined with the mounds and ridges formed during the winter, result in a complex hummocky surface, not always readily distinguished from a landslide accumulation.

BOULDERS OF WEATHERING OR RESIDUAL BOULDERS. Most massive rocks are traversed by one or more series of joints whereby they are divided into rhomboidal blocks of varying sizes. Even when not sufficiently developed to be conspicuous, such joints not infrequently exist as lines of weakness along which moisture and the accompanying agents of weathering make their way, gradually rounding the corners until there is left an oval mass, like the so-called *niggerheads* of the gabbro area about Baltimore. In nearly all such rocks the exfoliation and decomposition take place in the form of concentric layers, like the coatings on an onion. This holds true with the huge granitic bosses, as well as with the smaller joint blocks. If the block or mass is reasonably homogeneous, the agencies of decomposition penetrate nearly uniformly from all exposed surfaces, and the concentric structure results.

On top of the Rocky Mountains in the Pikes Peak region, as well as on the summit of the Laramie Range, the granite has weathered into large round boulders which resemble gigantic cannon balls piled up, tier upon tier.

*Sketched from U. S. Geological Survey Photographs*

ROCK STREAMS AND MUD FLOWS DUE TO SOLIFLUCTION

## SOLIFLUCTION OR SOIL CREEP

Solifluction, from *solum*, soil, and *fluere*, to flow, is a term applied to the slow imperceptible flowing from higher to lower ground of masses of rock and soil saturated with water. On mountain tops in humid regions this process can best be seen under way. Masses of rock mixed with soil and filled with an abundance of water, often from melting snow fields, gradually flow down the slopes, sometimes as extensive sheets of waste, and sometimes as "mud glaciers." It is simply creep of the soil aggravated by unusual conditions. Hills and slopes where this is going on show a very marked streakiness of surface, and in places the flow lines assume a crenulate or scalloped form resembling molten lava sheets.

Solifluction depends for its effectiveness upon certain climatic conditions. In regions characterized by a subglacial climate with heavy deposits of winter snow melting in summer, solifluction is the chief agent of destruction. In the polar and subpolar regions, where the ground is not covered with ice, this process is almost everywhere at work. The Falkland Islands, Spitzbergen, South Georgia Island, and northern Scandinavia are some of the localities where the effect of solifluction has been observed. In the same manner, the alpine tracts of lower latitudes are favorable for the development of this phenomenon. In Glacier Park the alps or uplands near the divides show clearly that the whole blanket of soil and rock is moving down the slopes en masse.

*Mud Flows.* In some valleys in the Rocky Mountains the creeping soil accumulates and slowly moves down the gulch as mud-rock glaciers. They are made up of rocks of every size from that of a pea to several feet in diameter. The abundance of melting snow upon the surrounding mountains accounts for their semifluid character. One of the best known is the Slumgullion Mud Flow on the headwaters of the Gunnison River in western Colorado. The damming of a valley into which this flow extended like a delta caused Lake San Cristobal, a sheet of water 2 miles long. The total length of the flow is about 4 miles, with a slope of about 2,500 feet in this distance. The thickness of mud filling the valley is probably 200 to 300 feet. The topographic features of the flow are very pronounced. It is bounded for nearly its entire length by two moraine-like lateral ridges of very sharp outline. Between these the flow is usually lower and characterized by furrows and trenches, knolls, and hollows of confused relations resembling those of modified landslide areas or of some glacial deposits.

EXAMPLES OF EXFOLIATION DOMES

## EXFOLIATION DOMES

Large homogeneous rock masses, when subjected to changes in temperature, spall off in flat or rounded slabs or flakes.

Two theories have been advanced to explain this phenomenon: First, it has been suggested that the separation of granite masses into curved plates is an original structure, derived perhaps from the once molten condition of the rock, possibly in the nature of flow lines. On the other hand, several lines of evidence favor the alternative theory that exfoliation is due to temperature changes. The dome structure does not extend downward and inward indefinitely. The curved plates, where they can be seen in cross section owing to the breaking apart of large rock masses, occupy a depth of usually not more than 50 feet. This suggests a surface phenomenon rather than a deep-seated inherent one. Moreover, the structure is not restricted to domes but occurs on the walls of canyons, the sides of ridges, and the bottoms of trough valleys. Since the forms of curvature are usually adjusted to the general shape of the topography, apparently their cause lies somewhere outside the body of the rock itself.

Another argument is that exfoliation or dome structure occurs in strongly bedded formations like conglomerate, cutting across the structure in some places at very sharp angles.

The three generally accepted explanations for exfoliation are (*a*) seasonal changes of temperature causing an expansion and contraction of the rock sufficient to bring about a peeling of the surface; (*b*) expansion of the surface due to hydration of the feldspars to form kaolin; (*c*) relief of internal pressure by erosion of surface masses.

The expansion and contraction of rock surfaces, based upon the known coefficient of expansion of gneiss, results in an increase in length of over ½ inch per 100 feet, if the temperature rise is 100°F. At depth it is considerably less than this because the range in temperature is so much less. At a foot beneath the surface the shrinkage or expansion is only one-twentieth as much, the temperature range there being only 5°. However, since surfaces a thousand feet in length may be exposed, the strain between the upper and lower portions must be very great.

The process of exfoliation is aided by other destructive agencies. The microscopic crevices first opened up by temperature strains furnish access for waters, gases, and perhaps minute plants. It is likely that slight changes of temperature at 15 feet are more potent than similar changes would be at the surface, for the rocks are there confined and not at liberty to seek relief in several directions. In certain regions, as in Brazil, where large flat surfaces of gneiss are openly exposed to the sun's rays, great flakes or shells are sometimes bulged and lifted away from the cooler mass below. Where roads pass over such places they give forth a hollow sound to the horses' hoofs.

*U. S. Forest Service*

"BALANCED ROCK," ROCKY MOUNTAIN UPLAND NEAR PALMER LAKE, PIKES PEAK REGION, COLO.

One of the numerous residual granite boulders which occur on the upland, rounded by exfoliation.

## REGIONS WHERE EXFOLIATION DOMES OCCUR

Exfoliation domes occur under varying climatic conditions: in the rain-swept mountains of Norway, the snowy Sierra Nevada of California, the hot, humid Brazilian highlands, the dry Kalahari desert. In each locality, however, the rock is massive granite or other igneous rock, usually of coarse texture. The joints are widely spaced and the blocks of mountainous proportions.

In Brazil, about Rio de Janeiro, these domes rise 1,000 feet and more from the water's edge. The rocks here are largely gneisses and porphyritic granites. In spite of the strong banding of the gneiss, exfoliation has developed rounded summits without relation to the structure. Some of the summits become clothed with vegetation but the steep faces still flake off in great sliding sheets.

In the granite areas of the Sierra Nevada, many summits have the form of domes. A few are symmetric, with approximately circular or oval bases, but most are one-sided or irregular. Associated with these domelike forms are closely related structures. The granite is divided into curved plates or sheets which wrap around the topographic forms. One of the domes in the Yosemite Park region, called *Half Dome*, presents a sheer vertical wall on one side, this being a relatively fresh joint surface.

Domelike mountains in desert lands are common, as are the smaller rounded blocks, ranging in size from a house to that of an egg. In the Sinai peninsula between Asia and Africa, in Egypt, and in the Nubian Desert these forms abound. As the blocks become smaller through exfoliation, they tend to split in half because of excessive heating on one side. If the heating on all sides is more or less uniform, further exfoliation occurs. Sudden showers of rain may also cause abrupt cracking along flat planes. Thus a mosaic of joints may be produced and the rock mass broken up into angular blocks, which upon further weathering become round and boulderlike. The desert surface is strewn with blocks of this sort, called "melons" by the Arabs.

Exfoliation domes occur in parts of the Adirondacks where homogeneous, medium-textured igneous rocks prevail. This area, largely syenite or granite, is broken by faulting into units like a gigantic checkerboard. The comparatively recent glaciation has left bold summits free of soil, a ready prey to changes of temperature. The summits are literally peeling or shelling off by the removal of exfoliation sheets of great size, some as much as 50 to 75 feet across and from 1 to 3 feet thick. Where the igneous rocks are gneissoid, the exfoliation disregards the direction of the gneissic structure and often great sheets come off at right angles to the foliation. This contrasts strongly with the schistose structure of the White Mountains of New Hampshire, where exfoliation never occurs, the rocks always parting along the cleavage planes of the schist.

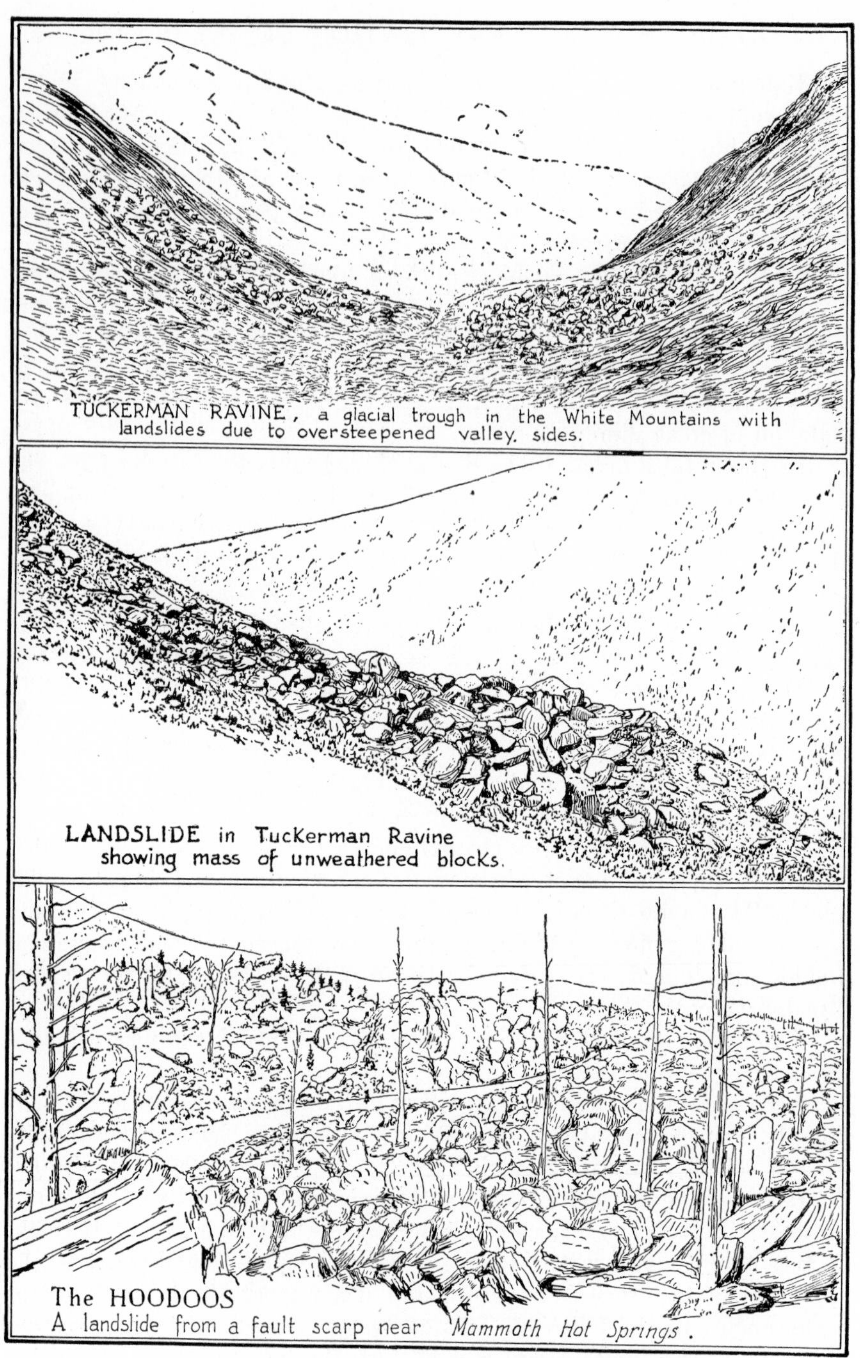

TUCKERMAN RAVINE, a glacial trough in the White Mountains with landslides due to oversteepened valley sides.

LANDSLIDE in Tuckerman Ravine showing mass of unweathered blocks.

The HOODOOS
A landslide from a fault scarp near *Mammoth Hot Springs*.

*From photographs by Douglas Johnson and U. S. Geological Survey*

LANDSLIDES AND ROCK STREAMS

## LANDSLIDES

Landslides and rock streams differ from creep of the soil and solifluction in that the movement is extremely sudden and involves usually the bedrock and not merely the soil cover. While talus cones are formed by the breaking away of a rock ledge, block by block, a landslide involves a great portion of the ledge itself in one sudden movement.

Two kinds of factors must be scrutinized in seeking the fundamental cause for landslides. First is the topography of the region; second is the geological structure, the kind of rocks and their physical characteristics. Of course, the immediate cause of the slide may be of interest. It may be due to an earthquake or the heavy rains which unduly saturate the ground, but these are incidental events rather than fundamental causes.

LOCATION OF LANDSLIDES. The first of the fundamental causes has to do with topography, and under this heading there appear to be four topographic situations which favor landslide development. One idea, however, lies at the root of each of the four situations. There is in all cases a steepness of slope greater than that which normally results from long-continued weathering in such a region. This oversteepened condition may be brought about in several ways but it is always the result of some relatively recent physiographic change in the region which has interrupted the orderly adjustment of the land surface to weathering conditions. For example, exceptionally steep valley walls are brought about by the local glaciation of a region. *Glacial troughs*, with their U-shaped profiles, have much steeper sides than did the V-shaped valleys which preceded them. The V-shaped valleys owed their more gentle slopes to the slow and long-continued process of weathering, which brought them into adjustment to the forces prevailing there. A certain equilibrium was thus established. Glaciation disturbs this condition by cutting a trough with more nearly vertical walls. For this reason, avalanches and landslides are common in alpine regions. These initial stages in the adjustment of the region to the forces of weathering are naturally the most violent.

A second situation conducive to landslides is along *fault scarps*. A fault scarp is a feature not at all in harmony with the forces of weathering and erosion. Equilibrium is established by active weathering which results in landslides. The fault blocks in the lava area of central Washington and in the Great Basin and the fault scarps in the Colorado Plateau reveal landslide features. Perhaps the most familiar landslide of this type is the region of the "Hoodoos" near Mammoth Hot Springs in Yellowstone Park. This jumbled mass of great blocks, through which the motor road winds its way, is situated on the dominant fault line controlling Mammoth Hot Springs and the Gardiner River Valley.

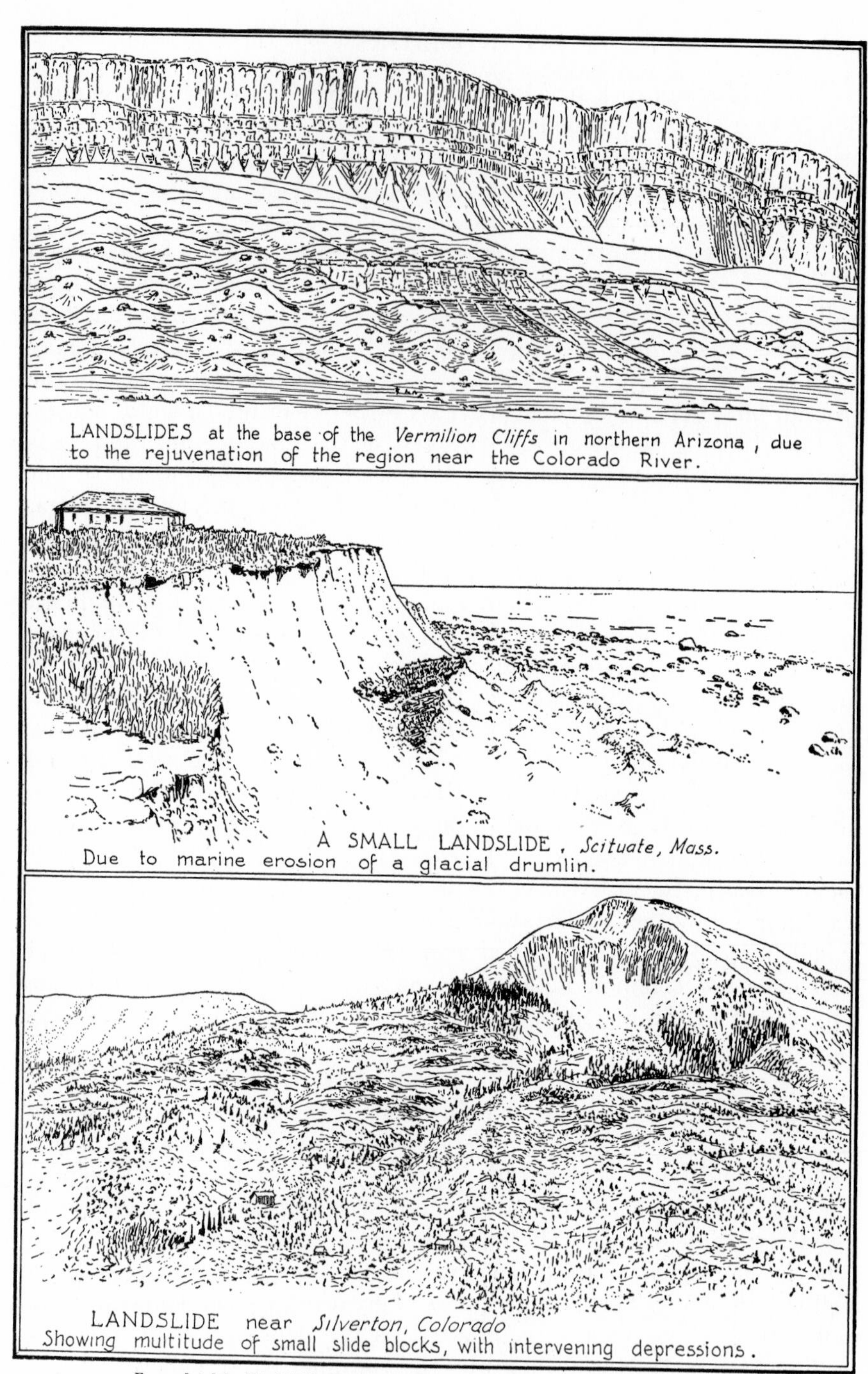


*From sketch by W. M. Davis and photographs by Douglas Johnson and U. S. Geological Survey*

TYPES OF LANDSLIDES

A third type of unadjusted slope results from the *rejuvenation* of a region. Streams resuming their erosional activity after a prolonged period of rest produce steep valley walls which are in unstable equilibrium, a ready prey to the forces of weathering. In fact, the presence of landslides and abundant talus is sometimes taken as evidence that rejuvenation has come about. For instance, in the Grand Canyon region a wonderful series of landslides occurs along the base of the Vermilion and Echo Cliffs. Professor Davis in studying the region observed that the slides occur relatively near the Colorado River, while farther away the cliffs have a much more mature profile entirely without slides. He concluded from this that, inasmuch as landslides are characteristic of the earlier and more energetic stages of a cycle of erosion, it seemed probable that these slides should be associated with the reviving activities in the youthful stage of a second cycle of erosion rather than with the fading activities in the advanced stage of a first cycle.

A fourth situation is that which obtains along *seacoasts* where waves have cut back into headlands to form steep cliffs. Landslides under such conditions are not uncommon, as, for example, along the Straits of Dover where the chalk cliffs, undermined by the sea, give way to great slumping masses of broken rock and soil.

In regard to the influence of the geological structure, it is sufficient to observe that two conditions are especially conducive to rock slips. First, a great mass of rock underlain by a weak shale, all dipping toward the valley, is, if saturated with water, likely to slide down the dip, the shale acting as a lubricating layer.

In the second place, any rock which is greatly shattered, or which is very deeply weathered so that at times of heavy rain it becomes a pasty mass, is liable to break away suddenly and flow, perhaps with violence, to lower levels.

In short, any geological structure or rock condition which permits complete saturation by ground water to lubricate the mass, and reduce internal friction to a minimum, favors landslide development. Any combination of these favorable conditions results in landslides, great in number and in size, as in the San Juan Mountain area of western Colorado.

Finally, the activities of man are occasionally conducive to landslides, as, for example, where oversteepening of slopes results from road building, excavations, and railroad cuts. Los Angeles and Seattle have both experienced destructive slides in recent years. Motorists who note the caution to "beware of fallen rocks" are observing landslides on a small scale. To prevent the sliding of large masses of water-soaked soil during construction operations, engineers drain the soil by means of pipes and tunnels or, by putting pipes through it, freeze it artificially.

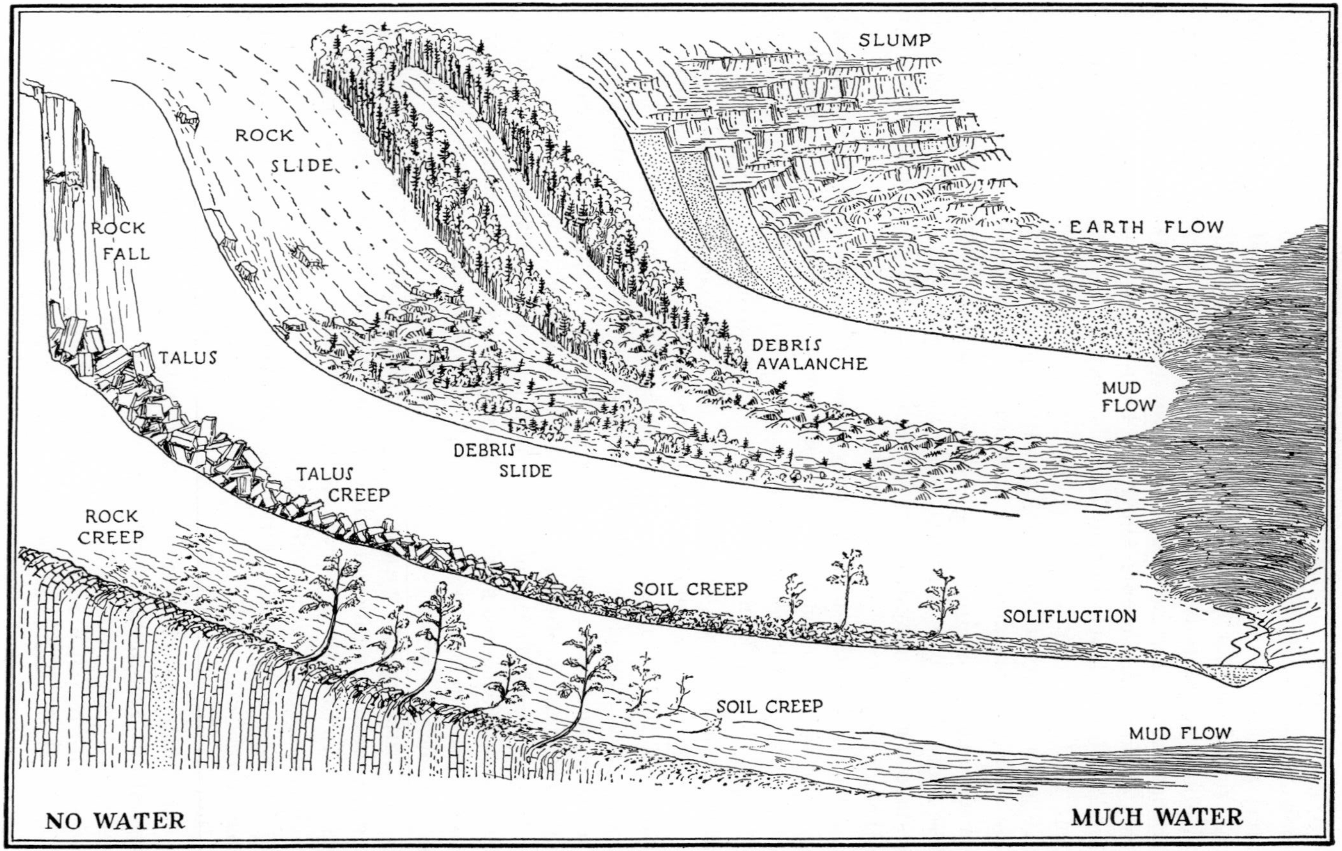

SCHEME SHOWING TYPES OF LANDSLIDES AND OTHER RELATED PHENOMENA

# CLASSIFICATION OF LANDSLIDES AND RELATED PHENOMENA

A recent study of mass movements of soil and rock by C. F. Stewart Sharpe has resulted in a systematic classification of many forms. They grade, in general, from rock falls and slides where the content of water is negligible through a series in which the amount of water becomes greater and greater. Near the end of the series are mud flows in which the soil is completely saturated. *Sheet wash* would represent a still more fluid condition, finally ending in flowing streams. It is apparent, therefore, that all of these types of soil movement are definitely parts of the drainage system of a region and that no sharp line separates one of the series from the next. The presence of much ice does not materially modify the scheme, although it introduces some additional terms, such as *rock glacier*.

Three classes of movement may be recognized, as follows:

*a.* Very rapid movement, consisting of sliding and falling. In these cases, the presence of water or ice as a lubricant is not necessary. The forms which result are called *rock falls*, *rock slides*, and *debris slides*.

*b.* Slow flowage occurs in material only partially saturated and produces *rock creep*, *soil creep*, and *solifluction*.

*c.* Rapid flowage results from complete saturation of the soil which causes *earth flows* and *mud flows* and the still more rapid movement of *sheet wash*.

It is clear, therefore, that on steep declivities—such as the walls of glacial troughs, along fault scarps, the walls of young valleys, and wave-cut cliffs—rock falls, slides, avalanches, and slumping may occur, even under relatively dry conditions. Soil creep and earth flow, however, demand a large moisture content as well as a fairly steep slope and, of course, are apt to occur in soil which has a plastic incoherent nature, such as clay and shale.

Mud flows result from a combination in a high degree of all the factors just enumerated: a large volume of water, a fair gradient, and especially a plastic slippery soil which permits the particles to move freely over each other.

The accompanying sketch illustrates a few of the many details of these forms, such as the backward-tilted terraces due to slumping, features which are sometimes called *catsteps;* the morainelike character of debris slides; and the interesting way in which trees regain their vertical attitude after being bent over from rock and soil creep.

## MAPS ILLUSTRATING WEATHERING

The small size of the topographic forms which result from weathering prevents showing them on most topographic maps. Large landslides can occasionally be represented. The *Red Rock, Wash.*, sheet clearly shows a landslide on the north side of Saddle Mountain with a distinct depression separating the downfallen material from the wall above it. The *Malaga, Wash.*, sheet also shows many landslides. On the *Silverton, Colo.*, quadrangle several landslides are depicted by the contours; an especially prominent one is at the head of Kendall Gulch, south of Silverton. The small scale of this map and the large contour interval causes these forms to be lost in the bold topography of this region. On the Silverton Folio, however, may be found several slides large enough to be shown by contours. The Slumgullion Mud Flow, with the dammed-up Lake San Cristobal is shown on the *San Cristobal, Colo.*, sheet. The easily obtained *Frank, Alberta,* sheet of the Canadian Geological Survey shows by 20-foot contours the great landslide of 1903 which tore down the side of Turtle Mountain and flowed a mile across the valley, obliterating everything in its path, including the Canadian Pacific Railway. In addition to the confused topography of the landslide debris, the map shows clearly the great scar left on the mountainside.

Differential Weathering and Rock Structure. The effect of weathering upon jointed and bedded formations is displayed in a unique way on the *Berne* and *Coxsackie, N. Y.*, sheets. The beds dip to the south and are cut by north-south and east-west joints. Although there are no great contrasts in the different beds, weathering has developed a series of steps in the topography.

The weathering of conchoidal fractures or curved joints is beautifully portrayed on the *Yosemite Valley, Calif.*, sheet (scale 1:24,000). The unusual skill displayed in representing the Royal Arches on this remarkable map incites the admiration of all topographers. The Yosemite Valley map shows also the great exfoliation domes which surmount the Sierra upland as well as the vertical cliffs along the joints in Half Dome and Liberty Cap.

Differential weathering of horizontally bedded rocks is superbly shown on the large-scale maps of the Grand Canyon, especially the *Bright Angel* and the *Vishnu, Ariz.*, sheets. Every little ledge where the resistant beds outcrop is revealed by the contours, as is the contrast between the plateau topography in the upper walls of the canyon and that produced by the crystalline rocks of the Granite Gorge.

On all maps showing various types of structures in both sedimentary and crystalline rocks, but especially in arid regions where the details are not concealed by soil and vegetation, the etching effect of differential weathering is abundantly displayed.

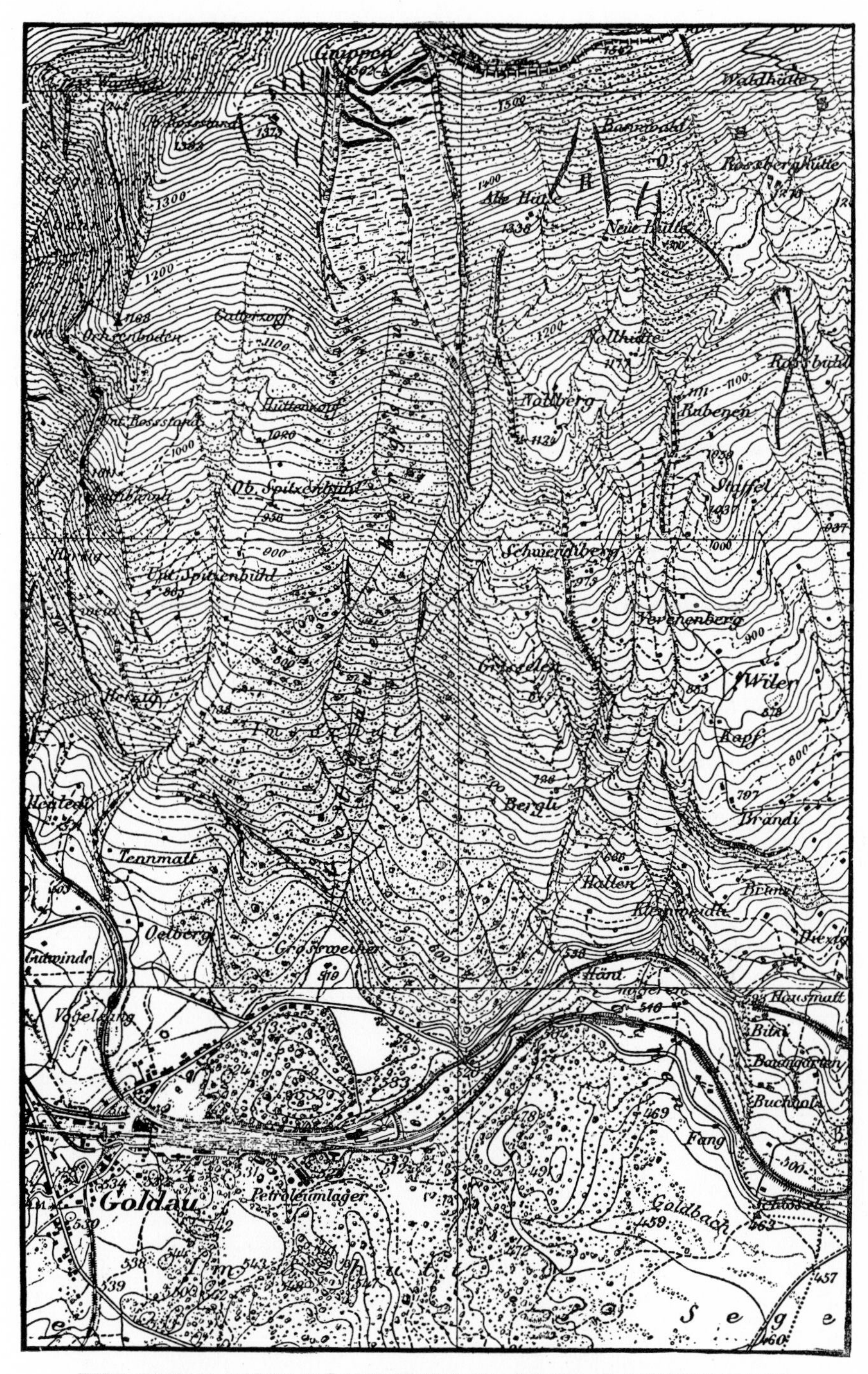


THE GREAT GOLDAU LANDSLIDE IN CENTRAL SWITZERLAND
Switzerland; *Arth* and *Lauerz* sheets, Nos. 207 and 209 (1:25,000).

## QUESTIONS

1. What is the source of the humus in chernozem or black-earth soils?
2. Why does rank vegetation develop in tropic jungles? Is the soil so good?
3. In what ways do the activities of man promote landslides?
4. What means can be taken to prevent sliding and slumping of the soil?
5. Why does sandstone in some cases weather rapidly and disintegrate into a loose sand?
6. In what ways do plants aid the process of weathering? How do animals aid weathering?
7. Does the rock structure affect landslides?
8. Are landslides more common in arid or humid regions?
9. Do landslides indicate rejuvenation in the geomorphic history of a region?
10. What determines the angle of slope of a talus cone?
11. Does the presence of vegetation, trees, etc., prevent or retard the development of soil creep?
12. Is glacial till as likely to slump as residual soil?
13. How can you distinguish between residual and transported soil?
14. Do you consider wind as one of the destructional forces or is it a part of the process of weathering?
15. What effective means can be taken to retard the process of weathering on buildings and monuments?
16. Would you rather have a tombstone of granite or marble?
17. What is kaolin?
18. If a large modern city were abandoned for a thousand years, what materials would be entirely destroyed in that time and what would be left? Would railway rails last? Would the copper roofs last? Or slate roofs? Would china and glassware last? Would concrete and cement and brick remain? Would anything be left of automobiles? Or of wooden furniture? Or of books? Or of clothes? Or of big bridges?
19. When a building goes to rack and ruin, what goes first? Then what happens?
20. Of the following ten chemical elements necessary to plant growth, which four are often present in insufficient amount and why is this? Potassium, nitrogen, phosphorus, calcium, oxygen, iron, aluminum, sodium, magnesium, and manganese. From what sources may potassium, nitrogen, phosphorus, and calcium be obtained?
21. What is meant by hypogene exfoliation?
22. Did you ever build a camp fire next to a granite boulder? What happened?
23. Which of the photographs at the beginning of this chapter illustrate arid regions and which illustrate humid regions?

## TOPICS FOR INVESTIGATION

1. Theories of origin of exfoliation.
2. Exfoliation domes. Massive curvilinear fractures. Relative importance of temperature changes and release of pressure in forming joints.
3. Soils. Soil types as related to climatic regions.
4. Chemical changes in weathering. Formation of kaolin from feldspar.
5. Landslides and avalanches. Causes and types.

## REFERENCES


GENERAL

LEVERETT, F. (1909) *Weathering and erosion as time measures.* Am. Jour. Sci., 4th ser., vol. 27, p. 349–368.

MERRILL, G. P. (1896) *Principles of rock weathering.* Jour. Geol., vol. 4, p. 704–724, 850–871.

MERRILL, G. P. (1906) *Rocks, rock-weathering and soils.* New York, 400 p. The most comprehensive book on the general subject of weathering.

MECHANICAL WEATHERING

BARTON, D. C. (1916) *Notes on the disintegration of granite in Egypt.* Jour. Geol., vol. 24, p. 382–393.

GEIKIE, A. (1882) *Geological sketches.* London. Rock-weathering measured by the decay of tombstones, p. 159–179.

LEONARD, R. J. (1927) *Pedestal rocks resulting from disintegration.* Jour. Geol., vol. 35, p. 469–474.

TABER, S. (1929) *Frost heaving.* Jour. Geol., vol. 37, p. 428–461.

WATSON, T. L. (1901) *Weathering of granitic rocks in Georgia.* Geol. Soc. Am., Bull. 12, p. 93–108.

CHEMICAL WEATHERING

GOLDICH, S. S. (1938) *A study of rock weathering.* Jour. Geol., vol. 46, p. 17–58.

HAYES, C. W. (1897) *Solution of silica under atmospheric conditions.* Geol. Soc. Am., Bull. 8, p. 213–220.

VAN HISE, C. R. (1904) *A treatise on metamorphism.* U. S. Geol. Surv., Mon. 47. The belt of weathering, p. 409–561.

WARD, F. (1930) *The rôle of solution in peneplanation.* Jour. Geol., vol. 38, p. 262–270.

EXFOLIATION

BLACKWELDER, E. (1925) *Exfoliation as a phase of rock weathering.* Jour. Geol., vol. 33, p. 793–806.

FARMIN, R. (1937) *Hypogene exfoliation in rock masses.* Jour. Geol., vol. 45, p. 625–635.

GILBERT, G. K. (1904) *Domes and dome structure of the high Sierra.* Geol. Soc. Am., Bull. 15, p. 29–36.

MILLER, W. J. (1911) *Exfoliation domes in Warren County, New York.* N. Y. State Mus., Bull. 149, p. 187–194.

SOILS

HILGARD, E. W. (1906) *Soils.* New York, 593 p. Particularly good on interrelations among climates, plants and soil.

RIDGLEY, D. C., and EKBLAW, S. E. (1938) *Influence of geography on our economic life.* New York, see p. 136–145.

WHITNEY, M. (1925) *Soil and civilization.* New York, 278 p.

LANDSLIDES AND RELATED PHENOMENA

The splendid contribution of C. F. S. Sharpe (1938) on *Landslides and related phenomena*, New York, 137 p., provides an extremely useful list of references on that subject, classified under such headings as *Soil flowage, Frost action and solifluction, Rock-glacier creep, Earthflow, Mudflow*, and *Landslides.* The following have been found especially useful:

ALDEN, W. C. (1928) *Landslide and flood at Gros Ventre, Wyoming.* Am. Inst. Min. Met. Eng., Trans., vol. 76, p. 347–361.

ANDERSON, J. G. (1906) *Solifluction, a component of subaërial denudation.* Jour. Geol., vol. 14, p. 91–112.

Behre, C. H., Jr. (1933) *Talus behavior above timber in the Rocky Mountains.* Jour. Geol., vol. 41, p. 622–635.

Blackwelder, E. (1912) *The Gros Ventre slide, an active earth-flow.* Geol. Soc. Am., Bull. 23, p. 487–492.

Blackwelder, E. (1928) *Mudflow as a geologic agent in semiarid mountains.* Geol. Soc. Am., Bull. 39, p. 465–480; discussion, p. 480–483.

Cairnes, L. D. (1912) *Differential erosion and equiplanation in portions of Yukon and Alaska.* Geol. Soc. Am., Bull. 23, p. 333–348.

Capps, S. R., Jr. (1910) *Rock glaciers in Alaska.* Jour. Geol., vol. 18, p. 359–375.

Dawson, G. M. (1899) *Remarkable landslip in Portneuf County, Quebec.* Geol. Soc. Am., Bull. 10, p. 484–490.

Hay, T. (1936) *Stone stripes.* Geog. Jour., vol. 87, p. 47–50.

Howe, E. (1909) *Landslides in the San Juan Mountains, Colorado.* U. S. Geol. Surv., Prof. Paper 67, 58 p.

McConnell, R. G., and Brock, R. W. (1904) *Report on the great landslide at Frank, Alberta.* Canadian Dept. Int., Ann. Rept. 1902–03, part 8, App., 17 p.

Patton, H. B. (1910) *Rock streams of Veta Peak, Colorado.* Geol. Soc. Am., Bull. 21, p. 663–676; discussion, p. 764–765.

Russell, R. J. (1933) *Alpine land forms of western United States.* Geol. Soc. Am., Bull. 44, p. 927–949.

Taber, S. (1929) *Frost heaving.* Jour. Geol., vol. 37, p. 428–461.

# IV
# UNDERGROUND WATER

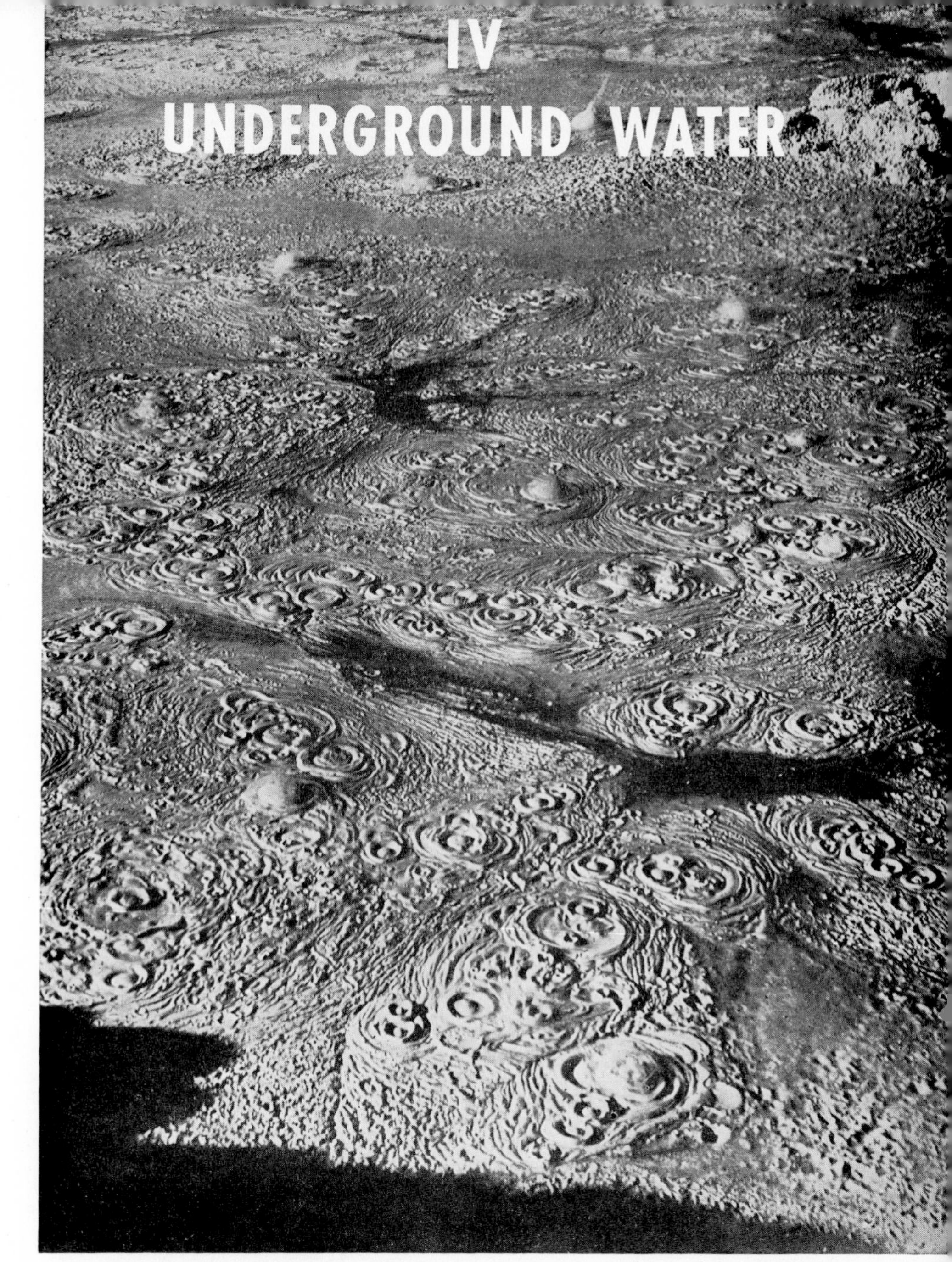

*Darton, U. S. Geological Survey*

PAINT POTS OR MUD VOLCANOES, YELLOWSTONE PARK
The "mud" is a very viscous clay resulting from the weathering of the igneous rocks through which the heated ground water rises.

*Jackson, U. S. Geological Survey*

LONE STAR GEYSER, YELLOWSTONE PARK

A geyser with an unusually high funnel formed of geyserite or siliceous sinter carried up in solution by the rising ground water which forms the geyser.

ISLAND SPOUTER, SARATOGA SPRINGS, N. Y.

An artesian spring in winter which has built up an ice funnel, strongly resembling a geyser cone. Fault-line scarp in rear.

*Saratoga Springs Authority*

*Ol. Magnusson, Reykjavik*

GEYSIR, ICELAND

The original "geyser" from which the word *geyser* comes. One of the many fine examples in Iceland, somewhat resembling Old Faithful in its low cone.

CASTLE GEYSER, YELLOWSTONE PARK

A geyser funnel surmounting a pediment, built up by violent eruptions in contrast with the rimmed pool in the foreground formed by a quiet steady flow.

*Jackson, U. S. Geological Survey*

*Jackson, U. S. Geological Survey*

TERRACES OF THE MAMMOTH HOT SPRINGS, YELLOWSTONE PARK, PHOTOGRAPHED IN 1870
Formed of calcium carbonate, not of siliceous sinter. The hot ground waters here rise through a thick limestone formation.

TUFA DOMES, PYRAMID LAKE, NEV.

These domes and pinnacles formed of calcium carbonate and other salts are built up by spray from the lake water which is almost saturated with mineral salts. It is a desert lake with no outlet.

*Russell, U. S. Geological Survey*

*Jackson, U. S. Geological Survey*

LIBERTY CAP, NEAR MAMMOTH HOT SPRINGS, YELLOWSTONE PARK

An unusually large extinct geyser cone formed of calcium carbonate, this being a limestone region.

EXTINCT GEYSER CONE, EAST FORK OF THE YELLOWSTONE RIVER, MONT.
Showing in places the many thin layers representing the numerous deposits of which it is built.

*Jackson, U. S. Geological Survey*

*U. S. Forest Service*

THE "SINKS," MONONGAHELA NATIONAL FOREST, W. VA.

Entrance to underground passageways in limestone plateau.



LOST RIVER ISSUING FROM UNDERGROUND CAVERN, WHITE MOUNTAIN NATIONAL FOREST, N. H.

An underground stream flowing not in a limestone region but under the rocks of a large landslide.

*U. S. Forest Service*

*Darton, U. S. Geological Survey*

SINK HOLE IN LIMESTONE, CAMBRIA, WYO.

An unusually good example of a sink with abrupt sides not modified by erosion.



SINK HOLES IN TERTIARY GRAVELS OF THE GREAT PLAINS

Alined along a subterranean water course. The sinking is due to the removal of the finer material, allowing the ground to settle.

*W. D. Johnson, U. S. Geological Survey*

*Veatch, U. S. Geological Survey*

MUD VOLCANOES, LONG ISLAND, N. Y.

Formed by ground water rising through the marshes. The "volcanoes" are built up because of the small explosions produced by the marsh gas.



MAORI WOMEN COOKING IN HOT POOL, ROTORUA, NORTH ISLAND NEW ZEALAND

The temperature of these boiling springs is 212°F. Many Yellowstone springs boil at less than 212°F. because of their higher altitude.

*New Zealand Tourist Bureau*

*U. S. Forest Service*

MARBLE CAVES, SISKIYOU NATIONAL FOREST, ORE.
Unusually fine examples of columns formed by the union of stalactites and stalagmites. In some cases the two have not yet joined.



CARLSBAD CAVERN, N. M.
Exceptionally rich display of stalactites, some of which have grown down to join the heavier and blunter stalagmites below.

*Lee, U. S. Geological Survey*

*Atchison, Topeka and Santa Fe Railway*

KINGS THRONE ROOM, CARLSBAD CAVERN, N. M.

Colossal pillars of dripstone, representing a complex type of stalagmite which has grown up to the roof of the cave 100 feet above.

*Atchison, Topeka and Santa Fe Railway*

KINGS THRONE ROOM, CARLSBAD CAVERN, N. M.

One of the most impressive rooms in any cave, renowned not only for its size but also for the rich lavishness of the cave deposits.

*Ewing Galloway*

LIMESTONE CLIFFS, DURANCE RIVER VALLEY, SOUTHERN FRANCE
Weathering along the joints has left these gigantic pillars whose sides are often scored with vertical grooves and ridges like the so-called *lapies* in the Karst.



ROAD CUT IN WIND-BLOWN CORAL LIMESTONE, BERMUDA
Showing residual limestone soil at the surface and vertical fluting where weathering has worked down along the joints.

*Bermuda Trade Development Board*

## UNDERGROUND WATER

Synopsis. The work of underground water is a part of the process of surface drainage. The self-same water may alternately be on the surface and underground at different times. Indeed many streams combine both surface and subsurface characteristics. During times of sufficient rainfall they are normal streams; during drier periods they maintain a circulation beneath the apparently dry gravels of their beds.

Underground water is effective mainly as a solvent rather than as an eroding agent. The fact that it moves slowly, that it is constantly present, and that it not merely is in contact with the surface of the ground but permeates a large volume of rock which has not already been altered by hydration, makes it possible for underground water to take up large amounts of mineral matter and bring them to the surface again in springs. That is why most spring water is mineralized, some of it to a very high degree.

The flow of underground water is subject to certain physical and mathematical laws which help to explain the occurrence of springs and seepage on hill slopes far above stream level. On the other hand, stream level may roughly be assumed to be ground-water level, below which the ground is saturated.

Rocks and soil may, however, consist of alternating layers of porous and nonporous beds with the resulting water tables "perched" above ground-water level.

Rocks with many joints, such as most massive sedimentary formations, permit the free circulation of subsurface water. Abundant springs occur where such beds outcrop along valley sides. In limestones, caverns develop along the more soluble beds and along the vertical joints. As passageways enlarge by solution, greater quantities of water flow through the rock and erosion becomes a factor. Reduction of base level of the cave streams, by the incision of the surface streams into which the cave streams flow, lowers ground-water level, and the upper parts of the cave system gradually become dry.

A cycle of development by underground drainage is recognized for both horizontal and contorted limestone beds. Davis has advocated the likelihood of two cycles in caves where dripstone is abundant: one cycle during which solution cavities are opened out by circulating ground water, and a second cycle following uplift, during which the cave passageways become dry and deposits are formed by evaporating water from the surface.

The surface topography of a region of excessive underground solution is characterized by vast numbers of depressions of all sizes, even several miles in extent, sometimes by great outcrops of fluted limestone ledges, and by an almost total lack of surface streams, the whole constituting *karst landscape.*

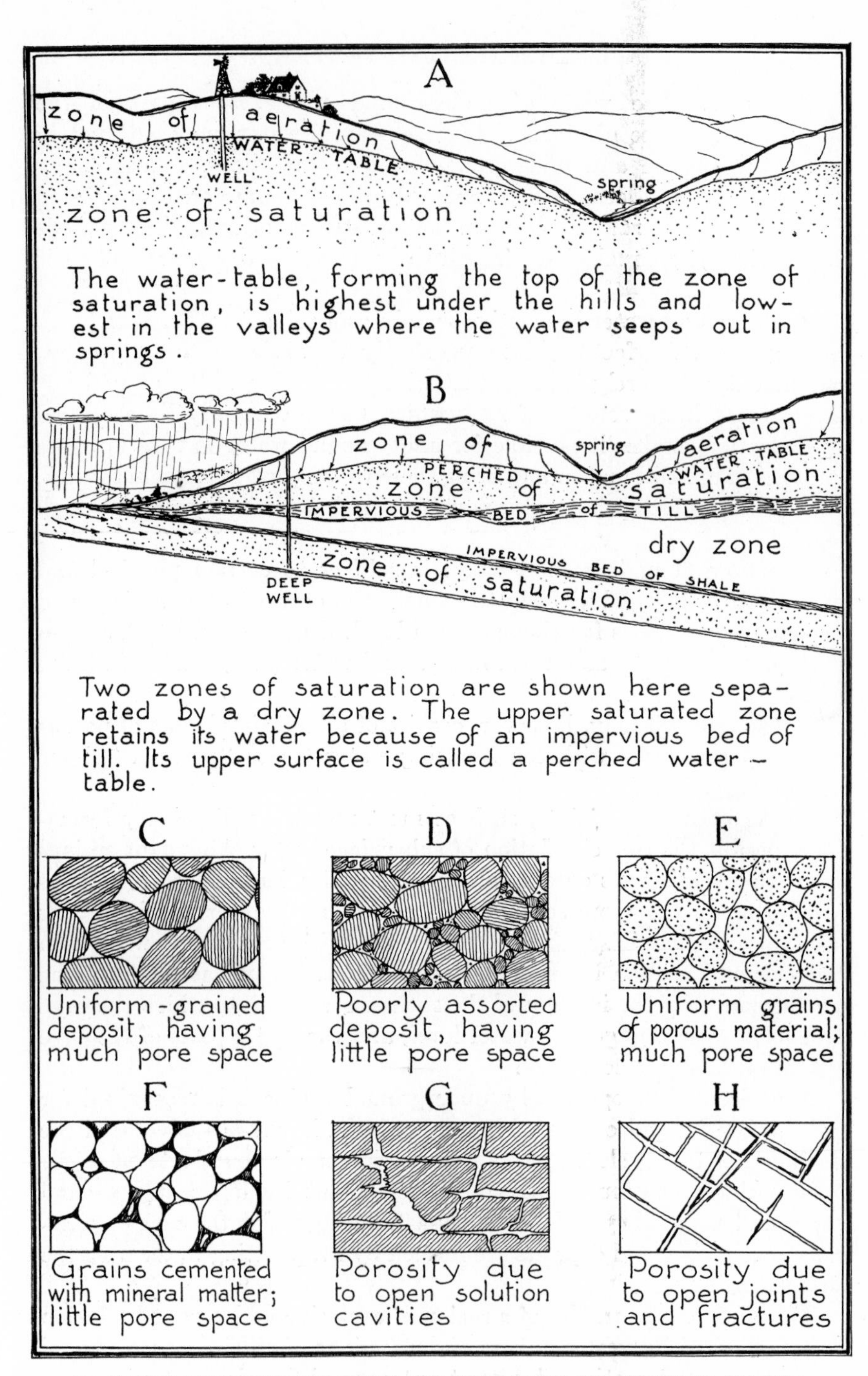


FACTORS INFLUENCING THE CIRCULATION OF GROUND WATER

General Conditions and Occurrence of Underground Water. A large amount of rain water soaks into the ground. Below a certain depth the soil and even the rocks are saturated. The top of the saturated zone is known as the *water table* (Fig. *A*). Occasionally several water tables are separated by zones of dry rock or soil. The surfaces of the upper water-bearing zones are then called *perched water tables.* The deeper or permanent water table derives its water from more remote sources (Fig. *B*). In porous sand-dune regions, and in limestone areas having extensive subterranean passageways, almost all of the water goes below ground immediately to form *ground water.* But in areas covered with an almost impermeable layer of compact glacial till or *hard pan,* practically all of the water immediately flows off the surface into streams or else accumulates as lakes or swamps. Nevertheless, the presence of water in the subsurface formations is the usual thing.

The occurrence of this water underground would be of little geological significance if it were stagnant in the rocks. But generally it is in motion, with the result that significant geological and topographic changes occur. Springs and the seepage of water along valley sides constitute the chief source of water for streams. They manifest clearly the constant movement of water underground, as do geysers, hot springs, and artesian wells.

Underground water works in two ways: first, as an eroding and dissolving agent and, second, as an agent of deposition.

The porosity of formations, or the amount of water which they can absorb, varies tremendously. The most porous materials are loose layers of sand and gravel, or of clay, in which the open spaces or interstices may equal 50 per cent of the volume. On the other hand, igneous rocks average less than 1 per cent of pore space.

In well-assorted deposits—formations made up of grains more or less uniform in size, like the gravel and sand deposits of outwash plains—the pore space is very great (Fig. *C*). If poorly assorted—so that small grains fill the spaces between the large ones, as in impure sandstones made up of sand and clay—the porosity is low (Fig. *D*). A gravel deposit of uniform pebbles of a porous sandstone is especially favorable as an aquifer or water carrier (Fig. *E*). But if the grains of the rock are not loose but cemented, forming a consolidated rock, then the porosity is greatly lowered (Fig. *F*). Compact sandstones and shales average 5 to 10 or 15 per cent pore space. Fine-grained limestone has a low porosity but its solubility often renders it very permeable (Fig. *G*). Igneous and metamorphic rocks absorb almost no water, unless they are much broken by joints or cracks (Fig. *H*). Lava formations, especially those which originally formed surface flows, are often *vesicular,* that is, having numerous cavities like congealed froth, and these openings may admit large volumes of water.

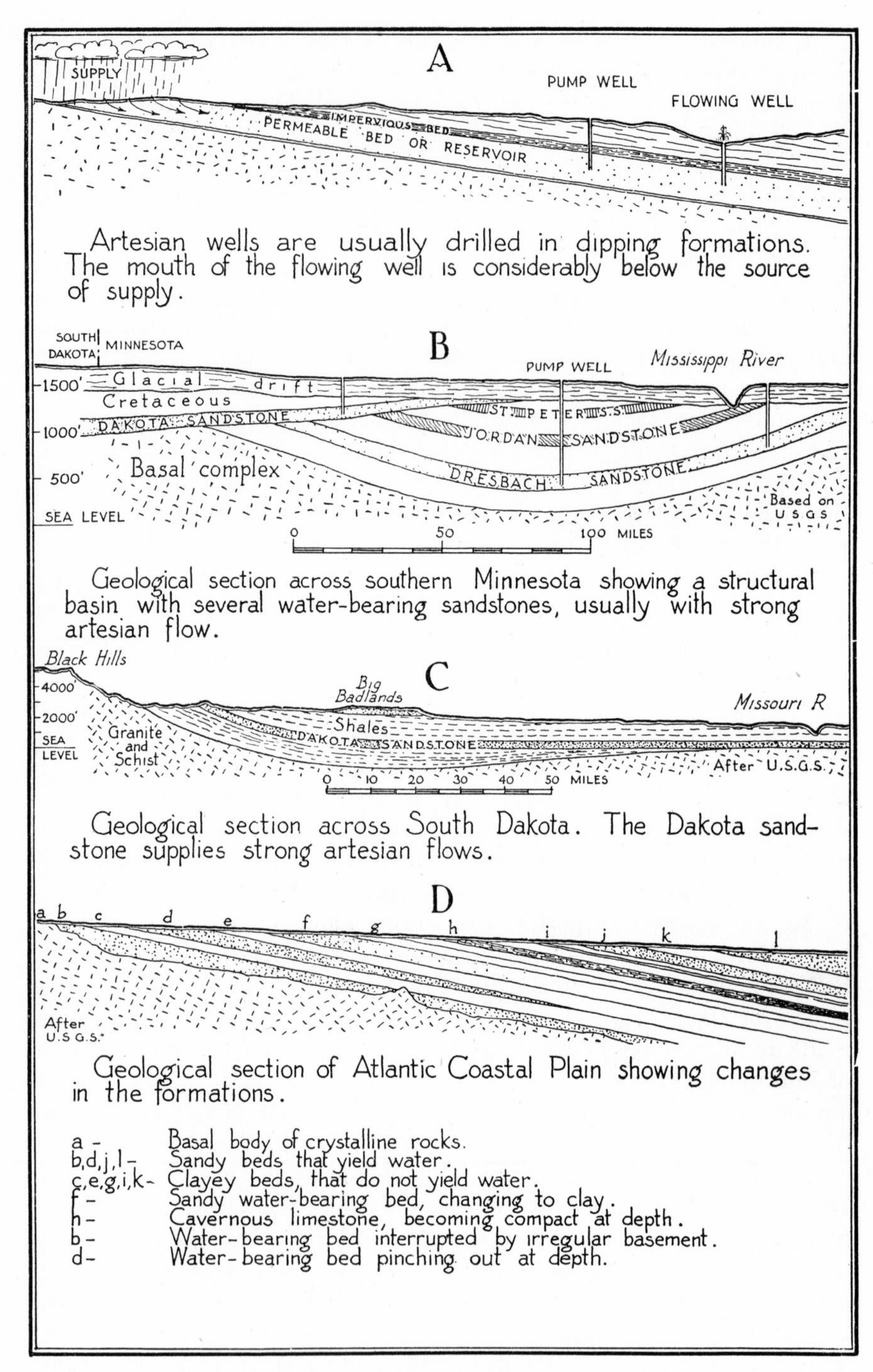


CONDITIONS INFLUENCING ARTESIAN WELLS IN DIFFERENT REGIONS

## WELLS

A well is an artificial boring or hole which penetrates the earth to the ground-water level. There are two classes of wells: shallow and deep. A shallow well reaches the zone of saturation at a moderate depth after passing through a zone of aeration. It may be only a few feet or in exceptional cases 200 feet below the surface. A deep well, however, may pass through various zones near the surface, possibly even perched water tables, but reaches a deep-lying zone of saturation where the water is held in by confining layers of impervious material. When this is punctured, the water may rise high above this impervious stratum because of the hydrostatic pressure behind it. If the top of the well is below the level at which the permeable bed or reservoir receives its supply at the surface of the ground, then there may be sufficient hydrostatic pressure, or head, to induce a constant flow of water above the mouth of the well. This constitutes a true *artesian well*, a name derived from the former province of Artois in northern France.

The usual geological conditions conducive to a natural flow of water from an artesian well are shown in Fig. *A*. There is a series of dipping formations, which includes a pervious or water-bearing stratum, with impervious beds above and below. The edge of the pervious bed outcrops and receives a supply of water from the surface. The water filling the pores of the pervious bed is therefore under pressure and, when this pressure is released by a well, the water, seeking its own level, rises to a height which may be almost equal to that of the outcrop of the pervious layer. Owing to the great friction encountered in the pores of the rock, this height is never actually attained.

Artesian Water Conditions in the United States. Lying east of the Rockies are the Great Plains. These are underlain by easterly dipping strata which include several important *aquifers* or water-bearing beds, particularly the Dakota sandstone formation, about 100 feet thick, overlain by a series of impervious shales in places over 2,000 feet thick (Fig. *C*). This Dakota sandstone contains abundant water and gives rise to strong-flowing wells in low-lying parts of North Dakota, South Dakota, Minnesota, Nebraska, Kansas, Wyoming, and Colorado. In the higher points of these states it supplies numerous pump wells. Other formations, *e.g.*, the Lakota, also yield abundant water, which everywhere in the Great Plains area is apt to be highly mineralized, because of the long journey it has made underground.

The deep wells of the Great Plains may penetrate down 1,000 feet and are of unusual importance because they provide water for a region markedly deficient in rainfall. Wells are also important in the Atlantic coastal plain, the Interior Lowland plains of Wisconsin and Illinois, and the alluvial-filled basins of the western United States, notably in California.

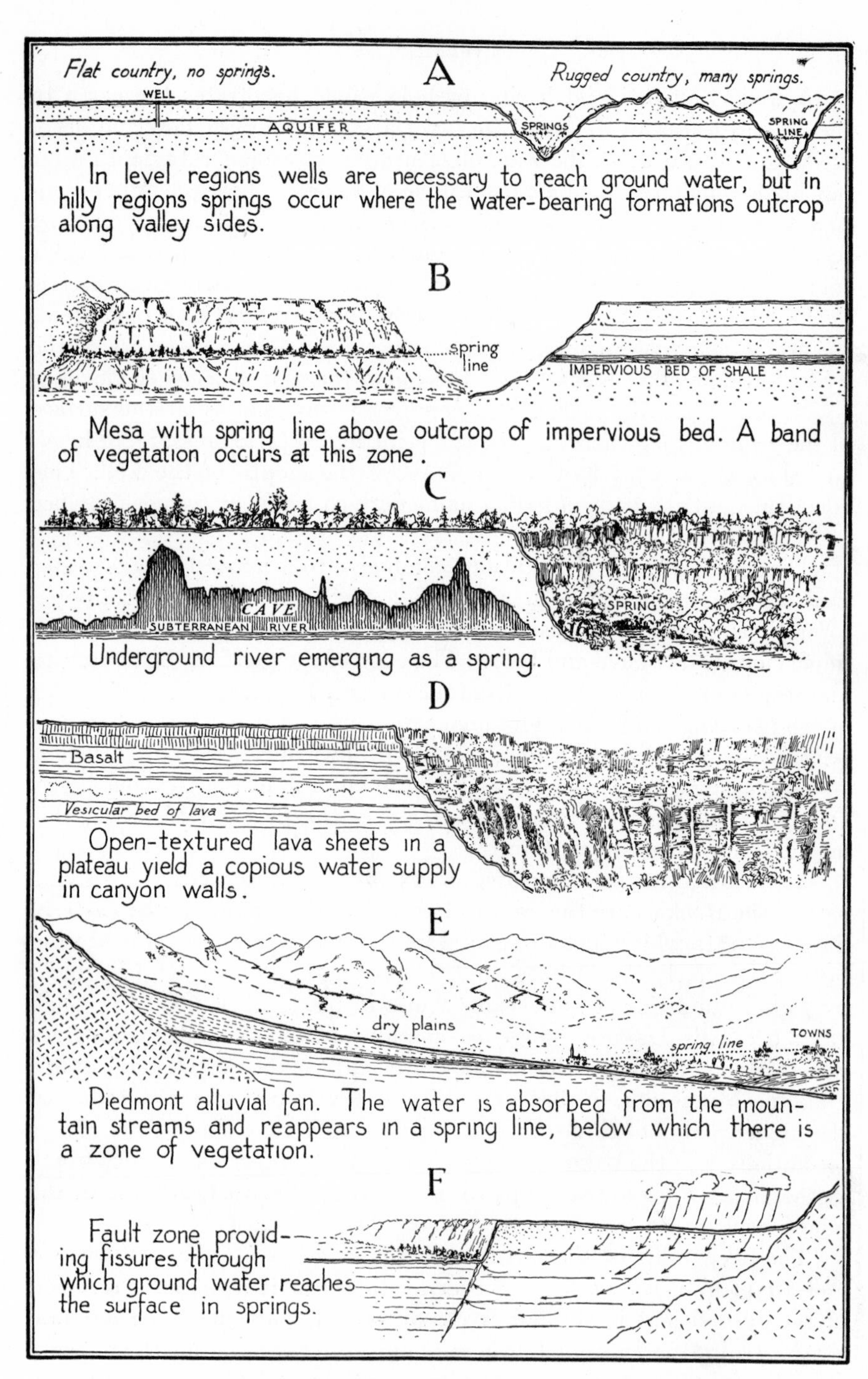


CONDITIONS INFLUENCING SPRINGS IN DIFFERENT REGIONS

## SPRINGS

In a flat region the porous formations become filled with water nearly to the surface of the ground. There is little movement of the ground water and springs are rare. If the formations contain much soluble matter, the water becomes highly mineralized because of its long-continued contact with the rocks. In a hilly or mountainous region, on the other hand, the water that seeps into the earth percolates rapidly downward and is likely soon to be discharged at a lower level, along the line of outcrop of some formation through which it has migrated.

This active flow of ground water results in effective leaching of the rocks and removal of the readily soluble material. Hence, the ground water from such rocks at the present time is generally soft.

In a flat region there may be good aquifers that yield large supplies, but these supplies can as a rule be recovered only by sinking wells; in a region of great relief, on the other hand, the ground water is more largely returned to the surface through springs, and wells are unnecessary. Thus in a gully or valley that has been eroded down to an aquifer, water will flow out of the aquifer, forming a spring. The gully or valley is not essentially different from a well dug to the aquifer (Fig. *A*).

In mountainous regions wells are not numerous, because ample supplies of good water are generally available from springs or from streams heading in springs. Indeed, much of the ground water is returned to the surface through springs.

LOCATION OF SPRINGS. Springs may be due to the following causes:

*a*. The outcropping of a porous bed along the side of a hill or in the walls of a valley. The porous bed may be underlain by an impervious layer which forces all of the water entering the ground above its level to follow it to its outcrop. This may cause a very copious flow. Conditions like this are found along the sides of mesas and plateaus where sandstone beds outcrop (Fig. *B*), as in Green Mountain near Denver; or where the underground water emerges from caves in limestone districts (Fig. *C*), as in Mammoth Cave; or where cavernous lava formations outcrop along the sides of canyons (Fig. *D*), as along the Snake River; or where gravel beds in alluvial fans lead the water which has seeped in near the apex of the fan to the margin of the fan, where it emerges in a spring line (Fig. *E*), as in northern Italy, and on the piedmont fans of southern California.

*b*. The presence of a fissure or crack, due to jointing or faulting, permits the ground water to rise from the reservoir, where it is confined under pressure, to the surface of the ground. Springs of this type occur at Saratoga Springs and elsewhere in New York; along the Balcones scarp in Texas; Owens Valley, California; and numerous places in the Basin and Range region of Nevada and Utah. Springs usually come from deep-seated sources and frequently contain much dissolved mineral matter. Most hot springs and geysers owe their location to fissures which extend to great depth.

Springs occurring along hillsides where small streams have cut gullies or valleys below ground-water level are apt to be small and intermittent, fluctuating with the rainfall. Springs due to the outcropping of deep-seated aquifers are much more reliable. Those due to deep fissures show little or no fluctuation with the seasons. Some of them possibly derive their water from magmatic sources.

## HOT SPRINGS AND GEYSERS

TEMPERATURE. There is no essential difference between hot springs or geysers, on the one hand, and cold springs, on the other. Nevertheless the activities of hot springs and geysers are sometimes so striking as to merit them a separate description. The source of the heat and the cause of flow are the chief considerations. The temperature of hot springs may vary from that of the rock through which the water rises, up to the boiling point of water, which varies for different altitudes. In Yellowstone Park the temperature in the Mammoth Hot Springs is 40°F. below the boiling point. Old Faithful Geyser has a temperature of 200°F., which is almost exactly the boiling point of water at that altitude, namely, 7,500 feet above sea level. The Giant and the Giantess have similar temperatures. The paint pots, spouting springs of mud, also reach the boiling point. Many geysers, however, run 10° or so below this figure, and the hot springs and pools are still lower.

The flow in the springs which have temperatures below the boiling point is supposed to be caused by convection, that is, by unequal heating of the water as in a hot-water heating system; by hydrostatic pressure, as in an artesian well; and possibly also by the pressure of confined gas, or gas developed in small quantities by chemical reactions. Under conditions of low temperature the flow is apt to be steady. But in a geyser the flow is usually spasmodic.

SOURCE OF WATER. The water supplying the flow of geysers and hot-springs is practically all meteoric in origin. That is, it comes from the rainfall of the region. A small amount may be magmatic in origin: it may be given off by the molten rocks beneath. This, however, is true only in a few localities, as, for example, in the high-temperature fumaroles of Katmai in Alaska, reaching 650°C., and in the hot springs of southern Idaho, many of which have no artesian source of water. In that locality the hotter springs are the smaller, which suggests that dilution with ground water from moderate depths is responsible for the low temperature and the greater volume of the larger springs.

One of the chief lines of evidence for the meteoric origin of the water in hot springs and geysers is the fact that practically all the elements brought to the surface in solution by such thermal waters occur in the rocks through which the waters have passed. Waters coming from magmatic sources contain some of the rarer minerals such as arsenic and boron. Moreover, it seems to be an established fact that fluctuations in rainfall are reflected in the behavior of the springs and geysers.

As to the depth from which the heat is derived by the descending ground water, very careful studies of thermal gradients, that is, changes of temperature with depth, made in deep wells and springs, indicates that the Yellowstone Park waters, for example, bring their heat from depths between 3,400 and 8,000 feet.

The ground water seeping through joints and cracks in the bedrock penetrates to depths of several thousand feet where it comes in contact with heated rocks. Hydrostatic pressure from behind forces it to ascend by circuitous courses through seams and cracks in unaltered rock which slowly widen under the disintegrating influences of igneous vapors. Finally, the thermal waters, following these cracks, issue at the surface as hot springs and pools. Underground reservoirs are thus excavated and become sources of hot springs and, under favorable conditions, of geysers. The geyser itself is simply a stage in the development of geological processes. In time geysers themselves become extinct. New geysers break out and, given the essential physical conditions, may develop eruptions quite as fine as any in action at the present time. Geologically speaking, the final stage of thermal activity is a hot spring. The tendency of a geyser is to develop a hot surface pool. It is frequently stated that some geyser has ceased to be active, and that this indicates the slow dissipation of the original source of heat. This is an error. The change is more probably due to a shifting of the channel of the ascending waters. New geysers originate by the opening of new waterways along fissure planes in the rock, and such new orifices of overflow are continually forming to compensate the diminution of activity of older vents.

DEPOSITS. Water pouring from hot springs and geysers is highly charged with gases and mineral matter. The mineral matter in solution is derived from the rock through which the water passes. If this rock is rhyolite or other siliceous igneous rock like that which abounds in the Yellowstone Park region, then silica forms the chief mineral constituent. If, however, the waters rise through beds of limestone as they do in the Mammoth Hot Springs area, then calcium carbonate is carried in large quantities. Several causes combine to bring about the rapid deposition of these minerals: (*a*) The water evaporates; (*b*) it cools and thus loses some of its solvent power; (*c*) the gases escape, rendering it less solvent; (*d*) the pressure is reduced; (*e*) algae growing in the hot pools and streams remove some of the mineral content.

The deposition of siliceous sinter or geyserite around the mouths of geysers builds up tubes or chimneys sometimes 5 to 10 feet in height, and the constant addition of material left by the water pouring away on all sides develops a geyser cone, a characteristic topographical form associated with many geysers. At the Mammoth Hot Springs a splendid series of terraces has been formed of calcium carbonate. Wherever the water collects in pools or basins and overflows, the rim at the point of discharge is built up most rapidly because the overflowing water cools most quickly as it passes over this edge and deposits there its mineral matter.

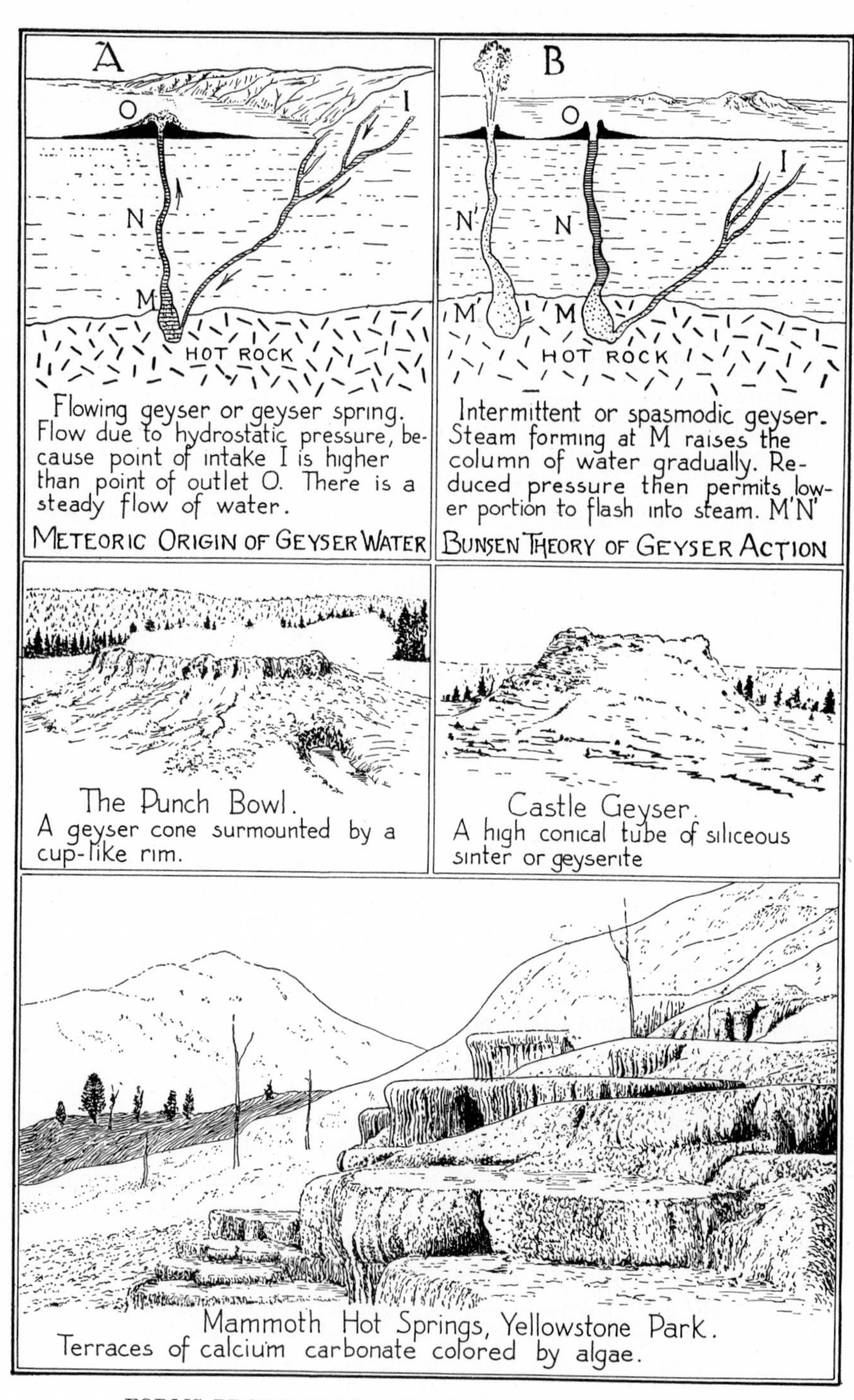


FORMS PRODUCED BY GEYSERS AND HOT SPRINGS

## GEYSER ACTION

There are two *types of geysers:* those which discharge violently at intervals, and those which pour over continuously. Old Faithful represents the first type. Excelsior geyser in Yellowstone Park, better called a *geyser spring*, represents the second.

THE BUNSEN THEORY. The principle of geyser eruption depends upon the fact that the boiling temperature of water increases with increased pressure. A higher temperature is required to boil water at the bottom of a vertical tube than higher up in the tube where the weight of the column of water is less.

The usual conditions are shown in Fig. *A*. Water of meteoric origin—rain water and melting snow—enters the ground at *I* and percolates down until it encounters a mass of heated rock. The geyser tube *N* is likewise filled with ground water. The only difference between tube *N* and the other tubes and crevices is that *N* is the most open and permits the escape of water to the surface most easily. If the point *I* is higher than the geyser mouth *O*, there will be a constant flowing out of water at *O* because of the hydrostatic pressure. This is the *Excelsior* type of geyser, simply a *flowing spring*. Its waters are violently boiling as they pour in vast volumes into the pond below.

If the column of water *OM* (Fig. *B*) is higher than the column of water *IM*, there is no natural and steady flow of water because there is insufficient hydrostatic pressure. The water at *M* remains in constant contact with the heated mass of rock until steam is formed. This slowly raises the whole column of water *OM* until it pours over at *O*. A quiet outflowing customarily occurs at Old Faithful and other geysers before an eruption. As the hot water at *M* is raised in the tube, as at *N*, it eventually, because of the reduced pressure, flashes into steam. The whole column of water is thus violently ejected. The display at Old Faithful lasts about 4 minutes, during which time it is estimated 3,000 barrels of water are discharged. The empty tube then gradually fills up again with ground water. After about an hour this water has reached the necessary temperature and the outburst is repeated.

Complications in the conduit system account for the irregular behavior of many geysers. In general, geysers which are irregular in their eruptions have continuously overflowing vents; and the most regular geysers have confined waters, which overflow only during eruption. This is explained by the fact that the overflowing vents are under hydrostatic pressure, cooler water from lateral ducts is continually replacing that which flows off, and the ebullition necessary to produce eruption is thus prevented. Where the water is confined and the supply of heat is constant, cooler water rushes in only after each eruption, and a definite interval is required to bring it to the boiling point at the base of the column.

## THE WORLD'S HOT SPRINGS AND GEYSER REGIONS

DISTRIBUTION. Thermal springs, or those whose mean annual temperature exceeds that of the locality in which they are found, are almost universal in their distribution. No continent is without them, with the exception of Australia, and even here they may be said to exist in a fossil state, for sinters and siliceous deposits are found in New South Wales, in a basaltic and trachytic region, indicating the former presence of hot springs and possibly of geysers.

Latitude has no effect upon their distribution, for we find them equally hot in the Arctic regions and under the equator. They are found in the frozen fields of Siberia and on the islands of Alaska, while the Andes have boiling springs from one end to the other. Venezuela and Patagonia, at the extremes of South America, both have hot springs.

When we come to geysers, we find them somewhat more limited in their occurrence, and yet even they are confined to no particular quarter of the globe, for each continent appears to have its geyser region. North America has the geysers of the Yellowstone National Park; Asia, a geyser region in Tibet; while the Iceland geysers may be considered as belonging to Europe, and the New Zealand field to Australia. Africa and South America seem to be left out, and yet the comparatively unimportant geyser area of the Azores can perhaps be considered the African representative; while in the boiling lake of Dominica and the water volcano of Guatemala, at least Central America may be said to have fair representatives of geyser action.

ICELAND. Three of these areas deserve especial mention because of the great size and number of their geysers, namely, Iceland, Yellowstone National Park, and New Zealand. In Iceland, geysers and hot springs occur over an area 5,000 square miles in extent. The Great Geysir, an Icelandic word meaning a *gusher* or *spouter*, has been known for many centuries and is the source of our present word applied to all similar phenomena. Another famous geyser in Iceland is known as *Strokr*, the churn, and there are others whose names indicate their behavior at the time of eruption. The Icelandic geysers, like those in Yellowstone, stand on low mounds resembling inverted saucers 70 feet or more in diameter. Many abandoned geyser mounds are to be seen in this locality. Numerous hot springs colored with algae still contribute vast volumes of water to the adjacent streams. The water flowing from the Icelandic geysers and springs is all meteoric in origin; that is, it comes from the rain and snow which now falls there. Great quantities of water derived from melting snow and ice penetrate the rocks along fissures down to zones of high temperature, rising along other channelways to the geysers and springs.

Yellowstone Park. In Yellowstone Park the geysers are largely concentrated in three distinct basins, known as Upper Geyser Basin, Lower Geyser Basin, and Norris Geyser Basin. These basins are structurally grabens or down-faulted blocks of the Yellowstone Plateau. The rimming walls of the basins are more or less abrupt fault scarps. The geysers and springs derive their water from the surrounding plateaus. The water seeps underground and rises again along the faults which have so completely shattered the floors of the basins. Many geysers and springs occur along other major lines of faulting, which have broken the plateau into large blocks, as shown in the preceding figure. The Mammoth Hot Springs, for instance, start at the northern end of the zone of prominent fractures which determine the geyser basins farther south. It is noteworthy, too, that the Yellowstone River, for part of its course, follows a similar fracture zone. Visitors to the Grand Canyon of the Yellowstone may occasionally observe jets of steam rising from the bottom walls of the canyon where hot springs occur. In fact, the brilliantly colored rocks of the canyon are due to the strong weathering of the plateau rhyolite under the chemical action of hot-spring vapors concentrated along this zone. The location of the canyon is to some extent determined by the weakness thus produced in the rocks.

In Yellowstone Park over two thousand springs have been enumerated and mapped, and among them are seventy-one geysers, of which twenty are known to spout to a height of not less than 50 feet. The stream from Old Faithful rises 150 feet but there are several which throw their water still higher.

New Zealand. The famous geyser area of New Zealand is situated along a zone of volcanoes in the central part of North Island. The largest geyser ever known was Waimangu, which during its greatest activity threw a column of water estimated at 1,500 feet in height and once threw a boulder weighing 150 pounds a quarter of a mile. However, it plays no more, after a short life of only two years. It came into being in 1901 following a terrific explosion during which it ejected great quantities of mud and rocks. Being a new geyser, it had not yet lined its conduit with siliceous sinter and its waters were always muddy and dark instead of clear, as is the case with geysers that have been longer active.

Fumaroles. In some districts *fumaroles* or steam jets take the place of geysers and hot springs. This, to a large extent, is true in the Valley of Ten Thousand Smokes of southern Alaska. These jets of steam, occurring almost without number on the floor of the valley, follow very definite fissures which run either around the margin of the valley or crisscross in all directions. Some of these clefts or chasms are 10 feet wide at the surface and extend to seemingly bottomless depths. In other cases the fumaroles are situated not on cracks but on domes or craters and seem to have been at one time explosive. The temperature of most fumaroles

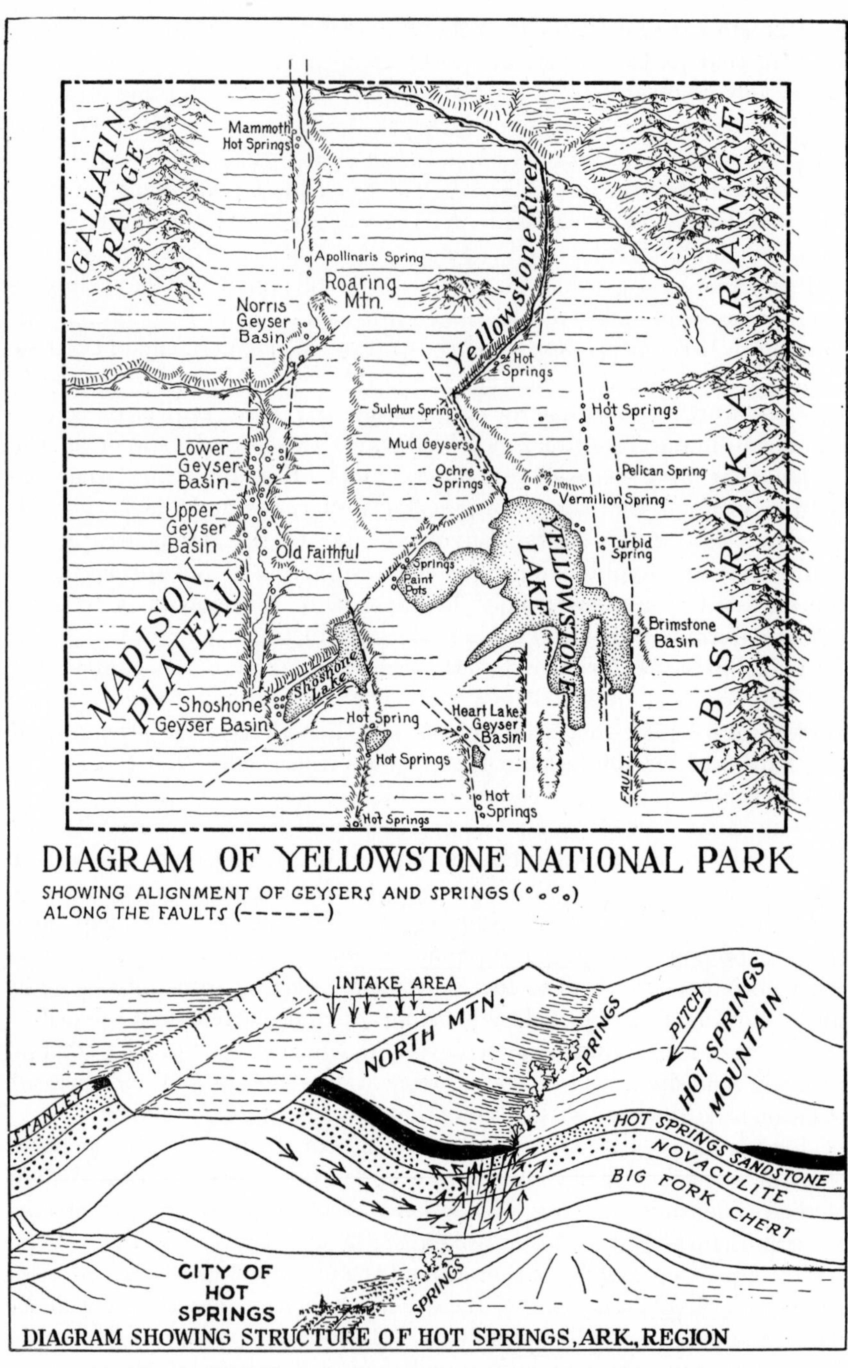


CONDITIONS INFLUENCING THE LOCATION OF GEYSERS AND SPRINGS IN YELLOWSTONE PARK AND HOT SPRINGS, ARK., REGIONS

is much greater than that of hot springs. The superheated steam is quite invisible where it first rushes out, with a temperature of 645°C., or 1200°F. Sticks and shavings of wood rapidly burst into flame when plunged into this hot vapor. Steam, the predominant constituent in all the fumaroles, varies from 98.4 to 99.99 per cent by volume. Other gases are hydrochloric acid, carbon dioxide, hydrogen sulphide, nitrogen, hydrofluoric acid, and some oxygen and ammonia. It is believed that part of the steam is undoubtedly derived from the vaporization of surface drainage, but also that much of it must be of volcanic origin. Practically all the drainage from the watershed tributary to the valley is vaporized by the heat and returned to the air. This, however, is only a small part of the total volume emitted by the fumaroles. Practically no seasonal change is to be noted in their behavior, in spite of the fact that during the winter period no run-off is contributed to the valley.

ARKANSAS. A unique hot-spring region, contrasting somewhat with the hot-spring regions already described, is in central Arkansas, around which the city of Hot Springs has been built. The contrast lies both in the fact that these springs do not lie in a volcanic area and also in the fact that they do not seem to be controlled in position primarily by any major fault zone. Instead, they come to the surface along the outcrop of a porous sandstone. The structural conditions are represented in the opposite figure. According to one interpretation, the water enters the Bigfork chert along the valley just west of North Mountain. Seeping underground beneath North Mountain, it reappears in the valley between Hot Springs Mountain and North Mountain. Owing to the pitch of North Mountain anticline, however, the greatest flow of water is at the tip end of the anticline, this point being slightly lower in level than the intake area to the north. Moreover, the broken character of the Arkansas novaculite, due to the anticlinal structure of Hot Springs Mountain, permits the water along that zone to rise through it and emerge in the Hot Springs sandstone and at the base of the Stanley shale, which acts as a capping layer. The heat is thought to be derived from contact with masses of hot rocks, not exposed anywhere at the surface. Some investigators have suggested that the mere movement of the water through the small pores of the rock is sufficient to raise its temperature, which in most of these springs is about 140°F.

It should be particularly noted, however, that some students of this region believe that there are deep-seated fractures which control the flow of water, but which do not seem to have any surface expression.

It is only fair to add in conclusion that, in practically all the regions of mineral springs and especially in the regions of hot springs the world over, faults seem always to determine the position of the springs. Of the many thermal springs of this type, those of the Great Basin may be mentioned as a final example.

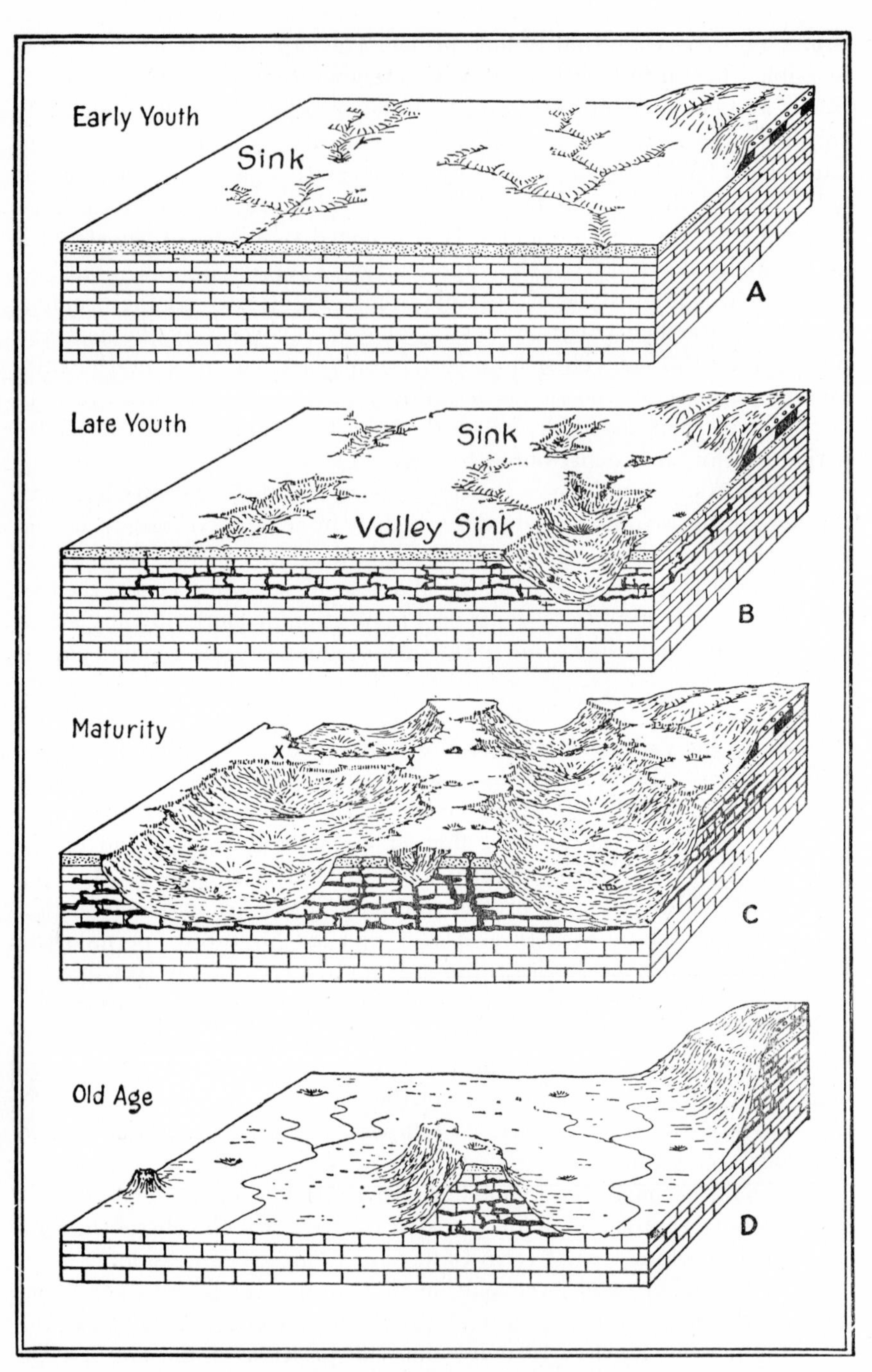


FOUR STAGES IN THE LIFE CYCLE OF A PLATEAU HAVING UNDERGROUND DRAINAGE

The plateau here represented is made up largely of limestone. Except for the earliest and latest stages, surface drainage is almost entirely absent.

## LIFE HISTORY OF PLATEAU REGIONS HAVING UNDERGROUND DRAINAGE

The geological structure of limestone regions having underground drainage may be highly complicated or relatively simple. The famous Karst district lying east of the Adriatic Sea is strongly folded. The Causses of southern France likewise lie in a much disturbed and faulted plateau. But in Indiana, Tennessee, Kentucky, and the Carlsbad Cavern region of New Mexico, the formations are virtually horizontal. Here the stages of development are relatively simple. Limestone beds may constitute the entire surface of the country or beds of shale or sandstone may be included in the limestone series. There must, however, be some massive or relatively pure limestone if characteristic underground-drainage features are to develop.

At first (Fig. *A*), there is the usual dendritic pattern of surface streams. If the limestone outcrops everywhere, these streams will lose their water underground. The figure shows a sandstone bed overlying the massive limestones. This maintains surface drainage for a certain period. The sandstone is gradually worn through by the streams and its open joints also allow the surface water to get underground. The limestone is soon attacked and subterranean solution channels are developed, followed by surface depressions known as *sink holes*. They are usually conical pits only a few feet across or as much as an acre in extent. The surface water drains into these depressions and disappears underground. With the advancement of youth the sink holes become numerous and grow larger as the water pours into them.

In maturity very little of the original plateau surface remains. Sink holes abound. Some have developed to great size and are known as *valley sinks*. They contain no surface streams and their floors are undulating and pitted. All of the water goes underground immediately. There are no perennial surface streams. Springs may occur around the margin of the valley sinks, but the water soon disappears again.

On a plain the development proceeds as on a plateau except that deep valley sinks are not formed. The mature stage shows numerous small sinks. On both plains and plateaus, the openings in the bottoms of the sinks may become clogged, and ponds may result. These constitute the chief source of drinking water for the cattle of the region. Frequently the sinks have sharp sides and open directly into caves.

In old age the country assumes a flat aspect with scattered mesas and buttes. Surface drainage reappears. In some regions this is due to the removal of all the limestone down to the water table. The surface water is unable to penetrate into the saturated bedrocks. Elsewhere it is because deeper lying rocks other than limestone have now been exposed to erosion. After this final stage, some regions enter upon a new cycle of development as the uplifted plain with lowered water table becomes pitted with a new generation of depressions.

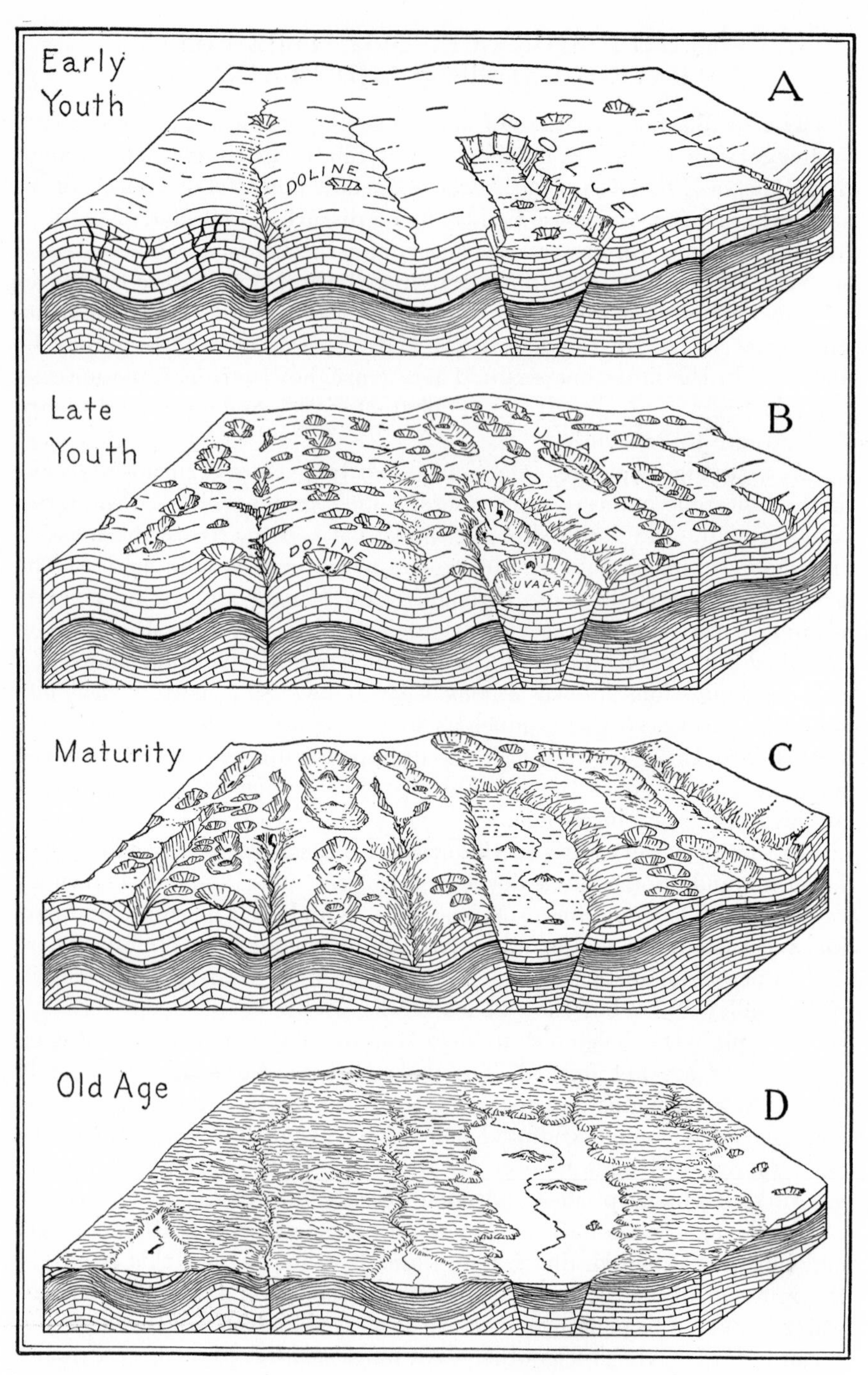


FOUR STAGES IN THE LIFE CYCLE OF A FOLDED AND FAULTED REGION HAVING UNDERGROUND DRAINAGE

## THE EVOLUTION OF KARST TOPOGRAPHY IN COMPLICATED STRUCTURE

The accompanying figures illustrate several stages in the development of karst topography, in the typical karst region of the world lying east of the Adriatic. In this region the rocks are strongly folded and faulted. Figure *A* represents the initial stage, just after a few sinks have formed. Surface drainage, however, is the usual thing. There are occasional long steep-walled depressions of tectonic origin. Some may be true grabens. These are called *poljes*. They are features peculiar to deformed limestone regions and are not found in undisturbed plateaus like those of Kentucky. The normal drainage of the region follows the zones of faulting or downfolding.

In stage *B*, late youth, the *dolines** or funnel-shaped depressions have increased greatly in number, so much so as to consume almost all the initial surface of the area. The surface drainage has given way almost completely to subterranean channels. Some of the dolines have greatly increased in size, partly by the erosion of their margins and partly by the collapse of caves. Several may in that manner be joined. The elongated depressions thus formed resemble the valley sinks of the Kentucky karst region and are called *uvalas*. On their floors, streams may for a short distance appear at the surface, emerging from caves and again entering caves. Natural bridges may also occur there.

During maturity, shown in *C*, an extremely rugged condition prevails everywhere, the original surface of the land having entirely disappeared. Some of the earlier dolines have been destroyed where the land is reduced to lower levels. Valleys again appear and the underlying shales or other impermeable beds are in many places laid bare. Wherever this happens, the drainage can no longer flow underground, and surface streams appear. At first they are not aboveground for any long distance, occurring most frequently where the edge of the limestone mass is eaten away or undermined by the extension of subterranean caverns. The surface of the ground where the limestone still remains is apt to be extremely irregular, being made up of a maze of sharp pinnacles and points and fluted ridges or *lapies*—the result of ages of weathering and solution along the joints. Such features range from a few inches to 15 feet or more in height and render a country almost impassable to man and beast. The floors of the poljes have become more expansive and are quite flat except for the isolated hills of limestone called *hums*. In the Causses of France such remnants are called *buttes temoines* and in Porto Rico and other islands of the West Indies similar features are called *haystack* or *pepino hills*. They are mere shells honeycombed with caverns.

Old age again shows a normal system of surface streams in possession of the land. The impermeable rocks outcrop everywhere over extensive areas, above which rise occasional isolated hums.

* English *doline* or *dolina;* Austrian, *Doline* (sing.); *Dolinen* (plural), from the Russian, *dolina*, a little dale.

THE INTERIOR OF A CAVE. A DOME IN COLOSSAL CAVERN, KY.
The vertical channeling is due to running water which has come through from the ground above. The horizontal lines are due to the bedding planes of the limestone.

## CAVES AND CAVE PHENOMENA

THE DEVELOPMENT OF CAVES. Caves are found by no means only in limestone regions. Lava flows, for example, often have extensive underground openings resulting from the fact that the surface of the lava cooled and hardened and the part still fluid beneath ran out, leaving a tunnel covered by a roof. Waves frequently produce caves along the coast by their constant pounding and washing away the weaker rocks. Sandstones and other types of sedimentary formations often have vertical cracks or joints, which gradually widen through weathering or by slight movement of rock masses.

The most extensive caves, however, occur in regions of thick and comparatively pure limestone formations. The solubility of the limestone is the prime reason for the presence of the caves. Dense and massive limestones tend to a maximum development of caves because such beds are practically impervious to water except along fissures and bedding planes. By the concentration of solution in certain places, large cavities result. Comparatively porous limestones, on the other hand, which offer free passage for the water in all directions, do not usually develop caves or sink holes, except where the circulation of the water is concentrated locally.

The chief constituent of limestone is calcium carbonate, which is only slightly soluble in pure water. Approximately 75,000 parts of water are required to take into solution 1 part of calcium carbonate. Rain water, however, is not strictly pure but contains some carbon dioxide absorbed from the air. This carbonated water acts upon the limestone and produces calcium bicarbonate, thus:

$$\underset{\text{calcium carbonate}}{CaCO_3} + \underset{\text{carbon dioxide}}{CO_2} + \underset{\text{water}}{H_2O} = \underset{\text{calcium bicarbonate}}{Ca(HCO_3)_2}$$

The resulting bicarbonate is thirty times as soluble as the calcium carbonate, even in pure water. But, if the water still contains some carbon dioxide, it is even more effective as a solvent. It is evident, therefore, that the presence of carbon dioxide in water, absorbed either from the air or from the soil, renders it highly capable of dissolving away the limestone. In fact, the amount of water which falls as rain upon one acre of land in the Mammoth Cave region in the course of a single year is capable of dissolving some 25 or more cubic feet of rock. Expressed differently, the rain falling on this region is easily capable of removing in solution 1 foot thickness of limestone from the entire region every 2,000 years. Inasmuch as this solvent activity is concentrated definitely in certain places, it is not difficult to appreciate the method by which so many caves of such large size have been formed.

Concentration of water in massive limestones is effected in two ways: first, along the joints or fissures and, second, along the bedding planes.

Nearly all stratified rocks are cut by cracks or joints, roughly vertical. These often occur in two sets or systems which intersect each other at nearly right angles. Such joints offer avenues whereby the water which has soaked into the soil may penetrate downward to greater depths. As the water descends along the joints, solution takes place, the joints are enlarged and crevices are formed. The intersection of two joints which cross each other is a particularly favorable spot for the penetration of water into the limestone. In such situations more or less circular openings are dissolved, instead of the long narrow crevices produced along single joints.

Water which finds its way into the joints of the rock descends until its downward progress is brought to a stop. This may be the result of one of two causes: First, a bed such as shale or a massive limestone layer, which is impervious to water, may be encountered; or, second, the ground-water table, or the depth at which all the openings in the rock are filled with water, may be reached. In either case, when the downward movement of the water is brought to a stop, lateral movement begins and channels are dissolved out along the bedding planes or in porous layers in the limestone.

Erosion in Caves. Cave channels and passageways are developed not only by the solvent action of water but also by its erosive force, in the same manner as on the surface of the ground. This is aided by the fact that the underground streams often contain a great deal of silt which has been carried into them from the surface of the region above. When it is realized that all the material which formerly filled the spaces now occupied by the sinks and depressions has been carried underground through various swallow holes in the floors of the sinks, it is easy to understand that much sand and silt must thus be carried into the caves. This sand and silt comes in part from the disintegration of any overlying sandstone or shale beds and in part is the residual soil left after the limestone is dissolved away. The residual limestone soil is usually red or orange in color. Accumulations of it may be seen in practically all caves.

Cave Silt. Cave silt is the source of the nitrate deposits for which many caves were exploited during the War of 1812. Nitrates of calcium, sodium, and potassium are present in this earth and are probably derived from the excretions of bats which formerly were numerous in many caves. In some of the caves many bats are still found. The nitrates were leached out from this "peter dirt," as it was called, and the resulting saltpeter was used in the manufacture of gunpowder. During the Civil War many of the caves of Tennessee provided a similar source. Water was conducted into the cave through a wooden pipe and the concentrated solution pumped back to be evaporated outside the cave.

Cave Passageways. Cave passageways in general are of two types: those which follow the joints and those which follow the bedding planes. The former are apt to be high and comparatively narrow, and the latter wide and comparatively low. However, both types, after they have been produced, are much modified by the constant cracking off of fragments from the walls and roof. This material, falling on the floor of the cave, is ground up, dissolved, and washed away by streams of water. Sometimes gigantic "halls" result, having a height of 100 feet or more and encompassing an area of an acre or so. When the ceiling of a cave consists of a firm and massive layer of limestone, it is less subject to destruction in this manner. Some such ceilings extend over thousands of square feet of cave, as flat and uniform as an artificial structure.

The natural tendency, however, is to develop the form of an arch, inasmuch as this provides the greatest strength. Sometimes the material which has broken from the ceiling accumulates in such quantities on the floor of the cave as to close it off entirely, though usually there are little openings between the top of the pile and the ceiling through which a person can crawl. Explorers in caves are rarely daunted in following a large passageway to find it interrupted in this manner, knowing that beyond the obstruction it will open out again.

Besides the more or less horizontal passageways which follow the bedding planes of the most soluble limestone layers, there are in many caves deep "pits" and high "domes." These are excessively large openings developed along vertical joints. They are formed by running water coming through from sink holes above and washing down the face of the joint planes. Seen from above, these openings are known as *pits*. Seen from below they are called *domes*. In them cave deposits are rare because erosion and solution are the prominent activities and the water does not halt to deposit its load.

The corkscrew passageways, known to the tourist as "fat man's misery," are found in almost every cave and are merely tortuous ways of going through piles of large broken blocks of limestone. Sometimes they provide convenient means of getting from one level of the cave to another. It should not be understood from this that caves are liable to collapse. This rarely happens. Blocks frequently drop from the ceiling of caves until an arch is produced. It used to be thought that the falling in of caves resulted in the formation of natural bridges if a portion of the roof happened to remain in position. But practically no examples of this are known. In the Mammoth Cave region, for example, with its numerous large caverns, sink holes, and valley sinks there is not a single natural bridge. Even the famous Natural Bridge of Virginia is known to have been formed in a different way.

Limestone caves occur in virtually every state in the union and in all countries. For every cave which has become famous there are scores of others off the beaten path known only to their discoverer, and for each one of these there are undoubtedly hundreds not yet visited by man.

CAVE DEPOSITS

## CAVE DEPOSITS

The deposition of calcium carbonate in caves occurs when the water carrying it in solution evaporates or loses its carbon dioxide owing to agitation as it trickles over rough rock surfaces. The most common forms are *stalactites* and *stalagmites*. On the ceiling of the cave are the stalactites, hanging like icicles but having greater variety of form. Rising from the floor of the cave are the stalagmites.

Some of the many forms of deposits seen in caves are illustrated in Fig. I. Long, thin, hollow, tubular stalactites (1) result from a rapid flow of water. These change to sturdier forms (2) when the central passageway becomes clogged. There may be double forms (3) or clublike masses (4) or branching forms known as *helictites* (5), or baconlike sheets (6), or there may be rows of stalactites following the joint planes on the ceiling (7).

The stalagmites rise from the floor as the dripping water evaporates there. If the dripping is slow and evaporation rapid, the stalagmites rise like a column (8); but if considerable water falls, the resulting stalagmite will have a broad base and a flat top (9) or it may be built up of a succession of disks (10), with iciclelike forms hanging around the margin.

Stalagmites rise to meet the stalactites directly over them. If the water drips slowly from above so that most of it is evaporated before it falls, the stalactite lengthens rapidly while the stalagmite below rises slowly (11 and Fig. II,*A*) but the reverse may happen (12 and Fig. II,*B*). Columns are ultimately formed either singly (13 and Fig. II,*C*) or in rows (14); occasionally projecting ledges intercept the dripping water with the turnip-shaped result shown at 15 or the long fluted columns shown at 16. Stalagmites perched on loose blocks appear at 17, and also in Fig. II,*D*. At 18 (shown also in enlarged view in Fig. III) there is a stalagmite with a hole worn in its summit followed by the growth of a new stalagmite.

Travertine terraces are shown in 19 and also in Fig. IV, where the water forms pools among the limestone blocks and, overflowing the rims of the basins thus formed, builds up little walls or ridges resembling the terraces of the Mammoth Hot Springs in Yellowstone Park. Sometimes, as in 20, cauldron-shaped features are produced as the rim grows inward over the basin with the constant addition of material at the water's edge (Fig. V).

For most of the forms just described the term *onyx* is popularly used. Strictly speaking, however, onyx is a deposit of silica. *Travertine* is the correct name for these calcareous formations. The term *cave onyx* has also been suggested. Impurities are always present. If they are in the form of iron oxide, the travertine assumes a reddish or yellowish hue, often very pleasing in tone. Sometimes manganese dioxide is present and this imparts a deep-black color to the formations. The ceiling of many caves, as in Star Chamber of Mammoth Cave, is coated over with manganese dioxide, which often takes the form of jet-black patches. Crystals of calcite sparkling through the mass simulate the stars.

Gypsum or hydrated calcium sulphate is another common deposit on the walls of caves. Sometimes the gypsum deposits take the form of elaborate rosettes several inches in diameter and often of exquisite beauty. In other places, the gypsum occurs as long fibrous crystals on the walls or in the silt on the floor of the caves. Epsomite or magnesium sulphate occurs in some caves as masses of fine fibrous crystals coating the walls.

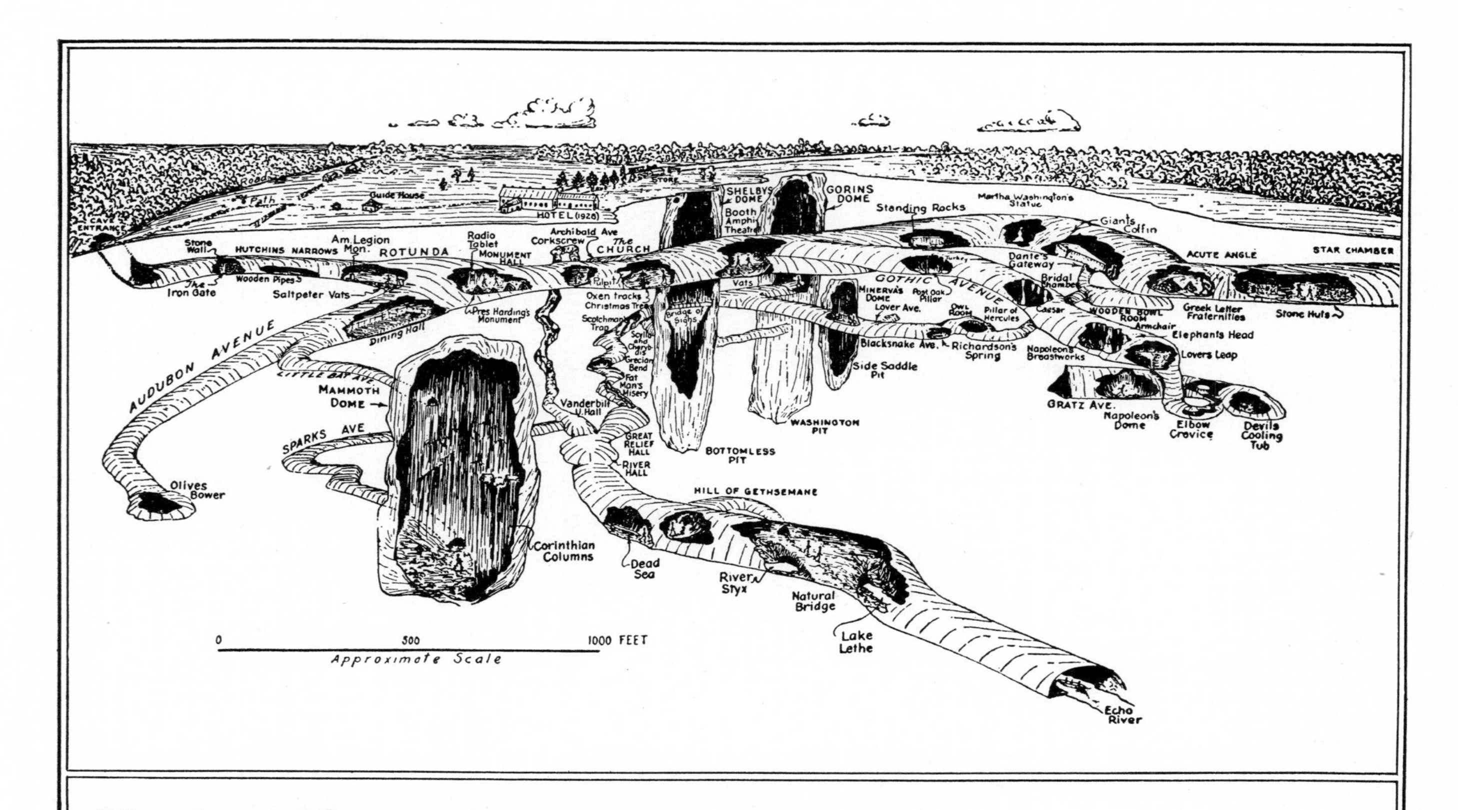


DIAGRAMMATIC REPRESENTATION OF MAMMOTH CAVE, KY.

Showing different "levels" of the cave which follow the horizontal bedding planes of the rock; and "domes" which follow the vertical joints.

Some notion of the extent and intricacy of the passageways in a typical cave may be gained from the accompanying figure which illustrates in a purely diagrammatic manner a part of Mammoth Cave. This represents the passageways followed by visitors on two of the shorter trips through the cave. Other longer routes extend beyond this view for several times the distance here shown. Moreover, there are scores of smaller byways and openings winding in and out in a maze of passages which are difficult of access and extremely confusing to the explorer. In short, the whole limestone mass is literally honeycombed with caverns, not unlike a gigantic sponge in texture. The various passageways and domes illustrated in the diagram, intricate as they seem, represent only a portion of the many caverns and openings known to exist under that part of the plateau.

The diagram shows more or less clearly a number of different cave levels. These correspond roughly with the more soluble bedding planes of the horizontal limestone layers. Entrance is gained on a hill slope where one of these openings has been exposed along the valley walls of Green River. The main avenue of the cave, sometimes known as *Broadway*, represents the second level. The first level is a bit above it and is represented by Gothic Avenue, entered by a stairway from Booth Amphitheater. The third level lies beneath the second level. It is represented by Blacksnake Avenue. The fourth level is reached from the third level by going through the Scotchman's Trap and Fat Man's Misery to Great Relief Hall. The fifth level is that of Echo River.

The domes and pits of the cave correspond with widened joint planes and have a vertical extent almost from the plateau surface to the river level. These have been formed by water seeping through from the surface of the ground. Washing down the walls of the joint surfaces, it dissolves away the limestone. In some places, as in the Corinthian Columns in Mammoth Dome, part of the burden of limestone thus removed has been redeposited.

In addition to the avenues which correspond with the bedding planes and to the domes and pits which represent the vertical joints, there are occasional smaller and much more devious ways from one level to another, formed by currents of water. Such channels are often narrow and show clearly the scouring effect of the moving water. Elbow Crevice at the end of Gothic Avenue is the best example. Fat Man's Misery is somewhat similar. Still another type of passageway is that represented by the Corkscrew, which provides a direct means of getting from the main cave, second level, down to the river level. It owes its presence to the breaking away of a great mass of limestone. Numerous blocks of all shapes and sizes became dislodged and it is through this maze of disordered rock that the Corkscrew makes its way.

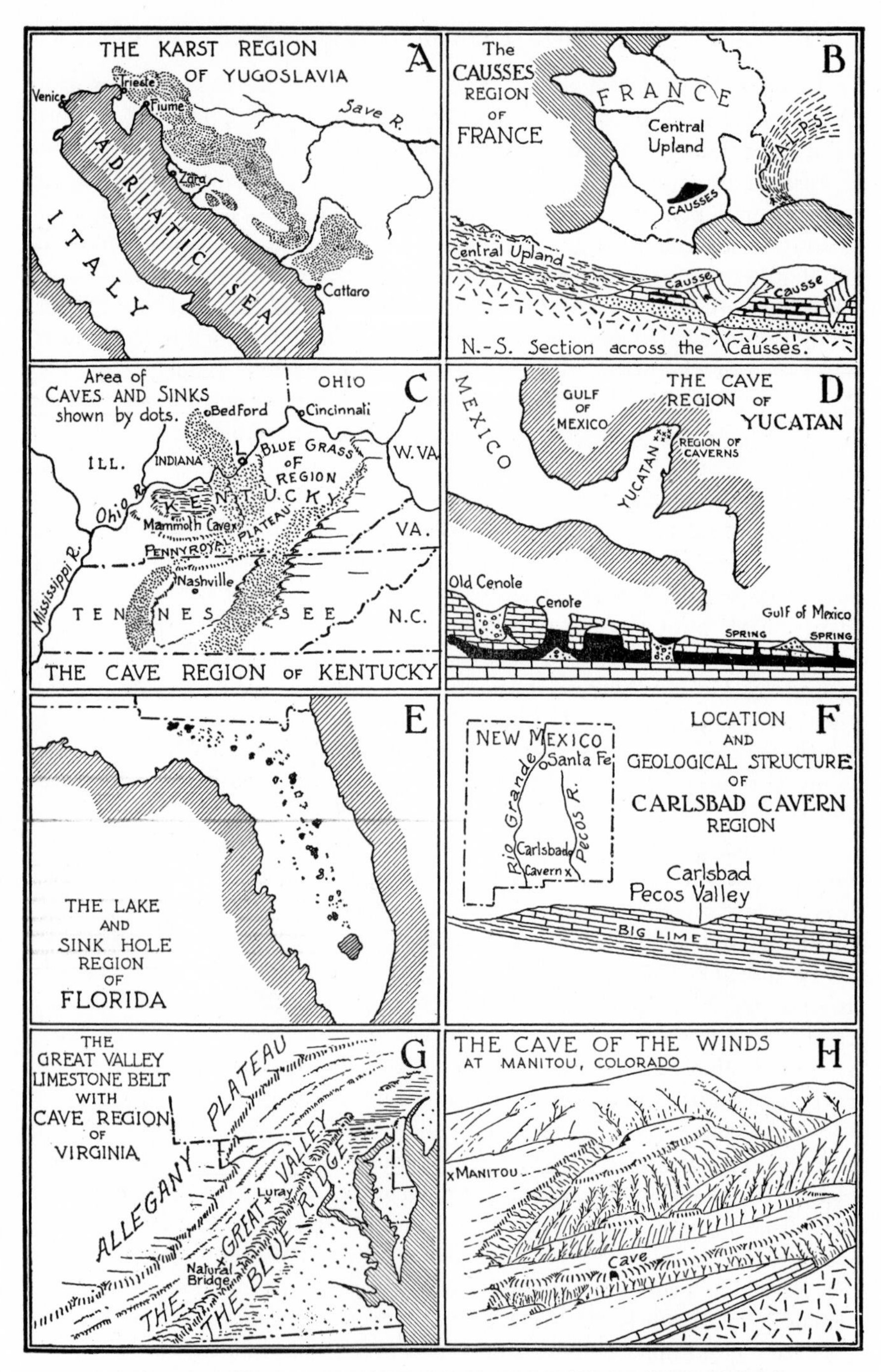


SOME OF THE IMPORTANT CAVE REGIONS OF THE WORLD

## SOME CAVE REGIONS OF THE WORLD

Limestone is so universally distributed that no large part of the world is without its famous caves. It is possible here to enumerate only a few of the outstanding localities where caves occur.

THE KARST REGION. Foremost of all perhaps is the famous Karst district of the Adriatic. This belt of country extends from Trieste over 450 miles southeastward to Montenegro. In places it is almost 100 miles wide, approximating New York state in area. This district consists chiefly of thick limestone rocks that are moderately folded. The surface of the limestone mountains or plateaus has become rough and uneven under the solvent action of the fiercest rainfalls of Europe. In some places the limestone is fretted by irregularly bordered furrows between which project rougher sharp-edged ribs of rock. But more peculiar to the Karst than these stretches of *karren* are the *dolines* or rounded funnels or pans of varied dimensions which break the smoothness of the surface. Even in regions overgrown with woods, the rocky substratum reveals itself in innumerable sharp edges of rock. Most of the Karst, however, in its barrenness is forbidding—a labyrinth of irregular forms that recur monotonously over wide expanses.

The stubborn intractability of the Karst to vegetation is due less to the absence of soil than to the lack of water. Rich as is the rainfall, the water quickly disappears into the clefts of the fretted rock and transfers its circulation and a part of its geological action to the interior of the mountain mass. There are large areas with no series of open descending valleys; some valleys, indeed, run in a deep furrow for a few miles, to be stopped short by a wall of rock; their streams disappear into its caverns, reappearing again at a considerable distance.

A particularly striking feature of these half-hidden, half-open water-courses is furnished by the so-called *polje*, large valleys often measuring a hundred square miles or more, and as a rule drained only underground through sink holes. The soil being fertile and abundantly watered, these depressions amid arid limestone ridges are often populous and industrious centers. The belt of Karst country is one of the most formidable barriers in Europe, cutting off the fertile plains of the Danube basin from access to the sea.

THE CAUSSES. A second famous region of karst topography is in southern France. It is known as the Causses, a term believed to relate to *calx*, meaning lime. Here on the southern slopes of the Central Upland is a region of limestone plateaus separated from each other by deep gorges. The surface of each block is a weird wilderness pitted with sink holes, some of which reach a depth of 700 feet. Below, it is honeycombed with caves and tunneled by a network of rivers. In these caves are the relics and carvings of the early cave dwellers, the first of mankind to inhabit Europe. In the course of about 30 miles the Tarn River alone receives thirty subterranean tributaries and not one on the surface. Sheep and mules are typical products of this district, so unsuited is it to agriculture and to grazing. Roquefort cheese is matured in the caves.

MAMMOTH CAVE REGION. A third region is the well-known limestone area of Kentucky, Tennessee, and southern Indiana. In Kentucky alone there are over 9,000 square miles with karst topography and underground drainage and probably as much more in the adjacent states. In the Pennyroyal or Mississippian plateau of central Kentucky there are estimated to be 60,000 to 70,000 different sink holes. The Mammoth Cave topographic map alone shows almost 3,000, an average of 10 to 15 to the square mile. The largest of the known sinks covers nearly 5 square miles but many are not more than an acre in extent. The great caves of the region lie underneath a plateau. Its margin, the Dripping Springs Escarpment, overlooks the undulating plain of the Pennyroyal. Whereas the Pennyroyal area is a land of sink holes, the plateau of the Dripping Springs Escarpment is the land of caverns and extensive valley sinks more like the typical karst region of Europe than any other portion of the United States.

YUCATAN. Northern Yucatan presents some karst features of unique interest. The limestone plain there slopes gently northward at a rate of about 1 foot per mile, and here one may look for miles with almost unobstructed view across the enormous plantations of henequen. In its more rugged and higher southern portion the surface is irregular because of the numerous channels or arroyos which interrupt it. Nowhere are there any permanent surface streams. The chief source of water is in deep caverns and sink holes. Most of these are of a peculiar chimneylike structure and are known as *cenotes.*

A cross section of the Yucatan plain is given in Fig. *D*. A typical cenote is shown; also a dome-shaped cavern the roof of which will later collapse to form a cenote or open well. This in turn will become filled with rubble as its steep walls break down and cease to serve as a supply of water. Away from the coast the cenotes are deep, the water level being 65 or 70 feet below the plateau. Near the coast the water level is very near to the surface of the plateau. In the lagoons fresh-water springs occur just offshore and there are actually springs of fresh water rising in the waters of the Gulf beyond the coastal reef.

FLORIDA. Somewhat similar to the Yucatan area is the sink-hole region of Florida. The surface of the interior of Florida is dotted with sink holes of all sizes from a few inches to several rods in diameter and often of great depth. When first formed, the typical sink is an opening leading from the surface through the superficial deposits into the limestone below. Many of these sinks, especially the smaller ones, are perfectly cylindrical, not funnel shaped, resembling the cenote of Yucatan. As a result of the subsequent caving of the banks, the bottom usually becomes clogged and the sides sloping. The formation of these sinks is practically instantaneous and is thought to result from a sudden caving of the earth. So frequent is their formation in certain sections that one must be on the lookout in driving through the country for newly formed sinks.

Sink holes are characteristic of that part of the state in which soluble limestone lies at or near the surface. If the limestone is covered by too great a thickness of clays or other impervious formations, sink holes do not form. Sinks, after being formed, tend to fill up by the caving of the sides, and as a result of the debris washed and blown into them. All ages of sinks, from the new steep-walled pits to the old and almost obliterated ones, are to be observed. Surface water collects in the sinks and forms ponds and lakes. This is especially true of the larger and shallower depressions whose floor has been covered with an impermeable layer of soil. Very large sinks or areas of depression are known as *solution basins*. When dry, they are known as *prairies;* when filled with water, they become lakes. Most of the lakes of central Florida are of this type. Above their floors rise tree-covered mounds called *hammocks*. Solution depressions occur in many other parts of the coastal plain. In the Carolinas they are known as *bays* and *savannas*.

CARLSBAD CAVERN. The conditions at Carlsbad Cavern, New Mexico, may next be mentioned. This gigantic cave occurs in a limestone bed which is over 1,000 feet in thickness. There are several other limestone, salt, and gypsum beds, all of which succumb to solution. Numerous large springs in the region and many extensive deposits of travertine, some at flowing springs and others at springs now extinct, attest the general honeycombed condition of the rocks. Sink holes are numerous wherever beds of salt and gypsum lie near the surface. The ranchmen fence them to prevent the cowboys from riding into them.

Carlsbad Cavern extends downward to a depth 1,000 feet beneath the present surface of the ground. Some of the rooms are of extraordinary dimensions, the ceiling of one being 250 feet high. The largest room thus far explored is called the "Big Room." It is more than half a mile in length and averages more than 200 feet in width. The floor lies about 700 feet below the surface of the ground and apparently the rock cover is little more than a shell. The profusion of cave formations and their majestic size render Carlsbad the most remarkable cavern in North America, if not in the world.

Luray Cavern in Virginia is unique in that it underlies a hill like a haystack, a honeycombed remnant of the vast mass of limestone which has elsewhere been removed from the valley.

Another unique cave is the Cave of the Winds near Manitou, Colo. It is in a small bed of limestone, perched on top of an isolated hill of crystalline rock. This explains its present dry condition. When it was forming it had not yet been separated by erosion from the limestone plateau which covered the region.

## THE GEOGRAPHICAL SIGNIFICANCE OF CAVES

In the earlier days of the human race, caves exerted a much greater influence upon the activities of man than they do at present. Almost all of the evidence of early man is to be found now in the caves which he inhabited. Caves were ready-made domiciles, a perfect protection from the weather. They were also fortresses into which he could retreat and gain protection from man and beast. Of course, the reason that the remains of early man and his way of living are found more abundantly in caves than elsewhere is because the conditions of preservation in caves are better than in most other situations. Caves constitute an ideal environment for the preservation of bones and implements. They are admirably protected from the ravages of the weather. Wind and storm do not beat upon them. Often, too, there is no moisture and hence less likelihood of decay. The uniform temperature conditions help also. In many caves the range of temperature is hardly more than a degree or two between the different seasons. Visitors to caves find them cool in the summer and warm in the winter. Neither animals nor man are likely to disturb objects left in caves, especially if they become only slightly buried in soil or ashes. By no means has all the evidence of early man been found in caves but it is not the purpose here to review the beginnings of the human race. When we reach the period of man's activities represented by paleolithic man, we find abundant evidence of his ability to make tools and weapons and to execute drawings and other artistic devices. Almost all of this kind of art is to be found in caves, notably in France and Spain. The walls of caves provided the first opportunity for mural decorations, and the artists of that time left many interesting outline sketches of the wild animals which they hunted. Most of the drawings are deeply cut in the weak limestone rock of the cavern wall or roof. Some are in bas-relief and a few are paintings made by daubing colored clay, charcoal, or ocher upon a smoothed rock surface. The fact that many of the animals thus depicted are now extinct lends particular interest to these relics, which throw a great deal of light upon the way man lived in the dawn of history.

Many of the early indications of human culture are found in the Pleistocene caves of Europe, which were inhabited by so-called *Paleolithic* man. These early cavemen ranged over middle Europe as far south as the Pyrenees and the Alps and inhabited the caverns of Belgium and Germany, Hungary, Switzerland, and southern France. They used huts as well as caves for habitation. They lived by hunting and fishing; they were fire users and lit up the darkness of their caves with stone lamps filled with fat. They were clad in skins sewn together with sinews of reindeer. They developed a marvelous facility for drawing animal figures. They possessed no domestic animals, nor were they acquainted with spinning or with the potter's art.

The later inhabitants of caves were distinguished from Paleolithic man by their use of domestic animals and by the fact that the wild animals living contemporaneously with them belong to existing species. The Neolithic, Bronze, and Iron Ages are represented successively in the culture of these later people, all of whom, however, belonged to prehistoric time.

The Neolithic caves are widely spread throughout Europe and have been used as the habitations and tombs of the early races who invaded Europe from the east with their flocks and herds. The Neolithic cave dwellers have been proved to be identical in physique with the builders of the cairns and tumuli which lie scattered over the fens of Great Britain and Ireland. These caves have been found in Wales, France, and Spain. The human remains found therein indicate that Neolithic man was long-headed and is represented at the present day by a part of the population inhabiting the Basque provinces of northern Spain.

The extreme rarity of articles of bronze in the European caves implies that they were rarely used by the Bronze folk for habitation or burial. Caves containing articles of iron and therefore belonging to that division of the prehistoric age are even less important. As man increased in civilization, he preferred to live in houses of his own building and he no longer buried his dead in the natural sepulchers provided for him in the rock.

In recent times the chief importance of caves centers around their use as a source of guano from which nitrates can readily be extracted to serve in the making of explosives and fertilizers. Guano is the excrement of bats or birds and consists largely of nitrates, which are soluble salts of nitrogen. Because of their solubility, they are valuable as plant foods. However, this very solubility causes nitrates to disappear from the surface of the earth in humid and rainy regions. Therefore nitrates of this kind are found only in arid regions or in caves. Bird guano occurs in the desert sections of Peru. Bat guano occurs in caves in many parts of the world, notably in New Zealand, South Africa, and the West Indies.

During the War of 1812 the accumulations of bat guano in Mammoth Cave, Kentucky, and in other caves served as a source of saltpeter for the manufacture of gunpowder. The wooden pipes and settling tanks may still be seen in the cave.

The response of organic life to a continuous existence in caves reveals some developments of exceptional interest, as, for example, the blind, wingless katydids, with extremely long and sensitive antennae, the blind, colorless crayfish, and blind and viviparous fish. Most cave-dwelling animals have evolved an extreme sensitivity to impressions other than those of sight. Bats, for example, are able to avoid obstacles in their flight, apparently from some sensory impressions not perceived by other animals.

## NATURAL BRIDGES

Natural bridges may be formed in a great variety of ways. The largest known are formed by the cutting action of streams. But most natural bridges occur in limestone regions and result from solution along joints and bedding planes.

The figure illustrates three stages in the development of this type of bridge. At first, there is a stream flowing over a limestone plateau. Next, at some part of its course the stream loses by seepage a part of its volume, which penetrates cracks and thence follows bedding planes at greater or lesser depth beneath the surface. Finally, the removal of most of the plateau mass by erosion and solution leaves a remnant in the form of a bridge, an arch, or a tunnel.

It is occasionally stated that natural bridges result when a cave collapses and only a small part of the roof remains. Very rarely do caves

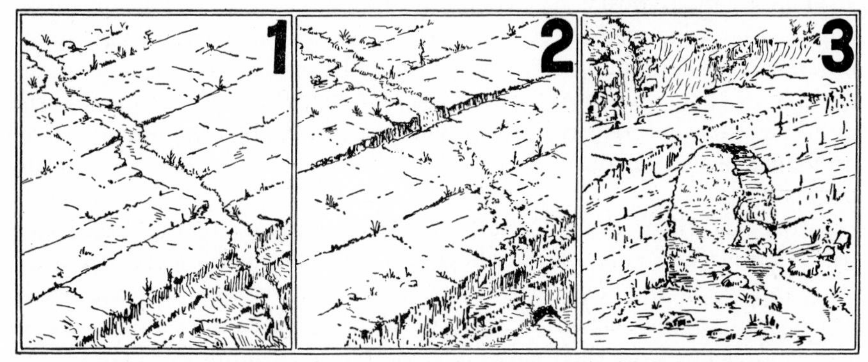

STAGES IN THE DEVELOPMENT OF NATURAL BRIDGE, VA.

collapse. In the Mammoth Cave region of Kentucky there is no evidence that any large cave has ever collapsed.

The famous Natural Bridge of Virginia was formed by seepage of water through a joint or fissure athwart a stream, thence along a bedding plane, until it emerged under a fall or rapid farther downstream. The channel thus formed was gradually enlarged until all the water of the stream was diverted from the stream bed below the point of ingress, leaving a bridge.

The height of the arch of a natural bridge above a stream will naturally depend upon the amount of cutting subsequent to the formation of the bridge, and to the weathering of the underside of the arch. Bridges formed in this way can readily be distinguished from the remnants of cavern roofs by the fact that the top of the bridge of the former was plainly at one time the bottom of the valley.

## MAPS ILLUSTRATING THE WORK OF UNDERGROUND WATER

Sink holes occur in many physiographic provinces of the United States. On the coastal plain the finest development of large sinks is in Florida. The *Arredondo, Fla.*, sheet is especially striking. It illustrates several rivers disappearing in sinks. Other good maps from Florida are the *Williston, Citra, Tsala Apopka*, and *Ocala* sheets. Large springs, whose flow is sufficient to produce headward sapping of the valleys they have formed, are shown on the *DeFuniak, Holt*, and *Niceville, Fla.*, sheets. Sinks occur on the unconsolidated deposits of South Carolina and New Jersey. The so-called "bays" or supposed meteor craters of South Carolina have been explained by some investigators as modified sink holes. Sinks of this character are shown on the *Peeples* and *Shirley, S. Car.-Ga.*, sheets and on the *Eutawville, Olar, Williston*, and *Allendale, S. Car.*, maps. The various sizes and irregular branching shapes of these sinks make it difficult to conceive for them a meteoritic origin. Still farther north sinks are shown on the *Smyrna, Del.-N. J.*, sheet where the limestone is evidently not very extensive. There the sinks may be due not to the solution of limestone but to the removal, by circulating ground water, of the finer sand from the gravel deposits, thus allowing the surface of the ground to settle.

The great sink-hole region of the country is in the limestone area of Kentucky and Indiana. The *Big Clifty, Bowling Green*, and *Munfordville, Ky.*, and the *Bloomington, Ind.*, maps show abundant sinks. On the *Mammoth Cave, Ky.*, map some of the sinks displayed are so large as to deserve the name *valley sinks*. For a limestone level with sinks and a higher insoluble formation without sinks, see the *Brownsville, Golconda*, and *Princeton, Ky.*, maps and especially the *Byrdstown, Ky.-Tenn.*, sheet. The *Renault* and *Cahokia, Ill.-Mo.*, maps show sinks on limestone which apparently has a slight dip. In the Folded Appalachians sinks are shown in the limestone valleys on the *Lexington, Va.*, and the *White Sulphur Springs, W. Va.-Va.*, sheets. On the *Falling Spring, Va.-W. Va.*, map there is an unusual occurrence of sinks on the crest of a high ridge.

On the plains of Texas, there are widespread areas of sinks, shown on the *Armstrong, La Sal Vieja, La Feria*, and *Saltillo Ranch, Texas*, sheets. As many of these are associated with sand dunes, some of the large depressions may be due to wind action. Of particular interest is the *Humble, Texas*, sheet upon which many small sinks are shown.

An interesting little group of sinks is shown near Thompsons Lake on the *Berne, N. Y.*, quadrangle. This lake has no surface outlet, as it occupies a sink hole.

Springs, both hot and cold, are shown on many of the maps of the Great Basin such as the *Ivanpah, Calif.-Nev.*, and the *Camp Mohave, Ariz.-Nev.-Calif.*, sheets. The relation of geyser basins and hot springs to faults is shown on the *Gallatin, Wyo.*, map.

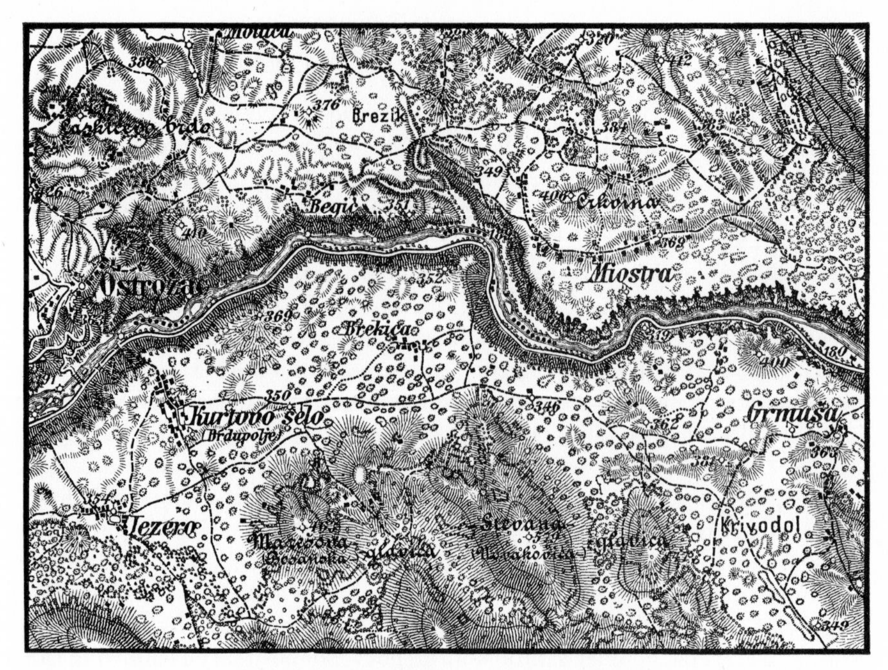


THE KARST REGION OF DALMATIA
Austria; *Bihać and Krupa* sheet, Zone 26, Col. XIV (1:75,000).

THE CAUSSE REGION OF SOUTHERN FRANCE
France; *Sévérac* sheet, No. 208 (1:80,000).

## QUESTIONS

1. Why are artesian-water supplies often hard, whereas water derived from streams is usually soft?
2. Can you account for the presence of hot springs in the bottom of the Yellowstone Canyon, where jets of steam may be seen actually bubbling up through the river?
3. Would you expect to find springs on top of a mountain?
4. The water supply of Ogden, Utah, is derived from great gravel deposits (originally alluvial fans spread out into a basin and then covered with lake deposits). Explain why this is favorable for artesian water. What function do the lake deposits have? This water actually flows. Prepare a section to show the conditions.
5. Why are some geysers intermittent whereas others sputter more or less continually?
6. What is the difference between juvenile and vadose water? How might these differ in chemical composition?
7. What is the explanation for the practice of flooding old oil wells with water in order to recover oil from oil-bearing sands which are practically exhausted?
8. Why do lakes occur so commonly in the sink holes of Florida but not in those of Kentucky? *Ans.* Because in Florida the ground-water level of the region is very near the surface. In Kentucky it lies 200 to 300 feet below, at the level of the rivers which dissect the region.
9. What is travertine? *Ans.* Travertine is a calcareous deposit left by water carrying lime in solution, called also *calcareous tufa.*
10. Is *onyx* the correct term to be applied to some of the colored and banded deposits found in caves? *Ans.* No, the term *onyx* is correctly used only with reference to siliceous deposits and not to calcareous deposits, such as occur in caves. The term *cave onyx* may perhaps be excusable.
11. Are all natural bridges due to collapse of caves? Are any of them due to this?
12. What general term is used for all kinds of cave deposits?
13. Which is the better solvent of limestone: perfectly pure water or that which contains carbon dioxide gas? Why?
14. What effect do dipping beds have upon the direction of cave growth? For example, on one side of a valley, limestone beds dip toward the valley; on the other side of the valley, they dip away from it. On which side of the valley will caves be most apt to develop?
15. What is your opinion concerning the effect of the proposed sea-level canal across Florida upon the artesian-water supply of that state?

## TOPICS FOR INVESTIGATION

1. Contrast between cycle of erosion in normal regions and in limestone regions.
2. Cave deposits: kinds; reason for deposition; relation to two-cycle theory of cave origin.
3. Artesian-well regions of the United States.
4. Karst; doline; lapies. The meaning of these terms.
5. Geysers; their relation to underground structure.
6. The two-cycle theory of cave formation (see Davis).
7. Underground water and its effect upon ore bodies.

## REFERENCES

Numerous studies of the occurrence and behavior of underground water in wells and springs have been made by the U. S. Geological Survey and the results published in the Water-Supply Papers. A general and readable discussion of the whole subject is given by O. E. Meinzer (1923) and is entitled *The occurrence of ground water in the United States*, in U. S. Geol. Surv., W.-S. Paper 489, 321 p. M. L. Fuller (1908) gives a *Summary of the controlling factors of artesian flows* in U. S. Geol. Surv., Bull. 319, 44 p.

A few years ago the American Geophysical Union held a symposium on the temperatures of hot springs and the sources of their heat and water supply. A number of papers on this subject were published in the Journal of Geology. Special mention may be made of R. B. Sosman (1924), *General summary of the symposium on hot springs*. Jour. Geol., vol. 32, p. 464–471.

An important paper explaining the behavior of geysers is T. A. Jaggar, Jr. (1898) *Some conditions affecting geyser eruption*. Am. Jour. Sci., 4th ser., vol. 5, p. 323–333.

The recent presidential address by A. L. Day of the Geological Society of America suggests that the Bunsen theory is not entirely adequate to explain the variable periods and time intervals of geyser eruptions. See A. L. Day (1939) *The hot-spring problem*. Geol. Soc. Am., Bull. 50, p. 317–336.

The long treatise on caves by W. M. Davis (1930) *Origin of limestone caverns*. Geol. Soc. Am., Bull. 41, p. 475–628, is the most important general discussion of that subject.

A classified list of references follows:


### General

Gregory, J. W. (1929) *Water divining*. Smithsonian Inst., Ann. Rept. 1928, p. 325–348.

Meinzer, O. E. (1923) *The occurrence of ground water in the United States with a discussion of principles*. U. S. Geol. Surv., W.-S. Paper 489, 321 p.

Meinzer, O. E. (1923) *Outlines of ground-water hydrology, with definitions*. U. S. Geol. Surv., W.-S. Paper 494, 71 p.

Russell, W. L. (1928) *The origin of artesian pressure*. Econ. Geol., vol. 23, p. 132–157.

Slichter, C. S. (1902) *The motions of underground water*. U. S. Geol. Surv., W.-S. Paper 67, 106 p.

Thompson, D. G. (1929) *The origin of artesian pressure*. Econ. Geol., vol. 24, p. 758–771.

Tolman, C. F. (1937) *Ground water*. New York, 593 p.

Versluys, J. (1931) *Subterranean water conditions in the coastal regions of the Netherlands*. Econ. Geol., vol. 26, p. 65–95.

### Springs and Wells

Bryan, K. (1919) *Classification of springs*. Jour. Geol., vol. 27, p. 522–561.

Bryan, K. (1925) *The Papago country, Arizona*. U. S. Geol. Surv., W.-S. Paper 499, 436 p.

Fuller, M. L. (1908) *Summary of the controlling factors of artesian flows*. U. S. Geol. Surv., Bull. 319, 44 p.

Hopkins, T. C. (1910) *Changes produced on springs by a sinking water table*. Geol. Soc. Am., Bull. 21, p. 774.

Meinzer, O. E. (1927) *Large springs in the United States*. U. S. Geol. Surv., W.-S. Paper 557, 94 p.

Mendenhall, W. C. (1909) *Some desert watering places in southeastern California and southwestern Nevada*. U. S. Geol. Surv., W.-S. Paper 224, 98 p.

### Hot Springs and Geysers

Allen, E. T., and Day, A. L. (1935) *Hot springs of the Yellowstone National Park*. Carnegie Inst. Washington, Publ. No. 466, 525 p.

Bryan, K. (1924) *The hot springs of Arkansas*. Jour. Geol., vol. 32, p. 449–459.

Davis, B. M. (1897) *The vegetation of the hot springs of Yellowstone Park*. Science, new ser., vol. 6, p. 145–157.

GRAHAM, J. C. (1893) *Some experiments with an artificial geyser.* Am. Jour. Sci., 3d ser., vol. 45, p. 54–60.
HAGUE, A. (1889) *Soaping geysers.* Am. Inst. Min. Eng., Trans. 17, p. 546–553.
HAGUE, A. (1911) *Origin of the thermal waters in the Yellowstone National Park.* Science, new ser., vol. 33, p. 553–568; Geol. Soc. Am., Bull. 22, p. 103–122.
JAGGAR, T. A., JR. (1898) *Some conditions affecting geyser eruption.* Am. Jour. Sci., 4th ser., vol. 5, p. 323–333.
SHERZER, W. H. (1933) *An interpretation of Bunsen's geyser theory.* Jour. Geol., vol. 41, p. 501–512.
SOSMAN, R. B. (1924) *General summary of the symposium on hot springs.* Jour. Geol., vol. 32, p. 464–471.
VAN ORSTRAND, C. E. (1924) *Temperatures in some springs and geysers in Yellowstone National Park.* Jour. Geol., vol. 32, p. 194–225.
WATSON, T. L. (1924) *Thermal springs of the southeastern Atlantic States.* Jour. Geol., vol. 32, p. 373–384.
WEED, W. H. (1912) *Geysers.* U. S. Dept. Interior, 29 p.
ZIES, E. G. (1924) *Hot springs of the Valley of Ten Thousand Smokes.* Jour. Geol., vol. 32, p. 303–310.

GEYSERS—LOCALITIES DESCRIBED

ALLEN, E. T., and DAY, A. L. (1924) *The sources of the heat and the source of the water in the hot springs of the Lassen National Park.* Jour. Geol., vol. 32, p. 178–190.
HAGUE, A., and WEED, W. H. (1891) *Hot springs and geysers of Yellowstone National Park.* 5th Intern. Geol. Cong., C.r., p. 346–363.
HOLMES, W. H., and PEALE, A. C. (1883) *Yellowstone National Park; geology, thermal springs, topography.* U. S. Geol. and Geog. Surv. Terr. (Hayden), 12th Ann. Rept., pt. 2, 490 p.
NOLAN, T. B., and ANDERSON, G. H. (1934) *The geyser area near Beowawe, Eureka Co., Nevada.* Am. Jour. Sci., 5th ser., vol. 27, p. 215–299.
PEALE, A. C. (1884) *The world's geyser regions.* Pop. Sci. Monthly, vol. 25, p. 494–508.

CAVES AND KARSTLANDS

CVIJIĆ, J. (1893) *Das Karstphänomen.* Geog. Abhandlungen, vol. 5, p. 49–330.
CVIJIĆ, J. (1924) *The evolution of lapies. A study in karst physiography.* Geog. Rev., vol. 14, p. 26–49. Remarkable illustrations.
DAVIS, W. M. (1930) *Origin of limestone caverns.* Geol. Soc. Am., Bull. 41, p. 475–628. A comprehensive treatise, involving one-cycle and two-cycle caverns. Numerous references.
DICKEN, S. N. (1935) *Kentucky karst landscapes.* Jour. Geol., vol. 43, p. 708–728.
HENDERSON, J. (1932–33) *Caverns, ice caves, sink holes, and natural bridges.* Colo. Univ. Studies, vol. 19, p. 359–405; vol. 20, p. 115–158.
MELTON, F. A. (1934) *Linear and dendritic sink-hole patterns in southeast New Mexico.* Science, new ser., vol. 80, p. 123–124.
SANDERS, E. M. (1921) *The cycle of erosion in a karst region* (after Cvijić). Geog. Rev., vol. 11, p. 593–604. Remarkable diagrams.
SWINNERTON, A. C. (1932) *Origin of limestone caverns.* Geol. Soc. Am., Bull. 43, p. 663–693.
THORNBURY, W. D. (1931) *Two subterranean cut-offs in central Crawford Co., Indiana.* Ind. Acad. Sci., Proc. 40, p. 237–242.

CAVE DEPOSITS

DAVIDSON, S. C., and MCKINSTRY, H. E. (1931) *Cave pearls, oolites, and isolated inclusions in veins.* Econ. Geol., vol. 26, p. 289–294.
EDWARDS, H. M. (1932) *The growth of stalagmites.* Science, new ser., vol. 76, p. 367–368.
ELLIS, R. W. (1931) *Concerning the rate of formation of stalactites.* Science, new ser., vol. 73, p. 67–68.

RICHARDS, G. (1931–32) *The growth of stalactites.* Science, new ser., vol. 73, p. 393; vol. 75, p. 50.

CAVERNS—LOCALITIES DESCRIBED

BAILEY, T. L. (1918) *Report on the caves of the eastern Highland Rim and Cumberland Mountains.* Tenn. Geol. Surv., Resources of Tenn., vol. 8, p. 85–138. A description of 109 caves studied to determine their value as source of niter.

BLATCHLEY, W. S. (1897) *Indiana caves and their fauna.* Ind. Dept. Geol. and Nat. Res., Ann. Rept. 21, p. 121–212.

BRETZ, J H. (1938) *Caves in the Galena formation.* Jour. Geol., vol. 46, p. 828–841.

COLE, L. J. (1910) *The caverns and people of northern Yucatan.* Am. Geog. Soc., Bull. 42, p. 321–336.

LEE, W. T. (1924) *A visit to Carlsbad Cavern.* Natl. Geog. Mag., vol. 45, p. 1–40.

LEE, W. T. (1925) *Carlsbad Cavern, New Mexico.* Sci. Monthly, vol. 21, p. 186–190.

LOBECK, A. K. (1929) *The geology and physiography of the Mammoth Cave National Park.* Ky. Geol. Surv., ser. 6, vol. 31, p. 327–399.

MARTEL, E. A. (1893) *The land of the Causses.* Appalachia, vol. 7, p. 18–30, 130–149.

MCGILL, W. M. (1933) *Caverns of Virginia.* Va. Geol. Surv., Bull. 35, 187 p.

STONE, R. W. (1930) *Pennsylvania caves.* Penn. Geol. Surv., 4th ser., Bull. G3, 63 p.

SWINNERTON, A. C. (1929) *The caves of Bermuda.* Geol. Mag., vol. 66, p. 79–84.

WILLIAMS, I. A. (1920) *The Oregon caves of Josephine County.* Nat. Hist., vol. 20, p. 397–405.

NATURAL BRIDGES (see also under Young Streams)

CLELAND, H. F. (1911) *The formation of North American natural bridges.* Pop. Sci. Monthly, vol. 78, p. 417–427.

MALOTT, C. A., and SHROCK, R. R. (1930) *Origin and development of Natural Bridge, Virginia.* Am. Jour. Sci., 5th ser., vol. 19, p. 256–273.

WALCOTT, C. D. (1893) *The natural bridge of Virginia.* Natl. Geog. Mag., vol. 5, p. 59–62.

WOODWARD, H. P. (1936) *Natural Bridge and Natural Tunnel, Virginia.* Jour. Geol., vol. 44, p. 604–616.

GEOGRAPHICAL ASPECTS

BAGG, R. M. (1933) *Underground water in human affairs.* Am. Water Works Assn., Jour. 25, p. 1000–1006.

BARBOUR, E. H. (1899) *Wells and windmills in Nebraska.* U. S. Geol. Surv., W.-S. Paper 29, 85 p.

CARNEY, F. (1908) *Springs as a geographic influence in humid climates.* Pop. Sci. Monthly, vol. 72, p. 503–511.

CASTERET, NORBERT (1938) *Ten years under the earth.* New York, 283 p.

EIGENMANN, C. H. (1917) *The homes of blindfishes.* Geog. Rev., vol. 4, p. 171–182.

MEINZER, O. E. (1927) *Plants as indicators of ground water.* U. S. Geol. Surv., W.-S. Paper 577, 95 p.

# V
# STREAMS IN GENERAL

*U. S. Forest Service*

PEABODY RIVER AND MOUNT WASHINGTON, NEW HAMPSHIRE

A river in its low stage and part of its flood load.

*Howe, U. S. Geological Survey*

GULCHES AND ALLUVIAL CONES, SAN JUAN MOUNTAINS, COLO.
Two complete stream systems, illustrating erosion and deposition.

## STREAMS IN GENERAL

SYNOPSIS. The water falling upon the surface of the land either evaporates or runs off. The *evaporation* may be immediate, as in desert regions, or may be long postponed. From the leaves of plants much water long held in the soil is eventually transpired. The *run-off* also may be immediate or postponed. Much water seeps into the soil to reappear later in springs. The movement of water, soil, and rock—the surface wash; the temporary rills; the underground water percolating through the soil; the moist soil slumping down hillsides; the soil creep; the soil and rock, less moist, falling from cliffs—all constitutes a great integrated system of which the streams themselves represent only a small part. It is customary to distinguish between the work of streams in eroding and transporting material and the slower movement of the soil in solifluction and soil creep. The behavior of streams, however, is greatly influenced by the rate of weathering prevailing over their drainage areas.

Rivers in their life history pass first through a period of youth in which they actively erode. During this period, downcutting is prominent; waterfalls and cascades are common; the longitudinal profile is irregular; the whole drainage area of the stream is giving way actively to erosion; there is strong slumping and sliding on the steeper slopes and there may be landslides.

Gradually, with maturity, a condition of equilibrium is established. The stream acquires a graded profile or slope just sufficient to permit transportation of its load, a condition it constantly strives to maintain. Varying factors of load and volume necessitate sometimes erosion, sometimes deposition. A change in load or volume or gradient in any part of the course of a mature stream affects the entire system, so delicately is it balanced. Flood plains, meanders, oxbow lakes, braided channels, natural levees, and terraces indicate a graded or a once-graded condition. A graded stream gradually reduces its slope as the load contributed by its headwaters decreases. Actual slope, therefore, is not a criterion of stage of development. Streams completely graded throughout their entire system are old.*

Streams are studied from many other aspects. Genetic types are recognized by such terms as *consequent, subsequent, resequent, and insequent,* as well as *superimposed* and *antecedent.* There are also many kinds of stream patterns: *dendritic, trellis, radial, annular, rectangular,* and other less common types, all of which reflect some different structural control. The stages in the life history of a stream and those in the development of the region are not always the same; for example, maturely dissected regions may have young streams.

* Some investigators have suggested that a stream should be termed mature when the width of the meander belt is equal to the width of the flood plain; and that when the flood plain is wider than the meander belt, old age has been attained.

## GENERAL STATEMENTS ABOUT RIVERS

WHAT IS A STREAM? The forces of destruction which modify the surface of the earth cannot everywhere be sharply separated. Under the action of wind and rain and the various kinds of weathering, and the force of water in the form of streams, glaciers, and waves, the larger land masses are being constantly destroyed. But these forces so work together that it is often impossible to attribute a particular observed change to any one of them. Streams, for instance, receive credit for much more work than they probably accomplish unless the term *stream* is made to cover a wider range of phenomena than is usually included. A stream system involves not only the familiar stream flowing in its channel but the millions of little rills which have their birth with each rain storm and flow in countless numbers down each hillside. As a matter of fact, the river system at such times more nearly resembles an entire leaf than the veins thereof, with which it is usually compared. It involves the full surface of the country. Nor is it at such times easy to distinguish the work of the stream, including all of its branches, from the accompanying results of weathering. The mantle of waste which covers all slopes and the soil which is continuously creeping downhill is as truly a part of the load of the stream as the sand and pebbles carried on its bed. Gravity is the ultimate agency which is accountable for all of the results observed. The differences in the results are due to the varying amounts of water present. The more water there is mixed with the rock waste, the more readily it is carried and the more gentle the slopes which result. Dry rock waste produces steep talus slopes having angles of approximately 35°, or even more for coarse blocks, whereas streams bearing no load can flow with a gradient of much less than a foot to the mile.

THE WORK OF STREAMS. The work done by streams in any given region depends upon several factors:

*a.* In general, the greater the rainfall, the more effective is the stream work. Abundant precipitation results not only in larger streams but also in more numerous and more permanent streams.

The rivers in the eastern United States, for example, are much larger and more numerous than those in the western United States.

*b.* Deep porous soil and heavy plant growth tend to absorb much of the water which falls as rain and thus greatly reduce the immediate run-off. We note vast areas of sandy uplands in the coastal plain of southern Alabama and Mississippi, where in spite of the heavy rains the streams are infrequent.

*c.* Porous limestone rocks with their underground drainage operate against the development of many surface streams. The Karst region of Dalmatia, for example, is particularly devoid of streams although the rainfall there is almost the heaviest in Europe.

*d.* Arid conditions with resulting scarcity of vegetation favor the work of streams although the volume, the number, and the permanency of streams is minimized.

*e.* Impervious clays and glacial till increase run-off, minimize addition to ground water, and accelerate erosion.

Relation between Rainfall and Run-off. After precipitation and before it finally reaches well-defined stream channels, water is subjected to the action of numerous agencies, including evaporation from ground and water surfaces, transpiration in plant and tree growth, infiltration and absorption by the ground, passage through the shallow storage upon ill-drained lands, and surface run-off through systems of trickles and streamlets. The importance of the subtractions from precipitation before it becomes stream flow may be illustrated by the statement that, to produce 1 pound of dry vegetable substance of a growing plant, several hundred pounds of water is taken up from the soil by the roots of the plant, passes through the growing plant tissues, and then evaporates into the air from the leaf surfaces. Thus to obtain satisfactory crop yields, many inches of water falling upon the crop-producing area must be used in this way. It is evident, therefore, that in arid and semiarid regions, and indeed in humid regions in times of drought, the flow of surface streams constitutes a surprisingly small part of the water initially falling as rain.

In the Red River Basin of North Dakota where the total annual precipitation is about 20 inches, there is about 5 per cent run-off; whereas in the more humid New England area, where the precipitation is twice as great, the run-off is close to 50 per cent. To put it in a different way: an increase of 1 inch in rainfall in the drier Red River region is reflected by an increase of about $\frac{1}{3}$ inch in run-off; but an increase of 1 inch in rainfall in the humid east is reflected by an increase of $\frac{3}{4}$ inch in run-off, thus indicating that in humid regions the run-off is more immediate than in arid conditions. In fact, in arid regions additional precipitation is reflected by immediate greater evaporation instead of by run-off.*

Run-off and Stream Flow. Run-off, namely that portion of the precipitation which appears as flow in surface streams, occurs in two ways: (*a*) as *surface run-off*, or that part of the precipitation which reaches surface streams by flowing over the surface of the ground and into tributary streams; and (*b*) as *ground-water run-off*, or that part of the precipitation which before reaching surface streams has passed through the ground; hence often called *seepage flow*, or *sustained flow*.

If the greater part of the precipitation runs off the surface of the drainage basin, the resulting stream flow will be immediate but erratic and will continue for only short periods after rains. This occurs in regions of clay soil or bare rock surfaces broken by few joints. Run-off then becomes concentrated and erosion is active. On the other hand, if the greater part of the precipitation reaches the stream as seepage from ground water, as in sandy regions and in those covered with deep loose soil or humus, the stream flow resulting therefrom will be delayed, possibly for weeks and months, but will be well sustained through drought periods.

* For numerous statistics relating to this problem see U. S. Geol. Surv., W.-S. Paper 772, on the *Relations of rainfall and run-off in the United States*.

*U. S. Forest Service*

VALLEY OF YOUNG STREAM IN THE SAWTOOTH MOUNTAINS, IDAHO
Showing narrow V-shaped valley with steep gradient.



VALLEY OF A MATURE RIVER IN ALASKA
Showing wide valley floor and gentle stream gradient.

*U. S. Geological Survey*

## LIFE HISTORY OF A RIVER

As a region passes through the geomorphic cycle, the rivers also exhibit changes, going from youth, through maturity, to old age. The stage in the life of the river at any given moment is usually not the same as the stage of the development of the region. It has been shown that a region is young when most of its initial surface is intact; that it is mature when reduced largely to divides and hilltops; and old when worn down approximately to base-level.

The stages in the life of a river, however, depend upon the behavior of the river itself and only indirectly have anything to do with the land form.

A river is *young* when it is constantly able everywhere to erode its channel. This means its gradient is sufficiently steep for it to carry all the load brought to it by its tributaries, both the perennial streams and the wet-weather rills, and that there is energy to spare. Young streams, consequently, usually flow in narrow valleys which they have cut for themselves. The walls of these valleys are steeply sloping because weathering has not widened them to the extent that the stream is cutting down. Rock ledges therefore abound on the valley walls.

The young stream occupies the entire floor of the valley. There is no flood plain. Young rivers normally have waterfalls and rapids due to the presence of more resistant rock masses exposed by erosion or due to initial irregularities in the region. The gradient of young streams varies because of variations in the rock structure. Lakes due to initial depressions in the area may be present along the stream's course. A young stream usually has a swift current of apparently clear water. It is not sufficiently loaded with debris to be turbid. Potholes and rock channels are common in the bed of young streams and often accompany waterfalls and rapids.

A river is *mature* when it has reduced its gradient throughout its course so that its velocity is just sufficient to carry the debris brought to it from all sides. It is quite unable to erode its valley any deeper until its load is reduced. Young streams have an excess of ability over the amount of work to be performed. In the case of mature streams, an equality of these two quantities is brought about, and the river is said to be *graded*. It has attained a *profile of equilibrium*. A thoroughly mature river, therefore, has no irregularities in its profile, no rapids or waterfalls. During the time necessary to attain the graded condition, weathering has reduced the valley walls to gentle slopes. Rock ledges are infrequent. The valley floor has been widened by the lateral swinging of the stream and a flood plain results.

When the trunk streams are graded, early maturity is reached; when the side streams are also graded, maturity is far advanced; and when the wet-weather rills are graded, old age is attained.

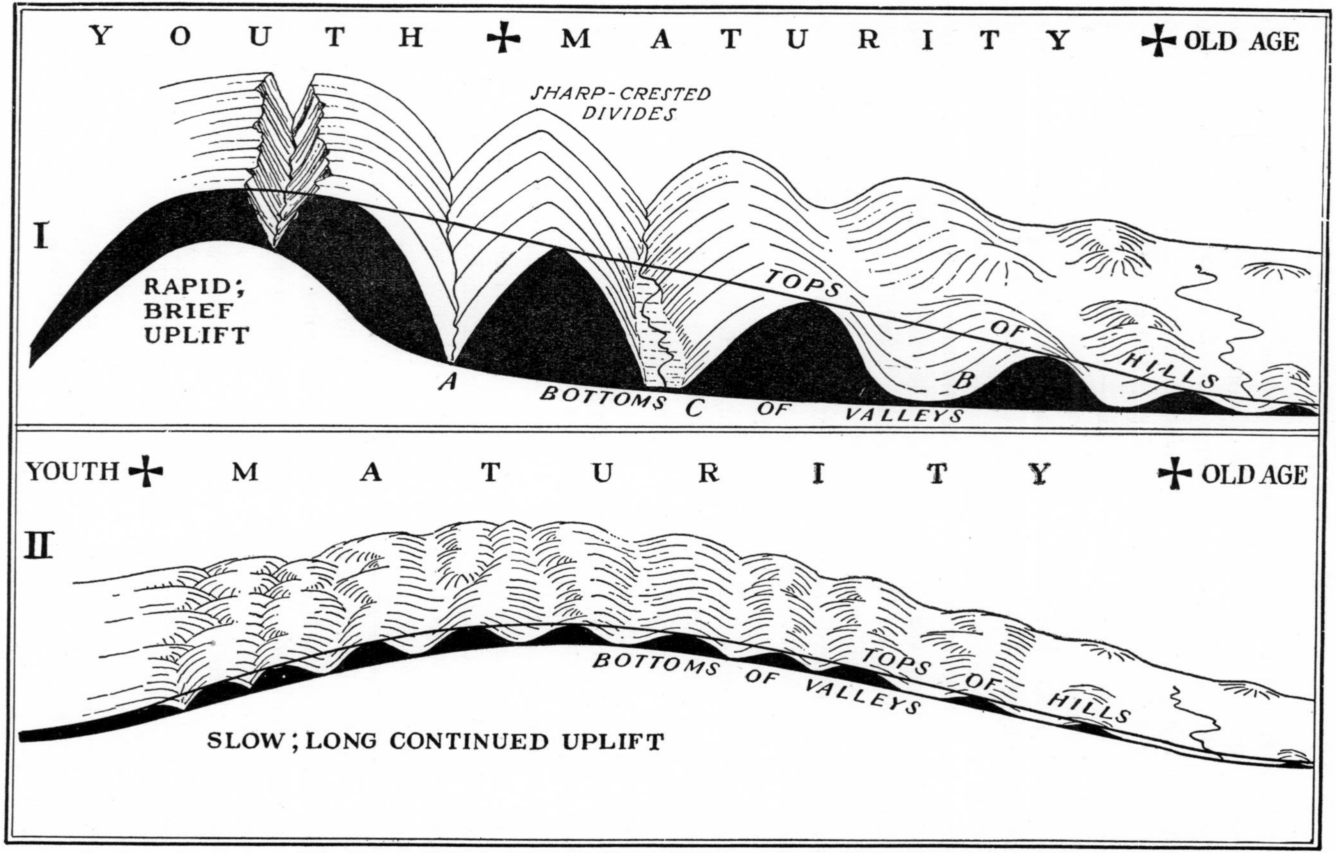

DIAGRAMMATIC REPRESENTATION OF EROSION CYCLES

## THE CYCLE OF EROSION

Distinction between Age of Streams and Erosional Stage of the Region. The cycle of erosion, sometimes called the *geographical* or *geomorphic cycle*, concerns the larger land mass rather than the streams. It is important to understand this, namely, that the cycle of erosion refers to the stages through which a land mass passes from the time of its uplift until peneplanation. These different stages are treated more fully under the several chapters on the larger land forms such as plains, plateaus, and the various kinds of mountains. One or two observations may, however, be introduced here.

Youthful Stage of Region after Rapid Uplift. If the uplift is rapid and brief in time (Fig. I), the streams begin their erosion upon an elevated mass and deeply dissect it so that at first the streams cut deeply and the divides are broad. This is youth in the stage of the erosion cycle for that land mass. Relatively soon, however, the streams have accomplished their greatest downward cutting. At the moment when the divides have narrowed down to sharp crests, maturity begins. This is the stage of greatest relief and is reached comparatively early in the history of most regions. The cross-valley profiles during the mature stage are apt to be curves convex upward, as at *A*.

Mature Stage of Region. As maturity advances, the slopes of the valley walls become gentle. The tops of the sharp-crested divides wear down faster than the streams cut down, and the relief thus becomes more subdued. The divides become rounded and the profiles are then concave upward, as at *B*.

Youth and Maturity in Region during Slow Uplift. The scheme of development just set forth was advanced by W. M. Davis. Penck, on the other hand, visualized several other possibilities. For example, if the uplift of a region is slow and long continued, as in Fig. II, the widening of the valleys will be relatively fast compared with the rate of downward cutting of the streams. This means that the period of youth is almost lacking and that maturity comes early in the history of the region. Davis also mentioned this possibility and pointed out the fact that valleys of open form, without flood plains, suggest slow uplift, whereas the presence of flood plains in the bottom of wide valleys with abrupt walls suggests rapid uplift, as at *C* in Fig. I opposite. Indeed, some students, notably Crickmay of England, believe that peneplanation is brought about more by the lateral erosion of streams, which pares away the divides and causes a coalescing of all the flood plains of a region to form what he terms a *panplane*, rather than to the wearing down of the summit areas. The development of peneplanes in this manner is believed to be much more rapid than peneplanation by weathering. This method, however, is not suggested in the illustrations opposite.

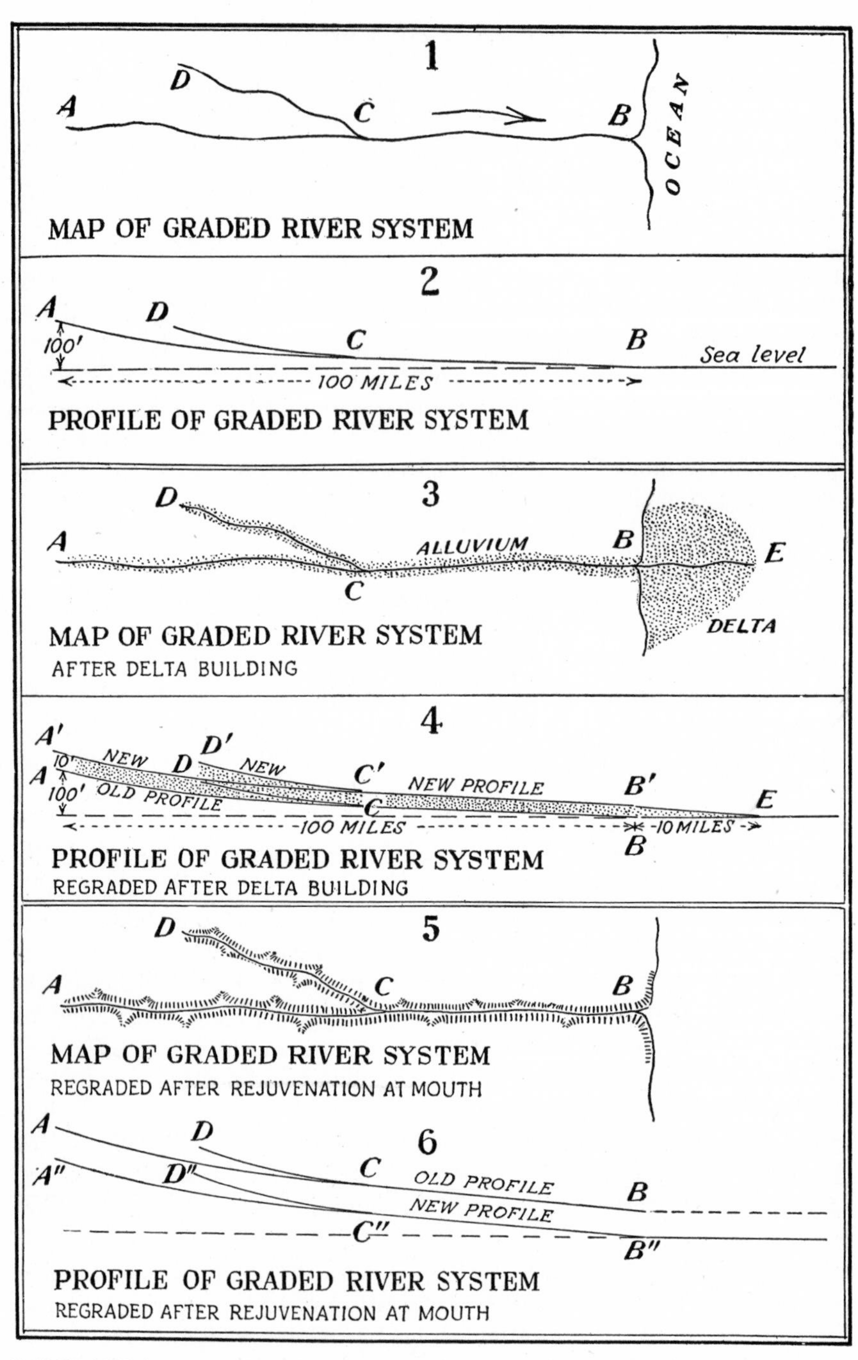


DIAGRAMS (1, 2, 3, 4) SHOWING EFFECT OF DEPOSITION AND (5, 6) EFFECT OF REJUVENATION AT THE MOUTH OF A GRADED RIVER

## GRADED RIVERS, I

A river system which in all its parts has acquired a graded profile is, theoretically, in a very delicate state of balance. The ability to carry a load is equaled by the amount of load to be carried. A perfectly graded river system neither deposits nor cuts down. In nature, this situation does not exist. Changing conditions intermittently alter the load or the volume of the stream. A so-called *mature* river is at one moment depositing material because of a temporary increase of load, or loss of volume, and at another moment cutting away what was previously deposited because of a reversal of conditions. The changing character of the load of a stream modifies its carrying ability and causes erosion or deposition. This constant struggle on the part of the stream to maintain a profile of equilibrium renders a mature river one of the most interesting phenomena in all nature. The delicate balance between all parts of the system is shown by the fact that a change in any part of the system is reflected by a readjustment of the entire system.

Depositional Changes at the Mouth of a Graded Stream. Delta Building. If a graded stream (*A-B*) enters a relatively quiet body of water (at *B*), its current is checked and it deposits its load as a delta. The building of a delta materially increases the length of the stream, which means a reduced gradient. A stream (Fig. 1, *A-B*) 100 miles long, with headwaters 100 feet above sea level, has an average gradient of 1 foot per mile (Fig. 2, *A-B*). If the stream is lengthened 10 miles by the building of a delta (Fig. 3, *B-E*), the gradient is reduced to 0.91 foot (100 ÷ 110) per mile. To reestablish the original gradient, the stream must everywhere deposit material along its course until it has raised its bed 10 feet higher (Fig. 4, *A′-C′-B′*). As long as the delta building continues, the river will raise its flood plain in order to maintain its gradient, this being one of the functions of the flood plain.

All mature tributaries (*e.g.*, *D-C*) keep pace with the main stream and build up their floors (to *D′-C′*) with flood plains. The only alternative to this is to form lakes.

Rejuvenation at the Mouth of a Graded Stream. If a mature stream (Fig. 6, *A-B*) has its gradient increased at its mouth by a lowering of sea level (from *B* to *B″*), it will reduce this gradient by cutting down its previously deposited flood plain until the former gradient is again established (*A″-C″-B″*). Any mature tributary (*D-C*) will do likewise, cutting down to *D″ C″*. The whole system is rejuvenated and alluvial terraces may result.

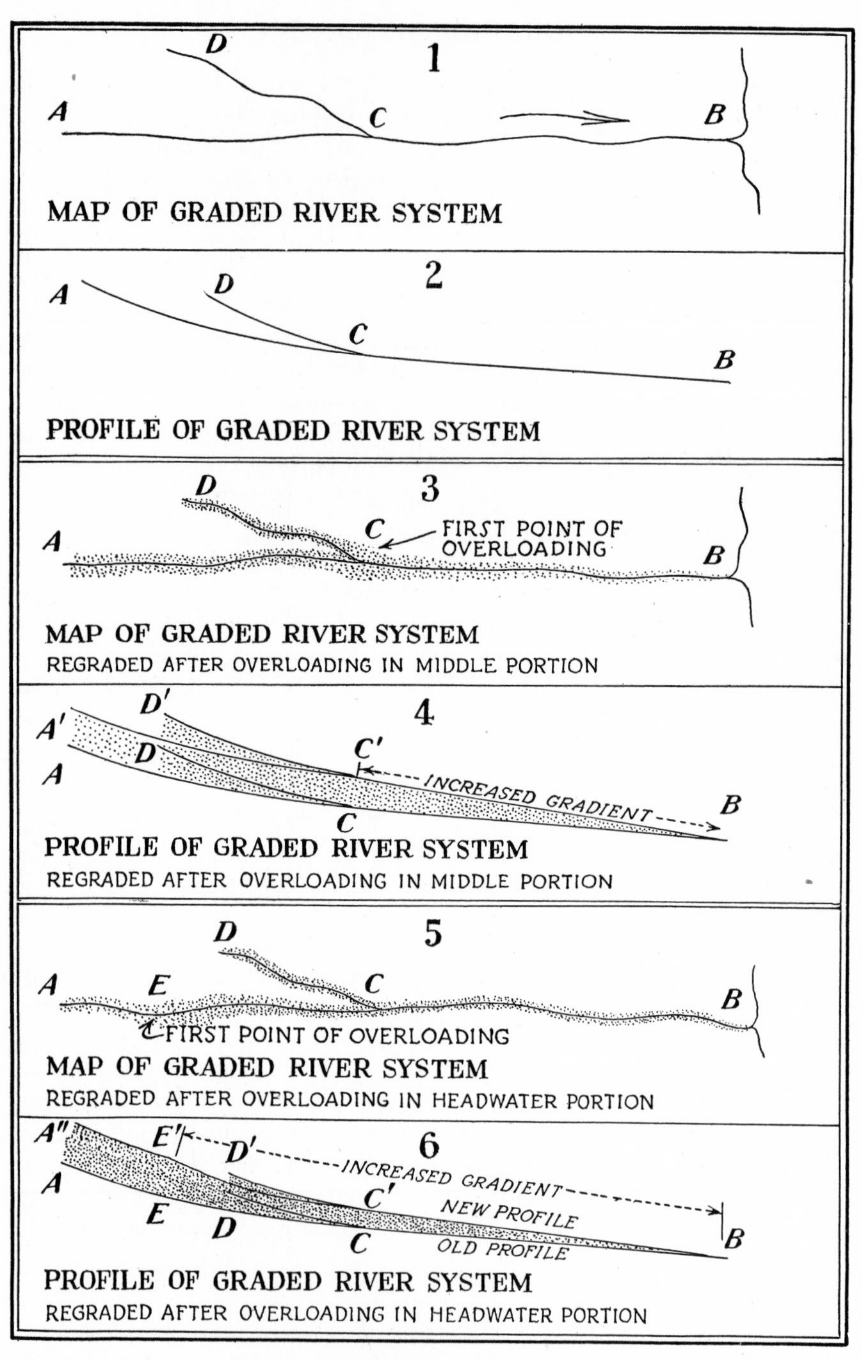


DIAGRAMS (1, 2, 3, 4) SHOWING EFFECT OF DEPOSITION BY TRIBUTARY ALONG THE MIDDLE COURSE AND (5, 6) ALONG THE HEADWATERS OF A GRADED STREAM

## GRADED RIVERS, II

DEPOSITIONAL CHANGES IN THE MIDDLE PORTION OF A GRADED STREAM. Assume again a graded river (*A-B*) with a length of 100 miles and headwaters 100 feet above the sea. As long as its load and its volume remain constant and as long as there is no delta building, these conditions are permanent. But suppose a more youthful tributary (*D-C*) joining its middle course starts to bring down an exceptional amount of material and continues to do so, a condition which would ensue if the tributary happened to advance headward into areas of weak and easily eroded rocks or if glaciation occurred in these headwaters. The master stream, not being able to handle any additional load, drops this material which eventually comes to be spread along the entire course of the stream below that point until a sufficiently steep gradient (Fig. 4, *C′-B*) is established to handle the greater load. With the upbuilding of the flood plain in its middle portion, the gradient above that point is thereby reduced. This, therefore, entails deposition of material above that point, until the former gradient is reestablished (as *A′-C′*). The flood plain is thus everywhere raised, part of it (the downstream portion) with increased slope (*C′-B*), the upstream part to the gradient which it had before. The increased gradient of the downstream portion is necessary to take care of the increased load below the point where the tributary enters. All graded tributaries entering the master streams are similarly affected. The elevation of the mouths of these streams by the elevation of the main stream results in the alluviation of the tributaries but with no increase in gradient. If the tributaries are not themselves graded streams, lakes will result because of their inability quickly to build up their channels. Thus the overloading of any part of a mature river system brings about deposition through the entire system both above and below that point.

DEPOSITIONAL CHANGES IN THE HEADWATERS OF A GRADED STREAM. Suppose next that the load in the headwater portion of a mature stream is augmented by an amount which remains constant for a long period. Then material will be deposited at that point (*E*) and continuously downstream in lesser amounts until a new and steeper gradient (Fig. 6, *E′-B*) is established, which will allow the stream to handle the new and greater load. The tributaries everywhere will have to build up their channels to keep pace with the main stream.

The three cases which have been considered, namely, deposition at the mouth, in the middle course, and in the headwater portion of a mature stream, all involve alluviation of the channel of the main stream but not everywhere an increase in gradient. Increase of gradient results only below that point where there has been an increase of load.

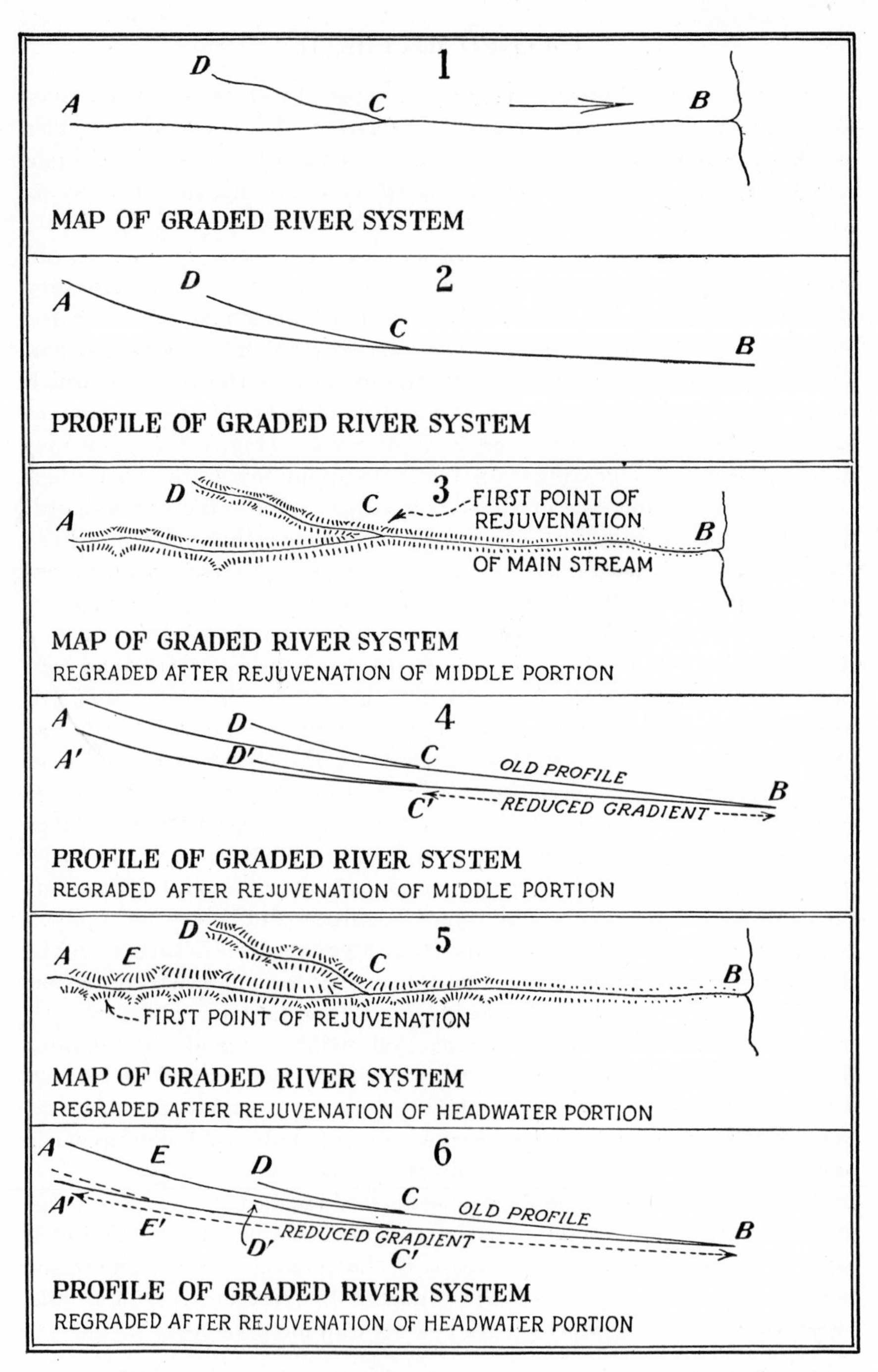


DIAGRAMS (1, 2, 3, 4) SHOWING EFFECT OF REJUVENATION ALONG MIDDLE COURSE OF STREAM, AND (5, 6) OF REJUVENATION OF HEADWATER PORTION

## GRADED RIVERS, III

Effect of Erosional Changes. Consideration may now be given to the effects produced upon a graded stream by the reduction in the load or by an increase in the relation of the carrying power to the load.

If (Fig. 1) a tributary (*D-C*) joining a mature stream (*A-B*) in its middle portion (*C*) ceases to contribute its usual load, the master stream below that point (*C-B*) is no longer loaded to capacity and cuts down its channel. It reduces its gradient to the state where it can just carry its load (Fig. 4, *C′-B*). Because the lower half of the stream's level is now reduced (*C′-B*), the gradient of the headwater portion is steeper (*A-C′*) than before. Downward cutting upstream results until the former slope of the headwater portion is reestablished (Fig. 4, *A′-C′*). The whole river channel has thus been lowered. As a result, the tributaries are able to lower their channels and reestablish their former gradients, if their loads are unchanged, or lower their gradients (*D-C* to *D′-C′*), as in this case, because *D-C* is carrying a smaller load than before. Rejuvenation, therefore, occurs everywhere in the river system. In this way terraces are formed. Obviously the cause of terrace formation at one point must sometimes be sought in some other part of a river system.

In Figs. 5 and 6, the load of the headwater portion (*A-E*) is reduced by a constant amount. This causes a rejuvenation of this headwater portion with establishment of the gentler gradient (*A-E* to *A′-E′*) and is accompanied by a lowering of the gradient of the main valley because of the reduced load throughout its length. Similarly, all the other headwaters have to reestablish their gradients because their mouths are cut down to keep pace with the main stream. They finally acquire the same gradients they had before because they have suffered no change in load, although they are now flowing at a lower level. Rejuvenation may be brought about at the mouth of the stream by reduction of the load introduced by the tributaries at this point, by a general uplift of the region, or by lowering of sea level as described previously.

It is apparent that a river is a highly sensitive organism. The building of a delta at the mouth of a stream may bring about the formation of flood plains in the uppermost tributaries. Stream capture by one tributary may result in rejuvenation of the whole system with resultant waterfalls and rapids and possibly other captures. The reduction of load in one part of a river system by the cessation of glaciation may account for the development of terraces in some region quite remote from the glaciated area. It is obvious that an explanation for any given set of phenomena is not to be sought always in the immediate neighborhood but that the behavior of the whole stream system must be understood.

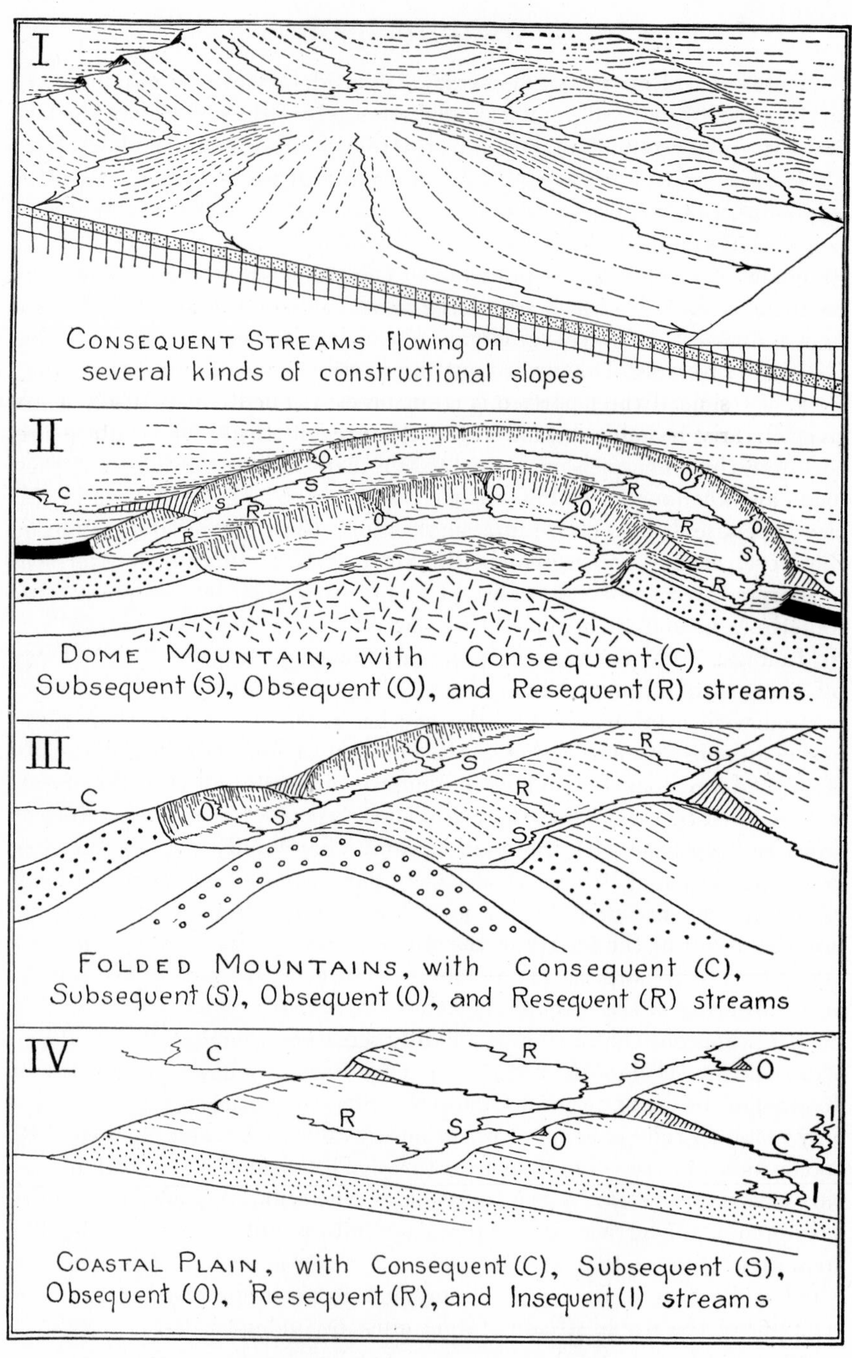


SEVERAL KINDS OF CONSTRUCTIONAL LAND FORMS, SHOWING THE DIFFERENT GENETIC TYPES OF STREAMS

## GENETIC TYPES OF STREAMS, I

A *consequent* stream is one whose position is the result of the initial slope of a land area. An uplifted dome, a newly raised block mountain, an elevated coastal plain acquires a drainage system made up at first entirely of consequent streams. They may flow in any direction and have any pattern. Their location is due solely to the original irregularities of the land surface. The little radial streams draining the surface of a Texas salt dome and the numerous relatively short streams of the Atlantic coastal plain are good examples of this type.

A *subsequent* stream is one which has developed a valley upon a belt of underlying weak rock. It is sometimes called a *strike* stream because it follows the strike of the formations. Such a stream is "adjusted" to the structure. It does not ordinarily cross resistant formations. The term subsequent refers not so much to the fact that its development is subsequent in time to that of the consequent streams but rather to the idea that it is working upon subjacent or underlying beds of less resistant rock. The Hudson River in its course between Albany and Newburgh occupies a subsequent valley, as does the Shenandoah in Virginia. Many of the streams in Pennsylvania follow the belts of weak rock in the Folded Appalachians. The term subsequent is applied also to those streams which follow joints and faults in crystalline-rock areas.

An *obsequent* stream is one which flows in a direction opposite to the dip of the formations* and opposite to that of the original consequent streams of the region. Obsequent streams are usually short, with steep gradients. They are often wet-weather rills cascading over escarpments. Most obsequent streams are tributary to subsequent streams. Kaaterskill Creek and Plattekill Creek flowing down the east face of the Catskills are streams of this type.

*Resequent* streams are those which flow down the dip of the formations in the same direction as the original consequent streams. But the resequent streams develop later and at a lower level on a stripped surface. The term *resequent* refers to the greater recency of their development and combines the two words *recent* and *consequent*. Resequent streams are frequently tributary to subsequent streams.

*Insequent* streams are those which are not controlled by any detectable cause. They do not follow the rock structure, nor do they flow down the dip of the beds. They flow in every conceivable direction, and the resulting pattern is dendritic. Millions of little streams tributary to the other types mentioned are termed insequent.

* Streams flowing down the face of obsequent fault-line scarps are also obsequent streams. Likewise, streams flowing down the face of a resequent fault-line scarp are resequent streams.

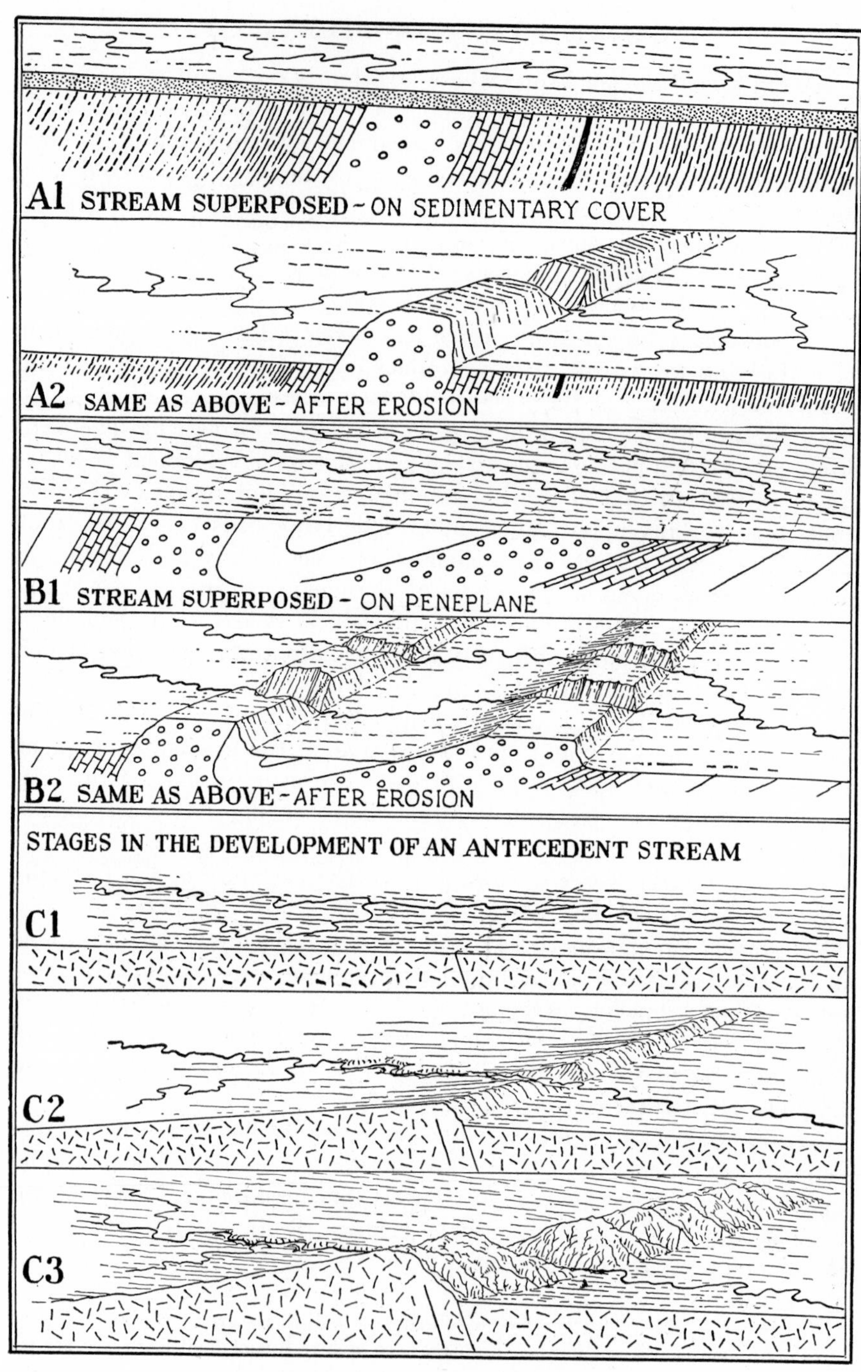


STAGES IN THE DEVELOPMENT OF SUPERPOSED AND ANTECEDENT STREAMS

## GENETIC TYPES OF STREAMS, II

A stream is said to be *superimposed* (or superposed) over crystalline or deformed rocks when it acquires a course upon flat-lying sedimentary or alluvial formations which conceal the underlying mass. Streams flowing on the thin veneer of alluvium or of detritus which covers a peneplane are superimposed above the rocks thus concealed. When rejuvenated, they may cut through the covering layer and transect the buried formations. Thus streams rejuvenated by the uplift of a peneplane become incised and develop courses without regard to the rock structure. Eventually the covering may be entirely removed or the peneplane dissected, and then only the physiographic pattern of the streams will suggest their having been let down from a superimposed position (Figs. *A2* and *B2*). The course of the lower Connecticut River from Middletown to Long Island Sound is due apparently to an earlier superimposed position upon a layer of coastal-plain sediments. The gorge of the Hudson River, the Delaware Water Gap, and the many other gaps which transect the Folded Appalachians are explained as having been formed by streams which earlier flowed down a coastal-plain cover above the present height of the ridge crest.

An *antecedent* stream is one which has maintained its course across an uplift which it antedates. This naturally presumes a very slow uplift. The Green River where it cuts across the Uinta Mountains through the Canyon of Ladore has often been cited as an antecedent stream. This is now considered doubtful as there is evidence that the Green River was blocked to form a lake whose outlet was superimposed at the point where the canyon now exists. A better example is the Sevier River across the Sevier Range in Utah. Otherwise it is impossible to account for this gorge cutting entirely across a block mountain.

The term *anaclinal* is applied to an antecedent stream flowing on a surface which has been slowly tilted in a direction opposite to the flow of the stream. If sufficiently vigorous, such a stream maintains its course. Davis cites the lower Raritan of New Jersey as an anaclinal stream.

*Reversed* streams are those which have been unable to maintain their course against the tilt of a region but change the direction of their flow to meet the conditions. The term *resurrected* has been suggested by McGee for those streams which resume courses where an earlier well-marked drainage system has become but slightly masked by a thin film of sediments as a result of brief submergence. The elevation of the masked surface permits the streams to follow lines essentially identical with the courses of their ancestors.

*Compound* streams drain areas of different geomorphic age. *Composite* streams drain areas of different geologic structure. Most large streams are both compound and composite, though some of the smaller Pennsylvania rivers like the Juniata are only composite.

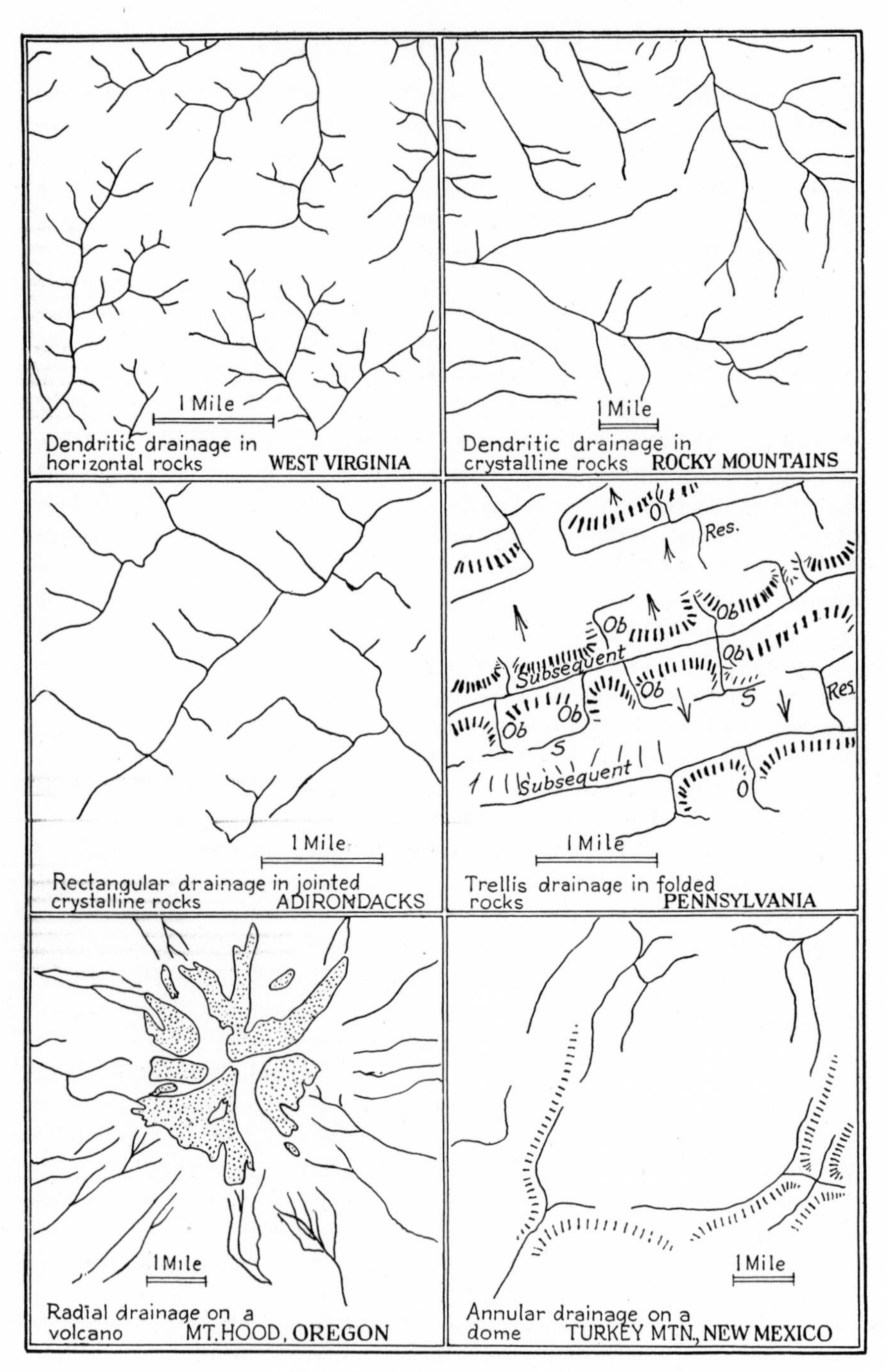


TYPES OF STREAM PATTERNS

## STREAM PATTERNS

DENDRITIC DRAINAGE PATTERN. In a region of homogeneous rock structure, such as the granite area of the Sierra Nevada, or the horizontal rock plains of the Middle West, the streams run in all directions like the branches of a tree; hence the term *dendritic*. The position of the streams is not influenced by rocks differing in resistance. The location of the many insequent streams is largely fortuitous.

RECTANGULAR DRAINAGE PATTERN. The tributaries to the Grand Canyon and other plateau streams have strongly angular courses because of the rectangular pattern of the joints which break up these rocks. Ausable Chasm, in the horizontally bedded Potsdam sandstone in northern New York, is strongly angular for the same reason. In the crystalline Adirondacks the mountain blocks are arranged in a regular checkerboard pattern by the rectangular plan of the valleys which follow joints and faults. Diagonal joints and faults produce an angular drainage pattern but not a rectangular one. In many regions of dendritic drainage pattern there is an obscure tendency to follow an angular or rectangular plan.

TRELLIS DRAINAGE PATTERN. This pattern is characteristic of strongly folded or dipping rocks. The plan of the drainage lines resembles the shape of a vine on a trellis. Three genetic types of streams combine to produce this plan: The longer streams following the outcrops of the weaker rocks are subsequent; tributaries from each side are obsequent and resequent, depending on whether they flow opposite to or in conformance with the dip of the strata. The transverse streams in such an area are probably due to superposition. Trellis drainage is conspicuous in the Folded Appalachian belt from New York to Alabama, in the Jura Mountains of France, and in the dipping formations of the coastal plain in southern England. In some regions of homoclinal dips the trellis pattern degenerates into a series of parallel streams.

RADIAL DRAINAGE PATTERN. Young dome mountains and volcanoes have radial drainage lines consequent in origin and centrifugal in direction. Structural basins also have a radial drainage pattern but centripetal in direction.

ANNULAR DRAINAGE PATTERN. Some of the streams which drain a maturely dissected dome follow circular paths around the dome in conformance with the outcrops of the weaker belts. These are subsequent streams. They have both obsequent and resequent tributaries. An annular drainage pattern is simply a variation or special form of the trellis pattern. The annular plans of the streams around the Black Hills and of the Weald of England clearly reflect the rock structure of these two eroded domes.

## PLAYFAIR'S LAW

Until the beginning of the nineteenth century many people believed that canyons and ravines were formed by some sudden splitting asunder of the earth's crust. These were the cataclysmists, or the catastrophists. It was of course quite natural to invoke the aid of some prodigious force to produce features of great prominence. But there was another group of people, the uniformitarians, who believed that the everyday forces of nature, which are everywhere slowly acting, are quite sufficient to produce mighty results if they can operate for a long enough period of time. The fact that little if anything in the way of changing landscapes can be observed over the span of one man's life made it difficult for those who believed in uniformitarianism to convince others. It was due largely to the writings of James Hutton of Scotland, a profound thinker about geology, that the newer views came to be accepted. Hutton's writings, however, were rather ponderous and obscure, but a young mathematical friend of his named John Playfair came to his rescue. In 1802, Playfair published his famous book called *Illustrations of the Huttonian Theory of the Earth,* which became celebrated for its charm and lucidity. Every student of geology and physiography knows by heart the following passage regarding the origin of valleys. Playfair says simply,

> Every river appears to consist of a main trunk, fed from a variety of branches, each running in a valley proportioned to its size, and all of them together forming a system of valleys, communicating with one another, and having such a nice adjustment of their declivities that none of them join the principal valley either on too high or too low a level; a circumstance which would be infinitely improbable if each of these valleys were not the work of the stream which flows in it.

The principle thus expressed is known as *Playfair's law.* It simply recognizes the fact that streams and their tributaries meet each other with accordant junctions. Those who have sought exceptions to this rule have examined great canyons in arid regions where it seemed most likely that the main stream could cut down more rapidly than its small and intermittent tributaries and leave them cascading down the valley walls. But even in the Grand Canyon this is not the case. The tributaries, in spite of their small size and intermittent character, seem to have kept pace with the main stream. It is easy to understand why this is apt to be the case. For if the Colorado River should cut below the mouths of its tributaries, this would greatly increase their gradients at those points; and in spite of their small volume, they would be able to deepen their valleys with greater vigor than that possessed by the Colorado itself. The competence of rivers to carve even the deepest valleys has thus been fully established.

It is true that Playfair's law, like most other laws, does indeed have some exceptions. But these are rare. In the Grand Canyon, for example, there are some streams which flow only after heavy thunderstorms and cascade for a brief hour or so over the brim of the canyon in waterfalls which vie with Niagara in height though not in volume. But these trivial exceptions are easily understood and by no means invalidate the general principle.

Playfair's law was only one of the ideas which Playfair gleaned from Hutton's observations upon the slow, persistent action of natural forces, but it called attention, as no other single expression had done, to the truth of uniformitarianism. Henceforth it was logical to see small forces producing great results if sufficient time were available. In short, Playfair paved the way to the modern belief in peneplanes. Stupendous as a canyon is, it is only the first step in stream erosion. The ultimate result is the complete wearing away of the land mass to a surface which practically coincides with sea level.

So compelling is this idea of uniformitarianism and so invincible its logic that of necessity the same principles are now applied to waves, glaciers, and winds. All of these forces are deemed capable of reducing land masses to base-level. Some students of geology have perhaps become too enthusiastic over this principle of uniformitarianism, for it is probable that some events take place on the earth's surface with cataclysmic suddenness. This is true of earthquakes and volcanic eruptions. Nevertheless, it is fairly certain that, looking back over all of geological time, we should see the same quiet, orderly processes of erosion, glaciation, wave action, and wind movement, as well as the slight shifting of land masses due to earthquakes and the very slight and virtually imperceptible changes in the relation between land and sea which we now observe going on from year to year. Ineffective as these changes seem to be, we know that throughout the long vista of the ages the cumulative effect is profound, and that by these means we can account for the present land features of the earth.

Other Observations of Hutton. Hutton's observations may now seem to us of small importance, but he drew from them conclusions which turned the whole trend of geological thought. The mere fact that rivers carved their valleys required proof in those days. Therefore Hutton called attention to the irregular spacing of tributary gorges on the two sides of certain rocky valleys, the north and south sides not matching in any way, and argued that this could hardly have been the case had the gorges been the effects of previous "concussions" of nature. He discussed the origin of alluvial plains, recognizing the gradation in the size of gravel. He understood the origin of sedimentary rocks as distinguished from igneous rocks and realized that earth movements occurred at different times to disturb their original horizontal position and that the oldest and deepest rocks were the most disturbed. He recognized the long period of time which "no doubt was required for the elevation of the strata." The origin of rocks and ores, the preservation of fossils, the development of landscapes, the origin of the earth—he saw as parts of one great continuous process of nature, all supporting the idea of uniformitarianism.

## STREAM DEFLECTION DUE TO ROTATION OF THE EARTH

One of the most remarkable observations made during the last century had to do with the effect of the earth's rotation upon flowing streams. It was well-known, from mathematical deduction, that bodies in motion in the Northern Hemisphere have a tendency to swerve to the right of what would be a straight path on the earth's surface. In the Southern Hemisphere the deflection is toward the left. This principle, known as *Ferrel's law*, was first applied by Ferrel to the behavior of winds and air currents on the earth's surface. The deflective force varies with the latitude, being much greater in high latitudes than in low. It is over 50 per cent greater at 60°N than at 30°N and is over ten times as great near the pole as it is a few degrees from the equator. On the equator, of course, there is no deflection. The deflective force at any given latitude is constant for any given rate of motion, regardless of the direction of movement. The deflective force, however, is proportional to the velocity. Hence swiftly moving objects at high latitudes are much more strongly affected than slowly moving bodies near the equator. A projectile fired by one of the long-range German guns during the war with a range of 75 miles was deflected about 1,500 feet in that distance. The bullet from a small rifle fired at a target a couple of kilometers away, that is, a little over a mile, in the latitude of New York is deflected a foot or so. This deflection, in the case of artillery fire, is usually too small to be taken into account.

One of the first localities where the deflective effect of the earth's rotation upon streams was observed was in southern Long Island. The south side of Long Island is a gently sloping outwash plain of remarkable evenness and homogeneous material. It is crossed by a number of small streams which have excavated shallow valleys in the homogeneous plain. Each of these little valleys is limited on the west, or right, side by a bluff from 10 to 20 feet high, while its gentle slope on the left side merges imperceptibly with the general plain. The stream in each case follows closely the bluff at the right. As the streams carve their valleys deeper, they are induced by rotation to excavate their right banks more than their left, gradually shifting their positions to the right, and maintaining stream cliffs on that side only.

The Yukon River and other Alaskan streams indicate a strong predominance of erosion on the right bank, revealed by asymmetry in the position of the river with respect to the flood plain and in the excess of bluffs on the right-hand side, and location of bars and islands relative to cut banks. In the last 600 miles of its course the Yukon flows close to the right side of its flood plain which is extremely wide. Of great significance also in the case of the Yukon is the behavior of driftwood and floating debris. Such material is almost entirely absent from the left bank but is plentiful on the right.

A study of the Missouri River showed that between Sioux City, Iowa, and Kansas City, Missouri, there is over four times as much area of flood plain on the left side of the river as on the right.

The rivers of Siberia present other examples. These streams flow over vast plains of unconsolidated homogeneous material. They display very much steeper banks on their eastern, or right-hand, sides. Von Baer was the first to point out this remarkable effect on the Russian rivers. Nansen, however, describes them very vividly in his book on Siberia. He affirms that it is quite possible to eliminate any effect due to tilted formations, inasmuch as the beds there are all horizontal and consist of loose sand and gravel deposits. On the west, or left-hand, side of the streams there are extensive plains of sand over which the rivers have migrated laterally. This renders it difficult or impossible for boats to land on the western banks of the streams. The towns and settlements are therefore mainly on the eastern side.

Still another interesting locality is the Lannemezan fan in southern France. This alluvial plain slopes north from the Pyrenees for many miles. From its apex radiate streams in almost all directions: east, north, and west. These streams flow in channels or ravines below the surface of the plain and in each case the ravine has a steep right-hand bank and a long gentle slope on the left, regardless of the direction in which the streams are flowing. It is difficult, therefore, to think that insolation, excessive precipitation, or dip of strata have been factors in determining this asymmetry.

In the case of meandering streams, the deflective effect in the Northern Hemisphere tends to cut off meanders on the left side of the river. This condition was noted by Eakin along the Missouri River. Between Fort Benton, Montana, and Sioux City, Iowa, five meander cutoffs were noted. Four of these are on the left side of the river and only one on the right. Even this exception was somewhat abnormal, for the river is now encroaching upon the meander instead of receding from it as is usually the case after a cutoff is effected.

The deflective effect of the earth's rotation may readily be demonstrated by the classic Foucault pendulum experiment, in which any swinging object, in the Northern Hemisphere, seems constantly to be diverted in a clockwise direction. The winds on the earth's surface are much more easily influenced than are streams of water. For that reason the trade winds in the Northern Hemisphere blow from northeast to southwest instead of directly toward the equator. In the Orient the southern trades blow from the southeast across the equator in the summer and then turn sharply to the right to produce the monsoons.

The counterclockwise movement of the winds in cyclonic storms in the Northern Hemisphere may be demonstrated as another example of deflection toward the right. Ocean currents also, the world over, are clearly under the influence of the earth's rotation.

## QUESTIONS

1. In what three ways does a stream carry its load? *Ans.* In suspension, dragging on bottom, in solution.
2. The Mississippi River each year transports material in each of these three ways. In one way it carries 340 million tons; in a second way it carries 40 million tons; and in a third way it carries 136 million tons. The figure 40 million is only estimated, as that method of transportation cannot be measured so readily as the other two. What method of transport does each of these figures represent?
3. Estimating the drainage basin of the Mississippi River at 1,000,000 square miles and a cubic foot of rock or soil as weighing about 200 pounds (*i.e.*, allowing a specific gravity of about 3) how much on the average is this whole drainage basin lowered in the course of a year; or rather how many years would it take to lower it 1 foot? And at the same rate how long would it take to lower it an average of 1,000 feet, that is, to peneplane it? *Ans.* 5,000,000 years.
4. A large mature stream is joined by a tributary, also mature, each stream carrying a heavy load. Is the gradient of the main stream apt to be greater above or below the point of junction?
5. Imagine a mature river system, young only in its most headward parts where erosion is still going on. Will the profile of this mature stream ever change or has it a slope which is fixed for all time? Explain what will happen as time goes on.
6. Study some of the topographic sheets showing the folded mountains of eastern Pennsylvania. Pick out examples of subsequent, resequent, obsequent, and possibly consequent streams, and determine their gradients in feet per mile. Which have the steepest gradient?
7. Will the deflective effect of the earth's rotation be noticeable more in young or mature streams? Is it more noticeable near the poles or near the equator?
8. A compound river is one which drains areas of different ages; a composite river is one which drains areas of different structures. What, therefore, is a simple river? Can you name examples of each?
9. What would be the effect of overgrazing upon the streams of a region?
10. What is meant by the dynamic cycle?
11. What are nickpoints? *Treppen?* (See Rich.)
12. What is the difference between base-level and peneplane?
13. What is meant by integrated and what is meant by nonintegrated drainage?
14. What are multiple-erosion surfaces and how are they explained?
15. What is saltation? Is coarse or fine material carried in this manner? What does a boulder of granite 1 foot in diameter, specific gravity 2.8, weigh under water?

## TOPICS FOR INVESTIGATION

1. Cycle of erosion. The exact meaning of this term. Varying usages.
2. Peneplanes. Various explanations for the development of peneplanes.
3. Stream deflection, influenced by earth's rotation. Description of examples and evidence from various parts of the world. (In this connection determine the effect of Ferrel's Law upon projectiles and other fast-moving objects.)
4. Typical river systems and their development (*e.g.*, Susquehanna, Rhine, Colorado).
5. Rivers and their geographical significance (*e.g.*, Rhine, Danube, Seine).
6. Drainage patterns. Varieties and significance.
7. The Huttonian theory. Problems considered in addition to those of streams.
8. Relation between rainfall and runoff under varying conditions.

## REFERENCES


General

Bissell, M. H. (1921) *On the use of the terms denudation, erosion, corrosion, and corrasion.* Science, new ser., vol. 53, p. 412–414.

Bonney, T. G. (1912) *The work of rain and rivers.* Cambridge, England, 144 p.

Brown, R. M. *Rivers and river valleys.* Am. Geog. Soc., Bull. 34, p. 371–383, 1902; Bull. 35, p. 8–16, 1903; Bull. 39, p. 147–158, 1907; Bull. 44, p. 645–657, 1912. Mississippi floods.

Campbell, M. R. (1929) *The river system; a study in the use of technical geographic terms.* Jour. Geog., vol. 28, p. 123–128.

Gilbert, G. K. (1877) *Geology of the Henry Mountains.* U. S. Geog. and Geol. Surv. Rocky Mt. Region (Powell). *Land sculpture,* p. 99–150. A classic work dealing with the principles of fluvial erosion.

Glock, W. S. (1931) *The development of drainage systems; a synoptic review.* Geog. Rev., vol. 21, p. 475–482.

Glock, W. S. (1931) *The development of drainage systems and the dynamic cycle.* Ohio Jour. Sci., vol. 31, p. 309–334.

Hoyt, W. G. (1913) *Effects of ice on stream flow.* U. S. Geol. Surv., W.-S. Paper 337, 77 p.

Johnson, D. W. (1922) *The scenery of American rivers.* Geog. Soc. Phila., Bull. 20, p. 22–27.

Johnson, D. W. (1932) *Streams and their significance.* Jour. Geol., vol. 40, p. 481–497.

Jukes, J. B., and Geikie, A. (1872) *Student's manual of geology.* Edinburgh, 778 p. (see Chaps. 25 and 26).

Playfair, Sir J. (1802) *Illustrations of the Huttonian theory of the earth.* Edinburgh (see p. 102, 350–371).

Russell, I. C. (1909) *River development* (*Rivers of North America*). London, 327 p.

Zernitz, E. R. (1932) *Drainage patterns and their significance.* Jour. Geol., vol. 40, p. 498–521.

Cycle of Erosion

Brigham, A. P. (1892) *Rivers and the evolution of geographic forms.* Am. Geog. Soc., Bull. 24, p. 23–43.

Davis, W. M. (1899) *The geographical cycle.* Geog. Jour., vol. 14, p. 481–504; Geog. Essays, p. 249–278.

Davis, W. M. (1923) *The cycle of erosion and the summit level of the Alps.* Jour. Geol., vol. 31, p. 1–41.

Johnson, D. W. (1933) *Development of drainage systems and the dynamic cycle.* Geog. Rev., vol. 23, p. 114–121.

Malott, C. A. (1928) *An analysis of erosion.* Ind. Acad. Sci., Proc. 37, p. 153–163.

Malott, C. A. (1928) *The valley form and its development.* Ind. Univ. Studies, vol. 15, p. 3–34.

Peneplanes

Ashley, G. H. (1930) *Age of the Appalachian peneplanes.* Geol. Soc. Am., Bull. 41, p. 695–700.

Campbell, M. R. (1897) *Erosion at baselevel.* Geol. Soc. Am., Bull. 8, p. 221–226.

Chamberlin, R. T. (1930) *The level of base-level.* Jour. Geol., vol. 38, p. 166–173.

Daly, R. A. (1905) *Accordance of summit levels among Alpine mountains.* Jour. Geol., vol. 13, p. 105–125.

Davis, W. M. (1909) *Geographical essays.* Boston, 777 p. Contains the following not elsewhere herein listed: *Baselevel, grade and peneplain; The peneplain* (also in Am. Geol., vol. 23, p. 207–239, 1899); *Plains of marine and subaerial denudation* (also in Geol. Soc. Am., Bull. 7, p. 377–398, 1896); *The geographical cycle in an arid climate* (also in Jour. Geol., vol. 13, p. 381–407, 1905).

Johnson, D. W. (1929) *Baselevel.* Jour. Geol., vol. 37, p. 775–782.

JOHNSON, D. W. (1916) *Plains, planes, and peneplanes.* Geog. Rev., vol. 1, p. 443–447.
LEE, W. T. (1922) *Peneplains of the Front Range and Rocky Mountain National Park, Colorado.* U. S. Geol. Surv., Bull. 730, p. 1–17.
MALOTT, C. A. (1928) *Base-level and its varieties.* Ind. Univ. Studies, vol. 15, p. 35–59.
MOSS, R. G. (1936) *Buried pre-Cambrian surface in the United States.* Geol. Soc. Am., Bull. 47, p. 935–966.
RICH, J. L. (1938) *Multiple erosion surfaces.* Geol. Soc. Am., Bull. 49, p. 1695–1722.
SHALER, N. S. (1899) *Spacing of rivers with reference to hypothesis of base-leveling.* Geol. Soc. Am., Bull. 10, p. 263–276.
SHARP, H. S. (1929) *The Fall Zone peneplain.* Science, new ser., vol. 69, p. 544–545.
SHARP, H. S. (1929) *A pre-Newark peneplain and its bearing on the origin of the lower Hudson River.* Am. Jour. Sci., 5th ser., vol. 18, p. 509–518.
TARR, R. S. (1898) *The peneplain.* Am. Geol., vol. 21, p. 351–370.
VAN HISE, C. R. (1896) *A central Wisconsin base-level.* Science, new ser., vol. 4, p. 57–59.
WILSON, A. W. G. (1903) *The Laurentian peneplain.* Jour. Geol., vol. 11, p. 615–669.

EVOLUTION OF CERTAIN RIVER SYSTEMS

BLACKWELDER, E. (1934) *Origin of the Colorado River.* Geol. Soc. Am., Bull. 45, p. 551–566. An integrated stream developed under arid conditions.
DAVIS, W. M. (1889) *The rivers and valleys of Pennsylvania.* Natl. Geog. Mag., vol. 1, p. 183–253; Geog. Essays, p. 413–484.
DAVIS, W. M. (1890) *The rivers of northern New Jersey.* Natl. Geog. Mag., vol. 2, p. 81–110; Geog. Essays, p. 485–513.
DAVIS, W. M. (1895) *The development of certain English rivers.* Geog. Jour., vol. 5, p. 127–146.
DAVIS, W. M. (1896) *The Seine, the Meuse, and the Moselle.* Natl. Geog. Mag., vol. 7, p. 180–202, 228–238; Geog. Essays, p. 587–616. Intrenchment of meanders.
JOHNSON, D. W. (1931) *Stream sculpture on the Atlantic slope.* New York, 142 p.
JOHNSON, D. W. (1931) *A theory of Appalachian geomorphic evolution.* Jour. Geol., vol. 39, p. 497–508.

STREAM DEFLECTION

EAKIN, H. M. (1910) *The influence of the earth's rotation upon the lateral erosion of streams.* Jour. Geol., vol. 18, p. 435–447.
GILBERT, G. K. (1884) *The sufficiency of terrestrial rotation for the deflection of streams.* Am. Jour. Sci., 3d ser., vol. 27, p. 427–432.

# VI
# YOUNG STREAMS

VERNAL FALLS, YOSEMITE NATIONAL PARK, CALIFORNIA
Stream flowing over flat side of joint plane. A position recently acquired, hence no notch has yet been formed.

*U. S. Engineers Office, Los Angeles*

SOUTHERN PACIFIC RAILWAY NEAR SANTA ANA, CALIF.

Showing transporting power of young stream. The flood waters of the Santa Ana River swept this track off the roadbed which formerly ran straight through the picture.

VIEW IN NORTH HOLLYWOOD TAKEN AFTER THE STORM OF MARCH 2, 1938

Showing erosion channel formed after a heavy rain in what was previously a city street.

*U. S. Engineers Office, Los Angeles*

*Lee, U. S. Geological Survey*

ONACHOMO NATURAL BRIDGE, UTAH

A bridge formed in massive horizontal sandstone, upheld by the strength of the flat tabular mass like a Greek lintel. Arch form almost lacking, because it is unnecessary. (*See also p.* 108.)

NATURAL BRIDGE, UTAH

A bridge formed in cross-bedded and much jointed rock, upheld by its almost perfect arch-like shape brought about by processes of weathering in a relatively weak mass. Tabular form lacking because it is structurally impossible.

*Darton, U. S. Geological Survey*

*U. S. Soil Conservation Service*

RECENT GULLYING, VENTURA COUNTY, CALIF.

Excellent example of branching system of gullies eating headward into cultivated fields.



GULLY NEAR SYRACUSE, NEB.

The headward growth of this gully is prevented by the ditch and diversion ridge, and wire check which spreads the water evenly over the pasture.

*U. S. Soil Conservation Service*

*U. S. Soil Conservation Service*

HALL COUNTY, TEX.

Pasture furrowing to prevent gullying. The furrows follow contour lines at irregular contour intervals, but close enough to prevent much accumulation of water in one place.



POTHOLES IN MASSIVE LAVA BETWEEN MEDFORD AND CRATER LAKE, ORE.

Characteristic details of a young stream bed. The joints in the lava cause the potholes to break up.

*A. K. Lobeck*

*Sharpe, U. S. Soil Conservation Service*

SPARTANBURG COUNTY, S. C.

A gully of long standing, constantly encroaching upon the adjacent fields by removing the deeply weathered residual soil of the crystalline rocks.



BADLANDS ON THE BLACK FORKS, WYO.

An advanced stage of gullying in a semiarid region of weak beds. Picture taken in 1870.

*Jackson, U. S. Geological Survey*

# YOUNG STREAMS

SYNOPSIS. The distinctive aspect of young streams is their ability to erode. This may be due to the steep slope of the land, as a result of initial uplift or later warping, to the volume of the stream, or to the small load which the stream is carrying. In any event the stream everywhere along its course is taking on more load and carrying it away. In its headwater portions this is expressed by rock falls, landslides, creep of the soil, mass earth movement of all sorts, springs, and sharply cut ravines and canyons. The main work of erosion, however, is concentrated in the bed of the stream where *corrasion*, *quarrying*, and *solution* are effective processes leading to an increase in load. Narrow steep-walled valleys with rocky ledges, falls, rapids, potholes, and natural bridges are characteristics of young streams. They all indicate active erosion and removal of material.

This load of material is transported in solution, in suspension, and by saltation (dragging or jumping along the bottom). A young stream may be loaded to capacity with one size of material but not with material of a smaller size which is more readily carried. Thus a mountain torrent after a cloudburst can move practically any size of boulder in its bed, even blocks as large as a small house. A stream slightly less vigorous can move large cobbles the size of footballs. In addition, many streams constantly drag a mass of small gravel along their beds but only occasionally move the larger blocks. Some young streams at times of low water find their ability to transport almost lacking. They may temporarily behave as mature streams, deposit sand bars, and build small flood plains. Indeed, most young streams have quiet reaches where mature conditions prevail.

An analysis of waterfalls shows that they indicate some interruption or disturbance in the orderly development of the stream or are due to the failure of the stream up to that time, because of varying rock resistance, to develop a graded condition.

Most young streams also show a distinct tendency to enlarge their drainage basins, by lateral or by headward erosion, usually at the expense of other streams. This causes stream capture or stream diversion with a concomitant series of minor details and adjustments. Numerous small diversions also undoubtedly occur during the process of stream adjustment to the rock structure. Streams following the strike of weak beds occasionally work laterally down the dip of the beds, a process which is termed *uniclinal shifting*. Streams following weaker belts, either softer beds or zones of jointing or faulting, have an advantage over their neighbors and gradually become the master streams of the region.

It is shown also in this chapter that incised meanders do not necessarily indicate rejuvenation from a former mature stage but may come about in the ordinary course of stream dissection.

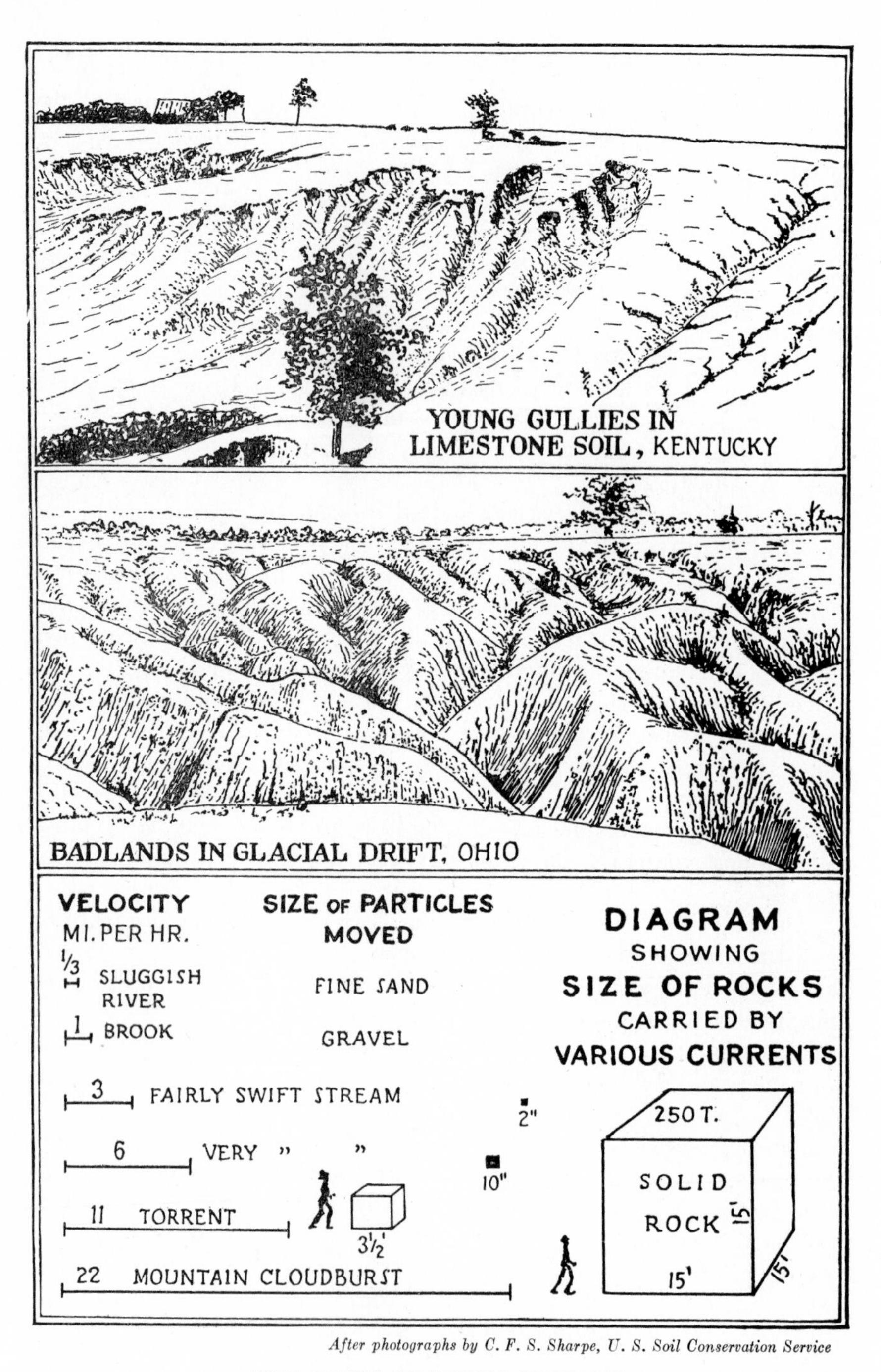


*After photographs by C. F. S. Sharpe, U. S. Soil Conservation Service*

THE WORK OF YOUNG STREAMS

## YOUNG STREAMS AND THEIR CHARACTERISTICS

Erosion. Young streams are defined as streams that are able, because of sufficient speed and volume, to carry their loads of sediment and at the same time further to erode their channels.

This is accomplished by (*a*) *corrasion*, or scraping and scratching away the bedrock; (*b*) *impact*, or the effect of definite blows on the bed of the stream by large boulders; (*c*) *quarrying*, due to the lifting effect of the water as it pushes into the cracks of rock; and (*d*) *solution*.

Corrasion and impact produce much fine material. Quarrying provides the larger blocks that are rolled along; and solution, especially in limestone regions like the Mississippi Valley, adds a vast amount of lime and soluble salts to the load. One-fourth of the load of the Mississippi River is carried in solution. All of these activities indicate excess energy, and the result appears as narrow gorges, falls and rapids, potholes, natural bridges, and rocky valley walls.

Young streams not only cut down their valleys but exhibit virility in the development of tributaries and in constantly adding to their drainage basins. Nor do young streams necessarily flow continuously. Some of the most active streams are the wet-weather gullies which develop along hillsides and rapidly by headward erosion denude the slopes. They wreak great damage where there is no protective cover of vegetation. The result is *badland* topography, characterized by multitudinous small ravines and arroyos only occasionally carrying a stream. The run-off from such regions after heavy bursts of rain is extremely rapid.

Transportation. The load transported by young streams may be very great because of their great velocity but is not so great as that of mature streams of similar size. Theoretically the carrying power of a stream is proportional to as much as the sixth power of the velocity (Gilbert's "sixth power law"). This means that, if the velocity is doubled, the size of the particles composing the load may be increased up to sixty-four times.

The character of the load, however, is a very important factor. Slow-moving streams can carry large quantities of small particles, but material a little larger cannot be moved at all. It is hard for many people to see boulder-strewn river beds and realize that such large pieces of rock can be moved by running water.

Streams may move their load by dragging or rolling it along the bottom (*traction*); by carrying it in small jumps (*saltation*); by carrying it in *suspension*; or in *solution*. The Mississippi, a mature stream, carries in suspension each day 1,000,000 tons of sediment, equivalent to a cube over 200 feet on a side, the size of a large building. Most young streams are not so heavily loaded with fine material as is the Mississippi and much of their work is done in time of flood.

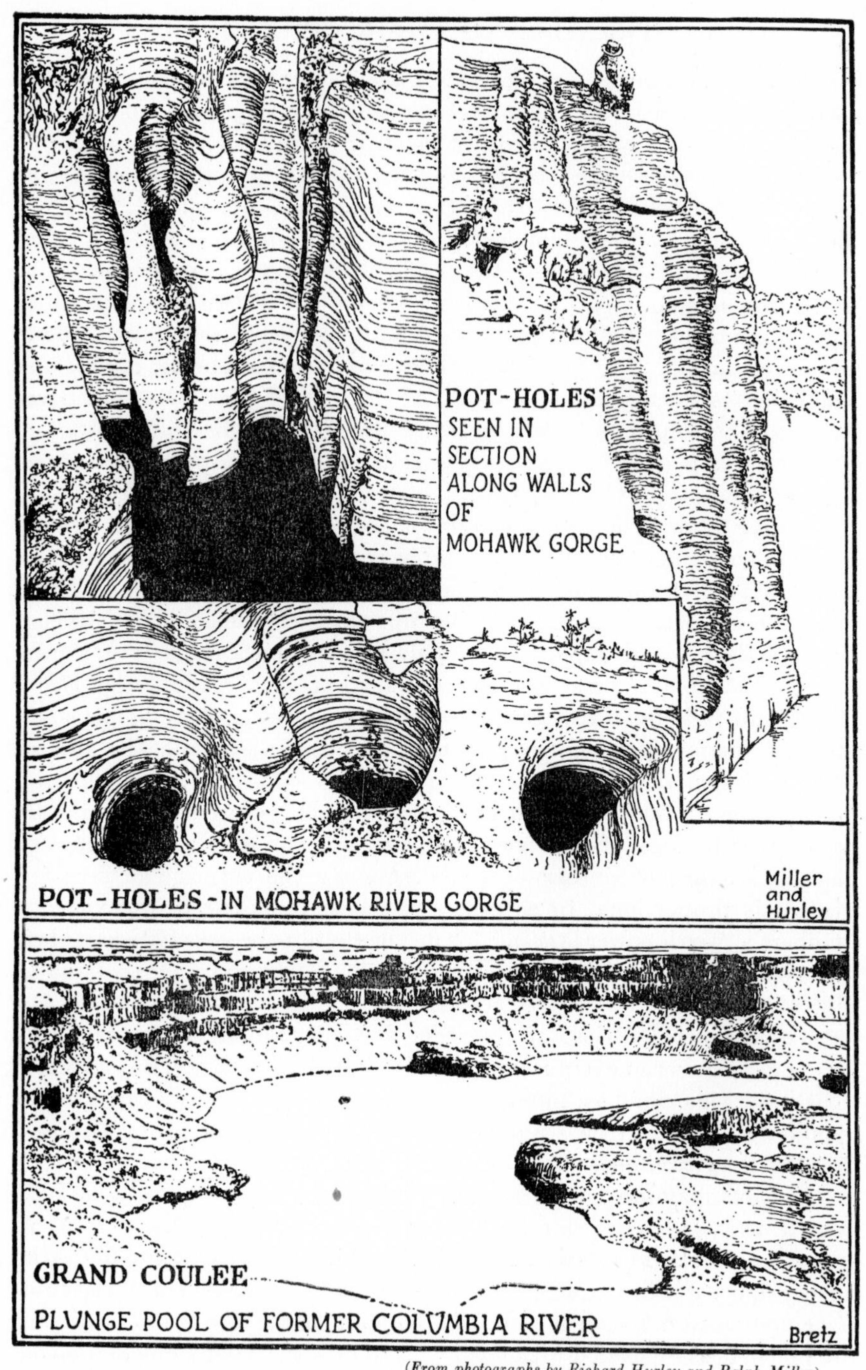


(*From photographs by Richard Hurley and Ralph Miller*)
(*From photograph by Frank Guilbert and J Harlan Bretz*)

POTHOLES IN MOHAWK RIVER GORGE AT LITTLE FALLS, N. Y.
PLUNGE POOL IN GRAND COULEE

## POTHOLES AND PLUNGE POOLS

There is continuous gradation from small hollows and depressions scoured out of the rock by swiftly flowing streams to giant potholes and great plunge pools at the base of waterfalls. These are not features of topographic importance, but they demonstrate the ability of streams to abrade their channels. This does not mean that the deepening of valleys by young streams is accomplished mainly by attrition of the stream bed. Undoubtedly more is accomplished by breaking off large blocks by hydraulic action along joint planes.

Nevertheless, potholes are interesting details which excite the curiosity of most people. Witness such names as are often bestowed upon them, such as "Jacob's Well," "The Devil's Punch Bowl," the "Witches' Caldron," "Aunt Sally's Basin," and "The Bathtub."

Potholes are undoubtedly formed most rapidly in weak rocks, like shale, but they are best preserved in massive rocks like basalt, granite, and quartzite. Examples almost without number can be cited along present-day rivers and there are many cases of abandoned potholes produced by glacial streams. Some potholes of the latter type are thought to have been formed beneath the ice sheet by water plunging into "moulins" or mills. Visitors to Ausable Chasm can see high up on the chasm walls potholes in various degrees of disruption formed when the stream flowed at a much higher level. There are several good examples of glacial potholes actually within the limits of New York City, notably those in Bronx Park, in Inwood Park, and along the cliffs above the Harlem Speedway.

Shallow potholes formed by currents moving constantly in one direction are apt to be elliptical in outline. Deep potholes with grooved sides are undoubtedly formed by spiraling currents which swirl boulders like a pestle in a mortar. Some large potholes are formed at the base of falls and cataracts and are more truly cylindrical in form. These, if very large, are termed *plunge pools.*

One of the largest potholes in the United States is at Archbald, Pennsylvania, just north of Scranton. It has a diameter and also a depth of about 40 feet and is cut out of shale. It is on a hilltop and, like others nearby, was formed by glacial waters.

Among the most interesting plunge pools, now abandoned, are those in the Grand Coulee, the former channel of the Columbia River in Washington. Formed by a stream of water falling over a cliff 400 feet high, this large basin has a depth of 80 feet and contains a perennial lake. Another abandoned plunge pool is at Jamesville in central New York State, formed by the waters from the Great Lakes which discharged here over a cataract 160 feet high and 800 feet across. The lake occupying this depression is now 60 feet deep. There are several other plunge basins along the course of this former stream.

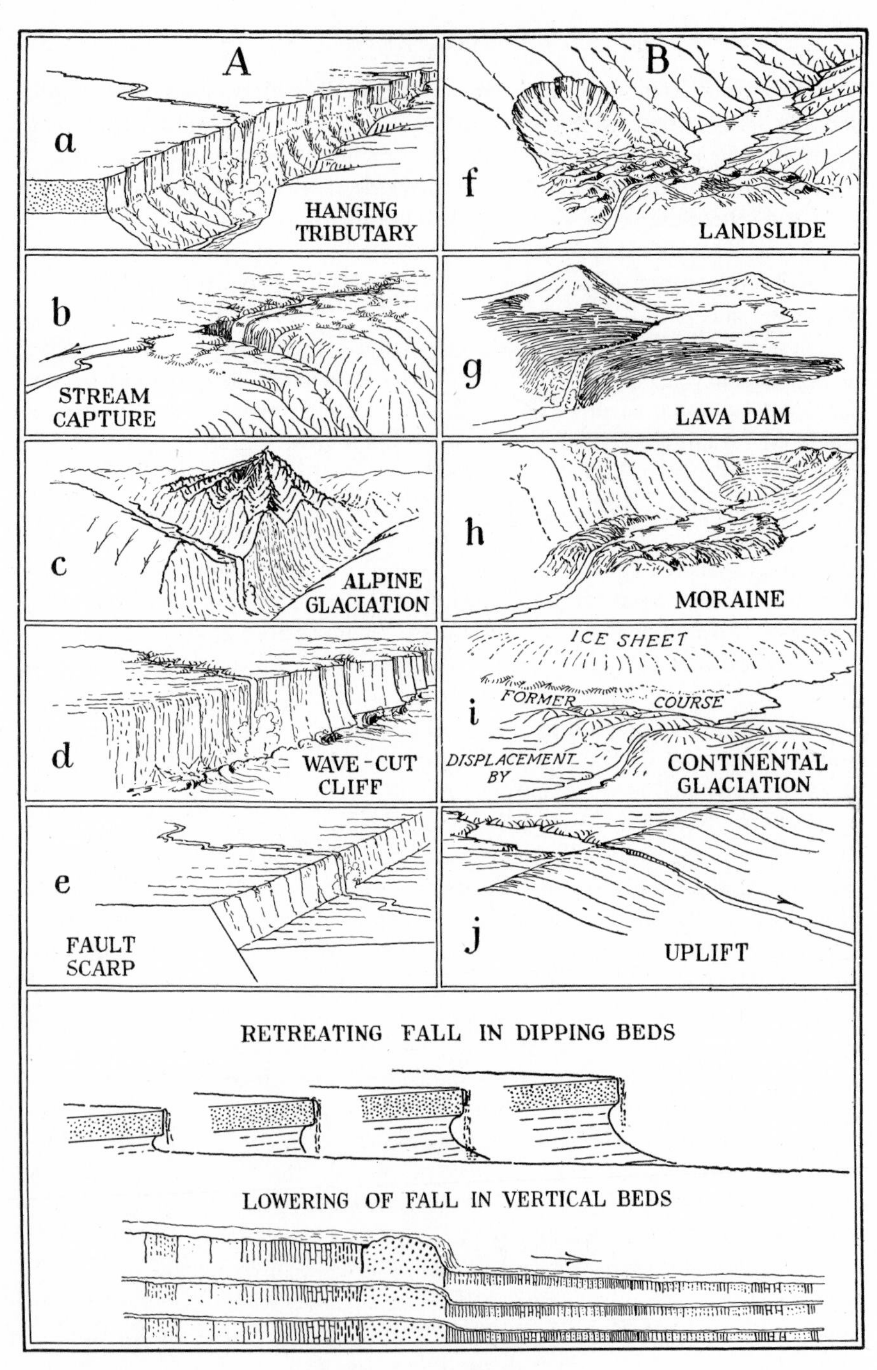


TYPES OF WATERFALLS

# WATERFALLS

Waterfalls and rapids are criteria of youth. They are of two kinds: first, those which develop in the normal life history of a river and indicate that the stream has not yet acquired a graded slope; and, second, those which result from some disturbance, accident, or interruption in the life of the stream, imposed upon it by some outside force.

1. The first type, which might be termed the *normal* type of waterfall, is due solely to the variation in the resistance of the rocks into which the stream is cutting. All streams in regions of complex rock structure erode unequally in different places, and rapids and falls result. Practically all streams, during the early stages of their life history, develop irregularities in their longitudinal profiles because of differential erosion. Many examples occur in the mountains of the Southern Appalachians, in the Adirondack Mountains, in the Appalachian Plateau and, indeed, in all regions where a variety of rocks occurs. The many falls of the fall zone between the piedmont and the coastal plain, and the rapids in such streams as the Susquehanna and the Potomac where they cut across the hard Appalachian ridges, result from the normal undisturbed development of young streams not yet completely graded. In all of the cases mentioned in the last sentence the streams appear to have been superimposed either from a coastal-plain cover or from a peneplane surface, but no outside force has marred their orderly development.

2. Disturbance of orderly stream development may come about because of the following factors, among the many which might be listed (the notation used here corresponding to that on the accompanying figure):

*A*. Lowering of stream's outlet.

*a*. Rapid downcutting of a main stream by sudden rejuvenation may leave its tributaries hanging.

*b*. Stream captures by other stream systems may result in strong differences in level.

*c*. Glaciation may produce discordance of valley junctions in mountainous regions.

*d*. Waves cutting against a coastline may leave the ends of streams hanging.

*e*. Faulting or warping may depress the lower course of a stream.

*B*. Temporary interruption in the life history of the stream by blocking its course in some way.

*f*. By landslide.

*g*. By lava dam.

*h*. By moraine.

*i*. By glacier forcing stream into a new position, causing superposition.

*j*. By uplift across course of stream as in case of dome or block mountain.

*k*. Any combination of circumstances which causes a stream to take up a new position or become superposed over irregular topography.

The actual pattern and behavior of the fall or rapids depends largely upon the attitude of the rocks causing it. Horizontal beds or beds dipping slightly upstream cause the falls to retreat upstream and usually to become lower. Falls coming under group *A*, not accounted for by differences in rock resistance, all retreat upstream as the lower course of the stream becomes graded. Falls due to vertical beds do not change their position but are simply reduced in height.

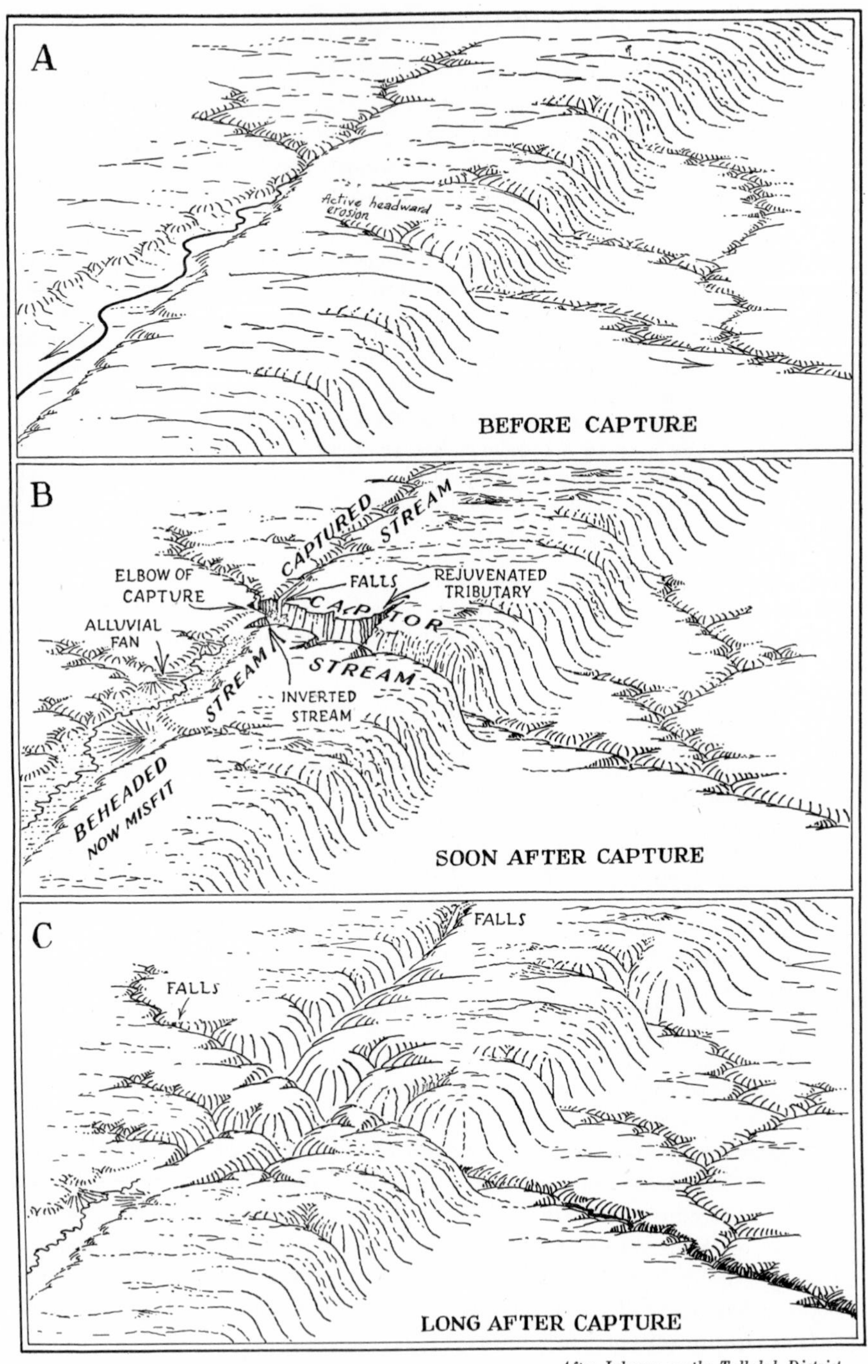


*After Johnson on the Tallulah District*

STAGES IN THE PROCESS OF STREAM CAPTURE

## STREAM CAPTURE

Stream capture results when one stream flowing in a lower region works headward and intercepts the headwaters of a stream draining a higher area. The stream flowing at the lower level always has the advantage. If there is a pronounced escarpment separating the two levels, the conditions for capture are especially favorable. Capture may also be effected by streams flowing upon weaker belts of rock. Such streams are able to cut down their valleys below the levels of streams which traverse resistant formations. In either case the capturing stream has the advantage of lower position.

DEDUCTIVE TREATMENT OF STREAM CAPTURE. Figure *A*, opposite, illustrates the conditions just before capture. The rivers heading in the escarpment can, because of their steep gradients, cut back rapidly into the drainage area of the streams flowing on the plateau above, even if the rocks of the region are all homogeneous and of equal resistance.

Figure *B* illustrates the conditions shortly after capture. The *captured stream* has been diverted by the *captor stream* and now turns sharply at the point of capture, known as the *elbow of capture.* The difference in level of the two streams results in a *waterfall.* The captor stream has its volume increased by the addition of the captured stream and begins to show signs of rejuvenation. Its gorge is deepened, and its tributaries on either side below the point of capture cut back rapidly to form other gorges. The *beheaded stream,* having lost much of its volume, acquires mature characteristics. It develops small meanders, not suited to the size of the valley. It becomes a *misfit,* or an *underfit,* stream. Its tributaries build *alluvial fans* on the valley floor because the beheaded stream in its shrunken condition can no longer transport the customary load. Lakes and marshes may thus be formed in the valley. Near the point of capture, some of the drainage may even turn back into the captor stream, thus forming an *inverted stream.*

In the third illustration (Fig. *C*) conditions are shown long after capture. The headwaters of the captor stream have all developed gorges. The falls at the point of capture have retreated upstream to the very head of the diverted stream or have been evened out into rapids. Other falls, resulting from the capture, have migrated upstream. Minor captures have been effected by some of the tributaries of the main captor stream as they eat their way headward into the upland.

Further development will eliminate all falls and rapids, the valley walls will become subdued, the streams on the upland will cut down, and the divide between the two systems will become stationary. Almost the only evidence of capture remaining will be the angular bend or elbow of capture and even this may be lost in the irregularity of drainage lines. Numerous stream captures thus bring about an adjustment of streams to the lands which they drain.

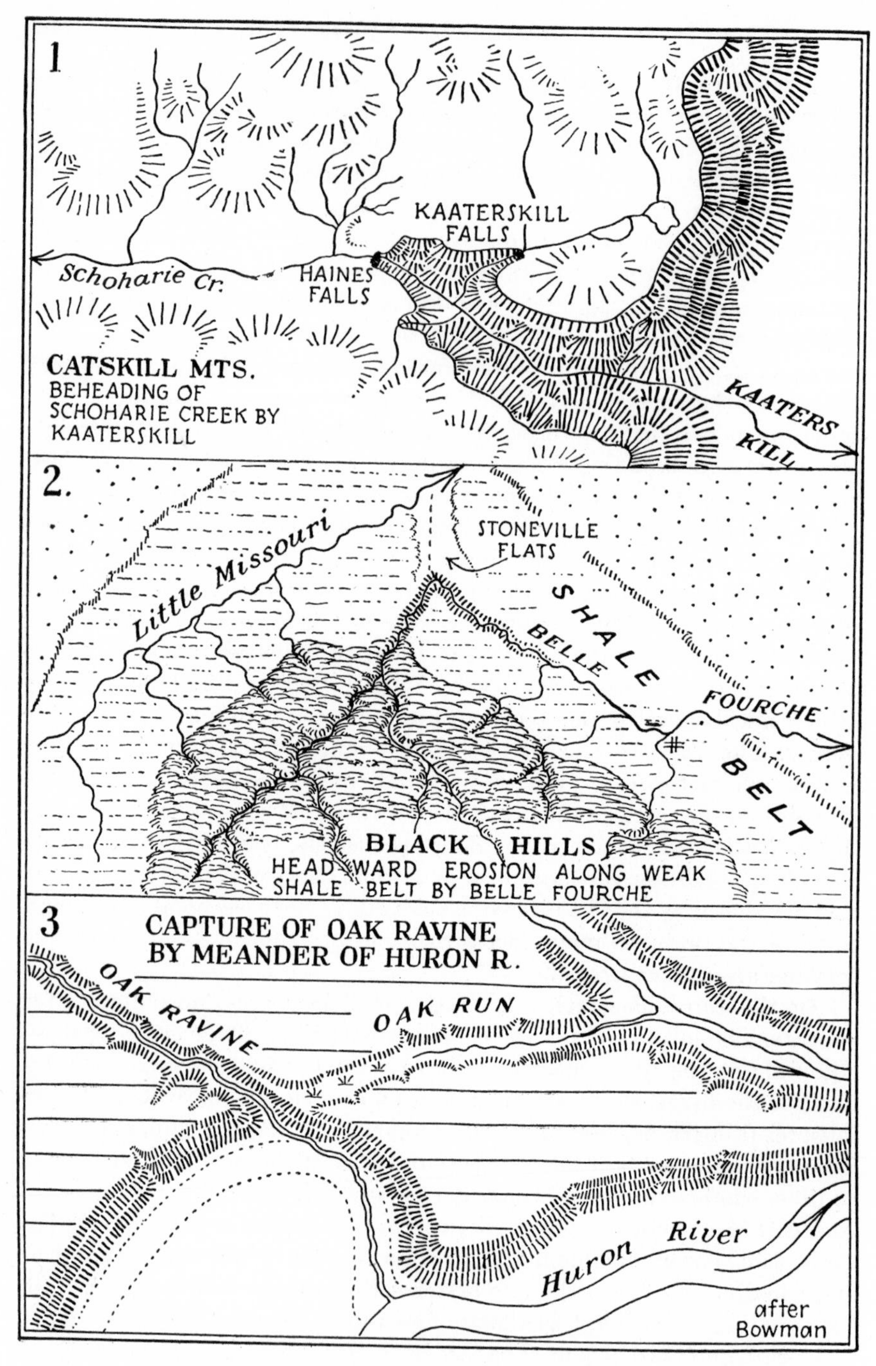


MAPS SHOWING THREE TYPES OF STREAM CAPTURE

## ILLUSTRATIONS OF THREE TYPES OF STREAM CAPTURE

The Capture of Upland Streams by Streams Flowing down the Face of an Escarpment. Kaaterskill Creek flowing down the eastern scarp of the Catskill Plateau has advanced its headwaters westward into the plateau and twice diverted some of the tributaries of Schoharie Creek. Schoharie Creek flows on top of the plateau, the waters going westward and northward into the Mohawk before eventually reaching the Hudson River. Between the foot of the escarpment and the top of the plateau there is a difference in altitude of 1,500 feet. The steep gradient of Kaaterskill compared with the gentle gradient of Schoharie explains the rapid incision which Kaaterskill has made headward into the drainage area of Schoharie. Two waterfalls have resulted from this capture: Haines Falls, at the head of the ravine, represents probably a reversal of drainage; the other, at Kaaterskill Falls, probably represents retreat of falls from the elbow of first capture. The extremely youthful character of Kaaterskill gorge is undoubtedly due to the rejuvenation resulting from these recent captures. This type of capture is represented along the face of the Blue Ridge Escarpment in North Carolina and also along the Allegheny Front which is analogous to the Catskill Escarpment.

The Capture of Consequent Streams by Subsequent Streams Working Headward along Belts of Weak Rock. The Belle Fourche River, a tributary of the Cheyenne, in the Black Hills has penetrated westward along a belt of weak shales until it has tapped one of the tributaries of the Little Missouri. The former course of the captured stream was through Stoneville Flats, a wide smooth-bottomed valley. The Belle Fourche at the great bend which represents the elbow of capture now occupies a new canyon cut about 100 feet below the floor of the old valley. This type of capture, so characteristic of the Folded Appalachians, is the explanation of many wind gaps. In eroded dome mountains and on coastal plains it is the normal method of stream adjustment.

By Sidewise Swinging of Mature Streams. This example was described by Isaiah Bowman. It occurred a few miles west of Detroit, Michigan, where the mature meandering Huron River impinged against the course of Oak Run and captured it. The captured portion which now flows in Oak Ravine was thus rejuvenated. The former channel is now a marsh. This type of capture has been termed *capture by stream intercision.*

The same explanation is offered for the junction of the Red River and the Mississippi. The Red River may have once run directly to the Gulf. Ordinarily the Red River discharges into the Mississippi, but in flood season it discharges partly into the Atchafalaya which runs directly into the Gulf. Similarly the Seine River cuts into the bluffs bordering the upland at Duclair and there has diverted the Ste. Austreberte.*

* This and similar cases on the Marne River were described by W. M. Davis.

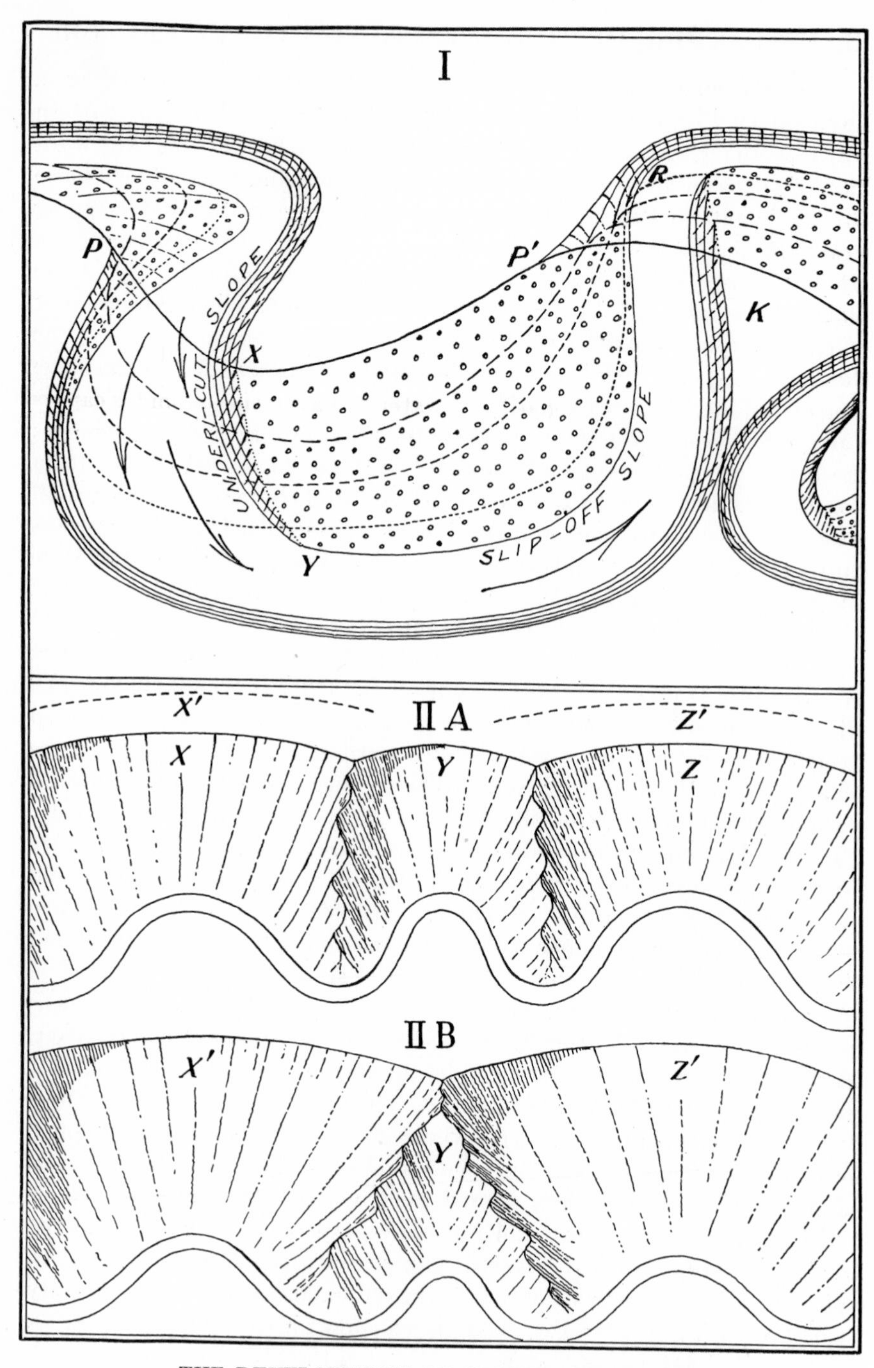


THE DEVELOPMENT OF INCISED MEANDERS

## INCISED MEANDERS

Incised meanders are evidences of youth but not necessarily of rejuvenation from a previous condition of maturity. That is, a stream of only moderate sinuosity may become incised and develop larger and larger curves. These in time simulate the meanders of a mature stream. Nor is it necessarily true that the meanders of one cycle are continued unchanged into the next cycle.

A young river not only cuts downward but it corrades laterally, sapping the outer bank of its curves. This enlargement of the curves increases their radii. At the same time the river saps the down-valley sections between the curves and this causes the whole system of curves to migrate slowly down the valley. As a result, the spurs become unsymmetrical. They come to have a steeper *undercut* slope on the up-valley side and a *slip-off* gravel-strewn slope on the down-valley side.

In Fig. I the line $P$-$P'$ represents the course of a stream on a plateau surface. The dotted lines represent successive positions of the stream as it cuts downward and enlarges its meanders, and as its meanders migrate down the valley. These dotted lines may be considered also as contour lines inasmuch as the gradient of the stream in the short distance represented is negligible.

It is to be noted that the undercut slopes on the up-valley sides of the spurs pass into the slip-off slopes around a pronounced salient where the contours turn at an angle, shown by the line $X$-$Y$. But farther downstream, where the slip-off slope passes ($P'$-$R$) into the next undercut slope, the transition is gradual.

In Fig. I the meanders have migrated only a short distance down valley, so that a large amount of upland appears on the spurs. A small amount of upland capping the spurs indicates in general a considerable down-valley movement of the meanders. Spurs may be so narrowed at their necks as to have a dovetail form with detached remnants of the upland on the spur heads, as at *K*.

The amount of upland preserved on the spurs depends not only on the amount of down-valley migration of the meanders but also upon the depth of incision. A stream which has cut down deeply has more actual meanders than the curves on the upland would indicate. In Fig. II *A* there are three meanders cutting into the upland, each showing a meander scarp on the upland surface. In Fig. II *B* the stream is more deeply incised and scarps *X* and *Z* have intersected each other. The smaller scarp *Y* between them has been obliterated on the upland. The scarps produced by the smaller meanders rise for only a short distance above the river before they are intersected by the higher scarps of the larger meanders.*

* Professor Davis's very suggestive paper on *Incised Meandering Valleys* (given in the list of references) should be consulted but it should be noted that his Figs. 3 and 6 have been turned upside-down by the printer.

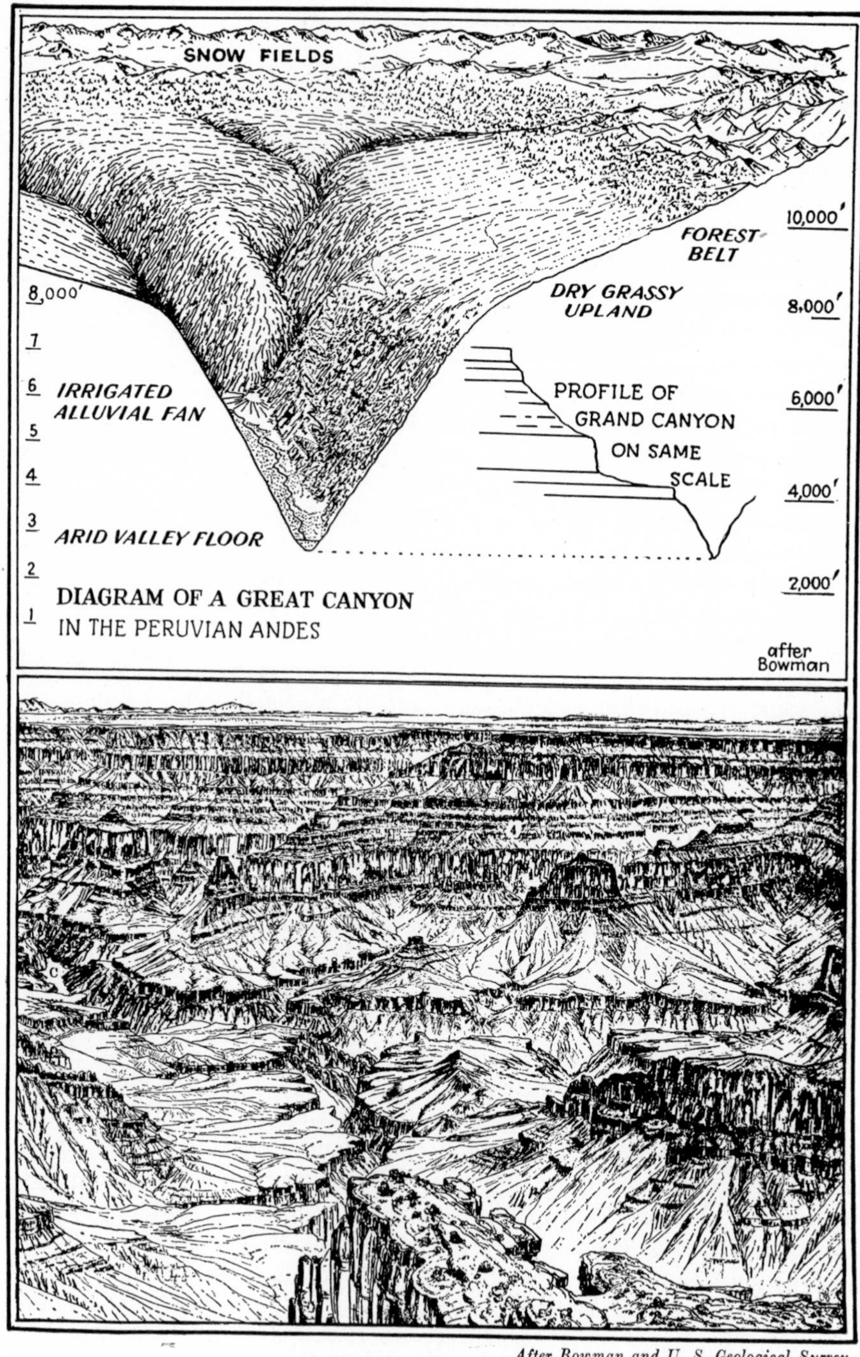


*After Bowman and U. S. Geological Survey*

EXAMPLES OF YOUNG VALLEYS: A GREAT CANYON IN THE PERUVIAN ANDES; AND THE GRAND CANYON OF THE COLORADO

## EXAMPLES OF YOUNG VALLEYS

THE CANYONS OF PERU. The Grand Canyon of the Colorado and the great canyons of the Andes are examples of young valleys, but in most respects they are quite unlike. In the Andes the streams have cut through masses of volcanic rock as deep as 7,000 feet. The canyon walls rise almost sheer for thousands of feet from the narrow valley floor, and with few side tributaries. In their upper parts, the Andean canyons flare out widely to form extensive sloping uplands. These are parts of earlier, more mature valleys below which the present steep-walled canyons have been incised. It is a "valley-in-valley" relationship. Above the valley rise rugged mountains culminating in snow-bedecked peaks at elevations of 20,000 feet and over.

The Peruvian canyons in their vertical range include a wide variety of climatic conditions. The high mountain tracts below snowline are forested but the expansive uplands are dry and support only grass. At lower levels still greater dryness prevails and the depth of the canyon is a veritable desert.

People inhabit all parts of this region. They irrigate the alluvial fans and the valley bottoms with tiny canals and raise sugar cane and tropical fruits. In the upland pastures sheep raising is carried on. Crops of corn, vegetables, and barley are also produced at these high altitudes, and highest of all, near the edge of the woodland, are fields of mountain potatoes.

THE GRAND CANYON. The Grand Canyon of the Colorado is cut into a plateau of sedimentary beds glowing with color. There is a quality of classic orderliness in the stretches of tier after tier of cliff and terrace. No great mountain masses rise above the plateau surface which extends as far as the eye can reach. But the variety within the canyon itself is overwhelming.

In actual depth the Grand Canyon is exceeded by many of the great Andean gorges and even by that of the Snake River Canyon in the United States. In the vertical range of 4,000 to 5,000 feet from its rim to the river level there are several zones of vegetation. The top of the Kaibab Plateau supports a luxuriant forest of giant trees and in the winter receives a heavy fall of snow. In the bottoms of the canyon, however, there is perpetual summer, where the Sonoran or Mexican type of vegetation exists.

The Grand Canyon, unlike the Andean gorges, never supported a population. There are remains of a few cliff dwellers near the upper rim but no dwellings in the depths of the canyon. To the scientist the Grand Canyon is interesting primarily because of the geological phenomena which it reveals by exposing rocks of many ages from the earliest to some of the most recent. The Andean gorges owe their fascination to the ways men have adapted their mode of living to conform with the topographic and climatic exigencies of the situation.

## SOME GEOGRAPHICAL ASPECTS OF YOUNG RIVERS

As Routes of Travel. Because of their narrow valleys and steep gradients, young rivers do not serve so admirably as routes for railway lines as do the valleys of mature streams. No railroad, for instance, runs along the bottom of the Grand Canyon. In the Royal Gorge, which is utilized by the Denver and Rio Grande Railroad, the right of way was constructed with great difficulty. In one place, where the gorge is extremely narrow, the tracks are actually suspended over the river. There is no room whatever for any highway through this same gorge. Even the telegraph wires are strung from point to point along the rocky walls.

These cases, however, are extreme. Most youthful valleys have small patches of flood plain here and there along the valley floor, with the result that little settlements find a foothold and railroad lines and roadways can follow the river. Tunnels are apt to be frequent where the route encounters projecting spurs of the valley wall. The gorge of the Kanawha River followed by the Chesapeake and Ohio Railway through the plateau of West Virginia offers facilities of this type. Similar to it are the valleys of the Allegheny and of the Monongahela Rivers in Pennsylvania. The valley of the Deerfield River in western Massachusetts is utilized by the Boston and Albany Railroad and by the Mohawk Trail, one of the most picturesque motor highways of New England.

The Hudson River Gorge is another example of a young valley which is followed by two great railways, one on either side, and an important motor highway, which winds its way far above the river around the highland spurs.

Although cities of large size can rarely find sufficient room on the bottom of a young valley, there are some notable exceptions—Pittsburgh, for example. This city occupies part of the flood plain of the Ohio, which here is not strictly young. Moreover, the city climbs far up the valley walls even to the top of the plateau.

For Water Power. Young rivers offer conditions most favorable for water-power development. The narrow valleys of young streams can be easily dammed to form reservoirs. The steep gradient makes a high head of water possible so that a large volume is not necessary. The occasional presence of lakes provides a steady and dependable flow. Counteracting these advantages is the fact that young streams are usually in mountainous regions, remote from the industrial centers where electric power is most needed.

The Boulder Dam near the Grand Canyon, the Shoshone Dam near Yellowstone Park in the Absaroka Mountains, the great Porjus power development in the Lappland mountains of northern Sweden, and the numerous water-power plants in the Alps, the Pyrenees, and the Southern Appalachians testify to the advantages of youthful valleys but at the same time illustrate the fact that they are apt to be distant from the consuming centers.

Youthful streams, with their natural falls and rapids, determined the location of early industrial towns in both Europe and the United States when the amount of power needed by any single mill or establishment was small. Scattered through New England are towns and cities situated on small streams which at one time supplied their power needs. Now, however, with the growth of industry the use of coal has become imperative.

As Political Boundaries. Young rivers, it might seem, should serve admirably as political boundaries, although not so well as mountain ranges do. One advantage that young rivers have over mountains is that the boundary line does not have to be surveyed. They are perfectly definite features. Nor do they require any markers. They can be described and defined in the simplest possible terms. The Snake River Canyon between Idaho and Oregon is a good example. It may be observed, however, that very few international boundaries follow young river valleys, but many of them do follow mountain ranges. The explanation for this is that, unless a young river valley is an actual canyon, it serves more to draw people together than to separate them and so is apt to be appropriated in its entirety by one country. The validity of this is emphasized by the fact that many political states and even entire countries coincide with river valleys. This is especially true in mountain regions where the streams are relatively small and easily bridged. The cantons of Switzerland are more or less coextensive with the different valleys. In arid mountain regions the identity of a district with a stream basin becomes still more pronounced, because here population must gather about the common water supply. Thus in Chinese Turkestan the several districts which comprise the country are identical with the different mountain tributaries of the Tarim River, whose basin in turn comprises almost the whole of Chinese Turkestan.

For Water Supply. Young rivers in mountainous regions are especially wholesome sources for the water supplies of large cities, even those lying at some distance away. Manchester, England, draws its supply from the Lake District of England, 60 miles distant; Birmingham, from the valleys of Wales, many miles to the west. New York taps the streams of the Catskills, 90 miles and more to the north. San Francisco looks far away to the Sierra with its numerous mountain streams and Los Angeles draws some of her most needed water from Owens Valley, which in turn is supplied by mountain streams, the water being carried 250 miles through the mountains.

As Biologic Boundaries. Although young canyons can in every instance be crossed by man, for certain animals they constitute impassable barriers more effective than an unbroken mountain range. The Grand Canyon separates two biologic provinces, having distinct species on either side. The snakes and the squirrels are notable examples of forms which have evolved different types on the north and south sides of the canyon.

## MAPS ILLUSTRATING YOUNG STREAMS

Young streams with steep gradients, and flowing in deep gorges or canyons with steep walls, are displayed on all maps of mountainous regions. An unusually fine example of a young gorge in a region of crystalline rocks is that of the Deerfield River shown on the *Hawley, Mass.-Vt.*, sheet; similar to this is the gorge of the Housatonic, shown on the *Derby, Conn.*, sheet. The *Cowee, N. Car.-S. Car.*, map also shows numerous young valleys. On the *Morgantown, W. Va.-Pa.*, sheet there is the splendid young gorge of the Cheat River incised over 1,000 feet through a series of sedimentary beds, although the horizontal structure of the plateau is not suggested by the contours. The *Canyon, Wyo.*, sheet shows the young gorge of the Yellowstone River cut into a lava plateau and formed by the headward retreat of the falls which are also depicted. On the *Niagara Falls, N. Y.*, sheet a similar though smaller gorge is represented, formed by the headward retreat of Niagara Falls, a retreat which is still proceeding at an average rate of 5 feet per year. Old plunge pools, called the "potholes," produced by former streams whose courses have been diverted, are shown on the *Quincy, Wash.*, sheet. In the glaciated areas there are many examples of falls produced by the superimposition of streams over rock ledges, such as the St. Croix Falls, shown on the *St. Croix Dalles, Wis.-Minn.*, sheet. The Grand Canyon, of course, is the superb example of a young stream, with an infinitude of young tributaries. The *Bright Angel, Shinumo*, and *Vishnu, Ariz.*, sheets are to be consulted in this connection.

The behavior of young and usually intermittent gullies in producing badland topography is illustrated on the *Rock Springs, Wyo.*, the *Craig, Colo.*, and the *Stoval, Ariz.*, sheets.

Incised meanders with meander spurs displaying undercut and slip-off slopes are shown on the *Cub Run, Frankfort*, and *Lockport, Ky.*, sheets. The *Shell Knob, Mo.*, the *Blacksville, W. Va.-Pa.*, the *Clearfield, Pa.*, and the *White Bluff, Tenn.*, all illustrate similar features. On the *Brownsville, Pa.*, sheet a high-level meander cutoff is shown.

Classic examples of stream capture are shown on the *Kaaterskill, N. Y.*, sheet, at the head of both Kaaterskill and Plattekill Cloves. Imminent and undoubtedly past stream capture is suggested along the Blue Ridge as it appears on the *Saluda, N. Car-S. Car.*, sheet. The *Waipio, Hawaii*, quadrangle exhibits numerous captures caused by the encroachment of deep canyons into an upland plain, resulting in striking elbows.

In the Folded Appalachians many captures are suggested by the presence of wind gaps, such as those shown on *Wind Gap, Pa.*, and on the *Harpers Ferry, Va.-W. Va.-Md.*, sheets.

Arches and natural bridges are usually the work of young streams but their form is rarely indicated by the contours. An unusual case is the natural arch shown on the *Calabasas, Calif.*, sheet, formed by a stream whose course has since been diverted.

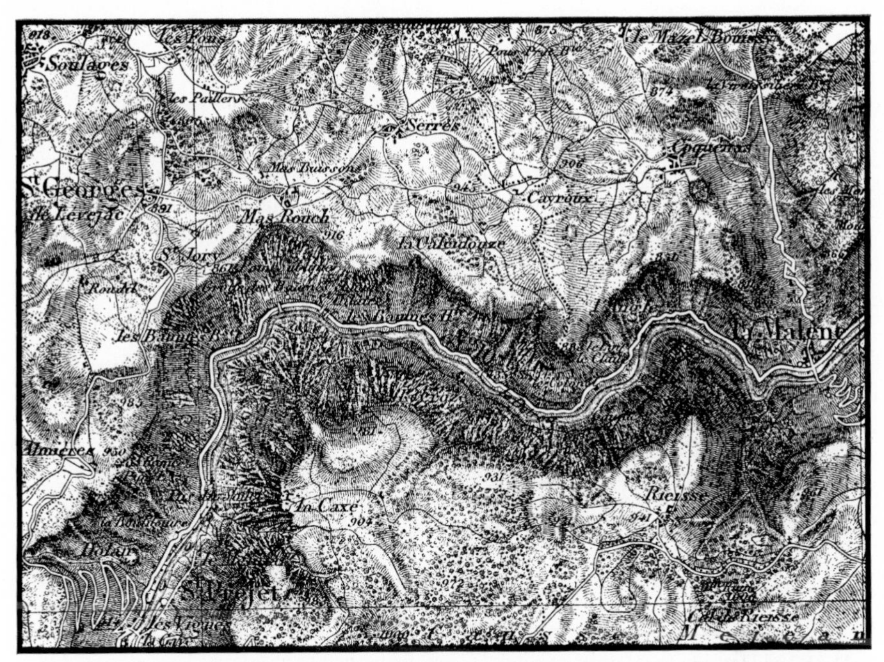


GORGE OF THE TARN RIVER, A YOUNG VALLEY IN LIMESTONE PLATEAU OF SOUTHERN FRANCE
France; *Sévérac* sheet (1:80,000).

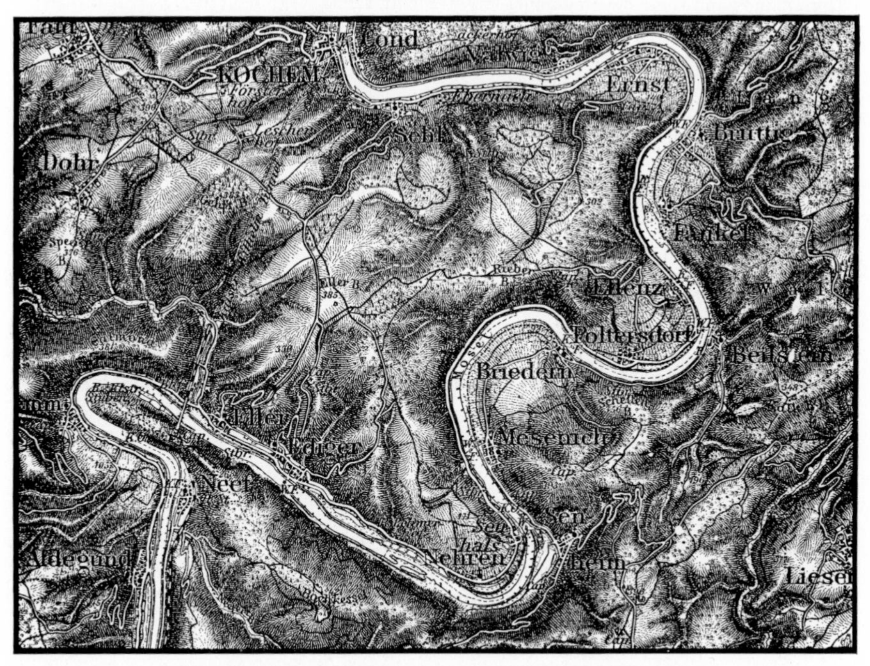


INCISED MEANDERS OF THE MOSELLE RIVER
German; *Kochem* sheet. No. 504 (1:100,000).

## QUESTIONS

1. What is the difference between corrasion and corrosion?
2. Draw a transverse profile of a valley (*a*) in which each valley wall is represented by a curve concave upward and (*b*) in which each valley wall is a curve convex upward. Which of these two probably represents the younger type of valley? Does the curve represented by *b* indicate rejuvenation?
3. Draw another transverse profile (*c*) in which each valley wall is made up of two curves, the upper part being convex upward, the lower part concave upward. What does a profile like this probably represent in the history of the stream?
4. Suppose a gorge or canyon is formed by the headward retreat of a waterfall. How will such a canyon differ from one formed by downward cutting along the whole length of the canyon at the same time?
5. During what periods in the life history of a river would you expect the most rapid adjustment to rock structure to take place?
6. In a region of slight rainfall are the streams more apt to be young or mature?
7. Do you think streams can cut below sea level?
8. Why are the heads of young canyons in arid regions amphitheater shaped, whereas those in humid regions are sharp and narrow? Try to formulate an explanation.
9. Name all the types of stream capture you can conceive of and deduce their consequences.
10. Study a number of contour maps illustrating young streams and determine the gradients of such streams. Plot accurately the longitudinal profile of a young stream from its headwaters to where it appears to have a graded condition. (Use vertical exaggeration.)
11. What is meant by stream surrender? (See Johnson, 1939.)

## TOPICS FOR INVESTIGATION

1. Badlands and their development. Favorable and unfavorable factors.
2. Stream load. Kinds of load. Effect on carrying power of streams.
3. Floods and their causes.
4. Stream capture. Varieties. Causes.
5. Waterfalls. Various types. Description and explanation of famous examples.
6. Natural bridges. Methods of origin.

## REFERENCES

EROSION


ALEXANDER, H. S. (1932) *Pothole erosion.* Jour. Geol., vol. 40, p. 305–337.

BENNETT, H. H. (1933) *The quantitative study of erosion technique and some preliminary results.* Geog. Rev., vol. 23, p. 423–432.

FORSLING, C. L. (1932) *Erosion on uncultivated lands in the intermountain region.* Sci. Monthly, vol. 34, p. 311–321.

JOHNSON, D. W. (1934) *How rivers cut gateways through mountains.* Sci. Monthly, vol. 38, p. 129–135.

MAXSON, J. H., and CAMPBELL, I. (1935) *Stream fluting and stream erosion.* Jour. Geol., vol. 43, p. 729–744.

O'HARA, C. C. (1910) *The badland formations of the Black Hills region.* S. Dak. Sch. Mines, Bull. 9, 152 p.

QUIRKE, T. T. (1925) *Potholes and certain features of glacial abrasion.* Ill. State Acad. Sci., Trans. 17, p. 194–198.

RAMSER, C. E. (1934) *Dynamics of erosion in controlled channels.* Am. Geoph. Un., 15th Ann. Meeting, p. 488–494.

REEDS, C. A. (1930) *Land erosion.* Nat. Hist., vol. 30, p. 131–149.

WENTWORTH, C. K. (1919) *A laboratory and field study of cobble abrasion.* Jour. Geol., vol. 27, p. 507–521.

TRANSPORTATION

GILBERT, G. K. (1914) *The transportation of debris by running water.* U. S. Geol. Surv., Prof. Paper 86, 263 p.

HAWKSWORTH, H. (1921) *The strange adventures of a pebble.* New York, 296 p.

LEIGHLY, J. (1934) *Turbulence and the transportation of rock debris by streams.* Geog. Rev., vol. 24, p. 453–464.

O'BRIEN, M. P., and RINDLAUB, B. D. (1934) *The transportation of bed load by streams.* Am. Geoph. Un., 15th Ann. Meeting, p. 593–603.

RUBEY, W. W. (1933) *Equilibrium conditions in debris-laden streams.* Am. Geoph. Un., 14th Ann. Meeting, p. 497–505.

STRAUB, L. G. (1932) *Hydraulic and sedimentary characteristics of rivers.* Am. Geoph. Un., 13th Ann. Meeting, p. 375–382.

FLOODS

BROWN, R. E. (1913) *The Ohio River flood of* 1913. Am. Geog. Soc., Bull. 45, p. 500–509. A report of the same flood was published by the U. S. Dept. of Agriculture in a *Special bulletin of the storms of Mar.* 22–27, on April 15, 1913.

GOLDTHWAIT, J. W. (1928) *The gathering of floods in the Connecticut River system.* Geog. Rev., vol. 18, p. 428–445.

IVES, R. L. (1936) *Desert floods in the Sonoyta Valley.* Am. Jour. Sci., 5th ser., vol. 32, p. 349–360.

McGEE, W J (1897) *Sheetflood erosion.* Geol. Soc. Am., Bull. 8, p. 87–112.

U. S. GEOL. SURVEY. *Floods.* W.-S. Papers 88, 92, 96, 147, 162, 234, 334, 487, 488; *Volume and load of rivers in the United States, ibid.*, Nos. 44, 93, 289.

WOLFF, J. E. (1927) *Cloudburst on San Gabriel Peak, Los Angeles County, California.* Geol. Soc. Am., Bull. 38, p. 443–450.

DIVERSION AND STREAM CAPTURE

BOWMAN, I. (1904) *Deflection of the Mississippi.* Science, new ser., vol. 20, p. 273–277.

BOWMAN, I. (1904) *A typical case of stream capture in Michigan.* Jour. Geol., vol. 12, p. 326–334.

CAMPBELL, M. R. (1896) *Drainage modifications and their interpretation.* Jour. Geol., vol. 4, p. 567–581, 657–678.

CLARK, H. (1911) *A case of preglacial stream diversion near St. Louisville, Ohio.* Denison Univ., Sc. Lab., Bull. 16, p. 339–346.

COBB, C. (1893) *A recapture from a river pirate.* Science, vol. 22, p. 195.

CROSBY, I. B. (1937) *Methods of stream piracy.* Jour. Geol., vol. 45, p. 465–486.

DARTON, N. H. (1896) *Examples of stream robbing in the Catskill Mountains.* Geol. Soc. Am., Bull. 7, p. 505–507.

DAVIS, W. M. (1903) *The stream contest along the Blue Ridge.* Geog. Soc. Phila., Bull. 3, p. 213–244.

FRIDLEY, H. M. (1933) *Drainage diversions of the Cheat River.* W. Va. Acad. Sci., Proc. 6, p. 85–88.

GOODE, J. P. (1899) *The piracy of the Yellowstone.* Jour. Geol., vol. 7, p. 261–271.

JOHNSON, D. W. (1907) *Drainage modifications in the Tallulah district.* Boston Soc. Nat. Hist., Proc. 23, p. 211–248.

JOHNSON, D. W. (1939) *Drainage modifications.* Jour. Geomorphology, vol. 2, p. 87–91.

LA GORCE, J. O. (1926) *Pirate rivers and their prizes.* Natl. Geog. Mag., vol. 50, p. 87–132.

MACKIN, J. H. (1936) *The capture of the Greybull River.* Am. Jour. Sci., 5th ser., vol. 31, p. 373–385.

MALOTT, C. A. (1921) *Planation stream piracy.* Ind. Acad. Sci., Proc. 1920, p. 249–260.

SCHOEWE, W. H. (1930) *Evidences of stream piracy on the Dakota hogback between Golden and Morrison, Colorado.* Kan. Acad. Sci., Trans. 31, p. 112–114.

Ver Steeg, K. (1930) *Wind gaps and water gaps of the northern Appalachians, their characteristics and significance.* N. Y. Acad. Sci., Annals 32, p. 87–220.

Wright, F. J. (1930) *Stream piracy near Asheville, North Carolina.* Denison Univ., Sci. Lab., Bull. 24, p. 401–406.

Waterfalls

Boyd, W. H. (1930) *The Niagara Falls survey of* 1927. Can. Geol. Surv., Mem. 164, 15 p.

Gilbert, G. K. (1895) *Niagara Falls and their history.* Natl. Geog. Mag., Mon. 1, p. 203–236.

Gilbert, G. K. (1907) *Rate of recession of Niagara Falls.* U. S. Geol. Surv., Bull. 306, p. 5–25.

Gregory, J. W. (1911) *Constructive waterfalls.* Scot. Geog. Mag., vol. 27, p. 537–546.

Lamplugh, G. W. (1908) *The gorge and basin of the Zambesi below the Victoria Falls, Rhodesia.* Geog. Jour., vol. 31, p. 133–152, 287–303.

Molyneux, A. J. C. (1905) *The physical history of the Victoria Falls.* Geog. Jour., vol. 25, p. 40–55.

Natural Bridges

Barnett, V. H. (1908) *A natural bridge due to stream meandering.* Jour. Geol., vol. 16, p. 73–75.

Barnett, V. H. (1912) *Some small natural bridges in eastern Wyoming.* Jour. Geol., vol. 20, p. 438–441.

Cleland, H. F. (1905) *The formation of natural bridges.* Am. Jour. Sci., 4th ser., vol. 20, p. 119–124.

Cleland, H. F. (1910–11) *The formation of North American natural bridges.* Pop. Sci. Monthly, vol. 78, p. 417–427; Geol. Soc. Am., Bull. 21, p. 313–338.

Cummings, B. (1910) *The great natural bridges of Utah.* Natl. Geog. Mag., vol. 21, p. 157–167.

Miser, H. D., Trimble, K. W., and Paige, S. (1923) *The Rainbow Bridge, Utah.* Geog. Rev., vol. 13, p. 518–531.

Nelson, W. A. (1915) *Two natural bridges of the Cumberland Mountains, Tennessee.* Tenn. Geol. Surv., Resources of Tenn., vol. 5, p. 76–80.

Geographical Aspects

Bennett, H. H. (1928) *The geographical relation of soil erosion to land productivity.* Geog. Rev., vol. 18, p. 579–605.

# VII
# MATURE STREAMS

*A. K. Lobeck*

THE LOUP RIVER IN EASTERN NEBRASKA
A mature stream with a braided channel, draining part of the Great Plains.

Darton, U. S. Geological Survey

THE PLATTE RIVER IN WESTERN NEBRASKA

Showing its braided channel and wide flood plain cut below the High Plains, which form the distant skyline.

THE VALLEY OF THE STIKINE RIVER, ALASKA

A mature stream on broad flood plain with natural levees and Yazoo type of tributary in the distance.

*U. S. Geological Survey*

*Norfolk and Western Railway*

THE SHENANDOAH VALLEY NEAR WOODSTOCK, VA.

A broad valley with meandering stream incised only slightly below the valley floor which, however, is not covered with alluvium.

A MEANDER ON THE WYE RIVER, ENGLAND

Showing meander spur with steep undercut and gentle slip-off slope.

*Fox Photos, Ltd.*

CYPRESS BEND
MONTEREY BEND
YELLOW BEND
BOLIVAR BEND
CHOCTAW BEND
ASHBROOK CUT
ASHBROOK NECK
MILLER BEND
BACHELOR BEND
POINT CHICOT
RTER
OINT
TAPLEY CUT-1935
LELAND CUT-OFF 1933
NECK
GREENVILLE
LELAND

*War Department, Corps of Engineers*

AIR VIEW OF GREENVILLE, MISS., REGION

Showing the three artificial cutoffs: the Ashbrook cut; the Tapley (preferably Tarpley) cut, and the Leland cut, also shown on the map on page 219.

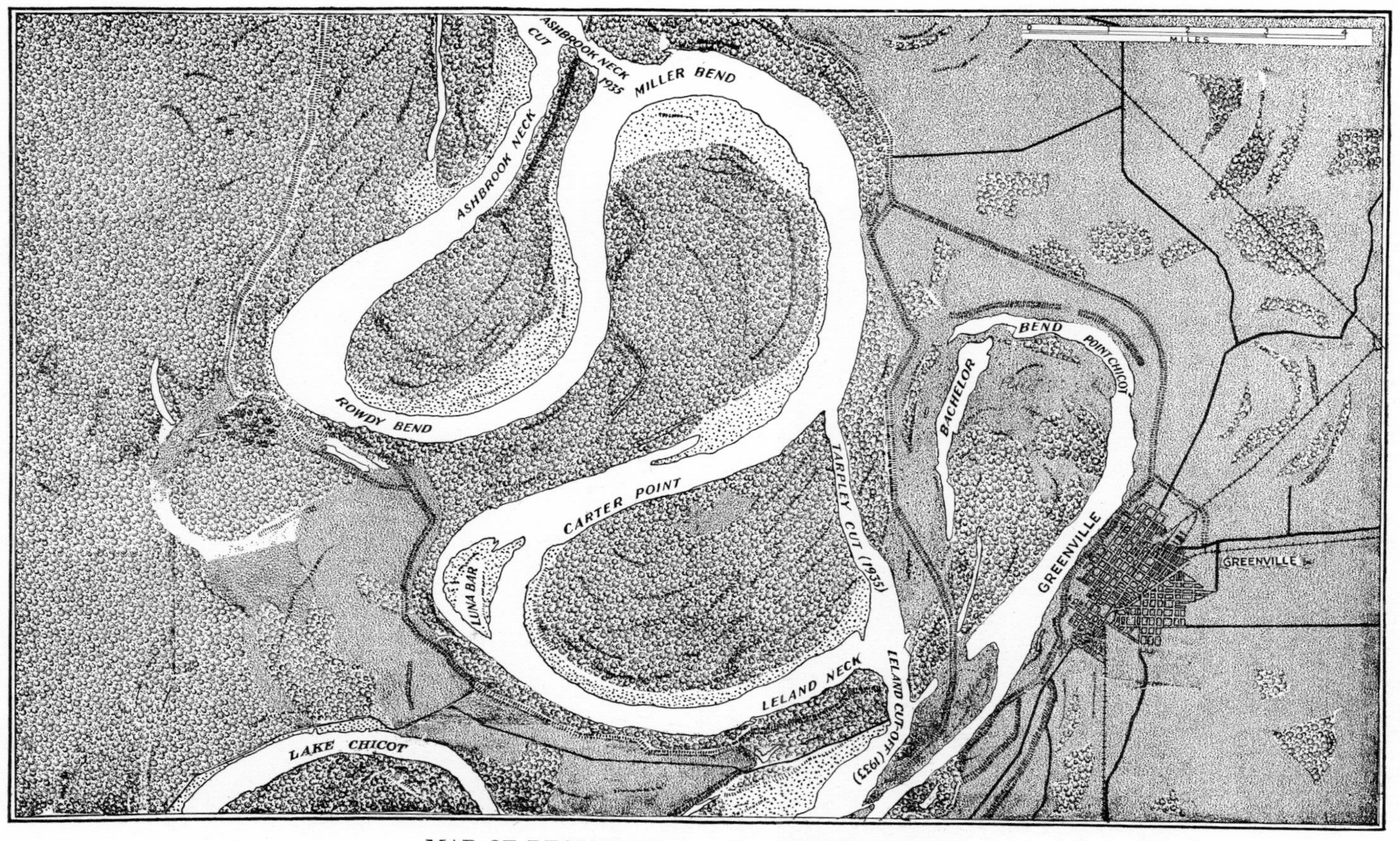


MAP OF REGION NEAR GREENVILLE, MISS.

Showing three cutoffs recently completed by the U. S. Engineers, shortening the river more than 30 miles in a total length of 40. From the *Refuge* and *Lamont*, *Ark.-Miss.* quadrangles, War Department Corps of Engineers.

*Geological Survey of Canada*

FLOOD PLAIN AND LEVEES OF THE KOOTENAY RIVER, BRITISH COLUMBIA

The flood plain is entirely flooded but the tops of the levees remain above water.



CANYON DIABLO, WEST OF COON BUTTE, ARIZ.

Sharply incised meanders of a mature stream. In the center foreground may be seen a supposed meteoric crater.

*Barnum Brown, American Museum of Natural History*

# MATURE STREAMS

Synopsis. A mature stream is one which has such a nicely adjusted slope that it neither erodes nor deposits. It is said to have a *profile of equilibrium.* Changing conditions, however, make it necessary for mature streams to change their slopes. A reduction in load, for example, permits erosion and hence a reduction in gradient. An increase in load causes deposition and demands an increased gradient which is supplied through deposition. For any given moment the profile may be not quite one of equilibrium. The stream is always oscillating one way or the other, like a tight-rope walker trying to maintain a balance.

The headwaters of a mature stream in time wear away the land and gradually give less and less load to the main stream to carry away. This means that the main stream keeps reducing its gradient. To all appearances it constantly has a profile of equilibrium but actually it keeps changing this profile as conditions are altered. This explains the varying slope of mature streams under various conditions.

An increase of load contributed by the headwaters, as, for example, during periods of glaciation, causes the main stream to deposit much alluvium and thus to build up its channel until the proper gradient is established to permit transportation of the increased load.

When this is followed by a reversal of conditions, erosion is resumed and terraces result.

The nice way in which a mature stream responds to change in any part of its system is one of the marvels of nature. A mature stream seems always to be doing exactly the right thing to improve conditions. But actually this is far from true. In developing meanders on a flood plain, for example, a mature stream lengthens its course, thereby decreasing its gradient. This aggravates still further the need to deposit material, and as a result of depositing material still more meanders are formed. So, in time, a continuous belt of meanders is developed. Associated with this are the formation of cutoffs and the production of oxbow lakes, and the deposition of natural levees, resulting in the so-called *deferred junctions* of tributaries of the Yazoo type.

The significance of delta building in the life history of mature rivers is a topic of great importance, for the effect is farreaching and is felt by the whole system. Rock pediments and alluvial fans, closely related features especially characteristic of arid regions, are found to be typical and normal features in the behavior of mature streams in certain localities.

Finally, it may be shown that an old stream is precisely the same as a mature stream as far as its profile of equilibrium is concerned. It is simply a master stream, all of whose children or tributaries are mature also.*

* See note at bottom of synopsis of Chapter V.

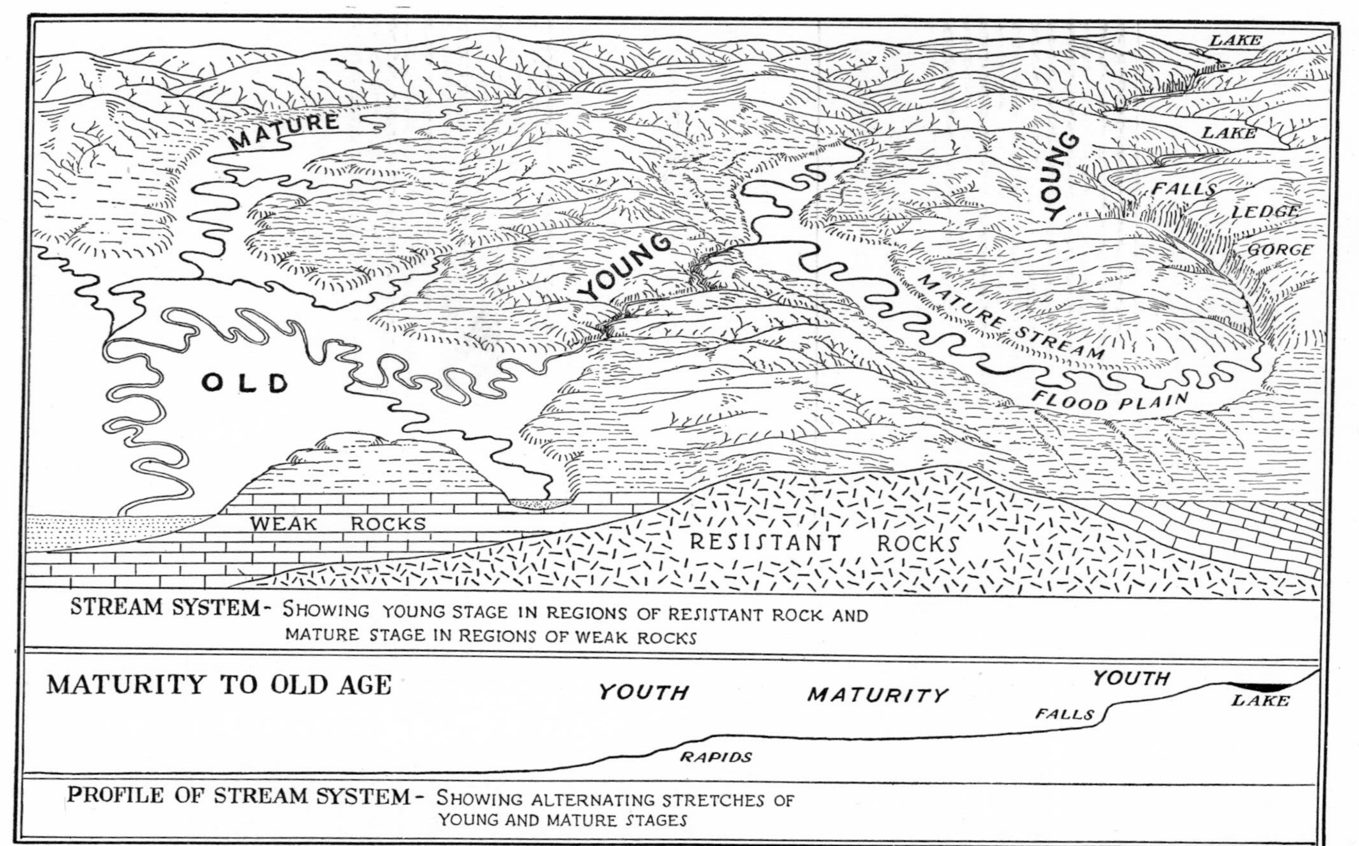


DIAGRAM AND PROFILE SHOWING DIFFERENT STAGES OF DEVELOPMENT IN DIFFERENT PARTS OF A STREAM SYSTEM

## MATURE STREAMS AND THEIR CHARACTERISTICS

MATURE STREAMS. A mature stream has been defined as a graded stream: one which is in balance, a stream whose energy is just sufficient to transport its load. It neither erodes nor deposits. No stream maintains this perfect profile of equilibrium. Temporary changes in load or volume cause it to deposit material or to erode material already deposited. There is a constant struggle to preserve just the right gradient. A stream acquires this profile of equilibrium by cutting down its valley, most rapidly in its upper portion, and constantly ironing out the irregularities in its channel. At the same time it widens its valley. Valley deepening at first may be more important than valley widening but only for a short time. A flat valley floor covered with a veneer of alluvium together with subdued valley walls usually indicates that a graded condition has been established.

It has been suggested that full maturity is attained when the width of the valley floor equals the width of the meander belt, although this may be long after the graded condition is acquired. Thus in maturity the stream can freely develop meanders appropriate to its volume. The width of the meander belt is usually about ten to twenty times the width of the river itself. Until the valley floor has attained that width, meanders cannot freely develop and the stream will constantly impinge upon the valley wall.

It is not clear that any definite relation exists between the width of a meander belt and the profile of a stream or its load-carrying ability. The observation above, nevertheless, recognizes an important period in the life of a stream, even though all graded streams do not meander.

OLD STREAMS. It has been suggested that any river whose members are all graded should be termed an *old river*. There is no critical change between maturity and old age, as there is between youth and maturity. Usually in old age the valley is several times the width of the meander belt because of the lateral migration of this belt across the flood plain.

Many streams exhibit alternating stretches of youth and maturity. Where resistant rocks prevail, the stream takes longer to attain a graded profile and in such regions narrow gorges may persist, while upstream and downstream it may widen out its valley and lower its profile in weaker formations. Such regions of resistant rock serve as temporary base-levels.

The following *observable characteristics of mature streams* may be taken to indicate that a graded profile has been established, although these features in themselves are not necessarily criteria of maturity and no particular one of them may be present in any given case:

(*a*) Flood plain, with natural levees; (*b*) meanders, with abandoned meander scrolls, cutoffs, and oxbow lakes; (*c*) width of valley equal to or greater than width of meander belt; (*d*) no rapids or falls; (*e*) slow-moving current of muddy water; (*f*) subdued valley walls, deeply soil covered, and few rock outcrops; (*g*) no lakes (except oxbow lakes).

Mature rivers build up the level of their flood plains by depositing some of their load. Naturally these deposits are greatest close to the river. When a stream overflows its banks at high water and is no longer confined to a definite channel, its velocity is checked because of its shallow depth, and it lays down some of its burden. This material forms *natural levees* which after a time may reach a height 10 to 20 feet above the level of the flood plain on either side of the river. With deposition taking place also on the bed of the river, the entire stream is raised and comes to flow on what is known as a *levee ridge*.

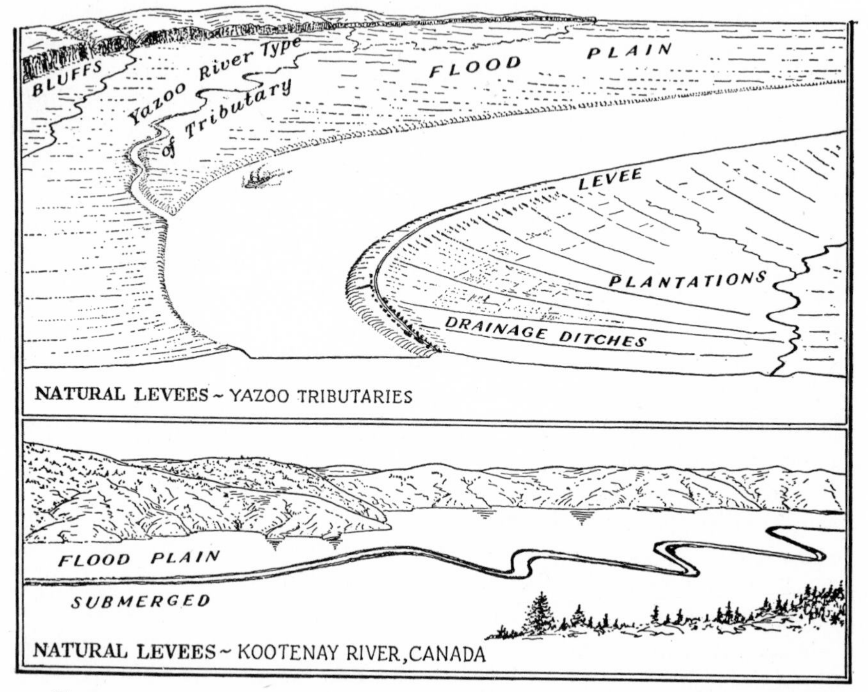


NATURAL LEVEES ~ YAZOO TRIBUTARIES

NATURAL LEVEES ~ KOOTENAY RIVER, CANADA

Even more pronounced are the ridges built by braided streams. Streams very heavily charged with gravel, like those emerging from mountains or from glaciers and flowing over alluvial fans, have numerous small interlacing channels. Such streams soon disappear beneath the surface of the alluvial fan and therefore drop their entire load. In this manner the stream channel is raised above the level of the fan and is not filled by the stream except in times of unusually abundant water. Braided channels of this type easily become clogged and the streams therefore change their positions frequently, usually splitting and forming distributaries on either side of the *alluvial dams* which they are constantly depositing.

The longitudinal slope of flood plains built by braided and therefore heavily loaded streams is greater than that of streams having ordinary meanders. It may be as much as 6 feet per mile as with the Platte River, whereas the average slope of the Missouri flood plain is 1 foot and the Mississippi less than $\frac{1}{10}$ foot per mile. The greater slope is due essentially to greater coarseness of the load.

When streams raise their courses above the level of their flood plains, their tributaries cannot join the main stream. In some cases such tributaries become blocked to form lakes but usually the tributaries flow along the side of the flood plain until they reach some point farther downstream where the main stream swings against the valley wall. Many of the tributaries of the Mississippi behave in this manner. Because the Yazoo River is a good case, it is taken as the type example, and tributaries which run for some distance parallel to the main stream are called *Yazoo rivers*.

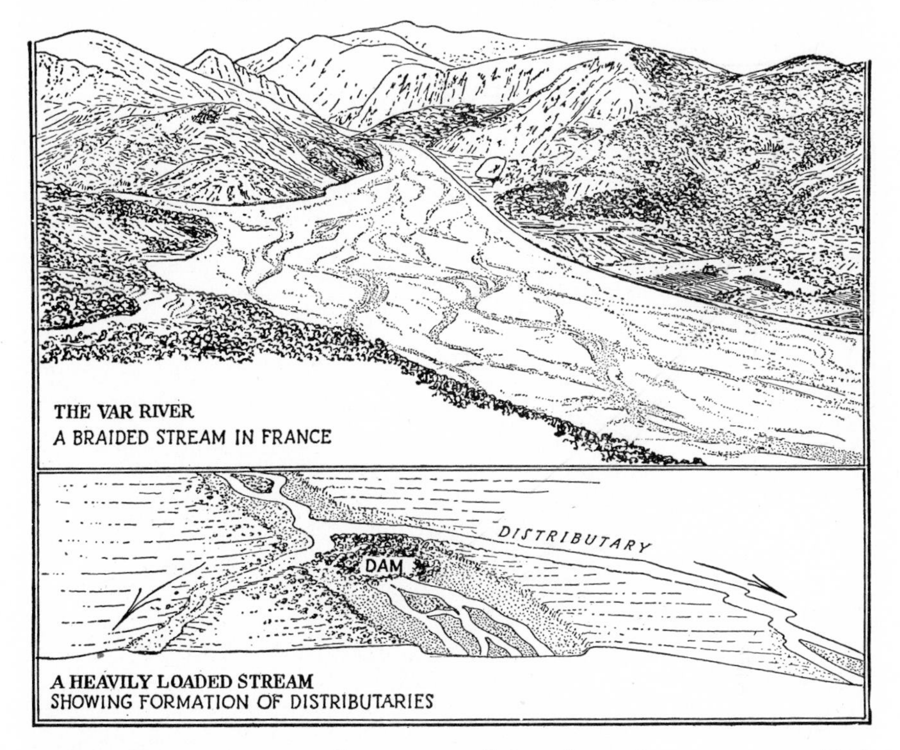

THE VAR RIVER
A BRAIDED STREAM IN FRANCE

A HEAVILY LOADED STREAM
SHOWING FORMATION OF DISTRIBUTARIES

Because of this condition some tributaries never do reach their main stream. This is true of the Atchafalaya system in Louisiana which enters the Gulf of Mexico independently of the Mississippi.

The drainage of flood plains is down the outer slope of the levees away from the river and into the "back country." In times of flood, levees become the safest places. A river may break through a levee but in doing so destroys only a small portion of it, whereas most of the back country or low part of the flood plain becomes inundated. Thousands of square miles of the lower Mississippi plain is subject to flood and this is true of the great rivers like the Ganges, the Hwang Ho, the Po, and on a smaller scale the Connecticut and other New England streams.

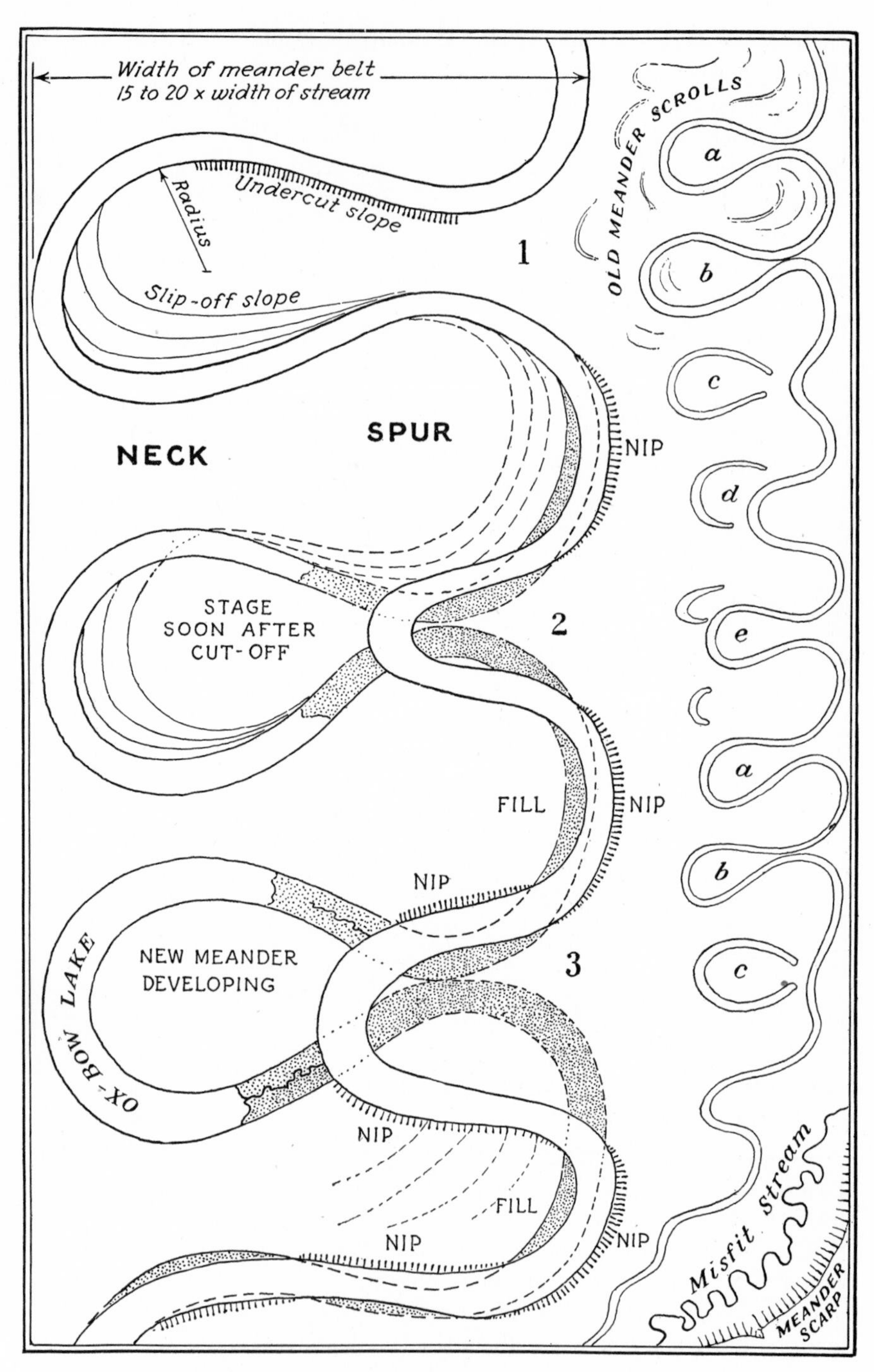


STAGES IN THE FORMATION OF CUTOFFS AND IN THE DEVELOPMENT OF NEW MEANDERS

## MEANDERING STREAMS

Most rivers which flow upon wide flat-floored flood plains have meandering courses. The word *meander* is from the name of a river in Asia Minor which has that habit.

Young rivers actively cutting downward do not meander. Neither do rivers, like the Platte, which are so excessively loaded that the entire force of the river is expended in the effort to keep its load moving, in which case the river may almost disappear from view, seeping slowly through its shifting sands in a braided channel. Meandering is the habit of only those mature rivers which have a load so adjusted that the river is not always depositing but is alternately cutting and filling its valley. The slight irregularities in the course of a river traversing a flood plain are slowly transformed into meanders by the deposition of sediment on the inside of the curve and the cutting away of the bank on the outside. If the load of the stream is so slight that no deposition takes place on the inner side of the curve, then the stream does not cut so actively against the outer bank and the tendency to meander is reduced.

In the process of meander development the whole meander migrates downstream. The stream attacks or undercuts the upstream side of a meander spur. The small scarp thus formed constitutes the *undercut slope.* The downstream side of the spur, on the other hand, presents a long gentle grade known as the *slip-off slope.*

The size of meanders is directly proportional to the size of the stream. The width of the belt of meanders is roughly eighteen times the mean width of the stream which produced them, the depth of water and the volume apparently being of negligible influence. For instance, many of the meanders on the Mississippi, which has a channel width of slightly over ½ mile, have a radius of about 4½ miles or a diameter of 9 miles.

It is to be noted that the development of meanders reduces the gradient of the stream by lengthening its course, a procedure which could not occur if the stream were fully loaded all the time.

During stages of high water, meandering streams cut across the neck of the meander spur and in this way shorten their courses. The arc of the meander, thus abandoned, is called an *oxbow* lake. Thus it is obvious that a mature stream maintains an approximately uniform length, as the reduction by cutoffs is in the long run equal to the added length produced by the enlargement of meanders. Each cutoff initiates a new meander which culminates in another cutoff. Therefore, all stages of meander growth may be represented at one time.

MISFIT MEANDERS. It is occasionally noted that small streams with small meanders occupy a flood plain apparently much larger than such a stream would make. The large size of the flood plain, however, should not be taken as evidence that a small stream did not make it; for, with sufficient time, a large flood plain formed by a small stream is possible. But if the radius of the arc of the meander swing scarps in the bluffs alongside the flood plain is much greater than the radius of the meanders now occupying the valley, it may be assumed that the meanders are the work of a larger stream. Meanders with small radius occupying a valley with meander swing scarps of large radius of curvature are said to be *misfit* or *underfit*, a condition due in some instances to an important change in the life of the river, which caused a reduction in its volume.

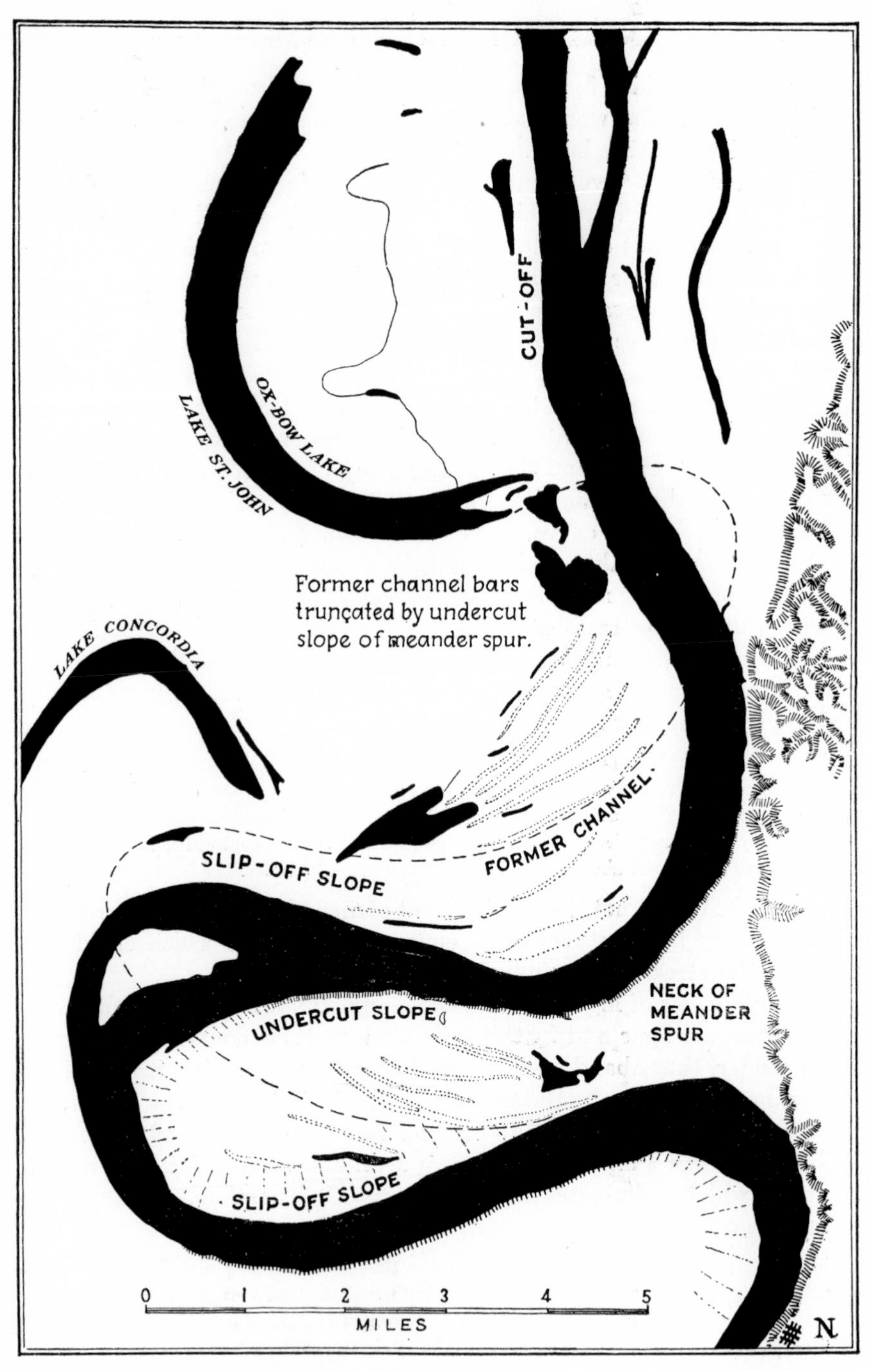


THE MISSISSIPPI RIVER ABOVE NATCHEZ (N) SHOWING MEANDERS AND MEANDER CUTOFFS

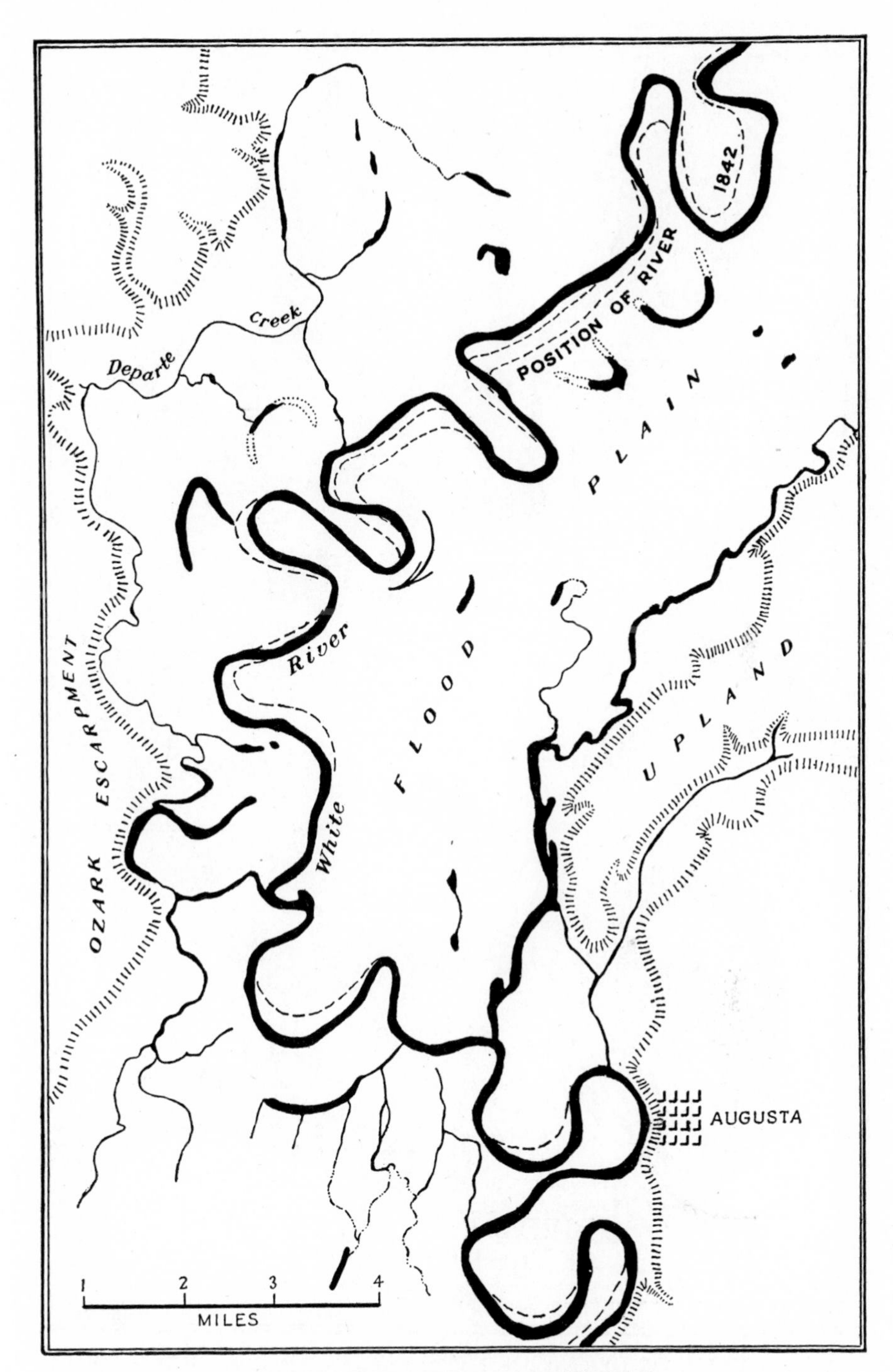


THE WHITE RIVER IN ARKANSAS
On the same scale as the Mississippi River opposite, and showing a much narrower meander belt.

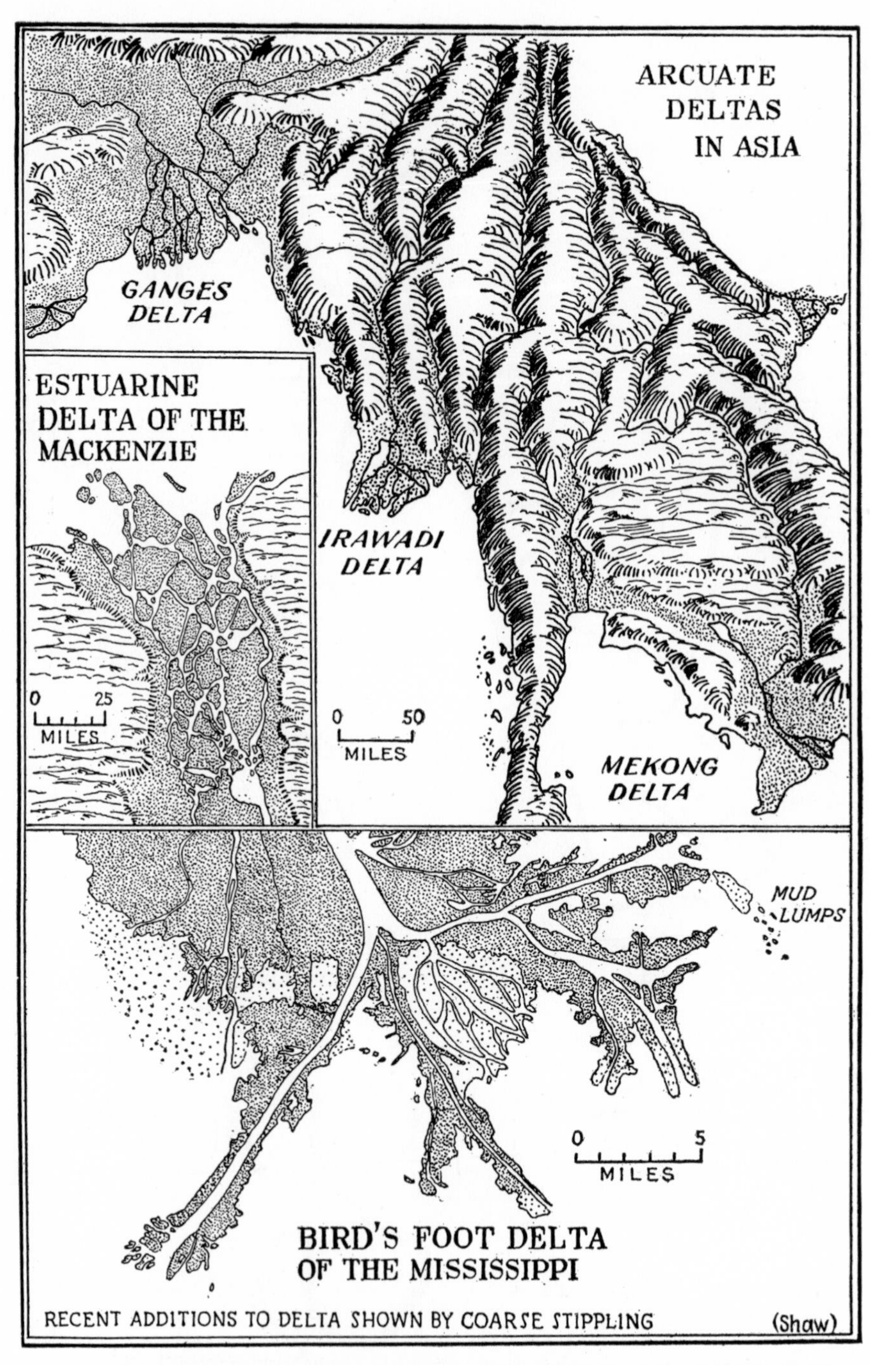


THREE TYPES OF DELTA; ARCUATE, ESTUARINE, AND BIRD'S-FOOT

## DELTAS

Just as a river deposits an alluvial fan on land when it can no longer carry its load, so also it deposits its entire load in the sea when it ceases to exist as a river. The form of a delta depends mostly upon two sets of factors, namely, those having to do with the river and its load, and those having to do with the body of water in which the delta is built. Not only is the quantity of load an important factor, but the character of the load is also significant.

CHARACTER OF DELTA MATERIAL. *a. Arcuate Deltas.* The load of a river may consist entirely of coarse gravel and sand, mainly quartz, and little material in solution, or, on the other hand, much of its load may be extremely fine colloidal material and large quantities of lime in solution. A delta built of coarse sand and gravel and silt is porous. Rivers flowing over such a deposit, on flood plains, alluvial fans, or deltas, tend to have braided channels and an extremely intricate system of anastomosing currents. Most of the channels are shallow and change their positions frequently during high water. Much water flows beneath the surface of the gravel. Desert climates aggravate this condition. Deltas built by streams with many distributaries, like those of the Nile, the Rhine, the Hwang Ho, the Niger, the Indus, the Irrawaddy, the Ganges, the Mekong, the Danube, the Po, the Rhone, the Ebro, the Volga, the Lena, and indeed most of the deltas of the world, are relatively simple in outline, more or less fan-shaped or arcuate, convex toward the sea.

*b. Estuarine Filling.* Other streams whose mouths have been and are still submerged, like the Mackenzie, the Elbe, the Vistula, the Oder, the Susquehanna, the Seine, the Loire, the Ouse, the Ob, and the Hudson, are depositing their loads in the form of long narrow estuarine filling which may constitute submerged bars or extensive flood plain or marshy land areas.

*c. Bird's-foot Deltas.* Still a third type of delta is built by streams carrying large amounts of extremely fine material and probably much lime in solution, like the Mississippi. Such streams maintain single channels instead of anastomosing systems. They do this because the very fine impervious material which they deposit does not permit any subterranean flow and, as a result, the water is concentrated in a few large channels. This combination is unusual and the Mississippi River is exceptional. No other large river drains so vast an area of limestone and other fine-grained sedimentary rocks. Three-quarters of the material deposited at the mouth of the Mississippi is ranked as silt or clay, the particles being less than $1/20$ mm. in diameter. Virtually all of the material in the delta is less than $1/10$ mm. in diameter.

The Mississippi River advances its delta by means of four distributaries or passes which radiate from a common point and give the unusual bird's-foot form to this delta. These channels are bordered by narrow banks of unyielding gray clay. The bays between the delta fingers or "passes" are deep and are being filled only very gradually by wave and tidal action. From time to time the main Mississippi or one of its distributaries breaks through the banks or levees in a crevasse and builds a minor delta into one of the bays, but the main channels are never abandoned. They are too deep.

Deltas built into lakes and inland seas are more perfect and less variable than those built into the ocean with its strong current. Some of the most perfect deltas known are those built into Lake Bonneville and now revealed by the disappearance of that lake. Waves and currents modify the shape of deltas, causing them to be cuspate, or fringed with bars, or snubbed in various ways by wave action.

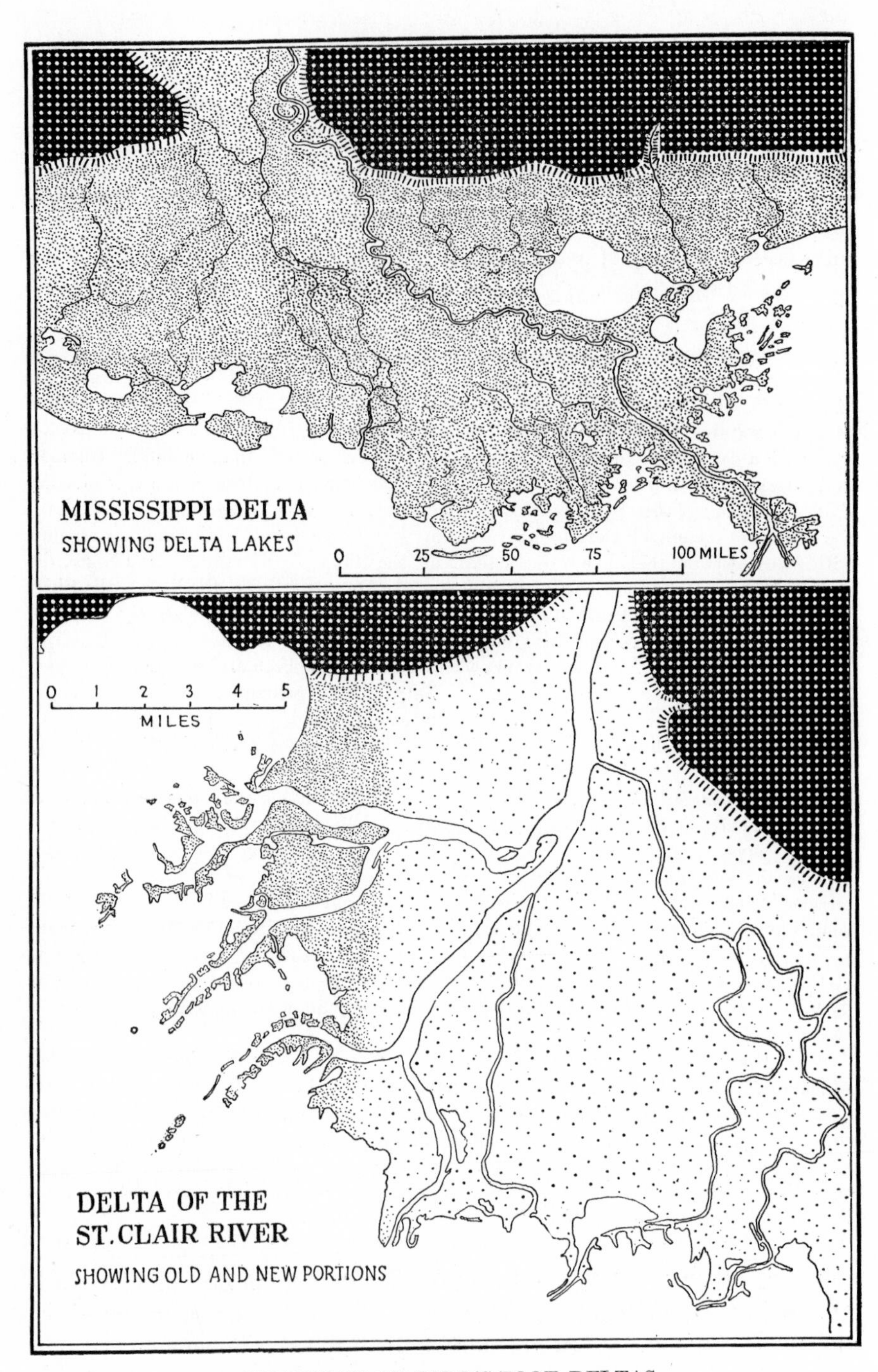


EXAMPLES OF BIRD'S-FOOT DELTAS

## DELTA CHARACTERISTICS

RATE OF GROWTH. Arcuate deltas advance seaward almost uniformly around the entire margin. The average increase of the Nile is about 12 feet per year. The Po delta has advanced into the Adriatic at rates varying from 80 to over 200 feet per year during the past 800 years. The town of Adria between the Po and Adige Rivers was a seaport before Christ but now stands 14 miles inland, an advance of 30 feet a year. The Terek River, flowing into the Caspian, pushes forward its delta at the rate of 1,000 feet a year. The Mississippi River advanced its various passes during the year 1908 at the average rate of 250 feet per year, but the bays between them remained unchanged. In 1912 after a break at the head of Garden Island Bay the land advanced 2,000 feet during a few months.

Deltas also become less extensive because of settling, or their margins may advance seaward by flowing bodily. Flowage of this kind at the mouths of the Mississippi has caused part of the submerged portion of the delta to be pushed up to the surface, forming flat-topped masses known as *mud lumps*. They are 5 to 10 feet high and have an extent of an acre or more. The different directions and rate of growth of the arms of the delta cause parts of the sea to be enclosed to form *delta lakes* like Pontchartrain near New Orleans, the Zuyder Zee on the Rhine delta, and Turtle Bay at the mouth of the Trinity River. Most delta lakes are formed along the delta margin where bars have been built across shallow embayments to form tidal lagoons.

DELTA STRUCTURE AND THICKNESS. Theoretically deltas consist of *bottom-set, fore-set,* and *top-set* beds. These are all well developed in the small deltas formed in glacial lakes and now exposed and dissected. But along the sea the almost horizontal fore-set and bottom-set beds merge together. The fore-set beds of young deltas of coarse material may slope at angles of 30 or 35° but the frontal slope of large marine deltas is much less, that of the Rhone being less than ½°.

The thickness of delta deposits depends largely upon the depth of the water body in which the delta is built. This is usually least near the head of the delta. It is stated that the mud of the Nile delta is not over 50 feet thick. Wells 2,000 feet deep have been sunk into the Mississippi delta without entirely penetrating the delta deposits.

Some deltas reveal by their form several stages in their development. The St. Clair delta has two parts, the old on the eastern side and the new on the western. Each of these represents a period of active delta building. After the old delta was built, the upper Great Lakes discharged by way of Georgian Bay through the Trent River outlet into the St. Lawrence and very little water flowed through the St. Clair River. With the uplift of the region to the north, the lakes have again renewed their former course of discharge and the present bird's-foot type of delta is being actively formed.

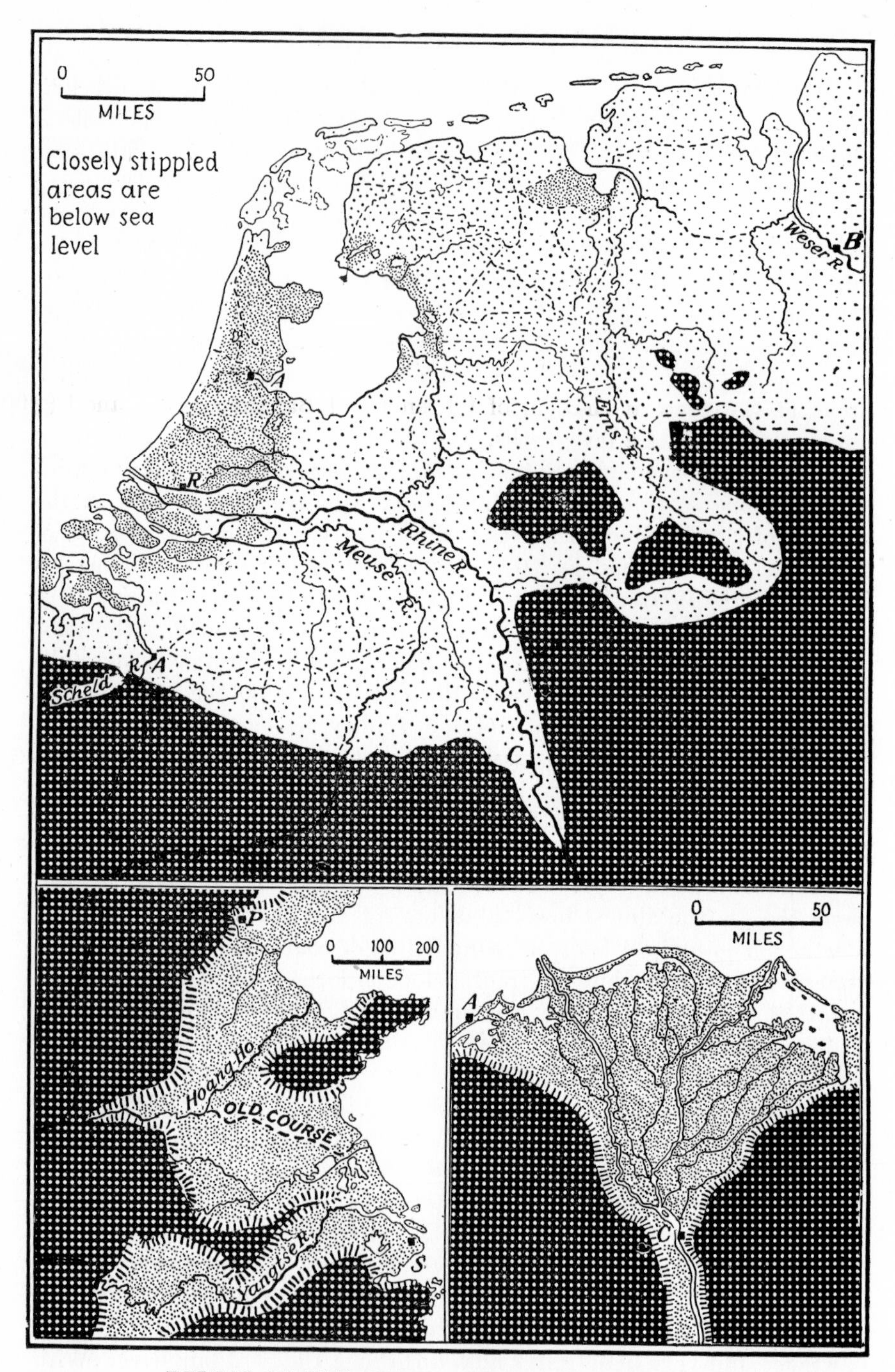


DELTAS OF THE RHINE, HWANG HO, AND NILE

The Rhine with its delta lakes and below-sea-level portions; the Hwang Ho surrounding the Shantung peninsula, formerly an island; and the Nile, the original Δ-shaped delta.

## SOME IMPORTANT DELTAS, I

THE RHINE DELTA. The region now occupied by much of Belgium, Holland, and the lowlands of western Germany constitutes the combined delta of the Rhine, the Meuse, the Scheldt, and the Ems. It comprises an area somewhat larger than the Nile delta and about the same shape. Its interior portion, called the *Geestland*, consists of coarse deposits of pebbles, gravels, and sand. This is a poor agricultural district, badly drained and filled with peat bogs. The seaward portion of the delta is low and is covered with fine silt. Here marine formations alternate with river deposits. Much of this low country, the polder land of Holland, has been reclaimed by diking, making it the site of the richest nation of central Europe.

THE HWANG HO DELTA. The Hwang Ho or Yellow River of China leaves the mountains about 300 miles from the present seashore, and over this distance it has built a very gently sloping, delta-shaped alluvial plain. The base lies along the present coast and extends for about 400 miles south from Peiping to the great plain of the Yangtze Kiang and surrounds the mountainous peninsula of Shantung, which was an island prior to the building of the delta. The head of the plain is only 400 feet above sea level; hence there is an average fall of only 1⅓ feet per mile, making the slope appear in all respects horizontal. The material of which this plain is built is mostly fine silt derived from the loess of the interior. It is the yellowish color of this material, due to the presence of hydrated iron, which has given the Yellow River its name. Because of the gentle slope, the river when swollen is easily diverted from its course. The main mouth of the stream has been repeatedly shifted as much as 200 miles. When the river breaks its banks, vast inundations result because the river in normal times is confined to a levee ridge 10 to 20 feet above the plain. In 1887 such a flood covered 50,000 square miles of immensely fertile and densely inhabited land.

THE NILE DELTA. This, the most famous delta of history, like the valley of the Tigris and the Euphrates, and indeed many of the great deltas of the world, has been the abode of large numbers of people. It is a true "delta," shaped like the Greek capital letter delta, and the first one to which this descriptive name was given. At Cairo, situated at its apex 100 miles from the coast, the Nile separates into two main branches and many smaller distributaries. The slope of the delta is less than 1 foot per mile, being about one-half that of the Hwang Ho, but three times that of the Mississippi delta. It is a region of intense cultivation and subject to few disastrous floods, because of the dams built along the Nile to conserve flood waters for irrigation. The streams flowing over the Nile delta are strongly impinging against their right-hand (or eastern) banks. As with many other large rivers in the Northern Hemisphere, this may be due to the influence of the earth's rotation as expressed in Ferrel's law.

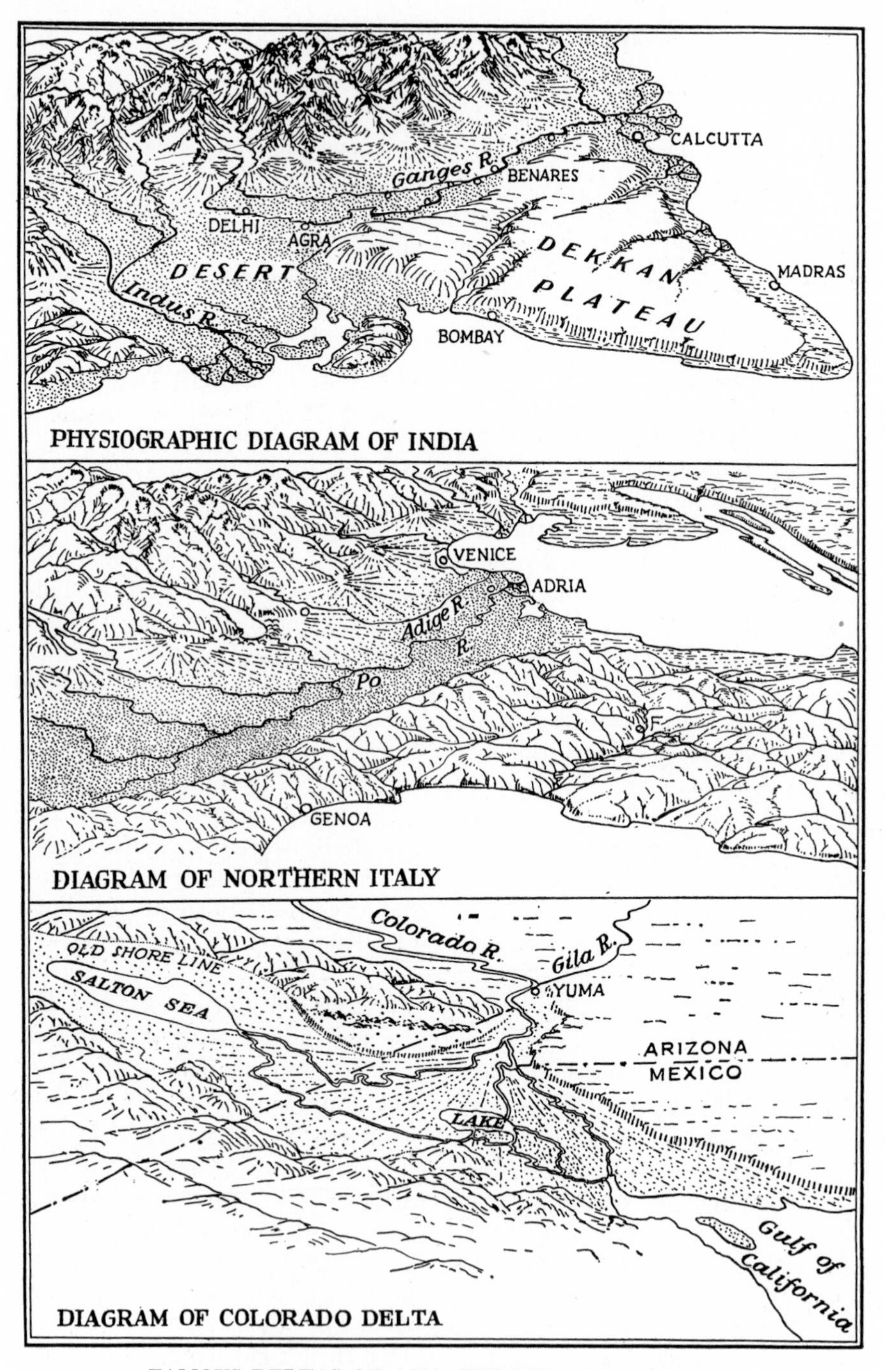


FAMOUS DELTAS OF ASIA, EUROPE, AND AMERICA

## SOME IMPORTANT DELTAS, II

THE INDUS AND THE GANGES. The great plain of northern India is drained by two major river systems, the Indus and the Ganges. Their tributaries have built numerous alluvial fans of coarse material and steep grade. Gravels are abundant at the foot of the mountains; but 20 or 30 miles from the hills the plain consists of assorted sands and clays. The highest portion of the plain is less than 1,000 feet above sea level and slopes about 1 foot per mile to the Indian Ocean. From Benares to the sea the slope is only 5 inches per mile. Its fertile area of 300,000 square miles supports an immense population.

Borings into this plain to nearly 1,000 feet below sea level show that the material is essentially the same throughout. This presumably means that this great trough parallel to the Himalayas has been subsiding ever since it first became loaded with deposits at the foot of the mountains and therefore exemplifies the principle of isostasy. The delta of the combined Ganges and Brahmaputra Rivers is a typical arcuate delta. The Ganges itself effects great changes in the country before it reaches its delta. It has been forced to flow along the southern side of its great valley by the fans built out from the mountains. Many decayed or ruined cities attest the frequent changes in the river bed. Nevertheless, numerous great cities line its course, although recently the river has left the city of Rajmabal high and dry, 7 miles from its banks.

THE PO DELTA. The Plain of Lombardy in north Italy is a long, narrow trough formed by subsidence between the Alps and the Apennines. It is covered largely by sediments brought down from the Alps. Waste material from the Apennines is negligible, as seen in the position of the Po, which has been pushed to the south by the affluents from the Alps, whereas those from the Apennines affect it very little. The Po contributes each year material for 334 acres of new land to its delta. The Brenta, the Piave, and the Tagliamento are all also encroaching upon the Adriatic.

THE COLORADO DELTA. The Colorado River, loaded with a vast amount of silt, has entirely separated the head of the Gulf of California from the main body of the gulf. Owing to the arid climate, the water has evaporated, leaving a depression nearly 275 feet below sea level. Traces of old shore lines 40 feet above sea level indicate that the Colorado River at times flowed into this depression and filled it to overflowing with fresh water. At present it is occupied by Salton Sea, a body of very saline water. In recent years (1891, 1905, 1906–07) the Colorado River has again found its way into this basin by enlarging irrigation ditches. The results were disastrous. Before these breaks occurred, broad belts around the lake shore were fringed with white crusts of salt. This entire basin, some 150 miles in length, is known as the Imperial Valley.

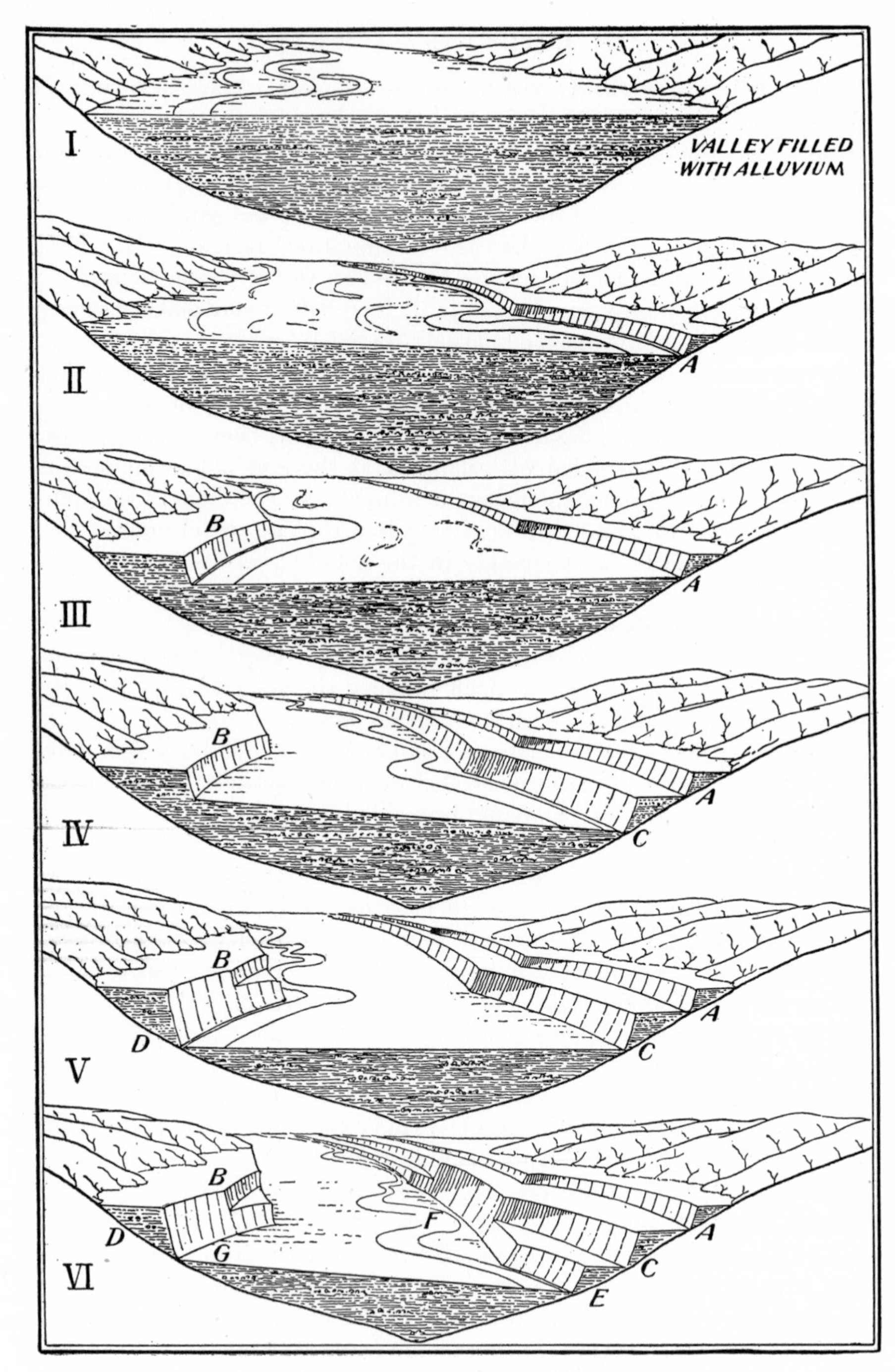


THE DEVELOPMENT OF RIVER TERRACES

## RIVER TERRACES, I

Alluvial terraces in river valleys result from the fact that alluvium formerly deposited in the stream valley has been partly removed by erosion. That is, terraces result from the rejuvenation of a stream which previously was more nearly mature. The rejuvenation may be brought about by increased gradient because of tilting, by increased volume, or by decrease of load. But the cause of rejuvenation has little, if anything, to do with the mode of formation of river terraces.

In a valley deeply aggraded and occupied by a meandering stream (Fig. I), if the entire load is immediately removed, and the stream thus enabled to cut down into its deposits, probably a deep gorge would result as the meanders become incised below the terrace plain. This rarely if ever happens. Instead, the stream loses its load gradually and continues to maintain most of the characteristics of a mature stream. It persists in its habit of swinging from one side of the flood plain to the other. The meanders migrate downstream. But, during the process, the stream constantly picks up some of the alluvium and reduces the level of the flood plain.

The resulting changes are depicted in the opposite illustration. The first diagram shows a valley well filled with alluvium. A mature stream, occupying only a small part of the flood plain, flows down one side of the valley. In the second view, the stream has swung across the valley and planed away a considerable thickness of alluvium. At the point $A$ its lateral migration has been checked by the buried rock ledge of the valley walls. A strip of the original flood-plain surface remains as a terrace. A terrace of this type is called a *rock-defended terrace.*

View III shows the stream again over at the left side of the picture and at a still lower level. The scarp of the terrace at $B$ is therefore higher than the one at $A$. $B$ is not rock-defended, and the stream does not come quite over to the valley walls before beginning its return swing across the valley.

In view IV the stream is back at the right and there are two rock-defended terraces, $A$ and $C$.

In view V the stream has swung left again until stopped by the buried valley wall. The terrace scarp at $B$ has been almost completely destroyed and replaced by the higher rock-defended terrace $D$.

Finally in view VI the stream is once more at the right with three rock-defended terraces, $A$, $C$, and $E$.

Thus we observe that any number of terraces may occur on either side of the river; that the terraces on opposite sides need not correspond in elevation, except the uppermost one on each side; that on one side of the valley we may find one terrace, as at $G$, or several terraces, depending on how far laterally the river may have cut at any particular point; and that terraces are not necessarily due to a progressive diminution in the volume of the stream or to successive uplifts of the land.

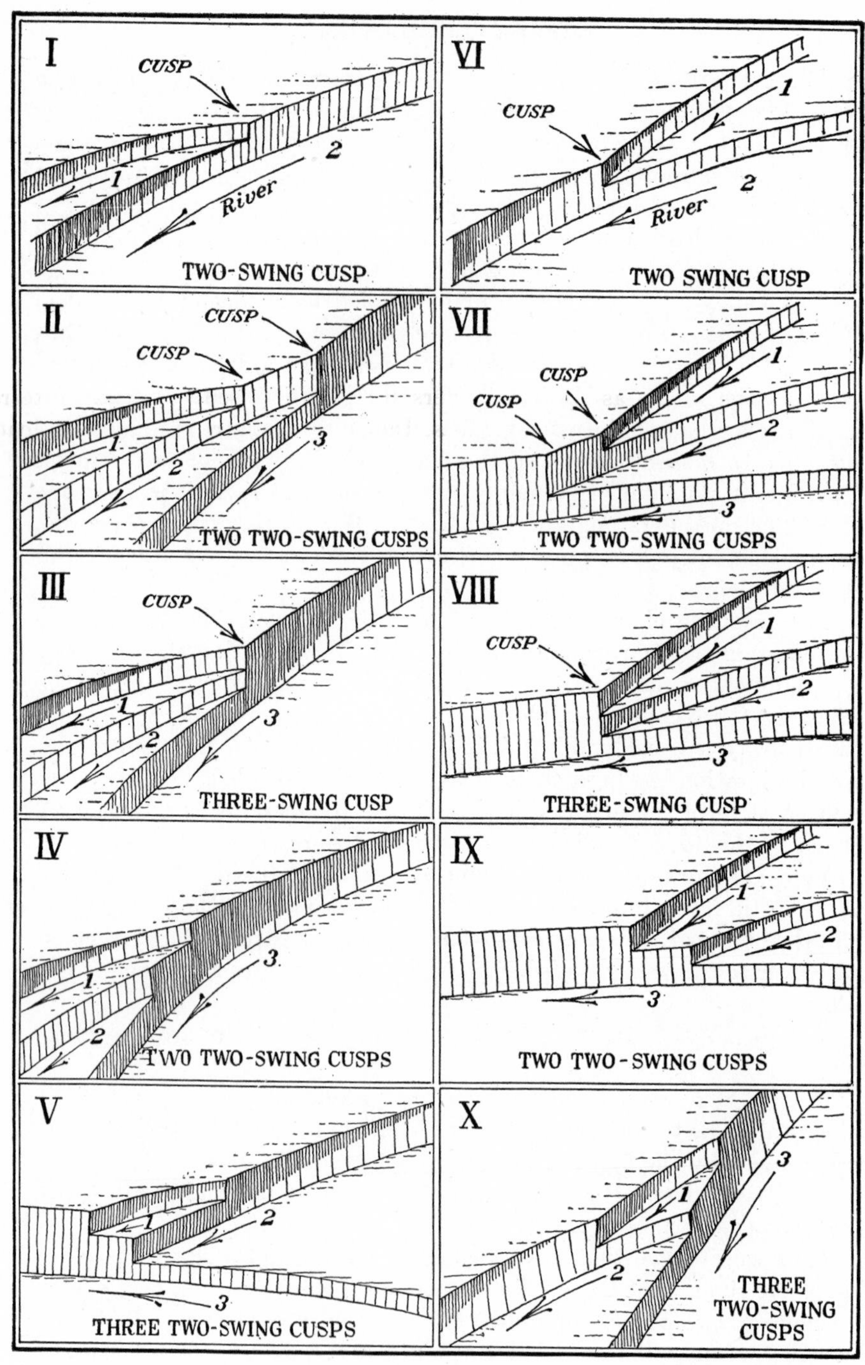


*After W. M. Davis*

FORMS OF RIVER TERRACES

## RIVER TERRACES, II

Two-swing Cusps. River terraces are almost infinite in the variety of their details. The patterns, however, may be resolved into two general groups. We may begin, as in Fig. I, with what is known as a *two-swing cusp* formed by two successive swings of a meander belt. The scarp produced by the first swing is intersected by the scarp of the second swing in such a way that a Y-shaped feature results. The handle of the Y is up-stream. In contrast with this is diagram VI, also a two-swing cusp but with the handle of the Y downstream. Corresponding figures symmetrical with these could be introduced for the other side of the valley.

These two initial forms may now be taken as the starting point for the development of several forms when a third swing of the river intersects either one or both of scarps 1 and 2.

In views II and VII the third scarp intersects the second one, thus introducing in each case another two-swing cusp.

In views III and VIII the third scarp has been pushed back until it just meets the point of the cusp formed by the first and second swings. This results, therefore, in a *three-swing cusp*.

When pushed back still farther, two two-swing cusps again result, as in views IV and IX, and the same is true in the last two figures of each group where the third scarp approaches the cusp from the other direction and therefore first intersects the two arms of the Y, producing in each case three two-swing cusps.

With a fourth swing, not illustrated, the pattern of terraces and cusps becomes still more involved.

Three-swing Cusps. Two-swing cusps are normal elements of all terraces and are to be expected. Three-swing cusps are rare. From the illustrations on the opposite page, it is evident that three-swing cusps are transitory features and are therefore not to be expected very frequently. Three-swing cusps, however, may remain as permanent features if the river is prevented from cutting back further at that point by a buried ledge of rock. A four-swing cusp might even be produced.

Two-sweep Cusps. A cusp formed by the sweep of two successive meanders migrating downstream while the river remains on the same side of the flood plain is termed a *two-sweep cusp*. The interval of time between the two sweeps is infinitely less than that between two swings of the whole river, and therefore the resulting scarps are very much lower. It is difficult, when the difference in height is not great, to determine definitely whether a cusp is a two-sweep or a two-swing cusp.

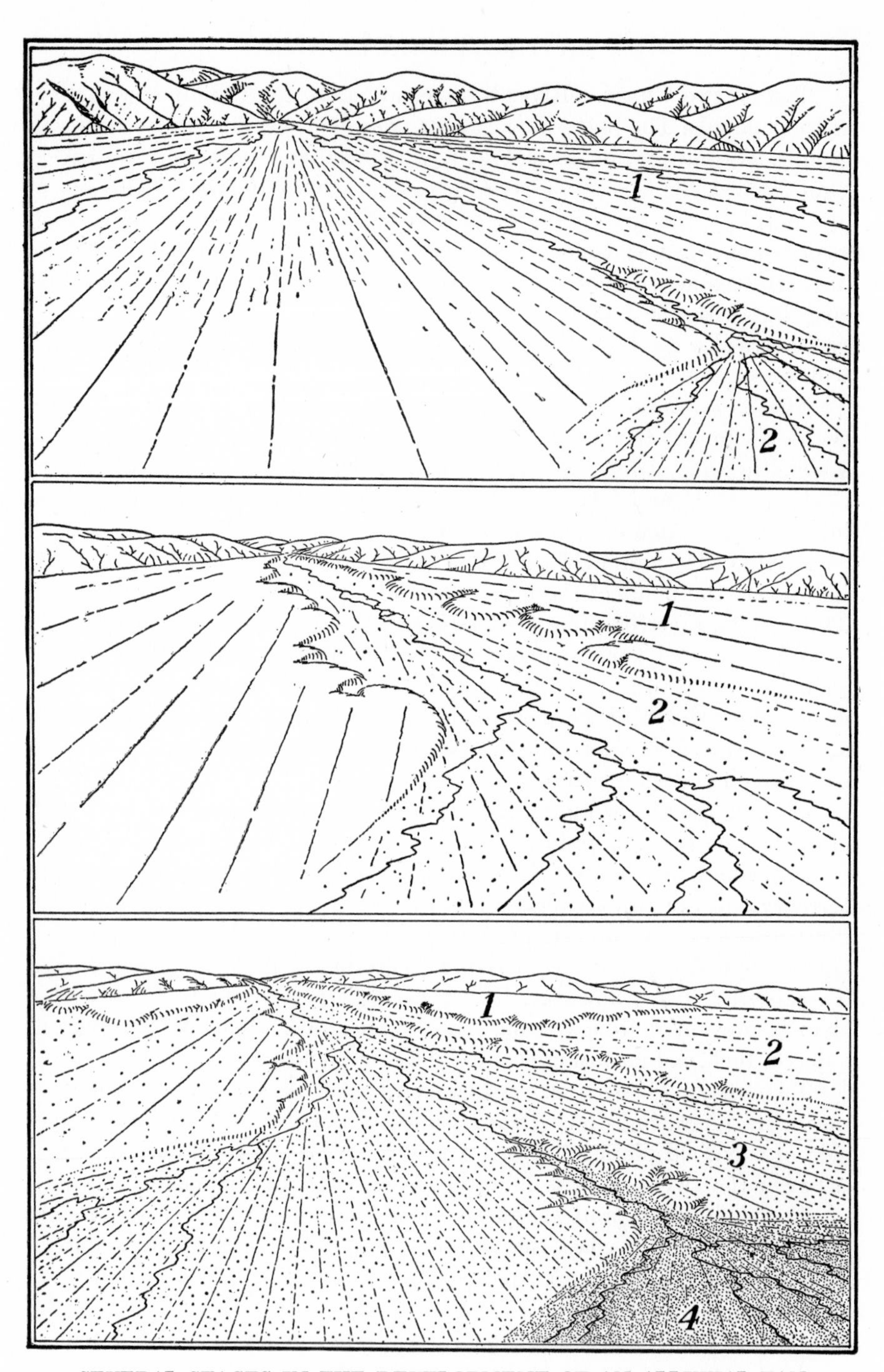


SEVERAL STAGES IN THE DEVELOPMENT OF AN ALLUVIAL FAN

## ALLUVIAL FANS

In nature there is every gradation from steep talus cones of coarse debris to plains of the finest alluvium sloping gently at 1 to 5°. They all, however, are due to the same cause, namely, that the streams which produced them are forced to deposit their load and thereby build up their gradients in order that the regular heavy burden of the stream can be transported. Thus it seems that, once the proper gradient is established, deposition will cease and the stream will proceed to carry its load down the steeper slope thus provided. That the problem is not quite so simple is apparent from the fact that deposition is always occurring at the margin of all alluvial fans and this necessitates the building up of the entire surface of the fan, a process which will go on until the load is actually reduced. The reduction in the load must some day be realized, for the headwaters of the stream will eventually wear down their drainage areas; this will lighten their loads and reduce their slopes. When this occurs, that part of the stream flowing across the alluvial fan will begin to cut down. Thus, without any uplift or change in slope, the stream becomes rejuvenated. As a result of the dissection of the main body of the fan, the stream at the periphery finds itself overburdened and deposits an amount greater than heretofore. This deposit may take the form of secondary fans, which in turn may become dissected, to be followed by a third set of fans of still smaller dimensions. A succession of alluvial fans, therefore, does not necessarily indicate abrupt changes in the physiographic history of the region.

During the process of fan building, the stream flowing over its surface may change its position many times. In fact, at the fan head the main stream may divide into several distributaries which may sink into the gravel to emerge as springs several miles below at the foot of the slope. The fan surface, therefore, is characterized by an interlacing network of braided channels, usually dry except immediately after a torrential downpour in the mountains above.

Alluvial fans do not occur either in valleys or on slopes which have resulted from the uninterrupted processes of normal erosion. Alluvial fans may develop at the foot of hillslopes where gullies are forming, usually as a result of the cultivation of the land; or along the front of block mountains where active mountain streams emerge upon a flat plain; or along the sides of glacial troughs where streams from hanging valleys and from higher mountain slopes debouch upon the flat-floored plain. Vast alluvial fans were built on the east side of the Rocky Mountains during Tertiary time owing probably to continual uplift of the mountain area, combined with a generally increased rainfall. In southern France, north of the Pyrenees, similar deposits were laid down; and in the plains of central India south of the Himalayas, in Patagonia east of the Andes, and on both the north and south sides of the Alps are extensive alluvial fans.

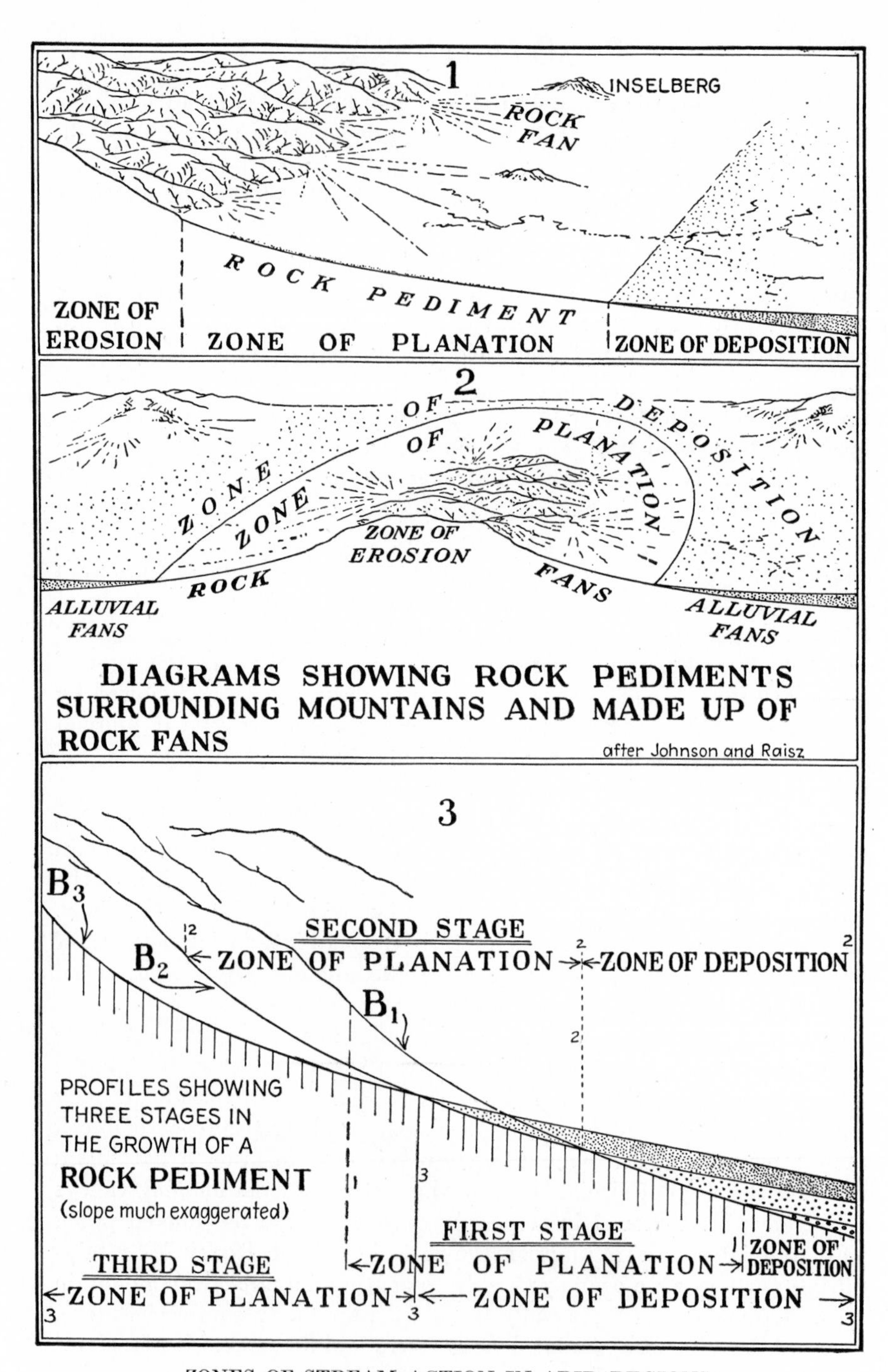


ZONES OF STREAM ACTION IN ARID REGIONS

# ROCK PEDIMENTS

At the base of many mountains and especially in arid regions, there is a flat zone of bedrock 1 mile to several miles in width, only slightly veneered with alluvium and which slopes away to the adjacent basins. Farther from the mountain front this rock plane is deeply buried under alluvium. The whole belt, however, presents the aspect of a continuous alluvial fan. Such a feature is known as a *rock pediment*, *rock plane*, or *conoplain* and may be thought of as a local peneplane beveling the rock structure. It has been shown that this belt represents a zone between the mountain area, where the streams are young and actively eroding, and the basin area, where the streams are actively depositing. This belt is called the *zone of planation*.

Where young streams emerge from the mountains on to a plain or basin area, they find their gradients much reduced; and in consequence they deposit their load of debris. If such a stream is heavily loaded, this deposition will take place at the very exit from the mountains, forming steep alluvial cones (or even talus cones). But, in general, just beyond the mountains the stream is still able to persist in carrying its burden. It has lost its ability to cut down, but it does not have to drop its load. It is in a state of equilibrium; that is, it is a mature stream. It swings laterally and planes off the irregularities of the country leaving isolated monadnocks, named *inselberge*. This beveling gradually reduces the mountain spurs and proceeds upstream, widening the pediment at the expense of the mountain area. In short, the stream develops a slope on the hard rock surface perfectly adjusted to its load so that, by and large, neither deposition nor incision takes place. But around the outer margin, the pediment is being buried under alluvium, if the stream is emptying into a closed basin. This raising of the outer or lower margin of the zone of planation by deposition means that, as this zone grows mountainward, it will have a higher elevation than it did earlier, if the successive profiles of the entire pediment at different times remain the same (Fig. 3).

Though the zone of planation lies in that part of the stream's course where the stream is in equilibrium and does not erode, this whole surface is veneered lightly with a load of coarse debris (like a heavy coat of mail) which is being moved down the slope and which consequently abrades the surface most effectively. There is no concentrated erosion along any given channel but there is erosion everywhere over this plane.

Just as a cone-shaped rock pediment is developed by the combined work of the several streams which emerge and flow away in different directions from a mountain mass, so each stream individually develops a rock fan or cone which is superposed upon the inner margin of the large pediment (Fig. 2). When the mountain mass is finally consumed and the streams become less and less loaded, downward cutting again prevails and the rock pediments as well as the alluvial fans may be dissected.

## SOME GEOGRAPHICAL ASPECTS OF MATURE RIVERS

Large sluggish rivers like the Danube, the Mississippi, the Hwang Ho, and the Amazon, as well as smaller mature streams such as the Ohio, the Missouri, the Connecticut, and the Seine, serve especially in the early development of a country as routes of travel. Such streams are usually navigable for great distances by small boats. In rare cases, rapids occur along such streams and there towns develop, as at Louisville on the Ohio.

The basins of all large navigable rivers, whether actually mature or not, tend to be geographical units. Rivers unite people rather than separate them. In spite of this, rivers have long served as political boundary lines. The Rhine, after the World War, again became the boundary between France and Germany although the two sides of the Rhine valley are similar and have many interests in common.

The inadequacy of mature rivers as boundary lines is due also to the fact that their channels are subject to change. This may result in important boundary disputes. The boundary between Texas and Oklahoma is determined partly by the Red River, a typical meandering stream with a wide flood plain. For many years the changes in the location of the river were of no consequence because few people lived there; but when oil was discovered in this valley, it became of paramount importance to each of the two states to command as much of the oil-producing land as possible. As long as the boundary line follows the main channel of the river, difficulties constantly arise as the channel changes. Now the boundary is established for all time as along the channel of a certain date. The same difficulty arose between the state of Mississippi on one side and the states of Arkansas and Louisiana on the other. A meander cutoff might in the course of a few hours throw an entire plantation from one state to another. It was possible in slavery days for the Mississippi River to transfer a parcel of land from Tennessee to Missouri and thus make a free man of some colored individual living in those parts. Maps of the Mississippi River show the state boundaries in many places following the former channels of the river which, since the establishment of the boundary, have been severed by meander cutoffs.

The fertile soil of river flood plains invites agricultural development and this results in a dense rural population. The Yazoo Basin of Mississippi is one part of the Mississippi River flood plain. Three or four others, all rich cotton lands, embrace 5,000 to 10,000 square miles each in the states of Arkansas and Louisiana. In times of spring floods, however, vast portions of these basins become inundated because these so-called *bottom lands* actually lie below the level of the main river.

Still more disastrous are the great floods of China. By means of the most gigantic diking system in the world along the Hwang Ho, a territory the size of England has been reclaimed for cultivation. Year by year the Yellow River mounts higher and higher on its silted bed above the surrounding lowlands. In 1887 it inundated an area of 50,000 square miles, wiped out of existence a million people, and left a still greater number to perish of famine.

From earliest days the Po and its tributaries have been diked. Since before the time of Christ the plains of Venetia have been drained and rendered tillable. The low plains of Holland are actually parts of the Rhine delta. Much of it is above sea level but a large part of it is polder land reclaimed by diking and pumping out the water. Zuyder Zee, like Lake Pontchartrain on the Mississippi delta, is a delta lake formed by partial detachment of a part of the sea by the outward and lateral growth of the delta.

Deltas are in many cases occupied by large populations owing to the superb fertility of the soil. Great cities grow up in spite of the difficulty of maintaining good ports in the shallow and ever-shifting channels of the rivers. Agricultural deltas are those of the Nile, the Yangtze, the Hwang Ho, the Rhine, the Rhone, the Mississippi, the Colorado, the Danube, the Po, and other streams of north Italy.

Mature rivers not naturally navigable have been rendered so by the building of dams, canals, and locks. The Ohio River from Pittsburgh to its junction with the Mississippi has several dams with a resulting depth of water of at least 15 feet everywhere along its course. In a similar manner the Rhine has been put under control, and this is true of other rivers of northern Germany. With the advent of railroads the flood plains of mature streams have come to serve as admirable rights of way, permitting the building of long, straight, level stretches of track through regions of relatively dense population.

Rarely do mature streams provide the proper conditions for water-power development. In Muscle Shoals on the Tennessee River in Alabama and in a similar plant on the Mississippi at Keokuk, Iowa, it was necessary to construct a long dam of relatively low height. Only the very great volume of water renders such a plant practicable.

Many young streams, emerging from mountains on to plains, acquire mature characteristics and develop flood plains and alluvial fans. These resemble deltas in the way they invite settlement. The alluvial fans covering the floors of the basins in southern California are famous for their fruit orchards. Similar alluvial deposits at the foot of the Wasatch Range in Utah are rich agricultural lands. In Algeria and other semidesert countries the alluvial plains constitute the chief tillable areas. They are the oases. They owe their fertility to underground water. Streams enter the fans at their heads and sink into the porous soil, reappearing farther down the slope as springs. The spring line often determines the location of towns, as in northern Italy where small villages mark the appearance of ground water around the margin of the alluvial plains which spread southward from the Alps.

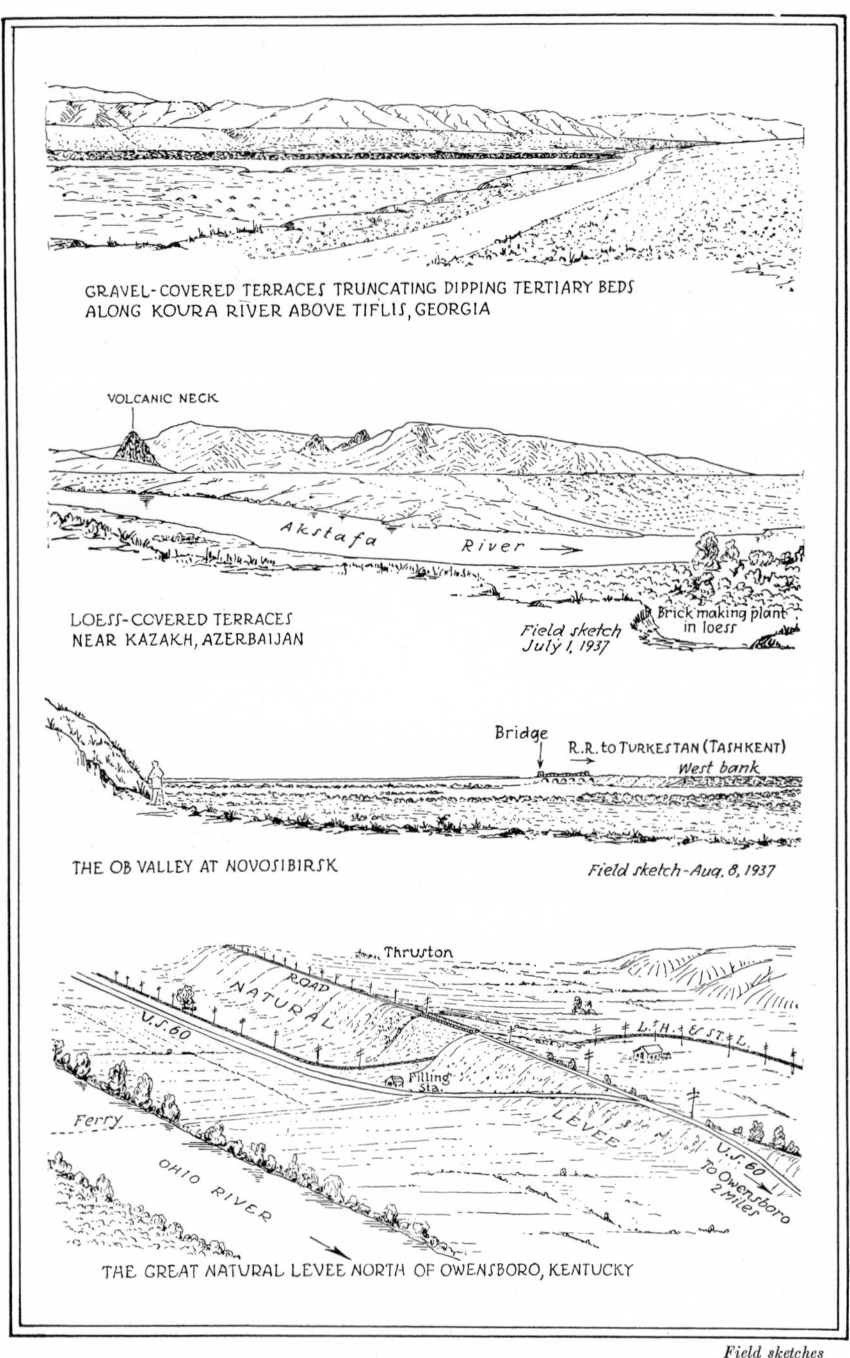


*Field sketches*

TERRACES, FLOOD PLAINS, AND LEVEES OF MATURE STREAMS

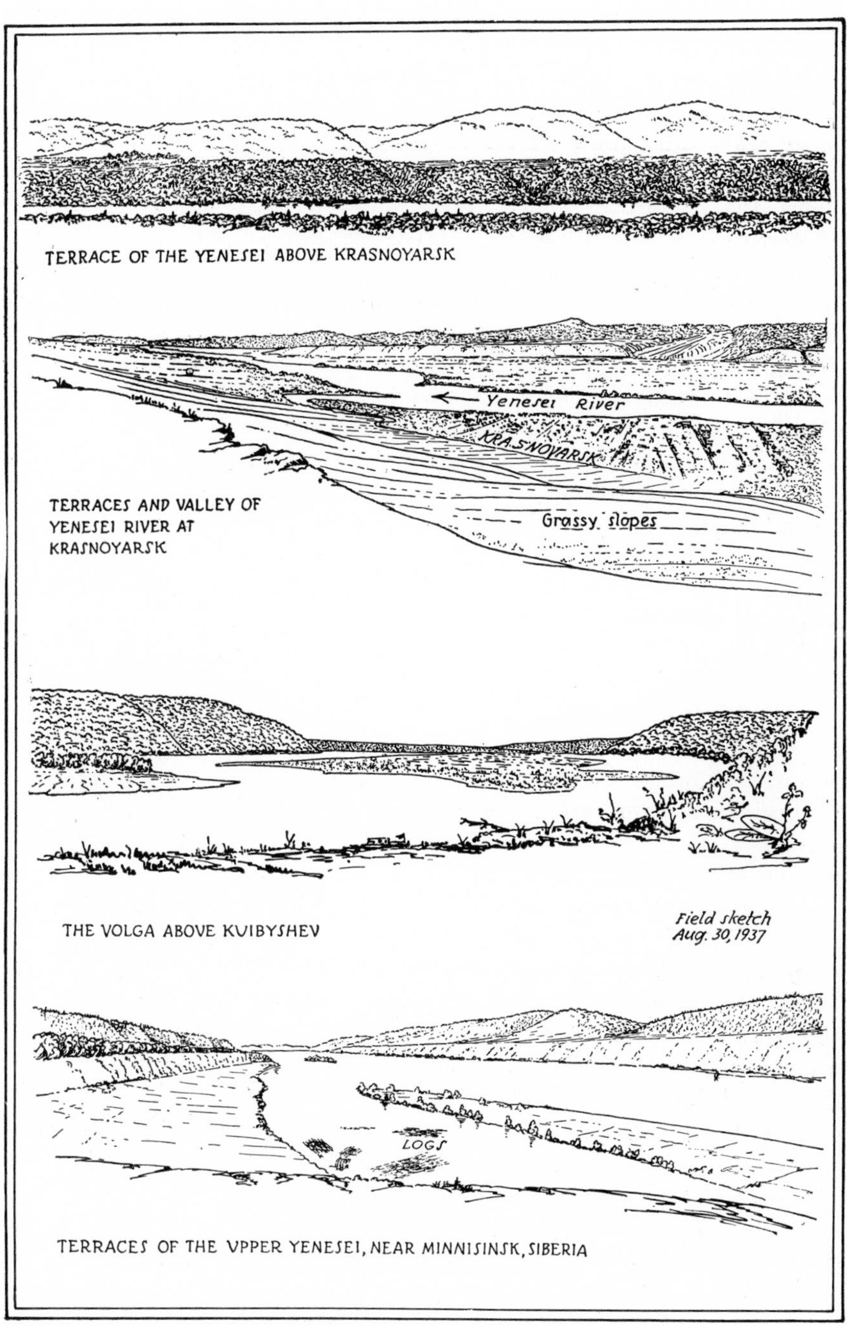


*Field sketches*

TERRACES AND FLOOD PLAINS OF THE YENESEI AND VOLGA RIVERS

## MAPS ILLUSTRATING MATURE STREAMS

Mature streams with flood plains so narrow that the stream is not able to develop completely formed meanders but impinges upon both valley walls are shown on the *Angelica, N. Y.*, the *Otway, Ohio*, the *Leavenworth, Kan.-Mo.*, and the *O'Fallon, Mo.-Ill.*, maps. Wide flood plains, that is, much wider than the meander belt, are shown on the *Chester, Ill.-Mo.*, the *Elk Point, S. Dak.-Neb.-Iowa*, and most of the maps of the Mississippi flood plain. Most flood plains exhibit numerous old meander scars, like those shown on the *Schlater*, *Coahoma*, and *Cleveland, Miss.*, sheets. Cutoff meanders forming oxbow lakes are shown on the *Moon Lake, Miss.*, the *Lake Providence*, *Bayou Sara*, and *Millikin, La.*, sheets. An unusual example is shown on the *Northampton, Mass.*, sheet where a meander migrating downstream has encountered an obstruction which brought about a cutoff. The varying width of meander belts is well shown on the *Elk Point, S. Dak.-Neb.-Iowa*, map which illustrates also a "Yazoo" type of tributary. Natural levees of large size appear on the *Donaldsonville, La.*, sheet, which shows also a former break through the levee, the Nita Crevasse. Smaller levees of natural origin are represented on the *Cairo, Ill.-Ky.-Mo.*, and the *Shawneetown, Ky.*, maps as well as on the *Mt. Airy, La.*, sheet. The *Quarantine, La.*, sheet shows several delta lakes as well as natural levees.

The *Cordova, Ill.*, sheet shows an interesting case of a river having alternating young and mature stretches. A bird's-foot delta, with mud lumps, is shown on the *East Delta, La.*, sheet.

Braided streams are represented on the *Waukon, Iowa-Wis.*, the *Edgington, Ill.-Iowa*, the *Anaheim, Calif.*, the *Lexington, Neb.*, and the *Winona, Wis.-Minn.*, sheets.

Alluvial fans, almost undissected, are remarkably well shown on the *Cucamonga* and *San Bernardino Calif.*, sheets and also upon the *Glendale, Calif.*, sheet, the city of Glendale occupying virtually the entire area of a fan. Dissected alluvial fans are well displayed on the *Stockton, Utah*, and the *Manitou, Colo.*, sheets. Rock pediments, grading downward into alluvial fans, appear on the *Ajo, Ariz.*, the *Hayes Ranch, Calif.*, and the *Needles, Ariz.-Calif.*, maps.

River terraces, with cusps, are shown on the *Lacon, Ill.*, map and also upon the *Hartford, Conn.*, *Springfield, Mass.*, *Fort Dodge, Iowa*, *Penacook, N. H.*, *Friant, Calif.*, and *Nyack, Mont.*, sheets. The new *Springfield North*, *Mt. Tom*, and *West Springfield, Mass.*, sheets (scale 1:31,680) with 10-foot contour interval are particularly good for river terraces.

Lakes formed in tributary valleys that have been dammed by extensive alluviation of the main valley are shown on the *Rathdrum, Idaho*, map.

Finally it may be interesting to note the way some mature streams have been completely straightened artificially as shown on the *Chandlerville, Ill.*, sheet.

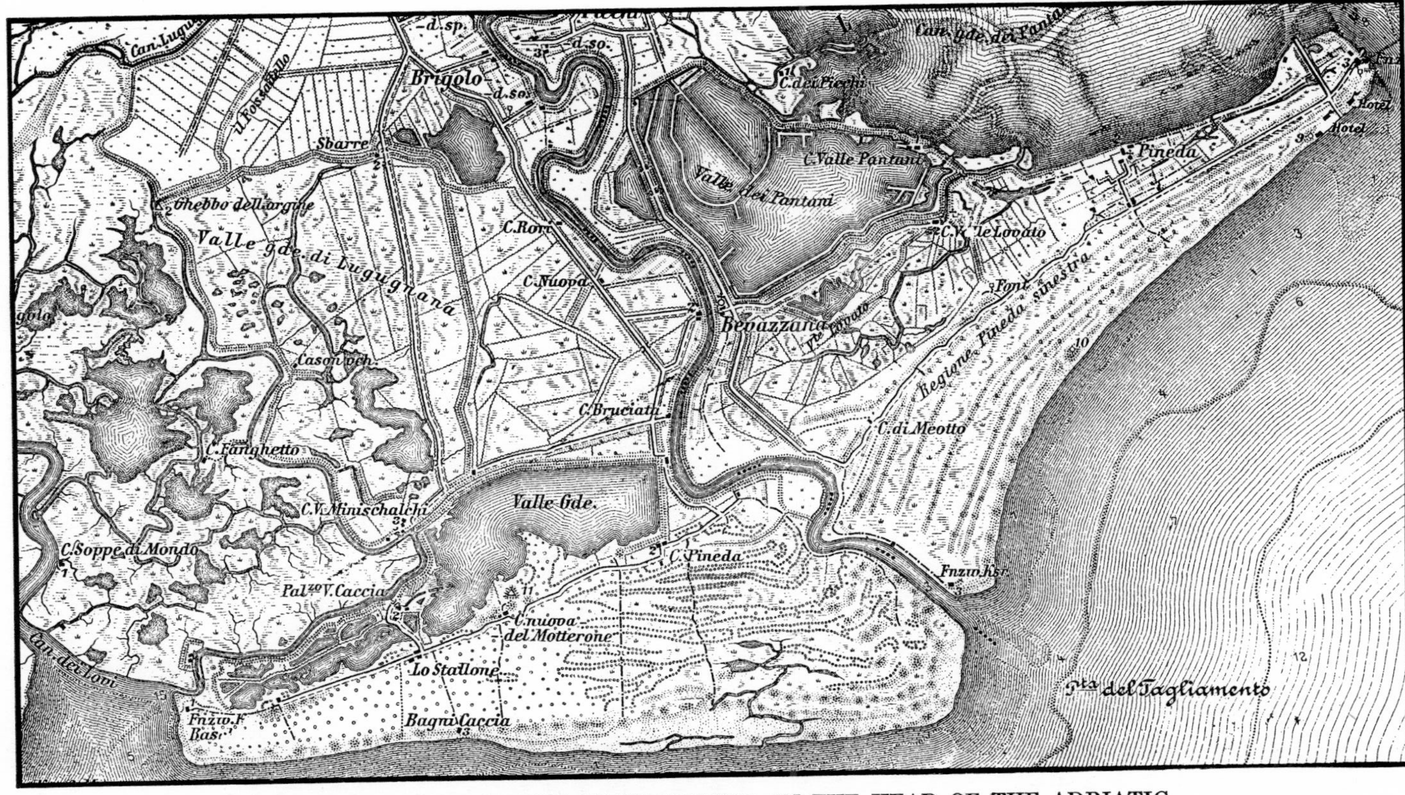


ARCUATE DELTA OF THE TAGLIAMENTO, IN THE HEAD OF THE ADRIATIC

Austria; *Porto Buso* sheet, Zone 23, Col. VIII (1:75,000).

## QUESTIONS

1. Name several different causes for local base-level in a stream.
2. Explain how a mature stream under certain conditions may have a gradient steeper than that of a young stream.
3. Study some of the recent Mississippi River Commission maps and determine how much the gradient of the river is increased as a result of either natural or artificial cutoffs (see, for example, the *Refuge* and *Lamont, Ark.-Miss.*, sheets).
4. Under what conditions may a mature stream capture a tributary?
5. Can you think of any conditions in which a stream would begin life in a mature condition? For instance, do you think it possible that some consequent streams on a very gently sloping coastal plain might be mature from the very start?
6. Are natural bridges indicative of young or of mature streams?
7. Does the dissection of an alluvial fan necessarily indicate uplift or some sudden cause of rejuvenation, or can it come about by the normal behavior of the stream? In other words, what might cause a stream to change from a régime of depositing to one of eroding on an alluvial fan?
8. Why does salt water cause silt to be deposited in deltas more rapidly than fresh water?
9. What is a *misfit* stream? Are all misfit streams *underfit* or can they be *overfit?*
10. Why in many parts of the world, notably in arid regions as in the Great Basin and on the southern side of Porto Rico in the lee of the trade winds, do the streams emerging from the mountains flow across a platform of rock sometimes covered with alluvium? What is such a platform called?
11. Why do meanders form on some flood plains, and braided channels on others?
12. How is the load of a mature stream measured?
13. Draw a contour map of a two-swing cusp; a three-swing cusp; a meander spur; a dissected alluvial fan; an entrenched meander.

## TOPICS FOR INVESTIGATION

1. Flood plains and their features: levees, oxbows, terraces.
2. Rock fans. Factors favoring their development.
3. River terraces. Patterns of terraces and their mode of origin.
4. Control of Mississippi floods by levees.
5. Mature rivers. World-wide examples and their geographical aspects.

## REFERENCES

FLOOD PLAINS

McGEE, W J (1891) *The flood plains of rivers.* Forum, vol. 11, p. 221–234.
MELTON, F. A. (1936) *An empirical classification of flood plain streams.* Geog. Rev., vol. 26, p. 593–609. Good aerial photos.

LEVEES

BROWN, R. M. (1906) *The protection of the alluvial basin of the Mississippi.* Pop. Sci. Monthly, vol. 69, p. 248–256.
JOHNSON, L. C. (1891) *The Nita crevasse.* Geol. Soc. Am., Bull. 2, p. 20–25.
SHAW, E. W. (1911) *Preliminary statement concerning a new system of Quaternary lakes in the Mississippi Basin.* Jour. Geol., vol. 19, p. 481–491. A new type of levee lake.

## Meanders

Campbell, M. R. (1927) *Meaning of meanders in tidal streams.* Geol. Soc. Am., Bull. 38, p. 537–555.

Cole, W. S. (1930) *The interpretation of intrenched meanders.* Jour. Geol., vol. 38, p. 423–436.

Davis, W. M. (1906) *Incised meandering valleys.* Geog. Soc. Phila., Bull. 4, p. 182–192.

Davis ,W. M. (1914) *Meandering valleys and underfit rivers.* Assn. Am. Geog., Annals 3, p. 3–28.

Eardley, A. J. (1938) *Yukon channel shifting.* Geol. Soc. Am., Bull. 49, p. 343–358.

Jefferson, M. (1902) *Limiting width of meander belts.* Natl. Geog. Mag., vol. 13, p. 373–384.

Leighly, J. (1936) *Meandering arroyos of the dry Southwest.* Geog. Rev., vol. 26, p. 270–282.

Macar, P. F. (1934) *Effects of cut-off meanders on the longitudinal profiles of rivers.* Jour. Geol., vol. 42, p. 523–536.

Moore, R. C. (1926) *Origin of enclosed meanders on streams of the Colorado Plateau.* Jour. Geol., Vol. 34, p. 29–57, 97–130.

Rich, J. L. (1914) *Certain types of stream valleys and their meaning.* Jour. Geol., vol. 22, p. 469–497. Interpretation of intrenched and ingrown meanders.

Tower, W. S. (1904) *Development of cut-off meanders.* Am. Geog. Soc., Bull. 36, p. 589–599.

## Alluvial Fans

Chawner, W. D. (1935) *Alluvial fan flooding. The Montrose, California, flood of* 1934. Geog. Rev., vol. 25, p. 255–263.

Eckis, R. (1928) *Alluvial fans of the Cucamonga district, southern California.* Jour. Geol., vol. 36, p. 225–247.

## Deltas

Cole, L. J. (1903) *The delta of the St. Clair River.* Geol. Surv. Mich., vol. 9, part 1, 28 p.

Credner, G. R. (1878) *Die Deltas, . . .* Petermanns Geog. Mitt., Erg'band 12, Erg'heft 56, 74 p.

Dryer, C. R. (1910) *Some features of delta formation.* Ind. Acad. Sci., Proc. 1909, p. 255–261.

McDougal, D. T. (1906) *The delta of the Rio Colorado.* Am. Geog. Soc., Bull. 38, p. 1–16.

Nevin, C. M. (1927) *Laboratory study in delta building.* Geol. Soc. Am., Bull. 38, p. 451–458.

Russell, R. J. (1936) *Physiography of lower Mississippi River delta.* Dept. Conserv., La. Geol. Surv., Geol. Bull. 8, p. 1–199.

Shaw, E. W. (1913) *Mud lumps at the mouths of the Mississippi.* U. S. Geol. Surv., Prof. Paper 85, 27 p.

Sykes, G. (1926) *The delta and estuary of the Colorado River.* Geog. Rev., vol. 16, p. 232–255.

Sykes, G. (1937) *The Colorado delta.* Am. Geog. Soc., Spec. Publ. 19, 193 p.

Trowbridge, A. C. (1930) *Building of Mississippi delta.* Am. Assoc. Petr. Geol., Bull. 14, p. 867–901.

## Rock Fans

Bryan, K., and McCann, F. T. (1936) *Successive pediments and terraces of the upper Rio Puerco in New Mexico.* Jour. Geol., vol. 44, p. 145–172.

Davis, W. M. (1930) *Rock floors in arid and in humid climates.* Jour. Geol., vol. 38, p. 1–27, 136–158.

Davis, W. M. (1938) *Sheetfloods and streamfloods.* Geol. Soc. Am., Bull. 49, p. 1337–1416. An important paper with many references.

Field, R. (1935) *Stream carved slopes and plains in desert mountains.* Am. Jour. Sci., 5th ser., vol. 29, p. 313–322.

Johnson, D. W. (1932) *Rock fans of arid regions.* Am. Jour. Sci., 5th ser., vol. 23, p. 389–420.

JOHNSON, D. W. (1932) *Rock planes of arid regions.* Geog. Rev., vol. 22, p. 656–665.
KOSCHMANN, A. H., and LOUGHLIN, G. F. (1934) *Dissected pediments in the Magdalena district, New Mexico.* Geol. Soc. Am., Bull. 45, p. 463–478.
RICH, J. L. (1935) *Origin and evolution of rock fans and pediments.* Geol. Soc. Am., Bull. 46, p. 999–1024. Review of the problem and many references.

TERRACES

DAVIS, W. M. (1902) *River terraces in New England.* Mus. Comp. Zool., Bull. 38, p. 281–346; Geog. Essays, p. 514–586.
DAVIS, W. M. (1902) *The terraces of the Westfield River, Massachusetts.* Am. Jour. Sci., 4th ser., vol. 14, p. 77–94.
DODGE, R. E. (1894) *The geographical development of alluvial river terraces.* Boston Soc. Nat. Hist., Proc. 26, p. 257–273.
FISHER, E. F. (1906) *Terraces of the West River, Vermont.* Boston Soc. Nat. Hist., Proc. 33, p. 9–42.

GEOGRAPHICAL ASPECTS

BOWMAN, I. (1923) *An American boundary dispute. Decision of the Supreme Court of the United States with respect to the Texas-Oklahoma boundary.* Geog. Rev., vol. 13, p. 161–189.
BROWN, R. M. (1902–03) *The Mississippi River from Cape Girardeau to the head of the Passes.* Am. Geog. Soc., Bull. 34, p. 371–383; Bull. 35, p. 8–16.
CLAPP, F. G. (1922) *The Hwang Ho, Yellow River.* Geog. Rev., vol. 12, p. 1–18.
EMERSON, F. V. (1912) *Life along a graded river.* Am. Geog. Soc., Bull. 44, p. 674–681; 761–768.
HILL, R. T. (1923) *Oklahoma-Texas boundary suit.* Texas Univ., Bull. 2327, p. 157–172.
LYONS, H. G. (1905) *On the Nile flood and its variation.* Geog. Jour., vol. 26, p. 395–421.
TWAIN, MARK. *Life on the Mississippi.*

# VIII
# ALPINE GLACIATION

*U. S. Forest Service*

THE TOE OF LYMAN GLACIER, CHELAN NATIONAL FOREST, WASHINGTON

Showing some of the debris carried on top of the glacier, and the river issuing from the melting ice.

*U. S. Geological Survey*

CRILLON GLACIER, LITUYA BAY, ALASKA
Showing rugged and broken moraine-covered ice. At the side is a glacial stream with delta

GLACIER BAY, ALASKA

Showing two entire glacier systems with numerous matterhorn peaks and hanging cirques.

*U. S. Geological Survey*

*Lee, U. S. Geological Survey*

HEAD OF SLATE RIVER VALLEY, A GLACIAL TROUGH, CRESTED BUTTE REGION, COLO.

Good example of rounded glacial trough with hanging valley at the right.



SWIFTCURRENT VALLEY, GLACIER NATIONAL PARK

Unusually fine example of rounded glacial trough with rock-basin lakes.

*Campbell, U. S. Geological Survey*

*U. S. Geological Survey*

ROCK STEPS AND ROCK-BASIN LAKES IN GLACIAL TROUGH ON BARANOF ISLAND, ALASKA

Post glacial erosion has only slightly reduced the outlets of the lakes.



SWANEA GULCH, NEAR SILVERTON, SAN JUAN MOUNTAINS, COLO.

A fine example of hanging cirque with post-glacial notch and alluvial fan. Reservoir in foreground.

*Cross, U. S. Geological Survey*

*Fairchild Aerial Surveys*

THE PRESIDENTIAL RANGE, WHITE MOUNTAINS, N. H.
Showing numerous glacial cirques indenting the eastern side of the range. A fair example of "biscuit-board" topography.

## ALPINE GLACIATION

Synopsis. Glaciers are rivers of ice; as such they behave in many respects like rivers of water. The resemblances and differences are as follows:

*a. The volume* of a glacier is vastly greater than that of a stream. Hence its channel is much larger. The channel of a glacier and the channel of a stream resemble each other in transverse profile, in that each may be represented as a catenary curve, as when a sagging string is supported at its two ends. The transverse stream profile is usually much flatter than the glacial profile. So-called *glaciated valleys* are described as rounded and troughlike; these forms are the channels formerly occupied by the glacier. The difference in volume between the main glacier and its tributaries accounts for the hanging relationship of the tributary troughs.

The volume of a glacier system may be so great as practically to fill the valleys. The projecting crests and divides are subjected to weathering in which alternate cold and heat play a large part and where plucking by numerous small hanging glaciers is important. Such erosion results in sharp-crested divides or arêtes instead of the rounded profiles of more normally eroded regions.

Cirques or amphitheaters develop at the heads of glaciated valleys. Instead of narrow clefts and ravines produced by stream action, the plucking action of the glacier takes place equally around the whole valley head, much as weathering produces amphitheaters of the valley heads in arid regions, with the difference that glacial cirques are scoured into true basins which may later be occupied by tarn lakes. The headward migration of cirques leads to the formation of the matterhorn peaks so characteristic of glaciated mountains.

*b. The depth* to which a glacier will lower its valley is greatest somewhere in its middle portion—not near its lower end as in the case of a stream. Hence the longitudinal profile of a glaciated valley is so flat indeed that, when reoccupied by streams, the gradient may be insufficient, and finger lakes result or deposition of alluvium takes place. This, in many cases, obscures the lower part of the U-shaped glacial trough. The steep, almost vertical walls of the trough suffer rapid weathering, and landslides result.

*c. The deposition* of material by glaciers occurs anywhere along their course. Terminal moraines blocking the ends of the valley may cause finger lakes; and lateral moraines along the valley walls, extending even into the cirques, are ordinary features.

*d. Postglacial erosion*, when stream cutting is resumed, adds postglacial notches in the lips of hanging valleys, and alluvial fans, and provides the waterfalls which abound in every glaciated region.

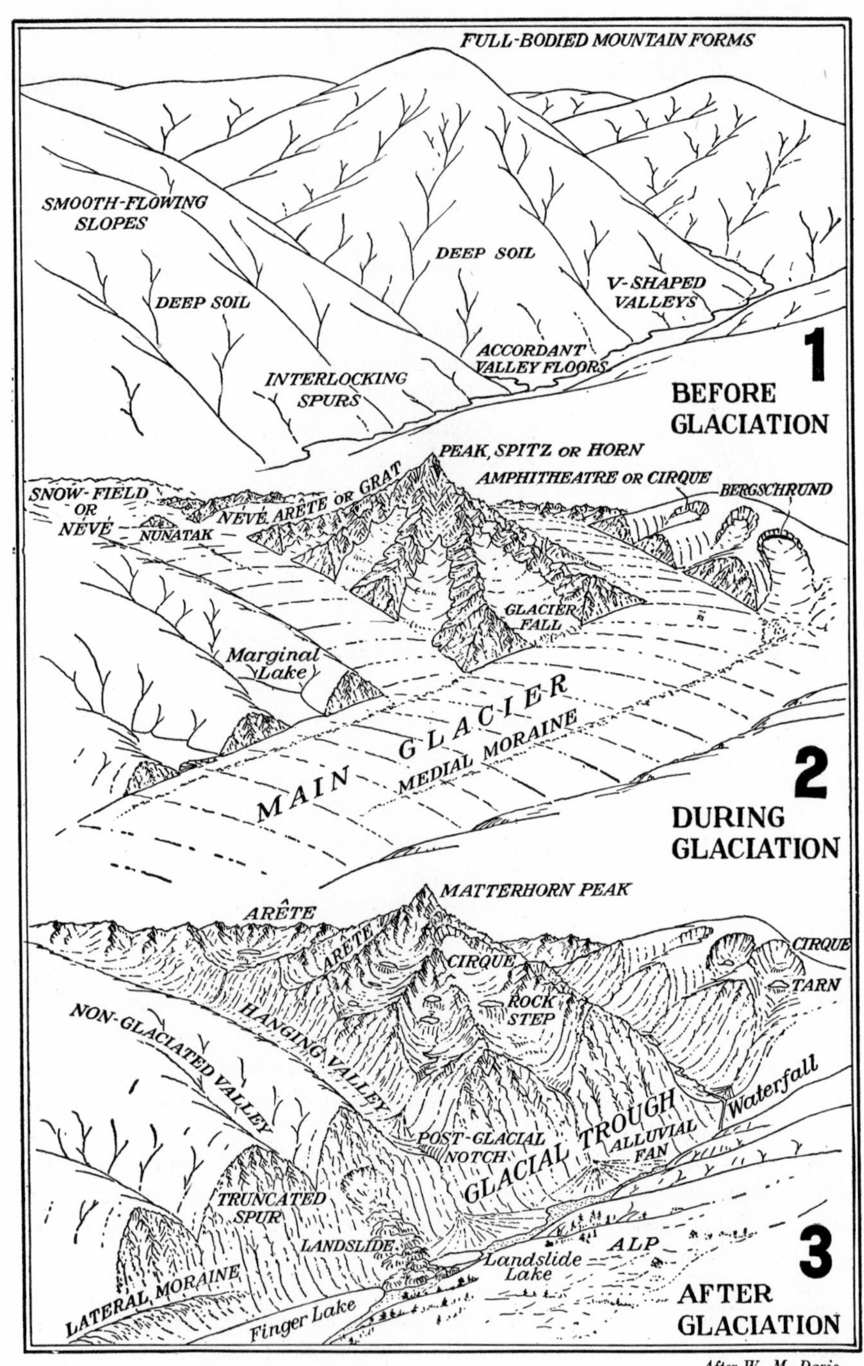


*After W. M. Davis*

A MOUNTAIN REGION BEFORE, DURING, AND AFTER GLACIATION

## ALPINE GLACIATION IN GENERAL

The terms *alpine, mountain, valley,* or *local glaciation* are applied when the glacier is in the form of a narrow tongue confined to a valley, in contrast with *piedmont glaciers* which spread out over plains, and *continental glaciers* or *ice sheets* which cover areas much more vast.

Alpine glaciers are rivers of ice. They differ from streams in the greater size of channel as compared with the total size of the valley. Moreover, they erode most actively in their mid-length, where two or more tributaries join, rather than at their lower ends where movement ceases and all power of erosion is lost. Hence somewhere in their middle courses glaciers deepen their channels and produce rock basins. The lower part of the glacier tongue may actually move uphill.

PREGLACIAL TOPOGRAPHY. Mountains produced solely by weathering and stream erosion (Fig. 1) are usually rounded and full bodied, with few, if any, sharp peaks. The valleys are typically narrow and V-shaped and are winding and irregular because of the overlapping mountain spurs. The valley floors at the points of junction are accordant; waterfalls are therefore rare. A fairly deep soil covers the mountain slopes and rock outcrops are uncommon.

DURING GLACIATION. With the advent of the glacier (Fig. 2) the former valley becomes deeply filled, most of it now serving as a channel to carry away the slowly moving ice in volumes equal to that previously transported by the quickly moving streams. Near the summits the divides or cols may be completely buried by snow, and glacier fields may fork and flow in several directions, in contrast with stream behavior. Glaciers show more erosive vigor at their heads than do streams of water. This results in the rapid lowering of the upper profile of the glacier and the formation of *amphitheaters* or *cirques.*

POSTGLACIAL FEATURES. The disappearance of the ice reveals more clearly the multitude of glacially produced features (Fig. 3). The former narrow ravinelike valley heads of stream erosion are now replaced by large amphitheaters or cirques. Sharp-pointed, pyramid-shaped matterhorn peaks dominate the mountain range. Sharp-crested divides or arêtes are the rule. The former V-shaped profile of the stream valley has been modified by glacial scour to a broad U-shaped or troughlike form. The tributary glacial troughs, not having been eroded so deeply as the main ones, now hang above them, and the streams descend in rapids or waterfalls. Lakes abound. Small *tarns* fill the cirques; larger *finger lakes* occupy the trough floors. The streams of the region, unable to transport their usual load of sediment, aggrade their valley floors. The valley bottoms are therefore flat with numerous alluvial fans. Rock outcrops and exceptionally steep slopes abound, resulting in landslides and avalanches.

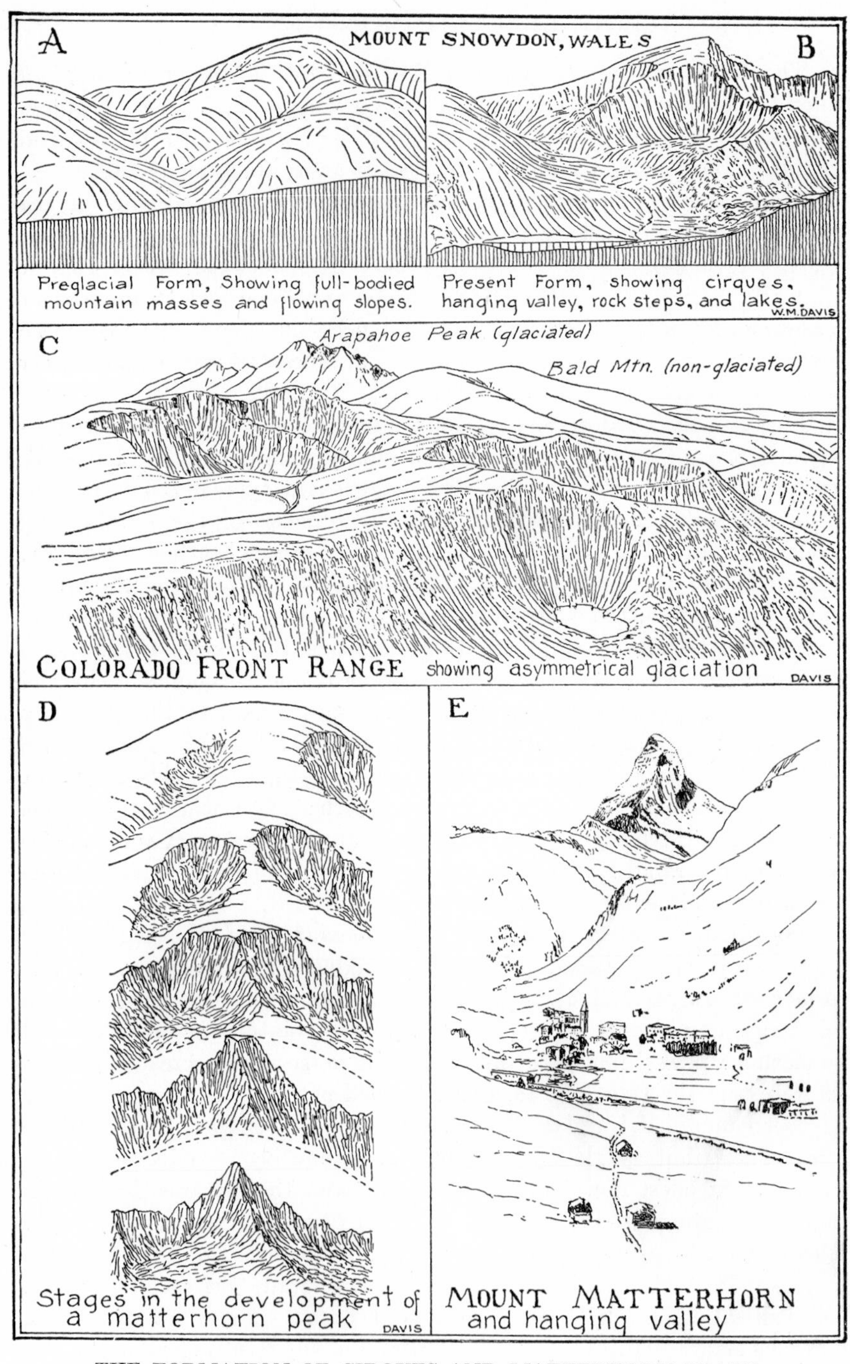


THE FORMATION OF CIRQUES AND MATTERHORN PEAKS

## GLACIAL CIRQUES

(*Corries*, Scotch; *cwms*, Welsh; *Karren*, German)

INCIPIENT GLACIATION; BISCUIT-BOARD TOPOGRAPHY. In low mountain ranges, like the Bighorn Range of Wyoming, the Harz of Germany, and the Mount Snowdon region of Wales, and in low-latitude ranges just within the limit of glaciation, like the Transylvanian Alps and the southern Rockies, only the uppermost parts of the valleys have been glaciated. The resulting cirques are small and the preglacial form of the mountain area is not greatly changed. A topography characterized by a rolling upland, out of which cirques have been cut like so many big bites, is known as *biscuit-board* topography (Fig. *B*). It represents an early stage, or only a partial completion, of the process of glaciation. Sharp-crested, pyramidal, or matterhorn peaks do not occur, for they characterize the later stages of glaciation after the cirques have encroached upon each other.

ASYMMETRICAL GLACIATION. When glaciation has been only slightly effective because of low altitudes or latitudes, it is likely to show the effect of those external influences which disturb or control glacial activity. For example, in the Rocky Mountains of Colorado (Fig. *C*) and in the White Mountains of New Hampshire, glaciation of the mountain summits has not been uniform on all sides. It has resulted in numerous cirques, but the upland surface of these ranges has not been entirely cut away. The cirques are largest and most numerous on the eastern and northern slopes and least developed on the southern and western. Two factors seem to be responsible: the amount of insolation or sunshine and the prevailing westerly winds. The southern and western slopes receive the heat of the sun during the warmest part of the day and, if the amount of snow is close to the margin of sufficiency for glaciation, the melting thus induced may be just enough to forstall glaciation on these exposures. Also the strong westerly winds which prevail at these latitudes may drift enough snow over the crest of the range to cause a sufficient accumulation for glaciation on the leeward or eastern sides, leaving the western slopes more or less bare. On the other hand, if the amount of snow is more than ample, as on the higher ranges of the Rockies and in Canada and the Alps, the relatively slight influence of wind or sunshine will not be enough to produce asymmetry of glaciation.

THE FORM AND POSITION OF CIRQUES. Most cirques are bowl shaped, being deeper in the center than at the rim They are true rock basins and consequently usually contain lakes after the glacier disappears (Figs. *B*, *C*). The steep sheer walls are due to the fact that the glacier formerly occupying the depression plucked away the cliff at its base. Some cirques do not continue downstream as valleys but are perched as tributaries alongside of larger troughs, constituting hanging cirques (Fig. *B*). In that position they are relatively small features, usually only a fraction of a mile across.

*U. S. Forest Service*

BRIDAL VEIL FALLS, YOSEMITE NATIONAL PARK

A perfect example of rounded hanging tributary valley with waterfall.

## HANGING VALLEYS

It has already been pointed out, in connection with Playfair's law, that hanging valleys are unusual features. Rarely do master streams cut down so rapidly as to cause their tributaries to cascade into them from the valley sides. When a block mountain is rapidly raised across a stream's course or when waves encroach rapidly upon a dissected coast, hanging valleys may result. But aside from these rather unusual instances, hanging valleys occur only in regions which have been glaciated. They are to be thought of not as actual *valleys* but rather as hanging *channels* of former glaciers.

In 1883 McGee recognized and explained the hanging valleys of the Sierra as the result of glacial erosion, but it was not until 1898 that there was a complete statement of the principles involved. In his study of Lake Chelan, a finger lake in the Cascades, in 1898, Henry Gannett observed the discrepancy in level between the lake and the valleys of the streams cascading into it. He inferred that the main glacier was at least 3,000 feet thick, since several of the smaller branches join the main valley at that height above its floor.

Glacial Deepening vs. Glacial Widening. The hanging character of tributary valleys is due either to the deepening or to the widening of the main valley. Studies by Douglas Johnson show that deepening by the main glacier is the chief cause. The amount of this deepening may be measured by the relations of the main and tributary valleys. If the present profile of a hanging tributary valley is extended to the center of the main valley, it comes far above the floor of the valley. This indicates much more erosion of the main valley by glaciation than of the tributary valley. This observable difference is only a part of the total amount of difference, which should include also the amount of alluvial filling on the floor of the main valley. The total overdeepening in the smaller glaciated valleys of the Central Upland of France is about 900 feet and in Yosemite Valley almost 2,000 feet. The latter valley is illustrated in the accompanying photograph; half of the overdeepening is concealed by alluvial filling which explains the flat valley floor.

Other Explanations for Hanging Valleys. It has been suggested that hanging valleys have resulted from their being occupied and protected by glaciers while the main valley was deepened by normal river erosion. This is an improbable explanation, because a series of independent glaciers would be unlikely to maintain their position long enough for the main valley to be greatly deepened and widened. An additional argument for glacial deepening is the fact that hanging valleys occur where only the main valley was occupied by a glacier.

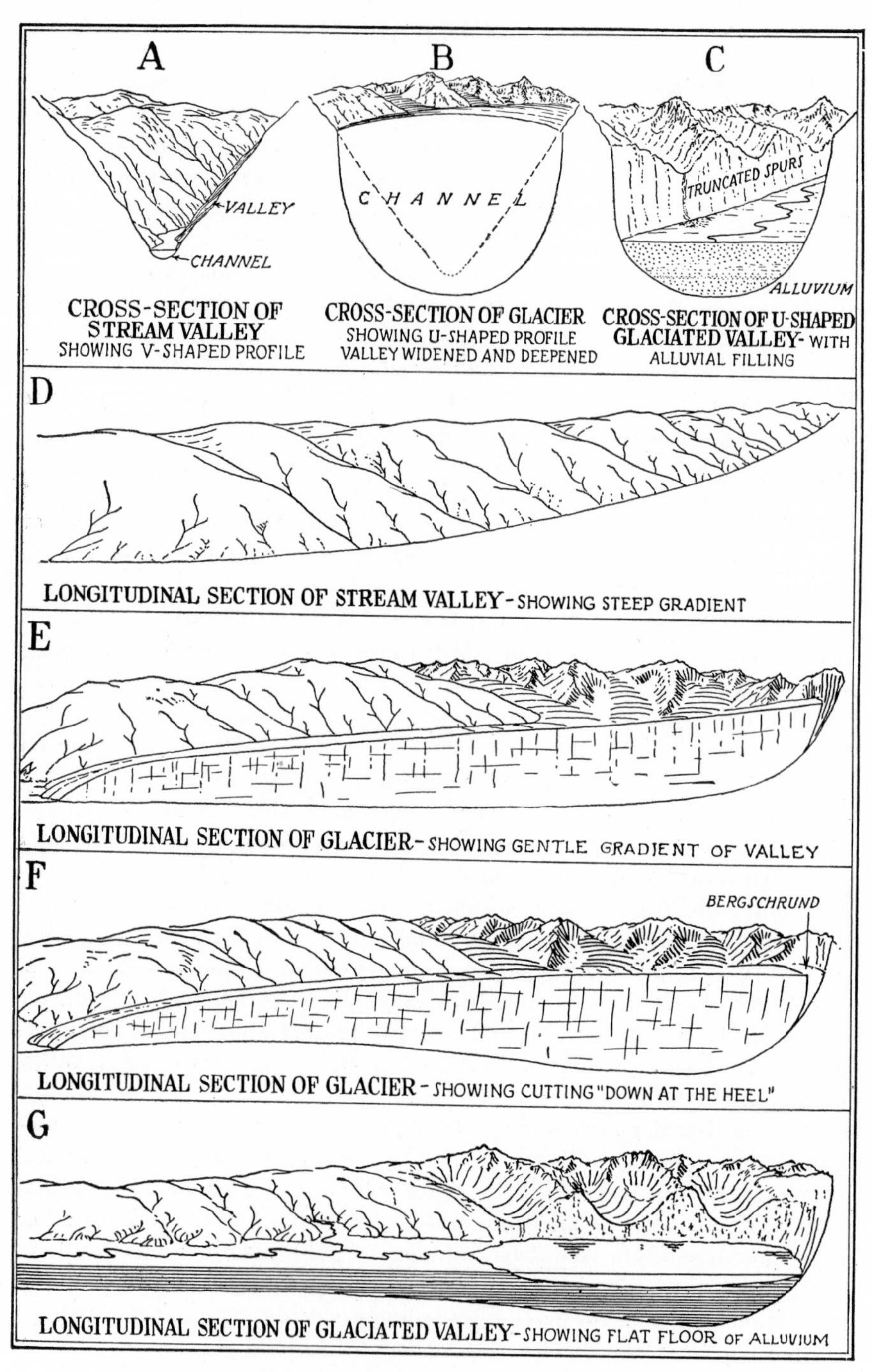


THE DEVELOPMENT OF GLACIAL TROUGHS

## GLACIAL TROUGHS

Glacial Channels vs. Glacial Valleys. It is commonly said that glaciated valleys are U-shaped. The fallacy of this statement lies in comparing the *channel* of a glacier with the *valley* of a stream. As a matter of fact, the channels of both streams and glaciers are U-shaped, much flatter, of course, in the case of streams than in the case of glaciers, but nevertheless displaying the form of a catenary curve. Streams occupy an infinitely small part of the cross section of the valley (Fig. *A*), but glaciers may occupy all of it (Fig. *B*) so that the channel becomes the dominating feature.

Glacial Deepening. In the development of a U-shaped trough from a preglacial V-shaped stream valley, a glacier deepens the valley and widens it (Fig. *C*). The deepening of the valley is not uniform throughout its length but apt to be greater in the middle and upper portions, especially where large tributaries unite. In the lower reaches erosion gradually gives way to deposition. This tendency results in a reduction of the gradient of the valley floor (Figs. *D*, *E*, *F*, *G*,) so that later when a stream comes to occupy the old glacial channel it finds the slope insufficient for the load with which it is burdened, and deposition takes place (Fig. *C*). Some glacial troughs, like Crawford Notch in the White Mountains, have never been filled with alluvium and still preserve their rounded profiles, but most glaciated valleys are flat floored or contain lakes. In Yosemite Valley the depth of the filling is estimated at 900 feet, much of it having been deposited in an old lake bed.

In many instances the overdeepening produces a "down-at-the-heel" condition, the middle part of the course being actually lower in elevation than that farther down the valley. The rock basins which result become filled with water and constitute finger lakes into which tributary streams debouch to form deltas and alluvial fans. The town of Interlaken in Switzerland stands on such a delta which has cut a finger lake in two.

Truncated Spurs. The widening of a stream valley by a glacier results in the truncation of the spurs which extend into it from the two sides. These truncated ends present triangular faces or walls of bare rock toward the valley, like El Capitan in Yosemite or "Jefferson's Knees" on the north side of Great Gulf in the White Mountains. The widening of the valley eliminates numerous minor curves so that extensive vistas are now to be had. The widening of the valleys results also in the oversteepening of the walls, occasionally almost to the vertical position. This unstable condition results in very violent erosion in early postglacial times and landslides are common features on the trough walls. Hermit Lake in Tuckermann Ravine in the White Mountains was impounded because of a landslide accumulation.

TWO EXAMPLES OF GLACIAL TROUGHS

**One rounded, the other flat-floored because it is buried under alluvium.**

*Above, from photograph by Douglas Johnson*

TWO EXAMPLES OF ALLUVIAL FANS BUILT BY TRIBUTARIES INTO LARGE GLACIATED VALLEYS

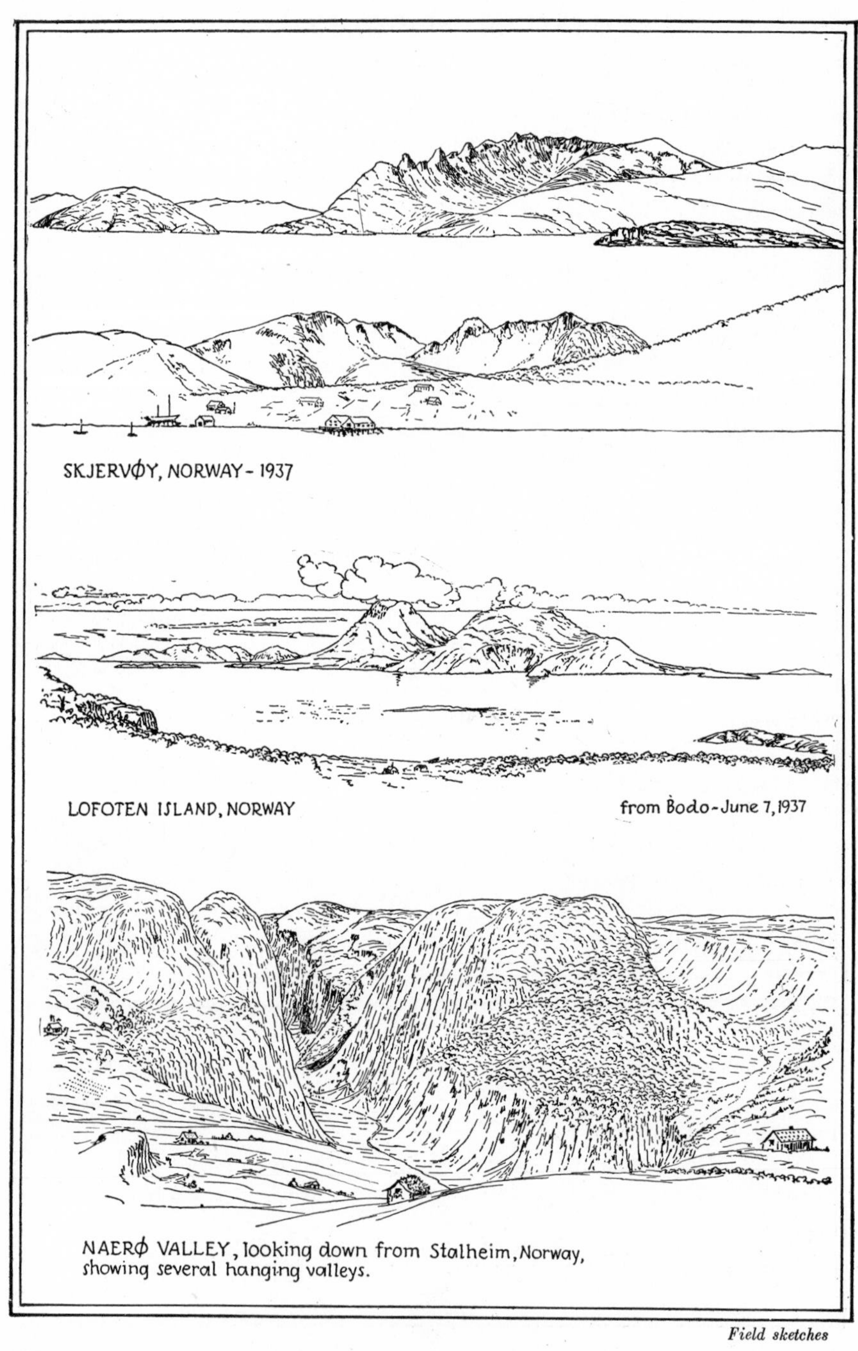


*Field sketches*

THE FIORDS OF NORWAY SHOWING EXAMPLES OF GLACIAL TROUGHS, CIRQUES AND HANGING VALLEYS

## FIORDS AND FINGER LAKES

EROSION BELOW SEA LEVEL. It has often been observed that fiords occur only in high latitudes and never in the tropics. Although some students of fiords have ascribed them to the submergence of valleys by subsidence of the land, the almost universal belief now is that fiords are simply glacial troughs eroded by the ice below sea level. Ice will not float until approximately seven-eighths of its volume has been submerged. This means that a glacier 1,000 feet thick will, upon entering the sea, continue to erode until it has worn its channel down to 800 feet or more below sea level. This explains the great depth of fiords. Moreover, when the varying depth of fiords adjacent to each other is taken into account, the difficulty of explaining them by a general sinking of the land becomes insurmountable. But it can readily be accounted for by glaciers of varying thickness.

THE SHALLOW THRESHOLD OF FIORDS. Fiords usually are deepest some distance above their mouths. This, of course, might be due to morainal accumulations at the end of the glacier. But in most cases this is clearly not the explanation, and the greater depth above the mouth is due to greater glacial erosion. At the very end of the glacier, where it is melting, glacial erosion ceases. The submarine contours of the Norwegian fiords, the Alaskan fiords, and the Chilean fiords all suggest that there has been strong glacial erosion beneath sea level, but that erosion did not extend to the toe of the glacier where it was melting and breaking off to form icebergs. This explains the shallow thresholds of most fiords. Measurements of the Rhone Glacier by Forel showed that its average movement downstream was 350 feet per year at a point some distance above its end, but close to the end it was only 15 feet per year. It is evident that the erosion of the glacial bed is at its maximum some distance upstream from the end of the glacier. The glacial channel must therefore become narrower and shallower as its end is neared. Finger lakes occupying such depressions are quite analogous to fiords. The "lochs" of Scotland are in some instances true fiords, but in other cases they are fresh-water lakes practically at sea level and barely separated from tide water by a low neck of land. The long finger lakes of central and northern Sweden on the eastern side of the Scandinavian Highland correspond to the fiords of Norway on the western side.

In low-latitude regions where glaciation was not active enough to produce rock basins by a sufficient overdeepening of the preglacial valley, as, for example, in the southern Rockies, the White Mountains, the Pyrenees, the Sierra Nevada of Spain, and the Transylvanian Alps, finger lakes are absent. But in the higher ranges and in more northern situations, like the Canadian Rockies, the Alps, and in Scandinavia, finger lakes or fiords are common. This is true also of the southernmost Andes and in New Zealand.

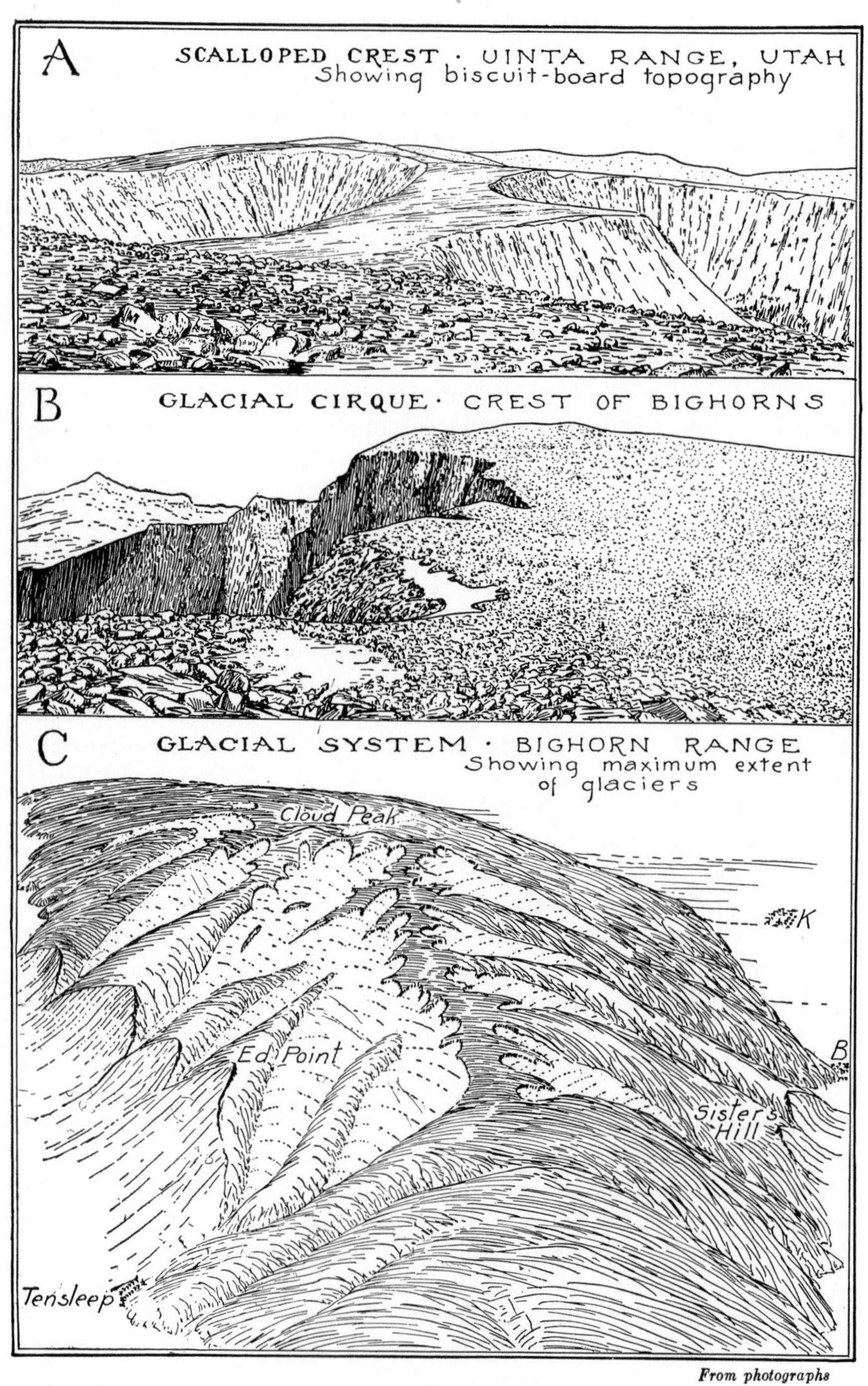


*From photographs*

EXAMPLES OF YOUNG ALPINE GLACIATION ILLUSTRATING BISCUIT-BOARD TOPOGRAPHY

# EXAMPLES OF YOUNG ALPINE GLACIATION

THE UINTA MOUNTAINS. The Uinta Mountains in northern Utah, in spite of their 13,000-foot crests, were not sufficiently glaciated to have their summit level completely obliterated. The biscuit-board topography there is remarkably well displayed (Fig. *A*). As in the Rocky Mountains, the glaciers in only a few instances advanced as much as 10 or 15 miles below the present summits and in no case actually debouched upon the surrounding plains. The divides between the glacial troughs are fairly wide and are rarely sharp crested.

THE BIGHORN MOUNTAINS. The Bighorn Range of eastern Wyoming has the form of a broad upland, seldom rising more than 10,000 feet above sea level (Figs. *B* and *C*). Into this upland numerous cirques and troughs have been cut. The cirques do not intersect each other, nor are the troughs wide enough to cover all the area. Hence a considerable part of the upland has not been obliterated, and the divides are wide and flat. Peaks of the matterhorn type are lacking and the higher summits are still full bodied in outline.

It is estimated that during the period of maximum glaciation in the Bighorn Mountains the amount of snow and ice was only two or three times as abundant as it is now for similar areas in Switzerland. This means only a mild amount of glaciation. The altitude necessary for the generation of glaciers in the Bighorn Mountains seems to have ranged from 9,500 to 11,500 feet, depending upon the exposure, the northern slopes being the more favorable.

The cirques of the Bighorn Mountains are of moderate size, averaging a third of a mile in diameter and having, as a maximum, cliffs 1,500 feet or so in height. The cliff walls are nearly vertical, usually sloping more than 75°. The heads of some of the glaciated valleys have been so little modified as not to have a cirquelike form. Most of the cirques have cleaned and polished rock floors with depressions containing lakes. Some small glaciers still occupy the valleys on the northern slopes. None of the former glaciers of the Bighorn Mountains was sufficient in size to reach beyond the mountain area into the plains (Fig. *C*).

THE COLORADO FRONT RANGE. The crest of the southern Rocky Mountains with an elevation of 10,000 to 12,000 feet, forming the Continental Divide 40 miles or so west of Denver and including Rocky Mountain National Park, is a flat-topped upland, into which remarkable cirques have been cut. Glaciation was most pronounced on the eastern side of the range. The glaciers, like those in the Bighorns, fell far short of reaching the plains. They advanced only 10 to 15 miles down their valleys where they left terminal moraines of striking grandeur.

THE TRANSYLVANIAN ALPS. This southern part of the Carpathian chain has been slightly glaciated, so that cirques occur in the upland. Unlike most of the ranges of the European Alpine system, glaciation never advanced beyond the mountain limit, and the long, beautiful finger lakes of Switzerland and Italy are here absent.

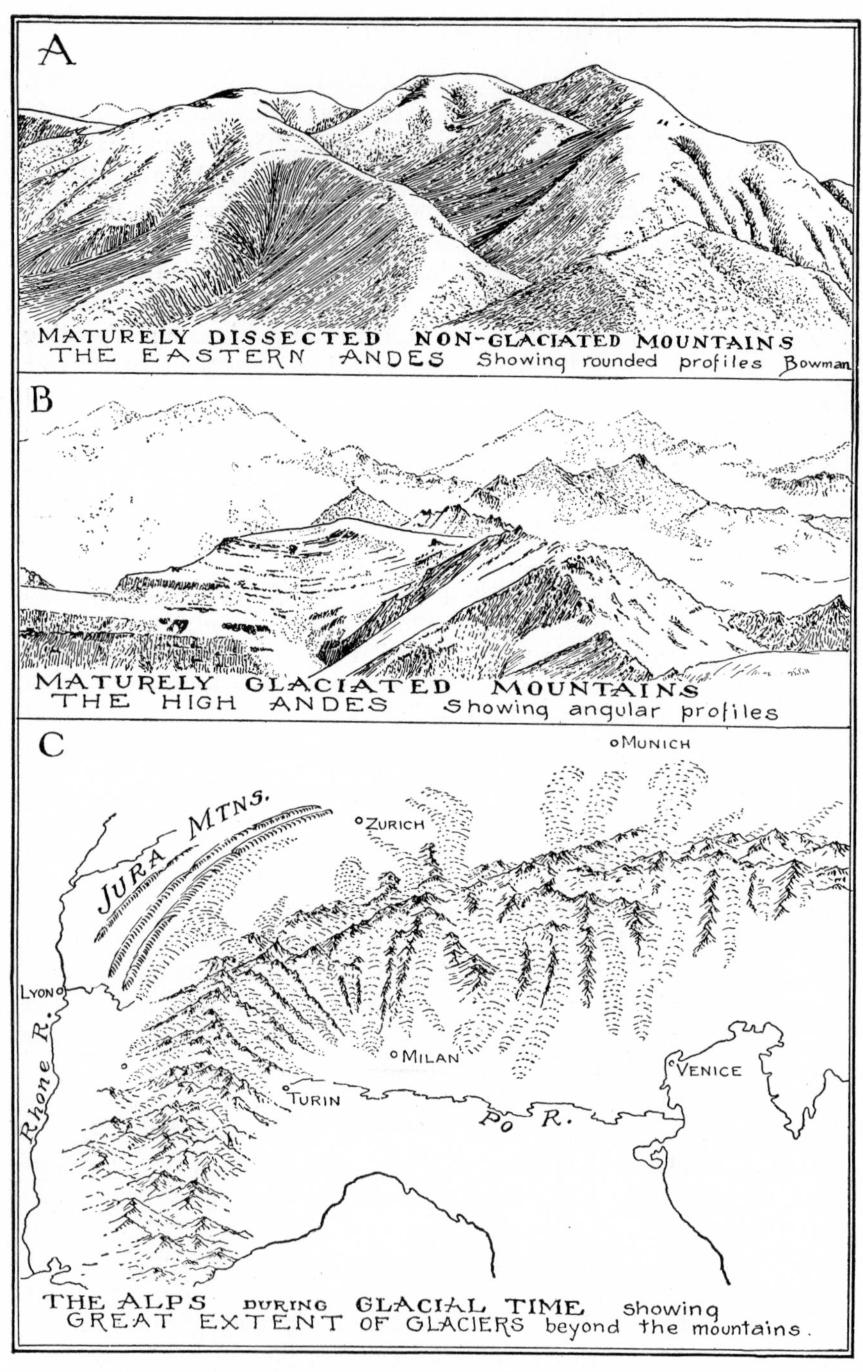


EXAMPLES OF MATURELY GLACIATED MOUNTAINS, CONTRASTED WITH NON-GLACIATED MOUNTAINS

## EXAMPLES OF MATURE ALPINE GLACIATION

Practically all the higher mountain ranges of the world in middle and higher latitudes in both hemispheres have suffered profound alpine glaciation with resulting serrate crests and sharp angular outlines. In the Northern Hemisphere this includes the Cordilleran ranges of Alaska, Canada, and the United States. British Columbia is famous for its alpine scenery and the present glaciers of Alaska are small remnants of those which formerly covered the mountains. The Sierra Nevada, culminating in the serrate crest of the High Sierra, the sharp spires of the Tetons, and that part of the northern Rockies now included in Glacier National Park, display perhaps better than anywhere else in the United States mountains maturely dissected by glacial processes. The high peaks of the Cascades, such as Mount Rainier and Mount Hood, although supporting the longest glaciers in the United States, have not reached the mature stage of dissection. Their initial volcanic form is still too well preserved. None of the ranges of the eastern United States is high enough or far enough north to have supported glaciers of sufficient size, or for a sufficient length of time, to have permitted the development of angular features. In the White Mountains of New Hampshire and Mount Katahdin of Maine the work of glaciers is only a beginning.

In Europe, most of the ranges of the alpine system have a serrate skyline, as in the Alps of Switzerland, Italy, and Austria, the Pyrenees, parts of the Carpathians, the High Tatra, and the Caucasus. The great Keel between Sweden and Norway likewise has a jagged crest line. The fiords on the western side and the Swedish lakes on the eastern side furnish evidence that the glaciers which occupied the region extended far beyond the present limit of the highland area. The large permanent snow fields or icecaps are evidence that active glaciation still exists. The vast extent of former glaciers in the Alps (Fig. *C*) is revealed not only by the long finger lakes of the Swiss and Bavarian Plateaus and the lakes of the northern Italian plain but also by the extensive deposits of morainal material, now far removed from the mountains, but having an alpine origin. As in Norway and in Alaska, glaciation in the Alps is still an active process of destruction. The high mountain ranges of Asia, notably the Himalayas, furnish perhaps the finest examples in the world of maturely glaciated mountains.

In South America the high southern Andes (Fig. *B*) of Chile and the Argentine present an aspect not unlike that of the Canadian Rockies, but the lower eastern Andes (Fig. *A*) have not been glaciated and resemble our southern Appalachians. Deep fiords indent the western side of the southern Andes, and finger lakes reach out eastward on to the plains of Patagonia. The glacier-clad Southern Alps of New Zealand furnish a beautiful example of mountain glaciation with spectacular peaks and steep-walled valleys, waterfalls, lakes, and fiords.

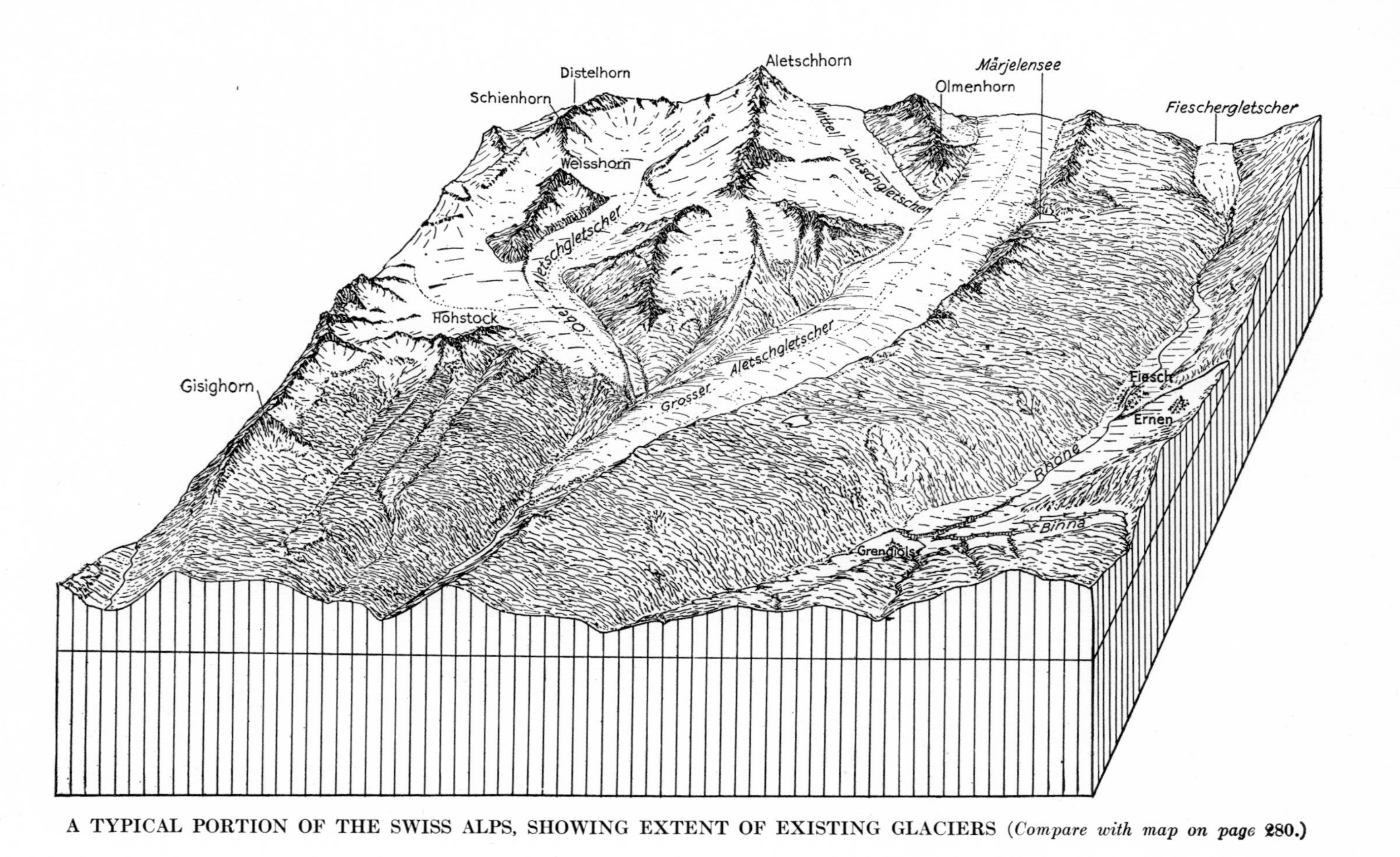


A TYPICAL PORTION OF THE SWISS ALPS, SHOWING EXTENT OF EXISTING GLACIERS *(Compare with map on page 280.)*

## GEOGRAPHICAL ASPECTS OF GLACIATED MOUNTAINS

Mountain regions, whether glaciated or not, are apt to harbor but a sparse population unless they are more or less surrounded by regions of dense settlement. This latter condition is true of the Alps and the Pyrenees. Glaciation, moreover, has rendered the valleys of these mountains fit for habitation by widening them and by causing their floors to become broad flat plains of alluvium. The protected nature of such valleys, both from undue harshness of climate and from inroads of unfriendly neighbors, is another asset of great advantage. These factors early in the history of Europe caused the Alps to become the abode of tribes who found there the conditions which favored, or at least made possible, peaceful pursuits. Above all, these people tended to develop a spirit of independence and self-reliance. The difficulty of moving from one great Alpine valley to another, and the fact that there was no inducement to do so because of the similarity of products and conditions in the different places, caused the different groups to grow up isolated from their neighbors.

In the Alps, dairying is fostered by the splendid grazing conditions which prevail on the high upland meadows or "alps." It is the custom there early in the summer to take the cattle to these high pastures where they remain in the open until winter. The milk is not shipped away but is turned into cheese by the one or two caretakers who live a somewhat lonely life in their little huts where they stay all summer far up on the mountainsides. Wood carving, the making of toys, and other domestic pursuits, especially during the winter months, play a large part in the lives of these people who have so little direct contact with the outside world.

A mountain environment favors the development of many small political groups, each of which inhabits a separate valley. The Republic of Andorra has for a thousand years preserved an independent existence in the protection and isolation of a high valley in the Pyrenees. Many of the cantons of Switzerland coincide roughly with a mountain-rimmed valley.

In their later history, the Swiss valleys have become routes of travel leading from the north to the south of Europe. The need of overcoming the obstacles of transportation imposed by hanging valleys and sheer precipices has caused Swiss engineers to resort to many ingenious combinations of bridges and tunnels. The abundant waterfalls have made possible the electrification of railroads in a region where coal is nonexistent. The lack of material resources, combined with abundance of water power, explains the prominence of countries like Norway and Switzerland in the development of electrochemical industries. The manufacture of nitrates from the air and the smelting of aluminum are examples of industries which require high temperatures that are obtained from electrical sources.

TOE OF GREAT ALETSCH GLACIER, SWITZERLAND
Switzerland; *Aletschgletscher* sheet, No. 493 (1:50,000).

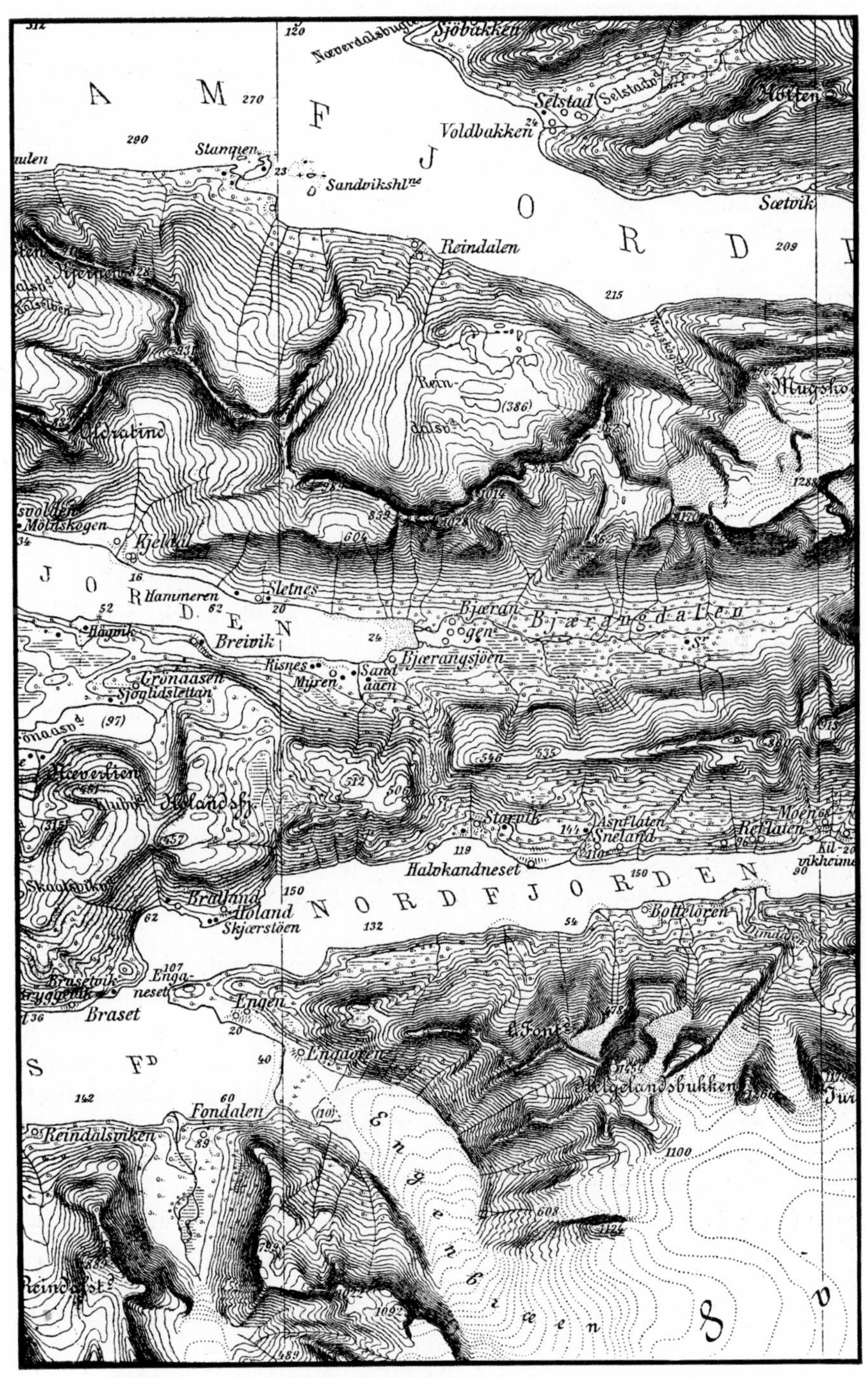


THE FIORD COAST OF NORWAY WITH HANGING GLACIERS
Norway; *Melöy* sheet, No. J 14 (1:100,000).

## MAPS ILLUSTRATING ALPINE GLACIATION

Incipient alpine glaciation resulting in biscuit-board topography is shown on the *Cloud Peak, Wyo.*, and less satisfactorily on the *Hayden Peak, Utah*, sheets. The *Katahdin, Maine*, the *Mt. Washington, N. H.*, and the *Manitou, Colo.*, sheets all show an advanced stage of cirque cutting and arêtes, without the development of sharp matterhorn peaks. They illustrate also the greater amount of glaciation on the northern and eastern sides of the mountain masses, a phenomenon also well shown on the *Central City, Colo.*, map. Cirques, glacial troughs, and hanging valleys are clearly shown on the *Anthracite, Leadville, Montezuma, Georgetown, Minturn*, and *Needle Mts., Colo.*, sheets. These maps show numerous tarns. Glaciation, however, has not reached the advanced stage exhibited farther north in Montana where the *Chief Mountain, Mont.*, sheet displays all features of alpine glaciation in a superb manner. Comparable with this map are several from the Sierra Nevada region, notably the *Mt. Lyell*, the *Mt. Whitney*, the *Mt. Goddard*, and the *Tehipite, Calif.*, sheets.

Terminal moraines of the horseshoe type are particularly well shown on the *Rocky Mountain National Park, Colo.*, sheet, in Glacier Basin, Moraine Park, and Horseshoe Park. On the *Leadville, Colo.*, sheet moraines enclose the lower ends of each of the Twin Lakes, and in a similar way on the *Grand Teton, Wyo.*, the *Snoqualmie, Wash.*, *Hamilton, Mont.*, and the *Custer, Idaho*, sheets moraines hold in several of the small lakes. The *Fremont Peak, Wyo.*, map shows several lakes (*e.g.*, New Fork Lake) divided by moraines.

Hanging valleys with waterfalls are represented on the *Yosemite Valley, Calif.*, map (scale 1:24,000) by Yosemite, Bridal Veil, and Sentinel Falls. This map shows also the flat floor of the glacial trough due to alluvial filling.

Rounded glacial troughs of large size are represented by Crawford Notch on the *Crawford Notch, N. H.*, sheet and by the Tuolumne valley on the *Sonora, Calif.*, sheet.

Actual glaciers are shown on the *Kotsina-Kuskulana District, Alaska*, and the *Chitina, Copper River Region, Alaska*, quadrangles. These maps show the melting ends of the glaciers with lateral, medial, and terminal moraines, outwash deposits, and terraces. The largest glaciers in the United States proper are shown on the *Mt. Rainier National Park, Wash.*, map (scale 1:62,500, contour interval 100 feet). Some of these magnificent glaciers, as, for example, Emmons Glacier, have lateral moraines large enough to be shown by the contours.

Finally, the following maps all show many features of alpine glaciation: *Kintla Lakes, Mont.* (finger lakes); *Lolo, Idaho-Mont.* (cirques in maturely complex mountain region); *Libby, Nyack*, and *Stryker, Mont.*, *Bidwell Bar, Bridgeport*, and *Dardanelles, Calif.* (numerous cirques and other details).

## QUESTIONS

1. When a mountain range, like the White Mountains, has been entirely covered by the continental ice sheet and has also been occupied by alpine glaciers, how is it possible to determine which came first?
2. Why do matterhorn peaks not occur in all mountains that have had alpine glaciers?
3. How would two periods of alpine glaciation be recognized in a mountain region?
4. In what part of the United States have tongues from the continental ice sheet produced features like those resulting from alpine glaciation?
5. Why are there more and larger cirques on the eastern side of Mount Washington than on the western side?
6. The profile of the divide between two opposing cirques is a hyperbola. Why is this?
7. Draw a contour map of a hanging valley.
8. Discuss the validity of the statement that maturely glaciated mountain regions closely resemble each other even though they differ widely in structure. Name some examples to bear this out.
9. Do you think it possible for one glacial system to capture part of another, the way streams do?
10. Can a glacier become overloaded with debris and be forced to deposit it, or does it carry its entire load until it melts?
11. In what ways does a glacier carry its load? Is much dragged along the bottom?
12. What causes a glacier to advance and retreat at intervals? How fast does a glacier move?
13. Do you think little pockets of snow which occupy many hollows in the hills in high latitudes late into the spring can produce cirquelike features? What is such a process called?
14. Do glaciers ever block up tributary valleys to form marginal lakes?
15. Can glaciers transport their load uphill?
16. What is meant by incipient glaciation?
17. Is it practicable to speak of young, mature, and old stages of glaciation? Do these terms refer to the valleys or to the land form as a whole?
18. What national parks in the United States exhibit features due to alpine glaciation?
19. What is a catenary curve? Name several examples. Is this form represented in any natural bridges? Explain.
20. Who first proposed the idea of "hanging" or "hung-up" valleys and where was this observation made?
21. What is meant by *nivation?*
22. Discuss the various kinds of avalanches.

## TOPICS FOR INVESTIGATION

1. Hanging valleys: their form, height, postglacial modification, and theories as to origin.
2. The glacial cycle. Comparison with stream cycle.
3. Cause for asymmetry of mountains in glaciated regions. Where is this to be seen?
4. The mechanics of ice flow. Rapidity of glacial advance and retreat.
5. Comparison of glacial features in various mountain regions as to the following: size of glaciers; stage of glaciation reached; moraines; evidence of multiple glaciation.
6. The origin of fiords.
7. Former extent of alpine glaciers in the mountains of the United States.
8. Correlation of periods of glaciation in the Basin ranges with the stages of Lake Bonneville.
9. The mechanics of glacier motion.

## REFERENCES


CYCLE OF GLACIATION

DAVIS, W. M. (1906) *The sculpture of mountains by glaciers.* Scot. Geog. Mag., vol. 22, p. 76–89; Geog. Essays, p. 617–634.

HOBBS, W. H. (1921) *Studies of the cycle of glaciation.* Jour. Geol., vol. 29, p. 370–386.

JOHNSON, W. D. (1904) *The profile of maturity in Alpine glacial erosion.* Jour. Geol., vol. 12, p. 569–578.

HANGING VALLEYS

BRANNER, J. C. (1903) *A topographic feature of the hanging valleys of the Yosemite.* Jour. Geol., vol. 11, p. 547–553.

CROSBY, W. O. (1903) *The hanging valleys of Georgetown, Colorado.* Am. Geol., vol. 32, p. 42–48.

GARWOOD, E. J. (1902) *On the origin of some hanging valleys in the Alps and Himalayas.* Geol. Soc. London, Quart. Jour., vol. 58, p. 703–718.

JOHNSON, D. W. (1909) *Hanging valleys.* Am. Geog. Soc., Bull. 41, p. 665–683.

JOHNSON, D. W. (1911) *Hanging valleys of the Yosemite.* Am. Geog. Soc., Bull. 43, p. 826–837, 890–903.

FEATURES OF GLACIATED REGIONS

ALLIX, A. (1924) *Avalanches.* Geog. Rev., vol. 14, p. 517–560.

GARWOOD, E. J. (1910) *Features of Alpine scenery due to glacial protection.* Geog. Jour., vol. 36, p. 310–339.

GILBERT, G. K. (1904) *Systematic asymmetry of crest lines in the high Sierra of California.* Jour. Geol., vol. 12, p. 579–588.

GILBERT, G. K. (1906) *Crescentic gouges on glaciated surfaces.* Geol. Soc. Am., Bull. 17, p. 303–316.

GREGORY, J. W. (1913) *The nature and origin of fiords.* London, 542 p.

HOBBS, W. H. (1911) *Characteristics of existing glaciers.* New York, 289 p. See p. 1–96.

MCGEE, W J (1894) *Glacial canyons.* Jour. Geol., vol. 2, p. 350–364.

RAY, L. L. (1935) *Some minor features of valley glaciers and valley glaciation.* Jour. Geol., vol. 43, p. 297–322.

GLACIER MOTION

AITKIN, J., MAIN, J. F., ET AL. *Glacier motion.* Am. Jour. Sci., 3d ser., vol. 5, p. 305–308, 1873; *ibid.*, vol. 34, p. 149, 1887; Nature, vol. 39, p. 203, 1888.

CASE, E. C. (1895) *Experiments in ice motion.* Jour. Geol., vol. 3, p. 918–934.

CHAMBERLIN, R. T. (1928) *Instrumental work on the nature of glacier motion.* Jour. Geol., vol. 36, p. 1–30.

GILBERT, G. K. (1906) *Moulin work under glaciers.* Geol. Soc. Am., Bull. 17, p. 317–320.

REID, H. F. (1896) *The mechanics of glaciers.* Jour. Geol., vol. 4, p. 912–928.

REID, H. F. (1895–1916) *Variations of glaciers.* Intern. Comm. on Glaciers, Jour. Geol., vol. 3, p. 278–288, 1895, and annually in the same Journal until 1916; Archives des Sciences Physiques et Naturelles, Geneva, vol. 2, p. 129–147, 1896, and annually up to 1905; Zeitschrift für Gletscherkunde, vol. 1, 1907, p. 161–181, and annually until 1914.

RUSSELL, I. C. (1895) *The influence of debris on the flow of glaciers.* Jour. Geol., vol. 3, p. 823–832.

VON ENGELN, O. D. (1938) *Glacial geomorphology and glacial motion.* Am. Jour. Sci., vol. 35, p. 426–440.

LOCALITIES DESCRIBED

ATWOOD, W. W. (1909) *Glaciation of the Uinta and Wasatch Mountains.* U. S. Geol. Surv., Prof. Paper 61, 96 p.

ATWOOD, W. W., JR. (1935) *The glacial history of an extinct volcano, Crater Lake National Park.* Jour. Geol., vol. 43, p. 142–168.

BLACKWELDER, E. (1931) *Pleistocene glaciation in the Sierra Nevada and Basin Ranges.* Geol. Soc. Am., Bull. 42, p. 865–922. Lateral moraines.

BOYD, L. A. (1935) *The fiord region of east Greenland.* Am. Geog. Soc., Special Publ. 18, 369 p.

CAMPBELL, M. R. (1914) *The Glacier National Park; a popular guide to its geology and scenery.* U. S. Geol. Surv., Bull. 600, 54 p.

DAVIS, W. M. (1900) *Glacial erosion in France, Switzerland, and Norway.* Boston Soc. Nat. Hist., Proc. 29, p. 273–322; Geog. Essays, p. 635–689.

DAVIS, W. M. (1909) *Glacial erosion in north Wales.* Geol. Soc. London, Quart. Jour., vol. 65, p. 281–350.

DAVIS, W. M. (1916) *The Mission Range, Montana.* Geog. Rev., vol. 2, p. 267–288.

FRYXELL, F. (1935) *Glaciers of the Grand Teton National Park of Wyoming.* Jour. Geol., vol. 43, p. 381–397.

GANNETT, H. (1898) *Lake Chelan, Washington.* Natl. Geog. Mag., vol. 9, p. 417–428.

GOLDTHWAIT, J. W. (1913) *Glacial cirques near Mt. Washington.* Am. Jour. Sci., 4th ser., vol. 35, p. 1–19.

HENDERSON, J. (1910) *Extinct and existing glaciers of Colorado.* Univ. Colo. Studies, vol. 8, p. 33–76.

JOHNSON, D. W. (1933) *Date of local glaciation in the White Mountains.* Am. Jour. Sci., 5th ser., vol. 25, p. 399–405.

KERR, F. A. (1936) *Quaternary glaciation in the Coast Range, northern British Columbia and Alaska.* Jour. Geol., vol. 44, p. 681–700.

MARTIN, L., and WILLIAMS, F. E. (1924) *An ice-eroded fiord. The mode of origin of Lynn Canal, Alaska.* Geog. Rev., vol. 14, p. 576–596.

MATTHES, F. E. (1900) *Glacial sculpture of the Bighorn Mountains, Wyoming.* U. S. Geol. Surv., 21st Ann. Rept., part 2, p. 167–190.

MATTHES, F. E. (1922) *The story of Yosemite Valley.* U. S. Natl. Park Service, Making of American scenery, No. 1, 4 p.

ODELL, N. E. (1933) *The mountains of northern Labrador.* Geog. Jour., vol. 82, p. 193–210, 315–325.

PEACOCK, M. A. (1935) *Fiord-land of British Columbia.* Geol. Soc. Am., Bull. 46, p. 633–696. Importance of joints in controlling glaciation.

REID, H. F. (1924) *Antarctic glaciers.* Geog. Rev., vol. 14, p. 603–614.

RICH, J. L. (1906) *Local glaciation in the Catskill Mountains.* Jour. Geol., vol. 14, p. 113–121.

RUSSELL, I. C. (1893) *Malaspina Glacier, Alaska.* Jour. Geol., vol. 1, p. 219–245.

RUSSELL, I. C. (1897) *Glaciers of North America.* Boston, 210 p.

RUSSELL, I. C. (1898) *Glaciers of Mount Rainier.* U. S. Geol. Surv., 18th Ann. Rept., part 2, p. 349–415.

RUSSELL, R. J. (1933) *Alpine land forms of the western United States.* Geol. Soc. Am., Bull. 44, p. 927–950.

SALISBURY, R. D. (1906) *Glacial geology of the Bighorn Mountains.* U. S. Geol. Surv., Prof. Paper 51, p. 71–90. See also Folios 141 and 142.

TARR, R. S., and MARTIN, L. (1914) *Alaskan glacier studies of the National Geographic Society in the Yakutat Bay, Prince William Sound and lower Copper River regions.* Washington, 498 p.

TARR, R. S. (1900) *Glaciation of Mount Katahdin, Maine.* Geol. Soc. Am., Bull. 11, p. 433–448.

TYNDALL, J. (1896) *The glaciers of the Alps.* New York, 442 p.

WENTWORTH, C. K., and RAY, L. L. (1936) *Studies of certain Alaskan glaciers in* 1931. Geol. Soc. Am., Bull. 47, p. 879–934.

WOOSTER, L. C. (1920) *Glacial moraines in the vicinity of Estes Park, Colorado.* Kan. Acad. Sci., Trans. 29, p. 91–94.

YAMASAKI, N. (1922) *Glaciation of the mountains of Japan.* Am. Jour. Sci., 5th ser., vol. 3, p. 131–137.

For *maps* showing former extent of mountain glaciation in America, see the following U. S. Geol. Surv. and other maps:

ALDEN: Geol. Soc. Am., Bull. 24, Pl. 13, op. p. 529.
ATWOOD: Prof. Paper 61, Pl. 4, 10.
ATWOOD: Jour. Geol., vol. 20, p. 390–398, Figs. 1–4.
BALL: Prof. Paper 63, Pl. 4 and 5.
BASTIN and BLACKWELDER: Prof. Paper 51, Pl. 28.
BASTIN and BLACKWELDER: Folios 141 and 142.
CALHOUN: Prof. Paper 50, Pl. 1.
CAPPS: Bull. 386, Pl. 1.
CROSS and HOWE: Folio 153.
HOLE: Jour. Geol., vol. 20, Pl. 1, p. 502.
LINDGREN: Folios 31 and 39.
RANSOME: Prof. Paper 75, Pl. 1.
RUSSELL: 8th Ann. Rept., pt. 1, Pl. 29.
RUSSELL: 20th Ann. Rept., pt. 2, Pl. 18.
TARR and MARTIN: Alaskan Glacier Studies, Map 1.
WEED: Bull. 104, Pl. 1.
WILLIS: Prof. Paper 19, Pl. 8.

GEOGRAPHICAL ASPECTS

ATWOOD, C. K., and ATWOOD, W. W., JR. (1937) *Land utilization in a glaciated mountain range (the Park Range).* Econ. Geog., vol. 13, p. 365–378.
BLANCHARD, R. (1921) *The natural regions of the French Alps.* Geog. Rev., vol. 11, p. 31–49.
HARSHBERGER, J. W. (1919) *Alpine fell-fields of eastern North America.* Geog. Rev., vol. 7, p. 233–255.

# IX
# CONTINENTAL GLACIATION

*Hillers, U. S. Geological Survey*

A BANK OF GLACIAL TILL
Showing beds of clayey till alternating with beds of bouldery till. Bottom beds of fluvio-glacial gravels.

*Hillers, U. S. Geological Survey*

A GLACIAL LIMESTONE BOULDER ABOUT NATURAL SIZE FROM NEAR NORWAY, IOWA
Showing flattened and polished surface with several sets of intersecting striae.

GIGANTIC GLACIAL GROOVES, FLATHEAD NATIONAL FOREST, MONT.
Produced by one of the lobes of the Cordilleran ice sheet.

*Northern Pacific Railway and U. S. Forest Service*

*Stose, U. S. Geological Survey*

THE TERMINAL MORAINE NEAR FRANKLIN FURNACE, N. J.
Showing hill and kettle topography, and in the foreground a lake ponded by the moraine.



THE GREAT ESKER AT PUNKAHARJU, FINLAND
With deltas and other outwash material in the Finnish lake country.

CREST LINE OF THE GREAT PUNKAHARJU ESKER, FINLAND
The summit of the esker varies from lake level to a height of 100 feet.

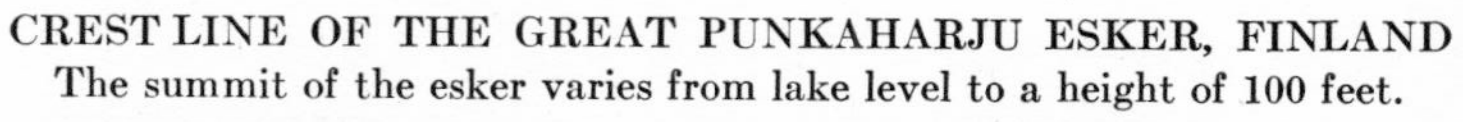

ESKER NEAR WHOLDAIA LAKE, NORTHWEST TERRITORY, CANADA

Showing also some of the vast area covered by glacial lakes; morainal topography in the foreground.

*Royal Canadian Air Force*

*Gilbert, U. S. Geological Survey*

DRUMLINS NEAR NEWARK, N. Y., LAKE ONTARIO REGION

Showing somewhat steeper ice-pushed ends at the right.



GLACIAL OUTWASH, NORTH SHORE OF LONG ISLAND, N. Y.

The dark layer is glacial till deposited when the ice overran the outwash previously laid down. Second outwash layer on top formed during retreat.

*A. K. Lobeck*

*A. K. Lobeck*

TERMINAL MORAINE AT HACKETTSTOWN, N. J.
Showing hill and kettle topography with small erratics. A region used mainly for grazing and woodland.



GLACIAL ERRATIC NEAR NORTH SHORE OF LONG ISLAND, N. Y.
A very large mass of crystalline rock transported from New England by the continental ice sheet.

*A. K. Lobeck*

*U. S. Forest Service*

GLACIAL OUTWASH OR VALLEY TRAIN COVERING VALLEY FLOOR NEAR WINONA, MINN.

Now being dissected to form terraces.



TERMINAL MORAINE BLOCKING THE END OF FLATHEAD LAKE, MONT.

Formed by a tongue of the Cordilleran ice sheet.

*U. S. Forest Service*

## CONTINENTAL GLACIATION

SYNOPSIS. The topographic effect of continental glaciation is due not so much to the action of the ice itself upon the land as to the influence it has had in disturbing, deterring, and otherwise influencing the drainage systems of streams.

*The erosive* work of continental ice sheets is topographically almost negligible. Because Canada is an old worn-down mountain region and because the ice sheet recently passed over the area, it should not be concluded that the ice sheet carried away the mountains. Polished surfaces, the smoothing of rôches moutonnées, glacial grooves, and striae constitute the visible effect of ice abrasion. However, vast quantities of residual soil were picked up by the ice and transported long distances. Where confined in narrow channels and even in wider valleys like those now occupied by the Great Lakes, glacial erosion was undoubtedly of great magnitude, resembling the erosive activity of alpine glaciers.

*The deposits* left by continental ice sheets are topographically inconspicuous. The moraines, for example, are infinitely smaller than those left by alpine glaciers, being rarely 100 feet high, as compared with deposits of the latter type which may be ten times as great in size. On the other hand, the deposits of continental ice sheets are very widespread and vary greatly in age and character. The oldest till plains are much dissected and are devoid of lakes. The youngest are almost intact and on them lakes abound.

The surface patterns of glacial deposits are most intricate and reveal the various forms of the glacial lobes at different times. The internal structure of glacial deposits indicates the many advances and retreats and serves as the chief means of learning about the duration of interglacial periods.

The variety of accumulations known as *till, drumlins, kames, eskers,* and *outwash,* their geographical distribution, as well as their composition, all are highly significant lines of evidence which must be studied not separately but as related to each other in order to understand properly the events and conditions of glacial time.

*The cause* of the glacial period remains almost as great an enigma as ever, except perhaps that one or two ill-founded theories have been permanently discarded.

The relative recency of the glacial period is a fact not always appreciated, nor is it always realized that glacial conditions did not preclude the occupation of a region by man and animals up to the very border of the ice sheet.

Finally, the conditions which brought about glaciation in higher latitudes had far-reaching effects in all other parts of the world as well. More humid conditions prevailed in what are now deserts, where there is evidence of many former lakes. The lowering of sea level because of glaciation is another matter of great significance, especially in relation to the growth of coral reefs.

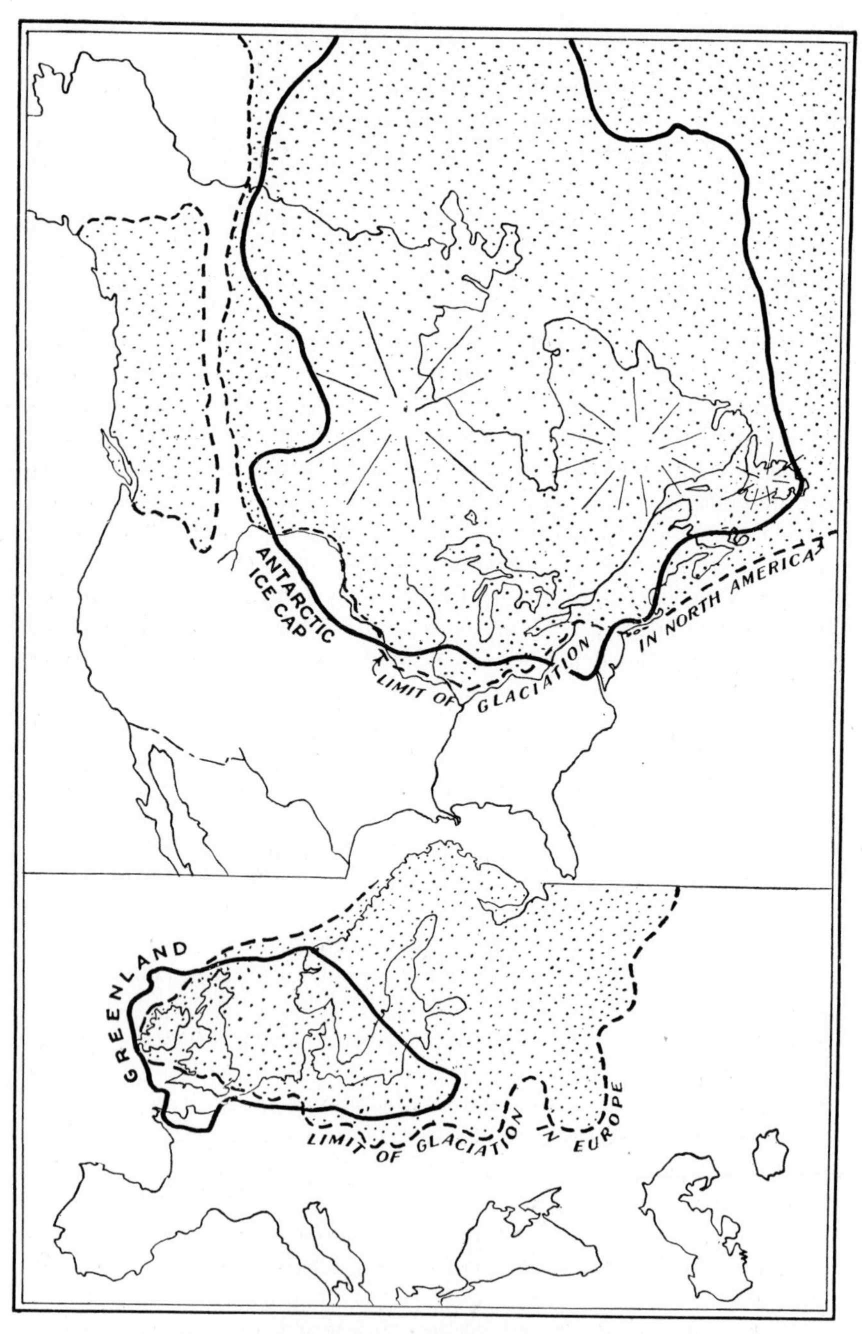


NORTH AMERICA AND EUROPE IN GLACIAL TIME
Continental ice sheets stippled. Present outlines of Antarctic and Greenland ice caps superposed for comparison.

# CONTINENTAL GLACIATION, I

General Characteristics. Continental glaciers erode, transport, and deposit material. To understand better how they do this, geologists have studied the icecaps of Greenland and Antarctica, the only present examples of great ice sheets. This study, however, has not explained the cause of glaciation but it has enabled us to visualize more accurately the thickness of an ice sheet, its rate of motion, its manner of eroding and transporting material, and the nature of its deposits.

Causes of Continental Glaciation. Widespread glaciation occurred several times during the geologic past. Extensive emergence of land areas brings about great extremes of climatic conditions. Oceans and other water bodies introduce uniformity. Early geological glaciation appears to be related to large land areas. However, the last great glaciation did not affect Siberia and northern Asia, the greatest land area on the globe and the one endowed with the most extreme climate; whereas in Europe glaciation came down to the sea and in many places glacial deposits and marine deposits are intercalated.

Elevation of the land brings about extremes of temperature and also abundant precipitation. This may account for Alpine glaciation but not for the great continental ice sheets of Europe and North America which accumulated in the low-lying country rather than the high.

Change in the direction and volume of ocean currents might easily modify the climate and has been suggested as a contributing factor in bringing about glaciation in Europe. There is, however, no proof that any such change occurred.

Certain atmospheric causes have also been suggested. Increase of carbon dioxide in the atmosphere produces milder climates because gas, like water vapor, acts as a blanket and prevents the radiation of heat from the earth. It has been suggested that extensive seas cause an increase of carbon dioxide because they permit the deposition of large volumes of limestone through organic activity and in doing so release large amounts of carbon dioxide to the atmosphere. The carbon dioxide in the water is the chief factor which enables the water to retain lime in solution and when the lime is removed the carbon dioxide is released. On the other hand, extensive land areas with their consequent vegetation cause a depletion of the carbon dioxide in the atmosphere. Thus extensive seas favor mild climates, and extensive land areas favor extremes.

Excess of volcanic ash in the atmosphere effectively prevents the sun's rays from warming the earth and temperatures have been distinctly reduced on that account. Several astronomic causes such as variability in the heat radiated by the sun, the precession of the equinoxes, and the displacement of the poles have been thought possible. At best these are only hypotheses and the true cause of glaciation appears still to be unknown.

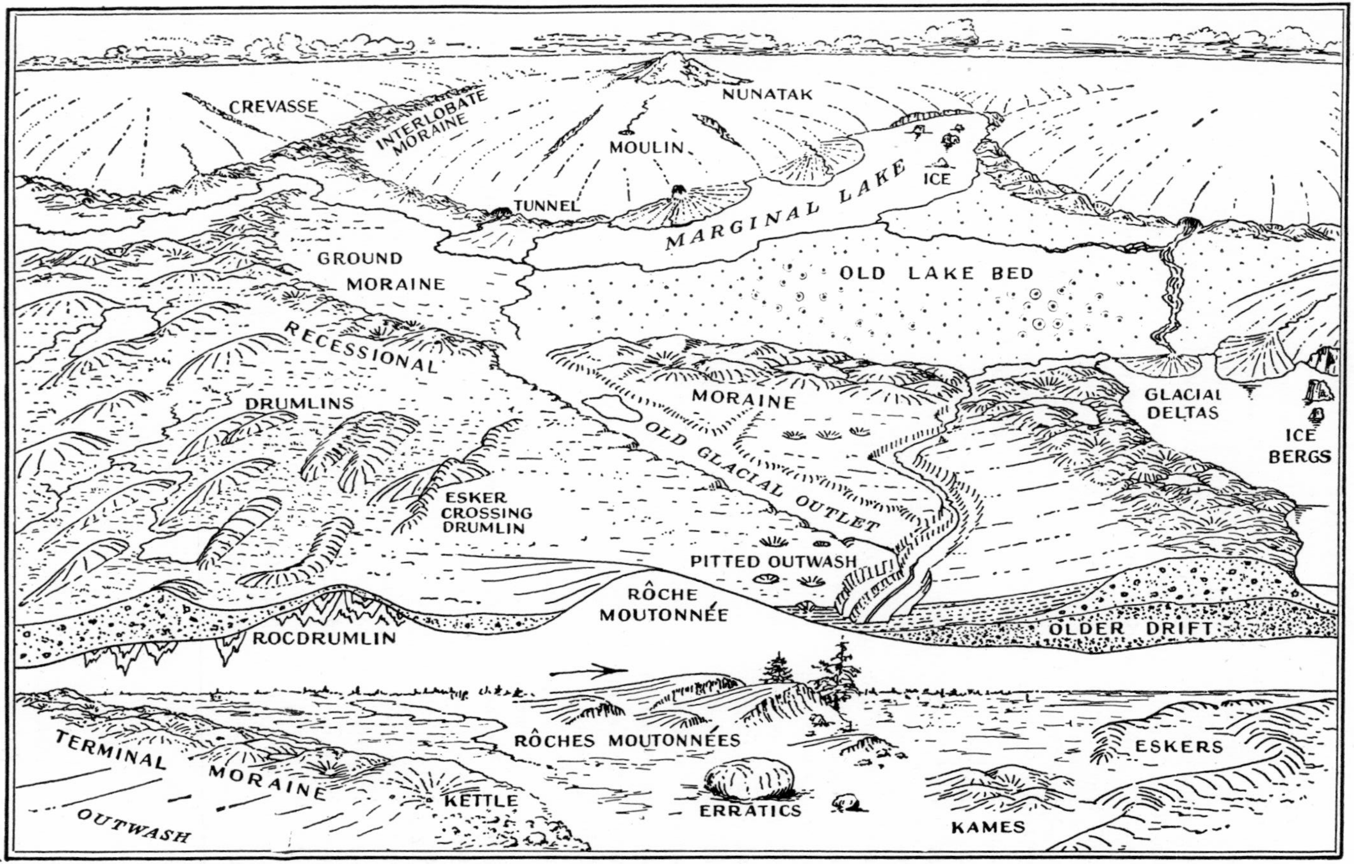

DIAGRAMS SHOWING VARIOUS TYPES OF GLACIAL DEPOSITS LEFT BY CONTINENTAL GLACIERS

# CONTINENTAL GLACIATION, II

Form and Behavior of Continental Glaciers. From our knowledge of the present icecaps of Greenland and Antarctica and from the inferences we can make concerning the ice sheets of North America and Europe, we know that a continental glacier is a low, almost flat-topped dome, probably not over 2 miles high and occupied most of the time by a permanent high-pressure area. As a result of this high pressure the winds blow outward, carrying snow toward the periphery of the ice mass. Here alone is there sufficient slope to cause movement; the degree of slope has little relation to the configuration of the underlying basement. This slope estimated from the terminal moraines in Wisconsin and Montana, as well as in the eastern United States, appears to be between 50 and 300 feet per mile. Icecaps spread radially partly because of the radiating wind movements. Precipitation results from centers of low pressure, which from time to time invade the heart of the region, and is most abundant along the margins.

Cracks and crevasses cannot extend to the bottom of a thick ice sheet and therefore the development of cirques and cliffs by sapping cannot occur. A great icecap smoothes a country, removes the preexisting soil, grinds the surface of the bedrock, and leaves deposits at the line of its farthest advance and also back of this line during its retreat by melting. It does not necessarily retain its forward movement until it disappears but may, through lack of nourishment, lapse into an immobile mass.

Icecaps constantly change their shape and extent, being guided by relatively small topographic features. In broad lowlands the ice front advances in tongues and lobes. When a balance between supply and melting is attained, movement stops and a terminal moraine is built up, but not along the whole ice front at any one time. Successive advances and retreats do not always coincide with previous ones, and this is reflected in the moraine pattern. Nowhere does a terminal moraine extend very far without being intersected by a later one. During melting, large portions of an icecap are isolated and die motionless. Great quantities of water-borne gravel, sand, and clay are deposited along the ice margin, down the valleys leading from the ice front, beneath the ice, as well as in the numerous pockets and hollows on top of its ragged surface. These form the *kames*, *eskers*, and various types of *outwash* which together are classed as *fluvioglacial* material.

The final withdrawal of the ice sheet leaves a country covered with lakes, marshes, bogs, badly disturbed drainage lines, as well as valleys and channelways temporarily used to carry off the abundant water. Some of the lakes fill rock basins scoured out by the ice but probably most of them are due to the blocking of valleys by glacial deposits. These are all in addition to the temporary marginal lakes which were held in by the ice itself.

The last great continental glaciation extended over a period of probably 1,000,000 years and was made up of several stages, both in Europe and in America. The interglacial stages were probably longer than the time which has elapsed since the withdrawal of the ice some 25,000 to 50,000 years ago.

The great volume of the continental ice sheets caused a lowering of sea level almost 300 feet throughout the world. Their weight caused a temporary subsidence of the earth's crust. Since the disappearance of the ice the preglacial surface is being restored by elevation or tilting. Areas to the north which had been most depressed have risen higher than those to the south which were less weighted down.

About 6,000,000 square miles of the earth's surface is still covered by ice. This is equivalent to twice the area of the United States and equal to one-half the total area covered by ice during the last glacial invasion. If this were melted the level of the ocean would be raised about 150 feet. Whether this will actually happen it is impossible to ascertain from present-day observations, but it is not unlikely in view of the fact that during several earlier geological periods the polar regions were far more temperate than they now are. Since the earliest earth history glaciation has occurred from time to time only to give way to milder conditions.

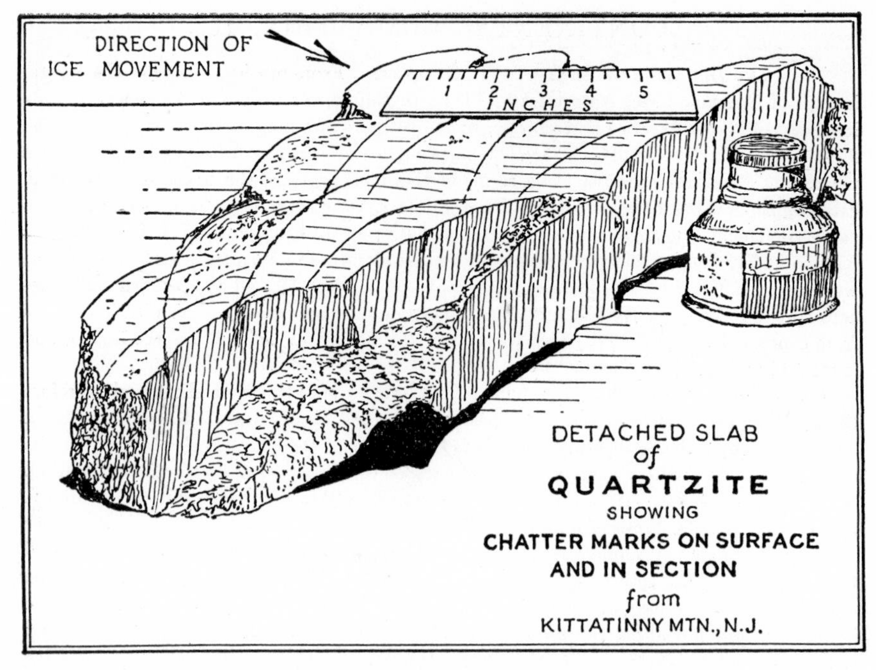


## GLACIAL EROSION

It is commonly believed that the continental ice sheet was a stupendous engine of erosion; that it stripped off the soil and wore away the rocks from the Canadian area and left this material in the form of thick deposits of drift in the United States. To some extent this is true, but it must be clear from what has already been said that only near its margin was the continental ice sheet in active motion. Of course the whole country which underlay the ice sheet was at one time or another under this moving marginal part of the ice and then only was it subjected to erosion.

That ice erosion did occur is shown by the presence of *glacial grooves, glacial striae;* polished rock surfaces with *chatter marks;* the absence of residual soil; glacial drift resting with a sharp contact upon fresh bedrock; an abundance of drift greater than the probable amount of preglacial residual soil; the presence in this drift of much fresh rock, both as boulders and as ground-up rock flour; rock basins in a position and of a form which could not have been produced by stream action; rounded hills known as *rôches moutonnées* with gentle slopes on the *stoss* side, which received the impact of the advancing ice, and steep declivities on the *leeward* side due to plucking.

Striae range in size from mere scratches to grooves a foot or more in depth. They may be formed on any rock surface but are preserved in the more massive rocks. Within the limits of New York City large glacial grooves cut in the Manhattan schist may be seen in many places, as, for example, along Riverside Drive and in Central Park. They are also to be seen at numerous localities on top of the Palisades, and elsewhere

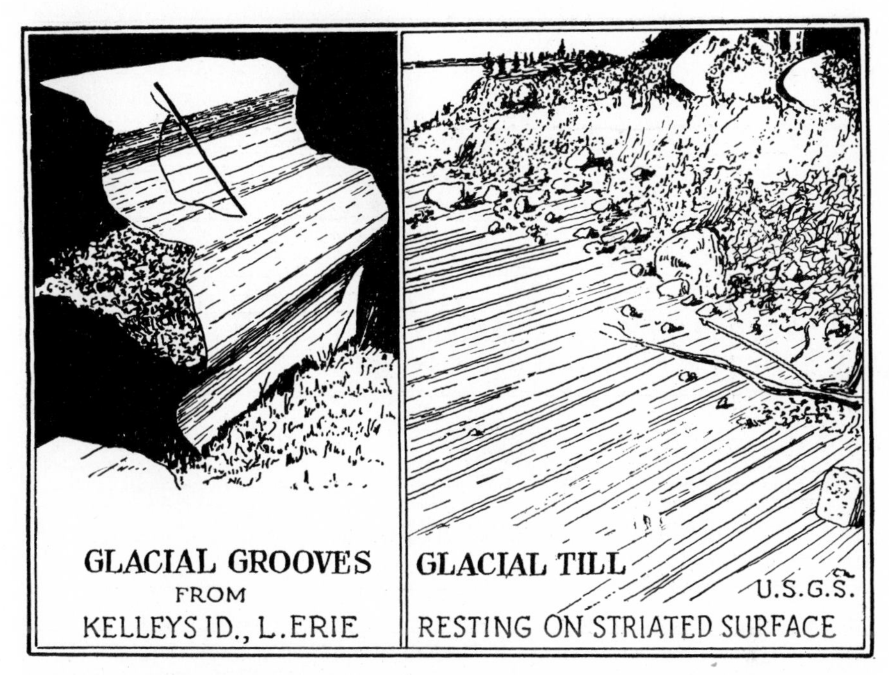

GLACIAL GROOVES FROM KELLEYS ID., L. ERIE

GLACIAL TILL RESTING ON STRIATED SURFACE U.S.G.S.

in the eastern United States they are very common phenomena. Rock surfaces polished like glass are characteristic of quartzites and conglomerates and may be seen on top of Shawangunk Mountain and Kittatinny Mountain where a massive conglomerate occurs. Here, too, chatter marks may be noted. These are crescent cracks, the horns of the crescent pointing in the direction of ice movement and away from the ice mass. They are formed by pressure, the crack resulting from tension. The principle is precisely the same as that used by Indians in prying off flakes from pieces of flint to make their arrowheads. *Crescentic gouges* pointing backward are also known and they can always be explained by the application of pressure opposite to that which produces chatter marks. *Faceted stones* (somewhat resembling *dreikanter* formed by wind action) occasionally occur in glacial drift and these often bear scratches as do also many of the boulders found in the drift.

Tongues of the continental ice sheet confined in narrow valleys behaved much like alpine glaciers in deepening the valley. This probably occurred in the Finger Lake region of New York and certainly in the gorge of the Hudson through the Highlands. There it was necessary to go 1,000 feet below the river to find bedrock in which to place the Catskill Aqueduct.

With evidence of this sort it is fair to conclude that the ice sheet acted like a mill, grinding the rock to the finest powder and producing glacial milk like that now to be observed pouring forth from Alpine glaciers. This very fine material, partly colloidal in character, makes much of the glacial lake deposits in which the minute laminae or *varves* are to be observed.

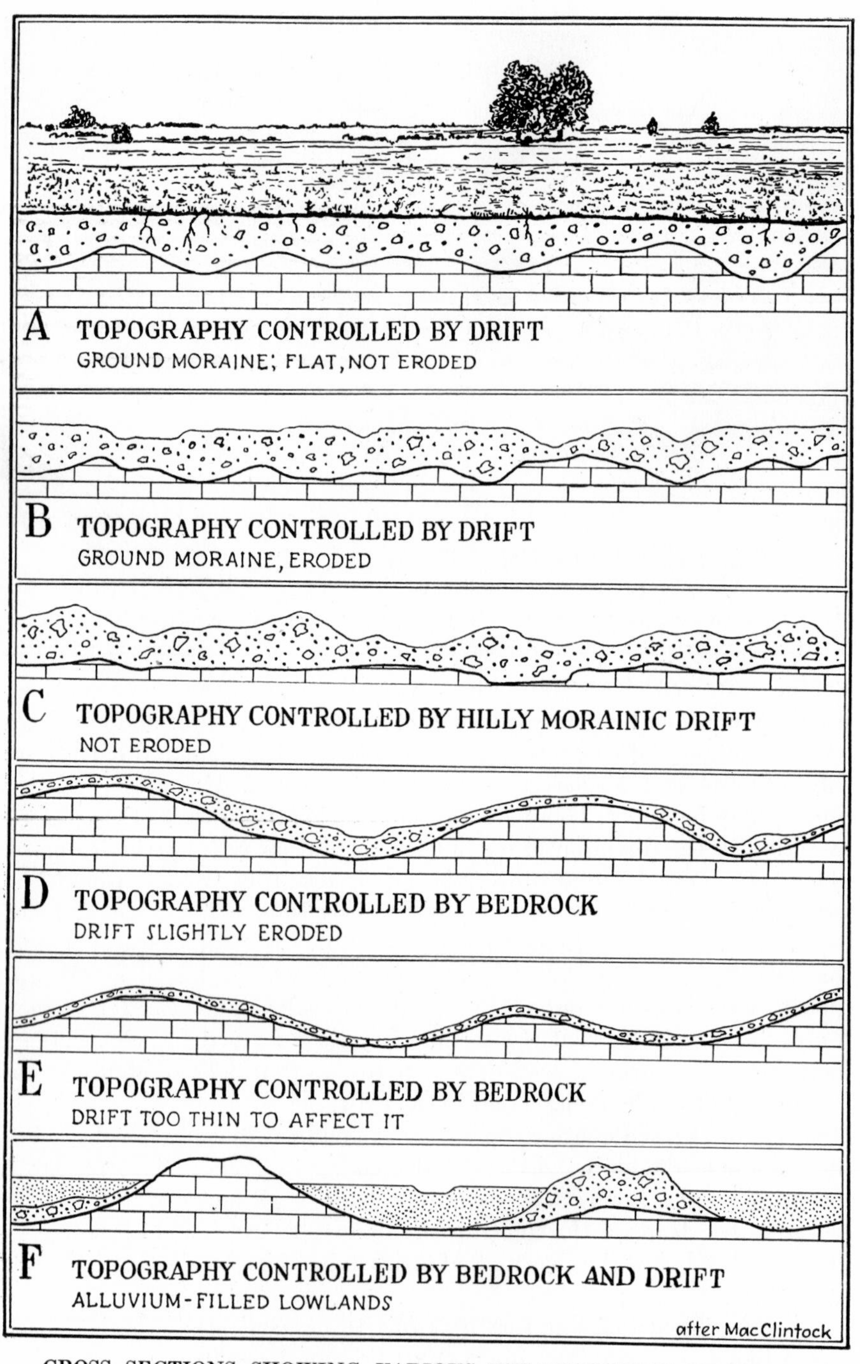


CROSS SECTIONS SHOWING VARIOUS RELATIONSHIPS BETWEEN BEDROCK AND GLACIAL DRIFT

## GLACIAL DEPOSITION: MORAINES

Moraines are designated according to their position with relation to the ice sheet as *terminal moraines, recessional moraines, interlobate moraines,* and *ground moraines.* Or they are designated according to the *material composing them,* as *till moraines, waterlaid moraines, delta moraines,* and *kame moraines.*

A *terminal,* or *end, moraine* is one formed at the outermost stand of a continental glacier. Moraines deposited during times when the ice border was stationary during the retreat of a moving icecap are called *recessional.* Most terminal moraines consist mainly of till, but as terminal moraines are laid down at the margin of the ice where melting is at its maximum, there is also much assorted or water-borne material.

MORAINAL TOPOGRAPHY. Terminal moraines grade from simple smooth ridges with very low slopes to the most complex aggregation of knobs and ridges interspersed with enclosed *kettles* or *pits.* This is sometimes termed *knob-and-basin* topography. The local relief within the moraine may exceed 100 feet and the total thickness several hundred feet. Moraines composed mainly of clay have low slopes and few kettles. Stony moraines have steep slopes, abundant kettles with lakes, ponds, and marshes; they present their steepest side toward the ice. In clay moraines the reverse is true; they are more gullied than stony ones.

Terminal moraines deposited on land include material deposited by running water originally in the form of alluvial fans and deltas, or as irregular fillings among ice blocks and drift hills. Hills of this material, usually conical in form and consisting of poorly stratified sand, clay, and gravel, are called *kames.* A moraine made up largely of such hills is termed a *kame moraine.* Moraines may also be deposited in deep water and have associated with them extensive delta deposits, whence the term *delta moraine.*

*Recessional moraines* may be formed at the time of a slight readvance of the retreating ice or when it halts temporarily during its retreat. Glacial geologists discriminate between the two kinds by the character of their deposits.

*Interlobate moraines* are formed in the angle between the margins of two distinct glacial lobes, a particularly favorable place for the formation of water-laid deposits. Such moraines include many kames and much outwash.

Terminal moraines frequently overlie older ones or rest upon outwash laid down before the advancing ice sheet. Outwash deposited during the melting back of the ice may also cover the moraine just formed.

*Ground moraine* includes the miscellaneous material left covering the region formerly occupied by the ice. If thick enough to cover the bedrock, it forms *drift plains* or *till plains;* if thin, it constitutes veneered hills and does not much affect the topography. Extremely flat till plains, largely of clay with almost no boulders, represent seas of glacial mud which flowed out over preexisting topography of probably slight relief.

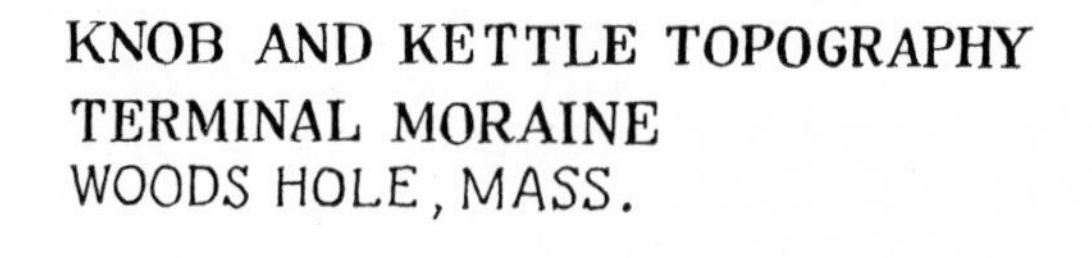

*From photographs by Douglas Johnson*

MORAINES

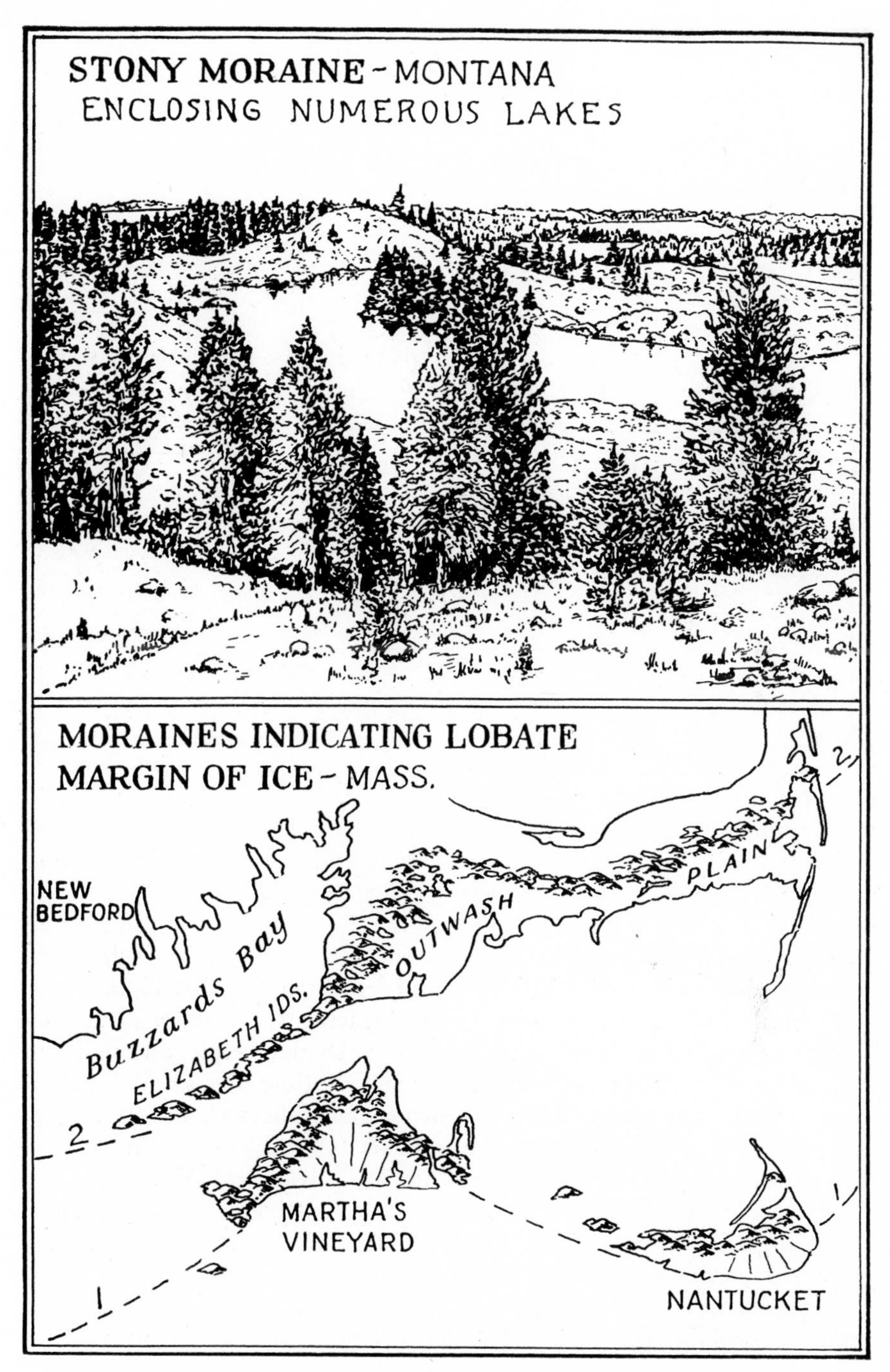
STONY MORAINE - MONTANA
ENCLOSING NUMEROUS LAKES
MORAINES INDICATING LOBATE
MARGIN OF ICE - MASS.
NEW
BEDFORD
Buzzards Bay
ELIZABETH IDS.
OUTWASH
PLAIN
MARTHA'S
VINEYARD
NANTUCKET
2
2
1
1

MORAINES

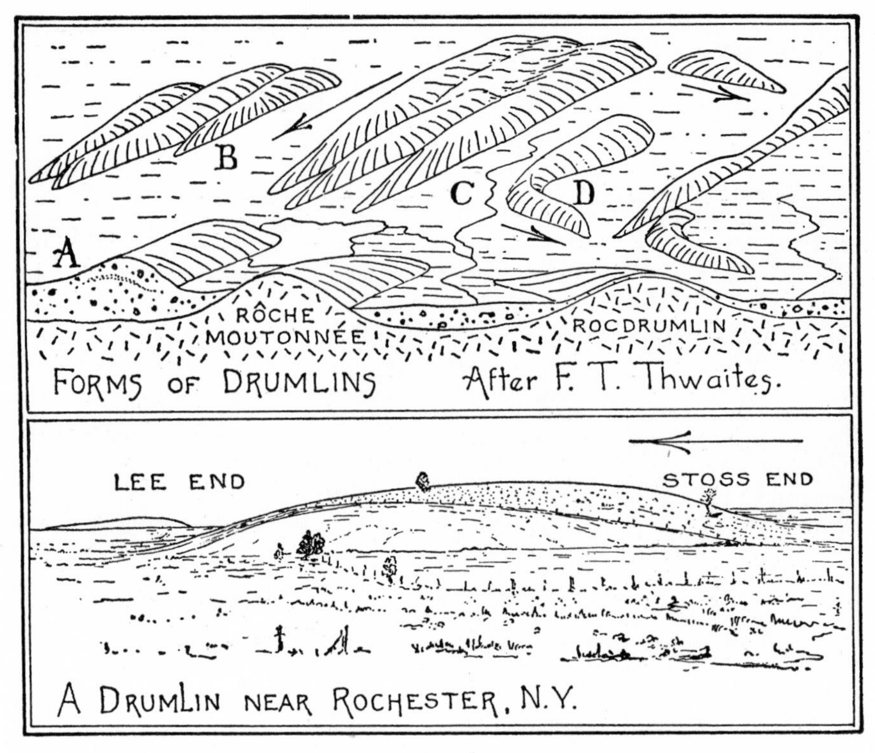


## GLACIAL DEPOSITION: DRUMLINS

Forms of Drumlins. Drumlins are smooth, oval hills composed mainly of till but sometimes including lenslike masses of gravel and sand (*A*). They have their long axes parallel to the direction of ice movement and usually occur in "swarms," more or less radiating in plan as if deposited by a lobe of ice moving outward from an axis. Most drumlins are less than half a mile in length and less than 100 feet high. Much larger ones are known. Their *stoss end*, facing the glacier, is usually blunter and steeper than the tail or *lee side*. Double, triple, and multiple drumlins occur together in all positions and relationships, small drumlins forming tails upon larger ones or benches alongside of them (*B*). Aggregations of drumlins form drumlin uplands (*C*). Some drumlinlike hills have rock cores with a thin veneer of till. These are called *rocdrumlins*.

Successive ice movements do not destroy previous drumlins but add to them, building tails and appendages often at an angle different from that of the original drumlins (*D*).

Origin of Drumlins. Several theories have been suggested for the origin of drumlins, one being that they are morainal deposits overridden

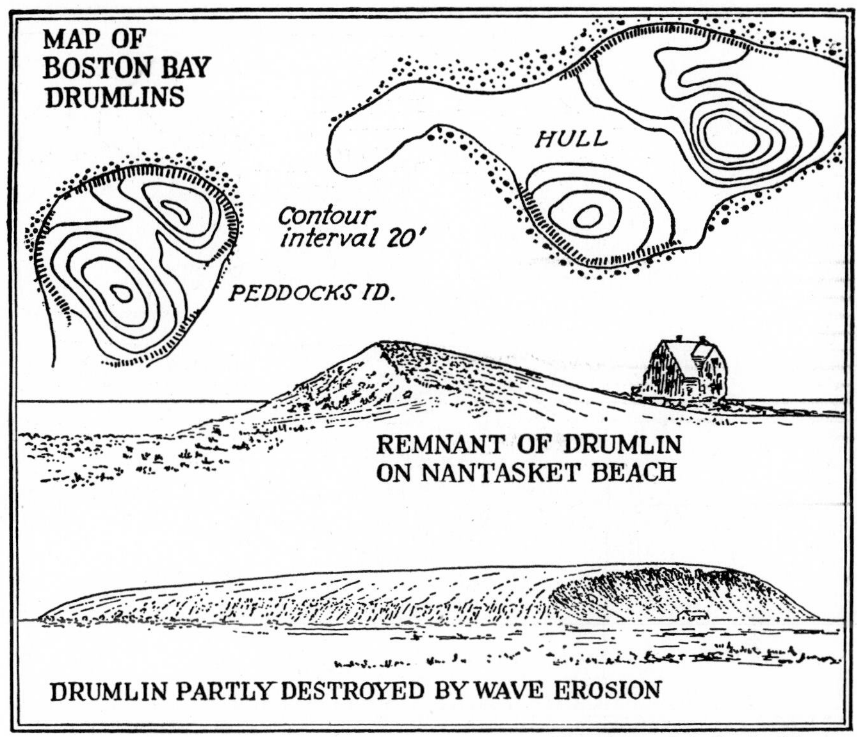


*From photographs by Douglas Johnson*

by later advances of the ice. On the other hand, most of the evidence suggests that drumlins are original deposits beneath heavily loaded ice, their arrangement being related to radiating fissures in the ice lobe through which much of the debris worked downward to the bottom of the ice sheet. Davis suggested that they are analogous to sand bars in a stream.

Occurrence of Drumlins. There are several well-known drumlin localities in the United States where drumlins are unusually abundant. In south-eastern Wisconsin there is a great swarm covering several counties of the state. Hundreds of them may be seen between Madison and Milwaukee, most of them cultivated and occupied by farms. Bunker Hill near Boston is a drumlin. And in Boston Bay most of the islands are of this character. In the outer harbor a group of six or eight drumlin islands has been dissected by the waves and tied together by bars to form the complex tombolo known as *Nantasket Beach.* Another large drumlin swarm occurs along the southern shore of Lake Ontario in northern New York and still others occur in northern Michigan. Rarely are drumlins prominent features in country of strong relief.

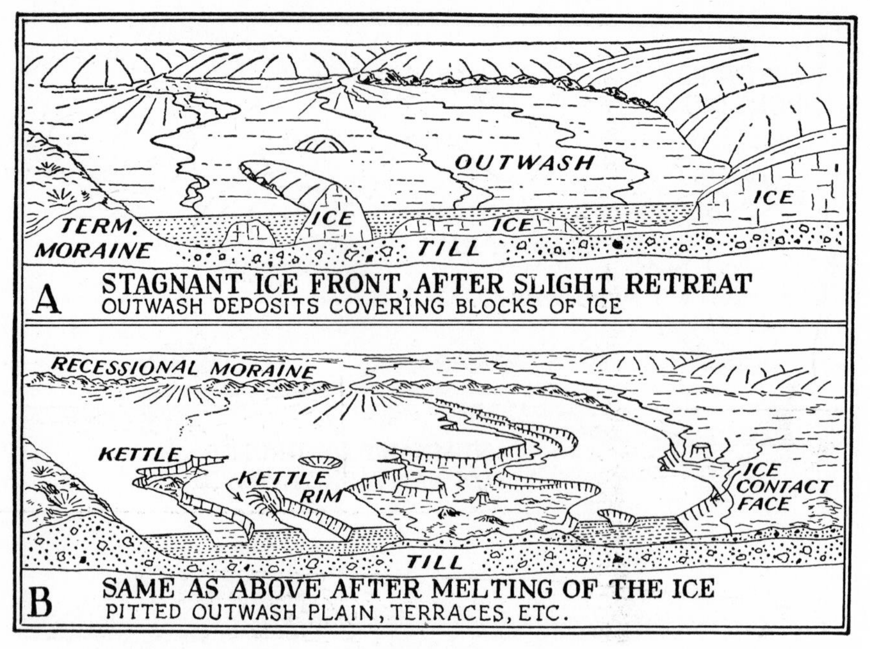


DIAGRAMS SHOWING DEVELOPMENT OF PITTED OUTWASH PLAIN

## FLUVIOGLACIAL DEPOSITS, I

*Outwash Plains and Related Features.* Streams emerging from melting glaciers carry gravel, sand, and silt. If the slope of the land away from the ice is sufficient, the streams continue to carry this load and do so until their velocity is checked. In most regions the deposition of the coarsest material takes place as soon as the streams emerge from the ice, but the finest material may be carried downstream many miles. The deposits thus formed constitute *outwash plains, alluvial fans, valley trains,* and *delta plains.*

Where outwash plains form narrow fillings in the bottoms of preexisting valleys, they are called *valley trains* and may extend scores of miles from the ice front into nonglaciated country. Elsewhere outwash plains may form coalescing *alluvial fans* which bury the entire country for many miles. The individual units of such fans head at breaks or low points in the terminal moraine, the slope of the plain being greatest near the moraine.

*Pitted outwash plains* are interrupted by *kettles* or pits, many of which contain lakes or ponds. These depressions vary from a few feet to many miles across and may be extremely irregular in shape. Some plains have very few pits, widely scattered. Others have so many that no vestige of the plain remains and the region has the aspect of a terminal moraine. Confused knobs, sags, and ridges occur, and in many of the larger kettles portions of the underlying drift topography, such as drumlins, eskers, ground moraine, or terminal moraine, are visible.

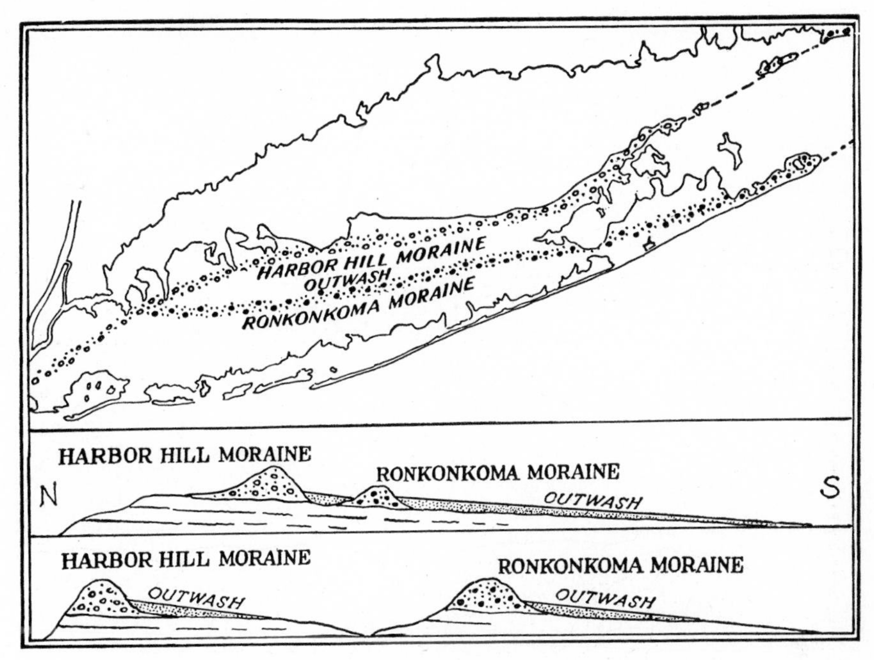


MORAINES AND OUTWASH PLAINS ON LONG ISLAND

Pitted outwash forms in places where outwash has been deposited on top of a more or less continuous sheet of stagnant ice or around isolated blocks of ice. In the latter case the debris contained in the block of ice is deposited as a *kettle* rim of till and boulders. *Kettle chains* or rows of kettles occur where ice has been buried in old drainage lines.

Some outwash plains, deposited during the advance of a growing ice sheet, have been overridden by the ice and covered with till. When this happens, the outwash in places shows much disturbance in its bedding, the laminae being folded and faulted and ploughed to pieces.

Some outwash plains do not slope gradually away from their sources but end abruptly in a steep face. These are delta plains, deposited in preglacial lakes. The water level of such lakes is marked by the break in slope between the flat top and the forward-facing slope.

Many outwash plains and valley trains now consist largely of steps or terraces formed by later stream erosion. This change of behavior from deposition to erosion may be due to one of several causes but perhaps most commonly to the fact that, with the retreat of the ice front, the streams crossing the outwash plains become less heavily loaded and consequently regain their erosive power. Increased gradient is another cause, the increase of gradient being due to the cutting away of barriers either of drift or rock farther downstream. When it is realized that outwash may be deposited not only in front of the ice but alongside of the ice mass and around great detached blocks of melting ice and under a wide variety of conditions, it is easy to see that terraces may result, not alone from erosion, but as initial features of deposition also.

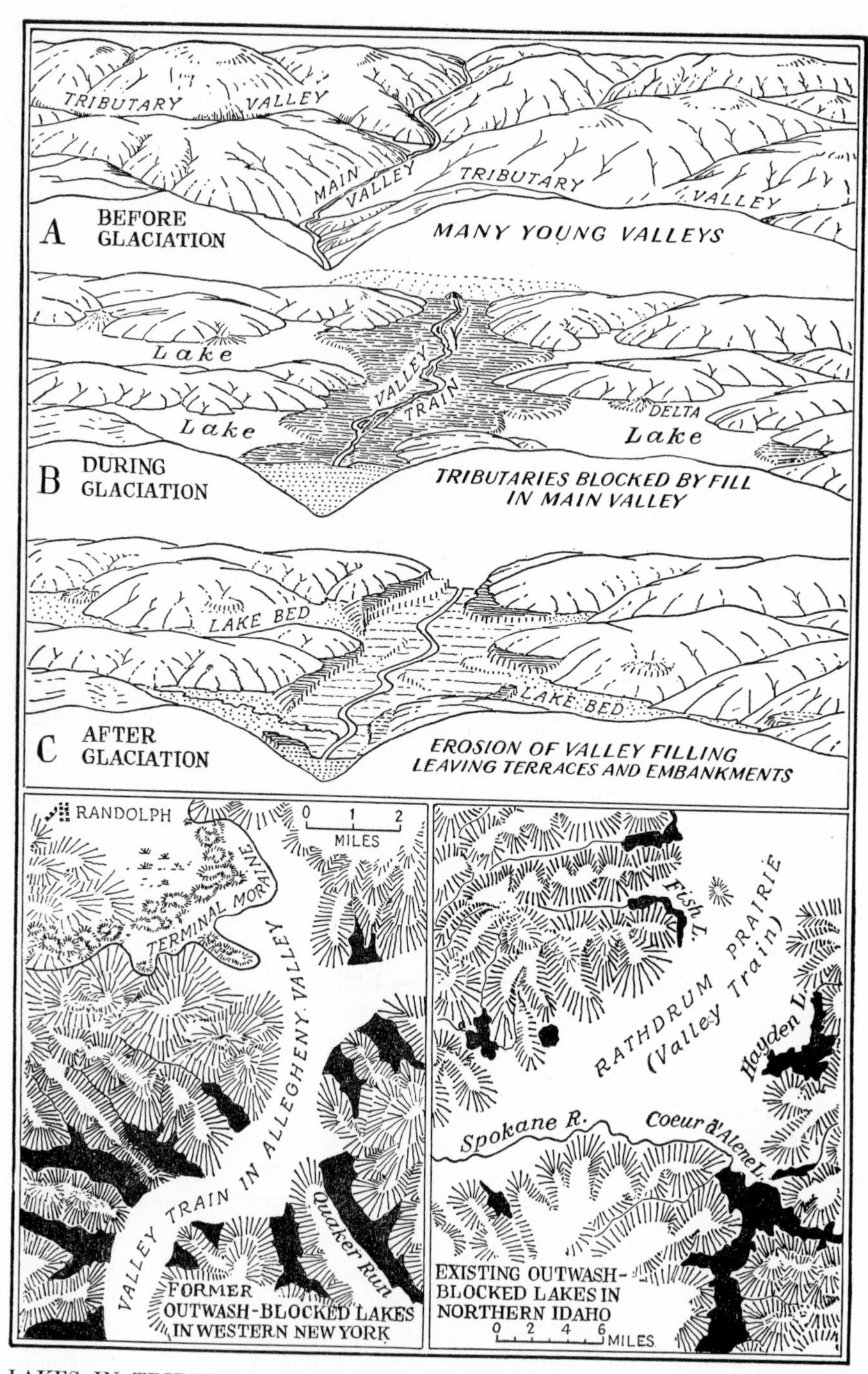


LAKES IN TRIBUTARY VALLEYS DAMMED BY VALLEY TRAIN IN MAIN VALLEY

# FLUVIOGLACIAL DEPOSITS, II

VALLEY TRAINS. Valley trains in hilly regions occasionally extend down the larger valleys far beyond the ice front and block the tributary streams to cause lakes.

In western New York State during glacial time tremendous quantities of outwash were poured into the Allegheny River, filling it with alluvium in places 300 to 400 feet. As the Allegheny filled its valley with gravel derived from the melting ice sheet, it constantly elevated its flood plain, and its tributaries became blocked. At first they deposited small amounts of sand and gravel in their channels but, not originating in glaciated country, their loads of debris were very small and they could not build up their beds so fast as the Allegheny did. Consequently, they soon ceased to flow as streams and became lakes. At the heads of the lakes where the streams entered them, small deltas were formed.

Practically every tributary valley was thus blocked. The resulting lakes are shown on the accompanying map. Since glacial time a large amount of the valley filling of the Allegheny has been removed, but remnants still exist along the valley sides to form terraces.

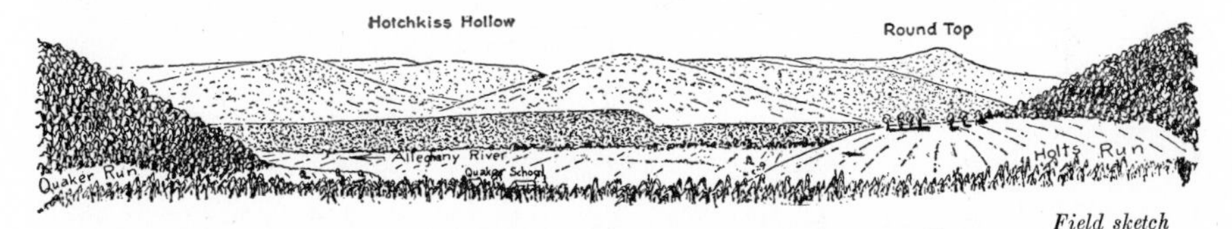


*Field sketch*

REMNANTS OF VALLEY TRAIN FORMING HIGH TERRACES ALONG THE ALLEGHENY RIVER IN WESTERN NEW YORK

This is by no means a unique case. In Idaho the outwash waters from the Cordilleran ice tongues so filled the headwaters of the Spokane River as to block many of the tributaries. This accounts for the lakes of that region, notably Coeur d'Alene, Hayden, and other smaller ones. The lakes of Idaho, unlike those in the Allegheny region, still exist. The gravel benches stretching from spur to spur, back of which these lakes lie enclosed, look like artificial embankments when seen from a distance. In one case the slope of the gravel bench or embankment into the tributary valley has the form of delta lobes, presumably built by small distributaries which flowed gently from the main aggrading ice-water river into the valley embayment.

Another interesting locality is in southern Illinois where the tributaries of the Ohio, far outside the region of glaciation, were blocked by alluvial accumulations in the Ohio River. The master-stream deposits of the Ohio grew more rapidly than those of the tributaries and the tributaries were thus dammed to form lakes. Similar lakes now extinct are known to have existed also in Kentucky due to the blocking of northward flowing tributaries of the Ohio River.

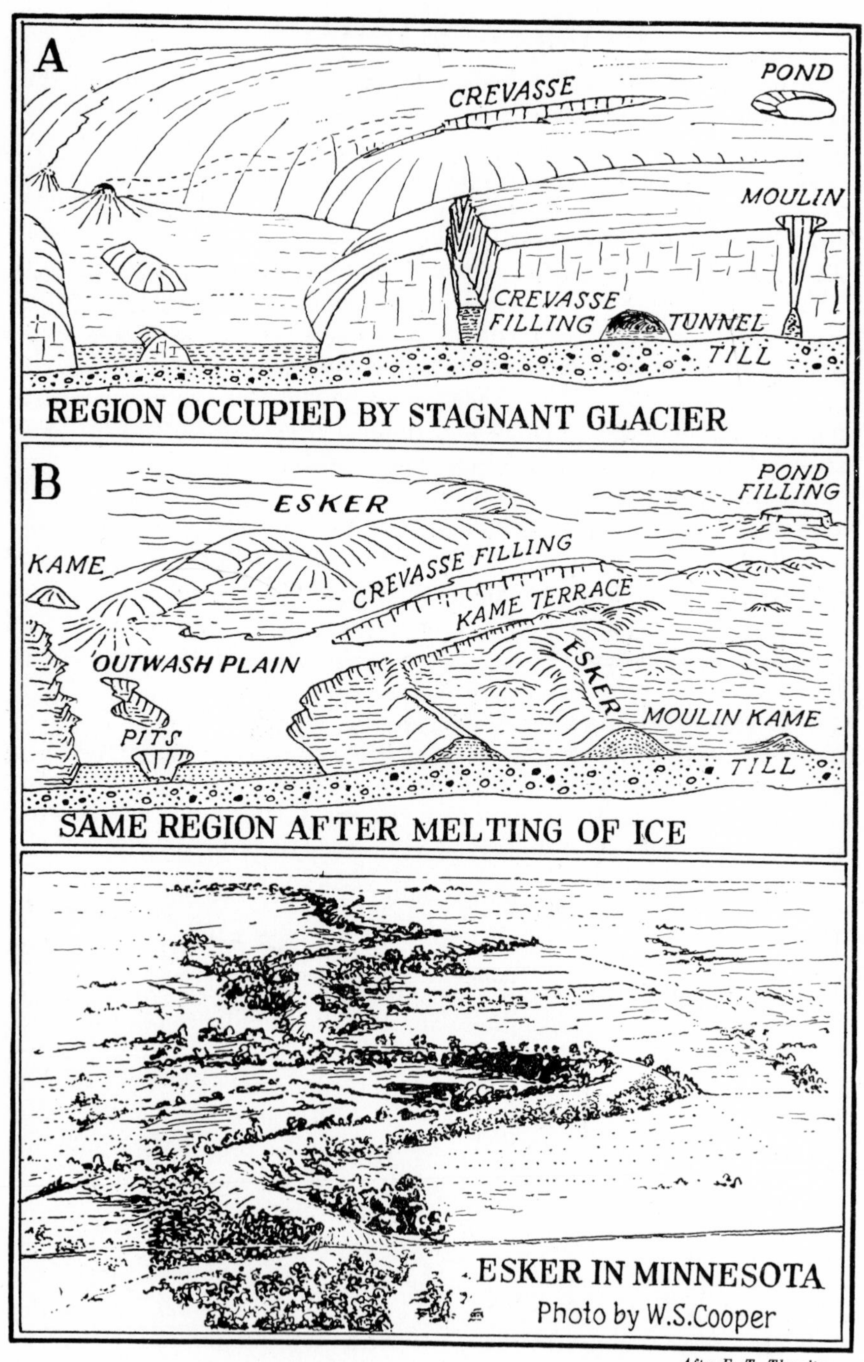


*After F. T. Thwaites*

THE ORIGIN OF ESKERS AND KAMES

# FLUVIOGLACIAL DEPOSITS, III

ESKERS, KAMES, AND CREVASSE FILLING. An *esker* (called also *Os*, plural *Osar*, from the Swedish) is a ridge deposited by a glacial stream in an ice tunnel. A *kame* is a more or less conical hill, usually of gravel or sand, deposited as a small delta cone or in a depression along the ice front or in a crack or hole within the ice border. A *crevasse filling* is a ridge of water-sorted material running in almost any direction, often associated with outwash or lake terraces, and deposited in a large crevasse.

Eskers and crevasse fillings cannot always be distinguished from each other. Both are narrow and steep sided, the side slopes being that of the angle of repose of sand or gravel, which is about 30°. Crevasse fillings are rarely over 1 mile long. Eskers, however, in series separated by relatively short gaps are known to extend for 150 miles. In height they range up to 150 feet. Most eskers are winding. They often branch and reunite like braided streams and are then termed *reticulate*.

Eskers are most common on low, swampy plains. The ground on one or both sides of an esker may form a pronounced depression, termed an *esker trough*. Eskers lie on all kinds of surfaces. They disregard the underlying topography and cross over or ascend hills several hundred feet high. Most eskers seem to have been buried by recessional moraines and not infrequently by outwash. They often lead into deltas.

The gravel (and some sand) composing an esker is but little sorted and usually very coarse, poorly rounded, and stony, with large open spaces due to too little sand to fill the voids. The layers of gravel are inclined in various directions, but never upstream. Owing to the slumping of the gravel beds when the ice walls melted, they dip away on either side from the axis of the esker.

Crevasse fillings usually contain layers of sand and silt, especially where associated with lake deposits. Transitional between eskers and crevasse fillings are isolated cones or kames deposited at the bottoms of *moulins* (mills) where water running over the ice plunges downward to the base of the ice sheet.

That eskers have been formed in tunnels under the ice and under hydrostatic pressure is now an accepted explanation of the fact that they go up and down over the irregularities of the underlying surface. Deposition in such tunnels on both up- and downgrades has been proved experimentally.

Eskers are known throughout the Mississippi Valley, in Wisconsin, Minnesota, Michigan, and Illinois. In the eastern states, owing to the far greater ruggedness of the country, they are less common. They are known, however, in central New York and in northern New Jersey. They are striking features at the head of Penobscot Bay, Maine. There they run for 15 miles or more as ridges 40 feet high above the adjacent tamarack swamps. Their branching habit suggests strongly the original stream pattern. They are known as "horsebacks," and roads follow the crests of most of them.

# THE GLACIAL PERIOD IN NORTH AMERICA AND EUROPE

Centers of Dispersion. The North American continental ice sheet moved outward from several centers of dispersion. It was as if Canada were covered with a snow field much as Greenland now is but with three or four instead of one center of excessive accumulation. The Keewatin center west of Hudson Bay provided the ice which inundated the central part of the continent, the Great Lakes region and the Middle West. From the Labrador center came the ice covering New England. Eastern Canada was invaded from Newfoundland. In the Cordillera there was still another center of dispersion. Probably the ice moved outward even toward the north.

The ice sought the paths of least resistance, which were the lowlands. Especially along the relatively thin ice border it was guided by the various lowlands, valleys, troughs, basins, depressions, and even gorges and passes, so that its margin became lobate, the tongues protruding farthest in the direction of the larger valleys and lowlands. The greatest depressions were those now occupied by the Great Lakes, which at that time were inner lowlands of a dissected ancient belted coastal plain. Thus were formed several lobes: Lake Superior Lobe, Green Bay Lobe, Lake Michigan Lobe, Saginaw Lobe, and Lake Erie Lobe. The ice moved southward far into the Mississippi Valley. The Superior and Michigan lobes remained well separated from each other as the ice flowed around the state of Wisconsin. During successive advances, first one of these lobes and then the other reached far south into Iowa or Illinois. But a section of southern Wisconsin, about as large as the state of New Jersey, was never glaciated and is known as the *Driftless Area.*

Stages of Glaciation. Broadly, the Glacial period can be divided into two parts, Early Glaciation and Late Glaciation. The earlier advances reached farther south than the later ones; in the Mississippi Valley older deposits occur as far south as Kansas. The later ones reached only into Iowa and Illinois. In the regions covered by later drift—most of Canada and the bordering states of the United States, as far south as the so-called *terminal moraine*—there are numerous lakes and marshes, peat bogs, and disturbed drainage. The streams flow aimlessly around the obstructions of drift with which the ice sheet blocked their former channels. Waterfalls abound because streams were forced to flow over buried rock ledges which they are now discovering. The glacial deposits are fresh and comparatively unweathered.

In the regions of older drift, notably in Iowa, Nebraska, Missouri, and Kansas, there are virtually no lakes. Earlier lakes have long ago been drained or filled up. The glacial deposits are profoundly weathered. Even the boulders have decayed and the mass of till, known as *gumbotil,* has gained a uniformity of character. The soil profiles indicate a very long period of adjustment to weathering conditions, whereas the recent glacial soils are immature. The stream systems are better organized as if the streams had had time to establish their courses and develop well-defined valleys and systems of tributaries.

The following names have been given to the several stages of glaciation: Nebraskan, Kansan, Illinoian, Iowan, and Wisconsin. The Wisconsin stage, which by some investigators is made to include the Iowan, corresponds with the period of Later Glaciation, the other stages representing Early Glaciation. The Wisconsin stage is itself made up of five different substages. Similar subdivisions are distinguishable in Europe. In all, probably a million years was represented. The world may now be in an interglacial period, to be followed by another glacial stage.

During the time of the continental ice sheet in eastern and northern United States, lakes covered parts of the Great Basin where before that time there had been only deserts, as there are now. Great Salt Lake is the remnant of one of those lakes. The deposits left by these lakes, as well as their shore lines, indicate that their history consisted of two distinct stages coinciding with the Early and Late periods of continental glaciation. Some of the lakes dried up completely between the two periods. Their greatest size, therefore, corresponds with the time of most profound continental as well as of mountain glaciation.

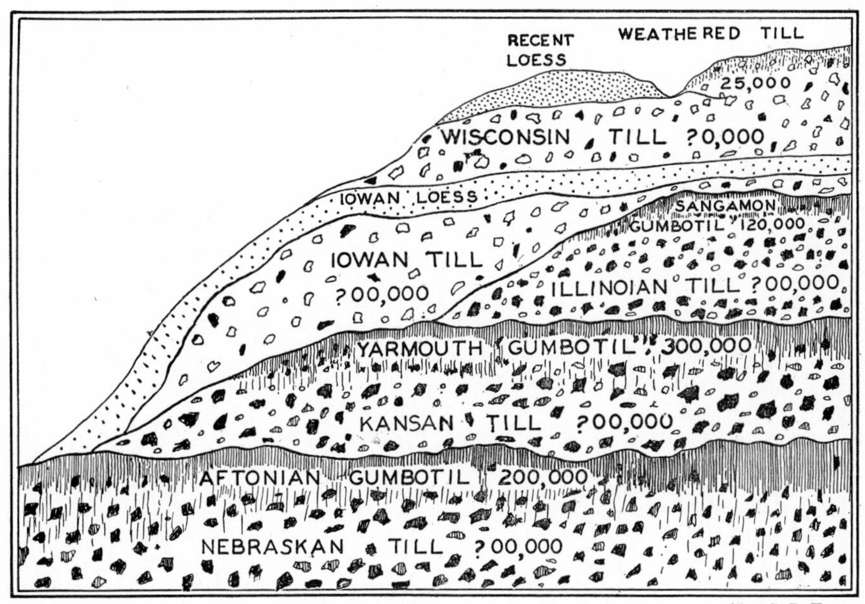


*After G. F. Kay*

FIVE GLACIAL TILL SHEETS
Separated by weathered gumbotil of interglacial times.

Marginal Lakes and Streams. As the ice advanced southward, interrupting the usual course of drainage, rivers were blocked or turned from their courses. Thus *marginal lakes* and *marginal streams* were formed. The present courses of the Ohio and Missouri Rivers represent the approximate southern limit of glaciation. As the ice withdrew to the north, lakes of great size occupied much of the country just evacuated. The Great Lakes of glacial time were somewhat larger than the present ones and discharged southward by several outlets. Lake Duluth, the predecessor of Lake Superior, found its way into the upper Mississippi. The waters of Lake Chicago, the ancestor of Lake Michigan, escaped through the Illinois River almost exactly along the route of the present Chicago Drainage Canal. Lake Maumee, preceding Lake Erie but covering much of northern Ohio, discharged through the present Wabash River valley across Indiana. From Lake Ontario the water escaped by way of the Mohawk Valley into the Hudson. With the retreat of the ice front northward, successively lower outlets were uncovered and the Great Lakes system came to have its present pattern. Niagara Falls was born when the waters of Lake Erie first flowed northward over the Niagara Escarpment into the Lake Ontario Basin. Because of this history there are many outlet channels of this former lake system in the places mentioned and many old beach ridges around the lake basins. The ridge road from Niagara Falls to Rochester is built along the crest of an ancient beach ridge of Lake Ontario. These old beaches rise constantly toward the north, as if the continent has been warped since glacial time. Indeed, it is practically proved that the release of the weight of the ice has permitted the earth's crust to rise several hundred feet in that part formerly most heavily loaded. This is shown in the Hudson-Champlain Estuary, in glacial time an arm of the sea, into which many rivers built deltas. These deposits now form terraces along the Hudson and in Vermont. They range in elevation from 80 feet above sea level at Croton Point, a few miles north of New York City, to almost 200 feet at Albany, and 300 feet along Lake Champlain. Still farther north in Canada this same plane, which once represented sea level, stands 600 feet above the sea.

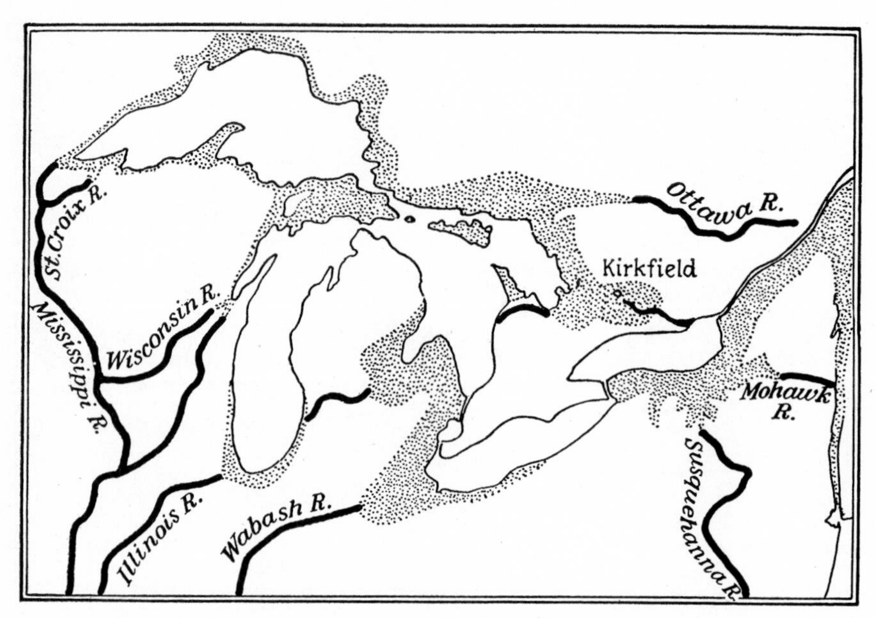


GLACIAL CHANNELS IN THE GREAT LAKES REGION
Used by glacial waters during retreat of the ice. Former lake areas stippled.

One of the largest marginal lakes occupied the basin of the Red River in Minnesota when that northward-flowing stream was blocked. It is called *Lake Agassiz* in honor of the man who first suggested the idea of continental glaciation. The rich alluvial soil deposited here now constitutes the fertile wheatlands of North Dakota, Minnesota, and Manitoba. At the time of its maximum extent Lake Agassiz discharged southward into the Minnesota River. Big Stone Lake and Lake Traverse now occupy the old channelway along the divide between the Red River and the Minnesota River.

In the eastern United States there were similar temporary lakes, some of them at times connected with the sea. Lake Passaic occupied the Passaic River basin back of the Watchung Ridges. The Berkshire Lowland and the Connecticut Valley also contained large lakes.

Drainage Changes. In many cases rivers were temporarily diverted from their courses. The Missouri River, for example, for a time cut a channel, the *Shonkin Sag*, around the northern side of the Highwood Mountains in Montana. On the Columbia Plateau, the diversion of the Columbia River brought about the Grand Coulee.

Abandoned Channels of Marginal Streams in Europe. Similar abandoned channels now containing only small streams occur on the Prussian Plains of North Germany. Northward-flowing streams such as the Oder and the Vistula, when blocked by the ice sheet, were forced to flow along the ice front westward into the Weser or the Elbe. Only small streams like the Spree, upon which Berlin stands, occupy these former channels, but their economic importance is great because they provide routes for the German canal system. Every city of note on the North German Plain stands either near the mouth of one of these northward-flowing rivers, as, for example, Bremen, Hamburg, Stettin, Danzig, and Königsberg, or else in one of the old connecting valleys, as, for example, Berlin and Warsaw.

Economic Effects of Glaciation. Whether the occupation of much of North America and Europe by the ice sheet made these continents more habitable or less so for man is a question difficult if not impossible to answer. It has been suggested that much of the energy and resourcefulness of the people of northern Europe has come about because of their enforced striving with adverse conditions of climate and topography.

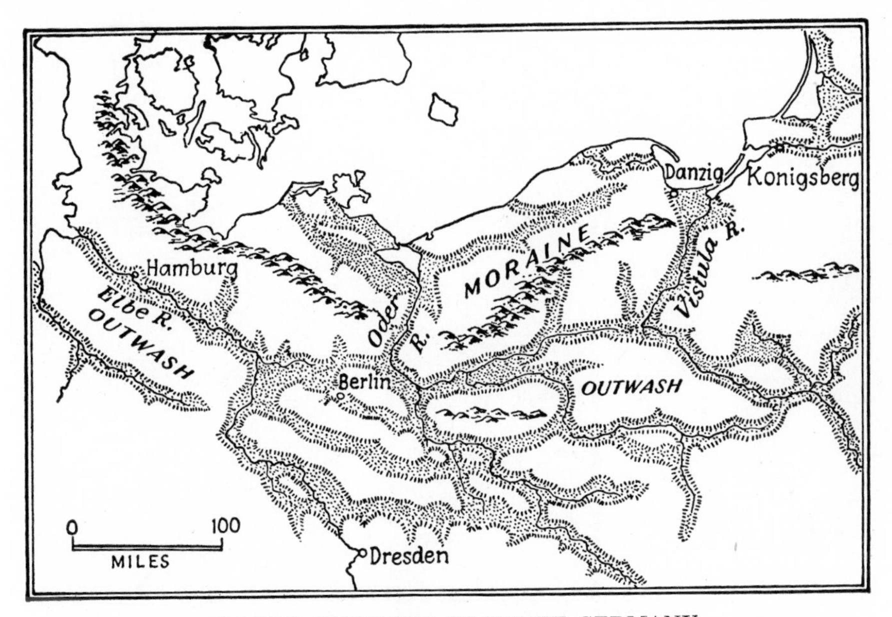


GLACIAL CHANNELS IN NORTH GERMANY
Showing routes used by marginal rivers at different times during retreat of the ice.

Whether the introduction of glacial till by the ice has been an advantage or otherwise is also not subject to a definite answer. No doubt in some parts of the United States, as in northern Ohio, Indiana, and Illinois, the level fertile till plains, containing almost no stones of appreciable size, are superior to the irregular and broken topography which is buried beneath them. Elsewhere, however, the stony moraines with their steep slopes and varied topography render the land unsuited to cultivation and best used only for the grazing of sheep and for woodlands. Nor perhaps can the extensive sand plains of Germany and Poland and also those of Aroostook County in Maine be considered very fertile. Nevertheless, they are not entirely unsuited to certain crops, like potatoes and rye.

Length of Postglacial Time. With the retreat of the ice many waterfalls came into existence. A number of these falls have been retreating upstream since that time and have left postglacial gorges on the way. The gorge below Niagara Falls, for example, is 7 miles long. During the past century these falls have been carefully observed, surveyed, and photographed. In that time they have worn back at the average rate of about 5 feet per year. If this rate were constant, it would indicate that the falls began 7,000 to 8,000 years ago. However, it is almost certain that the falls are now retreating much faster than they have at several times during the past. On two occasions, the water from Lake Erie only, representing about 15 per cent of the present volume, discharged over them; the water from the other great lakes went around through Canada via Georgian Bay and the Ottawa and other rivers. This fact greatly reduces the estimate for the rate of erosion and multiplies several times the time of recession. Therefore, it appears that 25,000 to 50,000 years is somewhere near the correct order of magnitude for the postglacial period.

Mountain Glaciation during Glacial Time. The presence of cirques and other glacial features in the White Mountains, on Mount Katahdin, in the Green Mountains and to a lesser degree in the Adirondacks and the Catskills indicates that these higher areas supported small ice fields and glaciers of their own, either before or after the glacial sheet had covered them. The existence of local moraines suggests that small glaciers persisted here after the main ice sheet disappeared.

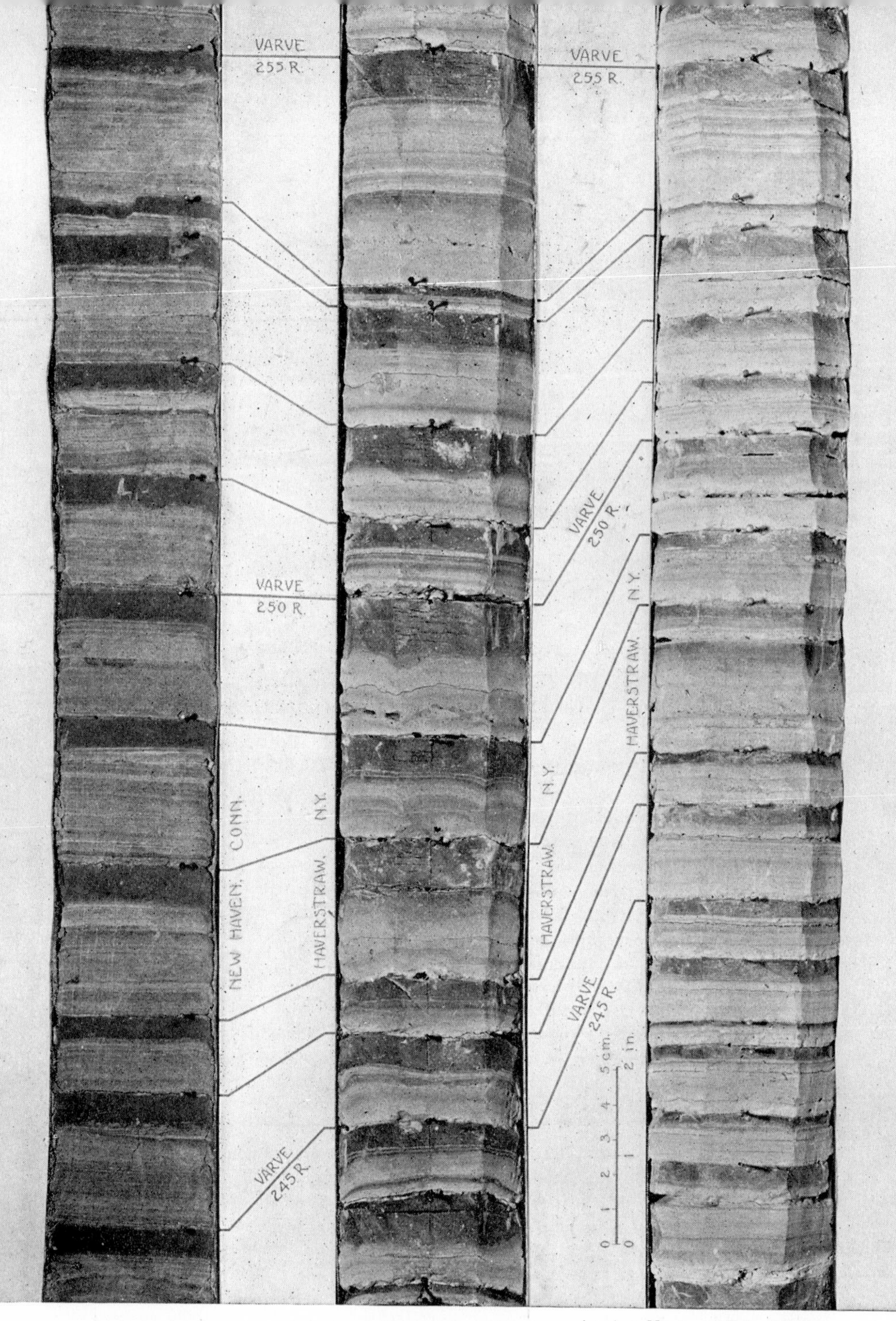


*American Museum of Natural History*

VARVED CLAY FROM NEW HAVEN, CONN., AND HAVERSTRAW, N. Y.
Showing correlation between three different lakes.

A varve is the yearly accumulation of sand, silt, and clay deposited in a lake basin by glacial waters. The coarser silt in each varve is at the bottom and represents that which settled during the maximum summer drainage. The fine clay at the top is that which settled out slowly during the winter. The thickness of a varve averages from ⅛ inch to ½ inch or even more and depends upon the prevailing conditions of rainfall and melting.

Varves occur in the lake deposits of most of the glacial lakes which temporarily existed along the front of the ice. It is evident, therefore, that if all the varves could be counted from the time the ice held its most advanced position to the time it occupied a known position farther north, the rate of retreat could be known and possibly some light be thrown on the number of years which have elapsed since the ice covered

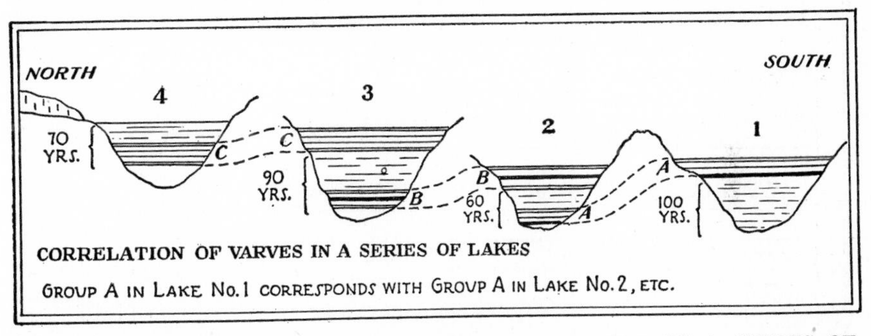


DIAGRAM SHOWING HOW VARVES ARE CORRELATED IN A SERIES OF LAKES

any given region. Inasmuch as any one lake covers a relatively small area, the problem of the geologist is to count the varves in that deposit; then go north and find a lake whose lower deposits correspond with the upper deposits of the first lake. This is done by correlating groups of varves by their thicknesses, the same succession of thick and thin layers representing seasonal changes being deposited simultaneously in each locality.

In the figure above, 320 years have elapsed since the deposition of the first varve in lake 1 to the time of the topmost varve in lake 4.

The technique of correlating varves was evolved in Scandinavia by Gerard De Geer, who determined a rate of retreat there of about 7 miles per century. In the United States similar studies by Ernest Antevs have shown that about 5,000 years elapsed after the ice occupied southern New England until it had retreated to the Canadian border, a total distance of some 250 miles. Correlation between the varves of the United States and Europe is probably not very reliable although it has been attempted. It is unlikely that similar seasonal conditions existed in both continents at the same time over an extended period of years.

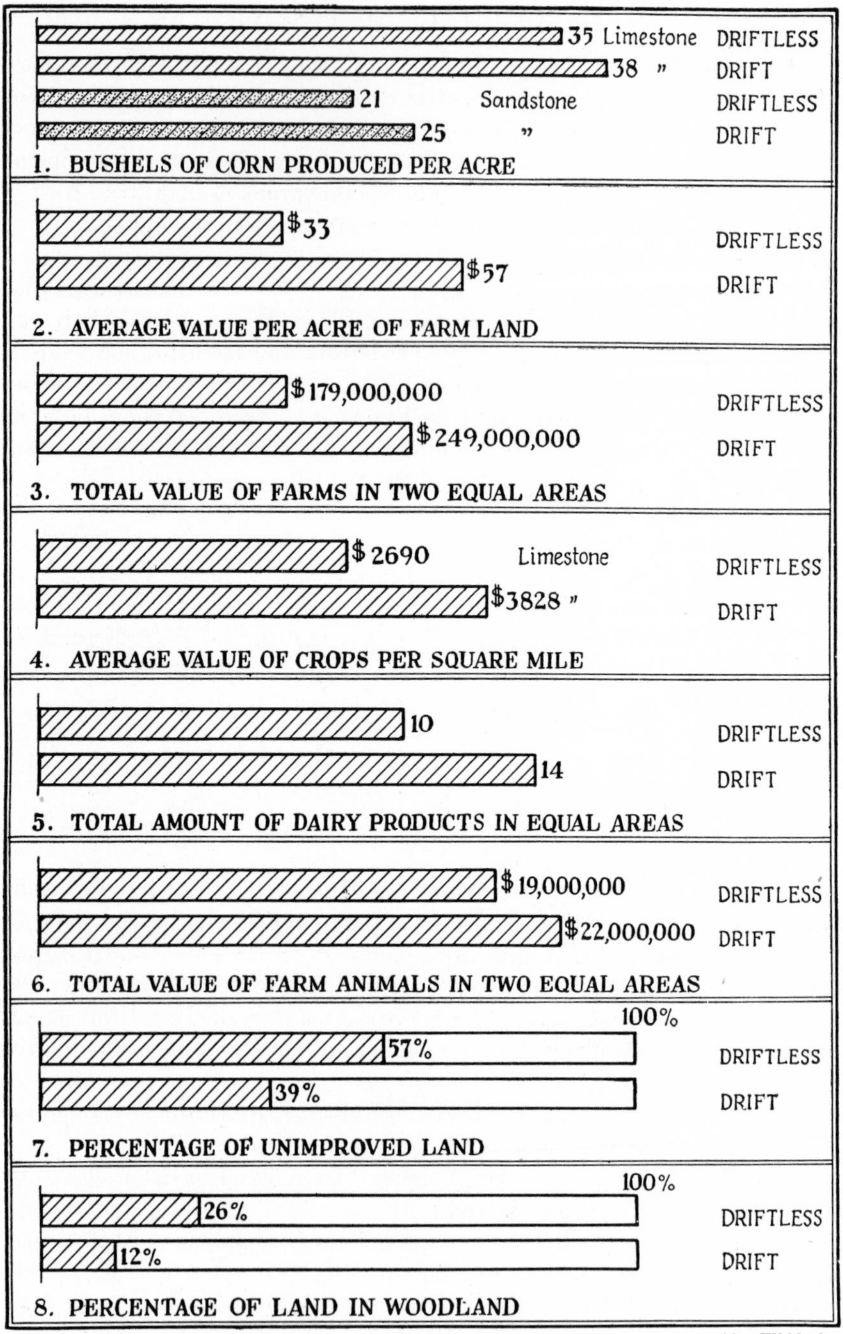

35 Limestone DRIFTLESS
38 " DRIFT
21 Sandstone DRIFTLESS
25 " DRIFT

1. BUSHELS OF CORN PRODUCED PER ACRE

$33 DRIFTLESS
$57 DRIFT

2. AVERAGE VALUE PER ACRE OF FARM LAND

$179,000,000 DRIFTLESS
$249,000,000 DRIFT

3. TOTAL VALUE OF FARMS IN TWO EQUAL AREAS

$2690 Limestone DRIFTLESS
$3828 " DRIFT

4. AVERAGE VALUE OF CROPS PER SQUARE MILE

10 DRIFTLESS
14 DRIFT

5. TOTAL AMOUNT OF DAIRY PRODUCTS IN EQUAL AREAS

$19,000,000 DRIFTLESS
$22,000,000 DRIFT

6. TOTAL VALUE OF FARM ANIMALS IN TWO EQUAL AREAS

100%
57% DRIFTLESS
39% DRIFT

7. PERCENTAGE OF UNIMPROVED LAND

100%
26% DRIFTLESS
12% DRIFT

8. PERCENTAGE OF LAND IN WOODLAND

*After Whitbeck*

STATISTICS SHOWING GREATER PRODUCTIVITY OF DRIFT-COVERED AREA AS COMPARED WITH DRIFTLESS AREA IN WISCONSIN

A comparison between contiguous glaciated and nonglaciated regions of Wisconsin made by Whitbeck indicates that, agriculturally, the glaciated regions are superior. The soils are more fertile and produce more per acre; the value of the individual farms is greater; the average value of all crops produced is greater; the percentage of improved land is greater; and the percentage of land not left in woodland is greater.

Graph 1 shows that the sandstone soils in the glaciated region produced almost 20 per cent more corn per acre than in the nonglaciated sandstone tract. The residual limestone soil of Wisconsin is inherently rich and is not so much improved by the addition of the drift; the residual sandstone is inherently sterile and was materially improved by the addition of the drift which happened to come from adjacent limestone regions. The average value per acre of farm land in the glaciated region is almost 50 per cent greater than in the driftless area (Graph 2).

Graph 3 indicates that the total value of farms, including houses and land, in an area of glaciated country is 40 per cent greater than in a similar equal area of driftless country. The average productivity of farms in a drift-covered limestone tract (Graph 4) is 40 per cent greater than in a driftless limestone area. Graph 1, and similar graphs for other crops not reproduced here, indicate only a slightly greater productivity of the land per acre in the drift-covered area. The greater difference in the total value of crops (Graph 4) is explained not by greater fertility alone but by the fact that glaciation improved the farming areas by making a much greater proportion of the land suitable to cultivation, probably by reducing the relief, in spite of the development of lakes and marshes in the glaciated area.

Not only in cultivated crops but also in grazing lands is the drift area superior. This is shown by the greater importance of dairying as well as the raising of other farm animals in drift regions as compared with driftless regions, as depicted by Graphs 5 and 6.

The last two graphs show that there is about 50 per cent more unimproved land and twice as much woodland per unit area in nonglaciated as compared with glaciated territory, the regions being contiguous and similar geologically, and presumably similar topographically before the advent of the ice sheet.

That the glacial lakes of the northern United States are a financial asset can readily be demonstrated. The increased value of property adjoining lake shores is considerably more than the average value of the land covered by lake waters.

The great development of water power is perhaps the most important result of glaciation the world over. While greater quantities of hydroelectric power have probably been developed by artificial damming of streams, this has been accomplished at far greater expense outside glaciated regions, where some natural damming had not already been brought about.

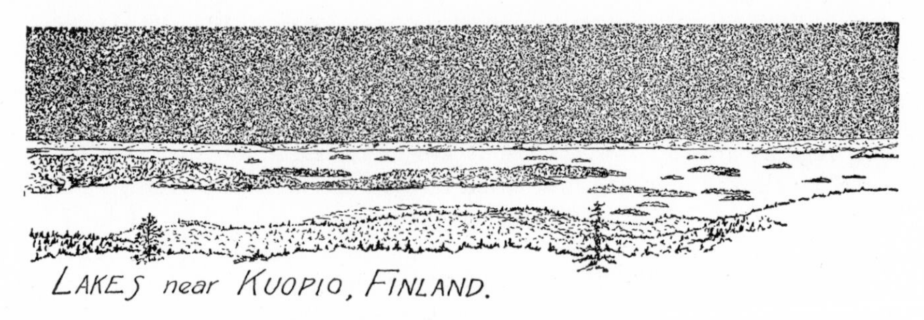


THE LAKE COUNTRY OF FINLAND
A vast area once covered by the continental ice sheet, now a maze of lakes and swamps.

LAKE SILJAN IN CENTRAL SWEDEN
One of the numerous glacial lakes occupying valleys in the Swedish upland.

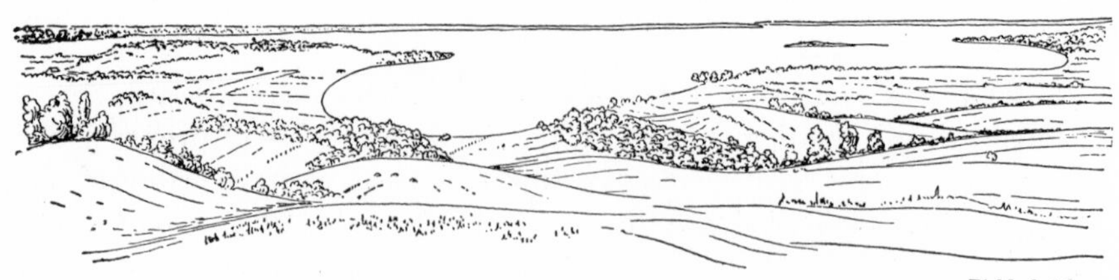

*Field sketches*

THE TERMINAL MORAINE IN CENTRAL DENMARK
With kames and large enclosed arms of the sea which correspond with the glacial lakes of other countries.

THE ICE CAP IN NORWAY ON THE HIGHLAND BETWEEN OSLO AND BERGEN
In several respects it resembles a continental ice cap.

LAKE MALAREN IN SOUTHERN SWEDEN
A glacial lake occupying a series of fault depressions.

*Field sketches*

TYPICAL ROLLING MORAINE IN DENMARK WITH KETTLE HOLES
Largely cultivated and used for dairying.

## MAPS ILLUSTRATING CONTINENTAL GLACIATION

The terminal moraine of the continental ice sheet is unusually well shown on the *Islip*, *Oyster Bay*, and *Riverhead*, *N. Y.*, sheets from Long Island. On all of these maps two separate and parallel moraines are shown, each fronted by an outwash plain. The *Barnstable*, *Falmouth*, *Marthas Vineyard*, *Nantucket*, and *Wellfleet*, *Mass.*, maps show a single belt of morainal topography with an outwash plain lying to the south. The *Block Island*, *R. I.*, sheet is entirely morainal in character except for the spits and bars. West of New York the moraine is less conspicuous but it can be traced on the *Passaic* and *Plainfield*, *N. J.*, sheets and on the *Staten Island*, *N. J.-N. Y.*, sheet where it forms the "Narrows" to New York harbor.

In the Middle Western states there are numerous moraines but usually no single topographic map shows so well defined a morainal belt as that shown on the *St. Croix Dalles*, *Wis.-Minn.*, sheet. Among the maps which show good morainal topography with numerous hummocks and kettles are the *Kongsberg*, *N. Dak.; Marshall*, *Stockbridge* and *Holly*, *Mich.;* and the *Vergas* and *Underwood*, *Minn.*, sheets. Many of the Middle Western maps show, in addition to the moraine, pitted outwash topography as, for example, the *Barrington*, *Ill.; Three Rivers* and *Niles*, *Mich.-Ind.;* and the *Battle Creek* and *Galesbury*, *Mich.*, sheets. The *Cut Bank*, *Mont.*, sheet shows part of the moraine in the Great Plains area. The *Grays Lake*, *Ill.-Wis.*, sheet shows a kame moraine probably interlobate in origin. Typical glaciated country with many lakes and swamps and disturbed drainage is shown on several of the New England and New York maps, as, for example, the *Burnham* and *Attean*, *Maine*, the *Wolfeboro*, and *Lovewell Mountain*, *N. H.*, the *Quinsigamond*, *Mass.-Conn.-R. I.*, and the *Cranberry Lake*, and *Old Forge*, *N. Y.*, sheets.

Among the maps showing drumlin swarms are the *Weedsport*, *Baldwinsville*, and *Clyde*, *N. Y.*, the *Sun Prairie*, *Waterloo*, and *Watertown*, *Wis.*, and the *Boston Bay*, *Mass.*, sheets. On the last-mentioned map the drumlins have been joined by bars to form the complex tombolo of Nantasket Beach.

Eskers of unusual size are represented on the *Passadumkeag*, *Great Pond*, *Nicatous Lake*, and *Boyd Lake*, *Maine*, sheets. As these eskers run through low country, many of them serve as the location for roads. Several of them show a branching form.

The old glacial outlet channel of Lake Agassiz is shown on the *Peever*, *Beardsley*, and *White Rock*, *S. Dak.-Minn.*, sheets. Smaller glacial outlets, now represented as belts of lakes, appear on the *Schoolcraft*, *Mich.*, *Battle Lake*, *Minn.*, and *Cooperstown*, *N. Y.*, sheets, the channel in the latter case being blocked by drift.

An unusual example of ice-push ridges is represented along the shore of Mille Lacs Lake on the *Deerwood* and *Wealthwood*, *Minn.*, sheets.

The bed of an old glacial lake, now a peat meadow drained for agricultural use, is shown on the *Goshen*, *N. Y.-N. J.*, sheet.

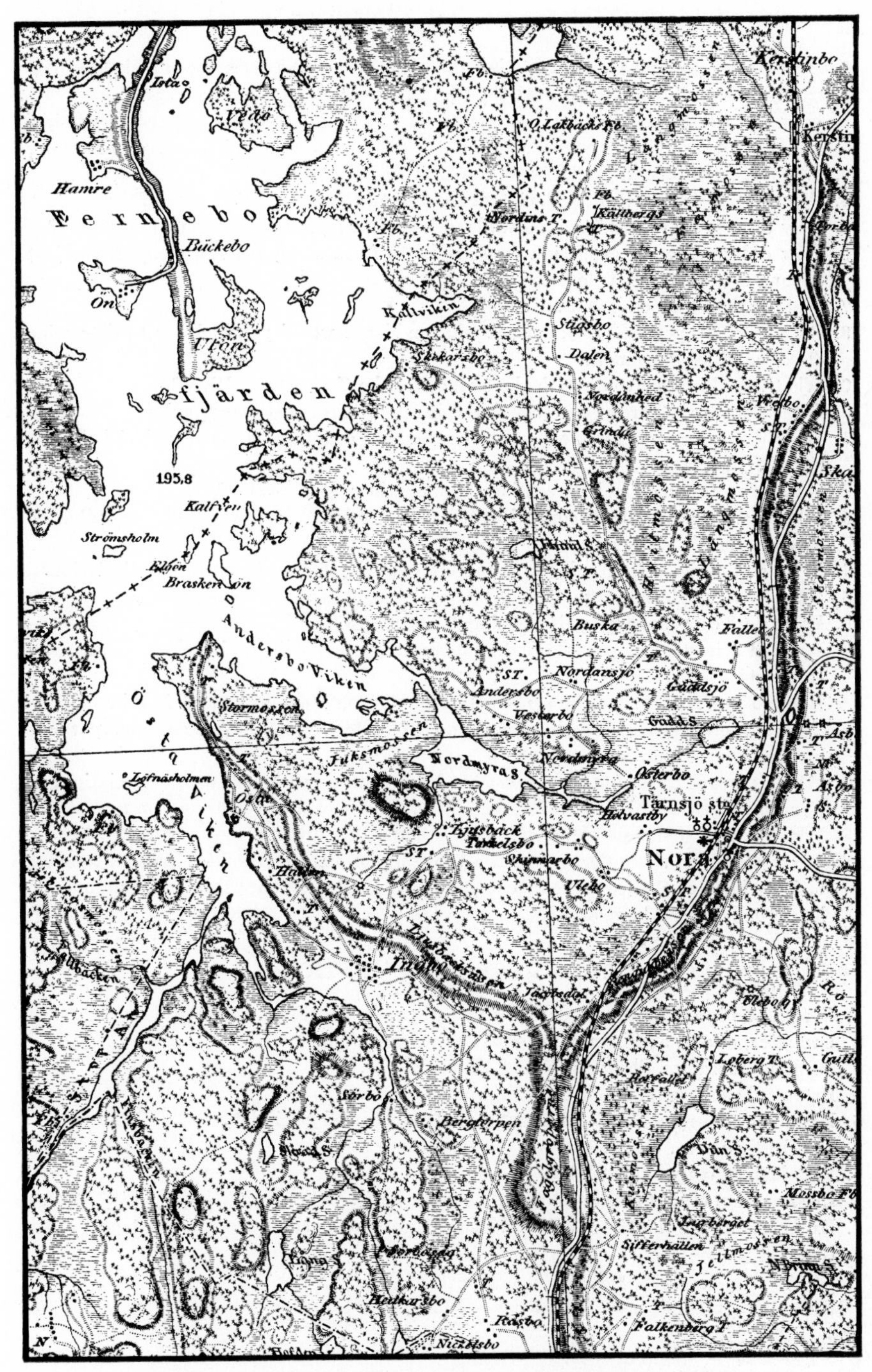

ESKERS AND GLACIAL LAKES IN SOUTHERN SWEDEN
Sweden; *Gysinge* sheet, No. 91 (1:100,000).

## QUESTIONS

1. How do you account for the fact that the glacial-till plains in northern Ohio are practically without boulders, whereas the till of New England is very stony?
2. Why did the continental ice sheet extend so much farther south in the Mississippi Valley than it did in the western part of the continent? Was this true also in Europe?
3. Was there continental glaciation in the Southern Hemisphere?
4. What effect did continental glaciation have upon the Great Basin region?
5. How can varves be used to indicate length of time since the ice occupied a region?
6. How did continental glaciation cause lakes to form along the Allegheny River just beyond the limit of the glaciated territory?
7. What is meant by crag and tail?
8. On a glacially polished rock surface what evidence would you look for to determine direction of ice movement?
9. What has glaciation to do with the terraces of the Connecticut River?
10. What evidence is there of postglacial warping in the Great Lakes region?
11. What difference would you expect in the chemical composition of early glacial till and late glacial till?
12. Is there any connection between loess and continental glaciation?
13. Who first proposed the theory of continental glaciation and where did he make his observations?
14. Is it practicable to speak of young, mature, and old stages of continental glaciation?
15. What types of glacial or fluvioglacial deposits are of economic importance?
16. What does osar mean? Is this the singular or plural form?
17. By what evidence is it possible to determine the centers of dispersion of the continental ice sheet?
18. How do you account for the small extent of the continental ice sheet in Siberia?
19. Do intersecting striae on rock surfaces indicate different stages of glaciation separated by long intervals of time?
20. What is meant by a hinge line and what has this to do with continental glaciation?
21. Can you suggest explanations for the different kinds of drift illustrated on page 287?
22. Which is the north and which is the south end of the drumlin shown on page 292?
23. What would be the difference in the topographic features resulting from a gradual melting back of the ice front as compared with a stagnant ice sheet melting more or less equally over a large area?

## TOPICS FOR INVESTIGATION

1. Glacial and interglacial periods. How and where distinguished.
2. Drumlins, eskers, and kames; theories as to their origin.
3. Terminal moraines. Show on map of United States.
4. Extent of continental ice sheet in Europe and Asia.
5. Economic advantages and disadvantages of continental glaciation.
6. The driftless area. Cause.
7. Postglacial changes in sea level.
8. Classification of all types of lakes on a genetic basis.
9. Explanations for the Glacial Period and for ancient glacial periods of past geological times.

## REFERENCES

The importance of glaciation in North American geology has resulted in a vast number of papers by T. C. Chamberlin (on the whole glacial period); Leverett and Taylor (on the Great Lakes and Mississippi Valley); Kay and Leighton (on the Mississippi Valley drift); Fairchild (on New York State); Antevs (on varves); MacClintock (in Illinois); Coleman (Canada), under whose names in the *Bibliography of North American Geology* specific references may readily be found. The following are among the more important:


GLACIAL PERIOD

ANTEVS, E. (1928) *The last glaciation, with special reference to the ice retreat in northeastern North America.* Am. Geog. Soc., Research ser., No. 17, 292 p.

COLEMAN, A. P. (1926) *Ice ages recent and ancient.* New York, 283 p. Glacial geology bibliography.

COLEMAN, A. P. (1933) *Ice ages and the drift of continents.* Jour. Geol., vol. 41, p. 409–417. Fixity of continents supported.

DALY, R. A. (1934) *The changing world of the ice age.* New Haven, 271 p.

GEIKIE, J. (1894) *The great ice age.* New York, 850 p.

HUNTINGTON, E. (1907) *Some characteristics of the glacial period in nonglaciated regions.* Geol. Soc. Am., Bull. 18, p. 351–388.

KAY, G. F. (1931) *Classification and duration of the Pleistocene period.* Geol. Soc. Am., Bull. 42, p. 425–466. One million years of glaciation.

LEVERETT, F. (1929) *Pleistocene glaciations of the Northern Hemisphere.* Geol. Soc. Am., Bull. 40, p. 745–760. Snow line and glacial stages discussed.

MCCABE, J. (1922) *Ice ages; the story of the earth's revolutions.* New York, 134 p.

THWAITES, F. T. (1935) *Outline of glacial geology.* Ann Arbor, 115 p. An excellent treatise, rich in detail, with many remarkable diagrams.

WRIGHT, G. F. (1889–1911) *Ice age in North America.* New York, 741 p.

WRIGHT, W. B. (1914–36) *The Quaternary ice age.* London, 478 p.

CAUSE OF GLACIAL PERIODS

CHAMBERLIN, T. C. (1899) *An attempt to frame a working hypothesis of the cause of glacial periods on an atmospheric basis.* Jour. Geol., vol. 7, p. 545–584, 667–685, 751–787. Maps.

UPHAM, W. (1890) *On the cause of the glacial period.* Am. Geol., vol. 6, p. 327–339.

GLACIAL STAGES

CLAPP, F. G. (1908) *Complexity of the glacial period in northeastern New England.* Geol. Soc. Am., Bull. 18, p. 505–556.

HOBBS, W. H. (1929) *Climatic zones and periods of glaciation.* Geol. Soc. Am., Bull. 40, p. 735–744. Man and climate.

EROSIONAL AND DEPOSITIONAL FEATURES

ALDEN, W. C. (1909) *Criteria for discrimination of the age of glacial drift sheets.* Jour. Geol., vol. 17, p. 694–709.

ALDEN, W. C. (1905) *The drumlins of southeastern Wisconsin.* U. S. Geol. Surv., Bull. 273, 46 p.

BROWN, T. C. (1931) *Kames and kame terraces of central Massachusetts.* Geol. Soc. Am., Bull. 42, p. 467–479.

BROWN, T. C. (1932) *Late Wisconsin ice movements in Massachusetts.* Am. Jour. Sci., 5th ser., vol. 23, p. 462–468.

CHADWICK, G. H. (1928) *Adirondack eskers.* Geol. Soc. Am., Bull. 39, p. 923–929.

CHAMBERLIN, T. C. (1883) *Terminal moraine of the second glacial epoch.* U. S. Geol. Surv., 3d Ann. Rept., p. 291–402.

CHAMBERLIN, T. C. (1888) *Rock-scorings of the great ice invasions.* U. S. Geol. Surv., 7th Ann. Rept., p. 147–248. Many illustrations of ice abrasion, accompanied by clear and concise explanations of the manner of their formation.

CHAMBERLIN, T. C. (1893) *The horizon of drumlin, osar, and kame formation.* Jour. Geol., vol. 1, p. 255–265.

CHAMBERLIN, T. C., and SALISBURY, R. D. (1885) *The driftless area of the upper Mississippi Valley.* U. S. Geol. Surv., 6th Ann. Rept., p. 199–322.

CROSBY, W. O. (1902) *Origin of eskers.* Am. Geol., vol. 30, p. 1–38.

DAVIS, W. M. (1890) *Structure and origin of glacial sand plains.* Geol. Soc. Am., Bull. 1, p. 195–202.

DAVIS, W. M. (1892) *The subglacial origin of certain eskers.* Boston Soc. Nat. Hist., Proc. 25, p. 477–499.

FAIRCHILD, H. L. (1907) *Drumlins of central western New York.* N. Y. State Museum, Bull. 3, p. 391–443.

FLINT, R. F. (1928) *Eskers and crevasse fillings.* Am. Jour. Sci., 5th ser., vol. 15, p. 410–416.

FLINT, R. F. (1930) *The origin of the Irish eskers.* Geog. Rev., vol. 20, p. 615–630.

LEWIS, H. C. (1884) *Report on the terminal moraine in Pennsylvania and western New York.* 2d Geol. Surv. Pa., Rept. Z, 299 p.

SCOTT, I. D. (1927) *Ice push on lakes.* Mich. Acad. Sci., Papers, vol. 7, p. 107–123.

STONE, G. H. (1899) *The glacial gravels of Maine, and their associated deposits.* U. S. Geol. Surv., Mon. 34, 499 p.

TARR, R. S. (1894) *The origin of drumlins.* Am. Geol., vol. 13, p. 393–407.

THWAITES, F. T. (1926) *The origin and significance of pitted outwash.* Jour. Geol., vol. 34, p. 308–319.

VARVES

ANDERSEN, S. A. (1931) *The waning of the last continental glacier in Denmark as illustrated by varved clay and eskers.* Jour. Geol., vol. 39, p. 609–624.

ANTEVS, E. (1922) *The recession of the last ice sheet in New England.* Am. Geog. Soc., Research ser., no. 11, 120 p.

COLEMAN, A. P. (1929) *Long-range correlation of varves.* Jour. Geol., vol. 37, p. 783–789.

REEDS, C. A. (1929) *Weather and glaciation.* Geol. Soc. Am., Bull. 40, p. 597–629. Correlation of varved clays.

REGIONS DESCRIBED

ANTEVS, E. (1925) *On the Pleistocene history of the Great Basin.* Carn. Inst. Wash., Publ. 352, p. 51–114.

ANTEVS, E. (1929) *Maps of the Pleistocene glaciations.* Geol. Soc. Am., Bull. 40, p. 631–720. Many references.

BELL, R. (1890) *On glacial phenomena in Canada.* Geol. Soc. Am., Bull. 1, p. 287–310.

BONNEY, T. G. (1910) *Some aspects of the glacial history of western Europe.* Scot. Geog. Mag., vol. 26, p. 505–532.

KRYNINE, P. D. (1937) *Pleistocene glaciation of Siberia.* Am. Jour. Sci., 5th ser., vol. 34, p. 389–398.

LEVERETT, F. (1921) *Outline of Pleistocene history of Mississippi Valley.* Jour. Geol., vol. 29, p. 615–626.

GEOGRAPHICAL ASPECTS

BELT, T. (1876) *Man and the glacial periods.* Quart. Jour. Sci., new ser. 6, vol. 13, p. 289–304.

KAY, G. F. (1939) *Pleistocene history and early man in America.* Geol. Soc. Am., Bull. 50, p. 453–464.

VON ENGELN, O. D. (1914) *Effects of continental glaciation on agriculture.* Am. Geog. Soc., Bull. 46, p. 241–264, 336–355.

WHITBECK, R. H. (1911) *Contrasts between the glaciated and the driftless portions of Wisconsin.* Geog. Soc. Phila., Bull. 9, p. 114–123.

WHITBECK, R. H. (1913) *Economic aspects of glaciation in Wisconsin.* Assn. Am. Geog., Annals 3, p. 62–87.

WRIGHT, G. F. (1892–99) *Man and the glacial period.* New York, 385 pp.

# X
# WAVES

*Canadian National Railways*

FLOWER POT ISLAND, GEORGIAN BAY, ONTARIO
A marine stack cut by waves and by shifting ice in horizontal sedimentary rocks, the remnant of an extensive plateau.

Fox Photos, Ltd.

WAVE ATTACK ON SCILLY ISLANDS, ENGLAND

A rocky headland on a shoreline of submergence pounded by tremendous seas after a 100 mile per hour gale.

SURF AT SAN JOSÉ, GUATEMALA

Cylindrical comber on gently shelving shoreline of emergence.

*Ewing Galloway*

*Burton Holmes from Ewing Galloway*

SEA-WORN ARCH, PINEY ISLANDS OF MATSUSHIMA, JAPAN

Characteristic features of a young shoreline of submergence, in a region of horizontal rocks

SOUTHERN CROSS CAPE, GRAND MANAN, NEW BRUNSWICK

Characteristic crenulate and minutely irregular young shoreline of submergence in volcanic rocks.

*Canadian National Railways*

CHESIL BEACH, PORTLAND, DORSET, ENGLAND
A long baymouth bar with smaller bar tying island to mainland to form tombolo.

THE LONG AYRE, TANKERNESS, ORKNEY ISLANDS, SCOTLAND

Bar of coarse cobble connecting island with mainland; wave-built deltas or wash-overs on leeward side.

*Crown Copyright*

*U. S. Geological Survey*

SERGIUS NARROWS, NEAR SITKA, ALASKA

Drowned coast of a rugged mainland, with numerous promontories and islands. No wave-cut cliffs.

CHICHAGOF ISLAND, SPASSKAIA BAY, NEAR SITKA, ALASKA

Drowned coast with double tombolo in the foreground.

*U. S. Geological Survey*

*A. K. Lobeck*

ELEVATED MARINE BENCH, DESECHEO ISLAND, PUERTO RICO

Truncating sharply upturned beds, and capped with recent horizontal deposits.



QUATERNARY WAVE-CUT BENCH, PORT HARFORD, CALIF.

With old stacks rising above it and new stacks now forming at its base.

*Stose, U. S. Geological Survey*

*A. K. Lobeck*

BEACH CUSPS, NORTH SHORE OF PUERTO RICO, WEST OF SAN JUAN
Formed in fine white limestone sand, resulting from the destruction of old consolidated dunes.



FRONT OF WASATCH MOUNTAINS NEAR LOGAN, UTAH
Showing several beach levels of the former Lake Bonneville, now partially dissected.

*U. S. Forest Service*

*Sketched from photographs in Columbia Univ., Lab. of Geomorphology*

FEATURES OF YOUNG COASTS

# WAVES AND CURRENTS

Synopsis. This chapter is essentially a discussion of the different kinds of shorelines and the way they are modified by marine action. Two types of shorelines are readily distinguished from each other, namely, those of emergence and those of submergence. Other types are essentially modifications of one of these. For example, delta and alluvial-plain shorelines closely resemble shorelines of emergence; whereas shores around volcanoes, lava flows, and drumlins and moraines deposited in the sea have many of the aspects of submerged coasts. In short, a coast line may be very simple, low, and regular in outline with shallow water offshore; or it may be high and irregular and deep offshore.

Shorelines of Emergence. Shorelines of emergence like those of coastal plains are reached only by small waves. Large waves, because of the shallow water, break far offshore and in doing so scour the bottom and throw up the sand to form a barrier bar. The submarine profile of the shore is thus changed by deepening. The quantity of sand used for bar building which is available from the bottom may be augmented by that which is brought from other places by longshore currents. In any event the continuity of the bar depends upon the available supply of material with the result that, the more remote it is from that point along the coast whence longshore currents get their supply, the more numerous are the tidal inlets. Tides washing through these inlets build tidal deltas both in the lagoon and on the seaward side. These inlets migrate in the direction of the longshore currents. In time the lagoon back of the bar is filled with silt brought down from the mainland and washed or blown into it from the bar. At the same time the bar is constantly pushed landward by the waves until it comes to rest upon the original shore. A new and deeper shore profile is thus established; the waves now reach the mainland; and maturity has been attained.

Shorelines of Submergence. Shorelines of submergence because of the depth of water are immediately attacked by all sizes of waves which truncate the ends of exposed promontories, producing cliffed and winged headlands with spits of all types. Bars known as *bayhead* bars, *midbay* bars, and *baymouth* bars, depending upon their location, are soon formed. Various types of beaches also result from wave action. Offshore islands may be tied to the mainland, to form tombolos, and to each other, to form complex tombolos. In time, if the coast remains stable, the islands and promontories are completely cut away until the bays no longer remain and the coast line becomes comparatively straight. This is the beginning of maturity.

In both shorelines of emergence and shorelines of submergence the mature stage is characterized by a shore profile having such a grade and depth that the waves move over it without much resistance and exert most of their force cutting into the mainland and not in wearing away the bottom.

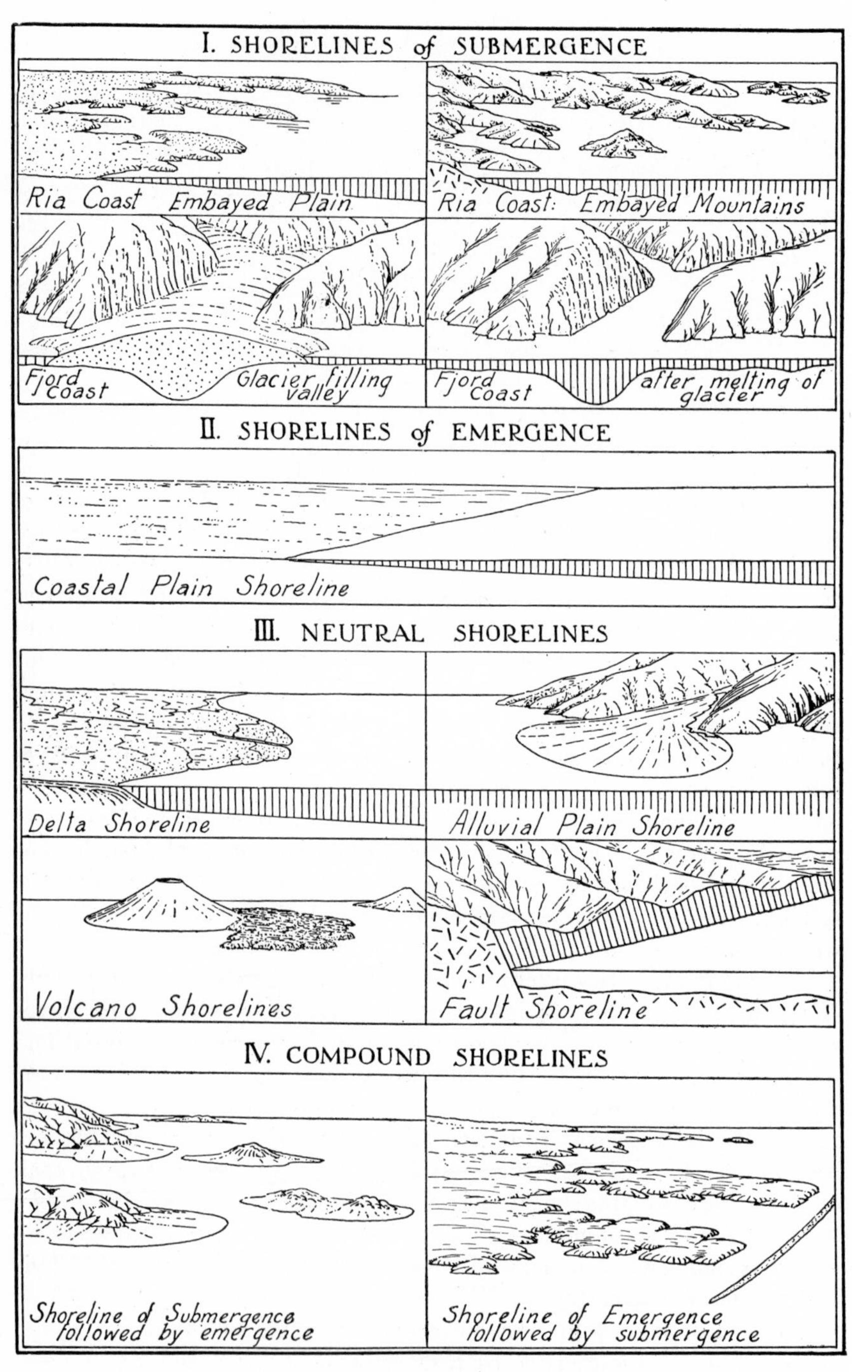


TYPES OF SHORELINES

## CLASSIFICATION OF SHORELINES

Any satisfactory classification of shorelines must be based upon the origin of the forms; that is, it must be *genetic*. Johnson recognizes four main classes in his well-known analysis of shorelines:

I. *Shorelines of submergence*, or those shorelines produced when the water surface comes to rest against a partially submerged land area.

II. *Shorelines of emergence*, or those resulting when the water surface comes to rest against a partially emerged sea or lake floor.

III. *Neutral shorelines*, or those whose essential features do not depend on either the submergence of a former land surface or the emergence of a former subaqueous surface.

IV. *Compound shorelines*, or those whose essential features combine elements of at least two of the preceding classes.

I. Shorelines of submergence may be further subdivided into *ria shorelines* and *fiord shorelines*. Ria shorelines result from the partial submergence of a land mass dissected by normal river valleys. There may be embayed plain or plateau ria shorelines, such as the Chesapeake Bay region; or embayed mountain shorelines of various structural types, such as the embayed folded-mountain shoreline of Dalmatia; or the embayed complex-mountain type of northwestern Spain, which is the type example of the ria coast. There may also be embayed volcano shorelines such as are found in the south Pacific, an evidence of the subsidence of the ocean floor.

Fiord shorelines are partially submerged glacial troughs. They do not necessarily imply a change in level between land and sea, for it is quite possible for glaciers coming to the sea to erode far below sea level so that, when the glacial ice eventually melts, the trough becomes flooded by the sea.

II. Shorelines of emergence result from the emergence of a submarine or a sublacustrine plain and in general constitute the coastal-plain shoreline. Rarely are coastal-plain shorelines strictly shorelines of emergence. Usually some later slight submergence has altered their simple aspects, so that they are compound.

III. Neutral shorelines include such types as (*a*) delta shorelines, of which several forms can be distinguished; (*b*) alluvial-plain shorelines; (*c*) outwash-plain shorelines; (*d*) volcano shorelines; (*e*) coral-reef shorelines; and (*f*) fault shorelines.

IV. Compound shorelines result when oscillations in the level of land and sea leave a shoreline with a variety of features, some of which resulted from submergence, others from emergence. The coast of Maine, for example, is essentially a shoreline of submergence, but recent emergence has resulted in the presence of small coastal-plain features around the islands and promontories. Similarly the originally upraised coast of North Carolina, and much of the Atlantic coastal plain, was later deeply embayed so that it is almost equally a shoreline of emergence and one of submergence.

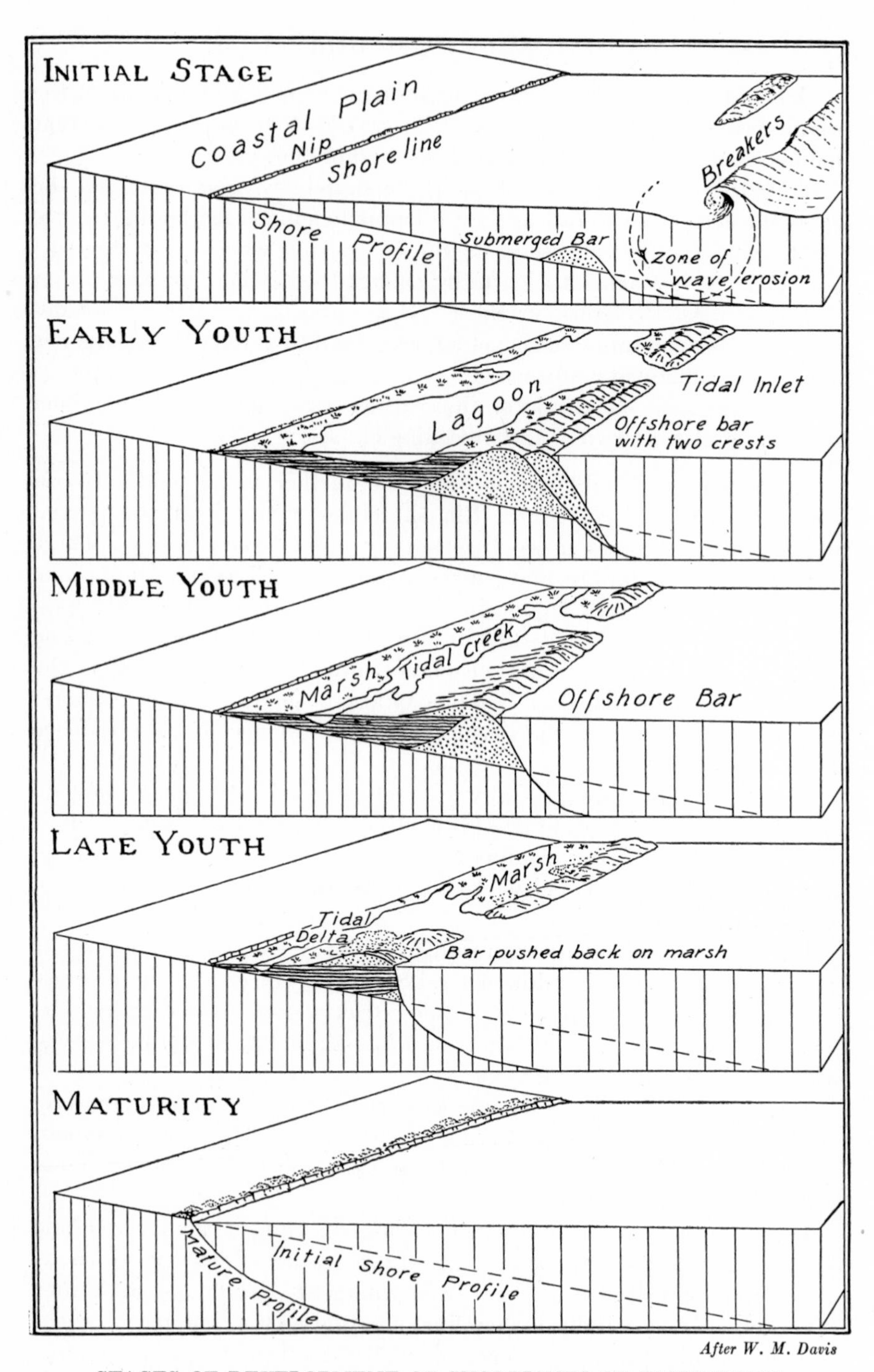


*After W. M. Davis*

STAGES OF DEVELOPMENT OF SHORELINES OF EMERGENCE

## SHORELINES OF EMERGENCE. STAGES OF DEVELOPMENT

INITIAL STAGE. The initial stage of a shoreline of emergence is characterized by a straight or nearly straight contour. This is due to the fact that the sea or lake floor is a smooth plain made up of material deposited under water. Unlike the shoreline of submergence, it does not have the many irregularities which result from the erosion of a land surface.

Some sea bottoms, it is true, represent rugged land areas recently depressed. The emergence of such regions gives an irregular shoreline. A coast of that type, therefore, has features essentially like those of a shoreline of submergence.

The offshore slope of a recently elevated sea floor is usually very gradual. This causes most waves to break offshore. The smaller waves advance landward and cut low cliffs in the weak material of which the land is usually composed. Thus a *nip* is produced. The larger waves breaking offshore cut into the sea bottom. Some of the material thus dislodged is thrown up to form a *submarine bar* parallel with the shoreline. This bar constantly grows in height with the accretion of more material, either cut from the sea bottom or drifted there by longshore currents. When the bar appears above the water surface, the shoreline is said to pass from the initial to the youthful stage.

YOUNG STAGE. The presence of an *offshore* or *barrier bar* with a *lagoon* behind it is the most striking feature of young shorelines of emergence. The material of which such bars are built comes from the sea floor just in front of the bar where the waves are deepening the sea bottom, and also from elsewhere along the coast. Undoubtedly much material drifted alongshore from places of active cliff erosion is added to the offshore bars forming in shallow water, where the waves have not yet succeeded in reaching the mainland. G. K. Gilbert thought this to be the case. He was the exponent of the "shore-drift" theory. De Beaumont, Davis, and Shaler, however, believed the material of the bar to be derived from the offshore bottom. These opposing views have been critically examined by Johnson who concludes that the de Beaumont theory is the more tenable because on the seaward face of the bar the ocean floor is overdeepened.

It is not uncommon to find two or even more than two bars developed parallel to the coast, the inner bar being first formed by a weaker set of waves which were able to advance farther landward in the shallow water.

MATURE STAGE. After the bar has been built and the waves are free to advance landward without encountering too much resistance on the sea floor, the bar itself is attacked and pushed back on to the lagoon or marsh.

When the bar has been forced back upon the mainland and the steepened submarine profile then provides the deep water required by large storm waves at the edge of the land, the shoreline of emergence is said to be *mature*.

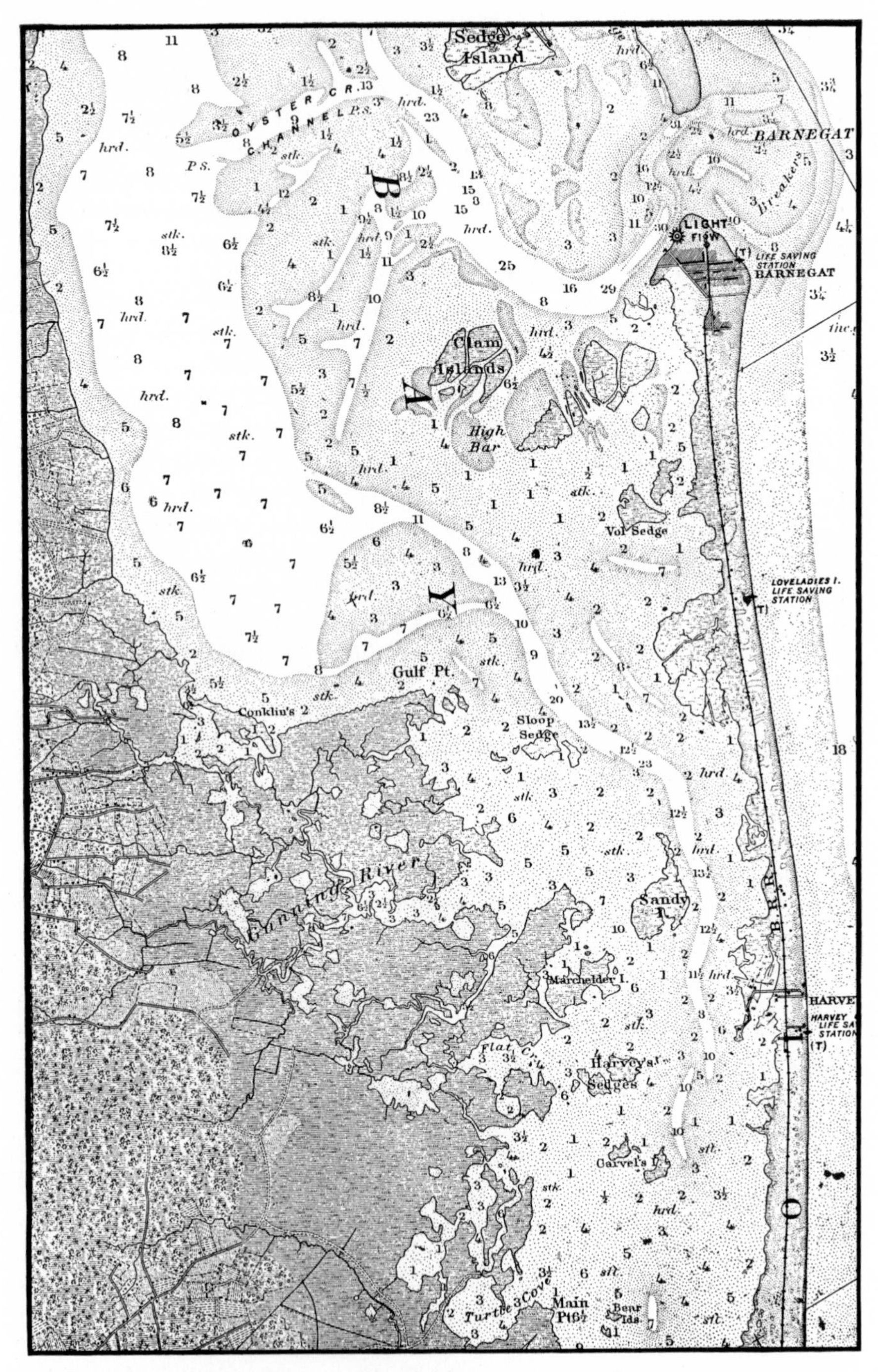


MAP OF OFFSHORE BAR, WITH LAGOON AND TIDAL INLET. COAST OF NEW JERSEY

U. S. Coast and Geodetic Survey Chart, No. 122 (1:80,000).

Offshore bars vary in size from a mere strip of sand appearing above sea level at low tide to a wide belt of sand two or three miles across surmounted by beach ridges and sand dunes. The beach ridges and their intervening swales are evidence that the bar has been formed by successive periods of bar building, the new additions being on the seaward side. Only after a bar attains appreciable size is there sufficient material for the building of large dunes by the wind.

TIDAL INLETS. Most offshore bars are not continuous features but are interrupted at intervals by transverse channels known as *tidal inlets*. These inlets permit the sea water to sweep in and out of the lagoon. Where the tidal range is very small, the tidal inlets are far apart. Thus, along the Texas coast, with a tidal range of but 1 or 2 feet, one offshore bar extends unbroken for about 100 miles; but along the Jersey coast where the range is from 4 to 5 feet, the inlets are more frequent.

The number and location of inlets bears an immediate relation to the direction of longshore currents because these currents are constantly providing material for the upbuilding of the bars.

The number of inlets gradually increases away from the source whence the currents derive their load. Not only are the waves less generously supplied with debris by the currents, but the bars themselves are smaller and more easily broken through by the waves and tides. Where the inlets are numerous, the lagoons are much more completely filled with marsh grass. This is because a uniform condition of salinity is maintained there from the numerous contacts with the ocean. Where the lagoons are large and more nearly isolated, the water is suited to neither fresh-water nor marine vegetation.

TIDAL DELTAS. The currents sweeping in and out of tidal inlets carry sand into the lagoon and also into the sea. Deltas are thus built. During storms waves break over a low portion of a bar and carry material back into the lagoon, depositing *wash-overs*. Almost every bar bears on its lagoon side a row of deltas thus formed. This causes the crenulate margin of the lagoon side in contrast with the simple seaward shore.

RETREAT OF BARS. The landward retreat of offshore bars causes them to rest upon the peat and muck deposits of the lagoon. This material is very compressible under the great weight of the bar thrown on top of it. As the bar is eroded on its seaward side, this depressed lagoon material may be exposed at low-tide level. This gives a false impression of subsidence, especially when stumps of trees are seen at that low level.

The entire bar, however, does not always come to rest upon lagoon filling because the various tidal inlets migrate along the coast in the direction of the longshore currents. This results in a constant scouring away of the lagoon material at the points where the inlets occur, so that when the bar again forms at those points, it rests upon the clean underlying sand of the inlet.

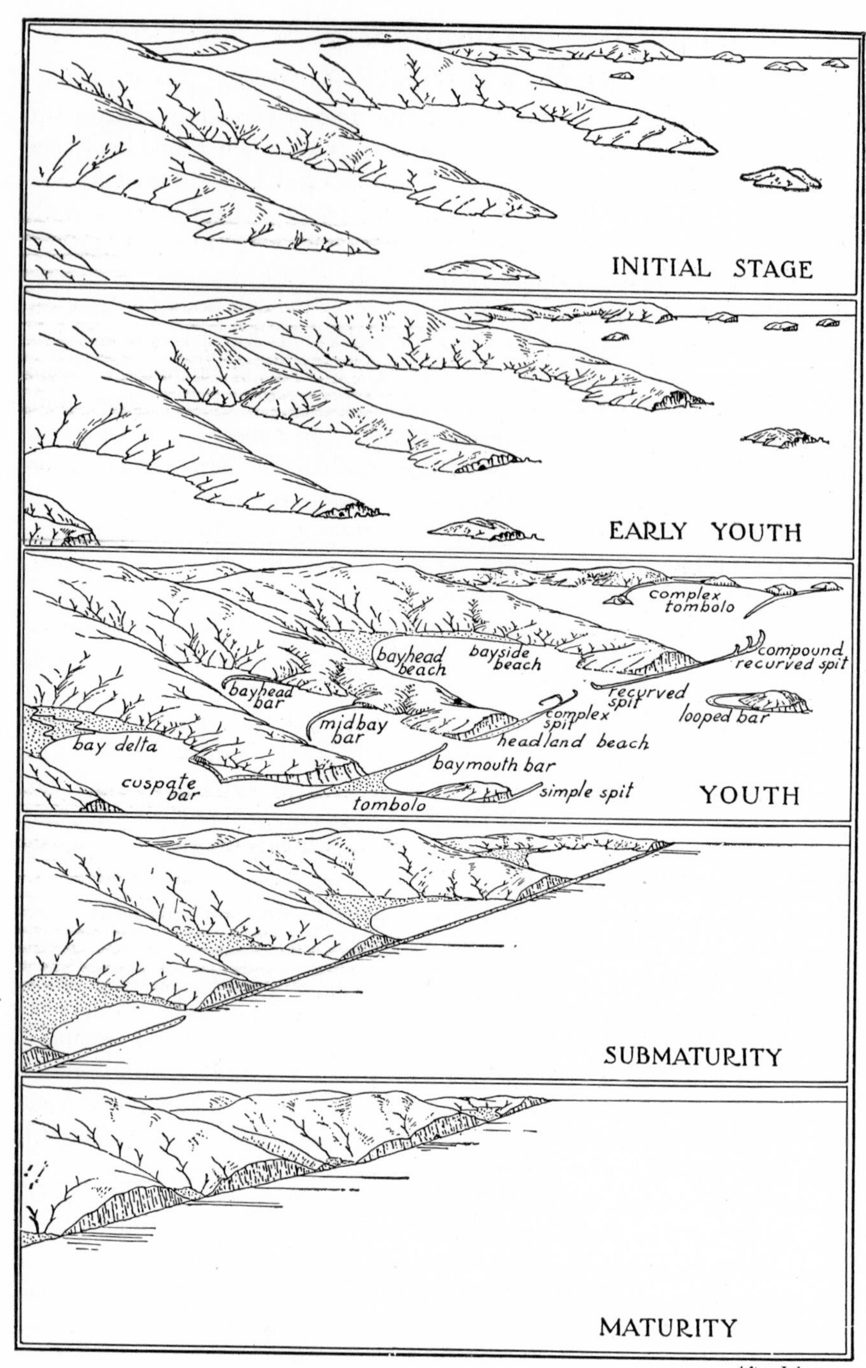


*After Johnson*

STAGES OF DEVELOPMENT OF SHORELINES OF SUBMERGENCE

## SHORELINES OF SUBMERGENCE. STAGES OF DEVELOPMENT

INITIAL STAGE. The initial stage of a shoreline of submergence is characterized by an exceedingly irregular shoreline; by numerous branching bays or drowned valleys; by many peninsulas; by a profusion of islands; and by an irregular sea bottom, whose inequalities represent the former hills and valleys of the land.

The initial stage of different shorelines of submergence may vary greatly, however, depending upon the structure of the region and the resulting topography developed prior to submergence.

YOUTH. During early youth, the waves beat upon the headlands and the exposed portions of islands and develop a minutely irregular or *crenulate* shoreline. This is due to the fact that every little joint and minor variation of structure influences the behavior of the waves so that the numerous small zones of weakness are etched out. The most picturesque features of cliff detail appear at this time. Pinnacles or isolated masses forming *chimneys* or *stacks* may be left standing in front of the main cliff. Weaker zones are excavated by the waves into *sea caves*. If a cave cuts through a projecting belt of rock, a *sea arch* is formed. Small and frequent landslides result from the rapid encroachment of the waves along the cliff base.

With the advance of youth there results the development of a great array of spits and beaches. The headlands and offshore islands suffer the most erosion. Cliffed headlands contribute the material which is carried by the currents. Beaches result from the accumulation of this material alongshore. Various names indicate the location of these beaches such as *headland beach*, *bayside beach*, and *bayhead beach*. If the debris carried in this way forms an embankment with one end attached to the mainland and the other terminating in open water, it is called a *spit*. Several forms of spits are distinguished, namely, the *simple spit*, the hooked or *recurved spit*, the *compound spit*, the *complex spit*. The term *bar* is also applied to spits built across from one headland to another. Such terms as *baymouth bar*, *midbay bar*, and *bayhead bar* distinguish their different locations. *Tombolos* result when offshore islands become attached to the mainland. Many other little forms are also associated with the youthful stage of development.

MATURITY. As youth advances into maturity, the numerous details are gradually lost. The headlands are cut back. Baymouth bars extend from one headland to another, leaving the bays cut off from the sea or filled in with detritus from the land.

Full maturity is attained when the shoreline has been pushed inland beyond the bayheads and lies against the mainland throughout its whole extent.

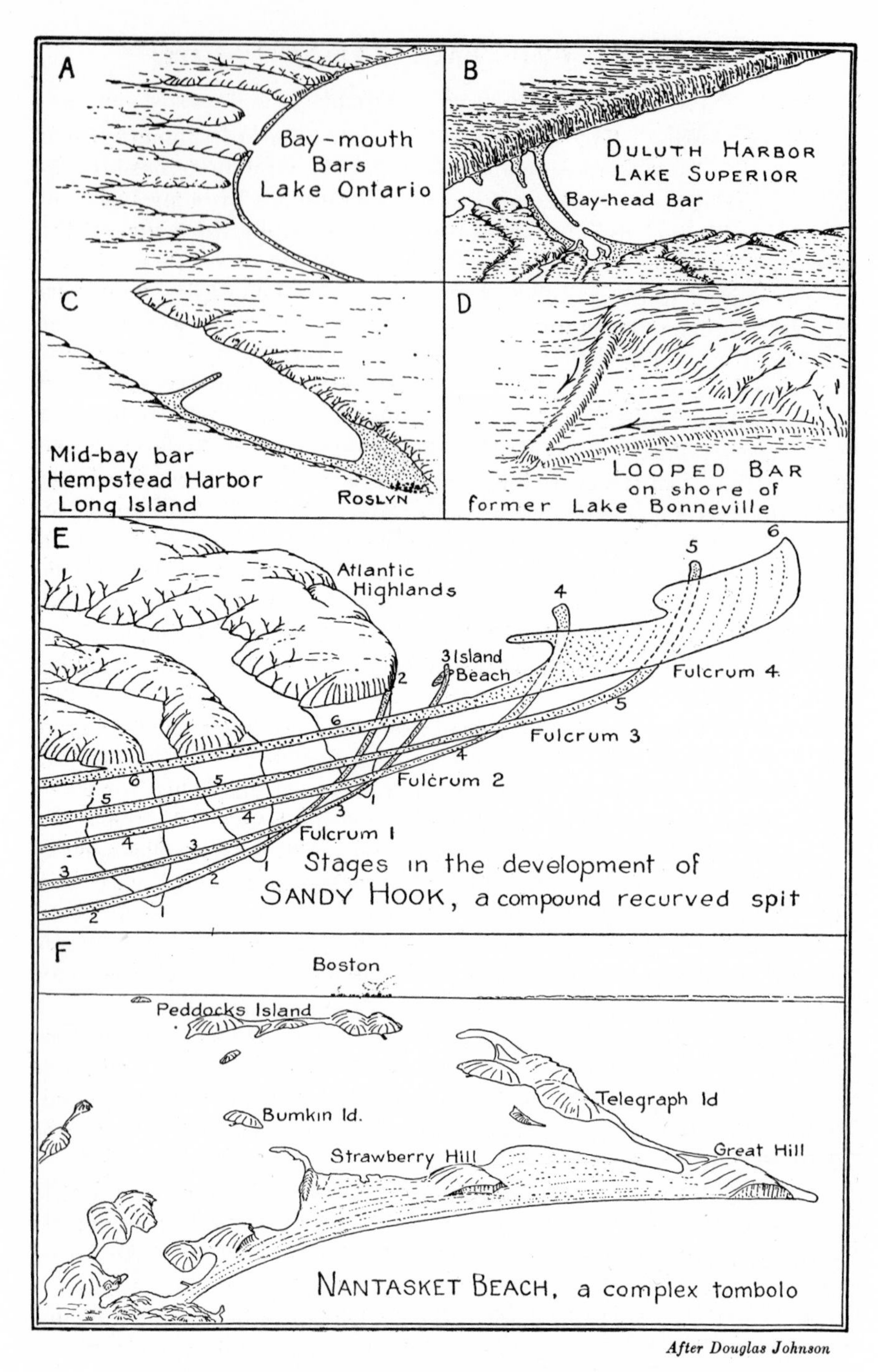


*After Douglas Johnson*

TYPES OF BARS, SPITS, AND TOMBOLOS

## BEACHES, SPITS, BARS, AND TOMBOLOS

Much of the debris eroded from headlands is not carried out into deeper water but is drifted by incoming tidal currents to the heads and sides of the bays where it forms bayhead and bayside *beaches*. In the initial stages this is especially apt to be true; but with the gradual shallowing of the bays and with the straightening of the shore due to the cutting off of the headlands, the longshore currents avoid the irregularities and sweep across the mouths of the bays. Under these conditions, the shore debris may be built out into the water in the form of a narrow embankment which grows by an excess of deposition at its seaward terminus. Such an embankment in time rises above water level and forms a *spit* or *bar*. Baymouth bars (Fig. *A*) run across from one headland to another. In some cases, however, the currents succeed in entering the bay to form bayhead or midbay bars. Usually the incoming tidal currents are stronger than the outgoing currents, with the result that the ends of most spits projecting from headlands are strongly recurved. A compound recurved spit (Fig. *E*) exhibits several recurved ends which represent successive stages in the development of the spit. In the later stages the end of the spit is usually not so strongly recurved as in the earlier stages. The successive points of intersection of each spit with the next succeeding stage are known as *fulcrums*, the latest fulcrums being nearest to the end of the spit. This principle of *migrating fulcrums* was first enunciated by Davis in his outline of Cape Cod and later was applied to Sandy Hook by Johnson. The successive embankments which are added to a growing compound spit give it a corrugated or ribbed aspect. They are called *beach ridges* and usually rise from 3 to as much as 20 feet above high-tide level and are separated from each other by corresponding depressions or *swales*.

*Complex spits* result from the development of minor or secondary spits like parasites on the ends or points of large ones. Cape Cod and Sandy Hook are both complex as well as compound.

*Cuspate bars* and *looped bars* constitute still other variations in the form of ridges built by currents. A *cuspate foreland* resembles a cuspate bar except that there is no lagoon, the entire deposit forming a continuous beach. Some cuspate forelands are very elaborate and exhibit series of ridges and swales occasionally intersecting each other at strong angles.

A tombolo is a bar together with an island which it connects with the mainland. Tombolos may be single or double or even triple or V-shaped when the island is attached to the mainland by two bars. *Complex tombolos* result when several islands are united with each other and with the mainland by a series of bars. Complicated beach patterns result when islands are destroyed and the material is swept inshore and used to prograde some of the bars previously built, as at Nantasket Beach.

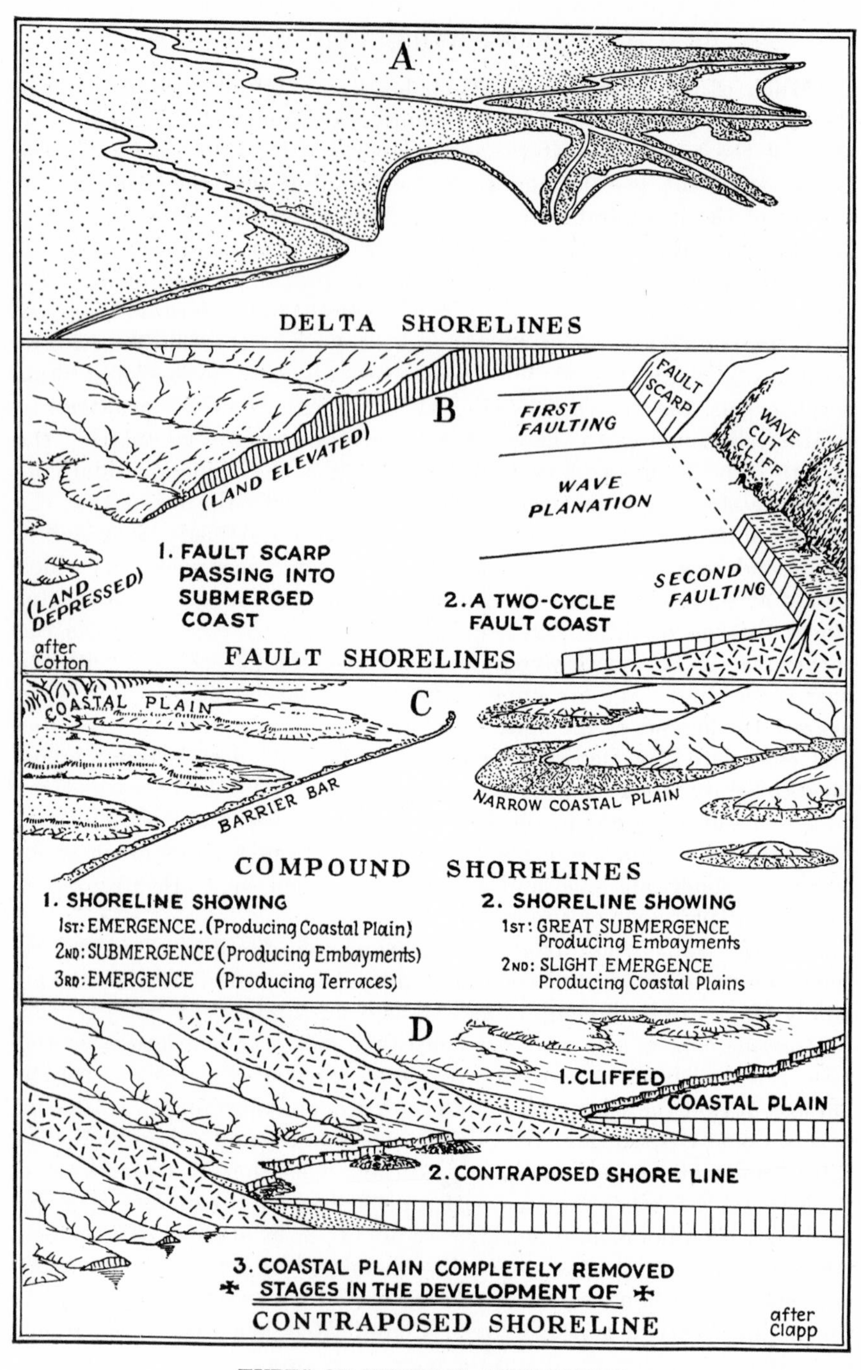


TYPES OF NEUTRAL SHORELINES

## NEUTRAL SHORELINES

DELTA SHORELINES, ALLUVIAL PLAIN, AND OUTWASH PLAIN SHORELINES. Shorelines formed by the deposition of alluvium in a body of water resemble in many respects shorelines of emergence. Their outline is relatively simple or gently sinuous. However, the water offshore usually deepens abruptly because the foreset beds of the delta or other deposit lie at a steep angle. This means that offshore bars are not so readily formed and the waves are able to come all the way to the shoreline before breaking.

Lobate or bird's-foot deltas may be very irregular in outline. In that case bars may be built across from one lobe to another. Eventually however, the lobes are cut back and a simple or arcuate shoreline results.

The youthful stages of all of these forms is generally characterized by rapid outward building. Maturity comes when the waves have destroyed the irregularities of youth and have produced a less devious shoreline. Perhaps in doing this, much of the delta has been removed. The various shapes of deltas are due both to the streams which produced them and to the waves which modified them later.

FAULT SHORELINES. Fault shorelines are of several types, depending upon the character of the region faulted and also depending upon whether the seaward block only has been depressed or whether both blocks have been altered in position relative to the sea.

If the seaward block only has been depressed and the landward region maintains its original position or is raised with respect to the sea, then an abrupt fault scarp will result and the streams from the mainland will cascade into the sea from the mouths of hanging valleys. Such a shoreline may resemble the mature stage of shorelines of submergence.

The contrast between youth and maturity in a fault shoreline is not so sharply defined as with other types. The more gentle marine cliff of the mature stage is in contrast with the steeper fault scarp of youth.

COMPOUND SHORELINES. Most shorelines result from both uplift and depression and therefore are compound. Several variations are shown in the accompanying illustration. *Contraposed shorelines* represent one type of compound shoreline. If the margin of a rugged oldland is bordered by a narrow coastal plain, the initial shoreline is developed upon the softer beds and will have some of the aspects of a shoreline of emergence. As erosion proceeds inland, the narrow coastal plain is destroyed, and the shoreline comes to rest upon the older resistant rocks. It may even change from a typical shoreline of emergence to one of submergence. The term *contraposed* is analogous to the term *superimposed* used for rivers which have been let down from an unconformable cover on to older rocks lying beneath.

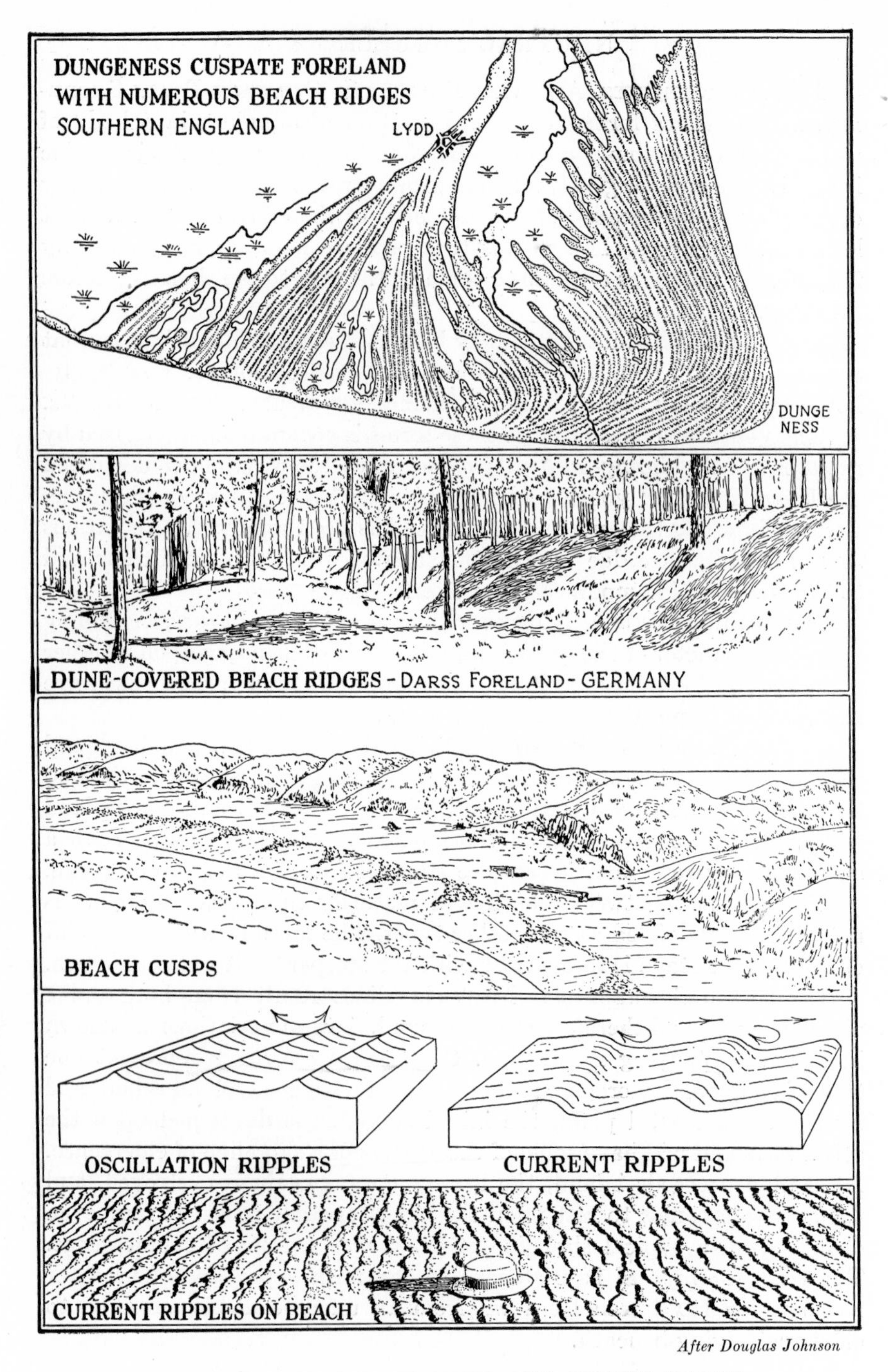


*After Douglas Johnson*

BEACH RIDGES, BEACH CUSPS, AND RIPPLE MARKS

## MINOR SHORE FEATURES

BEACH RIDGES. Beach ridges represent the successive positions of an advancing shoreline. The depressions between the ridges are known as *swales, slashes,* or *furrows.* Ridges and swales may occur on any beach or bar to which material has been added more or less regularly. Three methods for this have been suggested: (*a*) According to Gilbert, the debris of drifted sand is thrown up by the waves on the seaward side of the beach, each ridge being referable to some exceptional storm. (*b*) According to de Beaumont and Davis, the material is derived from the sea bottom, which is therefore deepened. Later ridges are due to successively more violent wave action. (*c*) A series of ridges may be formed at the end of a compound recurved spit by the addition of successive spits on the seaward side, the later ridges often truncating the earlier ones. But Johnson maintains that beach ridges cannot always be correlated with individual storms; that they are due more to fluctuations in the quantity of sand carried by longshore currents, which is controlled by the rate of wave erosion at other places.

Where abundant material is available, beach ridges may be added rapidly, especially at the ends of recurved spits. Within twenty-three years five ridges were formed at the end of Rockaway Beach near New York City, the end of the spit advancing at an annual rate of about 200 feet per year. Wide beaches like Cape Canaveral, Florida, and the Dungeness Foreland in England display dozens of parallel ridges, rising usually 3 to 6 feet above the intervening swales. The ridges stand about 200 feet apart from crest to crest.

BEACH CUSPS. Beach cusps are triangular accumulations of sand and other debris more or less regularly spaced along the shore, the long apex of the triangle pointing toward the water. Every gradation can be found. The cusps may be 1 to 30 feet from apex to base and as much as 100 feet apart measured from apex to apex. Johnson concludes that uniformly spaced cusps are not due to intersecting or oblique sets of waves impinging upon the shore but to the sweep of the water upon the small irregularities of a gently sloping surface. Absolute uniformity of slope and volume of water would result in perfect regularity and size.

RIPPLE MARKS AND OTHER DETAILS. Two types of ripples may be distinguished. Those due to waves, known as *oscillation ripples,* are symmetrical in profile; and those due to currents—*current ripples*—are asymmetrical. Oscillation ripples are infrequently seen because they are formed under deeper water. Current ripples may be observed on almost any tidal flat at low tide. They are relatively small, being 6 inches or less from crest to crest and an inch or so in height, but heights of 3 feet and distances of 10 to 30 feet from crest to crest are not uncommon. *Rill marks* are the miniature river systems formed on sand flats with the ebbing of the tide. *Swash marks* represent the innermost margin of wave advance. *Sand domes* are due to the expulsion of air from beneath the sand by advancing waves.

## THE GEOGRAPHICAL ASPECTS OF SHORELINES

Shorelines of emergence and shorelines of submergence the world over present contrasting types of human response. Shorelines of emergence usually have poor and widely spaced ports but a habitable hinterland often rich in agricultural and sometimes in mineral resources. Shorelines of submergence, on the other hand, have innumerable and excellent ports but a hinterland that is in many instances unsuited to a dense population and often quite deficient in agricultural possibilities. Compound shorelines, in some cases, combine the advantages of each type. Rarely do the disadvantages of each type occur on the same shoreline.

The Chesapeake Bay region of the Atlantic coastal plain is a compound shoreline presenting the advantages of a shoreline of emergence as well as those of a shoreline of submergence. This region has an extensive arable and fertile hinterland of low relief because it is an emerged coastal plain. It has also the advantages of good and protected harbors, such as Norfolk, Washington, and Baltimore.

Some shorelines of submergence, notably fiord shorelines, like those of Norway, Labrador, British Columbia, and southern Chile, are endowed with magnificent harbors but the rugged nature of the land renders such regions almost unfit for habitation. The people hug the coast, the houses of the farmer-fishermen occupying the few available patches of flat land or soil to be found near the water's edge. Along the fiords of Norway, and fringing many of the islands of that fretted coast, are narrow terraces of alluvium rarely 50 feet above the water's edge. These recently elevated parts of the sea floor are interrupted at numerous points where the bold headlands jut outward. The habitations are therefore widely scattered. Boats constitute the universal means of transportation. Great steamers ply several times a week from Stavanger in the south to Vardo and Vadso in the far north, a trip of well over 1,500 miles. They consume 7 to 8 days making the journey and stop only at the larger towns at the entrances to the fiords. Slightly smaller steamers serve the fiords and the islands lying offshore, making frequent stops at the numerous little villages. From these settlements, in turn, launches and motor boats run to all the nooks and corners of the interminable coast, linking up every inhabitant with the outside world. Everyone turns to the sea for a livelihood, and the sea in turn invites them to explore other parts of the world. The ancient Vikings and the Norsemen exemplified this tendency to wander far from their homes. What nations of the world are seafaring in their habits? Those which have rugged shorelines of submergence and an inhospitable hinterland. The land constantly repels while the sea beckons them on. The Greeks from their earliest history have been traders. Greek settlements fringe the shores of Asia Minor, even far east to the end of the Black Sea.

Scotland is a country of shipbuilding, not simply because she has the wherewithal to build ships but because she has the urge to use the sea for her livelihood. The French are a home-loving people because their land is rich and habitable. Whence, then, does she get sailors to man her vessels? From Brittany, the one part of France which has a rugged, embayed coast and which is a land of rocks and hills. Nine-tenths of the sailors in the French navy come from Brittany. And as for our own country, what part of the United States has turned its attention to shipbuilding, to fishing, to whaling? The answer again is that part which has the most irregular coast line—New England.

Turn now to the southern states, Virginia, the Carolinas, Georgia, Alabama, and Mississippi. Do we find the people there seeking a livelihood on the ocean? Not at all. The land satisfies them. They are interested in the sea only in that it provides an outlet for their resources. The result is that much of the trade of the south is carried in foreign ships. From Norfolk southward there are few good ports and this is due to the low shelving nature of the coastal plain. Ports like Savannah, Brunswick, and Jacksonville are river ports and they lack the spacious harbors which estuaries provide.

The population of a region having submerged shorelines and a rugged hinterland rarely shows a steady growth. It is apt to reach a certain figure and remain fixed or even to decline, as in New England. Some of the counties along the coast of Maine are experiencing an actual decrease in the number of permanent inhabitants. A stationary population is due to the exodus of some of the people each year to offset the natural increase.

In contrast with the countries previously mentioned, a peculiar response may result when the fertile and habitable part of a country is more or less cut off from the coast by a mountain or desert barrier. This is the condition in Yugoslavia. The interior valleys of the Drave and the Save Rivers harbor most of the inhabitants of the country. The rugged Dalmatian coast is isolated by the barrier of the almost impassable Dinaric Alps. The result is that it is open to settlement by seafaring people from other countries. Because of this the Dalmatian coast of Yugoslavia has many Italian settlements and is dominated by Italian cultural influences. So true is this that during peace negotiations after the World War, the Italians used this fact as a basis for their claim to possession of most of this coast.

An apparent exception to this explanation for prowess on the sea may be found in the case of Germany. The Germans were not forced to the sea by an uninhabitable land nor did they find the sea penetrating to the very doors of their households and so inviting them away. Quite the reverse. The Germans have always been a home-loving people, and long before they turned to commerce they devoted their energies to developing their own resources and capabilities at home. The British, Spanish, the Scandinavians, the French, all surpassed the Germans in seafaring throughout most of their history. Germany has taken to the sea only recently in order to find an outlet for the results of her great industrial development.

## MAPS ILLUSTRATING TYPES OF COASTS AND WORK OF WAVES AND CURRENTS

Coasts due to the submergence of rugged regions of complex rocks, are shown on the *Casco Bay, Penobscot Bay, Boothbay, Machias,* and *Bath, Maine,* sheets. All of these have numerous embayments and islands. The *Bar Harbor, Maine,* sheet (which is included on the *Acadia National Park, Maine,* map) is especially attractive and shows, in addition to features of submergence, many small coastal details such as bars, spits, and tombolos. The *Portland, Maine,* map illustrates a compound coast, as do several of the other maps, for it shows a narrow coastal plain of flat topography surrounding the more rugged hills of hard rock.

Coasts due to the submergence of a dissected low coastal plain are admirably represented by the *Kilmarnock, Va.,* the *Barnegat, N. J.,* the *Ocean City, Md.-Del.,* and the *Rehoboth, Del.,* maps, the last map showing by its terraces the effect of recent emergence.

Simple offshore bars, characteristic of young shorelines of emergence, are shown on numerous maps of which the following are particularly striking: the *Green Run, Md.-Va.; Barnegat, Atlantic City,* and *Long Beach, N. J.; Fire Island* and *Islip, N. Y.; Mary Esther, Villa Tasso, Fla.,* and *Lake Como, Texas,* the last map being on an unusually large scale. The *Mayport, Fla.,* sheet shows multiple offshore bars.

Multiple beach ridges are shown on the *Johnsons Bayou, Bayou Labauve, Grand Bayou,* and *Cameron, La.,* sheets and on several of the maps adjacent to the Great Lakes, such as the *Berea* and *Oberlin, Ohio,* maps, *Oak Orchard, N. Y.,* and the *Mt. Clemens* and *Bay City, Mich., Calumet City, Ill.-Ind., Toleston, Ind.,* and *Brockport, N. Y.,* sheets. Beach ridges along the old shore of Lake Agassiz are shown on the *Emerado* and *Larimore, N. Dak.,* maps. In the Great Basin, beach ridges of one of the old Quaternary lakes appear on the *Carson Sink, Nev.,* map. Raised beaches and wave-cut benches are shown on the *Capitola, Dume Point, Oceanside,* and *San Diego, Calif.,* sheets.

Splendid compound spits are shown on the *Erie, Pa., Provincetown, Mass.,* and *Sandy Hook, N. J.,* sheets. Tombolos appear on the *San Francisco, Calif., Oyster Bay, N. Y.-Conn., Biddeford, Maine,* and *Boston Bay, Mass.,* sheets. The following maps show many types of bars and other details: *Falmouth, Mass.* (baymouth bars); *Braddock Heights, N. Y.* (baymouth bars and spits); *Point Sur, Calif.* (young shore with stacks, clefts, etc.).

The following charts of the U. S. Coast and Geodetic Survey may be taken as examples of the numerous charts deserving of study: Nos. 103, 105, and 107, submerged rugged coast; 119, 121, 122, and 123, offshore bars; 145 and 147, cuspate bars; 110, remarkably beautiful map of the compound spit of Cape Cod with exquisite detail; 5126 and 5143, marine benches. The *Cabo Rojo, Puerto Rico,* sheet shows two superb examples of complex tombolos.

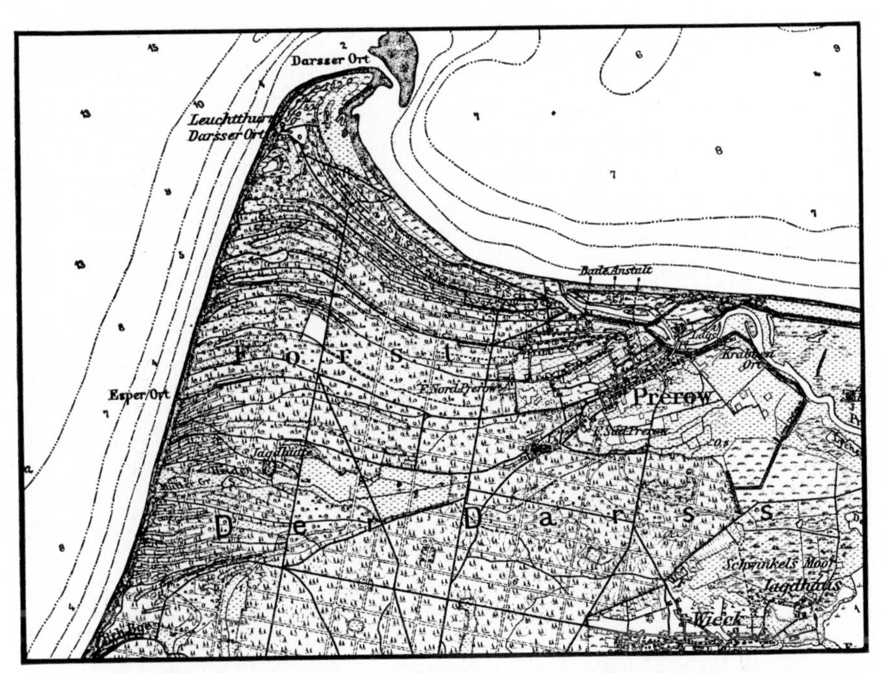


THE DARSS FORELAND, A CUSPATE BAR ON THE BALTIC COAST
Germany; *Barth* sheet, No. 62 (1:100,000).

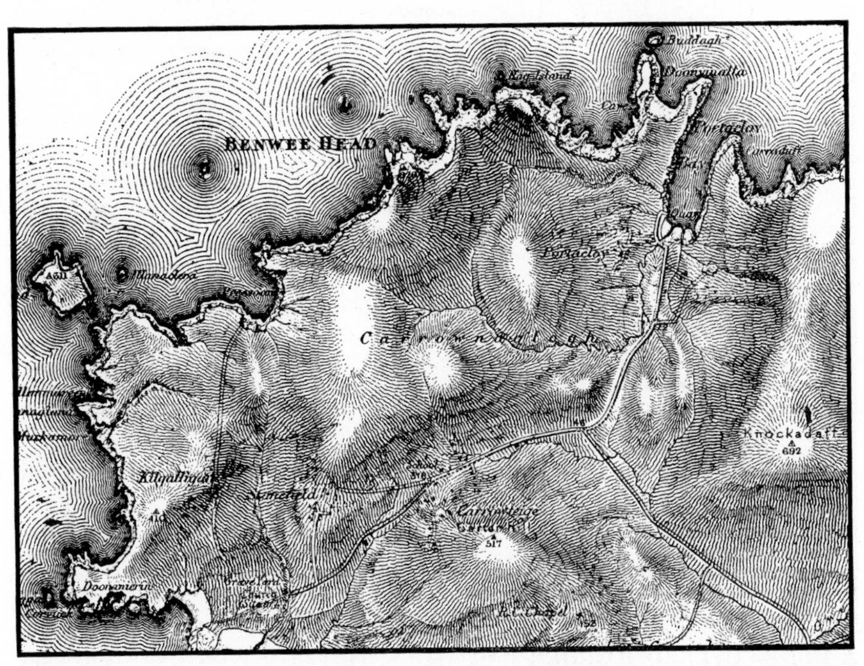


VERY YOUNG SUBMERGED COAST OF IRELAND WITH MANY FINE DETAILS
British Ordnance Survey sheet, No. 40 (1:63,360).

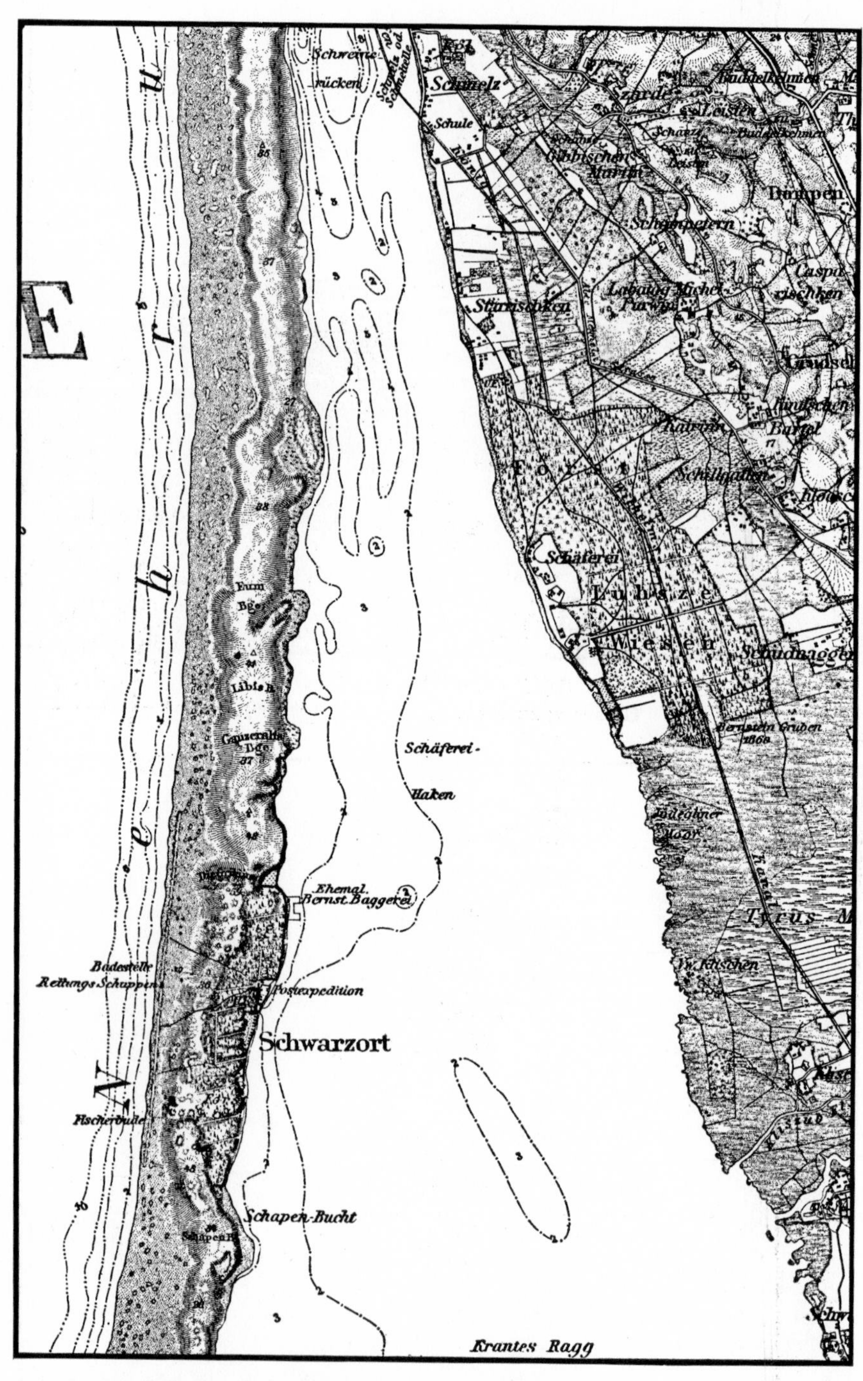


A DUNE-COVERED BARRIER BAR ALONG THE COAST OF EAST PRUSSIA
Germany; *Memel* sheet, No. 3 (1:100,000).

## QUESTIONS

1. Can a shoreline be both a shoreline of emergence and one of submergence at the same time?
2. Are barrier bars built around delta coasts?
3. What causes a spit to be recurved? What causes a spit to be compound? To be complex?
4. What is a contraposed shoreline?
5. What is a prograding shoreline and what causes it?
6. What is the difference between a beach and a bar?
7. What causes waves to stop cutting into a cliff and to build a beach in front of it?
8. What is meant by eustatic change of sea level?
9. Are ice-push ridges to be considered as shore-line features?
10. How high above sea level can waves build a barrier bar?
11. Does drowning of a coast line cause rivers entering the sea at that point to aggrade their valleys?
12. What events may cause the apparent subsidence of a lagoon marsh surface?
13. Can storm waves cut a terrace above sea level?
14. What is a *solution bench?* (see Wentworth).
15. Where do fault shorelines occur?
16. What is the difference between an oscillation ripple and a current ripple? Draw a profile of each.
17. Give several examples of shorelines of emergence? Of submergence? Of compound shorelines? Of neutral shorelines.
18. Why is the illustration on page 332 said to represent a shoreline of submergence when it is obvious that the beds were originally deposited in the sea and must have been exposed by emergence?
19. What evidence is there from the picture on page 334 that this is a compound spit? That it is a complex spit?
20. What evidence is there that the withdrawal of the lake waters from the area shown on page 335 took place slowly?
21. On page 348 label the barrier bar, lagoon, wash-overs and tidal delta. Does the Darss Foreland, mapped on page 361, preserve the complete cuspate form or has it been altered by wave erosion?
22. In what way are artesian conditions under the continental shelf related to the formation of submarine canyons? Is it possible that sub-oceanic springs along the seaward front of the continental shelf may account for these features?

## TOPICS FOR INVESTIGATION

1. Shorelines. Classification of the various types.
2. Changes of sea level. Causes. Extent.
3. Spits. Complex and compound. How formed.
4. Barrier beaches. Theories as to origin.
5. False evidence of subsidence.
6. The shore profile. What is a profile of maturity in relation to the load of the waves and currents?
7. Beach cusps and their origin.
8. Wave motion. Depth of wave erosion.
9. Marine *vs.* subaerial peneplanation.
10. Submarine canyons. Theories as to origin.
11. Fault shorelines. Criteria for their recognition. Their erosional development.

## REFERENCES


General

Cornish, V. (1910) *Waves of the sea and other water waves.* Chicago, 374 p.

Davis, W. M. (1912) *Erklärende Beschreibung der Landformen.* Leipzig. Der marine Zyklus, p. 463–554.

Fenneman, N. M. (1902) *Development of the profile of equilibrium of the subaqueous shore terrace.* Jour. Geol., vol. 10, p. 1–32.

Furneaux, W. S. (1911) *The sea shore.* New York, 436 p.

Gilbert, G. K. (1885) *Topographic features of lake shores.* U. S. Geol. Surv., 5th Ann. Rept., p. 69–123.

Gulliver, F. P. (1899) *Shoreline topography.* Am. Acad. Arts and Sci., Proc. 34, p. 149–258.

Johnson, D. W. (1919) *Shore processes and shoreline development.* New York, 584 p.

Reid, C., and Mathews, E. R. (1906) *Coast erosion.* Geog. Jour., vol. 28, p. 487–495.

Shaler, N. S. (1894) *Sea and land. Features of coasts and oceans.* New York, 252 p.

Shepard, F. P. (1937) *Revised classification of marine shorelines.* Jour. Geol., vol. 45, p. 602–624.

Changes in Sea Level, Marine Benches

Cooke, C. W. (1930) *Correlation of coastal terraces.* Jour. Geol., vol. 38, p. 577–589.

de Geer, G. (1893) *On Pleistocene changes of level in eastern North America.* Am. Geol., vol. 11, p. 22–44.

Johnson, D. W. (1932) *Principles of marine-level correlation.* Geog. Rev., vol. 22, p. 294–298.

Marmer, H. A. (1932) *Mean sea level as a geophysical datum.* Am. Jour. Sci., 5th ser., vol. 24, p. 35–45.

McGee, W J (1892) *The Gulf of Mexico as a measure of isostasy.* Am. Jour. Sci., 3d ser., vol. 44, p. 177–192. Evidences of depression.

Nansen, F. (1905) *Oscillations of shore lines.* Geog. Jour., vol. 26, p. 604–616.

Shaler, N. S. (1895) *Evidences as to change of sea level.* Geol. Soc. Am., Bull. 6, p. 141–166.

Sheppard, T. (1909) *Changes on the east coast of England within the historical period.* Geog. Jour., vol. 34, p. 500–513.

Tarr, R. S. (1897) *Changes of level in the Bermuda Islands.* Am. Geol., vol. 19, p. 293–303.

Wentworth, C. K. (1938) *Marine bench-forming processes: water-level weathering.* Jour. Geomorphology, vol. 1, p. 6–32. *Solution benching.* Same journal, vol. 2, p. 3–25.

Wentworth, C. K., and Palmer, H. S. (1925) *Eustatic bench of islands of the north Pacific.* Geol. Soc. Am., Bull. 36, p. 521–544.

Spits, Capes, Etc.

Abbe, C. (1895) *The cuspate capes of the Carolina coast.* Boston Soc. Nat. Hist., Proc. 26, p. 489–497.

Davis, W. M. (1896) *The outline of Cape Cod.* Am. Acad. Arts and Sci., Proc. 31, p. 303–332; Geog. Essays, p. 690–724.

Gulliver, F. P. (1896) *Cuspate forelands.* Geol. Soc. Am., Bull. 7, p. 399–422.

Johnson, D. W., and Reed, W. G. (1910) *The form of Nantasket Beach, Mass.* Jour. Geol., vol. 18, p. 162–189.

Tarr, R. S. (1898) *Wave-formed cuspate forelands.* Am. Geol., vol. 22, p. 1–12.

Wilson, A. W. G. (1904) *Cuspate forelands along the Bay of Quinte.* Jour. Geol., vol. 12, p. 106–132.

Barrier Beaches and Lagoons

Cornish, V. (1898) *Sea beaches and sand banks.* Geog. Jour., vol. 11, p. 528–543, 628–658.

Davis, C. A. (1910) *Salt marsh formation near Boston and its geological significance.* Econ. Geol., vol. 5, p. 623–639.

Hitchcock, C. B. (1934) *The evolution of tidal inlets.* Geog. Rev., vol. 24, p. 653–654.

Hite, M. P. (1924) *Some observations of storm effects on ocean inlets.* Am. Jour. Sci., 5th ser., vol. 7, p. 319–326.

Lucke, J. B. (1934) *A study of Barnegat Inlet, N. J., and related shore-line phenomena.* Shore and Beach, vol. 2, p. 45–93.

Lucke, J. B. (1934) *A theory of evolution of lagoon deposits on shorelines of emergence.* Jour. Geol., vol. 42, p. 561–584.

MacCarthy, G. R. (1933) *The rounding of beach sands.* Am. Jour. Sci., 5th ser., vol. 25, p. 205–224.

Patton, R. S. (1931) *Moriches Inlet: a problem in beach evolution.* Geog. Rev., vol. 21, p. 627–632.

Raisz, E. (1934) *Rounded lakes and lagoons of the coastal plains of Massachusetts.* Jour. Geol. vol. 42, p. 839–848.

Shaler, N. S. (1885) *Seacoast swamps of the eastern United States.* U. S. Geol. Surv., 6th Ann. Rept., p. 353–398.

Shaler, N. S. (1895) *Beaches and tidal marshes of the Atlantic coast.* Natl. Geog. Soc., Mon. 1, p. 137–168.

Stearns, H. T. (1935) *Shore benches on the island of Oahu, Hawaii.* Geol. Soc. Am., Bull. 46, p. 1467–1482. Pictures.

Thompson, W. O. (1937) *Original structures of beaches, bars, and dunes.* Geol. Soc. Am., Bull. 48, p. 723–751.

Features of Emergence

Bartrum, J. A. (1926) *Abnormal shore platforms (New Zealand).* Jour. Geol., vol. 34, p. 793–806.

Carney, F. (1911) *The abandoned shore lines of the Vermilion quadrangle, Ohio.* Denison Univ., Sci. Lab., Bull. 16, p. 233–244.

Carney, F. (1916) *The abandoned shore lines of the Ashtabula quadrangle, Ohio.* Denison Univ., Sci. Lab., Bull. 18, p. 362–369.

Carney, F. (1916) *The shore lines of glacial lakes Lundy, Wayne, and Arkona, of the Oberlin quadrangle, Ohio.* Denison Univ., Sci. Lab., Bull. 18, p. 356–361.

Gilbert, G. K. (1890) *Lake Bonneville.* U. S. Geol. Surv., Mon. 1, p. 23–187.

Goldthwait, J. W. (1907) *The abandoned shorelines of eastern Wisconsin.* Wisc. Geol. Surv., Bull. 17, 134 p.

Goldthwait, J. W. (1911) *The twenty-foot terrace and seacliff of the lower St. Lawrence.* Am. Jour. Sci., 4th ser., vol. 32, p. 291–317.

Ladd, H. S. (1930) *Vatu Lele, an elevated submarine bank.* Am. Jour. Sci., 5th ser., vol. 19, p. 435–450.

Putnam, W. C. (1937) *The marine cycle of erosion for a steeply sloping shoreline of emergence.* Jour. Geol., vol. 45, p. 844–850.

Smith, W. S. T. (1898) *A geological sketch of San Clemente Island.* U. S. Geol. Surv., 18th Ann. Rept., part 2, p. 459–496.

Smith, W. S. T. (1933) *Marine terraces on Santa Catalina Island.* Am. Jour. Sci., 5th ser., vol. 25, p. 123–136.

Subsidence and Assumed Subsidence

Cook, G. H. (1857) *On the subsidence of the land on the seacoast of New Jersey and Long Island.* Am. Jour. Sci., 2d ser., vol. 24, p. 341–355.

Dawson, J. W. (1855) *On a modern submerged forest at Fort Lawrence, Nova Scotia.* Geol. Soc. London, Quart. Jour., vol. 11, p. 119–122.

Johnson, D. W. (1910) *The supposed recent subsidence of the Massachusetts and New Jersey coasts.* Science, new ser., vol. 32, p. 721–723.

Johnson, D. W. (1913) *Botanical phenomena and the problem of recent coastal subsidence.* Botanical Gazette, vol. 56, p. 449–468.

Johnson, D. W., and Smith, W. S. (1914) *Recent storm effects on the northern New Jersey shoreline and their supposed relation to coastal subsidence.* N. J. Geol. Surv., Bull. 12, p. 27–44.

Johnson, D. W., and Winter, E. (1927) *Sea-level surfaces and the problem of coastal subsidence.* Am. Phil. Soc., Proc. 66, p. 465–496.

Lyon, C. J., and Goldthwait, J. W. (1932) *Study of drowned forests in New England and Nova Scotia.* Carn. Inst. Wash., Yearbook 31, p. 346–347.

Submarine Canyons

Daly, R. A. (1936) *Origin of submarine canyons.* Am. Jour. Sci., 5th ser., vol. 31, p. 401–420.

Johnson, D. W. (1938) *Origin of submarine canyons.* Jour. Geomorphology, vol. 1, p. 111–129, 230–243, 324–340; vol. 2, p. 42–58. Excellent bibliography of numerous recent papers. (This essay is continued in later numbers.)

Lindenkohl, A. (1885, 1891) *Submarine channel of the Hudson River.* Am. Jour. Sci., 3d ser., vol. 29, p. 475–480; vol. 41, p. 489–499.

Shepard, F. P. (1932) *Landslide modifications of submarine valleys.* Am. Geoph. Un., 13th Ann. Meeting, Trans., p. 226–230.

Shepard, F. P. (1933) *Submarine valleys.* Geog. Rev., vol. 23, p. 77–89. See also vol. 28, p. 439–451.

Spencer, J. W. (1905) *The submarine great canyon of the Hudson River.* Am. Jour. Sci., 4th ser., vol. 19, p. 1–15, 341–344.

Spencer, J. W. (1903) *Submarine valleys off the American coast and in the North Atlantic.* Geol. Soc. Am., Bull. 14, p. 207–226.

Regions Described

Cotton, C. A. (1916) *Fault coasts in New Zealand.* Geog. Rev., vol. 1, p. 20–47.

Davis, W. M. (1923) *The Halligs, vanishing islands of the North Sea.* Geog. Rev., vol. 13, p. 99–106.

Davis, W. M. (1926) *The Lesser Antilles.* New York, 207 p.

Davis, W. M. (1933) *Glacial epochs of the Santa Monica Mountains, California.* Geol. Soc. Am., Bull. 44, p. 1041–1133. Coastal evolution.

Fairbanks, H. W. (1897) *Oscillations of the coast of California during the Pliocene and Pleistocene.* Am. Geol., vol. 20, p. 213–245.

Gilbert, G. K. (1890) *Lake Bonneville.* U. S. Geol. Surv., Mon. 1, 438 p.

Gulliver, F. P. (1903–1910) *Nantucket shorelines.* Geol. Soc. Am., Bull. 14, p. 555; Bull. 15, p. 507–522; Bull. 20, p. 670.

Johnson, D. W. (1925) *The New England-Acadian shoreline.* New York, 608 p.

Lawson, A. C. (1893) *The post-Pliocene diastrophism of the coast of southern California.* Univ. Calif., Dept. Geol., Bull. 1, p. 115–160.

Russell, I. C. (1885) *Lake Lahontan.* U. S. Geol. Surv., Mon. 11, 288 p.

Sharp, H. S. (1929) *The physical history of the Connecticut shore line.* Conn. State Geol. and Nat. Hist. Surv., Bull. 46, 97 p.

Taber, S. (1934) *Sierra Maestra of Cuba.* Geol. Soc. Am., Bull. 45, p. 567–620. Fault scarps in marine benches.

Ward, E. M. (1922) *English coastal evolution.* London, 262 p.

Wilson, A. W. G. (1908) *Shoreline studies on Lakes Ontario and Erie.* Geol. Soc. Am., Bull. 19, p. 471–500.

Geographical Aspects

Matthews, E. R. (1913, 1918) *Coast erosion and protection.* London, 195 p.

# XI
# WIND

*U. S. Department of the Interior*

GYPSUM DUNES IN THE WHITE SANDS NATIONAL MONUMENT, NEW MEXICO

The prevailing direction of wind, toward the right, accounts for the barchane type of dune with the steep side away from the wind.

*Soil Conservation Service*

A RECENTLY FORMED CRESENT-SHAPED DUNE, DALLAM COUNTY, TEX.
The hard land around this dune has been listed (deeply plowed to turn up the subsoil) and planted to prevent further dune accumulation. This dune will gradually be dissipated.



BARCHANES AT BIGGS, ORE., ON THE COLUMBIA RIVER
These dunes are moving toward the right, the steep side being away from the wind.

*Gilbert, U. S. Geological Survey*

*U. S. Department of the Interior*

STEEP FACE OF ADVANCING DUNES OF GYPSUM SAND, WHITE SANDS NATIONAL MONUMENT, N. M.

These dunes are advancing over the level desert at an average rate of eight inches a year.



FRONT OF ADVANCING DUNES, SHORE OF LAKE MICHIGAN, DUNE PARK, IND.

The steep face of these dunes is about 35 degrees from the horizontal which is the angle of repose for material of this size.

*Stose, U. S. Geological Survey*

*U. S. Forest Service*

DUST WHIRLWIND, NEW MEXICO

Of local importance only, and rarely producing much destruction by the removal of soil.



APPROACH OF DUST STORM, UNION COUNTY, N. M., MAY 21, 1937

The dust cloud approached on a light wind and lasted 30 minutes, producing absolute darkness and filling even tightly closed rooms with fine choking dust.

*U. S. Soil Conservation Service*

*U. S. Soil Conservation Service*

DUST STORM, STANTON COUNTY, KAN., APRIL 14, 1936

As the dust storm passed over the region, complete darkness prevailed. Millions of tons of soil were removed by this storm, and transported hundreds of miles into more humid regions.



DURING A HEAVY DUST STORM, BEADLE COUNTY, S. D., OCTOBER 15, 1935

Human figure going south into the dust-laden wind which was blowing about 26 miles per hour. Typical scene in the "dust bowl" of America from which some of the dust is carried 2000 miles to the Atlantic coast.

*U. S. Soil Conservation Service*

*U. S. Soil Conservation Service*

DESERTED HOMESTEAD, DALLAM COUNTY, TEX., OCTOBER 1, 1937

Fences, farm machinery, and buildings buried under a five-foot accumulation of wind-blown dust.



FIELD IN OKLAHOMA PROTECTED FROM WIND EROSION BY LISTED FURROWS

These deeply plowed furrows turn up the heavy clods of subsoil which form a covering not so readily blown away. In the furrows wind-blown dust from unprotected fields accumulates.

*U. S. Soil Conservation Service*

*Jackson, U. S. Geological Survey*

CAVE ROCKS NEAR SIERRA LA SAL, DRY VALLEY, UTAH

Wind-swept plain and wind-scoured mesa in a region where the wind is a powerful eroding and transporting agent.



MEDICINE ROCKS NEAR BAKER, MONT.

A semiarid region where wind abrasion and wind scour are important factors in rapidly removing the products of rock disintegration.

*U. S. Forest Service*

*Walcott, U. S. Geological Survey*

WIND ERODED SANDSTONE IN THE ROCKY MOUNTAIN REGION

The sand grains have been sand-blasted away but the more durable binding silica, filling innumerable veins, has resisted the attack.

# THE WORK OF THE WIND

OUTLINE OF THE CHAPTER.

1. *Erosional Work of the Wind.* The wind erodes by deflation, producing hammadas or desert pavement in basins or bolsons; by abrasion, producing rock pedestals, desert lattices, wind-faceted pebbles or dreikanter.

2. *Transportation by the Wind.* The wind carries vast quantities of material, 10,000,000 to 100,000,000 tons in a single storm, for distances as great as 2,000 miles.

3. *Deposition by the Wind. Dunes.* Dunes or dune ridges may be transverse to the wind, if the strength of the wind be moderate; or parallel to the wind, if the wind be strong. Barchanes are intermediate in type.

4. *The Dune Regions of the World.* Dunes occur along river flood plains, as along the Columbia River; along the seacoast, as at Cape Cod, Cape Henry, France, Belgium, Holland, and Denmark; and in the interior of continents under desert conditions. Ancient dunes cover large parts of the Great Plains of Nebraska.

5. *Deposition by the Wind. Loess.* Loess is extremely fine, impalpable calcareous material transported by the wind from desert areas or from glacial outwash and river flood plains, deposited usually in humid or semihumid conditions.

6. *Erosional Features of Loess Desposits.* In their youthful stages loess deposits exhibit natural wells, gashes, tunnels, and sinks; in maturity they have sharply cut canyons with numerous side ravines which head in wall divides; in old age there are many spires and pinnacles standing on open flats.

7. *The Loess Regions of the World.* Great loess deposits occur in the Mississippi Valley derived largely from river flood plains; in central Europe derived largely from glacial outwash; in northern China derived from the Gobi Desert; and in northern Argentina derived from the deserts near the Andes.

8. *Economic Aspects of Wind-blown Deposits.* Houses are built of loess mud and are cut into loess cliffs; roads across loess uplands cut deep gashes or canyons; loess soil is extremely fertile. Sands take up moisture, concentrating it in basins to form oases. Soil removal by the wind and also the encroachment of sand upon farmland are disastrous to crops.

9. *Maps.*

10. *Questions and References.*

## EROSIONAL WORK OF THE WIND

Under the stimulus of aridity, wind scour becomes an erosive agent more constant than the rain, more potent than streams, and more persistent than the sea. The factors which encourage wind erosion are deficient rainfall, clear skies, high evaporation, great range of diurnal temperature, and sparse vegetation. The wind accomplishes its work in two ways: (*a*) by deflation, or the removal of material already produced by weathering and (*b*) by the actual abrasion of rocks after the manner of a sand blast.

DEFLATION. Desert areas containing rocks susceptible to weathering, and especially rocks apt to be broken up by changes of temperature, are easy prey to the wind. Some students of the great arid tracts of the American southwest, of the Kalahari Desert of South Africa, and of the Gobi Desert believe that desert leveling is due largely to the wind. Other investigators, however, believe that extensive beveled rock surfaces in desert regions have been formed by stream wash and that they are produced in the same manner as rock pediments.

BOLSON PLAINS. In southern New Mexico and in Texas the broad intermontane plains known as *bolsons* (from the Spanish word for purse) are largely wind-excavated depressions. Some of them contain deep alluvial accumulations washed into them from the surrounding mountains. But in many of the depressions the rock floor is exposed.

The desert plains of the Kalahari region, according to Passarge, resulted from wind planation; and he maintains the efficacy of the wind to produce peneplanes at different levels in different parts of a large region without any common base-level. Remnants standing above these peneplanes he calls *inselberge*, or island mountains.

HAMMADAS. It is well-known that many desert tracts are covered with only a thin veneer of sand and that bedrock is practically exposed at the surface. In some cases the wind removes the smaller particles, leaving the larger pebbles and rock fragments to form a *desert pavement* or *hammada*. This is the origin of the stony deserts of northern Africa, from which the sand has been largely removed. The remaining rock fragments tend to halt wind action. In many regions all of the weathered products are removed.

Sheet flood erosion in desert areas is effective in carrying material to the basin floors and into temporary playa lakes. The material carried in this manner is in many cases quickly removed by the wind down to bedrock.

The desert section of southern California reveals clearly the potency of the wind in removing such lake deposits. Many branching gullies formed by streams wind their way among small mesalike hills which represent the original floor of the lake. But these gullies converge toward a common depression at the lowest point in the basin. Since no outlet for the streams is possible there, it is clear that the work of excavation and the actual removal of the material brought there by the streams must have been accomplished by wind action.

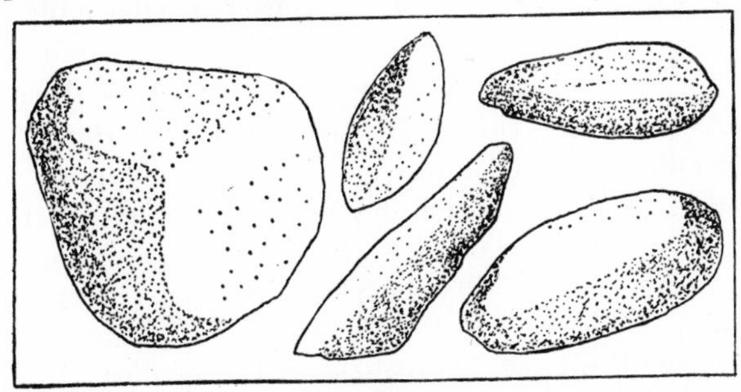

LAKE BASINS CREATED BY WIND EROSION. Even in less arid regions, the wind locally may excavate shallow, saucerlike hollows where the rocks rapidly disintegrate and where vegetation is scanty, due to the sterile condition of the resulting soil. Many basins of this type containing lakes were noted by Gilbert in parts of the Arkansas valley. These are often called *blowouts.*

WIND ABRASION. Travelers by automobile over the dusty plains of Wyoming know that a single driving dust storm may render their windshields completely nontransparent. The effect is that of a powerful sand blast. On the open plains, telegraph poles and fence posts are frequently cut down by the sand unless protected by a pile of stones around their base, or sheathed in metal.

Wind-faceted pebbles or ventifacts or *dreikanter* (three corners), shaped something like Brazil nuts, are common in many desert regions and along beaches where strong winds prevail. Their various forms have received much study and it appears that certain shapes are produced by constant winds and others by winds of variable direction.

Still other effects of wind abrasion are rock pedestals, mushroom rocks, rock lattices, and the many picturesque forms of desert landscapes.

## TRANSPORTATION BY THE WIND

Quantity of Material Transported. The effectiveness of the wind as a transporting medium depends not only upon its velocity but to a large extent upon the strength of the upward-rising currents of air and the height to which dust is carried. A current of air rising upward 1 meter per second will keep suspended a grain of quartz about 0.1 mm. in diameter. The diameter of the grain is roughly proportional to the velocity. An upward current moving 40 feet per second (about 30 miles an hour) will support grains over 1 mm. in diameter. This is about the size of sand. Observers declare that it is unlikely that dust larger than this is held suspended for any length of time. Much larger pieces, however, are moved along the ground and even for a time blown to considerable heights. Gravel the size of peas is readily drifted by the wind and there are many reliable accounts of rains of pebbles, lichens, and even fish and other animals. The actual amount of dust thus held suspended in the air may amount to thousands of tons over a square mile of wind-blown country. A cube of air 10 feet on a side may readily support 1 ounce of dust. This amounts to 4,000 tons per cubic mile. A thousand tons being carried over each square mile of country would not be unusual. Some dust storms are hundreds of miles in extent. A storm 300 to 400 miles in diameter would thus hold suspended over 100,000,000 tons of dust. This would represent, however, only the volume of material in a hill about 100 feet high and 2 miles across the base. If this dust were completely removed from the place where it originated and not replaced, it is easy to realize the rapidity with which desert tracts might be worn away by the wind.

The quantity of material transported during dust storms is readily determined if the dust falls upon clean snow. The snow is collected from a measured area, melted, filtered, and the residue weighed. In one of the dust storms which originated in the Colorado-Wyoming region and extended eastward to the Atlantic seaboard, it was found that in Wisconsin and Iowa the amount of dust in a square meter ranged from 5 to 10 grams. This is equivalent to a range of 15 to 30 tons per square mile. Even as far east as Pennsylvania there was an accumulation of 12 to 15 tons per square mile. This represents a total fall of at least 2,000,000 tons, and probably several times that amount. In another storm which reached as far east as New Hampshire, the amount of fall per unit of area was less than one-tenth as great as this amount. The smaller quantity was due largely to the remoteness from the place of origin of the dust.

The size of the dust particles appears also to bear a definite relation to the velocity of the wind during the storms, being much finer in the less violent storms. For any given storm the sizes of the dust particles vary over a wide range with a great preponderance of those of a given size. Analysis of old loess deposits indicates that the size of their constituents in many cases closely resembles that of recent dust storms, but not always.

DISTANCE. Observers have suggested that every square mile of the earth's surface contains pieces of dust from every other square mile. It is true that dust is transported for great distances. Recently falls of dust in the eastern cities of the United States have been identified as having their origin in the arid sections of the west, 2,000 miles away. Dust from Sahara occasionally falls in England 2,000 miles distant. Volcanic dust from Iceland has several times been recorded in Scandinavia. The dust formed by the eruption of Krakatoa in 1883 in the Dutch East Indies was projected so high in the air that it was carried around the world repeatedly before settling. Krakatoa ashes fell inches deep at distances of nearly 1,000 miles from the volcano, and small quantities fell even in Holland. Dust from Colima Volcano in Mexico in 1903 fell at points 200 miles north of the volcano and in Guatemala the eruption of Santa Maria produced a deposit of ash 8 to 10 inches thick 40 miles distant.

One of the chief factors in removing the dust from the air is the rain of humid regions. It is therefore apparent that much dust will not be transported far into humid localities before it is washed out of the air.

Nevertheless, the total accumulation of dust in humid regions over extended periods of time is considerable and doubtless much of the top soil of such localities is windblown. It has been estimated that during one storm 10,000 tons fell in England, and on another occasion 2,000,000 tons fell in Europe and as much more in northern Africa, much of it having been transported 2,500 miles. For the European area the fall averaged a layer ¼ mm. thick. If this much fell every 5 years, about 5 mm. would be deposited every century which would be enough to cover the earth several feet deep since the glacial period.

No one can realize the capacity of wind as a transporter of fine material who has not lived through at least one great storm on a desert. In such a simoom the atmosphere is filled with a driving mass of dust and sand, which hides the country under a mantle of impenetrable darkness and filters through every fabric; it often destroys life by suffocation and leaves in places a deposit several feet deep.

During periods of high winds, great dust storms prevail on the Missouri and Mississippi Rivers. Dust is everywhere. It sifts through the closed windows and doors of the houses, covering everything within. Outdoors all is yellow with an impalpably fine powder. The north and east sides of the river suffer more than the south and west sides. This is because the prevailing winds come from the southwest in the spring and summer when the silt bars are bare and dry. The amount of material thus brought out of the valley and deposited on the top of the bluff is often as much as $\frac{1}{100}$ inch in a day and averages about ¼ inch during the year.

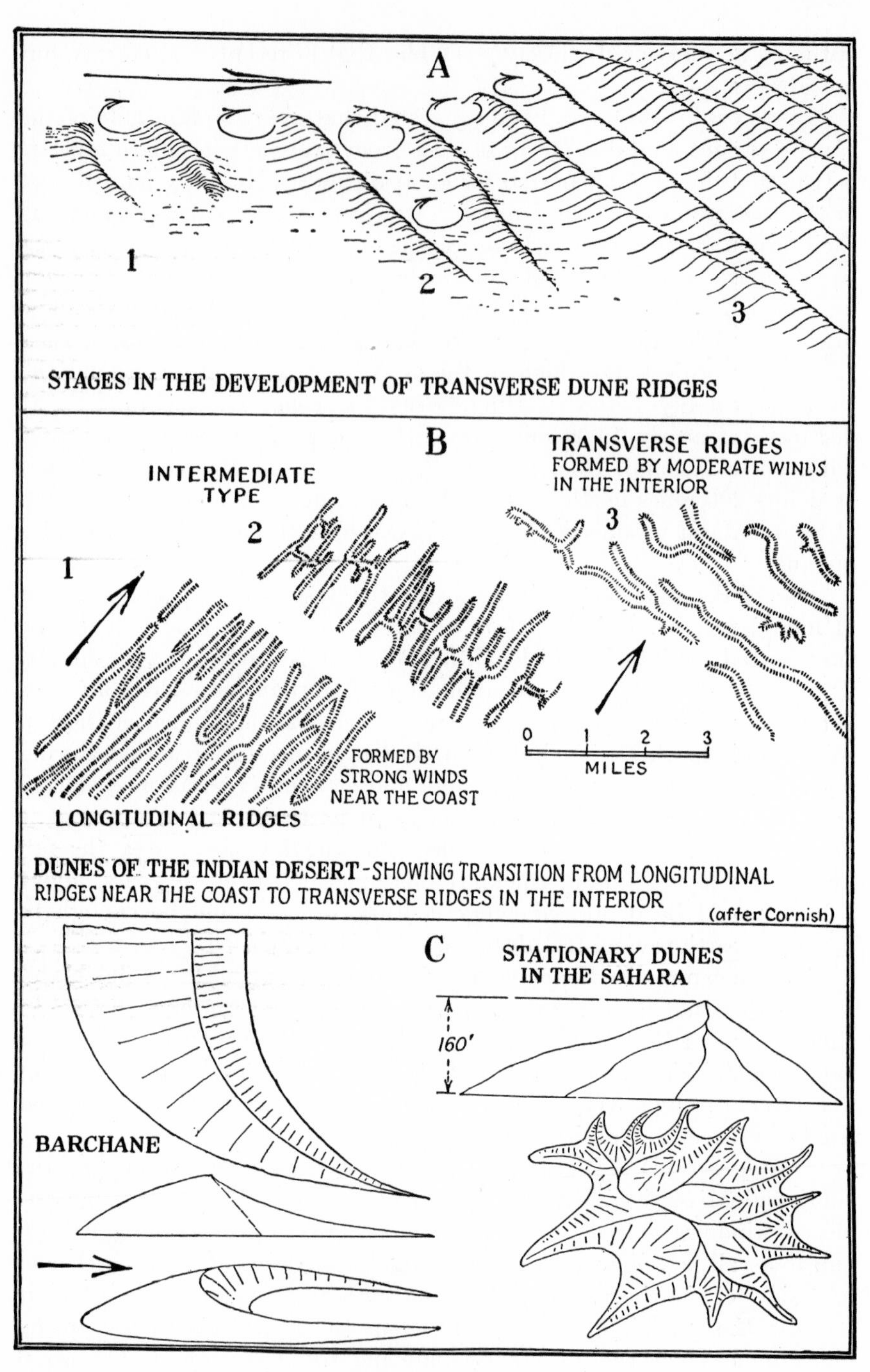


TYPES OF DUNES

## DEPOSITION BY THE WIND. DUNES

Wind blowing steadily from one direction over a sandy surface produces ridges which are either transverse or longitudinal, that is, at right angles to or in the direction of the wind. The first type is represented by ripples and transverse dune ridges, formed under moderate winds blowing from one direction; the second by longitudinal dunes and barchanes, formed under strong winds. Where the winds blow from all quarters, dunes may be of every shape.

Ripples and Transverse Dune Ridges. A moderate breeze is able to move only the finest sand particles and is obstructed by the larger ones. This sets up eddies close to the ground, which pick up the smallest grains and produce hollows. When a moderate wind blows upon deep sand, the conditions are favorable to the production of great transverse ridges. The eddy on the near side of the ridge, and to windward of it, burrows sideways along the base until the finer material is removed. A small hollow is thus elongated into a trough. The forward-flowing current passing over the top of the ridge has insufficient power to move the coarser grains of which the ridge is gradually becoming composed. The height of the transverse ridges and the distance between them depend upon the force of the wind and the amount of fine particles mixed with the sand.

Longitudinal Ridges and Barchanes. A strong wind behaves like a current and moves all the sand it strikes. It produces longitudinal stripes. The same effect is produced by a moderate wind blowing over a hard surface only slightly covered with sand, because eddies cannot excavate and transverse ridges cannot develop.

Winds of intermediate strength produce intermediate forms combining both transverse and longitudinal characteristics (Fig. *B*). In India the monsoon winds striking the coast produce longitudinal ridges. Farthest inland, where the winds are less intense, transverse ridges prevail. Intermediate types occur between these two.

Barchanes show both transverse and longitudinal characteristics. They occur neither where the sheet of wind has the complete mastery over the sand, nor where the drifting sand is too heavy for the carrying power of the wind. They dot the desert plain where the sheet of wind has, for the most part, the mastery of the sand, but where it drops its burden at certain points. The *horns*, or *cusps*, of the barchanes point to leeward, for the lowest parts of the dune travel fastest. A barchane exposed to winds from all quarters gradually changes to a stationary dune (Fig. *C*).

Coastal dunes exhibit similar forms. Where the sand supply is copious relative to the force of the wind, individual longitudinal coast dunes tend to merge in one another, forming a transverse ridge. Where the wind has more mastery over the sand, coast dunes show longitudinal development.

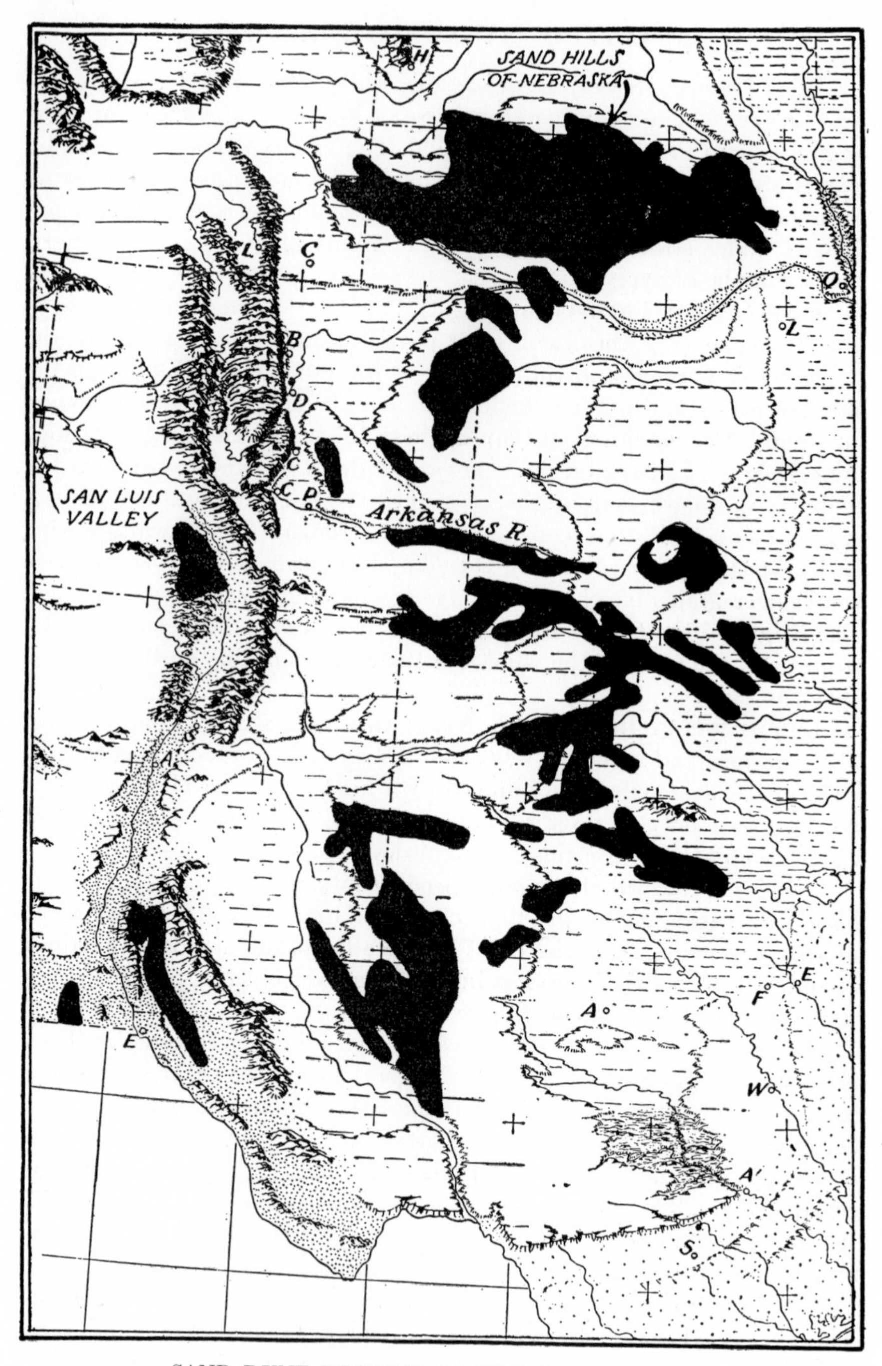


SAND DUNE REGIONS OF THE GREAT PLAINS

## SAND-DUNE REGIONS OF THE WORLD

Types of Occurrence. Sand dunes occur in three kinds of localities: as (*a*) shore dunes, (*b*) river-bed dunes, and (*c*) inland or desert dunes.

Shore Dunes. Shore dunes are typical of almost all shores, both ocean and lake, except those bordered by rocky headlands. From Provincetown on Cape Cod to Florida, the beaches and bars of the Atlantic coastal plain present an almost uninterrupted chain of dunes. Those of Cape Cod and those of Cape Henry, Virginia, are best known because of their unusual size and their picturesque setting. In some cases sand dunes are formed even on the top of near-by cliffs where these descend to sandy beaches, as along the Normandy coast and the coasts of Jutland and Schleswig-Holstein.

Along the Gulf of Biscay in France is a strip of coast, 5 miles wide, stretching for 150 miles and bearing many parallel rows of dunes up to 300 feet in height. The coast of Belgium and Netherlands supports a continuous dune belt which far exceeds in length that of the Biscayan coast but the dunes are rarely one-third as high. The dunes of the Danish coast are remarkable because of the rapidity with which they migrate over the country. Within 30 years the entire dune chain passed over a church. On the Baltic coast in East Prussia another important dune area supports great dunes, some reaching a height of 200 feet.

Lake-shore dunes are widely developed on the eastern shore of Lake Michigan. Extensive deposits of glacial sands, forming beaches of a past glacial lake larger than Lake Michigan, have been heaped into dunes by the prevailing westerly winds. Some of these are grassed over and wooded, but others are still advancing upon and burying the forests.

River-bed Dunes. Rivers flowing through broad open valleys, especially in semiarid regions, are often accompanied by an extensive development of border dunes. In the United States the rivers of the Great Plains, such as the Arkansas and its tributaries, have dunes bordering their eastern or leeward shores. The most extensive river-dune area in North America is that of the Columbia and Snake Rivers in Oregon and Washington. The railroads maintain their right of way against the drifting sand only by the constant use of plows. In this region the dunes assume a perfect crescent form, splendid examples of the barchane type.

In northern Europe as in the humid eastern United States, river dunes are seldom formed but in the dry regions of southern France and Spain they are abundant. The valley of the Gardou in Languedoc shows dunes 30 to 40 feet high, while dunes rise to almost 100 feet in the sandy desert region along the banks of the Guadalquivir in Andalusia in southern Spain. In southern Russia river dunes occur along the banks of the Dnieper, the Don, and the Donetz Rivers as well as along the Volga farther east. The river valleys of Asia have extensive dune deposits, many of them merging with those of the near-by deserts.

*Sketched from photographs*

TYPES OF SAND DUNES

Inland or Desert Dunes. The largest areas of dune deposits are in the deserts, notably those of Asia, Africa, and Australia. In the Sahara, dunes cover only one-ninth of the total area but this aggregates over 300,000 square miles. Sands form nearly one-third of the entire surface of Arabia, or not less than 400,000 square miles; and there are vast tracts of sandy desert in central Asia extending from Syria eastward through Iran, Baluchistan, northern India, and eastern Turkestan to Mongolia.

In North America extensive sand-dune areas of the continental type are found in the Great Basin in Nevada and in the Colorado and Mohave Deserts of southern California. One of the best known is in the Imperial Valley not far from Yuma, Ariz. There are also extensive dunes in the San Luis valley of southern Colorado between the Rockies and the San Juan Mountains.

One of the most unusual dune regions of the United States is in the White Sands National Monument near Alamogordo, N. M. An area of 500 square miles is covered with dunes of snow-white gypsum derived from disintegrating beds of gypsum exposed at the surface. Some of this gypsum sand is still shifting with the winds; but in many of the older dunes the mineral has crystallized until the grains are bound firmly together.

Fixed Dunes. The Sand Hill region of Nebraska and adjoining states comprises some 18,000 square miles. The region is now largely covered with vegetation, though many bare areas of drifting sand still occur. The sand is derived from the disintegration of the underlying Tertiary sandstone and is remarkably pure. The sand hills enclose depressions in size from mere basins to valleys a mile in width and many miles long. In these valleys occur small ponds and also lakes several miles in length. Where overgrazing is practiced or where fires have destroyed the vegetation, "blowouts" occur. These may attain a diameter of hundreds of feet and have a depth of 100 feet or more.

Another region of more or less fixed dunes is near Saratoga, N. Y., south of the Adirondack Mountains. This picturesque region contains many knolls now covered with scanty vegetation and evergreen trees. These are similar to dunes in parts of Europe formed at the end of glacial time, when extensive plains of loose alluvium invited wind action. Those north of Albany are formed on the ancient delta of the Hudson River.

In South America continental dunes cover much of the great Atacama Desert of Chile and Peru west of the Andes, and there are vast areas of dunes in the desert region of Australia.

Along the coasts of Bermuda, the West Indies, and certain Pacific Islands where lime sand has been blown up, the dunes have become firmly cemented to form a cross-bedded lime sandstone which can be cut into blocks for building purposes. In some cases only the exterior of the dunes is cemented; this forms a crust which serves to fix the dune in place.

## DEPOSITION BY THE WIND. LOESS

Loess Deposits. The term *loess* was first applied to the loose unconsolidated deposits which occur along the valley of the Rhine and extend eastward to the Black Sea. They lie just outside the glaciated area. In North America similar deposits occur along the Missouri and Mississippi Rivers; and vast deposits cover the plains of northern China. The loess consists of loosely arranged, angular grains of calcareous silt loam intermediate in fineness between sand and clay and of remarkably uniform mechanical composition. Normally loess is without stratification and breaks off in vertical slabs, forming perpendicular cliffs.

Source of Chinese Loess. The loess of China is made up primarily of the disintegrated rock material brought by the prevailing westerly winds from the Desert of Gobi in central Mongolia. In some places it is 1,000 feet thick. At first, it was thought to have been deposited under water but its aeolian origin is now universally accepted. However, the great rivers of China which flow through the loess region, namely, the Yellow River and Yangtze with their tributaries, have removed large quantities of this fine silt and have redeposited it again over wide expanses of their flood plains. These deposits, of course, have water-borne characteristics.

Source of American Loess. The contrast between some of the extensive loess areas of China and those of the United States may be expressed by saying that the Chinese deposits are primarily wind blown and secondarily stream deposited, whereas the American accumulations are primarily stream deposits and secondarily wind blown.

In the United States the loess deposits occur on the bluffs and uplands bordering the large rivers of the whole Mississippi Valley. Near the rivers the deposits are 100 feet or more thick but they thin out rapidly away from the stream courses. This material has been blown from the near-by flood plain at times of low water when large tracts of dry sand and silt were exposed to the strong winds of the region. Originally the streams derived the material from the glacial deposits farther north.

Some of the dust which has accumulated in the Mississippi Valley, however, is known to have come from the wind-blown plains of Colorado, Wyoming, and other western states. The dust which fell in the Mississippi Valley during the great dust storms of March, 1918, and March, 1920, was very similar in composition to some of the loess deposits of that area. This dust was readily traced to the High Plains of Colorado and Wyoming where wind deflation at that time was known to be unusually severe.

Source of European Loess. In Europe the loess deposits seem to have been derived from the adjacent glaciated areas to the north and are largely the work of the wind with little stream modification.

Composition of Loess. The chemical composition of wind-blown dust varies widely, depending upon the source from which it was derived. The dust falls of the Mississippi Valley contain about 67 per cent silica which represents about 35 per cent quartz, the rest of the silica being aluminum silicates or feldspars. This is much more siliceous than most of the dust storms of Europe which have probably brought their materials from the African deserts. Some of these have less than 10 per cent of quartz but high percentages of silicates, the feldspars running up to over 30 per cent, and an almost equal amount of kaolinite which is the weathered or hydrated form of feldspar. This is all equivalent to saying that the Mississippi Valley dust falls resemble far more closely some of the Mississippi loess deposits than they resemble the foreign dust falls. It seems clear, therefore, that much of our loess is solely wind-borne material.

Proofs of Aeolian Origin. Several other facts have been adduced to prove the aeolian origin of American loess. The size of the constituents indicates that 95 per cent of the material in dust falls consists of fragments finer than 0.05 mm. Samples of loess collected in Missouri, Iowa, and Nebraska were made up of 90 per cent of material of that size. This establishes a strong similarity between the two types of material. Organic remains found in the loess deposits of Mississippi are almost without exception land snails similar to those now existing. This is in contrast with some of the northern loess-like deposits which contain forms known to live in lakes and swamps. The lack of stratification in true loess seems to afford further evidence of aeolian origin. True loess deposits are characteristically penetrated by vertical tubes, calcite-filled hollows left by the decay of grasses and roots. These may reach great depth and indicate the steady accretion of wind-blown material while vegetation persisted.

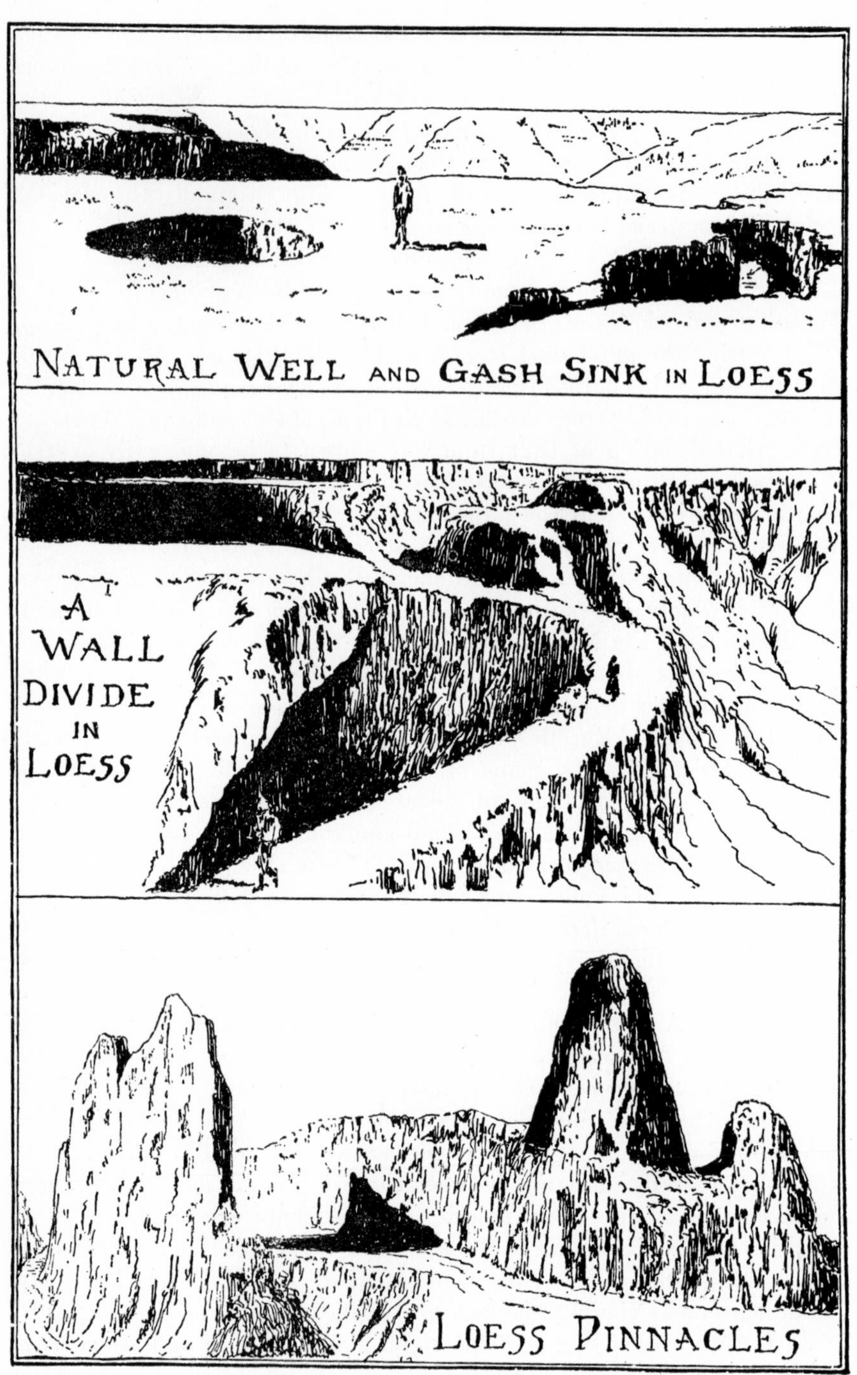


*Sketched from photographs taken by M. L. Fuller*

EROSIONAL FEATURES OF LOESS DEPOSITS

## EROSIONAL FEATURES OF LOESS DEPOSITS

Youth. The flat upland surfaces of loess plateaus frequently exhibit natural wells or pits. These usually occur near the rims of gullies, ravines, and canyons and are due to the settling of the soil consequent upon the mechanical removal of underlying material, rather than from its removal by chemical action as in ordinary limestone sinks. They result from seepage movement along bedded layers somewhat more porous than the surrounding material. The water movement along such lines results in some solution of the calcareous particles of the loess but leads first to increased porosity rather than to open passages. Soon, however, the openings along the line of movement become large enough to permit the mechanical transportation of the finer particles of the loess, and more or less tubular channels are formed. Eventually these may reach diameters of a foot or more. Caving of the roof due to downward percolation of surface water causes these cavities to enlarge upward and finally to reach the surface to form typical *loess wells.*

Between the wells the subterranean channel may so increase in size as to form a natural bridge or the tube may reach the side of a valley and appear there as an underground tunnel of large dimensions. Along the sides of canyons, loess pipes, tubes, wells, sinks, and bridges are characteristic features. The sinks in some cases form long gashes in the surface of the plateau. These features may all be termed youthful as they are associated with the active destruction of the loess upland.

Maturity. Maturity of dissection is reached when the loess upland is replaced by a multitudinous complex of ravines much like badland topography. The slopes of the valley walls, however, are more sharply cut and more nearly vertical than in badlands. The divides between the heads of opposing streams are like walls and are known as *loess dikes,* wall divides, or natural causeways. These serve the travelers of the region as means of passing from one plateau area to another and are maintained artifically because of their great convenience.

Old Age. In their later stages loess areas are reduced to numerous pinnacles which may be conical, sharp pointed, turretlike, or fin shaped, separated by wide flat-floored valleys. Such valleys result frequently from the erosion of roads and trails where the soft loess is quickly churned to dust and rapidly whirled away by the wind. There is reason to believe that many of the loess-canyon systems in China, although now showing no trace of roads, had their inception in the wear and wash of ancient highways.

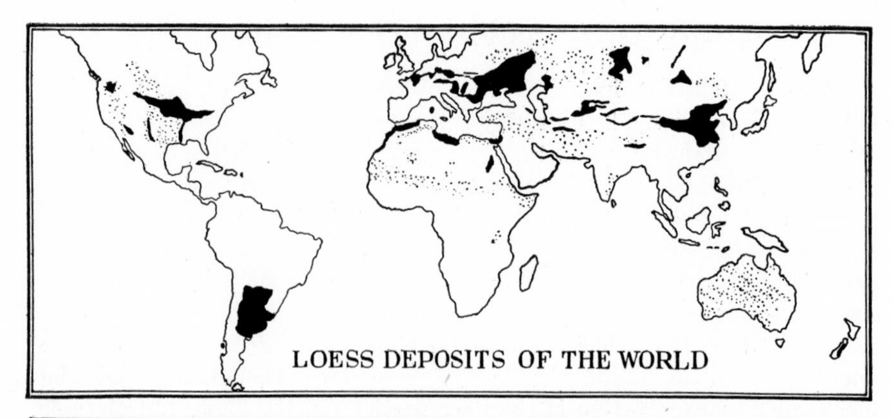


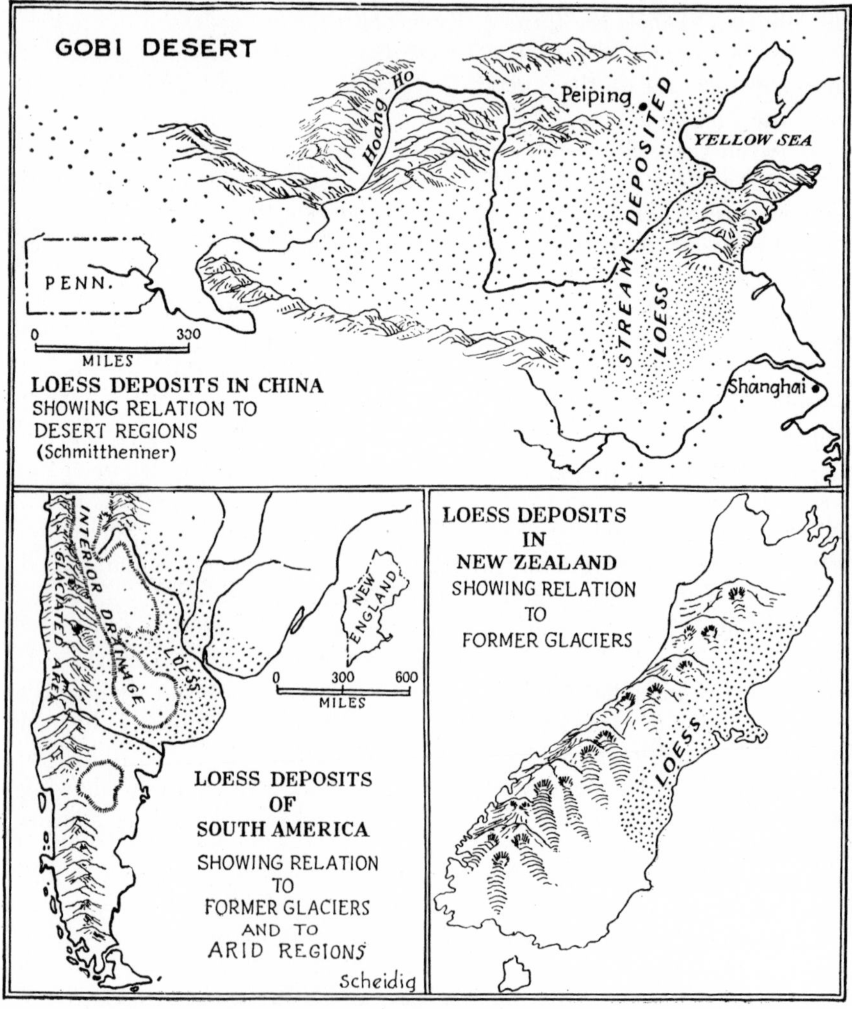


THE IMPORTANT LOESS DEPOSITS OF THE WORLD

## IMPORTANT LOESS DEPOSITS OF THE WORLD

THE UNITED STATES. The loess deposits of the United States come under three groups: (*a*) those capping the uplands adjacent to the large rivers of the Mississippi Valley, derived from their flood plains; (*b*) those covering much of western Kansas and Nebraska and extending into Iowa and Missouri, derived largely from the semiarid regions of Colorado, Wyoming, and other western states; (*c*) those constituting the Palouse area of Washington and Idaho, derived largely from glacial outwash plains.

The upland deposits are more extensive on the eastern sides of the great river valleys, owing to the prevailing westerly winds. This is shown by the long belt capping the eastern bluffs of the Mississippi almost to the Gulf of Mexico. In Illinois, Iowa, and Wisconsin, some of the loess deposits were probably laid down in old glacial lakes. They show laminations and the presence of aquatic molluscs not characteristic of true loess.

The loess deposits of Nebraska and Kansas are thicker than those of eastern Iowa. They are more siliceous, coarser, and laminated with sand. This is because the material was transported by the strong winds of the High Plains and was not derived from flood plains. The loess deposits of Iowa, probably from the adjacent river flood plains, are rarely over 10 or 12 feet thick, as in much of the Mississippi Valley. The Palouse deposits of Washington and Idaho are great stationary dunes of silt, forming deep, fertile soil. The 75-foot covering of the plateau between the Columbia River and the Bitter Root Mountains is being increased by material blown from the dry eastern slopes of the Cascades and the adjacent Columbia River valley.

EUROPE, SOUTH AMERICA, AND ASIA. Here the loess deposits lie just outside the regions of continental glaciation. Those of Europe constitute a rich farm-land belt across northeast France and Belgium, extending irregularly into Poland, Czechoslovakia, Rumania, and southern Russia.

In China the loess is strongly developed throughout the basin of the Yellow River and adjacent parts of North China. The term *loess*, however, is applied to many deposits somewhat similar in appearance to the true loess but differing vastly in age, composition, character, and mode of origin. The Chinese loess has been described as 1,000 to 2,000 feet in thickness but recent observers doubt if the true loess is much over 200 feet. It is clear that the Chinese loess has come from the Gobi Desert to the west but that some of it, especially along the river valleys, is from flood-plains.

The south American loess deposits cover much of the pampa of northern Argentina. The source of this material has been largely the arid regions farther west rather than the meager glacial deposits in the south. Like the Chinese deposits, the wind-blown material has occasionally been reworked and redeposited by streams.

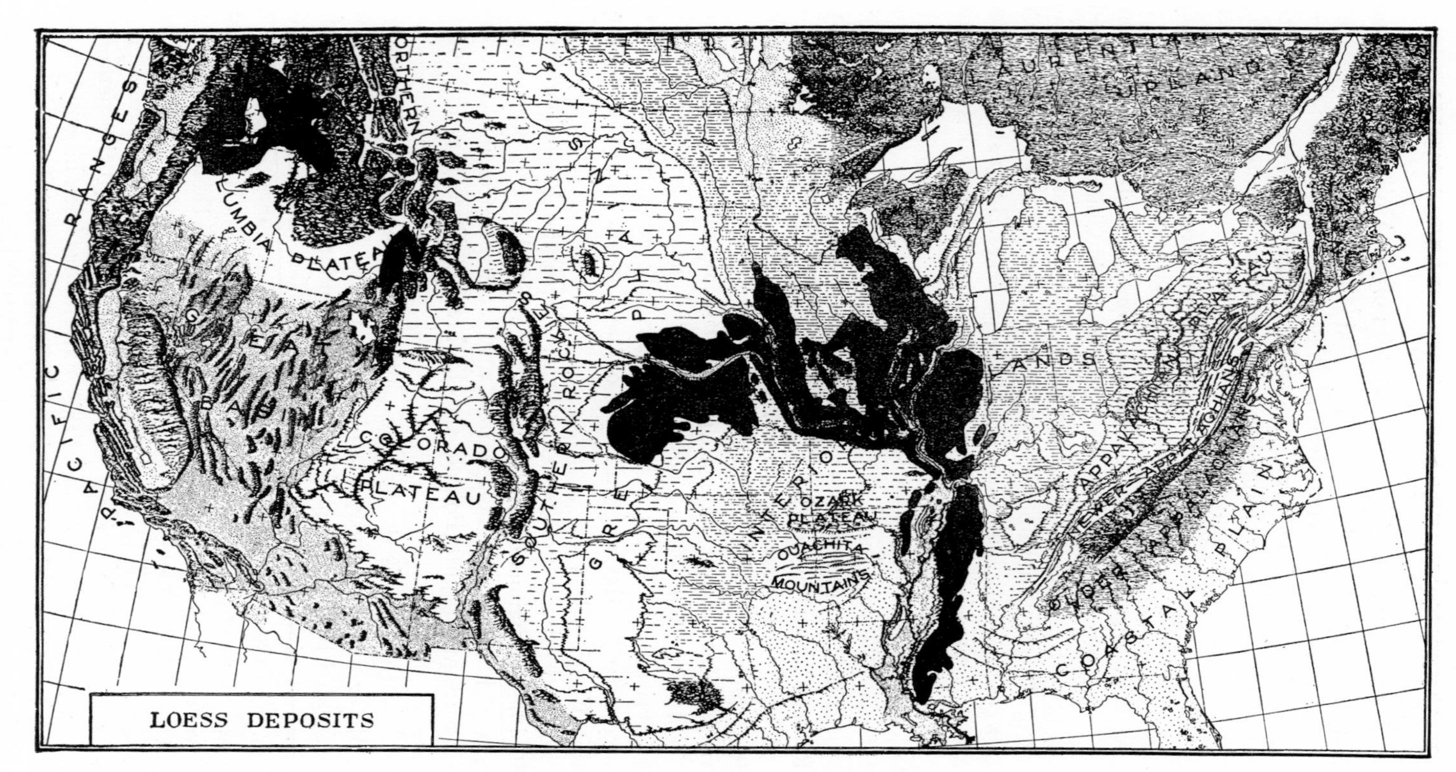


LOESS DEPOSITS IN THE UNITED STATES
Showing close relation to Mississippi River system and to the Great Plains.

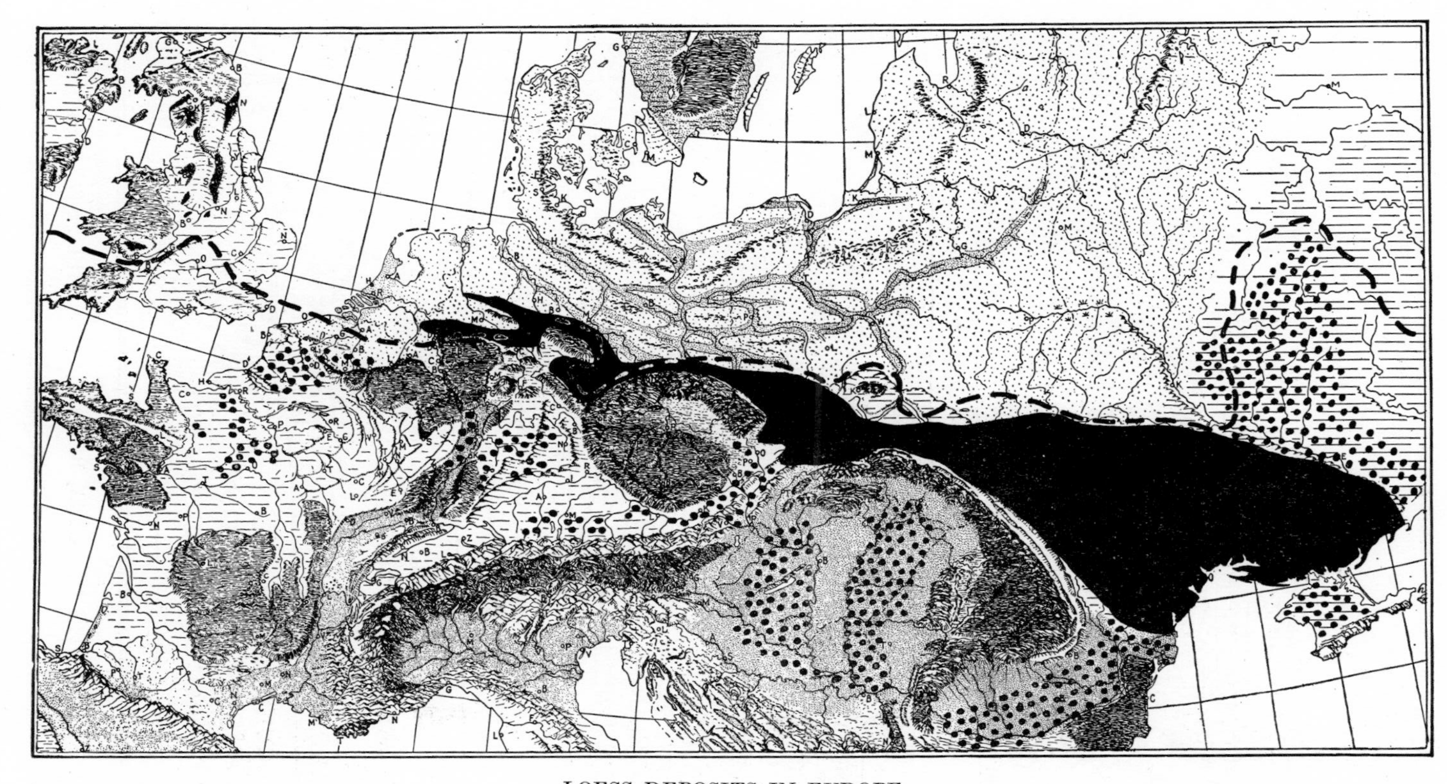

LOESS DEPOSITS IN EUROPE

Showing relation to glaciated area, indicated by broken line. Dotted areas represent discontinuous deposits.

*Sketched from photographs*

LOESS DEPOSITS IN CHINA, EUROPE, AND THE UNITED STATES

## SOME ECONOMIC ASPECTS OF WIND-BLOWN DEPOSITS

The geographer studying the loess-covered and dune-bedecked regions of the world will find himself giving attention to such topics as the character of the dwelling places, including houses constructed of loess mud as well as those carved out of the cliffs; roads and trails; sources of water supply; fertility of the soil; its removal from agricultural land, and its deposition and encroachment upon forests, habitations, and other useful areas.

In China, in Europe, and in the United States to a much lesser degree, the loess serves directly as the means for constructing dwellings. The dwelling is made of the plastic loess bound by straw, hair, grass, and sticks; it is built on top of the loess plateaus and plains. Its roof is of wood. Where loess is dissected, where cliffs abound, and where timber is lacking, rooms are carved out of the solid loess, usually when slightly moist. Water drawn to the fresh surfaces by capillarity evaporates and leaves a thin protective coating of lime, a form of casehardening. Chimneys and air vents from these underground rooms form an unusual sight in the tilled fields above. Owing to the fragility of the loess, collapse is common at times of earthquakes.

The adobe soils of the southwestern United States and of Mexico, used for making bricks, have some of the characteristics of loess. They are largely water deposited, that around Mexico City being old lake clays.

The fine impalpable character of loess and its calcareous nature render it an unusually fertile soil, especially where water is available for irrigation. The constant renewal and enrichment of the surface soil by wind and wash accounts for the cultivation of the grain regions of northern China for 4,000 years.

The rich chernozem or black-earth soils of southern Russia and similar soils of the Argentine pampa and the great plains of the United States are largely of loess origin, as are the wheat-raising Palouse soils of Washington.

Wind-blown regions of the western and southwestern United States have suffered severely from the removal of the soil at times of unusual drought. The encroachment of the sand upon cultivated tracts has been equally destructive. The great dunes at Cape Henry, Virginia, are constantly advancing upon an adjacent forest. In the Landes region of southwestern France the inland march of the extensive coast dunes is a constant menace.

Oases in deserts occur in regions of extensive dunes or *ergs*, the scanty rainfall of the region being quickly absorbed by the sand to reappear in the basins between the dunes. The Sand Hill section of Nebraska gives the impression of being uninhabited but almost every hollow or depression among the hills contains the homestead of a ranch surrounded by several acres of green fields and trees, while the rest of the country for miles about is fit only for grazing.

*U. S. Forest Service*

THE SAND HILL REGION OF NEBRASKA

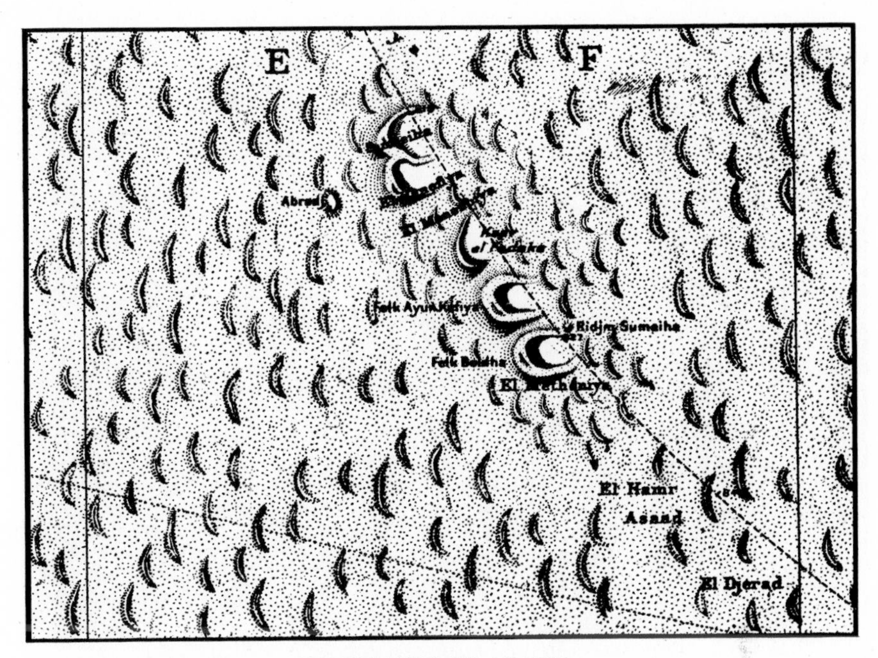


BARCHANES IN ARABIA
International Map of the World, *El Djauf* sheet, North H-37 (1:1,000,000).

## MAPS ILLUSTRATING WORK OF THE WIND

Large inland areas of active dunes are shown on the *Yuma, Calif.-Ariz.*, the *Arena, Buttonwillow, Casa Desierta, Fresno, Caruthers*, and *Holtville, Calif.*, sheets.

Inland areas of dunes, now largely grass covered, are admirably shown on the *Ogallala, Paxton, Chappell, North Platte, Camp Clarke*, and *Whistle Creek, Neb.*, maps.

Coastal dunes are unusually well shown on the *Cape Henry, Va.*, map and also on the *Rehoboth, Del., Mary Esther* and *Villa Tasso, Fla.*, the *Fernandina, Fla.-Ga.*, and the *Cumberland Island, Ga.*, sheets.

Lake-shore dunes appear on the *Three Oaks, Mich.-Ind., Lake Harbor*, and *Holland, Mich.*, sheets. Dunes along river flood plains occur on the *Syracuse, Larned, Great Bend, Garden City*, and *Kingsley, Kan.*, maps.

Wind-blown hollows are shown on the *Loveland, Colo., Holdrege, Neb., Laramie, Wyo.*, and *Hanford, Wash.*, sheets. Barchanes are unusually well shown on the *Moses Lake, Wash.*, sheet. On the *Reedsport, Ore.*, sheet large groups of coastal dunes have been swept inland to block the rivers and form lakes.

The *Albany, N. Y.*, sheet indicates the occurrence of dunes northwest of Albany on the old Mohawk Delta.

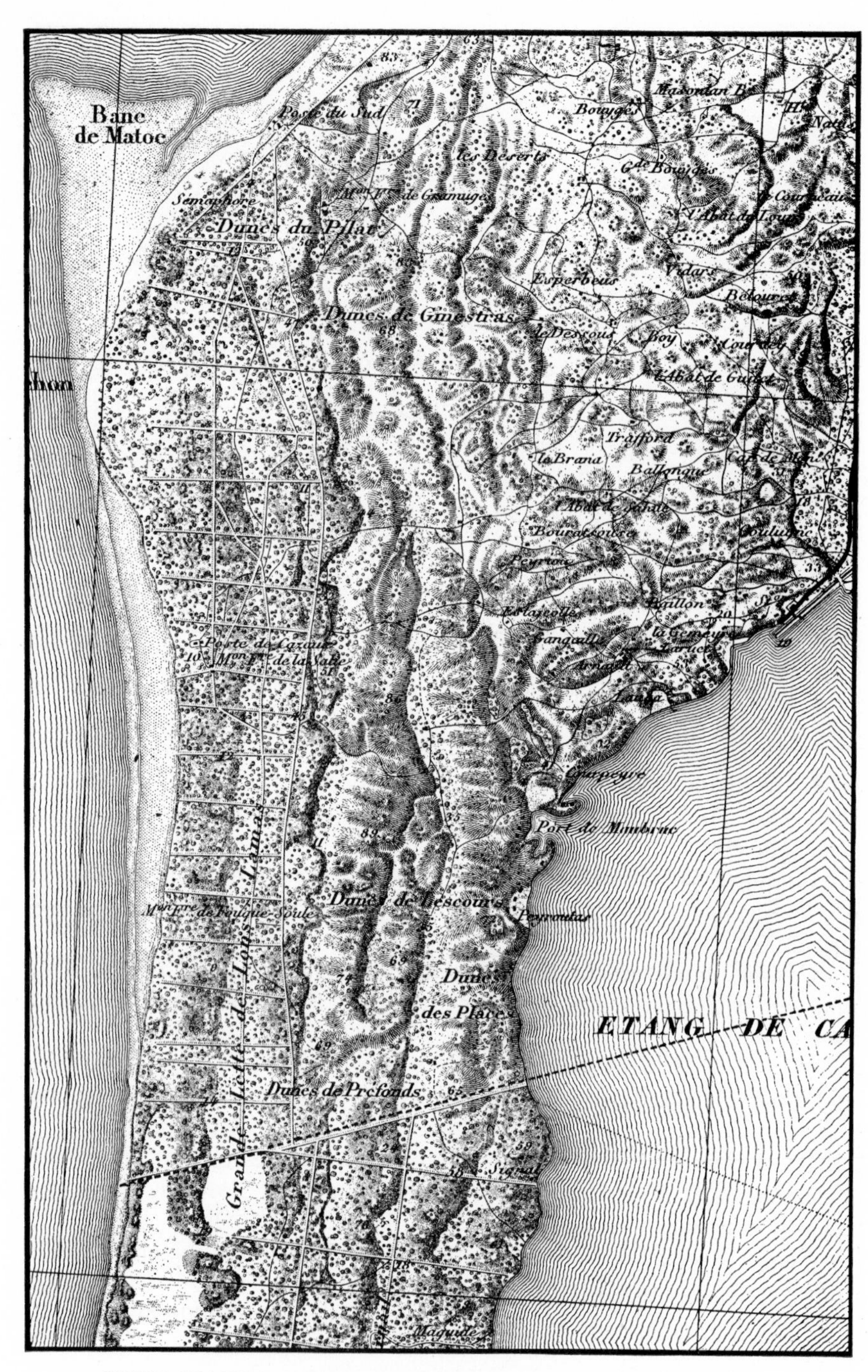


HIGH COASTAL DUNES OF THE LANDES REGION, FRANCE
France; *La Teste de Buch* sheet, No. 191 (1:80,000).

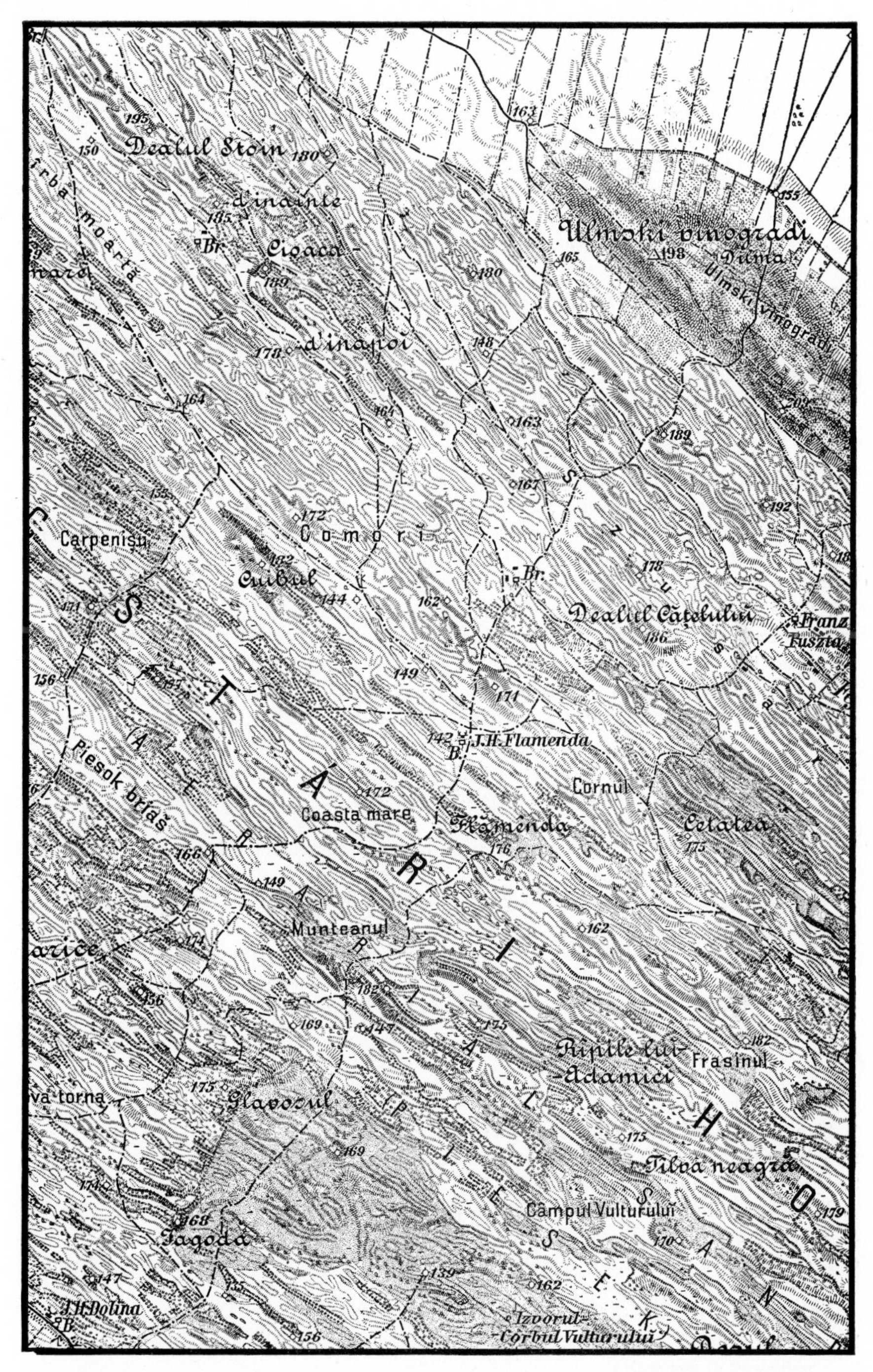


**PART OF THE GREAT SAND DUNE BELT OF CENTRAL HUNGARY**
Austria; *Bavaniste and Jaszenova* sheet, Zone 26, Col. XXIV (1:75,000).

## QUESTIONS

1. What are stony deserts? Ventifacts?
2. What is cross-bedding? How can wind cross-bedding and aqueous cross-bedding be distinguished? Are wind deposits ever laminated?
3. What are the physical and chemical properties of loess? Does loess produce a fertile soil? What is chernozem? Is it a loess deposit? What is palouse?
4. What human activities favor wind erosion?
5. What is the sand-dune region of Nebraska and when were these dunes built? Are dunes always built of quartz sand (*e.g.*, in New Mexico)?
6. Why should dunes develop sometimes as long ridges *transverse* to the direction of the wind and at other times as long ridges *parallel* to the direction of the wind?
7. What causes ripple marks to form? How do wind-formed ripples differ from water-formed ripples?
8. Which side of a sand dune is steep, the stoss or the lee side? What is the angle of repose of sand? What determines the angle of repose? Why do barchanes form?
9. In what respects do wind erosion, transportation, and deposition differ from stream erosion, transportation, and deposition? Does wind transportation correspond with suspension or with saltation in the case of water transportation?
10. Would wind activity be as prominent in the eastern United States as it is in the western United States, if the region were not humid?
11. How can loess be distinguished from gumbotil?
12. What is adobe soil? What quality makes it suitable for brick making?
13. In what parts of the eastern United States do sand dunes encroach upon forests?
14. What is molding sand? From what part of the eastern United States is it obtained? Is it pure quartz? To what other economic uses is sand put?
15. Where in the United States is drifting sand a serious problem? How rapidly do dunes migrate?
16. How are sand blasts used (*i.e.*, carving on granite; cleaning buildings)? How is wind action used in winnowing? And in what other ways?
17. Draw contour map of a barchane.
18. What method would you devise to determine the amount of dust falling on a given region during a certain dust storm?
19. Do you think wind abrasion along sandy coasts is ever sufficient to render lighthouse windows non-transparent?
20. Determine the validity of the figures given in the first paragraph on page 380. The hill mentioned may be considered as cone-shaped (the volume of a cone being equal to its area multiplied by one-third its altitude), and its specific gravity about 2. (One cubic foot of water weighs 62 pounds.)

## TOPICS FOR INVESTIGATION

1. Types and classification of sand dunes.
2. Wind transportation. Size of grains moved; distances; quantity.
3. Origin of loess in different regions.
4. Wind action in humid regions (*e.g.*, Scotland).
5. Abrasive action of the wind: economic importance.
6. Laboratory experiment. Determine the effect of the following factors upon dune formation: (*a*) strength of the wind; (*b*) size of sand grains; (*c*) quantity of material available; (*d*) steadiness of the wind; (*e*) uniformity of wind direction.
7. The dust bowl of America.
8. The deserts of the world.
9. Methods used to prevent drifting of sand and of snow.

## REFERENCES


### General

Davis, W. M. (1905) *The geographical cycle in an arid climate.* Jour. Geol., vol. 13, p. 381–407; Geog. Essays, p. 296–321.

Free, E. E. (1911) *The movement of soil material by wind, with a bibliography of eolian geology, by S. C. Stuntz and E. E. Free.* U. S. Dept. Agr., Bur. of Soils, Bull. 68, p. 1–173.

Grabau, A. W. (1924) *Principles of stratigraphy.* New York, p. 51–62.

Udden, J. A. (1894) *Erosion, transportation, and sedimentation performed by the atmosphere.* Jour. Geol., vol. 2, p. 318–331.

Walther, J. (1924) *Das Gesetz der Wüstenbildung. . . .* Berlin, 421 p.

### Wind Erosion: Abrasion

Blackwelder, E. (1934) *Yardangs.* Geol. Soc. Am., Bull. 45, p. 159–166. Chutes, troughs, and sharp ridges produced by wind.

Blake, W. P. (1885) *On the grooving and polishing of hard rocks and minerals by dry sand.* Am. Jour. Sci., 2d ser., vol. 20, p. 178–181.

Bryan, K. (1931) *Wind-worn stones or ventifacts.* Natl. Research Council, Reprint and Circ. Ser. 98, p. 29–50.

Davis, W. M. (1933) *Granite domes of the Mojave Desert, California.* San Diego Soc. Nat. Hist., Trans. 7, p. 211–258.

Laudermilk, J. D. (1931) *On the origin of desert varnish.* Am. Jour. Sci., 5th ser., vol. 21, p. 51–66.

Schoewe, W. H. (1932) *Experiments on the formation of wind-faceted pebbles.* Am. Jour. Sci., 5th ser., vol. 24, p. 111–134.

White, C. H. (1924) *Desert varnish.* Am. Jour. Sci., 5th ser., vol. 7, p. 413–420.

### Wind Erosion: Deflation

Blackwelder, E. (1931) *Desert plains.* Jour. Geol., vol. 39, p. 133–140. Good photographs.

Blackwelder, E. (1931) *The lowering of playas by deflation.* Am. Jour. Sci., 5th ser., vol. 21, p. 140–144.

Bryan, K. (1923) *Wind erosion near Lees Ferry, Arizona.* Am. Jour. Sci., 5th ser., vol. 6, p. 291–307.

Cross, C. W. (1908) *Wind erosion in the plateau country.* Geol. Soc. Am., Bull. 19, p. 53–62.

Gilbert, G. K. (1895) *Lake basins created by wind erosion.* Jour. Geol., vol. 3, p. 47–49.

Keyes, C. R. (1911) *Mid-continental eolation.* Geol. Soc. Am., Bull. 22, p. 687–714.

### Wind Transportation: Dust Storms

Hand, I. F. (1934) *The character and magnitude of the dense dust cloud which passed over Washington, D. C., May* 11, 1934. Monthly Weather Rev., vol. 62, p. 156–157.

Page, L. R., and Chapman, R. W. (1933) *The dust fall of Dec.* 15–16, 1933. Am. Jour. Sci., 5th ser., vol. 28, p. 288–297.

Sears, P. B. (1935) *Deserts on the march.* Norman, Okla., 231 p.

Udden, J. A. (1896) *Dust and sand storms in the west.* Pop. Sci. Monthly, vol. 49, p. 656–664.

Winchell, A. N. (1922) *The great dust fall of March* 19, 1920. Am. Jour. Sci., 5th ser., vol. 3, p. 349–364.

### Wind Deposition: Sand Dunes

Cornish, V. (1897) *On the formation of sand dunes.* Geog. Jour., vol. 9, p. 278–309.

Hitchcock, A. S. (1904) *Controlling sand dunes in the United States and Europe.* Natl. Geog. Mag., vol. 15, p. 43–47.

Holtenberger, M. (1913) *On a genetic system of sand dunes.* Am. Geog. Soc., Bull. 45, p. 513–515.

Olsson-Seffer, P. (1908) *Relation of wind to topography of coastal drift sands.* Jour. Geol.. vol. 16, p. 549–564.

Townsend, C. W. (1913) *Sand dunes and salt marshes.* Boston, 311 p.

WIND DEPOSITION: LOESS

CALL, R. E. (1882) *The loess of North America*. Am. Naturalist, vol. 16, p. 369–381, 542–549.

CAMPBELL, J. T. (1889) *Origin of the loess*. Am. Naturalist, vol. 23, p. 785–792.

CHAMBERLIN, T. C. (1897) *Supplementary hypothesis respecting the origin of the loess of the Mississippi Valley*. Jour. Geol., vol. 5, p. 795–802.

CRESSEY, G. B. (1934) *China's geographic foundations*. New York. Chapter on loess.

FULLER, M. L. (1922) *Some unusual erosion features in the loess of China*. Geog. Rev., vol. 12, p. 570–584. Good illustrations.

FULLER, M. L., and CLAPP, F. G. (1924) *Loess and rock dwellings of Shensi, China*. Geog. Rev., vol. 14, p. 215–226.

HILGARD, E. (1879) *The loess of the Mississippi Valley*. Am. Jour. Sci., 3d ser., vol. 18, p. 106–112.

MOYER, R. T. (1936) *Agricultural soils in a loess region of North China*. Geog. Rev., vol. 26, p. 414–425.

PUMPELLY, R. (1879) *Relations of secular rock disintegration to loess, glacial drift, and rock basins*. Am. Jour. Sci., 3d ser., vol. 17, p. 133–144.

SCHEIDIG, A. (1934) *Der Löss und seine geotechnischen Eigenschaften*. Dresden, 233 p.

SHIMEK, B. (1904) *Papers on the loess*. Iowa State Univ., Lab. Nat. Hist., Bull. 5, p. 298–381.

SMITH, H. T. U., and FRASER, H. J. (1935) *Loess in the vicinity of Boston, Massachusetts*. Am. Jour. Sci., 5th ser., vol. 30, p. 16–32.

WRIGHT, G. F. (1902) *Origin and distribution of the loess in northern China and central Asia*. Geol. Soc. Am., Bull. 13, p. 127–138.

REGIONS DESCRIBED

BAILEY, E. S. (1917) *The sand dunes of Indiana*. Chicago, 165 p.

BARBOUR, G. B. (1926) *The loess of China*. Smiths. Inst., Ann. Rept., p. 279–296.

BRADWELL, H. J. L. (1910) *The sand dunes of the Libyan Desert*. Geog. Jour., vol. 35, p. 379–395.

CHAMBERLIN, T. C., and SALISBURY, R. D. (1885) *The driftless area of the upper Mississippi Valley*. U. S. Geol. Surv., 6th Ann. Rept., p. 278–307. Dust carried out of the desert.

CHAPMAN, R. H. (1906–07) *The deserts of Nevada and the Death Valley*. Natl. Geog. Mag., vol. 17, p. 483–497; Sci. Am. Supp., vol. 63, p. 26126–26129.

CRESSEY, G. B. (1928) *The Indiana sand dunes and shore lines of the Lake Michigan Basin*. Geog. Soc. Chicago, Bull. 8, 80 p.

GAUTIER, E. F., tr. by MAYHEN, D. F. (1935) *Sahara, the great desert*. New York, 264 p. Foreword by D. W. Johnson.

HOBBS, W. H. (1918) *The peculiar weathering process of desert regions with illustrations from Egypt and the Soudan*. Mich. Acad. Sci., 20th Ann. Rept., p. 93–99.

HUNTINGTON, E. (1906) *The border belts of the Tarim Basin*. Am. Geog. Soc., Bull. 38, p. 91–96.

HUNTINGTON, E. (1907) *The pulse of Asia*. Boston, 415 p. The zone of the dwindling river, p. 210–222, 262–279.

HUNTINGTON, E. (1910) *The Libyan oasis of Kharga*. Am. Geog. Soc., Bull. 42, p. 641–661. War of dune and oasis.

MADIGAN, C. T. (1936) *The Australian sand-ridge deserts*. Geog. Rev., vol. 26, p. 205–227. References and pictures.

MACDOUGAL, D. T. (1907) *Desert basins of the Colorado delta*. Am. Geog. Soc., Bull. 39, p. 705–729.

MACDOUGAL, D. T. (1912) *North American deserts*. Geog. Jour., vol. 39, p. 105–123.

PASSARGE, S. (1904) *Die Kalahari. Versuch einer physisch-geographischen Darstellung der Sandfelder des südafrikanischen Bochens*. Berlin, 822 p.

PUMPELLY, R. *Explorations in Turkestan, expedition of* 1904. Carn. Inst., Publ. 73, vol. 1, p. 1–35. On war of dune and oasis.

PUMPELLY, R., DAVIS, W. M., and HUNTINGTON, E. (1905) *Explorations in Turkestan*. Carn. Inst., Publ. 26, p. 1–317. Views of barchanes.

# XII
# ORGANISMS

*American Museum of Natural History*

CORAL REEF GROUP IN THE AMERICAN MUSEUM OF NATURAL HISTORY

Showing the forms produced by several types of lime-depositing organisms which live on coral reefs.

*American Museum of Natural History*

GREAT BARRIER REEF, PORT DENISON, AUSTRALIA

A fringing reef lying several miles offshore and separated from the mainland by a wide lagoon.

*Pan-American Airways*

MIDWAY ISLAND IN THE PACIFIC
A low coral reef rising barely above sea level.

AN ATOLL IN THE PACIFIC
An almost perfect example of a circular atoll, viewed from the air, showing the surf breaking outside and, on the inside, the quiet protected waters of the lagoon.

*U. S. Navy*

*Charles E. Johnson, New York State College of Forestry*

BEAVER DAM AT END OF OSWEGO POND, IN THE ADIRONDACKS, NEW YORK

The water back of the dam is two to three feet higher than that in front and inundates the forest along the shore, causing the trees to die.



A RAFT OR LOG JAM ALONG THE RED RIVER, LOUISIANA

This raft is now breaking up and the trees are moving down stream.

*Veatch, U. S. Geological Survey*

*Veatch, U. S. Geological Survey*

FOREST FLOODED AS THE RESULT OF A LOG JAM ON THE RED RIVER, LOUISIANA

Hundreds of square miles of country were inundated at different times in this manner.



SITE OF FORMER LAKE DAMMED BY LOG JAM ON THE RED RIVER, LOUISIANA

The water in this lake was 5 to 10 feet or more in depth, and was navigable by river steamers.

*Veatch, U. S. Geological Survey*

*Chapin, American Museum of Natural History*

TERMITE NESTS, LUKOLELA, CONGO RIVER, AFRICA



TALL FLUTED NEST OF WHITE TERMITES NEAR PORT DARWIN, AUSTRALIA

*U. S. Department of Agriculture*

HILL OF TRUE ANTS, TOP OF WARRIOR RIDGE, CENTRAL PENNSYLVANIA

*A. K. Lobeck*

*American Museum of Natural History*

FLAMINGO NESTS. SOUTHERN ANDROS ISLAND, BAHAMAS

An example of mounds built by birds. The nests range from one to two feet in height, and are formed from the soft lime mud at low tide.

*American Museum of Natural History*

PRAIRIE DOG MOUNDS AND HOLES

In the Great Plains of the West there are millions of these openings connecting with underground burrows. A large amount of grass is carried into these holes and the humus content of the soil is thus enriched.

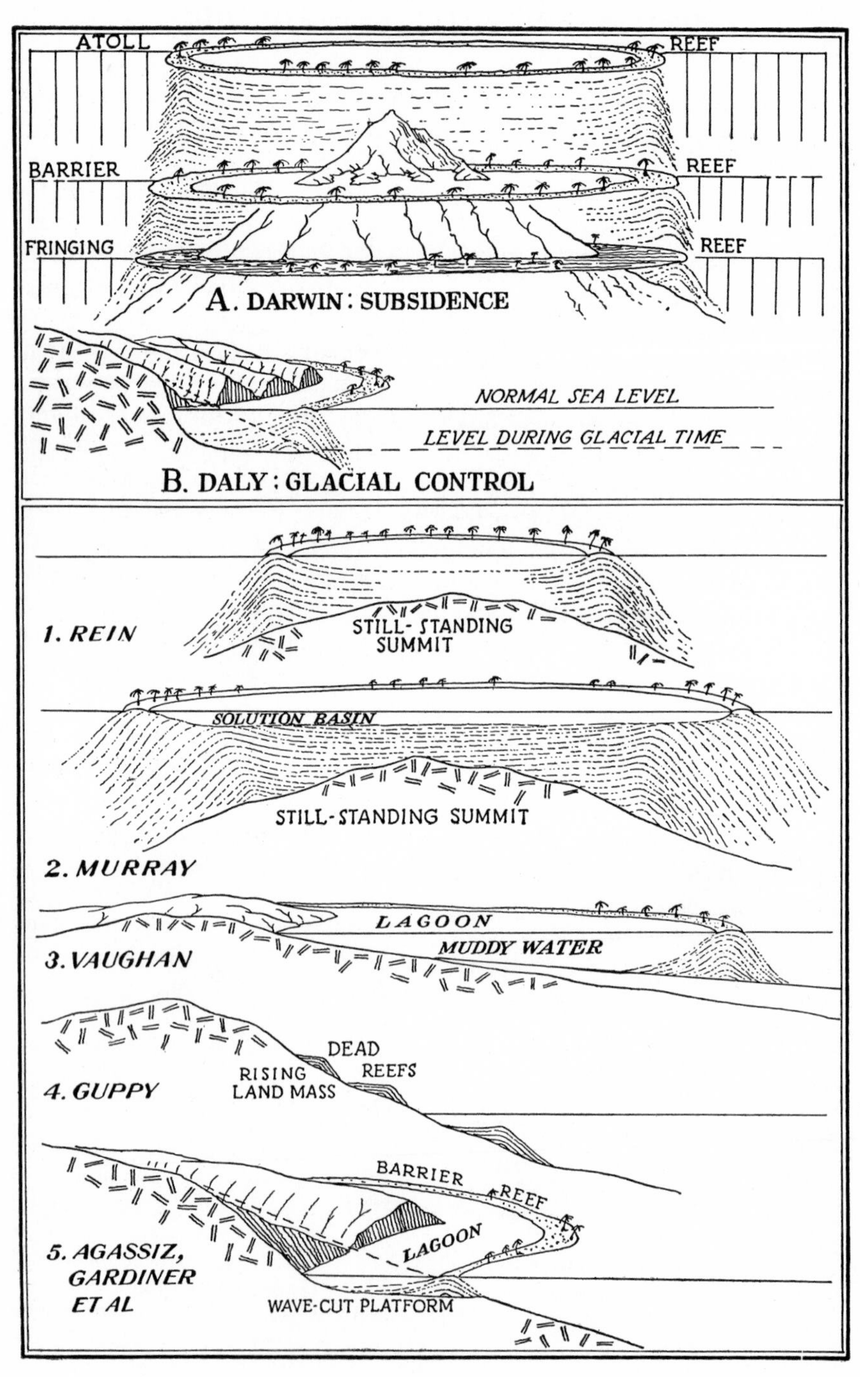


THEORIES EXPLAINING THE ORIGIN OF CORAL REEFS

# CORAL REEFS: OUTLINE OF THEORIES

Coral reefs occur as *fringing reefs, barrier reefs,* and *atoll reefs.*

*Fringing reefs* are closely attached to an island or to a continental shore. There is no lagoon or open water between the reef and the rocky land upon which the reef is attached.

*A barrier reef* lies some distance offshore from a few hundred feet to two or three miles or even more. There is a lagoon of open water between the rocky land and the reef. The barrier reef itself is a long narrow strip of coral rock and sand projecting above the water level. It may be roughly circular in shape if it encloses an island.

*An atoll* is also a roughly circular or elliptical reef. It is a narrow strip of coral rock and sand enclosing a relatively shallow lagoon of open water. An atoll is like a barrier reef but no island occupies the center of the lagoon.

DARWIN'S SUBSIDENCE THEORY. To explain the development of these three types of reefs, Charles Darwin in 1842 suggested that they grew up on a subsiding foundation. The subsidence in most cases was thought to have been intermittent and slow, and the reef grew up rapidly enough to keep at the surface. Only the outer edge of the reef supported actively growing corals. Hence, as the island or mainland sank, the reef stood farther and farther offshore until, with the disappearance of the enclosed island, an atoll was formed.

THE GLACIAL-CONTROL THEORY. The subsidence theory has recently been somewhat modified by the glacial-control theory as developed by Daly. According to this theory the sea level throughout the equatorial area was lowered during glacial time, permitting truncation of platforms by the waves, upon which reefs later grew when the water was again returned to the ocean.

OTHER THEORIES. Many investigators have contributed other suggestions, none of which, according to W. M. Davis, is so well suited to explaining the facts as the two just mentioned. For example,

*a.* Rein suggested that organic deposits accumulated on still-standing submarine summits eventually reaching sea level.

*b.* Murray believed a similar development and thought that the lagoons of atolls resulted from solution while the reef grew outward.

*c.* LeConte, Guppy, and Vaughan studied reefs on gently shelving shores and thought that barrier reefs might grow up from a shallow bottom at a considerable distance offshore; and that a lagoon would thus be enclosed where the water would be too muddy for coral growth.

*d.* Semper and Guppy thought some of the Pacific Ocean reefs grew up on rising foundations, inasmuch as they observed some dead reefs now entirely elevated above the sea.

*e.* Agassiz, Guppy, Wharton, and Gardiner all believed that reefs grow up on submarine platforms produced by wave planation.

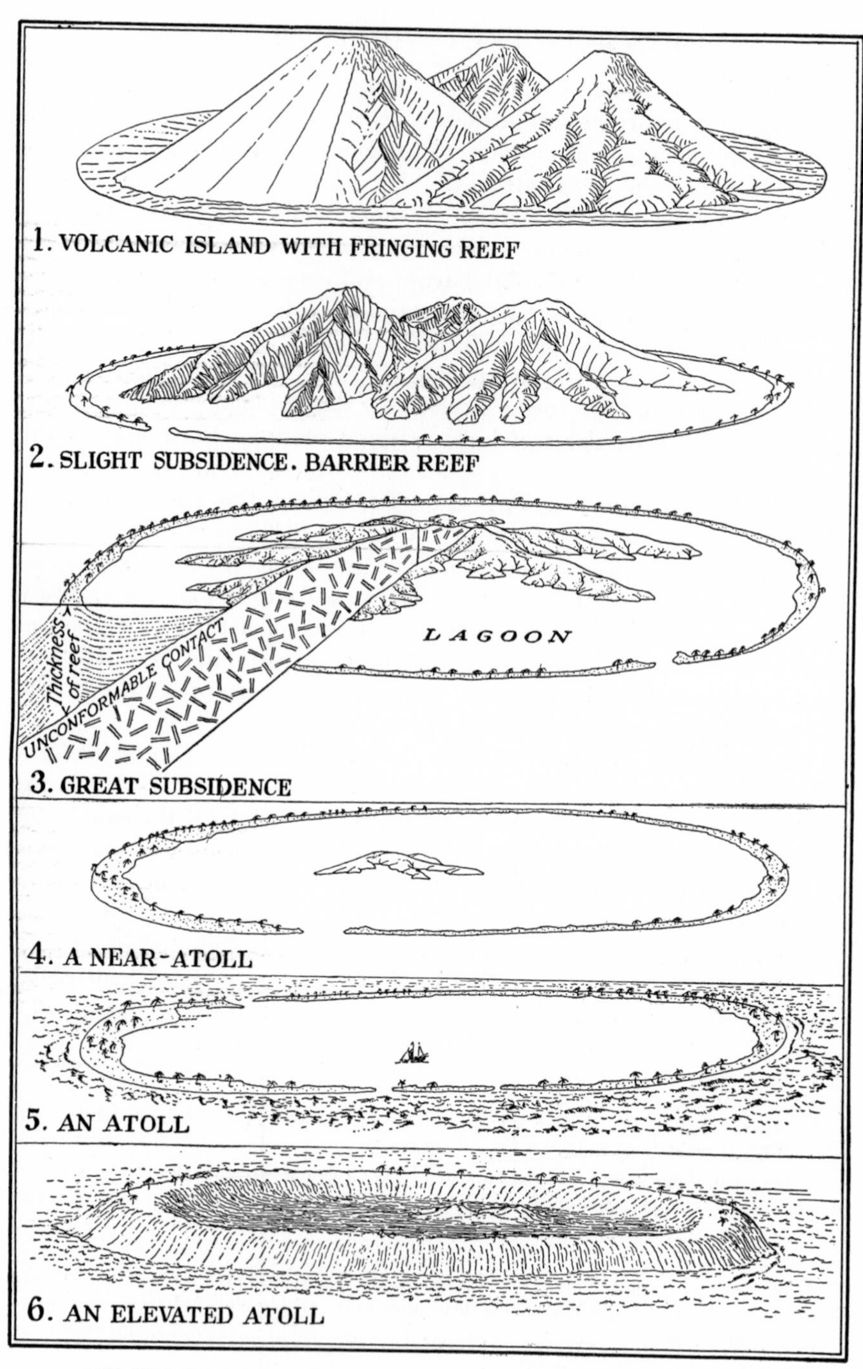


STAGES IN REEF DEVELOPMENT ON A SUBSIDING ISLAND

## CORAL REEFS: THE SUBSIDENCE THEORY

EVIDENCES OF SUBSIDENCE. The subsidence theory first propounded by Darwin is supported by several lines of evidence. First, shorelines bordered by barrier reefs are in numerous instances strongly embayed and the headlands rarely exhibit cliffs due to wave action. Such cliffs, if they ever existed, have disappeared through submergence. It is difficult to conceive of still-standing land masses in exposed positions without wave-formed cliffs of any kind. The bays are extensions of stream-cut valleys and were obviously formed by subaerial erosion. They were later drowned by submergence.

In the second place, regions in the Pacific which show many signs of having been elevated, in the form of raised beaches and marine deposits, are devoid of barrier reefs and atolls, though they may have very young fringing reefs closely skirting their shores. At the same time, regions with many barrier reefs and atolls usually exhibit no indication of having been elevated.

In the third place, the small islands surrounded by barriers, features which may be termed *almost atolls*, have pronounced slopes and in every respect resemble mountain summits. Their small size is due to the great amount of submergence which has occurred. If their small size had been brought about by long-continued stream erosion of still-standing masses, they would have very subdued slopes.

A fourth line of evidence is found in the so-called *unconformable contact* between the reefs and the land mass, as shown in Fig. 3. This means that there is an erosional surface beveling the rock structure upon which the reef rests, which indicates that the land mass subsided as the reef was formed.

In the fifth place, the great thickness of coral reefs indicates that they have formed on a subsiding foundation. Coral organisms do not grow at great depth. They flourish at depths not greater than 150 feet beneath the surface of the ocean. There are some atolls now elevated above the sea which have thicknesses at least 500 to 600 feet—that much being now visible above the present water level. It is almost necessary to conclude that the basement upon which the reef was built subsided gradually and that the reef was built up as rapidly as subsidence took place. The considerable thickness of reefs may also be inferred from the present slope of the mainland spurs beneath the sea. If the submarine profile is drawn at the same slope which the land has above sea level, the estimated depth of barrier-reef foundations may be as much as 1,200 or even 2,000 feet. In numerous instances this would seem to be the case. This depth is far too great for coral growth.

Finally, the sixth line of evidence is found in actually submerged barrier reefs and atolls. These have been discovered by sounding. From these examples it appears that submergence has been so rapid that the reefs have been "drowned" or killed and that upward growth has ceased.

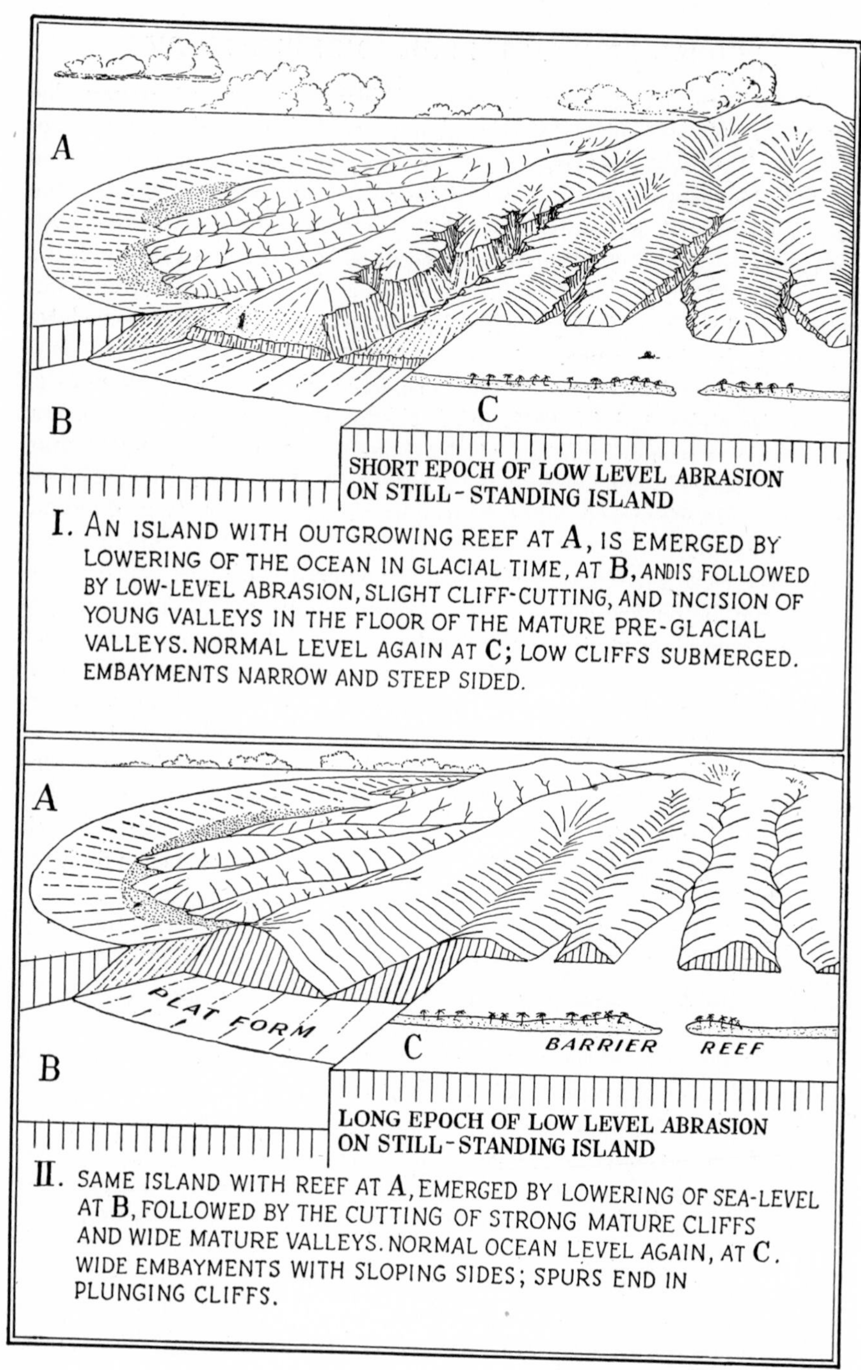


*After W. M. Davis*

EFFECTS OF GLACIAL LOWERING OF SEA LEVEL

## CORAL REEFS: THE GLACIAL-CONTROL THEORY, I

The glacial-control theory was advocated by Daly to explain the presence of barrier reefs and atolls which Darwin accounted for by subsidence. According to Daly, the ocean level in the tropic seas was lowered by 200 feet or more at the time of maximum glaciation. This was due mainly to the imprisonment of much water in the form of continental ice sheets, and also to the attraction which this large mass exerted upon the ocean near by.

EVIDENCE FOR GLACIAL-CONTROL THEORY. Daly's studies of reefs and atolls led him to conclude that the depth of lagoon floors is uniform, and that sublagoon rock platforms exist at a depth of 200 to 300 feet. He concluded also that these platforms were produced by wave erosion on stable islands during the glacial period. In addition, he believed that the chilling of the ocean waters at this time prevented the growth of reef-building organisms and thus permitted the previously formed reefs, then emerged, to be cut away and the islands behind them attacked. In this manner some of the older and much weathered volcanic islands would be completely truncated, while the younger islands would have benches cut at the same standard depth around their shores. On the margin of the platforms and benches thus produced, the now-existing atoll and barrier reefs would grow up and their lagoon floors would become more or less aggraded as the warming ocean rose to normal level in postglacial time. It should be emphasized that the reef foundations must remain stable throughout this period in order that the lagoon depths should be uniform.

OBJECTIONS TO THE GLACIAL-CONTROL THEORY. To this argument Davis raised the following objections:

*a.* He doubted, first of all, the uniformity of lagoon depths. Even within one atoll he finds the depth varying from 120 to 300 feet. Some small atolls have depths of 20 feet and others vary from 300 to 600 feet. Certain submarine banks are over 1,000 feet deep.

*b.* Daly emphasizes the stability of the regions involved, whereas Davis points out that, even in regions of known instability, reefs with normal and uniform depths occur. Therefore, he criticized Daly for using reefs with normal depths as evidence of stability in other places.

*c.* Daly measured the depths of bays and estimated a submergence of 270 feet at most. Davis criticizes this method because the bays were not measured at their most seaward ends (being covered by coral reefs). Davis thinks the submergence is far greater than 270 feet.

*d.* Davis points out that, during lowered sea level, wave erosion, if strong enough to develop platforms, would produce pronounced sea cliffs at the ends of the spurs. However, in very few cases within the coral seas are spurs cut off in plunging cliffs. Davis concludes, therefore, that the land was protected by reefs during this time.

*e.* As a corollary to the last argument, Davis makes the important deduction that in borderline regions just outside the coral seas, in what he calls the *marginal belts*, where no protecting reefs existed or where they were small, and also where they were killed by chilling, the waves during glacial time actually cut against the spurs and produced cliffs which are prominent features at the present day.

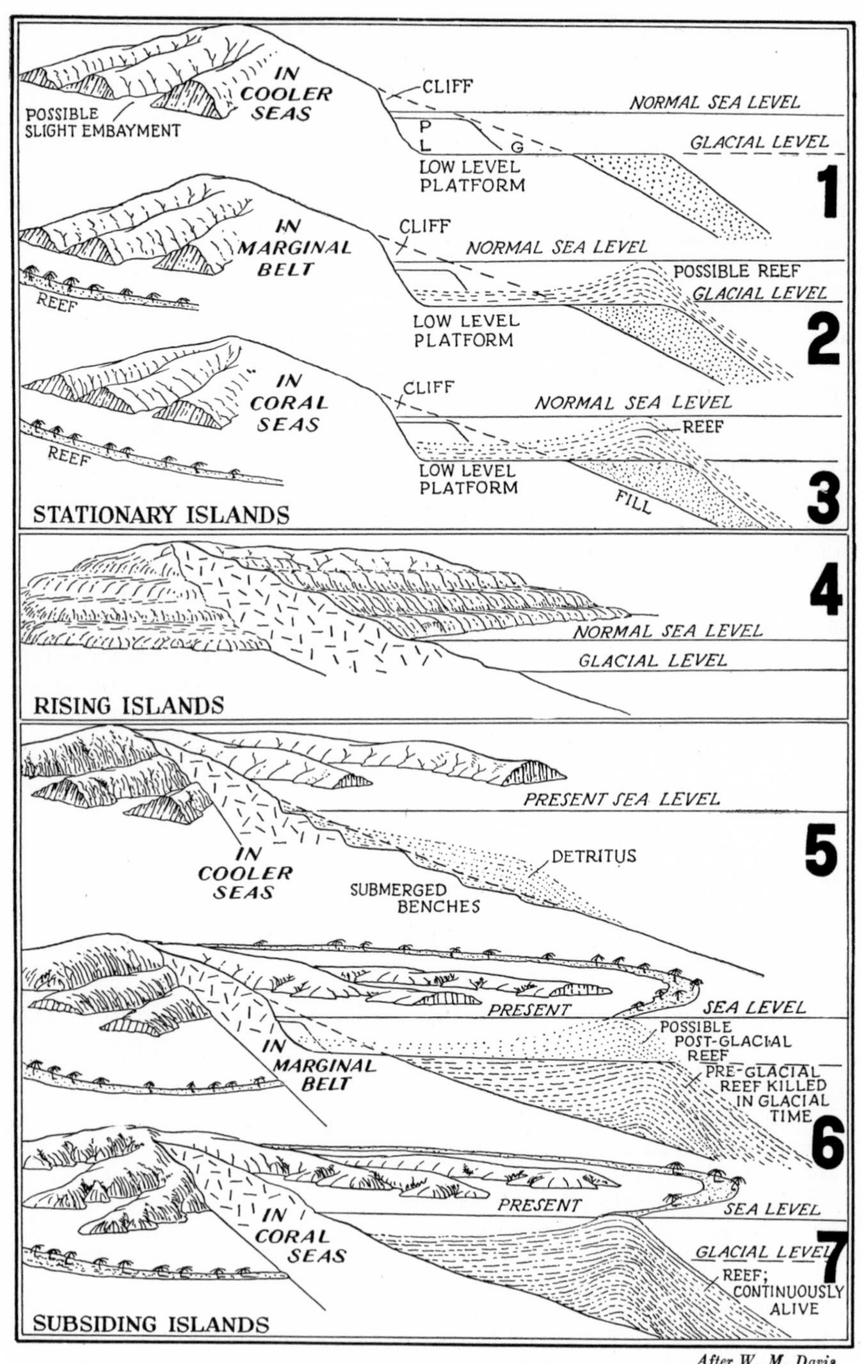


*After W. M. Davis*

EFFECTS OF GLACIAL FLUCTUATIONS OF SEA LEVEL UNDER VARYING CONDITIONS

# CORAL REEFS: GLACIAL CONTROL THEORY, II; DEDUCED FORMS

Brief Analysis of Forms. Three types of islands may be recognized: (*a*) stationary islands; (*b*) rising islands; and (*c*) subsiding islands. Each of these types may occur in the cooler seas where no corals exist; in the marginal belts where coral reefs grew in nonglacial periods; and in the coral seas where coral reefs have long flourished without interruption.

*Stationary islands* in the cooler seas (Fig. 1), if preglacial in origin, should have cliffs rising from a wave-cut platform (*P* in Fig. 1) around which a low-level platform (*G*) cut during glacial time is to be expected. The upper platform of normal marine abrasion level may in some cases (as at *L*) have been completely cut away. There may possibly be slight embayment of the valleys. No coral reefs would be found here.

In the marginal belt, stationary islands of preglacial origin (Fig. 2) should have cliffs much like those in regions of cooler seas, descending either to a low-level or a normal-level wave-cut platform, depending on whether the normal platform was completely destroyed by low-level erosion during glacial time. Reefs of postglacial origin may surmount the low-level platform, forming offshore barriers, or atolls if the central island has been completely destroyed by wave erosion.

Stationary islands in the coral seas (Fig. 3) should differ very little from those in the marginal belts. They should have cliffs descending to either high- or low-level platforms and should be encircled with narrow reefs of postglacial origin. Stationary islands in the coral seas, if they had not been subjected to changes in sea level by the glacial-control theory, would probably not have developed reefs of any kind. Fringing reefs, the type most to be expected, would constantly be buried in silt washed from the land.

*A rising island* (Fig. 4), in the cooler seas, the marginal belt, or the coral seas, will exhibit terraces at various levels. Glacial changes of ocean level will not significantly affect the character of the terraces although it may complicate them. Rising islands in coral seas and also in marginal belts will ordinarily be reefless, for their emerging shores will continually be cloaked with loose detritus. This, of course, takes no account of elevated reefs previously formed during periods of subsidence.

*Subsiding islands* in the cooler seas (Fig. 5) should have cliffs, and the shoreline should be embayed. The submarine profile should be more or less distinctly benched but the benches will be largely covered by detritus cut from higher benches as submergence proceeds. Glacial changes of ocean level may modify the succession of benches but not materially alter their character.

Subsiding islands in the marginal belt (Fig. 6) should be surrounded by barrier reefs and may show cliffs at the spur ends, due to marine erosion at times of lowered sea level when the cooler waters of the glacial period prevented coral growth and allowed wave planation. The coast line should be embayed. Reefs formed in postglacial time may form offshore barriers. Atolls would result from complete submergence.

Subsiding islands in coral seas (Fig. 7) would have embayed shorelines and would have barrier reefs but no cliffs, as they would remain reef protected during the lowered sea level of glacial time.

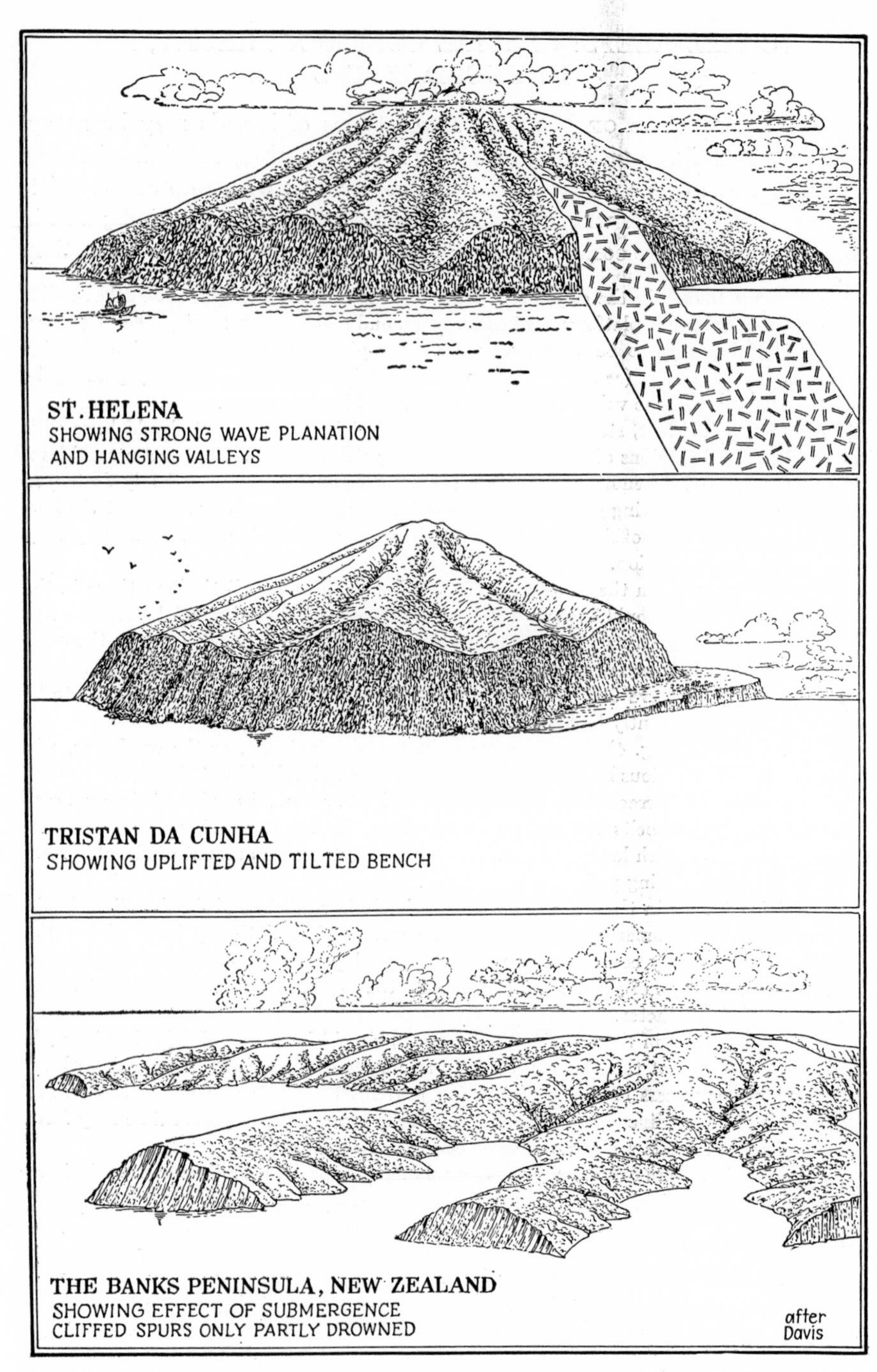


ISLANDS OF THE COOLER SEAS, WITHOUT REEFS

## CORAL REEFS: EXAMPLES. ISLANDS AND BANKS OF THE COOLER SEAS

W. M. Davis made a study of oceanic islands and banks well beyond the region of coral growth with the idea of discovering how effective wave erosion may have been in places where it was not interrupted by coral-reef growth at any time. He came to the conclusion that cliff recession by unhindered wave abrasion is more rapid than that of valley deepening by the short island streams and much more rapid than that of valley widening by the weather. This means that wave erosion is so rapid as to cut back the mouths of the streams and leave them hanging, a state of affairs clearly exhibited by the island of St. Helena. This island is believed to have remained stationary since it was first formed.

The island of Tristan da Cunha, 37° south, in the south Atlantic is much like St. Helena. Cliffs 1,000 feet high surround the island. It differs from St. Helena in that, since wave-cut platforms were formed around the island, it has been tilted so as to raise part of this platform above sea level, as shown in the sketch opposite. The main part of the island, rising 7,640 feet above the sea, still preserves its volcanic form, as it is very slightly dissected by streams. Wave erosion has been here much more rapid than stream erosion for the valley mouths hang high above sea level. This is in spite of the fact that the mountain slopes are steep and the rainfall abundant.

If strong submergence occurs after the condition reached by St. Helena or Tristan da Cunha is attained, then the valleys become embayed, while the spurs end in plunging cliffs. The "Banks" peninsula and islands off the New Zealand coast exhibit this condition.

The study of submarine banks was made especially to find out if there are many banks in the cooler seas which might have resulted from long-continued wave erosion at the present level or by low-level abrasion during glacial time. While few mid-ocean banks occur in either the north or south Pacific or in the south Indian Ocean, there appear to be a number in the Atlantic Ocean. These banks, however, vary greatly in depth; and if they represent truncated island masses, they indicate also considerable submergence. The absence of uplifted and benched islands in the cooler seas suggests that upheaval has been unusual. Submergence is thought to have occurred more commonly than emergence, but it has not been rapid enough to drown the mouths of the streams in the case of small islands like St. Helena and Tristan da Cunha.

The lack of banks in the Pacific is taken to indicate that after their development they have disappeared by subsidence rather than that they never existed.

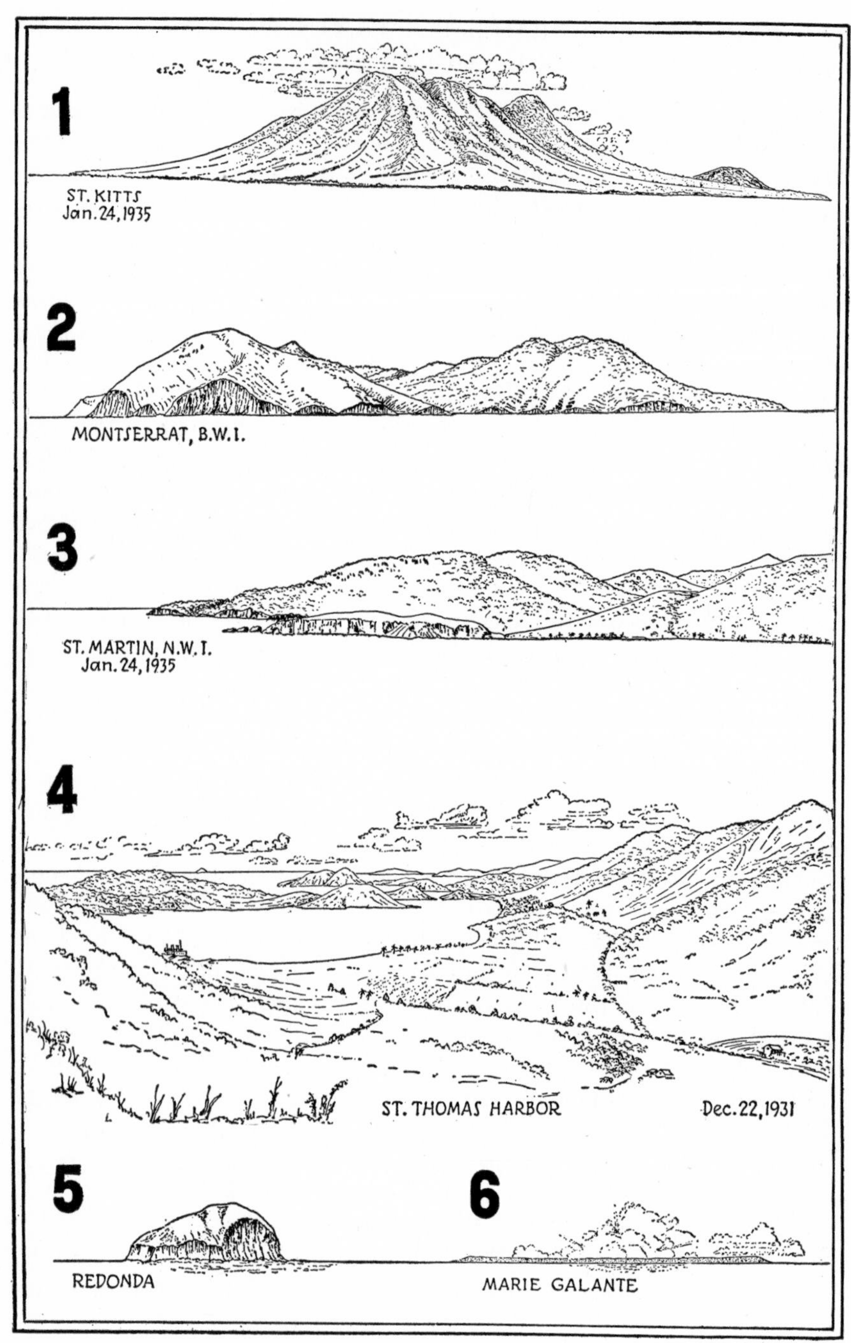


*Field sketches*

A SEQUENCE OF FORMS FROM THE WEST INDIES
Illustrating coastal development in the marginal belts of the coral seas.

## CORAL REEFS: EXAMPLES. THE MARGINAL BELT OF CORAL SEAS

The marginal belt of the coral seas in which coral growth was suspended during the glacial period is represented by the chain of the Hawaiian Islands in the Pacific, by the Lesser Antilles, and by scattered islands in the Indian Ocean and south Atlantic. These places display the features of subsiding islands in the marginal belt, namely:

*a.* Subsidence as revealed by an embayed coast.

*b.* Reef growth in preglacial time during subsidence.

*c.* Cessation of reef growth during glacial time with resultant low-level abrasion, producing platforms and cliffs.

*d.* Postglacial revival of reef upgrowth.

In the Lesser Antilles, Davis recognizes a regular sequence of forms embracing a whole series from young volcanic cones, standing at their original level and having no reefs, through those which have subsided more and more and have rimming reefs, until the series ends with the sea-level atoll type. Certain other examples are more complicated in that they include forms which, after passing through the cycle just mentioned, have been uplifted and are now beginning a second cycle of subsidence.

The sketches on the opposite page illustrate the series. First is St. Kitts, a volcanic group, still very young, showing no subsidence and no wave-cut cliffs. An even simpler case is represented by Saba Island (not illustrated). More advanced is Montserrat. Its hills are more rounded. It has lost its volcanic form. The valleys are more mature. Its cliffs, between 400 and 500 feet high, are younger than its valleys and may date back only to glacial time. No reefs occur around this island although it is surrounded by a platform 2 miles wide.

St. Martin is even more subdued in its topography, though its cliffs are not so bold as those of Montserrat. It shows strong evidence of submergence, as it consists of two or three islands tied together by benches of sand, enclosing lagoons. Discontinuous fringing reefs occur near the shore.

St. Thomas exhibits many features due to submergence. Its drowned valleys are thoroughly mature. The many islands and headlands are truncated by sea cliffs, like those of Montserrat, but inconspicuous in this general view. The cliffing is glacial and postglacial, whereas the valley cutting is preglacial, as is evidenced by a greater rock-bottom depth than can be accounted for by low-level erosion during glacial time.

Redonda (Fig. 5) represents a very late stage, the original land mass being almost completely submerged. The barrier reef, which is believed to have surrounded this island, was destroyed in glacial time.

The island of Marie Galante, an atoll recently elevated above the sea, represents the initial stage in the second cycle of development.

In all of the cases mentioned, the cliffing of the spurs, due to wave erosion in glacial time, is contrary to the conditions in the coral seas where the island masses were continuously protected against wave erosion. Moreover, these islands provide every evidence of instability, as advocated by Darwin's theory of subsidence. Change of sea level because of glacial control is important in this marginal belt where the dying of the reefs permitted wave erosion in glacial time; but it is not important in the coral seas where the reefs were never destroyed, and consequently where cliffs occur only in the early stages, before drowning took place.

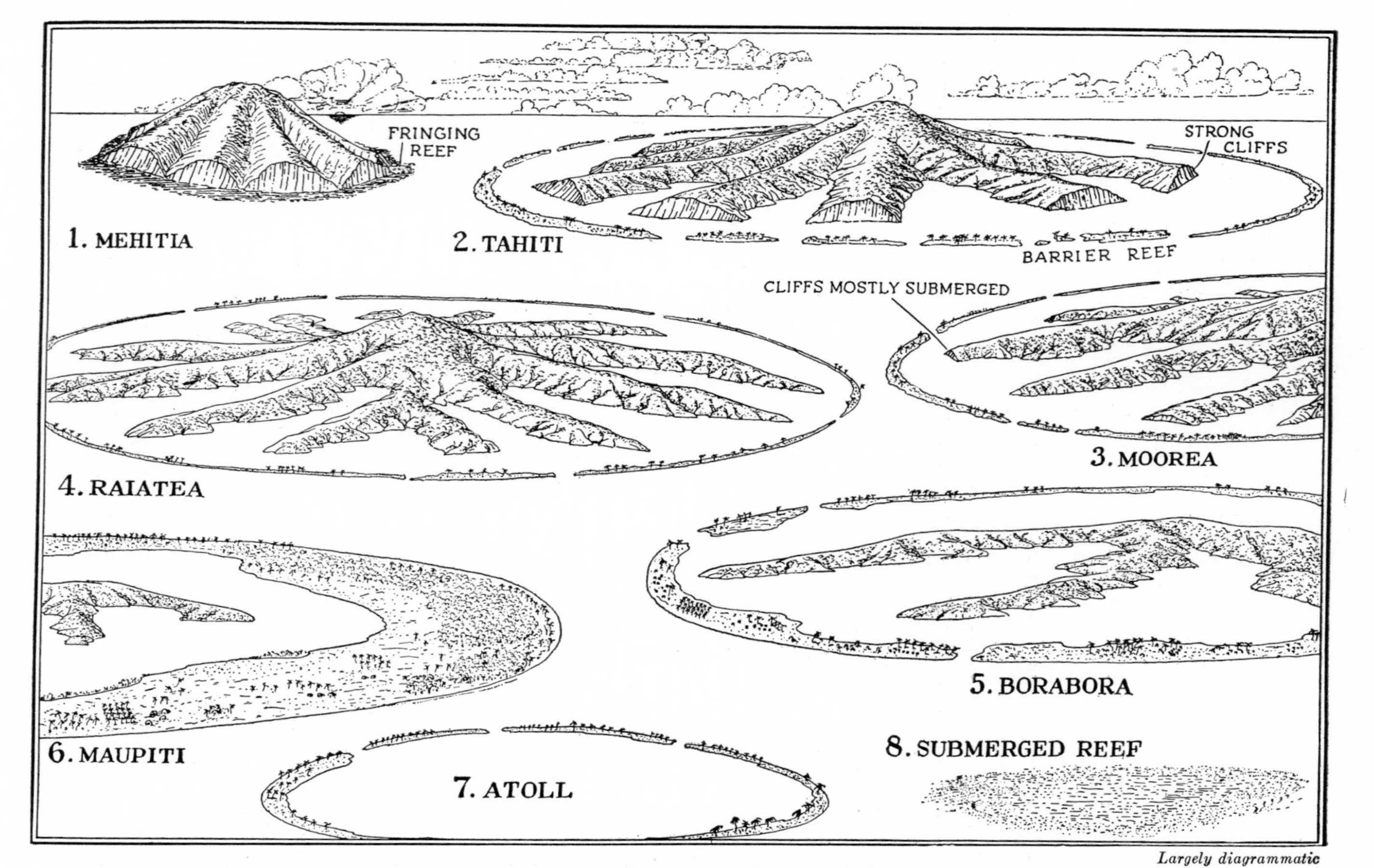


CORAL REEFS IN THE SOCIETY ISLANDS, SHOWING COMPLETE SEQUENCE OF FORMS

## CORAL REEFS: EXAMPLES. EMBAYED ISLANDS OF THE CORAL SEAS WITH BARRIER REEFS AND ATOLLS

These are the typical coral reefs of the world. Most of these islands are roughly circular and have a diameter of approximately 5 miles. They rise in height from 100 or 200 to 4,000 feet above the sea. Their shores are very irregular, the embayments extending into the interior of the mass, a mile or more in many cases, and occasionally as far as the very center of the island, suggesting a volcanic form with a breached crater. Ordinarily the spurs taper off gradually into the water of the lagoon. In some instances the spur ends are cliffed.

W. M. Davis suggests that the Society Group of islands exhibits a complete sequence of forms, starting with a young volcanic cone, called *Mehetia,* which is moderately cliffed. It has never suffered subsidence and does not have any barrier reefs. Next comes *Tahiti,* which is submaturely dissected. It was strongly cliffed in its reefless youth, just like Mehetia, then suffered subsidence and embayment, and consequently gained reef protection, although it has not yet subsided so far as to submerge completely its early-cut cliffs. Third is *Moorea* which has subsided so far as to submerge almost all of its cliffs, only here and there some remaining. The fourth in the series, represented by *Raiatea,* has noncliffed spurs because submergence has been very great. The shore lines are deeply embayed and there are well-developed barrier reefs. Fifth is *Borabora,* having a much denuded peak, worn-down spurs, and showing in its broad embayments evidence of deep submergence. It has a well-developed broad barrier reef. Next is *Maupiti,* a much smaller and lower island, with a very broad barrier reef and narrow lagoon. Maupiti has the appearance which Borabora would have if it subsided another 1,000 feet. Finally there are several small atolls suggesting complete disappearance of the central island, ending last of all with some reefs believed to represent submerged atolls.

In the case of Mehitia no reefs have developed because they are choked by the accumulation of mud and sediment around the shores of the land mass, due to erosion. Vigorous reefs do not form until after submergence begins; and once initiated, a fringing reef escapes being smothered by downwashed detritus only by continual subsidence of the island. This at the same time transforms it into a barrier reef.

The Society Group of islands stretches for 200 miles, in the order given, from east to west. It appears, therefore, that there has been a strong tilting of the sea floor, the western islands being those which show the greatest amount of submergence.

Life on coral reefs and atolls is distinguished by the scarcity of fresh water, the small variety of plants and animals which exist there, the remoteness from other lands, the lack of soil suitable for agriculture, and the constant exposure to wind and sun.

In the remote Pacific islands, fresh water is obtained from wells and shallows back of the beaches. During the rainy season the water accumulates in the hollows and, if the sand and coral rock be sufficiently compact, it does not readily mix with the sea water beneath, with which the whole mass is permeated. However, in dry seasons brackish water prevails. Islands like Bermuda, supporting large populations, find the water supply a difficult problem. Rain water is collected from all the roofs; and large cement collecting basins are constructed on the hillsides. Recently wells have been sunk into the coral rock to ground-water level, which is slightly above sea level. The rain water seeping through the rock accumulates just above the salt water and being of lighter specific gravity maintains this position.

On small sandy islands, where fresh water cannot accumulate, the cocoanut tree is used as a water trap. The leaves with their channeled midribs direct the water down the trunk where the natives catch it in hollow wooden vessels, often the hulls of old canoes. As a substitute for water, the oily milk of the green cocoanut is used everywhere, both for drinking and for making pudding and bread.

The lack of plants is more or less compensated for by the wide adaptability of the cocoanut to so many uses. This plant is found on every coral island. Its nut can float for six months or more at the mercy of the ocean without becoming waterlogged. It serves as the food and drink for man. Certain of the land crabs have pincers peculiarly adapted to tearing away the husk of the cocoanut and piercing its inner shell through one of the three "eyes." These crabs are of large size and many of them subsist entirely upon this food. In turn, the crabs are an important source of food to the natives and are useful in many ways.

The cocoanut palm thrives in brackish and salt water. It grows rapidly and profusely. The abundant husks from its fruit and, after death, from its decaying trunk serve to enrich the soil. Its leaves provide thatch for shelter and its husk yields fiber that can be spun into cloth or rope. The meat of the cocoanut is an important food; sugar is made from the sap. The nut takes only a year to ripen. There are over one hundred trees to the acre, and an average tree bears a hundred nuts a year. Ten to twenty acres of good cocoanut lands in an islet of twice this size provides a home for a community of one hundred people.

Lack of silica is made up by placing a basket of pumice or other volcanic rock around the roots of such trees, as the banana, which require this assistance. However, on most reefs these rocks are attainable only near some larger barrier-enclosed island. Fish remains and seaweed are also used to manure the land. Thus, breadfruit, papaya, pomegranate, tomatoes, and gourds can be grown. Seeds and plants from more richly endowed islands and lands are known to have been drifted 2,000 miles and more before taking a lodging on some remote land.

Occasionally boulders of igneous rock occur even on remote atolls, rafted there on the roots of trees. In fact, the natives search the roots of all drifting vegetation in order to find such stones which are highly prized in the making of tools and weapons. Indeed, on some islands such finds are deemed the property of the ruling "king."

The total population of the coral reefs of the Pacific at the time of their discovery was not more than that of a good-sized city, and that scattered over an area twice as great as the United States. All told, they inhabited only about 800 square miles of land, of which perhaps one-half was cultivated.

All visitors to coral lands remark on the contrast between the ever-pounding ocean on the outside and the serene emerald water of the lagoon which serves as a safe haven of refuge after long weeks at sea.

The villages are usually set on the lagoon shore. The seas there are calm and the sandy beach is convenient for hauling up the outrigger canoes. Here in the lagoon, the natives can travel for many miles, in the case of the larger atolls, and secure fish and sea turtles which, in places, occur in great abundance. The outer side of the reef, where attacked by the waves, is at low tide a platform of jagged and broken coral rock with pools filled with an abundance of marine life. Only the unremitting activity of millions of coral polyps at its outer edge, where no sand occurs and where the water is constantly agitated, keeps the reef from being destroyed. The coral polyp cannot grow at depths greater than 20 or 30 fathoms, for there sand and broken rock accumulate and suffocate them. Their chief enemy appears to be a certain species of fish whose bony jaws are adapted to stripping them from the coral rock upon which they grow.

Low coral atolls are always in great danger of being overwhelmed by earthquake waves. Certain of the highly uplifted atolls of the Pacific are spared this danger and, moreover, have much more varied topography, as well as a more diversified flora and fauna, than the sea-level islands.

Among the famous coral atolls of the world is Wake Island, in the mid-Pacific, one of the bases of the trans-Pacific air route. Its horseshoe shaped form encloses the small lagoon upon which the great air liners alight. Twice a year a supply vessel anchors offshore. Water is obtained in large concrete catchment basins.

## TIMBER RAFTS OF THE RED RIVER

Just before the advent of the early settlers into northern Louisiana, the Red River became blocked by a series of log jams. This came about naturally as the river, flowing through a forested flood plain, undercut its banks and caused whole trees with their interlocking branches to be carried downstream. By the year 1600 when the first explorers began to arrive the jam had worked upstream as far as Alexandria. For more than 200 years the jam kept increasing in size and advanced upstream until it reached the Arkansas border in the middle of the nineteenth century. Its rate of advance was almost a mile a year, on the average.

As the raft advanced, it blocked the outlets of the tributary streams and produced a series of lakes. The timber in these flooded areas soon died, and the exposed portions decayed, leaving extensive areas of the flood plain covered with the stumps of the former forest.

As the jam advanced upstream, the lower portion decayed and was gradually carried down the partly blocked channel. At the time of the early settlements the foot of the raft was near the present site of Natchitoches and was one of the important factors in determining the location of that town, which was thus placed at the head of ordinary navigation.

History records the growth of a number of the lakes, especially those around Shreveport which were formed during the close of the eighteenth century. The raft continued to grow until 1873, when its upper end was close to the Arkansas state line and produced the last of the Red River lakes, called *Poston Lake*, through which, for a time, steamboats passed on their way to upper Red River.

The raft was gradually removed in 1873 by means of river steamers with heavy cranes which lifted the logs on deck to be cut up by steam and hand saws.

Practically all of the lakes have now been drained and much of the land formerly covered by the lakes is under cultivation. Several of the streams now flowing over the old lake beds have developed rapids and waterfalls where they have been superposed over spurs of the older bedrock of the country. This has somewhat retarded the draining of the lakes.

The lakes of the Red River Valley are not unlike the lakes formed in valleys tributary to heavily silt-laden streams, as explained in the chapter on continental glaciation. There are, in fact, many small lakes, varying in area from 10 to 250 acres, along the Red River in Arkansas, upstream from the former Lake Poston. They have been formed by the silting up of the mouths of small streams and are much more permanent features than the "raft" lakes.

## BEAVERS

These industrious rodent engineers are responsible for surprising local modifications of topography, particularly in regions where the aspen is abundant. The most important beaver activity is ponding of streams. In the United States numerous "beaver meadows" with impenetrable willow growth have replaced valuable timber land.

The dam is generally placed where the least amount of construction will impound the most water. Normal dams are made of logs, branches, mud, and stones; exceptional ones of mud, coal, or stones, some of which weigh 50 pounds, as in the Bad Lands of North Dakota. In Montana, some beaver dams have become encrusted with lime and are termed *petrified dams*. The cementing mud is scooped from the pond bottom directly subjacent to the dam. Dams vary in length from a few feet to over 2,000 feet, and in height from less than 5 to more than 11 feet. A 260-foot dam may contain 7,000 cubic feet of material. Ponds vary from 50 to 800 feet or more in length.

As the pond bottom is silted up, the dam is raised and strengthened by the beavers, and subsidiary dams are thrown across sags in the advancing shore line. Stick and mud lodges are built either in the pond or on the bank and vary in diameter from 6 to 35 feet, and in height from 5 to 9 feet. These lodges or houses are entered from below water level. In the fall they are plastered on the outside with mud or sod which freezes to a hard protecting shell. Canals are dug into surrounding woodland, so green logs can easily be floated into the pond. Frequently ditches are cut across meander necks to shorten traveling distance to food supply or to divert the water of other streams into the pond. Not uncommonly the canals have one or more "locks."

Beaver dams require constant attention and repair and, when left to disintegrate, the pond is soon drained. A dozen years after a pond is abandoned by the beavers, the resulting meadows are quite firm. They are usually underlain with deep silt representing centuries of accumulation, topped by sand and gravel. A pond maintained by beavers for many years develops into a peat bog.

In a very few years a pair of beavers with its progeny rapidly changes the aspect of a landscape. Large areas of forest become inundated and killed, but at the same time a multitudinous variety of small animal life is introduced. Water-loving birds become more abundant as well as small animals.

Large numbers of beavers occupied North America before the advent of the white man. It is now believed that many meadows and swamps formerly considered beds of glacial lakes are more likely the work of long-forgotten beavers.

*(From material supplied by Dr. F. Eyolf Bronner)*

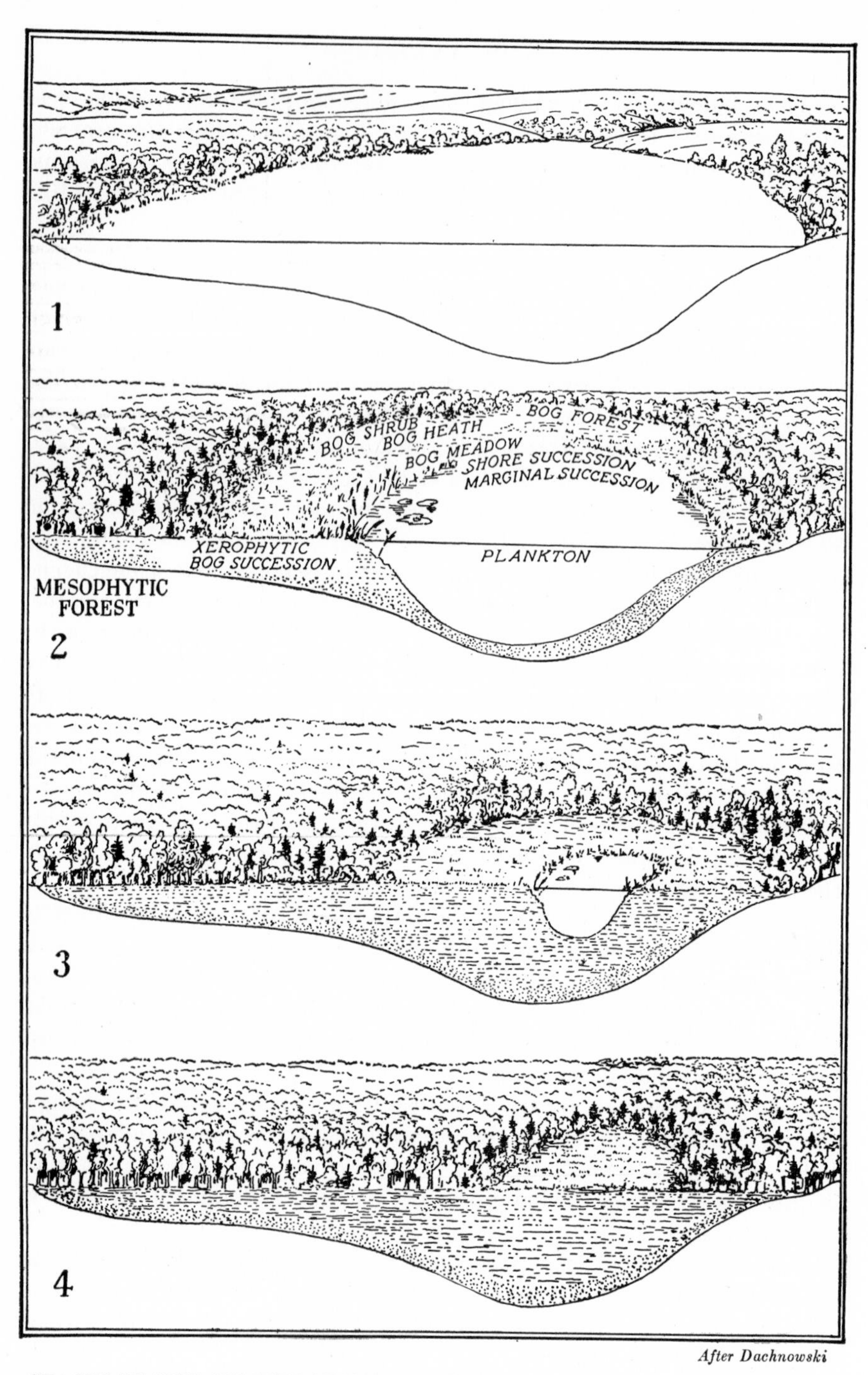


*After Dachnowski*

STAGES IN THE OBLITERATION OF A LAKE BY THE ENCROACHMENT OF VEGETATION UPON ITS SHORES

## PEAT MARSHES

Plants are effective in changing the shape of lakes and estuaries by the formation of peat. These changes are much more rapid than most geological changes. Lakes are known to be ephemeral but it is usually assumed that they are destroyed by being filled with sediment or by being drained when their outlets are cut down. Filling by plant growth is, especially in relatively shallow glacial lakes, much more important.

The filling of lakes by plant deposits proceeds in a very systematic way, somewhat as follows: Where the water is deepest, only floating plants develop, such as diatoms, one-celled algae, and duckweed. This is a plankton type of vegetation. The spores and the remains of these plants sink to the bottom and produce a very fine-grained peat. The bottom layers of most peat deposits indicate this kind of origin.

Near the shore of the lake are water lilies, pickerel weed, grasses, and reeds. These plants grow rapidly and die down each year. Their abundant remains accumulate on the bottom. Thus the shore is pushed outward and the area of open water reduced in size. With the accumulation of vegetable debris the soil becomes acid and a new type of vegetation prevails. This series, the *bog succession*, is classed as *xerophytic*, plants able to subsist under scanty water conditions or where the water is heavily charged with salts. As shown on the accompanying illustration, four types make up this xerophytic bog succession; (*a*) the *bog meadow*, containing sedges, cranberry, ferns, and sphagnum moss; (*b*) the *bog heath*, containing blueberry, wintergreen, and other heaths; (*c*) the *bog shrub*, containing alder, poplar, and willow; and (*d*) the *bog forest*, with tamarack, larch, yew, red maple, dogwood, and other common trees. Finally, as a thick accumulation of leaves produces a soil suitable to the usual flora of the country, a *mesophytic* vegetation appears. Most of our common trees belong in this type, intermediate between the xerophytic (dry) and the aquatic. In the northern United States and Europe there are lakes in every stage of being overgrown by peat-producing vegetation, as well as many flat tracts of peat land, occupying the sites of former lakes.

The types of vegetation that encroach upon salt marshes is quite different from the succession just outlined, because relatively few plants can thrive in salt water. Nevertheless, the process is similar and an accumulation of peat results.

The total area of peat land in the United States adaptable to economic use is equal to twice the area of New York State. Aside from the use of peat land for cultivation, peat is important as a fertilizer, stock feed, absorbent and disinfectant, fuel, packing material, and for paper making.

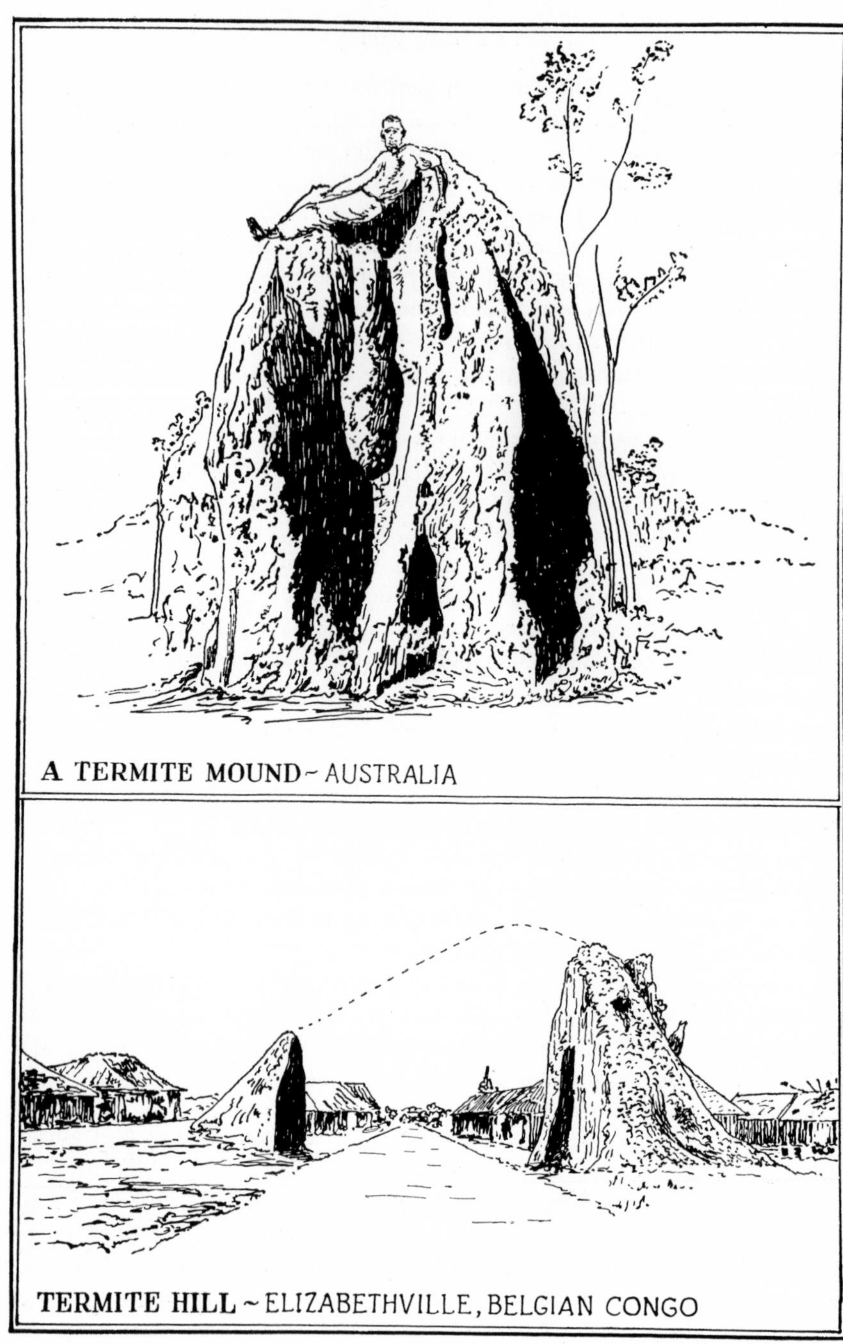


*From photographs*

TERMITARIA

## TERMITARIA

Termitaria, the large mound-shaped or pinnacle-shaped dwellings of the termites, attain heights of 25 feet. In equatorial Africa and in Australia, they not only attract attention because of their curious form and striking size but they must also be reckoned with as topographic obstacles where roads and villages are built. They are scattered over vast areas and in some instances form congested colonies a mile or more in extent. They vary widely in shape, some being sharp-pointed pinnacles, fluted down the sides. Others are solid mounds as big as haystacks. It is almost impossible to stand anywhere in the level grasslands in central Africa without seeing the mounds, the rounded bosses, or the towers of the dwellings of these so-called *white ants*. In fields, hillocks upon hillocks rise, as tall as a giraffe, bristling with grasses. The hillocks slope to the ground and seemingly end there, yet yards away run their tunnels in a miraculous, invisible network. Or in a wooded section may stretch a little group of cathedrals, buttressed with hard uneven ridges.

The termitaria are resistant masses, little affected by erosion. By the natives they are accepted as a permanent part of the landscape, enduring longer than the span of a human life. Cattle graze upon their summits. Man cuts through them with the use of dynamite. The material of which they are composed is extremely hard and brittle, being much like solidified plastic wood. In fact, the termites manufacture their homes by masticating wood and leaves to a soft pulp which is plastered, layer after layer, on both the inside and outside of their dwelling. A certain amount of clay also goes into the construction.

Termites are insects but not ants. Found most abundantly in warm climates, they are occasionally very destructive. Biologically they are of unusual interest because of their great colonies and the specialized activities of the different members.

ANTHILLS. The hills of sand built by real ants are quite unlike the nests of termites. Anthills are only the accumulations made by the ants when excavating their burrows beneath the ground. Anthills are built in much the same way as volcanic cinder cones. The tube or neck of the anthill may serve the ants as means of ingress and egress but usually there are other openings to the underground passageways, not so likely to become clogged up or destroyed.

Anthills of large size occur even in northern latitudes, being quite common in Finland and the northern United States, where they attain heights of 2 or 3 feet. When the ants cease to use them, the mounds become covered with a layer of leaves and mold but remain conspicuous for a long time.

## EARTHWORMS

Frequently overlooked in discussions of topographic forms is the Nibelung among animals—the earthworm. While abrasion, frost, and gravity are the mortar and pestle of rock weathering, and water, acids, and air the reagents, this retiring but diligent phylum acts collectively as a stirring rod that churns the soil in ceaseless slow turbulence. From sea level to elevations of many thousand feet, these segmented creatures unwittingly contribute to the production of fertile soil.

On the morning after a heavy rain worm castings, thrown up during the night, dot the surface of the ground. They consist of finely comminuted rock particles, which may be fairly loose or well agglutinated into a viscid mass of a few grams to more than 120 grams. The castings are larger on poor land than on rich, probably because the former contains a smaller amount of nutriment per unit consumed by the worms.

Each miniature pyramid marks the entrance to a burrow in which one worm lives, and near which it remains. The burrow is perpendicular to the surface, from 3 to 8 feet deep, terminating in a small chamber usually lined with pebbles and seeds. The walls of the tube are lined with leaves pulled into it by their apices, plastered with an earthy cement expelled from the worm's body in pellet form, and smeared by the passage of the worm's body. The excavated material actually passes through the worm's digestive tract, in which soft mineral particles are triturated and exposed to humic acid, while organic matter is extracted for food. In this manner worms keep the soil in constant motion, aerating it, and mixing it intimately with decaying vegetable matter on the surface.

Cave-in of many abandoned burrows necessarily causes slump, with attendant sinking of heavy objects, such as large stones. Simultaneously, castings brought up along the sides of the stone tend to be protected from dissipation, so that they eventually form an elevated rim standing above the general level of the ground. In a similar manner, cement garden walks and even building foundations are undermined, causing them to collapse or crumble. In this way worms may also be of prime importance in burying ancient buildings.

In one garden acre there are more than 50,000 worms, and in an acre of grainfield approximately one-half as many. These worms annually bring to the surface 0.083 to 0.22 inch of earth, depending directly on whether the soil is rich or barren. Careful measurements indicate that 16 tons of castings are annually deposited on every acre of grainfield in the temperate zone, and 7 tons on every acre of lawn. These castings are evenly spread over the ground, by flow while they still are wet, and by rolling when they are desiccated.

Thus all superficial soil passes through the bodies of worms every few years, and the earthworm performs a menial task which no other animal can do.

*(From material supplied by Dr. F. Eyolf Bronner)*

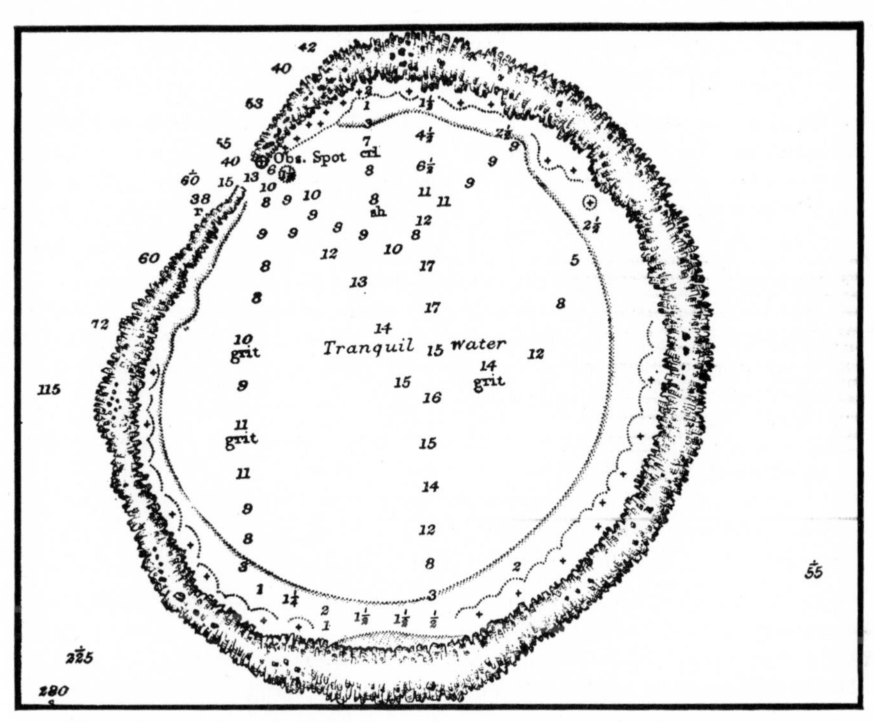


NORTH MINERVA REEF, AN ATOLL IN THE SOUTH PACIFIC
United States Hydrographic Office Chart, No. 2018 (1:73,761).

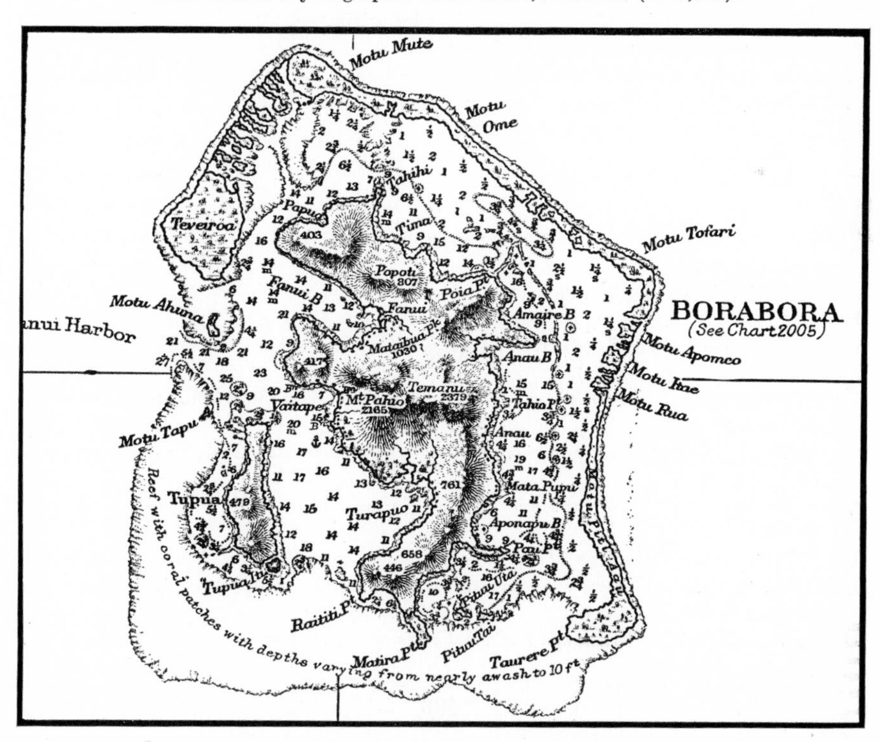


BORABORA, SUBMERGED ISLAND WITH BARRIER REEF, SOCIETY ISLANDS
United States Hydrographic Office Chart, No. 2023.

## QUESTIONS

1. What does a coral polyp look like? Is it an animal or plant? Does it thrive on a sandy sea bottom?
2. Suppose a reef is found to be 1,000 feet thick. Can this be taken as evidence of subsidence?
3. How do elevated reefs ever serve as evidence of subsidence?
4. Who studied the reefs along the Florida coast? How did he think they were formed?
5. Why do barrier reefs and atolls sometimes have a height of 15 to 20 feet above sea level, although elevation of the reef may not have occurred?
6. Why would it be reasonable to find fringing reefs common in one part of the ocean or coast line; and atolls or barrier reefs common in another part?
7. Give a summary of all the evidence which favors the subsidence theory.
8. If you were on a barrier reef surrounding an island, how would you determine the probable amount of subsidence? Where would you go and what would you do?
9. Do you think it likely that one might find somewhere a series of islands which would show more and more subsidence as one followed the chain from one end to the other? What would this mean?
10. What do you think of the theory that atolls grew up on the rims of submarine craters? What objections are there to this theory?
11. What are tepee buttes? Do you consider them forms due to organic activity? What are effigy mounds? In what classification do they belong?
12. For what purposes is peat used? How would you distinguish former glacial lake beds from those due to beavers?
13. Why are angle worms more abundant in gardens than in open fields?
14. If *xerophytic* means "loving dryness," why do xerophytic plants grow in bogs?

## TOPICS FOR INVESTIGATION

1. The Florida Keys: description; explanations for origin.
2. Bermuda: description; explanations for origin.
3. The Great Barrier Reef of Australia.
4. The coral reefs of the Pacific.
5. The coral reefs of the West Indies.
6. The glacial-control theory.
7. Vegetation as a geological agent.

## REFERENCES

Most of the important literature on coral reefs has been listed by W. M. Davis (1928) in his comprehensive treatise on *The coral reef problem*, Am. Geog. Soc., Spec. Publ. 9. Of the many other papers on coral reefs by W. M. Davis the following are particularly important:

DAVIS, W. M. (1914) *The home study of coral reefs*. Am. Geog. Soc., Bull. 46, p. 561–577, 641–654, 721–739.

DAVIS, W. M. (1915) *A Shaler Memorial study of coral reefs*. Am. Jour. Sci., 4th ser., vol. 40, p. 223–271.

DAVIS, W. M. (1918) *Coral reefs and submarine banks*. Jour. Geol., vol. 26, p. 198–223, 289–309, 385–411.

DAVIS, W. M. (1918) *Subsidence of reef-encircled islands*. Geol. Soc. Am., Bull. 29, p. 489–574.

DAVIS, W. M. (1918) *The reef-encircled islands of the Pacific*. Jour. Geog., vol. 17, p. 1–8, 58–68, 102–107.

DAVIS, W. M. (1920) *The islands and coral reefs of Fiji*. Geog. Jour., vol. 55, p. 34–45, 200–220, 377–388.

DAVIS, W. M. (1923) *The marginal belts of the coral seas.* Natl. Acad. Sci., Proc. 9, p. 292–296; also Am. Jour. Sci., 5th ser., vol. 6, p. 181–195.

C. R. Darwin's original contribution to this subject is still deserving of much study. It is entitled *The structure and distribution of coral reefs.* London, 344 p. 1st ed., 1842; 2d ed., 1874; 3d ed., edited by T. G. Bonney, New York, 1889.

R. A. Daly's extensive discussion of the glacial-control theory may be represented by the following:

DALY, R. A. (1910) *Pleistocene glaciation and the coral reef problem.* Am. Jour. Sci., 4th ser., vol. 30, p. 297–308.

DALY, R. A. (1915) *The glacial-control theory of coral reefs.* Am. Acad. Arts and Sci., Proc. 51, p. 155–251.

DALY, R. A. (1919) *The coral-reef zone during and after the Glacial period.* Am. Jour. Sci., 4th ser., vol. 48, p. 136–159.

DALY, R. A. (1920) *A recent world-wide sinking of ocean level.* Geol. Mag., vol. 57, p. 247–261.

Among J. D. Dana's important papers are the following:

DANA, J. D. (1853) *On coral reefs and islands.* New York, 143 p.

DANA, J. D. (1872) *Corals and coral islands.* New York, 398 p.; 2d ed., 1874; 3d ed., 1890.

DANA, J. D. (1885) *Origin of coral reefs and islands.* Am. Jour. Sci., 3d ser., vol. 30, p. 89–105, 169–191.

Gardiner's ideas are presented in his latest contribution:

GARDINER, J. S. (1931) *Coral reefs and atolls.* London, 181 p.

Note also:

GARDINER, J. S. (1903) *The origin of coral reefs as shown by the Maldives.* Am. Jour. Sci., 4th ser., vol. 16, p. 203–213.

The following readily accessible papers from other authors are selected from a long list:

AGASSIZ, A. (1883) *The Tortugas and Florida reefs.* Am. Acad. Arts and Sci., Mem. 11, p. 107–134.

AGASSIZ, A. (1889) *The coral reefs of the Hawaiian islands.* Mus. Comp. Zool., Bull. 17, p. 121–170.

AGASSIZ, A. (1898) *A visit to the Great Barrier Reef of Australia.* Mus. Comp. Zool., Bull. 28, p. 95–148.

AGASSIZ, A. (1899) *The islands and coral reefs of Fiji.* Mus. Comp. Zool., Bull. 33, p. 1–167.

AGASSIZ, A. (1903) *On the formation of barrier reefs and of the different types of atolls.* Royal Soc. London, Proc. 71, p. 412–414.

AGASSIZ, A. (1903) *The coral reefs of the tropical Pacific.* Mus. Comp. Zool., Mem. 28, p. 1–410.

GUPPY, H. B. (1888) *A criticism of the theory of subsidence as affecting coral reefs.* Scot. Geog. Mag., vol. 4, p. 121–137.

HUNT, E. B. (1863) *On the origin . . . of the Florida reef.* Am. Jour. Sci., 2d ser., vol. 35, p. 197–210.

JUKES-BROWNE, A. J., and HARRISON, J. B. (1891) *The geology of Barbados.* Part I: The coral-rocks of Barbados and other West Indian islands. Geol. Soc. London, Quart. Jour., vol. 47, p. 197–250.

MACCAUGHEY, V. (1918) *A survey of the Hawaiian coral reefs.* Am. Naturalist, vol. 52, p. 409–438.

MAYOR (MAYER), A. G. (1916) *Sub-marine solution of limestone in relation to the Murray-Agassiz theory of coral atolls.* Natl. Acad. Sci., Proc. 2, p. 28–30.

MAYOR (MAYER), A. G. (1917) *Coral reefs of Tutuila, with reference to the Murray-Agassiz solution theory.* Natl. Acad. Sci., Proc. 3, p. 522–526.

MURRAY, J. (1880) *On the structure and origin of coral reefs.* Royal Soc. Edinburgh, Proc. 10, p. 505–518.

MURRAY, J. (1887–89) *Structure, origin, and distribution of coral reefs and islands.* Royal Inst., Proc. 12, p. 251–262; also Nature, vol. 39, p. 424–428.

MURRAY, J., and IRVINE, R. (1889–90) *On coral reefs and other carbonate of lime formations in modern seas.* Royal Soc. Edinburgh, Proc. 17, p. 79–109.

SOLLAS, W. J. (1899) *Funafuti: The study of a coral atoll.* Nat. Sci., vol. 14, p. 17–37; Smiths. Inst., Ann. Rept. for 1899, p. 389–406.

VAUGHAN, T. W. (1919) *Corals and the formation of coral reefs.* Smithsonian Inst., Ann. Rept. for 1917, p. 189–276.

WHARTON, W. J. L. (1897) *Foundations of coral atolls.* Nature, vol. 55, p. 390–393.

WOOD-JONES, F. (1910) *Corals and atolls.* London.

YONGE, C. M. (1930) *A year on the Great Barrier Reef.* London, 245 p.

A recent beautifully presented description in English of reefs in the Netherlands East Indies is given by P. H. Kuenen (1933) on *Geology of coral reefs* in the report of the Snellius Expedition, Utrecht, 125 p. This author's observations strongly support Darwin's theory of subsidence and are contrary to the glacial-control theory.

References on organisms other than corals:

ANTS

ANDREWS, E. A. (1925) *Growth of ant mounds.* Psyche, vol. 32, p. 75–87.

BEAVER DAMS

HEGNER, R. W. (1935) *Parade of the animal kingdom.* New York, 675 p.

STONE, W., and CRAM, W. E. (1914) *American animals.* New York, 4th ed., 318 p.

WARREN, E. R. (1927) *The beaver.* Am. Soc. Mammal., Mon. 2; Baltimore, 177 p.

EARTHWORMS

DARWIN, C. F. (1881–97) *The formation of vegetable moulds through the action of worms.* London, 328 p.

PEAT

DACHNOWSKI, A. P. (1912) *Peat deposits of Ohio.* Ohio Geol. Surv., 4th ser., Bull. 16, p. 259–262.

DACHNOWSKI, A. P. (1926) *Selection of peat lands for different uses.* U. S. Dept. Agri., Bull. 1419. List of references.

SOPER, E. K., and OSBON, C. C. (1922) *The occurrence and uses of peat in the United States.* U. S. Geol. Surv., Bull. 728, 207 p.

TIMBER RAFTS

VEATCH, A. C. (1906) *Geology and underground water resources of northern Louisiana and southern Arkansas.* U. S. Geol. Surv., Prof. Paper 46, 422 p.

# XIII
# COASTAL PLAINS

*Holmes, U. S. Geological Survey*

A TIDAL CREEK NEAR MOUTH OF CAPE FEAR RIVER, NORTH CAROLINA
An estuary on a partially submerged coastal plain.

*U. S. Forest Service*

OCALA NATIONAL FOREST, FLA.

Very young, undissected coastal plain with lakes and solution depressions.

"HAYSTACK" HILLS, NEAR BAYAMON, PUERTO RICO

An old coastal plain surmounted by scattered limestone monadnocks honeycombed with caves.

OLDLAND PENEPLANE IN PUERTO RICO SLOPING NORTHWARD BENEATH COASTAL PLAIN

NORTHWARD DIPPING COASTAL PLAIN BEDS, FORMING TERRACES. MANATI VALLEY, PUERTO RICO

STRIPPED RUGGED OLDLAND OF SOUTHERN PUERTO RICO WITH INFACE OF CUESTA AT EXTREME RIGHT



OLD COASTAL PLAIN, PUERTO RICO, WITH MONANDNOCKS

*British Geological Survey, Crown Copyright*

CHALK ESCARPMENT, BETCHWORTH, SURREY, ENGLAND

Inface of cuesta overlooking clay lowland, English Coastal Plain.



LIMESTONE ESCARPMENT, STOKE-SUB-HAMDON, SOMERSET, ENGLAND

Inface of cuesta overlooking inner lowland, English Coastal Plain.

*British Geological Survey, Crown Copyright*

*U. S. Forest Service*

BLACK RIVER REGION, WISCONSIN

Rolling inner lowland, with inface of dissected limestone cuesta in the distance.



HICKORY RIDGE LOOKOUT, HOOSIER NATIONAL FOREST, SOUTHERN INDIANA

Maturely dissected former coastal plain with profile of cuesta in the distance.

*U. S. Forest Service*

*A. K. Lobeck*

PROFILE OF CUESTA, SOUTHERN NEW JERSEY, VIEW LOOKING SOUTH



STEEP INFACE OF MATURELY DISSECTED LIMESTONE CUESTA NEAR LARES, PUERTO RICO, VIEW LOOKING NORTH

*A. K. Lobeck*

# COASTAL PLAINS

SYNOPSIS. A coastal plain is an exposed part of the sea floor due to emergence. It may be very narrow or even fragmentary or it may be very broad. It may recently have emerged or it may have been formed in some earlier geological time and now be far inland from the sea.

The structure of a coastal plain may be simple and represent only a continuous sequence of beds. Or it may be complex as a result of several advances and retreats of the sea; in this case some of the beds may be separated by disconformities.

A coastal plain may be formed by simple regional uplift or it may be warped into low domes and basins. It may rest upon an oldland of simple structure or of complex structure; and upon a surface of low relief or of rugged topography.

A coastal plain may have been rapidly exposed to erosion; in this case all parts of it exhibit about the same stage of development. It may have been uplifted slowly, so that its inner portion is maturely dissected, while the outer part is young.

At first, a newly emerged coastal plain exhibits large initial depressions which contain shallow lakes or swamps. These are drained as an erosional system is gradually established.

Coastal-plain stream development passes through an orderly series of stages. At first, the coastal plain is drained by consequent and extended consequent streams flowing seaward, with numerous insequent tributaries. Then along the inner margin of the coastal plain the lower weaker beds become exposed and subsequent streams open out an inner lowland. This lowland consists in part of a stripped belt, where the coastal-plain strata have been removed down to the rock of the oldland, and in part it is cut from the coastal-plain strata. The edges of the more resistant members of the coastal plain form cuestas, with scarps facing inland on one side, and, on the other side, with long gentle slopes down the dip of the beds to the sea. Thus the coastal plain eventually comes to have a belted pattern, consisting of alternating strips of lowlands and cuestas. The short streams flowing down the cuesta face are obsequent streams; those flowing seaward down the back slope are consequent or in some cases resequent streams. The entire drainage pattern is trellislike, but simpler and more regular than the trellis pattern of folded rocks.

The fall line, inliers, outliers, the drowning of inner lowlands, the cessation or in some instances the union of cuestas, marine terraces, stream capture, undrained upland marshes, solution depressions, doming, submergence of coastal-plain streams, the economic uses of lowlands and cuestas—these are a few of the topics which are comprehended in a study of coastal plains.

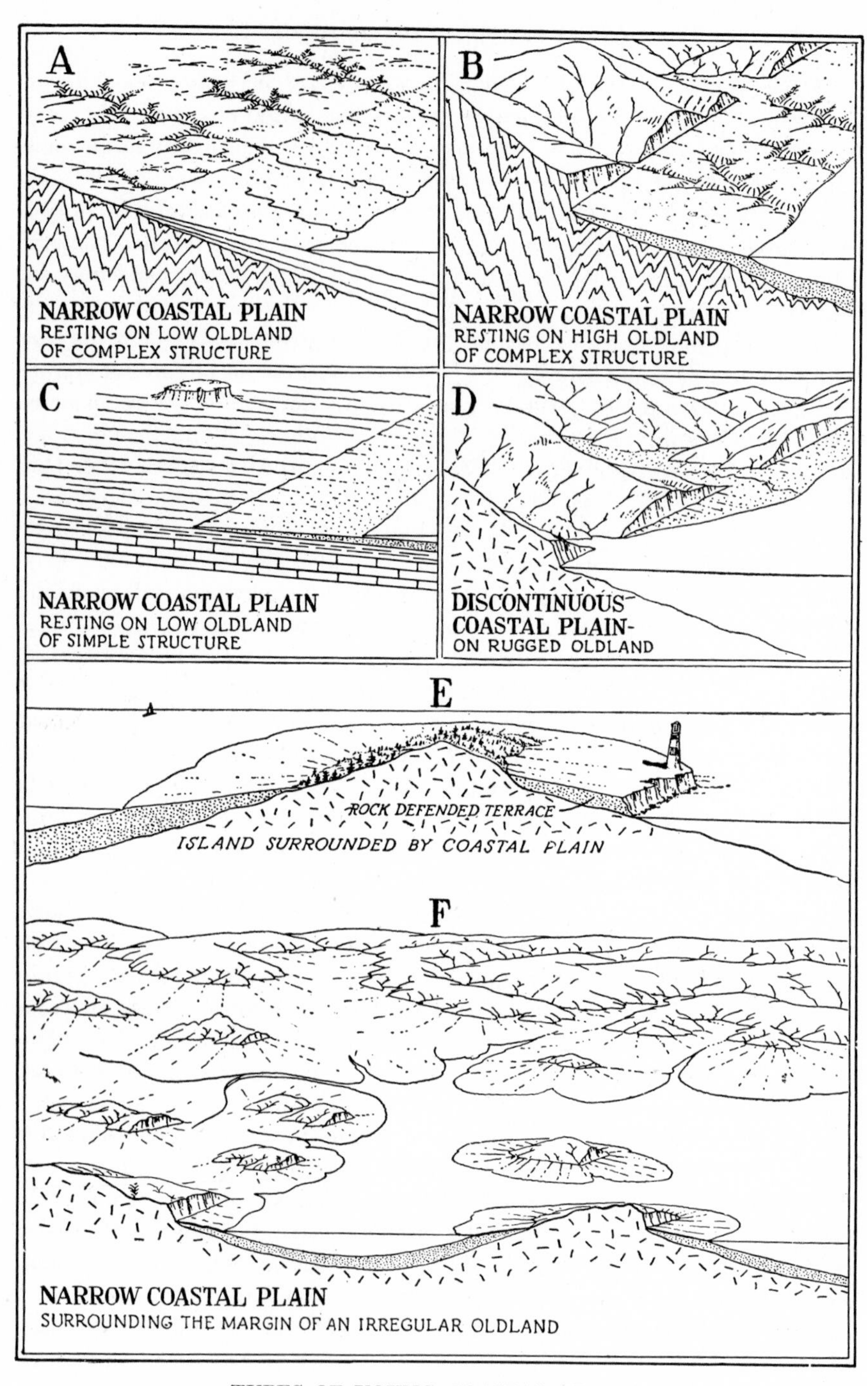


TYPES OF YOUNG COASTAL PLAINS

TOPOGRAPHY OF THE OLDLAND. A young coastal plain is a strip of the former sea floor exposed along the margin of an *oldland* area. It is usually narrow and built up of loose or weakly consolidated sands and clays. The oldland upon which it rests may be low (*A*) or high (*B*). It may be complex in structure (*A* and *B*) or very simple, as simple as the coastal plain itself (*C*). In fact, the oldland may be a former coastal plain. Narrow coastal plains are apt to be discontinuous, especially if the oldland mass drops abruptly into the sea (Fig. *D*). If the oldland shelves very gradually seaward, even a slight emergence is apt to expose a considerable belt of the sea floor (Figs. *A*, *B*, and *C*).

If the oldland is very irregular (*F*), the coastal plain will have a very irregular outline. What at one time were islands offshore become hills in the midst of the coastal plain. Such hills are called *mendips*, from the Mendip hills near Bristol, England.

The border of the oldland may or may not show the effect of vigorous wave erosion in the form of cliffs descending abruptly to the plain. And these cliffs may exhibit caves, stacks, arches, and platforms which until recently were undergoing wave attack. The rivers which dissect the oldland may be young (*A*) or mature (*B*), and they may cross the coastal plain in a direct manner or wind about among the mendip hills (*F*). The coastal plain may slope very gently to seaward or pitch at a strong angle. Its surface may be flat or it may be undulating with hollows which contain swamps or lakes. When gently sloping, the coastal plain will be attacked by the waves offshore and barrier bars result. Steeply sloping coastal plains are attacked near shore and are cut back to form terraces descending seaward in cliffs.

The loose nature of the coastal-plain sediments renders them an easy prey to wave attack, especially on the exposed sides of islands (Fig. *E*) and at the ends of promontories. The coastal plain is thus rapidly worn back until the waves encounter the buried oldland mass at the foot of the cliff cut in the coastal-plain sediments, as in *E*. The resulting terrace is said to be *rock defended* and resembles similar terraces along river valleys.

TYPES OF COASTAL-PLAIN RIVERS. The short and simple streams which originate upon the newly uplifted coastal plain are known as *consequent* streams. This is a genetic type of stream whose origin and position is determined by the initial slope of a newly formed land area. Other streams arising in the oldland and extending their courses across the coastal plain, as in both *A* and *B*, are called *extended consequents*.

Narrow coastal plains do not show much variety of rock structure or soil types and consequently do not have belts such as wider coastal plains have. Narrow coastal plains occur along the New England coast and the Scandinavian coast of Europe as the result of recent postglacial uplift and tilting.

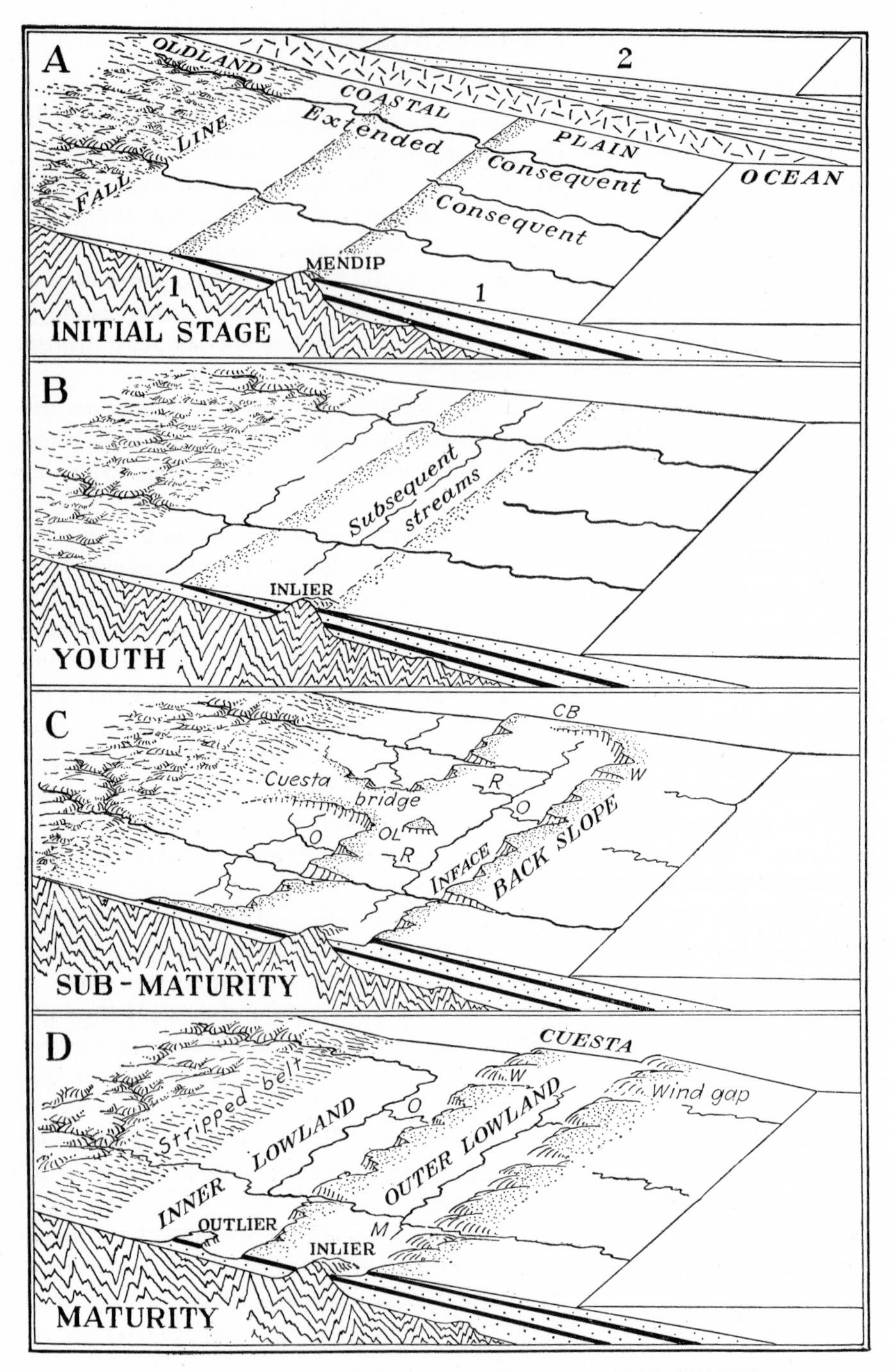


THE EROSIONAL DEVELOPMENT OF COASTAL PLAINS

# EROSIONAL DEVELOPMENT OF BROAD COASTAL PLAINS

STRUCTURE. The cross section of a broad coastal plain may be like either *A1* or *A2* or it may be a combination of both. If it is like *A1*, then it is due to *regressive overlap,* or *offlap,* the deposits having been laid down successively by the retreating sea. If like *A2*, then it is due to *progressive overlap,* the deposits having been laid down by an advancing sea. The withdrawal of the sea in the latter case was so rapid that very little sediment was deposited during the withdrawal. With only slight modifications the erosional development is similar in each case. In this presentation, type *A1* will be used.

TYPES OF STREAMS. Figure *A* shows the initial stage. There is a moderately irregular *oldland* of complex rocks, but any type of oldland might have been used. The initial streams originating upon the newly uplifted coastal plain are *consequent streams.* Streams which have extended their courses from the oldland across the coastal plain are *extended consequents.* Hills of the oldland which project through the coastal-plain strata are called *mendips* or *inliers.* The zone where the streams pass from their rocky channels in the oldland to their more gentle courses in the coastal plain is called the *fall line* or *fall zone.*

LOWLANDS AND CUESTAS. As erosion progresses during the youthful development of the region, the streams incise themselves below the surface of the plain and develop tributaries (Fig. *B*). The tributaries increase most extensively on the belts of weaker rocks, usually beds of clay. These tributaries are termed *subsequent streams.* The region becomes mature when these streams work headward along the weaker belts and erode lowlands. The innermost lowland, next to the fall line, is called the *inner lowland;* the other lowlands are called *outer lowlands,* as shown in Fig. *D*. The ridges or uplands between the lowlands are termed *cuestas.* Each cuesta has a steep *inface* and a gentle *back slope,* down the dip of the beds. The streams flowing down the steep *inface* are called *obsequent streams, O* (*i.e.,* opposite to the consequent streams in direction); those flowing down the stripped back slope are called *resequent streams, R* (*i.e.,* recently formed consequent streams). The flat divides between the headward ends of two opposing subsequent streams are called *cuesta bridges* (*CB*, Fig. *C*). Detached parts of the cuestas standing out in the lowlands are called *outliers* (*OL*, Figs. *C* and *D*).

Full maturity is reached when one of the major streams, *M*, becomes complete master of the region because its tributaries have worked headward and captured the territory of adjacent streams, thus diverting most of the drainage into one system. The presence of *wind gaps, W*, attests to some of these changes.

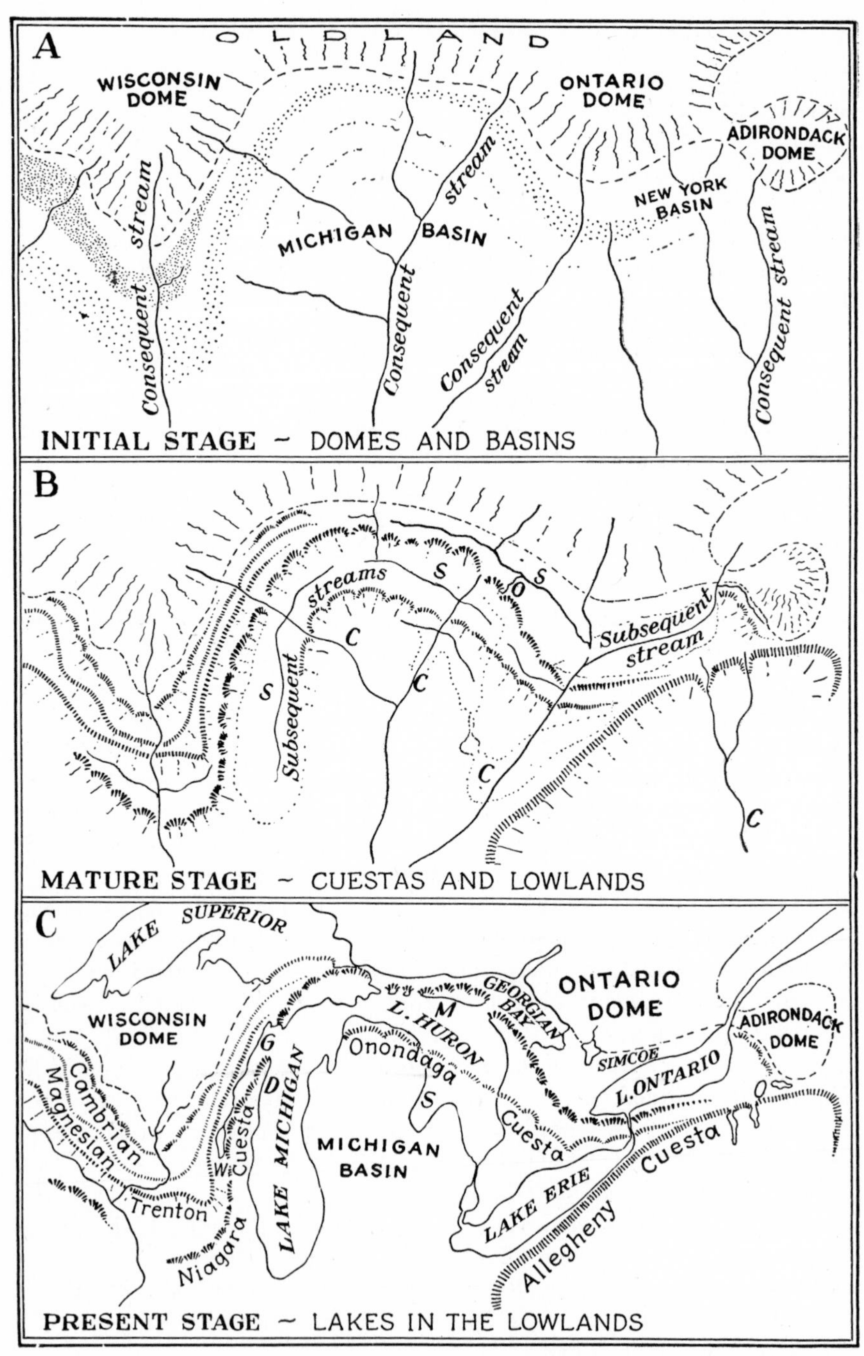


*After A. W. Grabau*

STAGES IN THE DEVELOPMENT OF THE GREAT LAKES

*A*. Young warped belted coastal plain. *B*. Mature belted coastal plain. *C*. Same after partial drowning of lowlands.

## WARPED COASTAL PLAINS

As an example of a warped coastal plain the region around the Great Lakes is unsurpassed. It is a Paleozoic coastal plain and is occasionally called an *ancient* coastal plain, thus distinguishing it from that along the Atlantic coast, which is modern or recent.

INITIAL WARPING AND DRAINAGE PATTERN. The Great Lakes region exhibits several domes and basins, formed by warping during uplift. The Wisconsin Dome, the Ontario Dome, and the Adirondack Dome alternate with down-warped areas or basins such as the Michigan Basin and the New York Basin. The oldland is represented by the ancient rocks of Canada, the coastal plain by the Paleozoic formations overlapping the old rocks from the south and ranging in age from Cambrian to the Pennsylvanian.

Upon this initial surface (Fig. *A*) after the withdrawal of the Paleozoic sea, there developed a system of consequent streams radiating in general from the centers of the domes and converging toward the basins.

CUESTAS AND LOWLANDS. As erosion progressed, these streams incised themselves below the coastal-plain surface and their tributaries etched out lowlands along the belts of weaker rock. The procedure was quite like that which is exemplified in all coastal plains but the belts, instead of being straight, were curved; the pattern of the subsequent streams was annular or ringlike, clearly shown in Fig. *B*.

The mature stage indicates the development of at least six different cuestas of varying heights and continuity. Stream capture must have occurred in several places, as suggested in the illustrations, but the exact stream pattern cannot, of course, be restored.

DROWNING OF THE LOWLANDS. Recently this region of complex ridges and lowlands was overridden by the continental ice sheet and the whole drainage pattern changed. Many of the lowlands were blocked by glacial deposits, and the Great Lakes, as we know them now, were formed (Fig. *C*). In only one other part of the world is any analogous series of water bodies to be found and that is in the Baltic Sea of Europe with the Gulf of Finland and related Russian lakes, such as Ladoga and Onega, and the White Sea, all occupying the inner lowland.

The most persistent cuesta in the Great Lakes region is the Niagara Escarpment. It may be traced from central New York westward past Niagara Falls across Ontario into the Indian Peninsula and Manitoulin Islands, *M*, westward into the Door Peninsula of Wisconsin. In southern Wisconsin it is buried under glacial drift but reappears in Iowa and then runs far northward for hundreds of miles into northern Canada. Among the bodies of water which occupy the lowland at the foot of this scarp are Lake Ontario, Georgian Bay, Green Bay, Lake Winnebago, and, in Canada, Lake Winnipeg. Lake Winnipegosis, Great Bear Lake, and Great Slave Lake lie in similar lowlands. Lake Huron is an unusual lake in that it lies in parts of two lowlands. The submerged Onondaga Cuesta runs across the middle of this lake and may readily be detected by the soundings on the lake charts. Saginaw Bay is probably part of the valley of a former consequent stream.

Unlike the four other great lakes, Lake Superior lies almost entirely within the Canadian oldland. Nor are the finger lakes of central New York analogous to the Great Lakes. They are in the same category with Saginaw Bay, being parts of the valleys of earlier consequent streams.

The great Mohawk Lowland, of inestimable economic importance, connecting the Hudson River valley with the Great Lake region, contains only relatively small lakes like Lake Oneida.

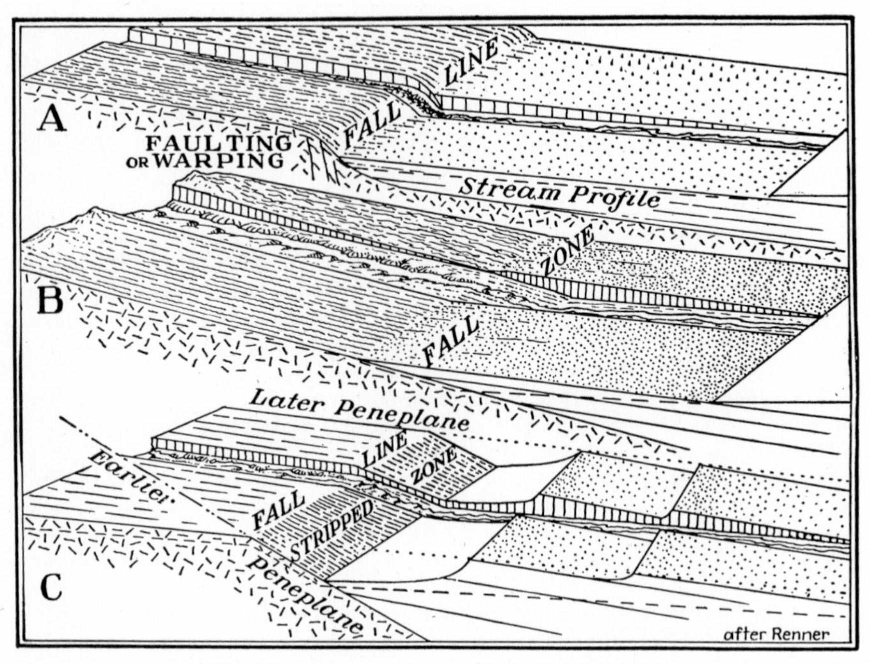


THREE INTERPRETATIONS OF THE FALL LINE

## THE ORIGIN OF THE FALL LINE

Studies of the fall line suggest several ways to account for the steeper gradient of streams along this zone. These are indicated in the three profiles above.

Profile *A* shows a zone of flexing or faulting. Profile *B* interprets the fall line as due to the difference between the steep gradient which the streams have on the crystalline oldland and the more gentle gradient which they have on the weaker coastal-plain beds. That is, on the coastal plain the streams are more nearly mature than they are on the oldland. Finally, profile *C* interprets the fall line as due to the intersection of two peneplanes. The peneplane upon which the coastal plain rests has been tilted and beveled across by a later peneplane which is now the surface of the oldland. The fall zone is a stripped part of the older peneplane. This interpretation conforms with the facts observed along the Atlantic coastal plain and also along the border of the ancient Paleozoic coastal plain in central Wisconsin.

In these places the slope of the buried peneplane beneath the coastal plain is distinctly greater than is the present slope of the oldland surface.

The intersection of two peneplanes in this manner has been termed a *morvan* by Davis, named after the Morvan region in central France. From the above it is apparent that the fall line is in general a *zone* of appreciable width rather than a line.

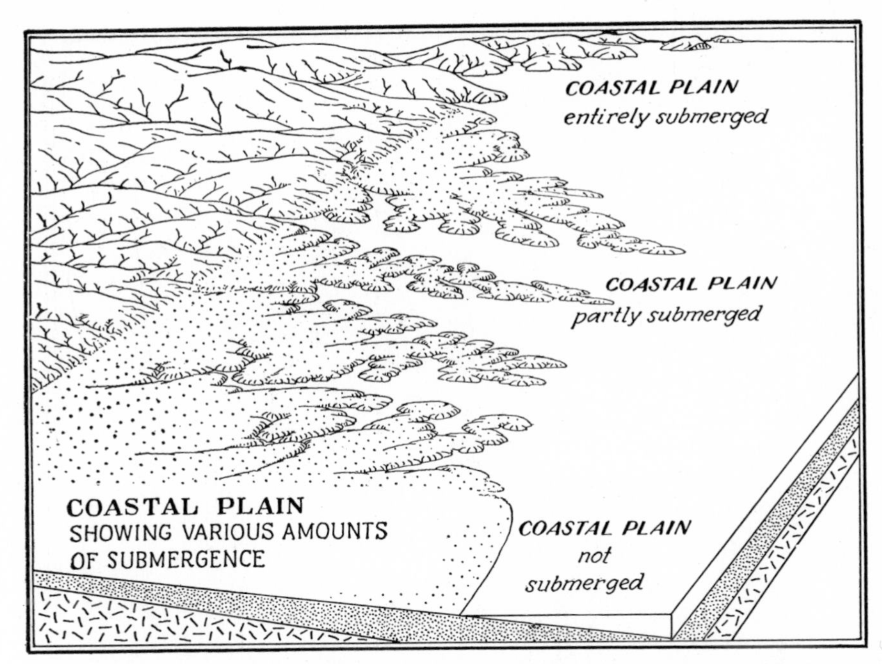


A COASTAL PLAIN SHOWING VARYING AMOUNTS OF SUBMERGENCE

## EMBAYED COASTAL PLAINS

Most modern coastal plains show in a greater or lesser degree some evidence of recent submergence due to the constantly changing relation between land and sea. The elevation of a coastal plain is in no sense a single event but consists of a series of uplifts and submergences.

Many of the early changes are recorded in the character of the coastal-plain deposits, but the later changes affect the present topography. Very slight changes in the relative level of sea and land, because of the low elevation of the coastal plain, make a great difference in the position of the new shore line. A coastal plain which has been only moderately dissected by streams with shallow valleys becomes deeply embayed with only a slight submergence. Coastal-plain streams having gradients of 5 to 6 feet to the mile are drowned for 10 miles up from their mouths when the vertical change is only a matter of 50 feet or so.

Embayed coastal plains rarely display the numerous islands which more rugged lands display when submergence takes place.

Some coastal plains are crossed by relatively few streams and these have only small tributaries. Such plains do not have many embayments when submergence occurs. This is the character of the Argentine coast where, between the estuaries of the Plata River and Bahia Blanca, there are hundreds of miles of low coast with no indentations, because few streams drain that arid region.

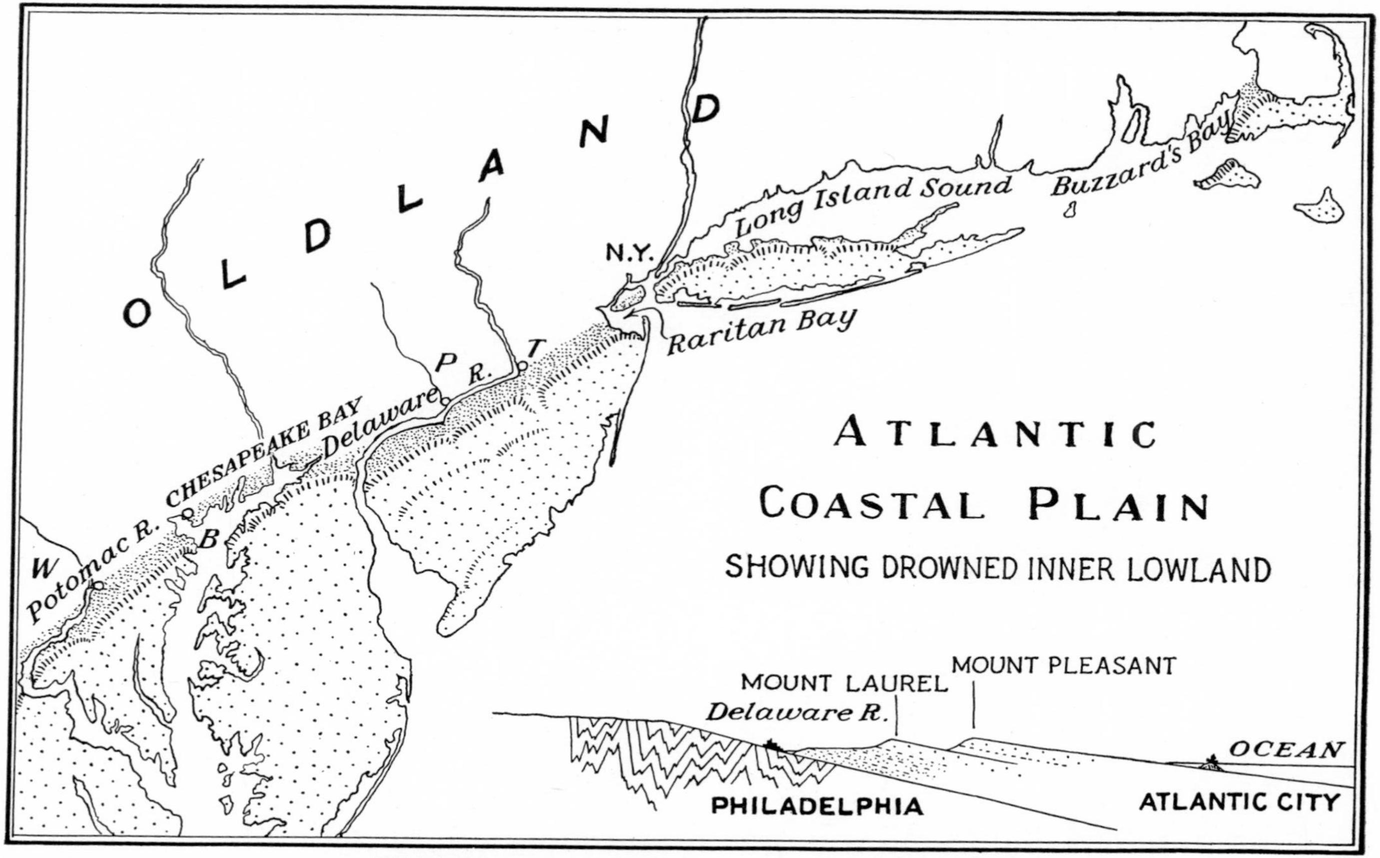

DIAGRAMMATIC MAP OF ATLANTIC COASTAL PLAIN

# ATLANTIC COASTAL PLAIN

Between Massachusetts and Virginia the Atlantic coastal plain has been submerged; many parts of the inner lowland belt constitute water bodies. Buzzards Bay almost separates the arm of Cape Cod from the oldland of New England. The Cape Cod Ship Canal cuts through the remaining unsubmerged portion of this lowland. Martha's Vineyard is part of the cuesta. So also are Block Island and Long Island, with Long Island Sound representing the inner lowland. On all these islands, as well as on Cape Cod, there are thick glacial deposits capping the cuesta.

Raritan Bay is a continuation of the inner lowland, south of which rises the steep inface of the cuesta forming the Atlantic Highlands. The narrow *waistline* of New Jersey comes next. Here run the main line of the Pennsylvania Railroad, the superhighway U. S. Route 1, and the Delaware and Raritan Canal. From Trenton (*T*) to Philadelphia (*P*) and beyond, this inner lowland is drowned and forms the head of Delaware Bay.

The cuesta traversing New Jersey southwest from the Atlantic Highlands is called *Beacon Hill*, *Pine Hills*, *Mount Holly*, *Mount Laurel*, and *Woodbury Heights*. On top of this cuesta near Trenton was the site of Camp Dix during the war. This location provided an extensive, flat, well-drained upland of sandy soil not apt to become muddy, with excellent artesian water supply, close to important lines of communication but a suitable distance from any large city. The adjacent lowland to the west with its richer clay soil was intensively cultivated, providing an abundant supply of fresh vegetables.

In line with the Delaware River to the southwest are the head of Chesapeake Bay and also part of the Potomac River. The form of the Delaware, Chesapeake, and the Potomac systems are almost identical. Each has three parts: (*a*) an uppermost portion in the oldland, which is consequent; (*b*) an intermediate portion in the inner lowland, now drowned, which is subsequent in origin; and (*c*) a lower portion reaching the sea, which is again consequent. Apparently this part of the coastal plain epitomizes coastal-plain development.

The five cities of New York, Trenton, Philadelphia, Baltimore (*B*), and Washington (*W*) are all fall-line cities; they are also ports due to the submergence of the coastal plain. In the Carolinas and Georgia, submergence was much less extensive and the fall-line cities are inland towns. There the coastal plain is also much broader than in New Jersey. Off the Maine coast, however, the coastal plain is completely submerged. The cuesta appears on the coastal charts as submerged banks, such as Georges Bank. The Gulf of Maine constitutes the drowned inner lowland.

The two interesting peninsulas which form Cape May and Cape Henry are not shore features produced by current action, like Sandy Hook, but apparently outer cuestas. Their form suggests Indian Peninsula in Lake Huron.

THE COASTAL PLAINS OF ENGLAND AND FRANCE

## COASTAL PLAINS OF ENGLAND AND FRANCE

The coastal plain comprising southeastern England and northern France is essentially a single unit, bisected by the English Channel.

In England two pronounced and almost parallel cuestas, called, respectively, the *Cotswold Hills* (the inner cuesta) and the *Chiltern Hills* (the outer cuesta), run north from the English Channel south of Bristol to the North Sea near Newcastle. The oldland is represented by the three highland tracts of older rocks which constitute the Cornwall peninsula, Wales, and the Pennine Range of central England. That the surface of the underlying oldland is hilly is attested by the oldland rocks projecting through the coastal plain, notably in the Mendip Hills immediately south of Bristol. The inner lowland, represented in part by the Midland Plain, contains important cities. Bristol (*B*), Gloucester (*G*), Nottingham (*N*), Lincoln (*L*), and York (*Y*) are indicated on the accompanying map. The Severn, the Avon, the Trent, and the Ouse drain this lowland as subsequent streams. Between the Cotswold and the Chiltern Hills is an outer lowland in which both Oxford (*O*) and Cambridge (*C*) lie, as well as several famous schools. Hence the expression "the Educational Lowland" of England.

The Chiltern Hills north of London are a wide tableland of chalk, sloping gently toward the south. The edge of this tableland descends northward more than 400 feet in several steps toward the Cambridge Lowland and the Valley of the Great Ouse. In Yorkshire this same scarp, swinging to the west, joins the Cotswold Hills to form the York Wolds; the intermediate lowland, because of the thinning out of the weak formation, disappears. The Weald is a dome extending across the channel into France, where it is called the *Boulonnais*, or Boulogne district.

The coastal plain of northern France is a great embayment, usually termed the *Paris Basin*. Like a pile of saucers, the layers of rock form circular escarpments facing outward, their back slopes dipping toward the center of the basin near Paris. The oldland, like the English Plain, is not continuous. It is represented by the Ardennes, the Vosges, the Central Upland of France, and Brittany. The outward-facing escarpments are the "natural fortifications" of Paris, directed eastward toward Germany. Those which became familiar to Americans during the war are the Côtes de Moselle, the Côtes de Meuse overlooking the Woëvre Lowland, the Forest of Argonne, and the Isle de France, the innermost of all, which looks down upon the broad plain of the Champagne wherein lies the city of Rheims. Many of the smaller cities of eastern France command gateways cut through these ridges by the tributaries of the Seine. The cuesta forming the southeastern limit of the Paris Basin where it descends into the Saône Lowland is called the *Côte d'Or*, Golden Slope, with Dijon at its base. In the western part of the Paris Basin the cuestas disappear. The plain does not have the belted characteristics so strongly developed on its eastern side.

Agriculture vs. Seafaring. People living along the coast tend to become seafaring only if the land is less hospitable than the sea. If the land is rugged and mountainous, if the soil is poor, if the climate is harsh, people turn to the sea for a living. But even under these conditions, the sea may still not attract them if no harbors are available.

A young coastal plain, not submerged after dissection, with good soil and a mild climate, is certain to be populated with an agricultural and land-loving people. The sea can offer them few inducements, as shown by our southern states.

If such a coastal plain is deeply embayed, the sea becomes more of a factor in the lives of the people, but still not necessarily a dominating one. Thus in Virginia and the Chesapeake Bay region early colonial enterprises first found root. The people quickly became agriculturally minded but maintained close ties and traded with countries across the seas. It was an even balance between the land and the sea. The contrast between the living conditions on the Atlantic coastal plain and those on the oldland of New England is striking. In New England the forbidding nature of the land, the severe climate, and the innumerable places of refuge along the coast enticed men to the sea. Fishing, whaling, shipbuilding, and commerce became leading industries.

In New England, Norway, Brittany, Galicia in northern Spain, Cornwall, and Scotland, people soon discovered that the sea, with its numerous estuaries, had advantages over the land. Such places have bred sailors and seafaring men from their earliest days. British Columbia and Alaska became the home of Indians who learned to fish, to build great sea-going canoes, and to travel far and wide.

Contrast also North China and Japan. China, with its extensive plains bordering the sea but with almost no harbors, is an agricultural country. Japan, with almost no plains but many bays and inlets, resorts to fishing and commerce. The irregular coast line and mountainous character of southern China leads to maritime activities. Thousands of small fishing boats may be seen offshore, and these provinces have sent people to populate other lands.

In South America a similar contrast exists between Chile and Argentina. Chile, with its embayed coast and mountainous mainland, is the maritime nation of South America. Argentina is agricultural. Only one or two large estuaries bring the sea far inland, as at Buenos Aires and Bahia Blanca.

Cuestas vs. Lowlands. Belted coastal plains in all parts of the world show contrasts between their less habitable ridges and their fertile lowlands. The ridges, because of their hilly nature and because they usually are regions of infertile sand and gravel or occasionally limestone, with a resulting very low ground-water level, are invariably less densely populated than are the intervening lowlands. In Alabama, Mississippi, Texas, and New Jersey, the cuestas are sandy uplands supporting extensive pine forests and providing the greatest tract of timber in the United States; the lowlands are fertile agricultural lands, like the Black Belt of Alabama, the Black Prairie of Texas, and the clay belt of New Jersey. In England the cuesta hills forming the Cotswolds and the Chilterns are devoted to sheep rearing in contrast with the cultivated belts on either side. To some extent this is true also of the cuestas in Wisconsin. In France several of the cuestas are still wooded, such as the Forest of Argonne.

Water Supply. Water supply in coastal plains is rarely a serious problem. The dipping strata with their sandy water-bearing beds render artesian supplies adequate for even large cities. Atlantic City thus meets her needs and so also to some extent does Brooklyn on Long Island. Throughout the Atlantic and Gulf coastal plain this condition holds true. Many of the cities in Wisconsin and Illinois are supplied by artesian wells drilled through the beds of the ancient Paleozoic coastal plain, some of them having depths of 1,000 feet.

The water derived from these sources, because it has remained so long in the rocks and has traveled so far, is apt to contain much dissolved mineral matter, especially calcium carbonate, and is therefore known as *hard water*. In some instances appreciable amounts of sulphur occur in the water in the form of hydrogen sulphide.

Mineral Products. The mineral products occurring in coastal plains are those associated with sedimentary rocks. Clay is abundant. Pottery making and brick manufacture are common industries, as at Trenton and Perth Amboy, New Jersey, and in the Carolinas. Lignite in Alabama, Mississippi, and Texas constitutes an important fuel, its woody characteristics still remaining because of the recency of deposition. Phosphate is a very important product of the coastal-plain beds of Florida, which produces about 80 per cent of that mined in the United States, representing close to $10,000,000 worth. Practically all of the sulphur produced in the United States, having a total value of about $30,000,000, is derived from the coastal plain of Louisiana and Texas. It occurs in the capping rocks overlying salt domes and is obtained by pumping superheated steam down through the pipes, melting the sulphur and drawing it to the surface. By far the most important mineral products of the coastal plain are petroleum and natural gas, the total value of which each year is some $700,000,000, representing about half the production of the United States. Vast quantities of salt occur in the salt domes of Louisiana, but the total production is small. Over 50 per cent of the United States production of salt comes from New York and Michigan from beds of Silurian age. These are also coastal-plain deposits but of an earlier geological time, and now remote from the sea.

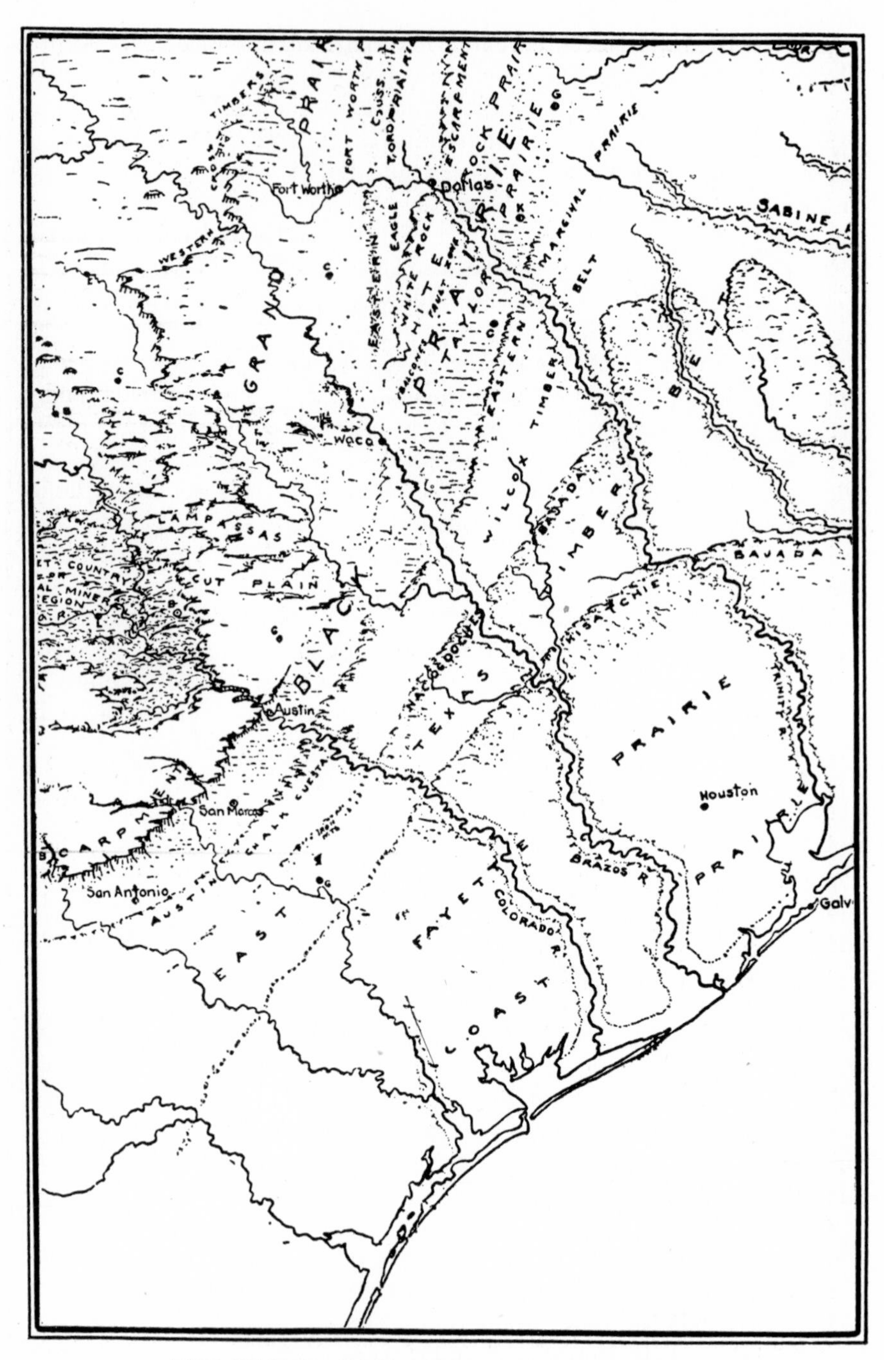


THE BELTED GULF COASTAL PLAIN OF TEXAS
Showing succession of cuestas and lowlands.

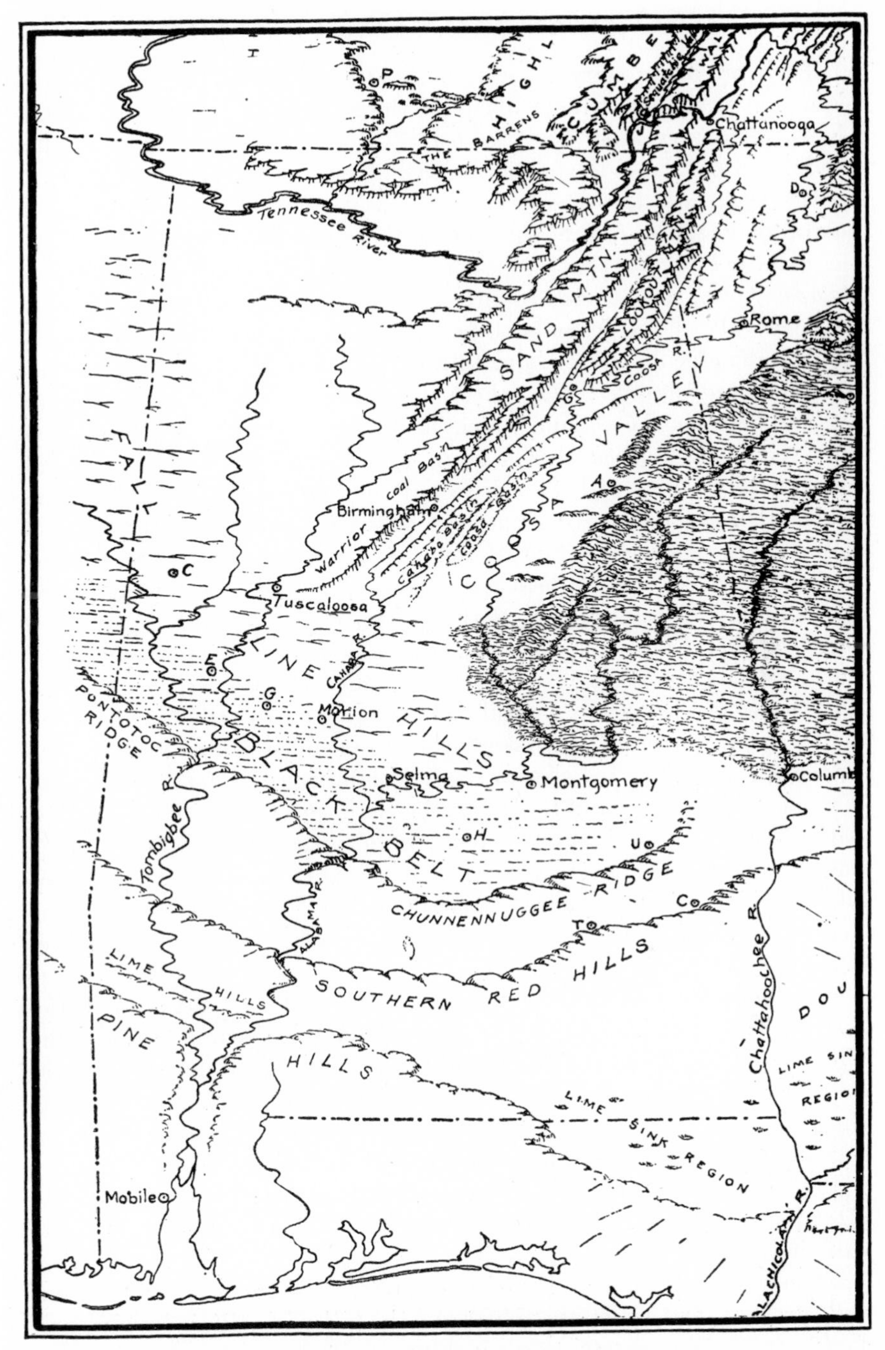


THE BELTED GULF COASTAL PLAIN OF ALABAMA
Showing the varied topography of the oldland in the north and the succession of cuestas and lowlands.

## MAPS ILLUSTRATING COASTAL PLAINS

There are many maps covering parts of the Atlantic and Gulf coastal plains but very few of them illustrate the significant features of coastal plains, such as cuestas. The *Camden, N. J.-Pa.-Del.,* sheet is unusually good because it shows the fairly rugged oldland, the inner lowland drained by the Delaware, the dissected cuesta, and all of the genetic types of coastal-plain streams. The *Navesink* and *Cassville, N. J.,* sheets illustrate the dissected cuesta, part of which forms the Atlantic Highlands. The *Patapsco, Md.,* sheet includes part of the oldland in the northwest, with its deep valleys and irregular topography, and in the southeast the recently submerged coastal plain. In fact, this map is situated on the fall line. On the *Tolchester, Md.,* sheet, Chesapeake Bay follows the inner lowland between the oldland which lies to the northwest and the coastal plain lying to the southeast. The *Epes, Ala.,* sheet shows the belted coastal plain of that state, with two dissected cuestas and two lowlands. The *Seale, Ala.,* sheet vaguely suggests a belted coastal plain which is also shown on the *Pelahatchee* and *Morton, Miss.,* sheets. On the *Warrenville, S. Car.-Ga.,* sheet there appears part of the dissected cuesta which forms the fall-line hills of that part of the coastal plain. As more topographic sheets appear from time to time, especially from Alabama, it will be possible to have further examples of unusually fine cuestas.

Cuestas of the ancient coastal plain forming part of the Interior Lowland Province in the Great Lakes region are shown on many maps, such as the *Winnebago, Wis.,* special sheet (comprising the *Fond DuLac* and *Neenah, Wis.,* sheets) and the *Niagara, N. Y.,* sheet, both of which show part of the Niagara cuesta overlooking an inner lowland. The *Kendall* and *Mauston, Wis.,* sheets show a much dissected cuesta with outliers. The *Berne, N. Y.,* sheet shows two cuestas in the Helderberg region facing northward toward the Mohawk Valley, which represents the inner lowland.

Maps illustrating the more recently elevated and therefore little-dissected seaward portion of the coastal plain are represented by the *Trent River, N. Car.,* sheet which still has swamps on the upland, because an effective drainage system has not yet been established. The *Bamberg, S. Car.,* sheet illustrates a maturely dissected part of the coastal plain much like that shown on the *Patuxent, Md.-D. C.,* sheet.

The *Heathsville, Va.-Md.,* sheet shows part of the coastal plain which, after partial submergence, has recently been slightly elevated, so as to exhibit marine terraces.

The solution depressions of the coastal plain, known as bays, savannas, prairies, and sinks are shown on the *Allendale, Olar, Peeples,* and *Williston, S. Car.,* sheets, as well as on the *Williston, Citra, Arredondo,* and *Ocala, Florida,* maps.

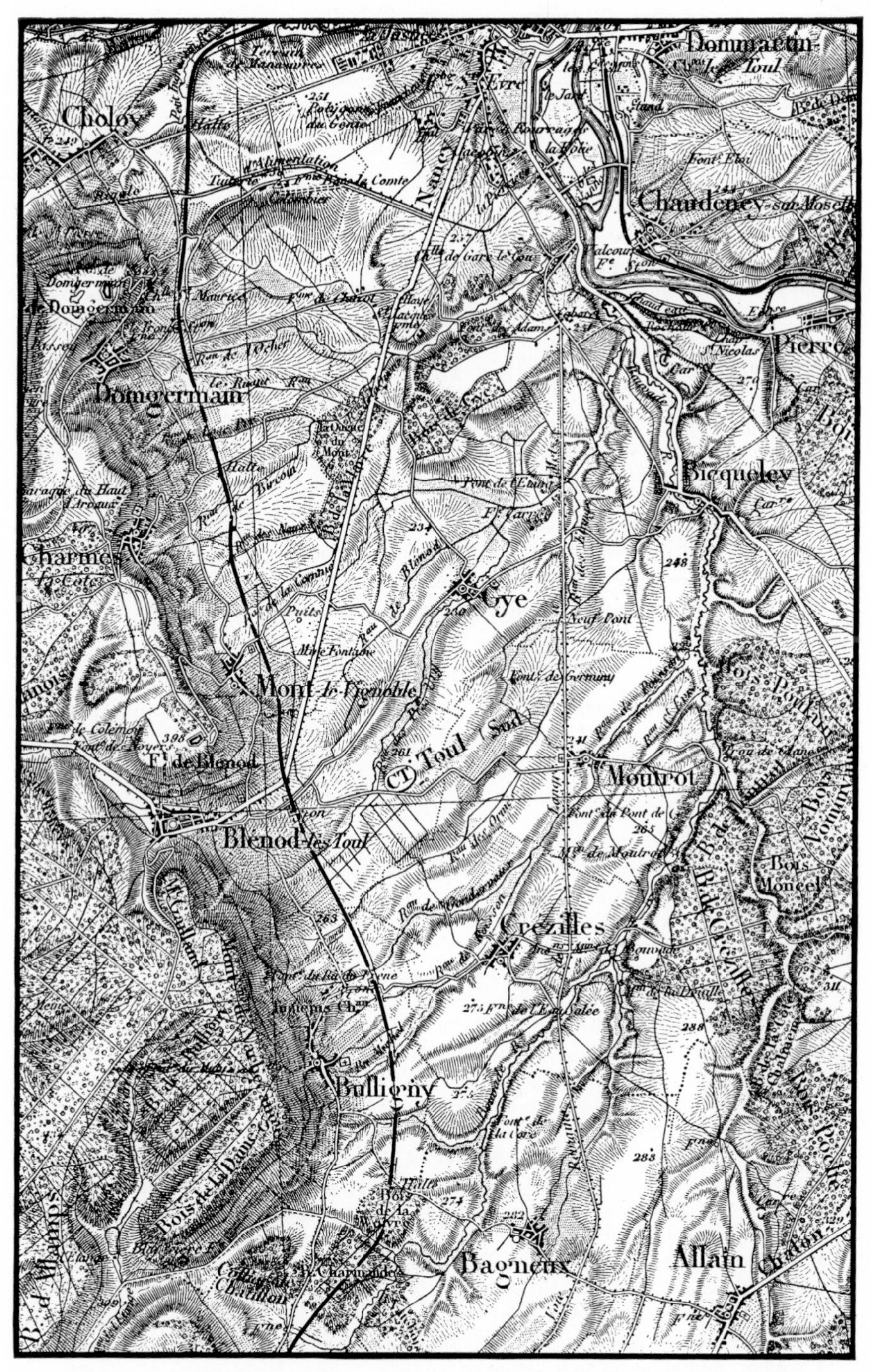


CUESTA WITH BENCHES, AND LOWLAND, NEAR TOUL, FRANCE
France; *Nancy* sheet, No. 69 (1:80,000).

## QUESTIONS

1. Name all possible ways in which a cuesta can come to an end or disappear.
2. What is meant by a fall-zone peneplane?
3. Suppose a coastal plain were maturely dissected and then completely submerged; how would this fact be known?
4. What is a mendip? Is it an inlier? Or an outlier?
5. Draw a map showing only the stream system of a belted coastal plain, representing consequent, subsequent, obsequent, resequent, and insequent streams, as well as indications of stream capture.
6. Suppose a coastal-plain system of valleys were covered again by the sea and buried under sediment, but that upon retreat of the sea the valleys were reoccupied by streams, what name might be given to these streams? (See McGee.)
   Note: The renewal of buried drainage systems in the eastern Gulf section of the coastal plain is apparent in many instances. Several times the sculptured surface has been submerged and mantled with sediments, only to rise and resume more or less fully its old aspect under the influence of waterways following the old lines. Such resurrected or *palingenetic* drainage is characteristic of much of Mississippi.
7. How can the thickness of sediments on the continental shelf be determined?
8. What evidence is there of a submerged coastal plain off the coast of Maine?
9. Camp Dix in New Jersey is situated on top of a sandy cuesta, not far from Trenton. What advantages can you ascribe to such a site, as regards water supply, sanitation, etc? What other camps were situated on the Coastal plain? (*e.g.*, Yaphank).

## TOPICS FOR INVESTIGATION

1. Geological structure of coastal plains.
2. The fall-line cities of the United States and their early development.
3. Coastal plains in various parts of the world that is, India, Puerto Rico, Mexico, Argentina, Siberia.
4. Former coastal plains, for example, Bavaria, southern Russia, France, Great Lakes region.
5. Narrow coastal plains of the world, for example, Scotland, Norway, Maine.
6. Former landward extent of Atlantic coastal plain.
7. Artesian water supply of coastal plains.
   Note: Artesian conditions on the Atlantic coastal plain have a direct bearing upon the origin of the submarine canyons which occur along the seaward margin of the continental shelf. Some of these canyons appear to be opposite the mouths of large rivers, but many of them bear no relation to land drainage systems. Because they are thousands of feet beneath sea level, it is difficult to explain them as erosional features formed when the continent stood higher with relation to the sea, for there is no evidence that so great an elevation of the land existed in recent geological times. After reviewing many theories for the origin of submarine canyons Johnson presents strong evidence in support of the theory that they were formed as a result of *sapping by submerged artesian springs*. Water-bearing beds of the coastal plain are believed to continue beneath the continental shelf and provide conduits through which water can pass seaward all the way from the zones of intake on the land to the springs on the relatively steep outer slope of the continental shelf, far beneath the surface of the ocean. The great distance to the edge of the shelf is not of great moment, since there are instances in which a single artesian horizon has been traced and found productive many hundreds of miles from its surface outcrop.
8. Solution depressions: the Carolina bays or savannas, the Florida prairies and lakes, sink-hole regions of Florida and other coastal plain states. (Consult references on p. 718.)
9. Phosphate and other mineral resources.

## REFERENCES


### Drainage Development of Coastal Plains

Davis, W. M. (1895) *The development of certain English rivers.* Geog. Jour., vol. 5, p. 127–146.

Davis, W. M. (1898) *Physical geography.* Boston. *Plains and plateaus,* p. 113–158.

Davis, W. M. (1912) *Erklärende Beschreibung der Landformen.* Leipzig. Küstenebenen, Ebenen, und Hochebenen, p. 197–245.

Grabau, A. W. (1908) *Preglacial drainage in central western New York.* Science, new ser., vol. 28, p. 527–534.

Grabau, A. W. (1913) *Principles of stratigraphy.* New York. *The coastal plain,* p. 830–840.

Grabau, A. W. (1920) *Textbook of geology.* New York. The erosion cycle on a simple coastal plain, part 1, p. 709–722.

Smith, Laurence L. (1931) *Solution depressions in sandy sediments of the coastal plain in South Carolina.* Jour. Geol., vol. 39, p. 641–652.

### Cuestas

Cleland, H. F. (1920) *The Black Belt of Alabama.* Geog. Rev., vol. 10, p. 375–387.

Davis, W. M. (1899) *The drainage of cuestas.* Geol. Assn. London, Proc. 16, p. 75–93.

Dicken, S. N. (1935) *A Kentucky solution cuesta.* Jour. Geol., vol. 43, p. 539–544.

### Fall Line

Dietz, E. A. (1905) *The fall line.* Jour. Geog., vol. 4, p. 244–248.

Renner, G. T., Jr. (1927) *The physiographic interpretation of the fall line.* Geog. Rev., vol. 17, p. 278–286.

### Recent Changes of Level

Goldthwait, R. P. (1935) *The Damariscotta shell heaps and coastal stability.* Am. Jour. Sci., 5th ser., vol. 30, p. 1–13.

Grabau, A. W. (1920) *The Niagara cuesta from a new viewpoint.* Geog. Rev., vol. 9, p. 264–276.

Johnson, D. W., and Stolfus, M. A. (1924) *The submerged coastal plain and oldland of New England.* Science, new ser., vol. 59, p. 291–293.

McCallie, S. W. (1928) *Georgian coastal plains terranes.* Pan.-Am. Geol., vol. 49, p. 167–178.

Raisz, E. (1934) *Rounded lakes and lagoons of the coastal plains of Massachusetts.* Jour. Geol., vol. 42, p. 839–848. Similar to South Carolina "bays."

Stephenson, L. W. (1928) *Major marine transgressions and regressions and structural features of the Gulf coastal plain.* Am. Jour. Sci., 5th ser., vol. 16, p. 281–298.

### Regions Described

Abbe, C., Jr. (1899) *A general report of the physiography of Maryland.* Md. State Weather Service, vol. 1. *Costal plain,* p. 74–114.

Clark, W. B., and Mathews, E. B. (1906) *The physical features of Maryland.* Md. Geol. Surv., vol. 6, p. 29–92.

Cobb, C. (1897) *North Carolina.* Jour. School Geog., vol. 1, p. 257–266, 300–308.

Davis, W. M. (1918) *Handbook of northern France.* Cambridge, England, 174 p.

Dryden, L. (1935) *Structure of the coastal plain of southern Maryland.* Am. Jour. Sci., 5th ser., vol. 30, p. 321–342.

Glenn, L. C. (1898) *South Carolina.* Jour. School Geog., vol. 2, p. 9–15, 85–92.

Hill, R. T. (1900) *Physical geography of the Texas region.* U. S. Geol. Surv., Folio 3.

Hill, R. T. (1901) *Geography and geology of the Black and Grand prairies, Texas.* U. S. Geol. Surv., 21st Ann. Rept., part 7, 666 p.

KINDLE, E. M. (1925) *The James Bay Coastal plain. Notes on a journey.* Geog. Rev., vol. 15, p. 226–236.

MACKINDER, H. J. (1902) *Britain and the British Isles.* New York, 377 p.

MARTIN, L. (1916) *The physical geography of Wisconsin.* Wis. Geol. Surv., Bull. 36, 549 p.

MATSON, G. C., and CLAPP, F. G. (1909) *Geology of Florida.* Fla. Geol. Surv., 2d Ann. Rept., p. 25–49.

MCGEE W J (1891) *The Lafayette formation.* U. S. Geol. Surv., 12th Ann. Rept., part 1, p. 347–521. Includes good account of the Atlantic and Gulf coastal plain. Introduces the term *palingenetic.*

PARTSCH, J. (1905) *Central Europe.* London, 358 p. Descriptions of Netherlands-Germanic Plain, and the Danubian plains.

REUSCH, H. (1894) *The Norwegian coast plain.* Jour. Geol., vol. 2, p. 347–349.

SALISBURY, R. D. (1898) *The physical geography of New Jersey.* N. J. Geol. Surv., vol. 4, p. 1–170.

SANFORD, S. (1909) *The topography and geology of southern Florida.* Fla. Geol. Surv., 2d Ann. Rept., p. 175–231.

SELLARDS, E. H. (1919) *Geology of Florida.* Jour. Geol., vol. 27, p. 286–302.

SHALER, N. S. (1890) *The Dismal Swamp district.* U. S. Geol. Surv., 10th Ann. Rept., p. 261–339.

SMITH, E. A., ET AL. (1894) *Report on the geology of the coastal plain of Alabama.* Geol. Surv. Ala., 759 p.

STEPHENSON, L. W. (1928) *Structural features of the Atlantic and Gulf coastal plain.* Geol. Soc. Am., Bull. 39, p. 887–899.

VEATCH, A. C. (1906) *Outlines of the geology of Long Island.* U. S. Geol. Surv., Prof. Paper 44, p. 28–32.

VEATCH, A. C. (1906) *Geology and underground water resources of northern Louisiana and southern Arkansas.* U. S. Geol. Surv., Prof. Paper 46, p. 14–69.

GEOGRAPHICAL ASPECTS

BLANCHARD, R. (1917) *Flanders.* Geog. Rev., vol. 4, p. 417–433.

CLELAND, H. F. (1920) *The Black Belt of Alabama.* Geog. Rev., vol. 10, p. 375–387.

GABRIEL, R. H. (1921) *The evolution of Long Island.* New Haven, 194 p.

WENTWORTH, C. K. (1930) *Sand and gravel resources of the coastal plain of Virginia.* Va. Geol. Surv., Bull. 32, 146 p.

# XIV
# PLAINS AND PLATEAUS

*Hillers, U. S. Geological Survey*

MARBLE CANYON, COLORADO RIVER, ARIZ.
Young plateau of strongly jointed rocks in second cycle of erosion, suggested by distant mesa standing on peneplaned plateau surface.

*Barnum Brown, American Museum of Natural History*

CANYON OF LITTLE COLORADO RIVER IN MARBLE PLATFORM, ARIZ.
Young canyon in young plateau. Strong angularity due to jointing.

CATSKILL MOUNTAINS, N. Y., FROM THE EAST

A maturely dissected plateau of coarse texture in a humid region. Horizontal bedding clearly shown. The Great Valley in the foreground is a region of folded rocks.

*Fairchild Aerial Surveys*

American Museum of Natural History

MODEL OF THE GRAND CANYON IN THE AMERICAN MUSEUM OF NATURAL HISTORY
Showing geological section at the left. A maturely dissected plateau in an arid region.

BRYCE CANYON NATIONAL PARK, UTAH

A maturely dissected plateau in weak rocks with numerous vertical joints. An undissected mesa in the distance.

*U. S. Department of the Interior*

*Barnum Brown, American Museum of Natural History*

BORDER OF TYENDE MESA NEAR AGATHEA PEAK, COLORADO PLATEAU, NORTHERN ARIZONA

An old plateau with sandstone-capped mesa monadnocks.

THE TOTEM POLE, MONUMENT VALLEY, ARIZ.
An old plateau with monadnocks of sandstone, forming pinnacles and buttes.

*Monsen*

*U. S. Forest Service*

OZARK MOUNTAINS, ARK.

A maturely dissected plateau in a humid region. Rock benches not apparent.



HAYDENS CATHEDRAL, UINTA MOUNTAINS, UTAH

A maturely dissected plateau of coarse texture. Horizontal rock structure well exposed.

*Jackson, U. S. Geological Survey*

*Campbell, U. S. Geological Survey*

CANYON OF NEW RIVER, ALLEGHENY PLATEAU, W. VA.
A young valley in a maturely dissected plateau. Horizontal structure apparent.



OHIO RIVER, SOUTHERN INDIANA
A mature valley in a maturely dissected plateau.

*U. S. Forest Service*

*U. S. Forest Service*

THE LLANO ESTACADO, THE STAKED PLAINS OF EASTERN NEW MEXICO

A young, totally undissected plain subject to strong wind action.



PRAIRIE PLAINS OF SOUTHERN INDIANA

Maturely dissected plains of slight relief, and therefore well cultivated.

*U. S. Forest Service*

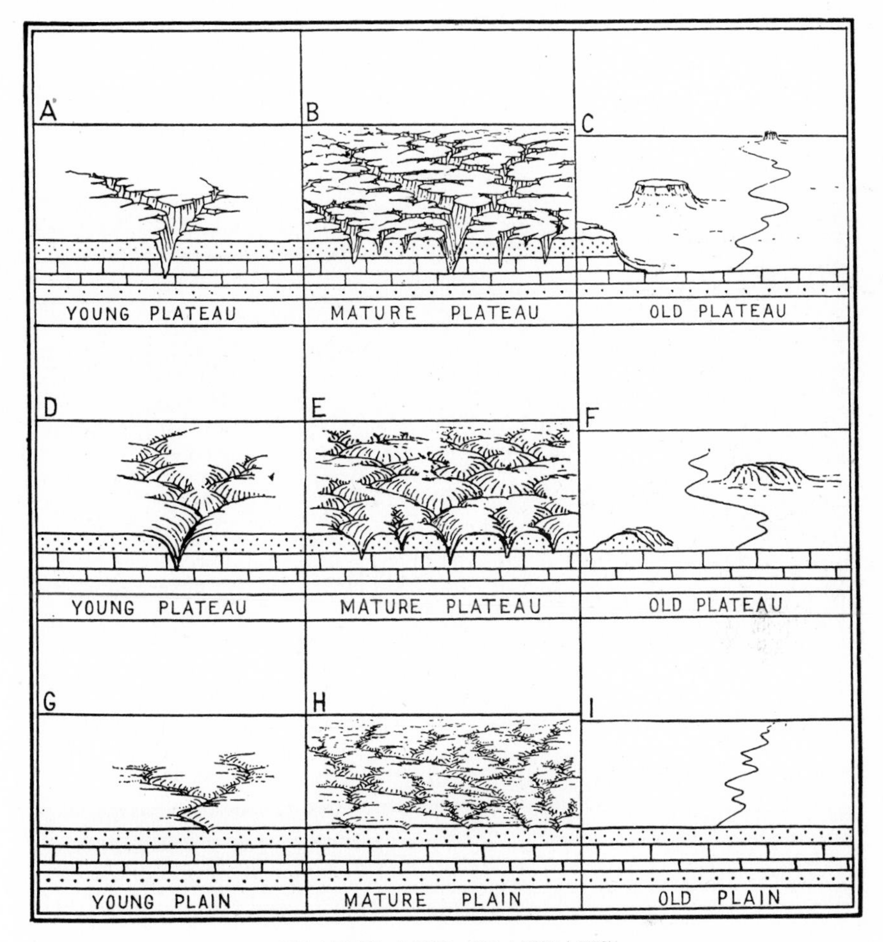


## PLAINS AND PLATEAUS

Synopsis. Plains and plateaus are regions of horizontal structure. Plateaus have high relief: the valleys and canyons are deep. Plains have low relief: the valleys and canyons are shallow.

Young plains and plateaus have few streams widely spaced. Mature plains and plateaus have many streams close together. Such country is hilly or mountainous. Old plains or plateaus have been worn down to a peneplane with scattered monadnocks of mesas and buttes.

There are many *constructional* types of plains, such as coastal plains, interior plains, lake plains, lava plains and till plains; and there are certain *destructional* plains, like delta plains, flood and outwash plains.

Most plateaus and plains are warped and even broken by faults. A study of plains and plateaus involves the character of the dissection, whether fine or coarse and whether it is controlled by humid or dry conditions; as well as the effect which the rock structures, such as warping, doming, and faulting, have upon the topography.

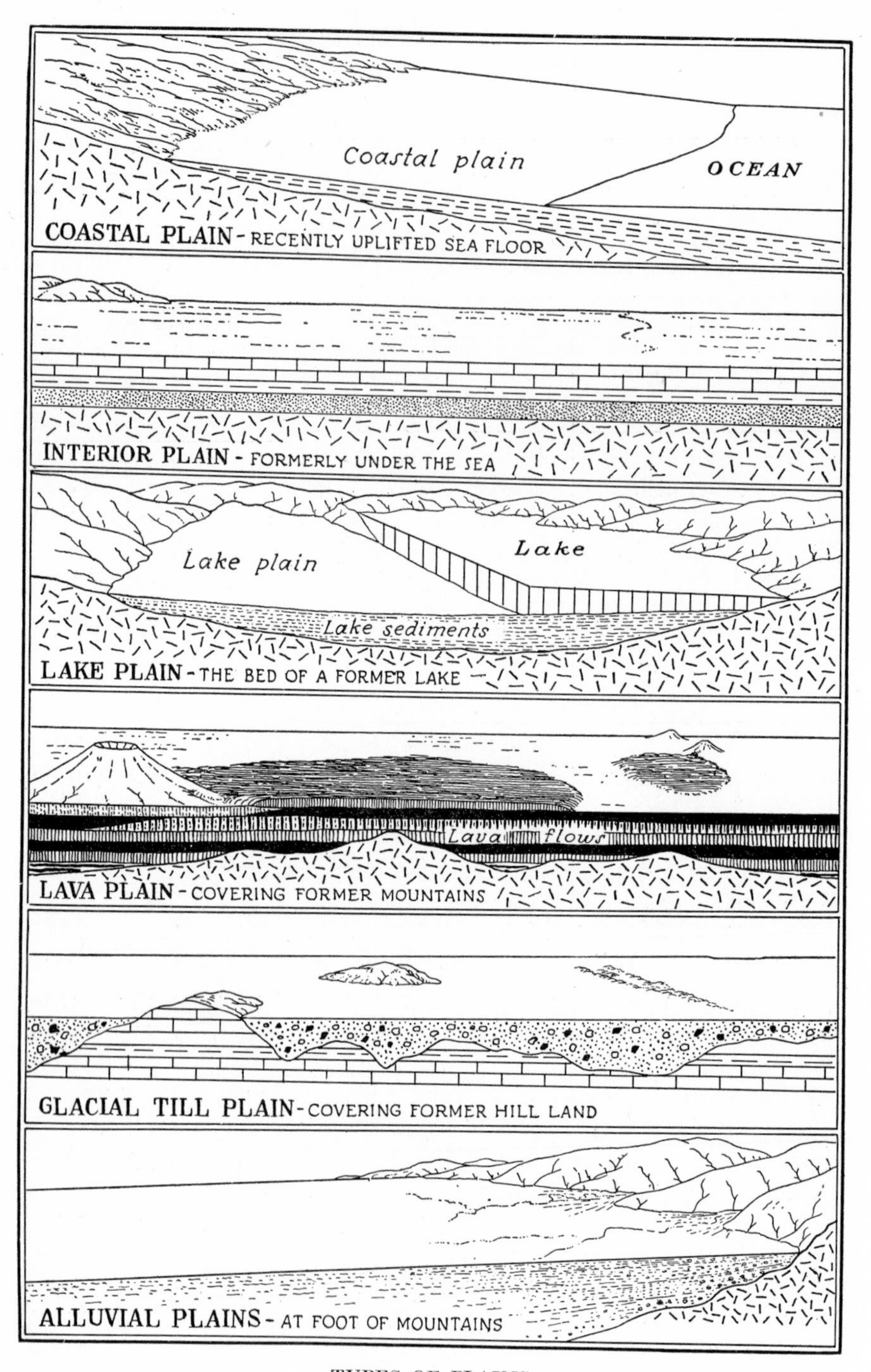


TYPES OF PLAINS

## TYPES OF PLAINS AND PLATEAUS

There are six types of plains which are of wide extent.

a. *Coastal plains,* formed by the emergence of the sea floor. These have been treated in the preceding chapter.

b. *Interior plains,* which originally constituted part of the sea floor but are now remote from the sea.

c. *Lake plains,* formed by the emergence of a lake bed by the draining of the lake. Lake plains thus have an origin similar to coastal plains.

d. *Lava plains* and plateaus, built up by successive flows of lava.

e. *Till plains,* formed of glacial till covering an older irregular topography.

f. *Alluvial plains,* formed by the building of alluvial fans which may extend for vast distances from the base of mountains.

Lava plains rightfully belong with volcanic features, but in most essential respects they resemble plains made up of sedimentary beds. Till plains are features of glacial origin and alluvial plains are due to stream work but, because of their occasional vast extent, some of them deserve to be included with plains of the constructional type.

Other types of plains, all produced by destructional forces, such as delta plains, flood plains, outwash plains, and peneplanes, have received consideration under the chapters on streams.

The rocks which underlie plains may be either loose or consolidated. Gravel or pebble beds and layers of sand, clay, and marl constitute the most common types of loose material. These occur on coastal plains. Alluvial plains consist of sand, gravel, and clay, varying in thickness from place to place. Lake plains are made up largely of finely laminated clays. Interior plains are underlain by the consolidated types of sediments, shales and limestones being the most common, with sandstone and conglomerate less frequent.

Like the term *plain*, *plateau* is also occasionally applied to regions which are not underlain by horizontal layers of hard rock. The ordinary nontechnical conception of a plateau is that of an upland standing above the surrounding country, regardless of what the structure of the upland may be. For instance, the so-called plateau of Guiana is a region of complex rocks; and the Piedmont Plateau and the Laurentian Plateau are regions of disturbed crystalline rocks. To all of these the term *peneplane*, or upland, is more properly applied, because they are not regions of horizontal structure.

SCARPED PLAINS OR TILTED PLAINS. The Osage Plains of Kansas and the Triassic Lowland of northern New Jersey are plains in which the beds have a strong dip, a dip of 5 or 10 degrees. The resulting topography is characterized by many parallel ridges or intermittent hills where the more resistant beds and minor layers come to the surface. These ridges or scarps are somewhat analogous to the cuestas on a coastal plain but usually do not bear any definite relation to an oldland. Such plains with their ribbed topography are known as *scarped* or *tilted* plains.

*John H. Maxson*

THE PLAINS OF THE UKRAINE, SOUTHERN RUSSIA
Extremely young undissected region of horizontal rocks.

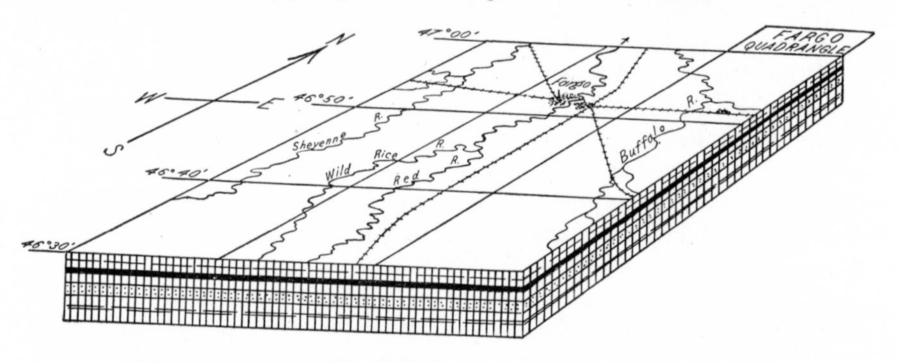


BLOCK DIAGRAM OF FARGO, N. D.-MINN. QUADRANGLE
A young lake plain dissected by mature streams now slightly rejuvenated.

## YOUNG INTERIOR PLAINS

Young interior plains are regions of horizontal rock structure remote from the sea. There are few streams, since the drainage is not yet fully developed, and hence most of the original surface of the land remains. Large portions of interior plains do not have a belted topography such as usually occurs on coastal plains. This is because of their great size and also because the oldland with its bordering lowlands and cuestas may be no longer existant. Interior plains, therefore, resemble those portions of extensive coastal plains which are most distant from the oldland.

Interior plains may be made up of rock formations originally laid down under the sea and these may be limestones, shales, or even sandstones. They may also consist of unconsolidated clays and sands. Old lake beds and even glacial-till plains, when of large extent, are often classed as interior plains, although they are actually depositional features due to destructive processes.

*George W. Bain*

DISSECTED PLAINS OF SOUTHERN SIBERIA
Flood plain of the mature Yenesei River in foreground.

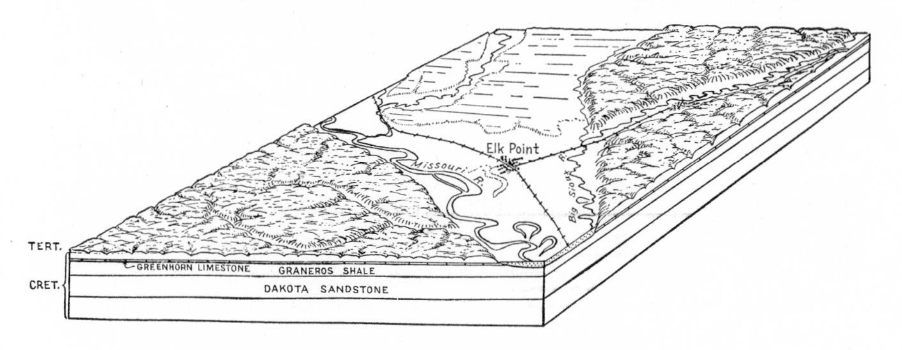


BLOCK DIAGRAM OF ELK POINT, S. D., QUADRANGLE
A mature plain, dissected by mature streams.

Most young plains are not only flat but also very level. This means that they are poorly drained. Broad level tracts of land between the streams have little or no run-off. Some regions, like the old bed of Lake Agassiz of North Dakota and Minnesota, are veritable mud lakes. The few streams, like the Red River and its tributaries, cross these plains in meandering channels. Small streams are rare. Roads and railroads traverse these plains for many miles without a curve. Distant buildings, from whatever direction they are seen, reveal only their topmost parts like ships at sea, thus attesting the spherical form of the earth's surface. On the broad divides between the streams, shallow lakes and marshes occur, the land near the streams being better drained.

Young plains are dissected by both young streams and mature streams. With advancing age of the plain, the streams increase in number and the plain passes through the period of late youth to full maturity.

*U. S. Forest Service*

A maturely dissected plain in southern Indiana of moderately strong relief and hence only slightly cultivated.

## MATURE PLAINS

Mature plains are described by most people as rolling hills. But rolling hills are by no means always mature plains. A region of horizontal rocks everywhere drained by a well-developed stream system, with only slight to moderate relief and having little if any flat upland surfaces, is a mature plain. Some regions of complex rock structure, worn to old age and then dissected, have about the same appearance as mature plains, but such regions should be termed *maturely dissected complex mountains in the second cycle of erosion*. This is an important distinction, to both the student of geology and of human geography. The nature of the surface topography may be less important than the nature of the underlying rocks with all of their associated phenomena.

The stream pattern of maturely dissected plains is typically dendritic; most of the streams are youthful and incised only 50 to 100 feet below the surface of the plain. All the land drains to some valley and, under humid conditions, the slopes are gentle with few rock outcrops. In regions of massive rocks, like thick sandstones and limestones, the streams may develop a right-angled pattern, due to joint systems, and the topography may have a fairly coarse texture; but in regions of weak rock, like shales and clays, the streams flow uniformly in all directions, and the topography is apt to be fine textured or even have badland characteristics.

In maturely dissected plains the larger streams may be mature. The age of the streams and the stage of a land form do not always agree with each other.

Plains underlain by massive and pure limestones do not develop normal dendritic stream patterns because of the subterranean drainage lines. Such regions, in the state of maturity, have numerous short inter-

*U. S. Forest Service*

A mature plain (or low plateau) of strong relief in northern Kentucky with almost mature valleys.

mittent streams disappearing in sink holes. In unusually humid regions lakes may abound, as in the lake district of Florida. Glaciation may also completely modify the normal stream pattern on plains, as on any other land form.

THE PRAIRIE PLAINS. The Prairie Plains of the Middle West, in Iowa, Illinois, and adjacent states, are good examples of mature plains. The topography is mildly rolling, the total relief being rarely more than 100 to 200 feet. The slopes are so gentle that the valley sides are cultivated with only slight danger of soil erosion. The valleys and narrower ravines support groves of trees but the upland surface is open and windswept. These uplands are the natural prairies so enthusiastically described by the early settlers who found them each spring richly bedecked with a profusion of flowers of great variety.

Where covered by glacial till, especially in Illinois and Ohio, the rolling topography is quite concealed and the level almost undissected surface of the till plains extends unbroken for great distances. Numerous well borings, however, reveal the character of the buried landscape and show it to have been maturely dissected by a well-defined dendritic system of streams. Over much of the area, notably in Iowa, the drift cover is relatively thin and, in the Driftless Area, is quite lacking. In brief, then, the Prairie Plains may be considered as maturely dissected constructional plains of almost horizontal beds with varying degrees of glacial covering, up to complete burial.

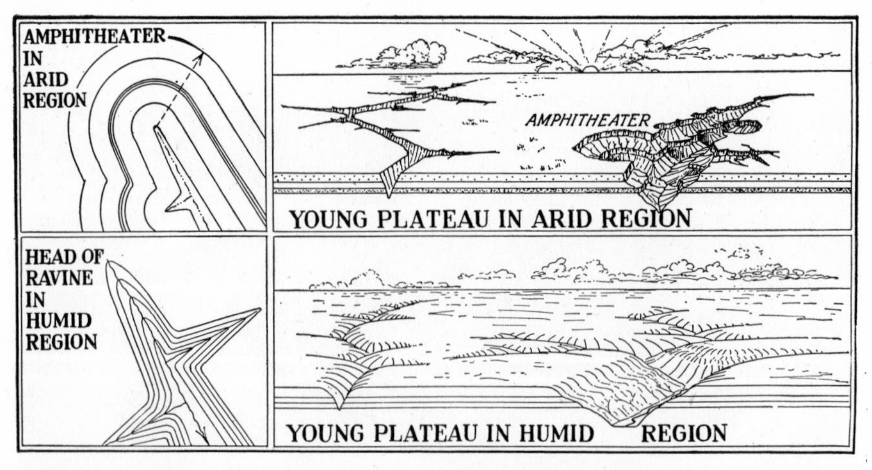


YOUNG PLATEAUS IN ARID AND HUMID REGIONS SHOWING CONTRAST BETWEEN VALLEY HEADS

## YOUNG PLATEAUS

Young plateaus are regions of horizontal rock structure, often remote from the sea and dissected by a few deeply cut streams. The relief is great and that distinguishes them from plains. The region may be higher than the surrounding country and be bordered by scarps, as in the case of the Colorado Plateau which descends to the Great Basin. Or it may be lower than the adjacent country, which may have high mountains overlooking the plateau, as is the case in Idaho where the towering mass of the Salmon River Mountains surmounts the Snake River Plateau.

In arid or semiarid regions the valleys in a plateau are canyonlike. That is, they conform with the usual conception of a canyon and have steep or vertical walls, with bare rock exposures. The presence of vertical cliffs in canyon walls is due to the presence of resistant formations in the plateau structure. In arid regions limestone is resistant to weathering. Many of the canyons in the western United States exhibit cliffs hundreds of feet in height where the streams have cut through massive limestones. Shales, in general, are weak and usually produce long gentle slopes or rolling surfaces covering platforms of more resistant rock.

The heads of most gullies in arid plateaus are wide amphitheaters, instead of sharp-pointed ravines as they are in humid regions. This is because the gully is formed mainly by weathering and not by direct stream erosion. Where streams are constant, the head of the gully is subjected to continuous erosion and it therefore wears back faster than the sides do where weathering alone is the agent. A narrow ravine

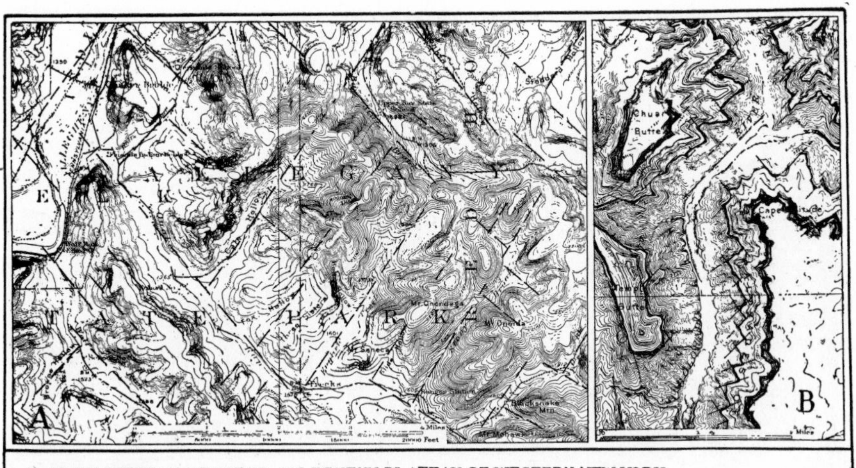


A. JOINT SYSTEMS OF HUMID ALLEGHENY PLATEAU OF WESTERN NEW YORK
B. JOINT SYSTEMS IN ARID COLORADO PLATEAU OF ARIZONA

RECTANGULAR FEATURES IN STRONGLY JOINTED PLATEAUS

results. But in arid regions where streams are intermittent, the walls of the gully weather back equally in all directions from a central point, and a circular form or amphitheater results. Gully development of this kind is shown in the Colorado Plateau.

Valleys in humid regions, even where the rock structure includes massive and resistant beds, have less precipitous profiles than in arid districts. The soil produced by weathering is not so quickly removed but is held in place by vegetation. Talus is abundant. Wind action is negligible. Soil has an opportunity to accumulate. Tributaries to the main canyon wear back in narrow ravines owing to continuous stream erosion.

Due to the many vertical joints, canyon walls in arid regions are sharply angular in plan. In fact, the system of ravines, alcoves, and indentations, alternating with projecting spurs, points, promontories, detached pinnacles, buttes, and mesalike remnants of the plateau, give to the canyon rim an extreme rectangularity because of weathering along intersecting joint systems. This sharp angularity is rare in humid regions, although many plateaus, both in humid and in arid regions, exhibit rectangular stream patterns.

With the increase in the width of the canyons or valleys and in their number and perhaps also by the added action of alpine glaciation, a plateau becomes extremely rugged and mountainous in aspect. It passes thus from the condition of youth through late youth and submaturity to the fully mature stage.

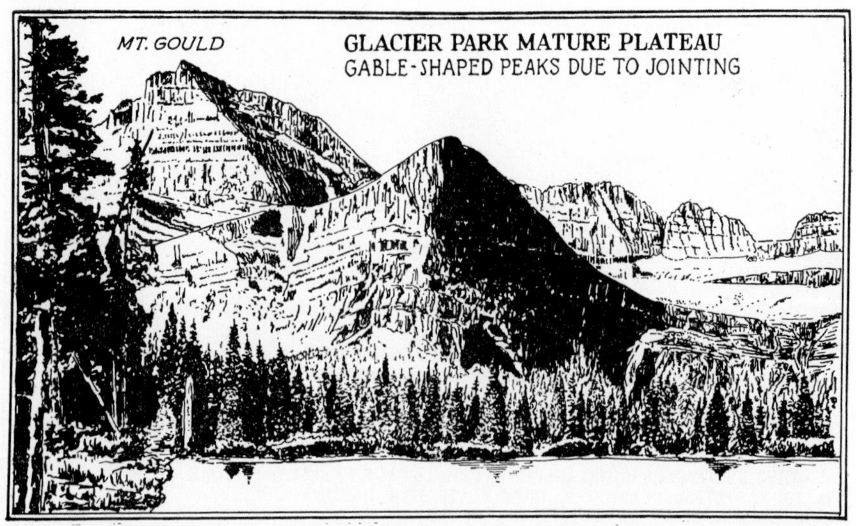


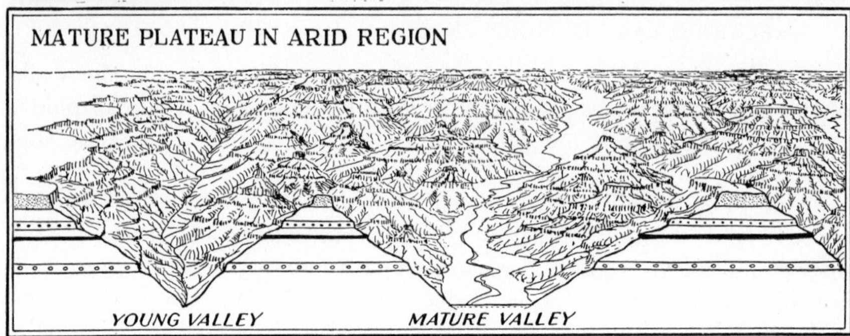


## MATURE PLATEAUS

Mature plateaus are called *mountains* by most people. We have noted, however, a regular transition from young undissected plateaus to maturely dissected ones. Hence the logic of maintaining the term *plateau* even for rugged areas of horizontal rock. This is the approved geological usage.

MATURE PLATEAUS IN ARID REGIONS. In arid regions, dissected plateaus are characterized by sharp and angular peaks with vertical walls and rock terraces. The effect of vertical jointing is everywhere evident. Where earth stresses have introduced a system of dipping joints, the mountain summits weather along these diagonal planes and have a gablelike form, a shape very common in Glacier Park and in the Canadian Rockies.

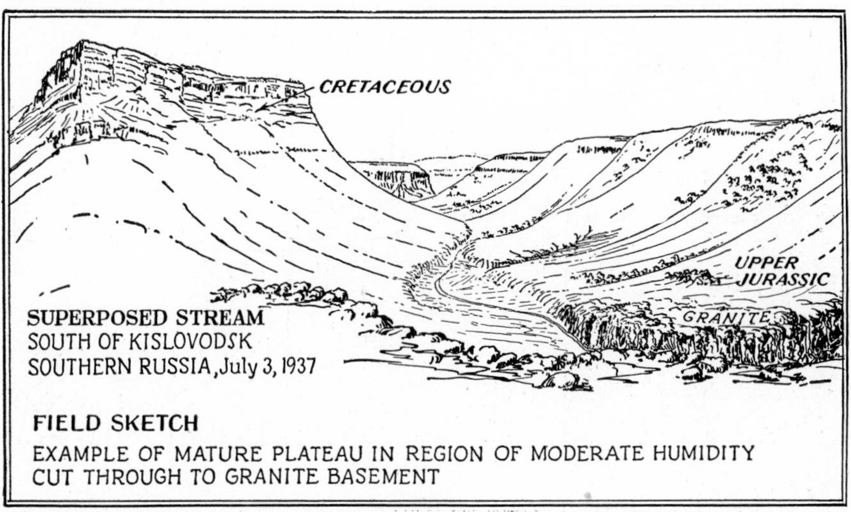


FIELD SKETCH
EXAMPLE OF MATURE PLATEAU IN REGION OF MODERATE HUMIDITY CUT THROUGH TO GRANITE BASEMENT

MATURE PLATEAU IN HUMID REGION

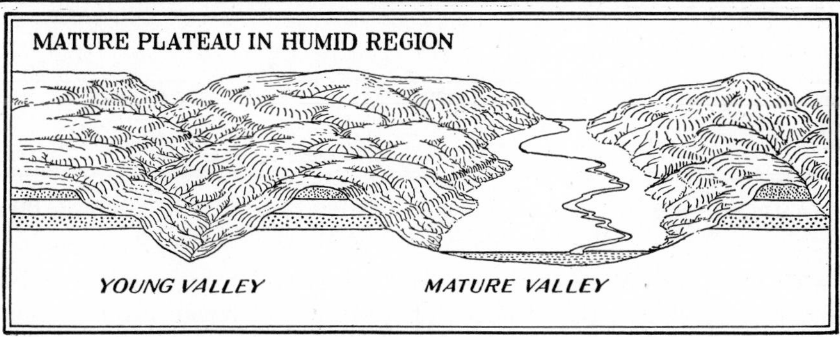


Mature Plateaus in Humid Regions. Mature plateaus in humid regions have rounded full-bodied forms. Nevertheless, in spite of heavy vegetation and deep soil, the horizontal structure is revealed by flat-topped summits and terraces or benches along the mountainsides. In the Catskills this is a noteworthy feature. In some plateaus the differential erosion in valleys has caused the removal of weak rock layers above some hard, resistant bed, and a cirquelike valley head or pocket results. Such valleys may have the appearance of hanging above the valleys to which they are tributary.

In some humid regions the heavy forest cover completely obscures the minor terraces and benches which are due to the horizontal rock structure. But the character of the vegetation may reveal the different kinds of slopes. The tops of the benches, being soil-covered, and also often being seepage zones, support deciduous or broad-leaved trees, usually light green in appearance. The steeper declivities support the coniferous trees, which are dark green. A horizontal banding of the vegetation may therefore be noted.

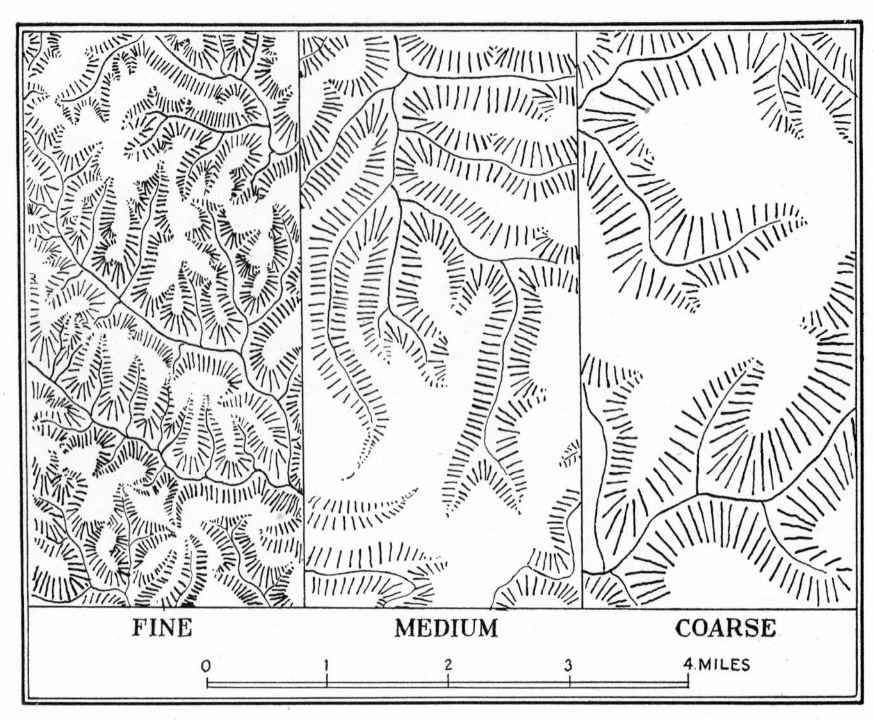


EXAMPLES OF FINE, MEDIUM, AND COARSE-TEXTURED TOPOGRAPHY FROM THE PLATEAU OF WEST VIRGINIA

Coarse- and Fine-textured Topography. In regions of massive and resistant rocks the elements of a dissected plateau are large and bold, the rivers being far apart. This constitutes *coarse topography.* The Catskills are of this type. In regions of weaker rocks the streams are numerous and the topography is *fine textured.* Such is the Kanawha Plateau in West Virginia. If the texture is extremely fine, as it is in clays and weak shales, true *badland topography* results. This is the case in the Goshen Hole country of western Nebraska as well as in Bryce Canyon, Utah.

Even if the formations which make up a plateau do not vary enough in resistance to produce pronounced scarps and benches, nevertheless the layers of rock are revealed by the belts of vegetation or variations in the soil. In maturely dissected plateaus, where all the land is in slopes, these belts appear like contour lines around the hillsides and are especially apparent when viewed from the air.

Maturely dissected plateaus of high relief have in some instances supported local glaciers, as in the rugged Absarokas near Yellowstone Park and in Glacier National Park. "Flat-topped" mountains and rock steps due to the horizontal bedding are common features, as are also high alpine meadows and steplike divides.

THE ROCK CITY AT SALAMANCA, N. Y.

Showing the main ledge at the right and large blocks at the left breaking off along joint planes.

When overridden by a continental ice sheet, a maturely dissected plateau is but little altered in appearance. Moraines, if any, are insignificant features among the much higher hills of the plateau. Lakes, if formed by damming of valleys, soon disappear by the erosion of the barrier. Only occasionally are old outlet channels preserved across low divides on the plateau surface.

In many plateaus there is a suggestion of right-angled drainage lines due to the prevailing joint systems. Usually, however, the pattern appears to be typically dendritic.

Rock Cities. When the joints are not more than a few feet apart, and the plateau is capped with an unusually resistant rock formation, the rim of the plateau adjacent to valleys breaks off in blocks, cubical or tabular in shape. The blocks gradually slump down the valley side, thus widening the joints until they appear as ditches, 10 to 20 feet in depth. Such an array of blocks resembling houses, separated by streets, is called a *rock city*. Well-known rock cities occur at Olean in western New York and at other places in the vicinity of Allegany State Park.

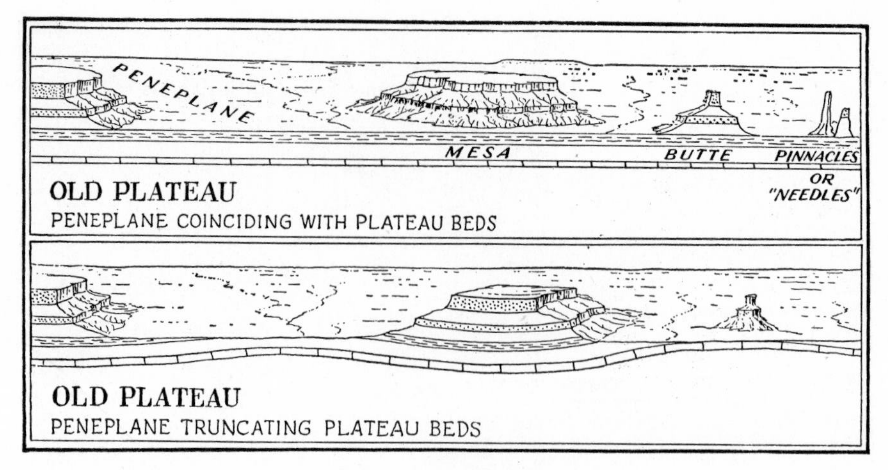


## OLD PLATEAUS

Old plateaus are regions of horizontal or almost horizontal structure, which have been worn down to a peneplane. Remnants of the former plateau mass remain as outliers, mesas, or buttes. Over the old plateau surface flow mature streams in shallow valleys, unless rejuvenation has occurred and caused them to be incised. At one side the plateau may rise in steps to higher levels still but slightly dissected.

Peneplanes and Bedding Planes. Whether the surface of an old plateau represents a base level of stream erosion, or whether it is a differential erosion surface is difficult to determine. In most plateaus the beds are somewhat warped. A plane of erosion, due to base-leveling, therefore truncates the structure, and different formations outcrop on the old plateau surface. If due to differential erosion, that is, to different erosional effects upon different kinds of beds, then the surface has a warped shape conforming with whatever warping the beds themselves have.

Mesas. One of the famous plateau remnants is the *Enchanted Mesa*, with towering vertical cliffs 400 feet high. Such mesas are occasionally inhabited by Indian tribes, a well-known example being Acoma in New Mexico. It is a favorite haunt for tourists who are willing to climb the steep trail to its summit.

A portion of central Wisconsin, the vast inner lowland of a former coastal plain, is actually an old plateau. This extensive worn-down area is even more impressive than the Grand Canyon. It is easy to realize that a prodigious amount of erosion has taken place, for many scattered mesas and buttes still rise above its surface.

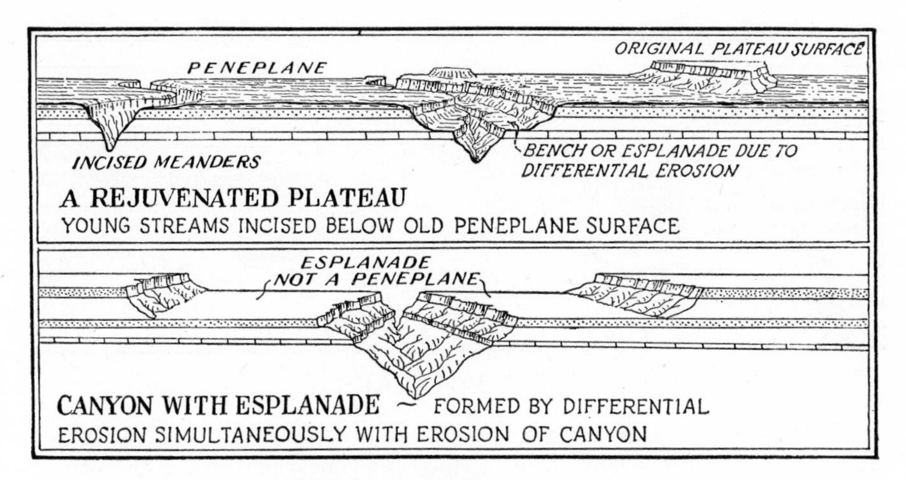

A REJUVENATED PLATEAU
YOUNG STREAMS INCISED BELOW OLD PENEPLANE SURFACE

CANYON WITH ESPLANADE ~ FORMED BY DIFFERENTIAL EROSION SIMULTANEOUSLY WITH EROSION OF CANYON

## REJUVENATED PLATEAUS

A plateau, having reached old age (usually evidenced by the presence of mesas rising above its surface) and then having been rejuvenated, is considered to have passed through one or possibly more cycles of development and to have begun a new cycle. Such a region is said to be now in its $n + 1$ cycle, $n$ being any number of cycles preceding the present one.

The presence of incised meanders is not necessarily a proof of rejuvenation from an old-age condition. Streams with irregular courses tend to do away with their irregularities and to develop large sweeping curves, which resemble meanders developed by mature streams.

One of the most extensive old plateau areas of the United States is in eastern Montana and Wyoming and the western Dakotas. This is called the *Missouri Plateau.* It is characterized by mesas and buttes, and by broad terraces, probably in part due to differential erosion. The Missouri Plateau peneplane has apparently been rejuvenated since it was formed, as the Missouri River and its tributaries now flow below its surface. Along the bluffs of these rivers badlands are common.

Much of northern New Mexico is also an old but rejuvenated plateau. Mesas and buttes rise above its surface and rivers are now cutting canyons below the plateau level.

Canyons in plateaus occasionally have broad benches or terraces known as *esplanades.* These might at first be taken as local peneplanes once formed at that level. However, they conform strictly with the rock structure and they have an almost equal width on each side of the present canyon, clearly suggesting differential erosion simultaneous with the cutting of the canyon.

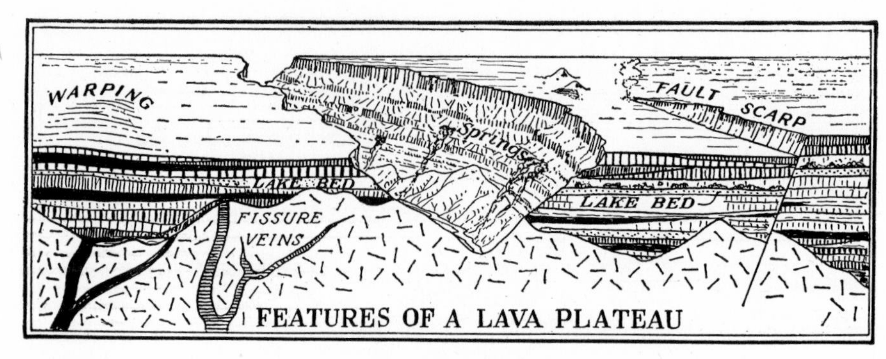

FEATURES OF A LAVA PLATEAU

## LAVA PLATEAUS

Lava plateaus constitute a special type of plateau. They consist of numerous lava flows laid down at different times, one above the other, and frequently alternating with lake beds, stream deposits, layers of volcanic ash, and weathered lava. No single flow is coextensive with the plateau itself. The flows interfinger with each other.

Most lava is highly porous because of its vesicular structure. Therefore surface water readily penetrates it and finds its way down to impervious layers, usually lake beds, which were formed on what was at one time the plateau surface. For this reason water tables are common. The ground water migrates along these levels until it flows out along the walls of canyons to form springs—indeed veritable rivers in some cases. It is not surprising, therefore, in lava plateaus which have undergone dissection, to find caves and natural bridges.

Extensive lava plateaus are rarely surmounted by volcanoes, though occasionally young cinder cones of later date may grow upon their surface. Some lava areas, like the Cascades, have been greatly elevated and are now surmounted by high volcanoes. It is not uncommon, also, to find hot springs and geysers in lava plateaus, where faults have permitted ground water to reach great depths and to rise again to the surface.

Ordinarily, canyons cut into lava plateaus have steep walls, due to the columnar structure of the lava, but the various cliffs are rarely so persistent and so uniform in height nor do they follow so nearly a level plane as do the cliffs in plateaus of sedimentary rocks.

In some instances the lavas are deeply weathered and altered by fumarolic action, and then, instead of somber vertical-walled cliffs, the canyon sides are glorious with many delicate hues of pink and yellow. Such is the canyon of the Yellowstone River.

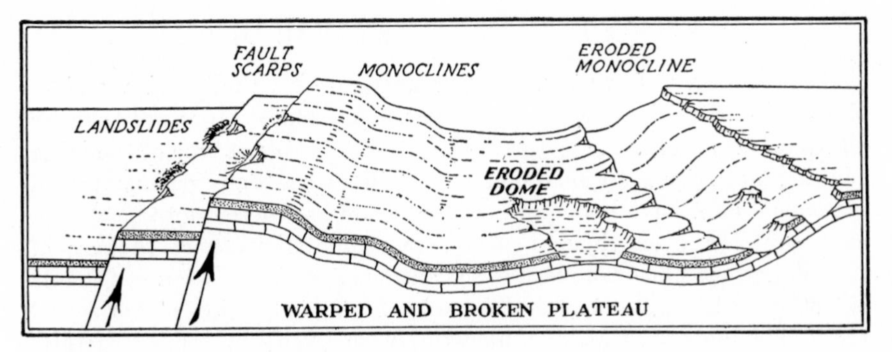


## WARPED AND BROKEN PLATEAUS

Rarely is the structure of extensive plateaus not disturbed by faults or folds. The Colorado Plateau in the Grand Canyon district and the Columbia Plateau in central Washington are both strongly faulted and warped with great monoclines.

Like gigantic steps, the plateau blocks rise from the Great Basin on the west to the top of the Kaibab Plateau on the east, ascending in all 5,000 or 6,000 feet to an elevation 9,000 feet above the sea. The *fault scarps* thus produced run for many miles straight across the country. The western winds, impinging upon these high cliffs, produce rain in an otherwise dry region. Landslides are common at the foot of these cliffs, whence comes the name *Grand Wash* cliffs applied to this escarpment. Similarly, the large escarpments in central Washington, not far from Ellensburg, have at their bases great accumulations of irregular landslide deposits, containing in some places landslide lakes. Landslides usually indicate recent faulting and suggest that these scarps are fault scarps rather than fault-line scarps formed by erosion. However, landslides and renewed erosion may occur along fault-line cliffs if rejuvenation has occurred and streams have increased activity.

In plateaus, scarps of three kinds may occur: (*a*) fault scarps, (*b*) fault-line scarps, and (*c*) scarps of differential erosion not on fault lines but due to the normal retreat of resistant beds by weathering. The erosion of domes and warped parts of plateaus results in circular cliff patterns, like the Circle Cliffs of Utah. Eroded monoclines like the San Rafael Swell in the Colorado Plateau exhibit strongly dipping hogbacks and peculiarly sloping buttes and mesas.

The Allegheny Plateau in western Pennsylvania is warped to form a series of parallel undulations. Some of the anticlines still exist as topographic features, like Chestnut Ridge and Laurel Hill. Along other structural arches oil and gas occur. The eroded synclines are represented by rows of mesa-like hills and by shallow synclincal mountains capping the plateau.

Plateaus vs. Uplands. Larger areas of North America and of Europe are underlain by rocks of simple horizontal structure than by all kinds of disturbed structures combined. This is not true of the other continents; Africa is an upland of crystalline rocks. Many regions termed plateaus do not have horizontal structure and are therefore to be classed as mountains even though they have a plainlike aspect because of peneplanation. For example, the so-called *Laurentian Plateau* of Canada is not a plateau, nor is the Piedmont Plateau a region of horizontal rocks. The plateau of Mexico, likewise, is one of mountainous structure.

For these highlands the term *upland* is better. The term *meseta* applied to the upland portion of Spain is something of a misnomer also. Parts of the meseta are covered with flat-lying sedimentary beds but most of it consists of strongly folded and metamorphosed rocks.

The Bolivian and Peruvian Plateau should be conceived not as a region of flat-lying beds but as an elevated part of the complex Andean system not so rugged as the rest. In Bolivia, Mexico, and Tibet the plateaus are partly intermontane basins filled with thick deposits from the surrounding ranges. These parts are true plateaus.

North America and Europe. The two greatest plains areas in the world are the central parts of the North American continent, between the Appalachians and the western Cordillera, and a still larger area in Europe and Asia encompassing most of European Russia, much of northern Siberia, and countries bordering on the Baltic and North Seas. Both of these regions for long geological ages were under the sea. The bedrock therefore is of sedimentary origin, consisting of shales and limestones, with clay and sand where most recently elevated. The rocks have been warped into low domes and basins, and erosion has slightly beveled the rock formations. But the actual dip of the beds in most places is hardly 1°.

There is a great contrast between regions of this type and the steppe lands of southern Siberia, in spite of a superficial similarity. The steppe lands, like the plains of Russia and the United States, are open rolling areas suited to grazing and certain forms of agriculture. But the bedrock structure consists of formations intricately folded and bearing many igneous intrusions.

South America. The Argentine pampa is a remarkably level plain covering some 250,000 square miles, with a surficial cover of terrestrial sediments of alluvial and aeolian origin. Beneath this veneer is a series of horizontal strata upon a granitic base. The drainage is largely underground. The peculiarly uniform relief of the pampean plain has long been a source of wonder, for few areas in the world resemble the pampa in flatness. Originally this was a great expanse of grass but man's activities have left a pattern of fields and farms and occasional shade trees. Still the endless plain is the impressive feature of the landscape.

## ECONOMIC ASPECTS OF PLAINS AND PLATEAUS

MINERALS. The mineral wealth of plains and plateaus consists largely of the nonmetallic minerals usually associated with sedimentary rocks. These include clay for pottery making; sand for glass manufacture; phosphate for fertilizer; salt, gypsum, and sulphur; limestones and other building stones; lignite and coal, usually of bituminous type; oil and gas; and artesian water. Among the common metallic minerals are lead, zinc, and iron.

WATER SUPPLY. The water supply in young plains and plateaus is largely artesian. In maturely dissected plateaus, with mountainous topography, surface streams are generally used. In most plains the slightly dipping rock formations cause water-bearing beds to slope to considerable depth. A dip as small as 1° means a difference in elevation of about 1,000 feet in a distance of 10 miles. Artesian water derived from limestone strata is hard; that is, it contains much mineral matter in solution. This is the character of the artesian water throughout Wisconsin, Illinois, Michigan, and Indiana. On the coastal plain, because of its sandy composition, the artesian water is much softer.

HUMAN LIFE ON PLAINS AND PLATEAUS. In vast areas of level, almost undissected plains, the uniformity of soil, the monotony of the topography, the sameness of the resources, a similarity of climate over large areas, often combined with deficient rainfall, encourage migration and wandering. Nomadic life, temporary villages, movable herds and flocks are the rule. Thus live the people of the steppe lands of Europe and Asia, the plains Indians of America, and the wandering tribes of Australia. A sedentary life in such regions has come about only with settled forms of agriculture and the introduction of means of overcoming the handicap of isolation.

Dissected plateaus, like all rugged areas, tend to a permanent way of life. On young plateaus and plains people inhabit the broad uplands and avoid the narrow valleys or canyons. They go freely from place to place. In maturely dissected regions with small, detached upland areas, the people live and travel in the valleys. The actual stage of dissection of a plain or plateau and the width of its valleys influence the mode of activity of the people. Plateaus with open mature valleys having broad flood plains may support large numbers of people. Plateaus with young narrow valleys unsuited to agriculture remain sparsely populated and are likely to be regions of lumbering and hunting.

The mining methods in young plains or plateaus contrast with those employed in maturely dissected areas. Deep coal beds in an undissected plain can be reached only by vertical shafts because there are usually no outcrops, as in much of the Illinois and the Ukraine coal fields. In dissected regions, however, the coal outcrops on the valley walls and can be worked by horizontal drifts, as in the Appalachian field. Shallow beds are mined by open pits.

## MAPS ILLUSTRATING PLAINS AND PLATEAUS

The *Fargo, N. Dak.*, sheet illustrates a very young plain, in this case a lake plain. The region is extremely flat. The streams meander over its surface, being incised a trifle. They began their existence as meandering streams but now exhibit slight indications of youth. The *Olivet, S. Dak.*, sheet also shows a very young plain with valleys, like that of the James River, which are mature. The *Wahpeton, N. Dak.-Minn.*, sheet illustrates another part of this plain, also very young, its monotony being broken only by the moraine in its southwestern corner. Another very young lake plain is shown on the *Chatom Ranch* and *El Rico Ranch, Calif.*, sheets, which include part of the former bed of Tulare Lake. Parts of the coastal plain are extremely young, such as that portion shown on the *Glennville, Ga.*, and the *Trent River, N. Car.*, sheets. A more advanced stage of dissection is represented on the *Wilson* and the *Falkland, N. Car.*, sheets which are in middle or late youth. A young till plain, with very flat relief but dissected by wide mature valleys is shown on the *Avon, Ill.*, sheet.

Maturely dissected plains are shown on the *Millen* and *Oliver, Ga.*, and the *Bamberg, S. Car.*, sheets. Many of the sheets of the Middle West provide good examples of mature plains, as, for example, the *Eureka Springs* and *Fayetteville, Ark.-Mo.*, and the *Kahoka, Mo.-Iowa-Ill.*, sheets.

As an example of old plains, parts of eastern Kansas may be taken as fairly representative, such as those portions shown on the *Burlingame, Burden, Fredonia,* and *El Dorado, Kan.*, sheets.

The *Soda Canyon*, and *Mesa de Maya, Colo.*, the *La Sal, Utah-Colo.*, and the *Watrous, N. Mex.*, maps illustrate young plateaus with sharply incised young canyons whose terraced walls indicate the horizontal structure of the several regions. The *Hollow Springs, Tenn.*, sheet displays a very young plateau surface bordered by a wonderfully dissected scarp. The *Kaibab, Ariz.*, sheet illustrates a warped plateau.

Maturely dissected plateaus in humid regions are represented on the *Hurley, Va.-Ky.*, and *Iaeger, W. Va.-Va.*, sheets which are fine-textured; by the *Andes, N. Y.*, sheet, which shows medium texture, and by the *Walton* and *Kaaterskill, N. Y.*, sheets, which have coarse-textured features. The *Kaaterskill* sheet also shows good rock benches along the mountain front. The *Buckhorn, Ky.*, region is a mature plateau with young streams, whereas the *Falmouth, Ky.*, area is a mature plateau with mature streams.

The *Highmore* and *Parachute Creek, Colo.*, and the *Bright Angel* and *Vishnu, Ariz.*, sheets illustrate in a superb manner maturely dissected plateaus in arid regions where beds of the plateau are revealed in the canyon walls. The *Roan Creek, Colo.*, and *Salina, Kan.*, maps illustrate a postmaturely dissected plateau.

Finally, the *Abilene, Texas*, sheet may be taken to illustrate an old plateau with mesas and buttes.

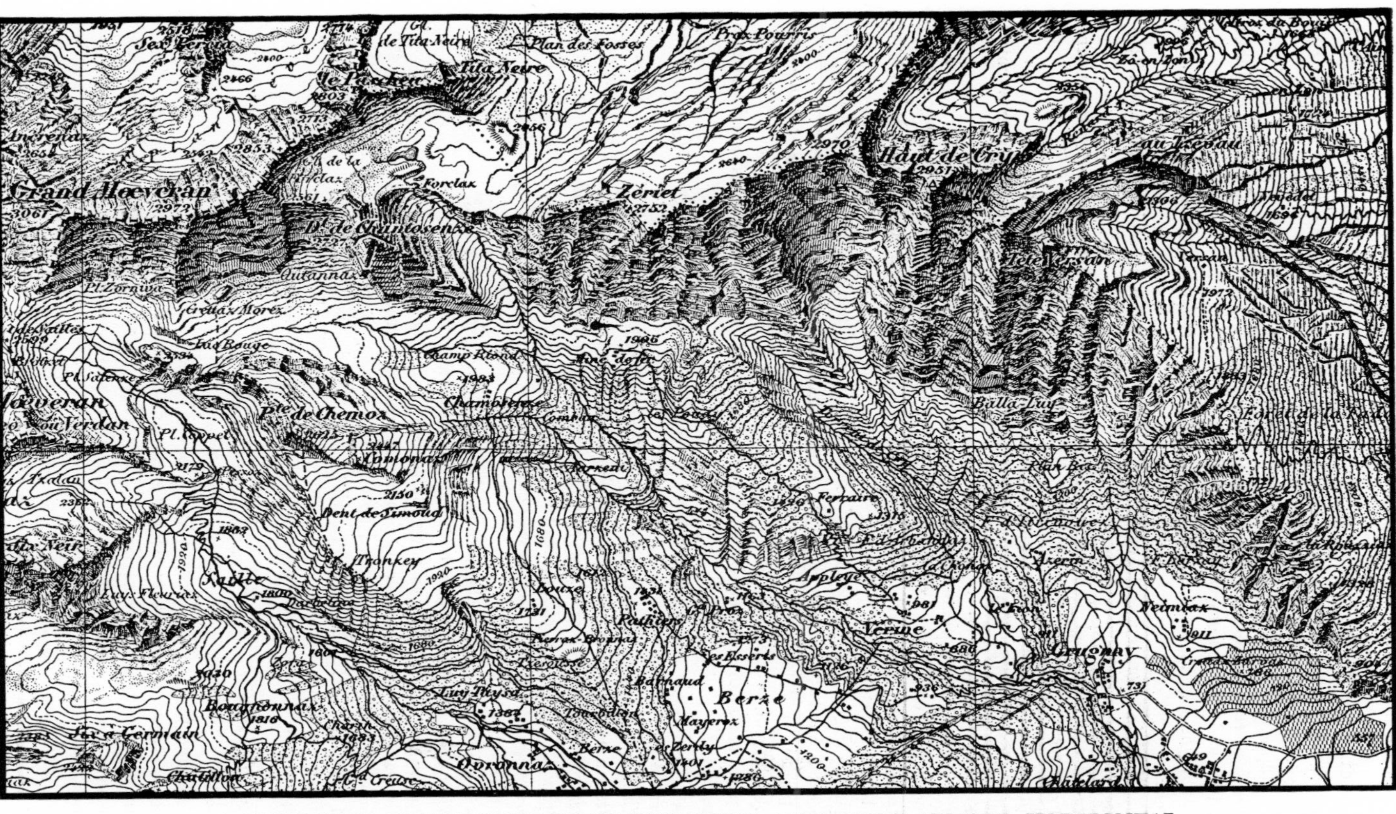


A DISSECTED PLATEAU SCARP IN THE SWISS ALPS. BEDS ALMOST HORIZONTAL
Switzerland; *Saxon* sheet, No. 485 (1:50,000).

## QUESTIONS

1. How can a young plain be distinguished from an old plain which has no monadnocks?
2. On a submaturely dissected plain, where will the highways and villages be, on the divides or in the valleys?
3. Are steppe lands always plains? What is a prairie plain?
4. What difference is there in the erosional history of a plain by normal surface drainage and by underground drainage?
5. Would you expect to find subsequent streams on plains? Explain how they might occur.
6. Is a till plain a constructional or destructional type of plain, according to the distinction used in this book?
7. Are all constructional plains necessarily coastal plains, either present-day or ancient in origin?
8. What is the difference between an ancient and an old coastal plain?
9. Draw a contour map of a mesa standing upon an old plain. Show that it is surmounted by a cap of hard rock.
10. In dissected plains, where would you expect to find springs?
11. Under what conditions would glacial troughs occur in a plain or a plateau?
12. What physiographic divisions of the United States are plains or plateaus? What type is the Triassic Lowland?
13. Why do the massive joint blocks of rock cities, when slumping down a valley wall, come to tilt toward the plateau rather than toward the valley?
14. A plain is dissected by streams which are for the most part mature. What is the probable stage of development of the plain?
15. A plain is dissected largely by young streams. What is the probable erosional stage of the plain?
16. Can a plain be gradually transformed into a plateau? Can a plateau be gradually transformed into a plain?
17. What criticism would you make of question 9 above?
18. Draw a contour map illustrating badland topography.
19. Draw a cross section of a maturely dissected plain buried under a till plain.
20. Draw a cross section of a plateau with coal beds outcropping on the valley sides.
21. Draw contour map of canyon wall cut in slightly dipping, alternating weak and resistant, sedimentary beds.
22. Is the interpretation of the picture on page 469 necessarily correct?
23. What type of rock is illustrated in Bryce Canyon on page 473?
24. Do the two illustrations on page 483 represent similar kinds of country?
25. Do you think the numerous horizontal benches depicted on the Swiss map, page 499, could be so clearly represented by contours as they have been? What is this method of showing relief? Hachures? In which direction do these beds dip?

## TOPICS FOR INVESTIGATION

1. A classification of plains. On some logical basis.
2. The Colorado Plateau. Variations in structure.
3. The Grand Canyon region. Its physiographic history.
4. Cycles of erosion in plateaus. How are they distinguished?
5. Important plateau regions: the Catskills; the Allegheny Plateau; the Cumberland Plateau; the Columbia Plateau. Physiographic and geographic aspects.
6. Block diagrams. Convert some of the photographs at the beginning of this chapter into block diagrams. Show geological structure along the sides of the blocks.

# REFERENCES


General

Davis, W. M. (1898) *Physical geography.* Boston. *Plains and plateaus,* p. 113–158.

Davis, W. M. (1912) *Die erklärende Beschreibung der Landformen.* Leipzig, Chap. V, *Plains.*

Erosion Cycles

Fridley, H. M., and Nölting, J. P., Jr. (1931) *Peneplains of the Appalachian Plateau.* Jour. Geol., vol. 39, p. 749–755.

Sharp, H. S. (1932) *The geomorphic development of central Ohio (part* 1). Denison Univ., Sci. Lab., Jour. 27, p. 1–46.

Ver Steeg, K. (1931) *Erosion surface of eastern Ohio.* Pan-Am. Geol., vol. 55, p. 93–102, 181–192.

Ver Steeg, K. (1931) *Erosion surfaces of the Appalachians.* Pan-Am. Geol., vol. 56, p. 267–284.

Ver Steeg, K. (1932) *Erosion surfaces of the Appalachian Plateau.* Pan-Am. Geol., vol. 58, p. 31–44.

Water Resources

Darton, N. H. (1898) *Underground waters of a portion of southeastern Nebraska.* U. S. Geol. Surv., W-S. Paper 12, 56 p.

Darton, N. H. (1899) *Geology and water resources of Nebraska west of the one hundred and third meridian.* U. S. Geol. Surv., 19th Ann. Rept., part 4, p. 719–785.

U. S. Geol. Surv., Water-supply Papers, 227, 317, 319, 335, 341, 428, 518, 539, 598, and many others.

Prairie Plains

Barton, T. F. (1936) *The Great Plains tree shelterbelt project.* Jour. Geog., vol. 35, p. 125–135.

Elias, M. K. (1935) *Tertiary grasses and other prairie vegetation from High Plains of North America.* Am. Jour. Sci., 5th ser., vol. 29, p. 24–33.

Lesquereux, L. (1866) *On the origin and formation of the prairies.* Worthen's Geol. Surv. Illinois, vol. 1, p. 238–254.

Shimek, B. (1911) *The prairies.* Iowa Univ. Lab. Nat. Hist., Bull. 6, p. 169–240. Contains a beautiful description of the prairies as first seen by the settlers.

Visher, S. S. (1916) *The biogeography of the northern Great Plains.* Geog. Rev., vol. 2, p. 89–115.

Willard, D. E. (1902, 1923) *The story of the prairies.* Chicago, 375 p.

Regions Described

Bowman, I. (1911) *Forest physiography.* New York. *Appalachian Plateaus,* p. 685–706; *Atlantic and Gulf Coastal Plain,* p. 498–542; *Colorado Plateaus,* p. 256–297; *Columbia Plateaus,* p. 192–207; *Great Plains,* p. 405–450; *Prairie Plains,* p. 460–497.

Campbell, M. R., and Mendenhall, A. C. (1896) *Geologic section along the New and Kanawha rivers in West Virginia.* U. S. Geol. Surv., 17th Ann. Rept., part 2, p. 473–511.

Cross, C. W. (1896) *Geology of the Denver Basin.* U. S. Geol. Surv., Mon. 27, p. 285–316. Table mountains.

Davis, W. M. (1901) *Excursion to the Grand Canyon of the Colorado.* Mus. Comp. Zool., Bull. 38, p. 107–201.

Davis, W. M. (1903) *Excursion to the plateau province of Utah and Arizona.* Mus. Comp. Zool., Bull. 42, p. 1–50.

Dutton, C. E. (1880) *Geology of the high plateaus of Utah.* U. S. Geog. and Geol. Surv. Rocky Mt. Region (Powell), 307 p.

DUTTON, C. E. (1882) *Tertiary history of the Grand Canyon district.* U. S. Geol. Surv., Mon. 2, 264 p. and atlas.

FENNEMAN, N. M. (1931) *Physiography of western United States.* New York, 534 p. Chapters on the Great Plains and Colorado Plateau.

FENNEMAN, N. M. (1938) *Physiography of eastern United States.* New York, 714 p.

GREGORY, H. E., and MOORE, R. C. (1931) *The Kaiparowits region: a geographic and geologic reconnaissance of parts of Utah and Arizona.* U. S. Geol. Surv., Prof. Paper 164, 161 p.

GUYOT, A. (1880) *Physical structure and hypsometry of the Catskill Mountain region.* Am. Jour. Sci., 3d ser., vol. 19, p. 429–451.

HEILPRIN, A. (1907) *The Catskill Mountains.* Am. Geog. Soc., Bull. 39, p. 193–199.

POWELL, J. W. (1875) *Exploration of the Colorado River of the west, and its tributaries.* Washington, 291 p.

SMITH, J. R. (1905) *Plateaus in tropic America.* 8th Intern. Geog. Cong., p. 829–835.

VAN TUYL, F. M., and COKE, J. M. (1932) *The late Tertiary physiographic history of the High Plains of Colorado and New Mexico.* Colo. Sci. Soc., Proc. 13, p. 19–25.

GEOGRAPHICAL ASPECTS

BARROWS, H. H. (1910) *Geography of the middle Illinois valley.* Ill. Geol. Surv., Bull. 15, 128 p.

CONDRA, G. E. (1906) *Geography of Nebraska.* Lincoln, Nebr., 192 p.

DRYER, C. R. (1897) *Studies in Indiana geography.* Terre Haute, Ind., 113 p.

HODGE, F. W. (1897) *The enchanted mesa.* Natl. Geog. Mag., vol. 8, p. 273–284.

JOHNSON, W. D. (1901–02) *The High Plains and their utilization.* U. S. Geol. Surv., 21st Ann. Rept., part 4, p. 601–741; 22d Ann. Rept., part 4, p. 631–669.

TOWER, W. S. (1906) *Regional and economic geography of Pennsylvania. Plateau Province.* Geog. Soc. Phila., Bull. 4, p. 204–217, 271–281.

VAN VALKENBURG, S. (1939) *Elements of political geography.* New York, 401 p. See chapter on relief, pp. 127–138. Many references listed.

# XV
# DOME MOUNTAINS

*Chamber of Commerce, Colorado Springs*

VERTICAL HOGBACK RIDGES, GARDEN OF THE GODS, COLO. Sedimentary beds upturned along the front of the Rocky Mountain dome, partly visible in the distance.

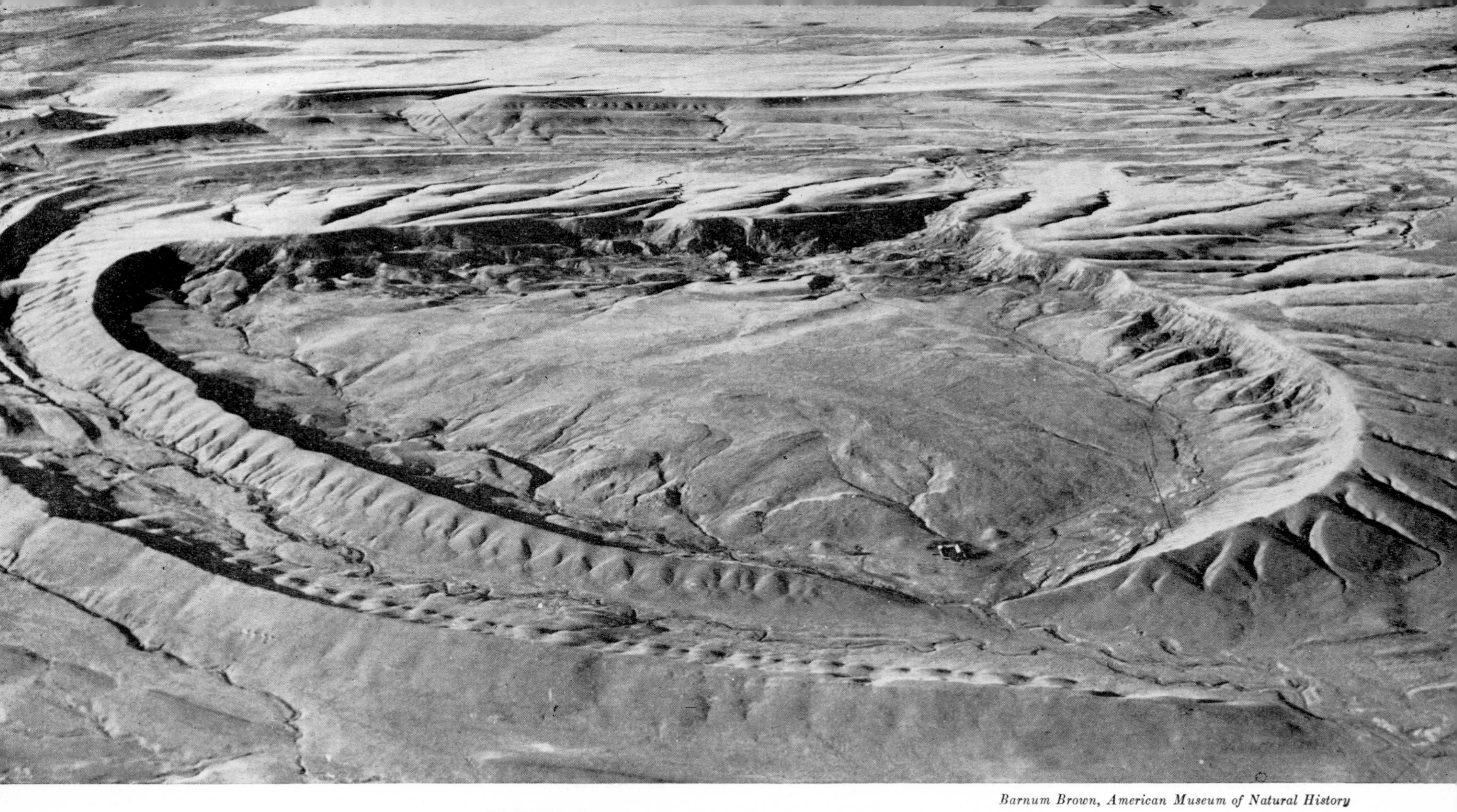

*Barnum Brown, American Museum of Natural History*

MIDDLE DOME, SOUTHEAST OF HARLOWTON, MONT.

A maturely dissected low dome with encircling hogbacks, some having a triangular form.

SHEEP MOUNTAIN, NEAR GREYBULL, WYO.

A strongly uplifted and maturely dissected dome encircled by many hogback ridges of various heights, shapes, and colors.

*Barnum Brown, American Museum of Natural History*

*Palace Studios, Boulder, Colorado*

FLATIRONS, BOULDER, COLO.

Resistant sandstone beds upturned against the front of the crystalline mass of the Rockies.

BEAR BUTTE IN THE BLACK HILLS, S. D.

A small dome of igneous rock, showing hogbacks of strongly upturned sedimentary beds around its base.

*Darton, U. S. Geological Survey*

*Edgar Tobin Aerial Surveys*

AVERY ISLAND, NEAR NEW IBERIA, LA.

A low young salt dome, two miles in diameter and 100 feet high, with many radial drainage lines. (Mapped on *De Rouen* sheet, Miss. River Comm.)

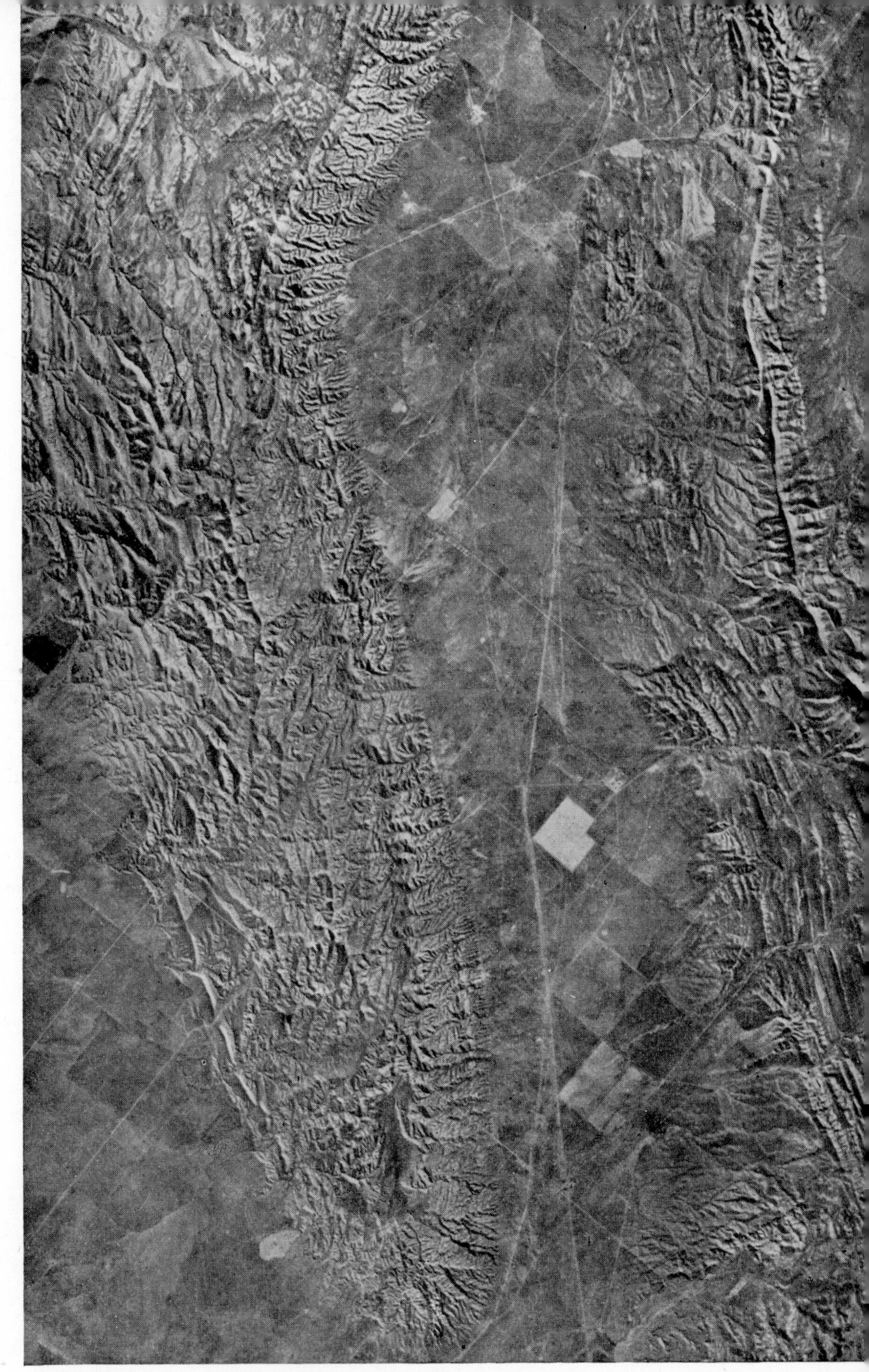

*Fairchild Aerial Surveys*

KETTLEMAN HILLS, NEAR COALINGA, CALIF.

An elongated dome, maturely dissected, the central area rising 1000 feet above the surrounding plains.

*Douglas Johnson*

THE DAKOTA HOGBACK, NEAR CANYON CITY, COLO.

Beds dipping eastward (to the right) from the Rocky Mountain uplift, with other low parallel hogback ridges of higher, but thinner formations.



THE DAKOTA HOGBACK, ZUNI UPLIFT, N. M.

Part of the dipping sandstone ridge encircling the Zuni dome, which lies to the right.

*Darton, U. S. Geological Survey*

# DOME MOUNTAINS

SYNOPSIS. Domes and basins together make up large areas of the earth's surface, even though to the unaided eye the rocks appear to be horizontal. Where the warping is inconsiderable, the region is termed a *plain* or *plateau;* where the uplift is great, a dome mountain results. There is every gradation from low domes, like the Cincinnati uplift where the dip of the beds is 1° or less, to the bold mountain masses, like the Black Hills and the Bighorns around which the beds in places have been bent up to a vertical position. In size, domes vary from a fraction of a mile to several hundred miles in extent. There is little relation between the area involved and the sharpness of uplift.

There are many causes for doming. Salt domes, due to concentration and crystallization of masses of salt, are low, small, and inconspicuous. Laccolithic domes, due to intrusions, are often high, but small in area, and are always distinct landmarks. Batholithic domes are large in area, high in elevation, and constitute true dome mountains.

No young domes of large size and height are known to exist. Most large domes are maturely dissected. This means that dome uplift must be relatively slow and that maturity of form is acquired at the outset. Many examples of domes eroded to a peneplane are known. In other cases, as in the Weald, it appears that, after peneplanation, domes have been buried and the present drainage system is *epigenetic;* that is to say, it has come much later than the doming. It is superposed.

Every conceivable kind of structural disturbance may be found around the flanks of a dome. The beds may dip away gently; they may be strongly upturned; they may be overturned. The beds may be broken off sharply by faulting; they may be bent up and faulted at the same time. A fault in one place may pass gradually into a monoclinal fold along the flank of the dome.

Every conceivable shape of hogback may be encountered. Some hogbacks are long, narrow, even-crested ridges, interrupted only by stream gaps and wind gaps. Other hogbacks are cuesta shaped with extensive surface areas on top. Still others are triangular masses, adhering to the ends of the spurs or rising up to the crest of the range, to form so-called *flatirons.*

Domes and basins grade into anticlines and synclines of folded regions, on the one hand; or into much complicated, uplifted, and disturbed complex mountain areas, on the other. Erosion may produce only basins along their axes or it may expose a resistant core of crystalline rocks which constitutes the highest parts of the mass. In some instances, as in the central Texas uplift, the crystalline core may be weak and a basin may result.

Domes may also be compound, large domes being surmounted by smaller ones, each type revealing its own peculiar erosional forms as determined by its own peculiar origin and structure.

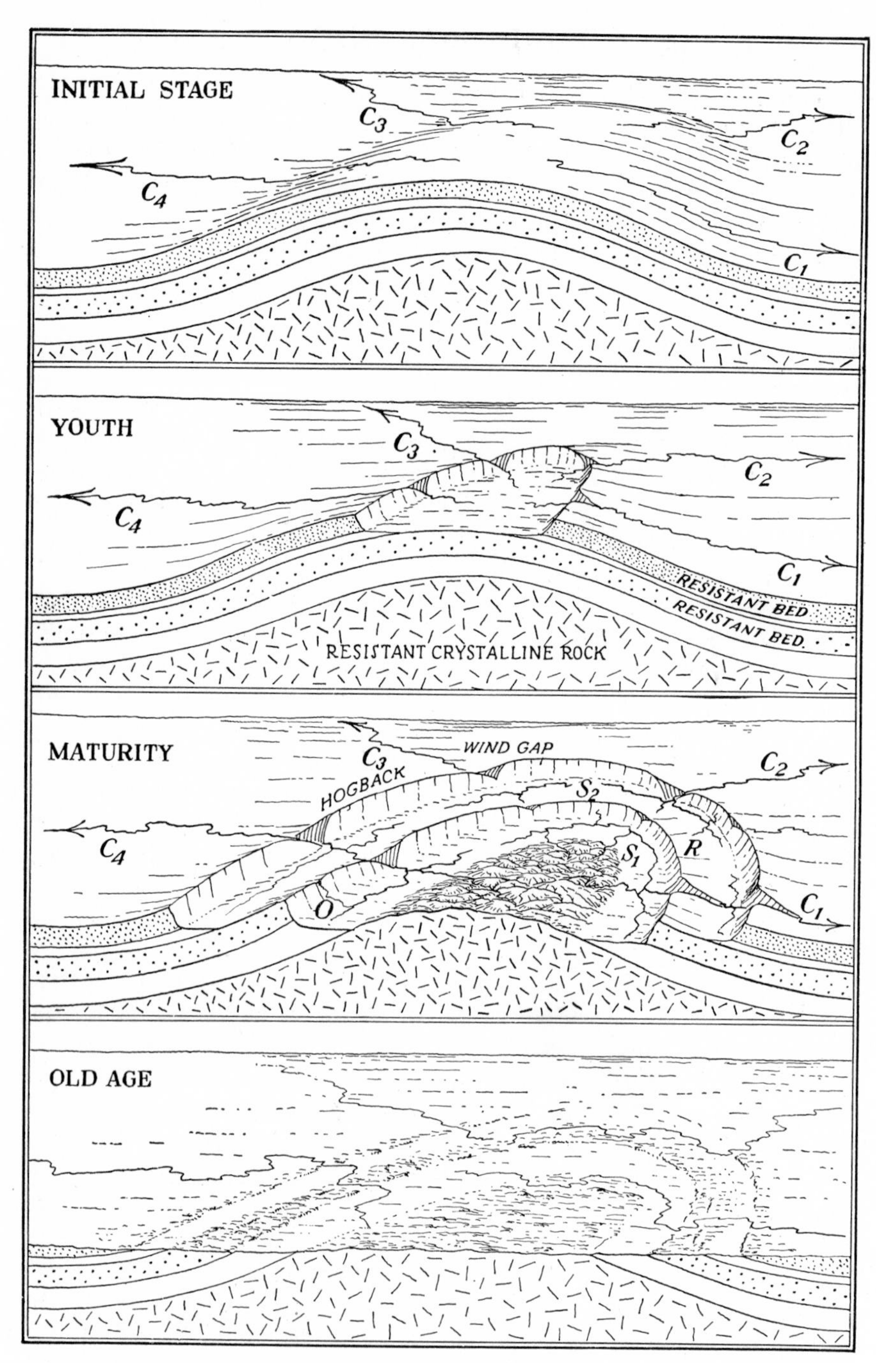


STAGES IN THE EROSIONAL DEVELOPMENT OF DOME MOUNTAINS

## THE DEVELOPMENT OF DOME MOUNTAINS

INITIAL STAGE. In the initial stage dome mountains are drained by consequent streams radiating from the center of the uplift. A *radial drainage pattern* is characteristic of very young and undissected domes. Some of the low young salt domes on the Gulf coastal plain can be detected from an airplane by the radial drainage lines when they are so low and subdued as to be almost unrecognizable as domes.

YOUTH. The first striking change from the erosion of a dome is the development of a basin along its crest. The several consequent streams work headward and concentrate upon the summit of the uplift until it is breached. The underlying weaker beds are removed as rapidly as the streams cut down their gaps in the resistant upper layer. The scarp formed by the topmost resistant bed retreats down the slope and the basin is thus enlarged. Lower and lower beds are successively exposed at the summit until perhaps a hard core of crystalline rock is uncovered.

MATURITY. Maturity shows the greatest variety of detail and the strongest relief. In the accompanying figure, consequent stream *C*1 has developed tributaries. One of them, *S*1, has worked headward along a belt of less resistant rock and tapped the headwaters of *C*2, leaving a wind gap on the crest of the innermost ridge. Several other similar captures may be noted. Subsequent stream *S*2 has captured the head of consequent stream *C*3. The lengthening of the subsequent streams by headward advance around the belts of weak rock develops an *annular drainage pattern* characteristic of the mature stage. The ridges formed by the upturned beds are termed *hogbacks*. The hogbacks usually present a steeper face toward the dome unless the formation is turned up at an angle of 45° or more. In this case the two sides of the hogback slope about equally. As the hogback is worn farther back where the dip of the beds is less, it comes to have a plateaulike aspect with an infacing escarpment. The short inner-slope streams of the hog-backs are *obsequent streams* (*O*). Those flowing down the back slopes, following the dip of the beds, are *resequent* (*R*). Many wind gaps interrupt the crests of the hogbacks.

The erosion of a pronounced uplift with crystalline core results, in the mature stage, in a mountain area which occupies the center of the dome. This is the case in the Black Hills. If the dome is a low one, then the entire central part of the mature dome is a broad basin, as in the Weald of England. Many domes are asymmetrical; in some instances the lowermost or oldest beds are very resistant and these may form a broad plateau over the center of the area; small parasitic domes may occupy the flanks of the main uplift; these departures from the simple scheme represented cause great variety in the actual aspect of domes.

OLD AGE. The final figure illustrates the conditions when erosion has beveled across the structures and reduced the region to a peneplane. The annular drainage pattern is almost lost.

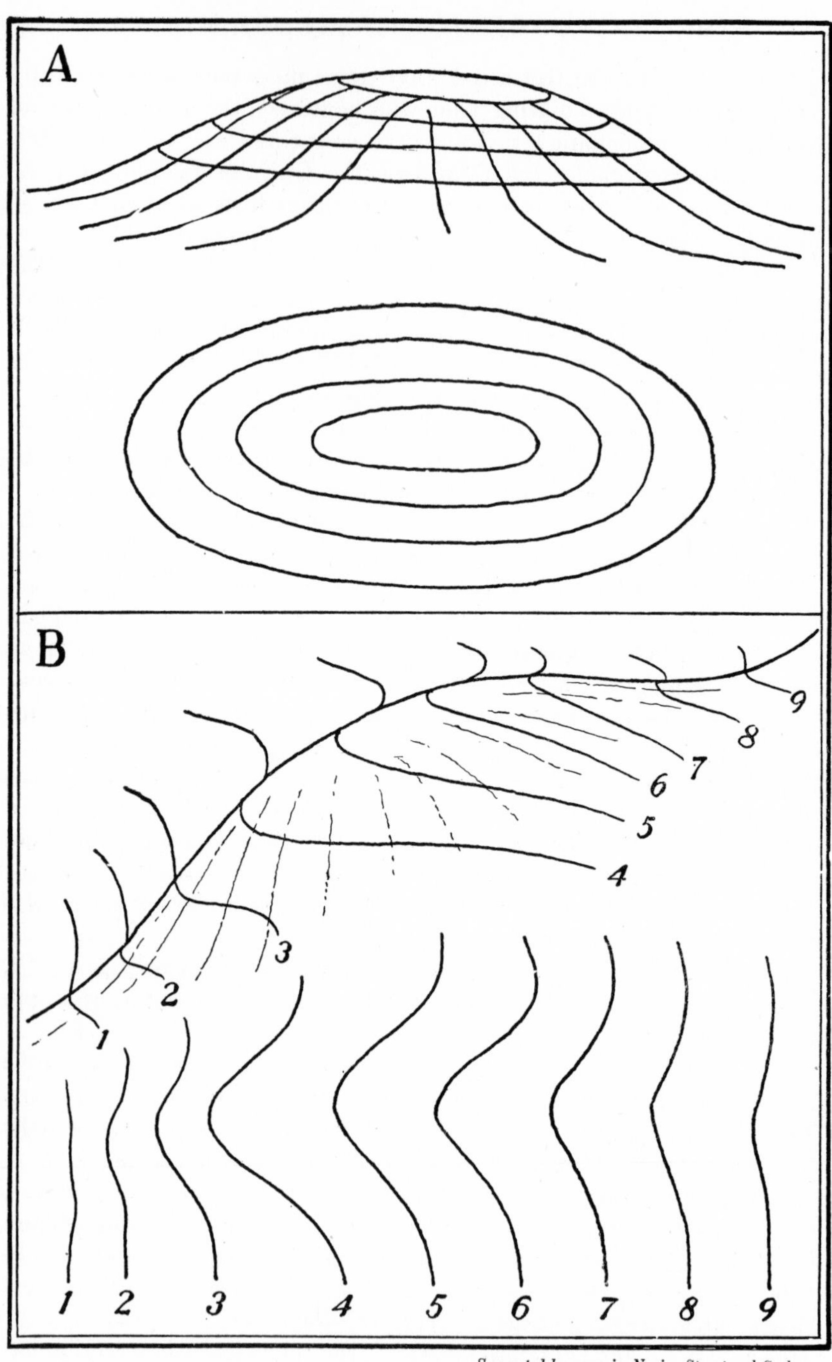


*Suggested by map in Nevin, Structural Geology*

*A*. DOME STANDING ON HORIZONTAL PLANE, AND CONTOUR MAP
*B*. DOME FORMING A NOSE ON TILTED PLANE, AND CONTOUR MAP

## SOME STRUCTURAL CHARACTERISTICS OF DOMES

Domes due to actual warping of sedimentary regions are usually low but may be of great extent. They may be linear and may resemble folds, several of them running parallel with each other.

Very large and high domes result from the uplift of extensive areas by the intrusion of large batholiths. Such domes may be almost round or broadly elliptical and may rise several thousand feet above the surrounding country. Domes may result also from the intrusion of lavas between beds, as sheets or laccoliths. Domes of this type may be limited in area but may be very symmetrical.

Salt domes are probably the smallest of all domes, being in some cases only a fraction of a mile across. They are due to the expansion of salt and gypsum beneath the surface.

Swarms of small domes may occur on the flanks of large ones where laccoliths are associated with batholithic intrusions.

Not all domes are due to an actual change in the initial dip of the beds, produced by either warping or intrusion. Some dome structures are explained as deposits covering original irregularities in the sea floor. There may, under those conditions, be a strong initial dip.

Domes due to warping and to regional earth movements may be associated with intervening broad basins, as in the eastern United States or in western Europe. This is not so true of the batholithic and laccolithic domes because the effect of their uplift is not widely felt.

The dips around the flank of a dome may vary greatly, so much so that the dome is strongly asymmetrical. If the dip is very steep, the flexure may pass into a fault and many actual displacements, constituting normal faults, are common around the margin of a dome. Domes due to volcanic activity, and especially those accompanied by violent outbursts, have many radiating fractures and dikes.

DOMES AND "NOSES." A dome on an otherwise flat surface is readily recognized as such and ordinarily is represented by closed contours on a map.

In contrast with this, however, a similar dome occurring in a region of dipping rocks offers a distinctly different appearance and is usually not recognized as a dome because it has the aspect of a pitching anticline. The actual form of the dome in the two cases may be identical.

*A* on the opposite page represents an almost circular dome standing on a horizontal plane. Its form is shown by the closed-contour pattern.

*B* shows the same dome as it would appear on a gently dipping surface. The contour map of this dome is in marked contrast with the other contour map. It shows the dome as a "nose" or pitching anticline. In areas of strong regional dip, noses of this type are common, whereas in regions of flat dip domes are common. Actually both represent the same shaped structure.

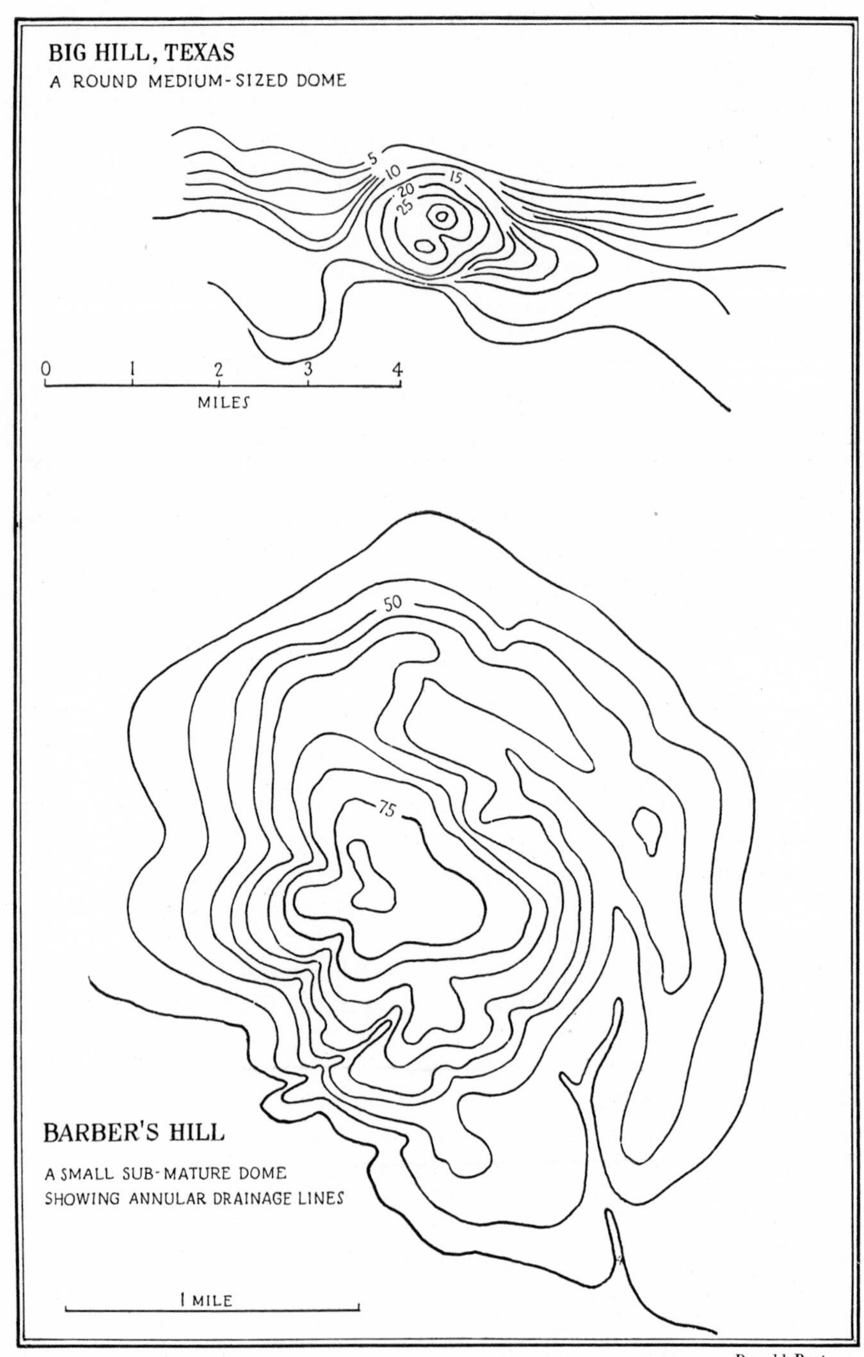


*Donald Barton*

CONTOUR MAPS OF SMALL ALMOST UNDISSECTED SALT DOMES ON THE GULF COASTAL PLAIN

## SALT DOMES

Many of the salt domes of the Gulf coastal plain in Louisiana and Texas are examples of very young almost undissected domes. When first studied, these features were believed to be erosion remnants, that is, small monadnocks rising above a peneplane. Their occurrence, however, in a region of more or less poorly organized drainage of a consequent pattern and their position immediately above and coincident with areas of uplift soon led to the recognition of their deformational origin.

Domes occur in all stages of dissection. In general, those lying farthest inland are maturely dissected and those nearest the coast are in early youth. Most of the domes have a circular form but they may be somewhat longer than they are wide. In the inland group the symmetry is expressed by one or more circular rings of hills around a central depression or around a central hill, or by a circular-radial drainage pattern, while in the coastal group some sort of topographic mound rises above the general level.

YOUNG SALT DOMES. The smallest and youngest salt-dome mounds rise only 3 to 10 feet. The rise of the mound from the plain is so gradual that there is no break in slope between the slopes of the mound and the level prairie. Such mounds as these would usually not be noticed by a person used to even slightly hilly country but they can be detected by an eye which is accustomed to the level country of the coastal plain.

The medium-sized mounds average 23 feet in height and 1.4 miles in diameter and show a marked break in slope. The large mounds average over 100 feet in height and only 1.9 miles in diameter, with much steeper slopes. The smaller mounds usually have so small a drainage area as to have developed no particular drainage characteristics, but the larger mounds have a clearly defined radial pattern of minor gullies. Some of the larger mounds have small "pimple" mounds rising above their convex surface, giving it an undulating character. Sink holes occasionally abound where the underlying salt has been dissolved away. There may even be a large central depression, due apparently to the same cause.

MATURE SALT DOMES. The mature salt domes lie far inland from the coast, where, in general, the coastal plain is in a mature stage of dissection. These domes are characterized by radial and circular drainage, with or without a central depression.

The West Point dome is an interesting example. A central circular hill, about 1½ miles in diameter and almost 100 feet high, is surrounded by a circular "racetrack," which in turn is surrounded by a ring of "hogbacks." Similar domes have a central depression instead of a hill.

OLD SALT DOMES. Some of the mounds of the Gulf coast seem to have been completely obliterated by erosion. Geophysical methods of surveying have detected the presence of salt cores and this is often the only indication that a mound ever existed. Some of the mounds appear to have been beveled off by antecedent streams during their uplift.

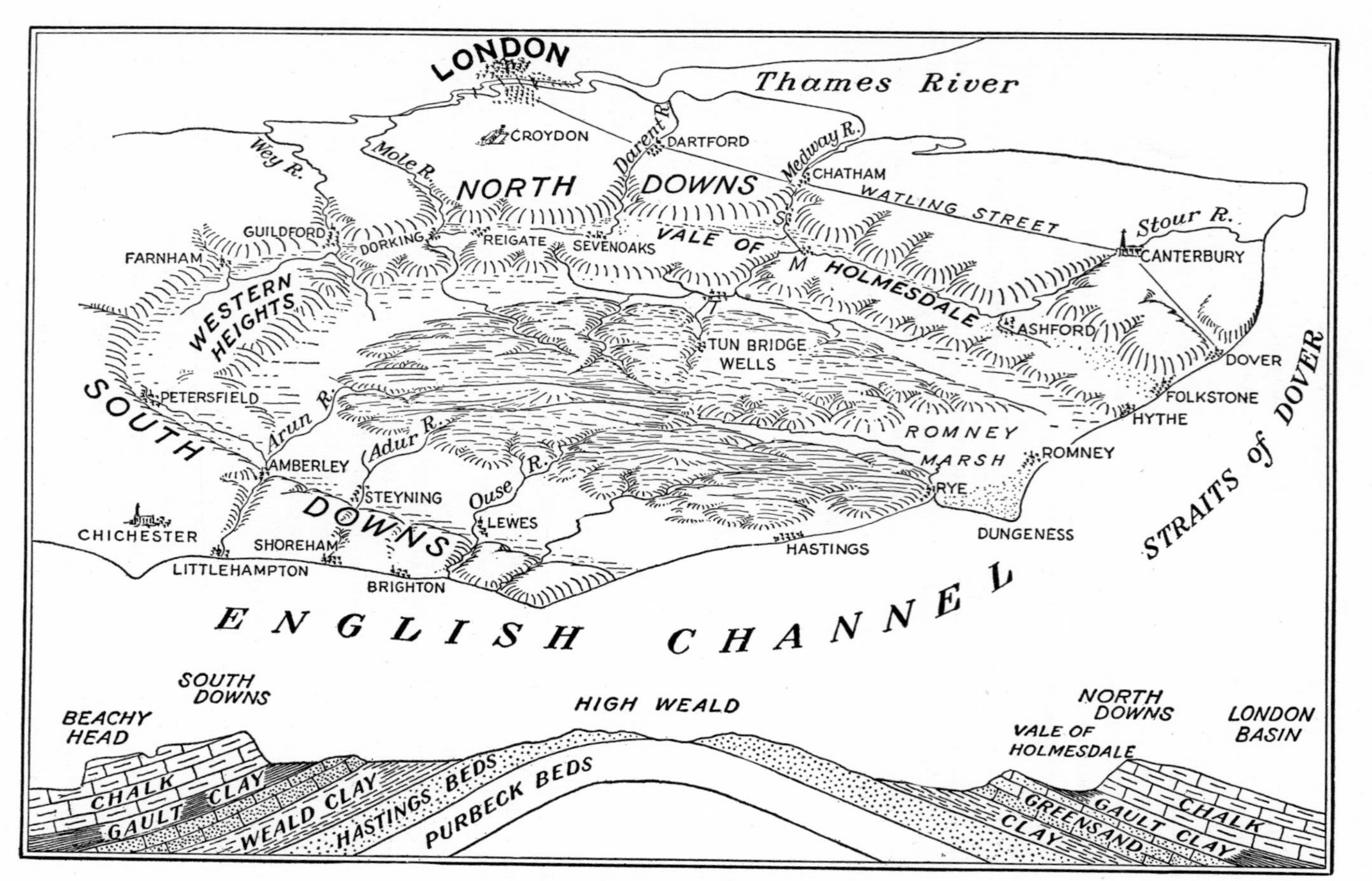


BLOCK DIAGRAM OF THE WEALD

## THE WEALD. PRESENT FORM

The Weald, a low, maturely eroded dome, comprises most of that part of southeastern England lying between London and the English Channel. It is a symmetrical arch, one end of which is transected by the Straits of Dover; the easternmost tip of the arch is represented by the Boulognnais region of France. The English portion of this uplift is about 90 miles long and half that wide.

The Weald is essentially a core of older sedimentary rocks surrounded by successive outcrops of newer formations, the strata dipping gently to the north and south, and giving rise to a series of scarped ridges and alternate vales roughly parallel to the anticlinal axis.

Rimming the Weald on the outside is an almost continuous chalk plateau or gently sloping hogback with a steep inner face, overlooking a very distinct lowland, known on the northern side of the Weald as the Vale of Holmesdale. The chalk scarp is designated on the north the North Downs and on the south the South Downs. It comes to the sea in the cliffs of Dover and again at Beachy Head.

A second circle of cliffs, facing toward the axis of the dome, is separated from the Chalk Downs by the Vale just mentioned. This inner hogback, made up of the Greensand, as shown in the geological section opposite, is not so continuous as the outer belt of cliffs but at the western end of the Weald it broadens out to form the Western Heights, which rise 1,000 feet above the sea and constitute the highest part of the entire region.

The central part of the Weald is a region of high hills, at one time an iron-mining region, and in general too rugged for easy agriculture. It is this region which has always supported the extensive woods to which the name *weald* was originally applied, the name having a common origin with the German word *Wald*, a wood or forest.

The diagram suggests the intimate control which the topography has upon the location of towns, roads, and railroads. The site of villages since prehistoric time has been determined largely by the presence of good water. The most favored locality is the narrow belt at the foot of the chalk escarpment. Here the water, which readily finds its way through the porous chalk, is arrested by the impervious underlying bed of clay, called the *Gault*, and springs result. The chalk uplands are practically devoid of streams.

Coal is mined in several localities between Canterbury and Dover. Deep workings penetrate the Carboniferous beds that occur here at less depth than at any place between this region and central England where the beds actually come to the surface.

Farming in the Weald is influenced by the nature of the rocks and the available water. Because of their porous soil the chalk uplands do not produce a rich pasture and the raising of sheep is one of the chief industries.

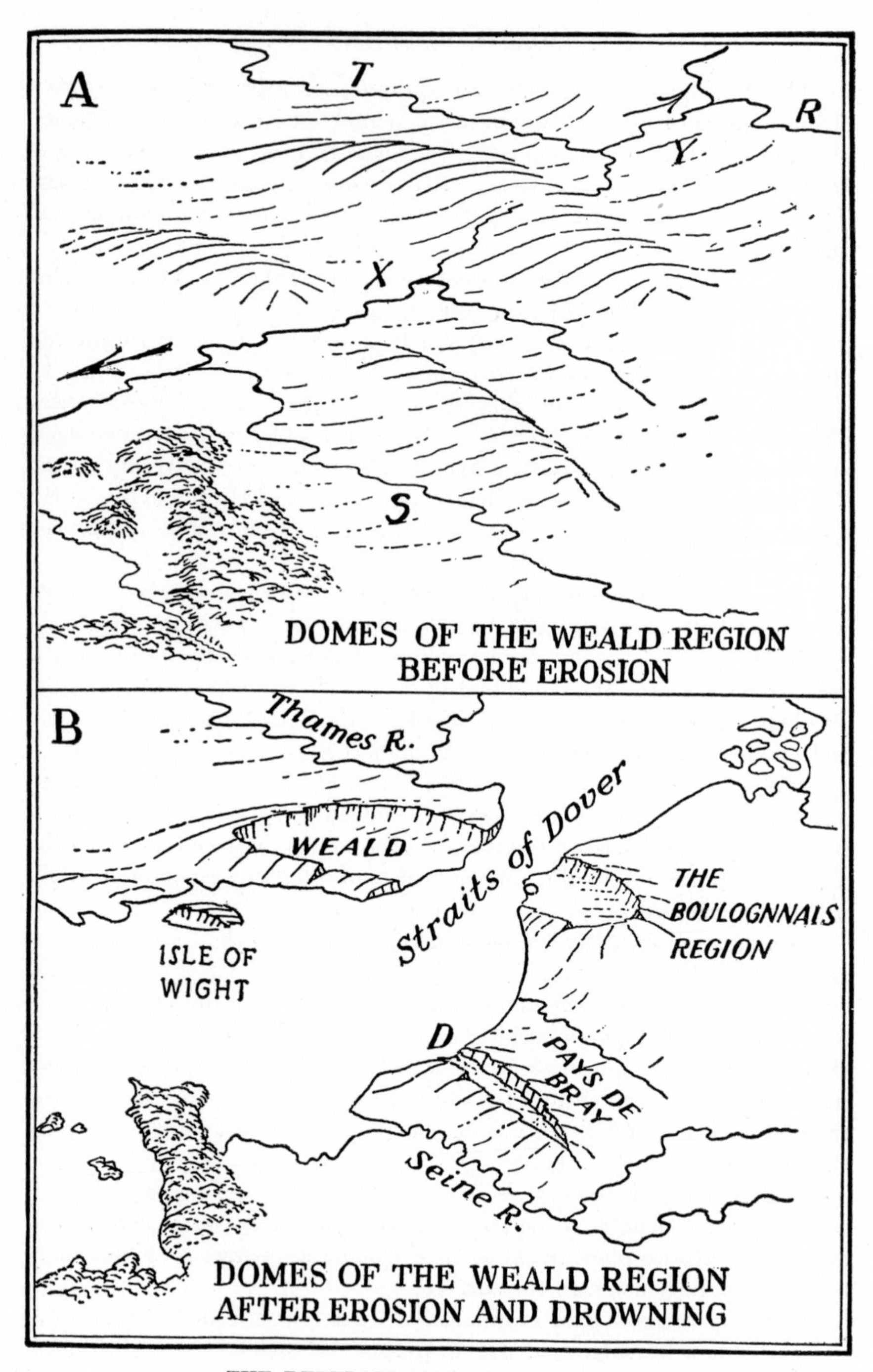


THE DEVELOPMENT OF THE WEALD

## THE WEALD. PHYSIOGRAPHIC DEVELOPMENT

ONE-CYCLE EXPLANATION OF THE WEALD. The Weald is one of several parallel arches, trending northwest-southeast in southern England and northern France. *A* suggests the initial form of these folds, the tops of which must have been 3,000 feet higher than the present altitude of the region. From the crest of the fold now forming the Weald two consequent streams, *X* and *Y*, flowed in opposite directions, one to the southwest, the other to the northeast. They received as tributaries other consequent streams from the synclinal valleys on the north and south sides. Three of these tributary streams (*S*, *T*, and *R*) will be recognized as the predecessors of the Seine, the Thames, and the Rhine. Erosion eventually breached the summit of the Weald as well as the crests of the smaller arches. By a succession of stream captures, the drainage lines were shifted to the axes of the anticlines, exactly as in the development of drainage in folded mountains. Wind gaps still remain transecting the ridges which rim the different basins, attesting the capture of earlier consequent streams by the later subsequent ones.

The present configuration of the land is the result of recent submergence. Were the submergence only slightly less, the English Channel would be an estuary like Bristol Channel. The Straits of Dover, the narrowest part of the English Channel, are situated exactly on the axis of the Weald uplift; here the water is so shallow that any English cathedral standing on the sea bottom would reveal most of its steeple to passing ships.

The Isle of Wight, of the same chalk formation as the South Downs, is the north limb of one of the arches. Between the Isle of Wight and the Weald is Salisbury Plain, a synclinal basin similar to London Basin north of the Weald.

In France the so-called *Pays de Bray* is an anticlinal valley, drained by a small river which emerges at the coast at Dieppe (*D*). Boulogne, farther east, occupies a similar location on the eroded French portion of the Weald, called here the *Boulognnais*.

On both sides of the English Channel the erosion of the various arches has, in general, not resulted in the formation of valleys below the level of the earlier synclines. As a result, the later drowning has submerged only the synclines to form such estuaries as the Seine, the Solent at Portsmouth, and the Thames.

TWO-CYCLE EXPLANATION OF THE WEALD. The explanation given above suggests that stream erosion of the Weald went on continuously from the time of uplift until the present-day forms resulted. Close study of the region suggests that it was peneplaned and partially covered by a sheet of marine sediments, upon the surface of which the present drainage lines took their courses as consequent streams. This suggestion is closely similar to that proposed for the development of drainage systems in the folded Appalachians of Pennsylvania, namely, superposition from a marine cover.

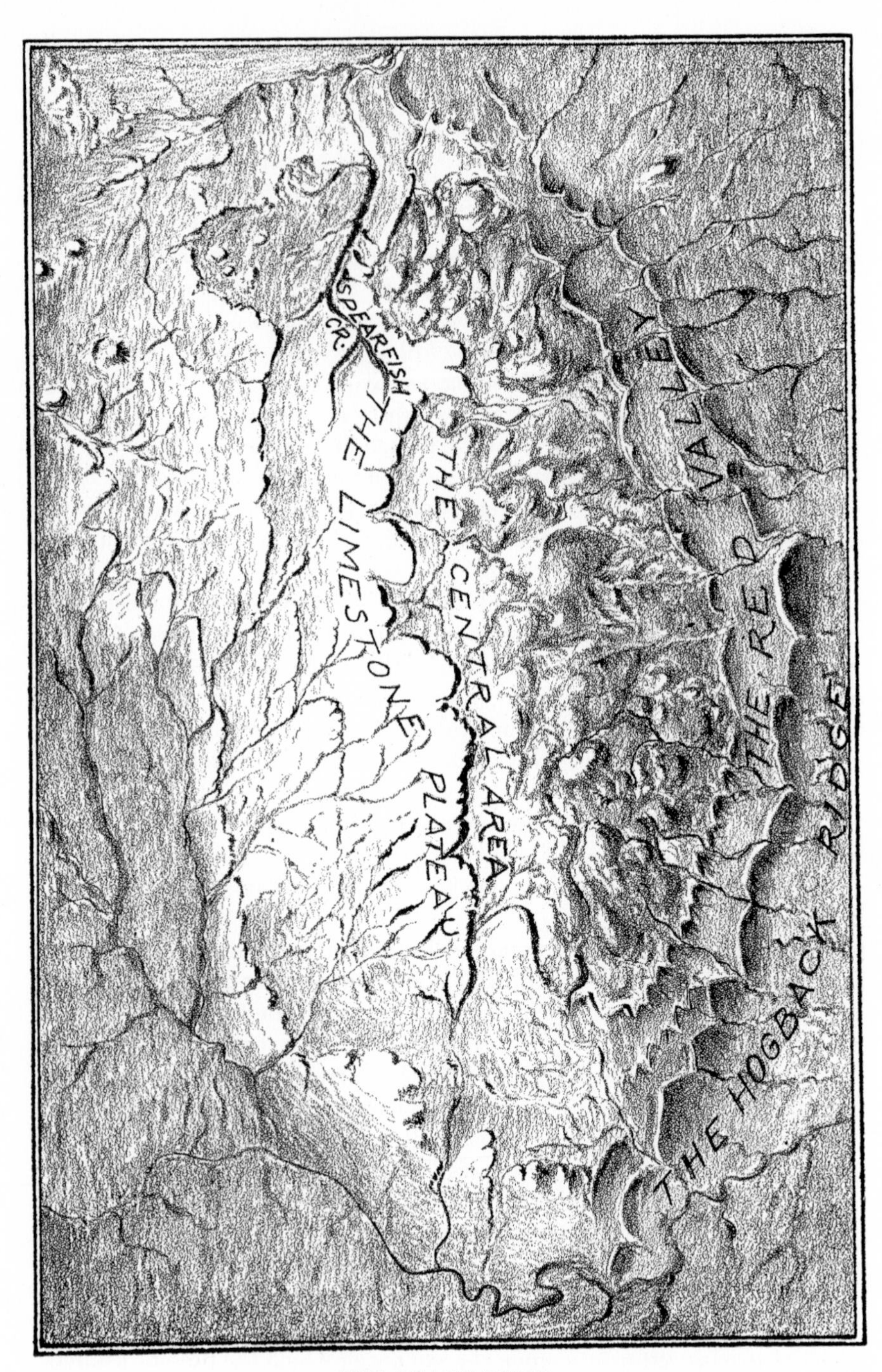


THE BLACK HILLS

## THE BLACK HILLS, A MATURELY DISSECTED DOME

The Black Hills uplift is probably the finest known example of a maturely dissected dome. Lying on the border between South Dakota and Wyoming, it has a length of over 100 miles and a width of about half as much. It is almost perfectly elliptical in shape.

The eroded mass now exhibits in the so-called *central area* a core of granite and other crystalline rocks. This constitutes the highest part of the whole region, rising 4,000 feet above the plains surrounding the dome.

THE LIMESTONE PLATEAU. Immediately rimming the granite core is a platform of limestone which is actually higher than most of the granite area except for certain outstanding summits of the granite area like Harney Peak. On the western side of the dome the limestone plateau is 15 to 20 miles in width and is almost a level upland except along its westernmost margin where it dips down with the dip of the rocks on the western flank of the dome. The escarpment marking the eastern edge of the plateau rises 800 feet above the parklike valleys of the crystalline area. The streams rising on the surface of the plateau have eroded deep canyons, with precipitous walls hundreds of feet high, notably the canyon of Spearfish Creek which is more than 1,000 feet deep. The corresponding limestone belt on the eastern side of the dome narrows to a ridge or hogback with a steep westerly facing scarp and a more gentle slope eastward down the dip. Near the outer margin of the limestone plateau is a low ridge or cuesta formed by the next higher resistant bed (the Minnekahta limestone) which is marked by an inward-facing escarpment 50 to 60 feet high. Both of the limestone escarpments or hogbacks are intersected by many stream gaps with canyonlike aspects.

THE RED VALLEY. Almost continuously surrounding the Black Hills is the Red Valley, bounded by the limestone slopes on the inner side toward the dome and by a very steep-sided hogback ridge on the outer side. It is one of the most conspicuous features of the Black Hills, a wide valley of bright red earth, bearing few trees and known to the Indians as the "Racetrack."

THE HOGBACK RIDGE. The Hogback Ridge, forming the outer rim of the Black Hills, is a single-crested ridge of hard sandstone, the Dakota sandstone, usually steeply dipping but in places flattening out into a gently sloping plateau. Its crest stands 400 to 500 feet above the Red Valley and is a dominant feature of the Black Hills landscape. Many streams have cut their way across this prominent hogback on their way to the surrounding plains.

Beyond the hogback ridge the rocks of the plains are gently inclined. A few of the strong strata produce inward-facing, cuesta-like escarpments encircling the uplift and separated from each other by annular valleys.

VOLCANIC MOUNTAINS. Numerous small laccolithic hills or mountains surmount the northern flanks of the Black Hills dome. Where maturely eroded they repeat in miniature the features that the Black Hills display as a whole.

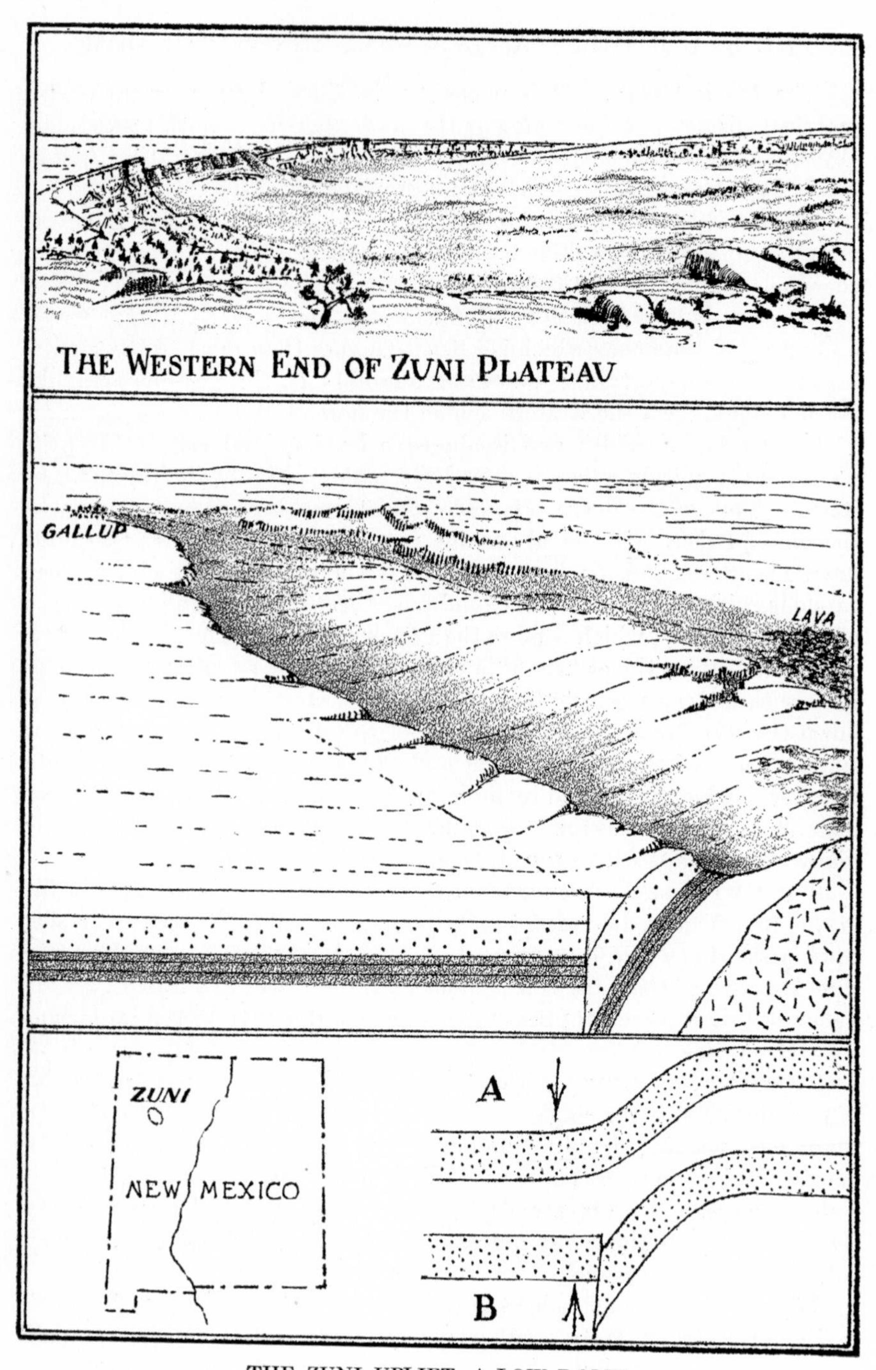


THE ZUNI UPLIFT, A LOW DOME

## THE ZUNI UPLIFT

In northwestern New Mexico, within the limits of the Colorado Plateau, there is a tract of country covering about 1,500 square miles which has been bowed up above the level of the surrounding regions. Like the Black Hills, it is elliptical in shape, being 60 miles or so in length and half that wide. The uplift appears to have been caused by an intrusion of granite, for a large area of granite is now exposed in the center, and within the granite are chunks and masses of the beds through which the granite made its way. Moreover, the sandstone beds now in contact with the granite have in places been much metamorphosed so as to be indistinguishable from a quartz porphyry of igneous origin. There are, in fact, all shades of metamorphism.

AMOUNT OF UPLIFT. FAULTING. When first formed, the dome must have risen several thousand feet above the surrounding country but erosion has stripped away a large volume of the capping rocks and at present the difference in relief is not much more than a couple of thousand feet.

On the northern side the beds dip very gently at an angle of 5° or less away from the axis of the dome, but on the southwestern side the beds are sharply upturned with dips of 75° so that they appear to stand out almost vertically. Where the dip is very steep, too, faulting in some sections seems also to have occurred.

An unusual feature about this fault is that the beds on the outside of the dome seem to have been uplifted. It would appear that the monocline was first formed with the beds strongly bent, as shown in Fig. *A*. This was followed by faulting in the reverse direction (Fig. *B*). What is now the upthrown side of the fault was formerly the downthrow of a monocline of older age.

CONCENTRIC BELTS. Because erosion has removed so extensive an area of the uppermost sedimentary beds, thus exposing the lower strata, the formations now outcrop in concentric belts, much as they do in the Black Hills. Here and there outliers of the higher beds are scattered around the central part of the area. This central portion, being considerably elevated and being also dissected by deep canyons, has the aspect of a plateau and is usually called, therefore, the *Zuni Plateau.* The broad valley 4 to 6 miles in width, lying between the pine-covered plateau and the imposing red and variegated cliffs forming the cuestas (or hogbacks) to the north, serves to determine the location of the railroad which runs northwestward to Gallup, much as do some of the valleys rimming the Black Hills. In contrast with the valleys in the Black Hills, there is a large portion of this great valley which is covered over by an extensive lava field of very recent age.

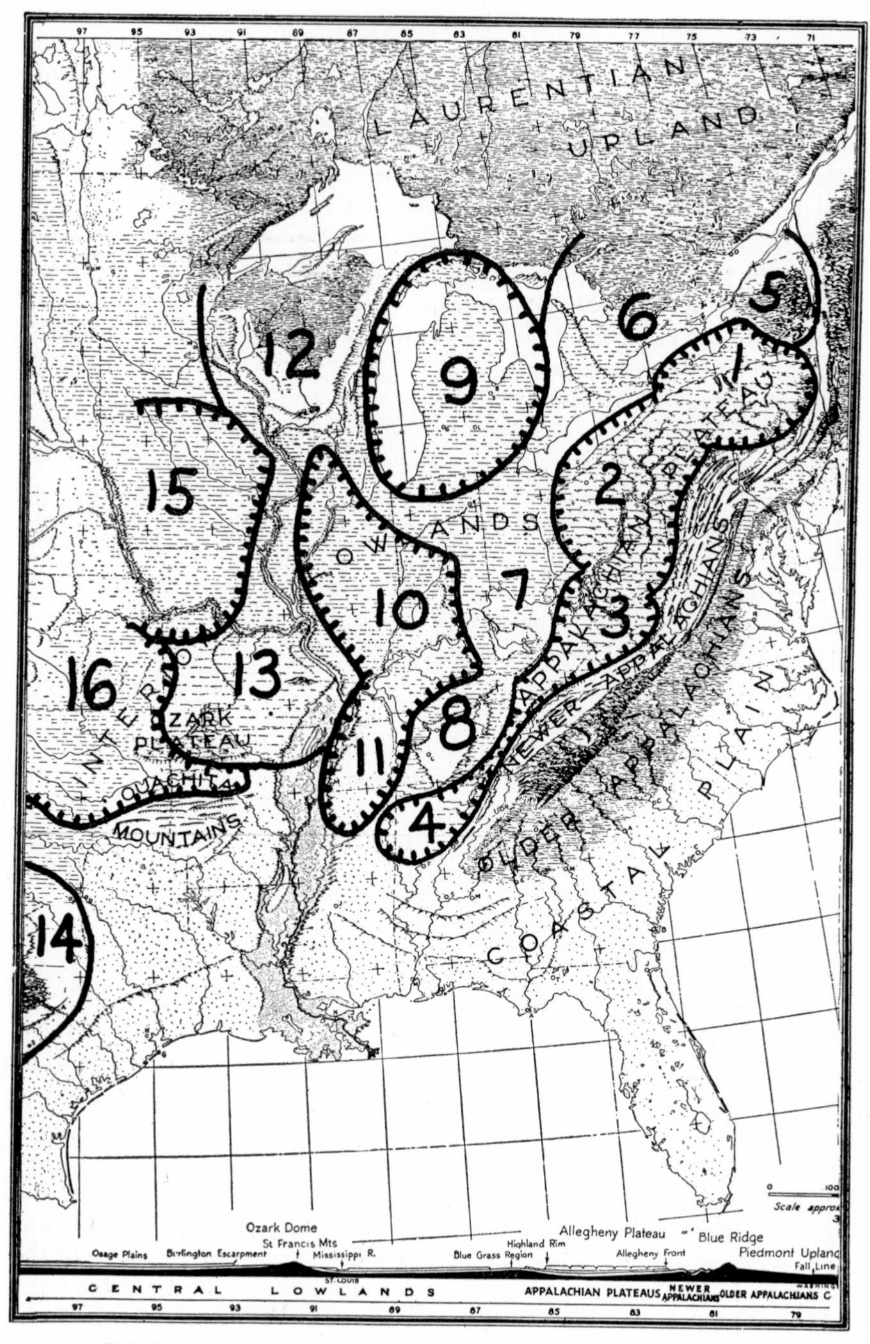


DOMES AND BASINS OF THE EASTERN UNITED STATES

## DOMES AND BASINS OF THE EASTERN UNITED STATES

The eastern United States is marked by a number of distinct basins and domes forming five recognizable belts, trending in general north and south. On the accompanying map the basins are shown by hachures. The easternmost belt is a row of basins coinciding with the Appalachian Plateau. A second belt of basins lies east of the Mississippi River. The third belt is the one farthest west on the border of the Great Plains. Alternating with the belts of basins are belts of low domes or narrow anticlines. All these domes and basins owe their present character to the Appalachian folding, but some had their beginnings in ancient geological time.

In accordance with the numbering on the map the basins and domes may be described as follows:

1. *The New York Basin* embraces the northern part of the Appalachian Plateau, which includes along its eastern border the Helderberg, Catskill, and Pocono Mountains, and along its northern edge the Allegheny escarpment.

2. *The Allegheny Basin* comprises most of western Pennsylvania and West Virginia and involves the great Allegheny bituminous-coal basin.

3. *The Cumberland Basin* of eastern Kentucky and Tennessee includes the eastern Kentucky coal field. Its southern end forms Walden Ridge.

4. *The Alabama-Mississippi Basin* is the southernmost portion of the Appalachian Plateau and includes several important coal fields such as the Warrior Coal Basin.

5. *The Adirondack Dome* is a highly complicated structure with a core of high-standing crystalline rocks, surrounded by sedimentary beds dipping away in all directions, and in places with strongly faulted margins.

6. *The Ontario Dome.* Around the southern margin the sedimentary beds curve in big arcs, producing distinct escarpments like the Niagara Cuesta.

7. *The Cincinnati Arch* is a very gentle dome eroded at its summit to expose the underlying limestone beds in the Blue Grass District of Kentucky.

8. *The Nashville Dome* is a similar gentle uplift.

9. *The Michigan Basin* is a mildly down-warped structure, preserving coal beds in its center.

10. *The Illinois Basin* is another important coal field which includes the coal basin of western Kentucky.

11. *The Western Tennessee Basin* lies between the Nashville Dome and Ozark Uplift.

12. *The Wisconsin Dome* is much like the Ontario Dome. Crystalline rocks are exposed in its central portion and cuestas curve around its southern end.

13. *The Ozark Dome* is a moderately strong uplift with crystalline rocks exposed in the St. Francis Mountains near its eastern side. The Mississippi River follows the uplift which connects the Wisconsin and Ozark Domes, erosion having here reduced the arches to a lower topographic level than the structural basins on either side.

14. *The Central Texas Dome* has an eroded basin in its center where crystalline rocks appear at the surface.

15. *The Iowa-Missouri Basin.*

16. *The Oklahoma Basin.* These last two basins represent the Western Interior Bituminous Coal Field.

DOMES AND BASINS OF THE WESTERN UNITED STATES

## DOMES AND BASINS OF THE WESTERN UNITED STATES

The Rocky Mountain belt is a region of many domes and basins. The domes are of considerable height and form true dome mountains, with strongly upturned sedimentary beds around their flanks and surrounding the much-dissected crystalline core.

Within the basin areas, where the beds dip in general toward the center away from the surrounding uplifts, there are many minor and low domes which are partially eroded.

A glance at the accompanying map reveals, in a rather crude arrangement, three belts trending in a north-south direction: two belts of basins and one belt of domes. The easternmost belt is a series of basins comprising the Great Plains, with certain extensions running into the mountains. West of this belt is the series of domes making up the Southern Rockies of Colorado and the so-called *Middle Rockies* of Wyoming. Still farther to the west is the plateau region, which again is essentially a series of basins, interrupted by several domes. These various basins and domes may be listed in order as follows:

1. *The Northern Great Plains* or *Missouri Plateau.*
2. *The Yellowstone Basin.*
3. *The Bighorn Basin.*
4. *The Powder River Basin.* These last three are extensions of the Missouri Plateau.
5. *The High Plains,* a long broad syncline not interrupted by any uplifts.

6. *The Black Hills Dome,* an outlier of the Rocky Mountains with which it is connected by
7. *The Hartville Uplift,* a low structural arch.
8. *The Front Range* of the Colorado Rockies, the highest of all the Rocky Mountain domes, branching northward to form the Laramie Range and the Medicine Bow Range.
9. *The Sawatch Range,* ending at the north in the Park Range. This is more or less separated from the Front Range by a series of basins, notably North Park, Middle Park, and South Park, which are too small to be numbered on the map.
10. *The Sangre de Cristo Range,* a long uplift similar to the Sawatch.

11. *The Wyoming Basin,* broken up by minor domes, like the Rock Springs Uplift, into several parts, notably the Washakie and Bridger Basins.
12. *The Uinta Basin.*
13. *The Colorado Plateau* which contains several minor domes.

14. *The Little Rocky Mountains,* one of several smaller domes isolated from the main Rocky Mountain area.
15. *The Big Belt Mountains.*
16. *The Bighorn Range,* a continuation northward of the Colorado Rockies.
17. *The Wind River Range.*
18. *The Uinta Range.*
19. *The Henry Mountains.*
20. *The Kaibab Arch.*
21. *The Zuni Uplift.*

*U. S. Geological Survey*

BEAR BUTTE, BLACK HILLS, S. D., A SMALL LACCOLITE SHOWING UPTURNED BEDS

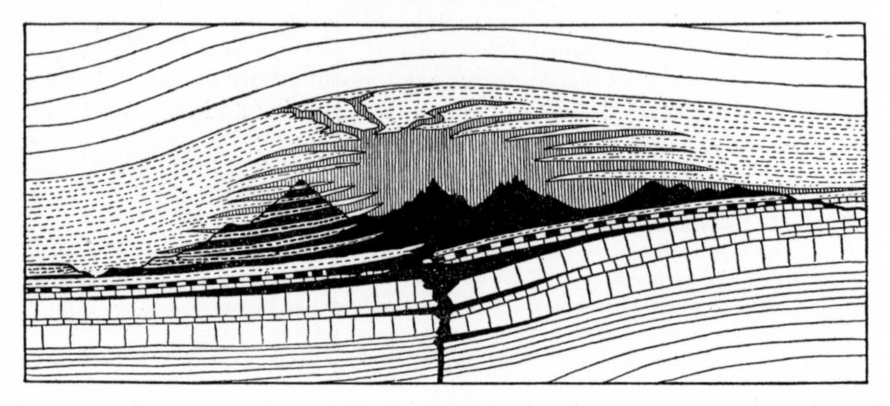

SECTION OF LA PLATA MOUNTAINS, COLO., A LACCOLITE WITH MANY SILLS

## LACCOLITHIC DOMES

Among the unusual types of domes are those produced by laccolithic intrusions. These are apt to be very small affairs, possibly not more than a mile or two in diameter and ranging up to 5 or 6 miles across. The smaller ones, after erosion, rise 1,000 feet or so above the adjacent country; the larger ones may be several times that high.

Laccoliths are frequently accompanied by intruded sheets. When erosion occurs, these layers of igneous rock, because of their greater hardness, stand in relief and produce hogbacks or ridges.

Laccoliths have a vast variety of shape and mode of occurrence, some being thin, almost like sills, others being thick and stocklike, and cutting across the intruded beds.

Many small laccolithic domes, or laccolites, occur in the northern portion of the Black Hills. Others of larger size occur on the Colorado

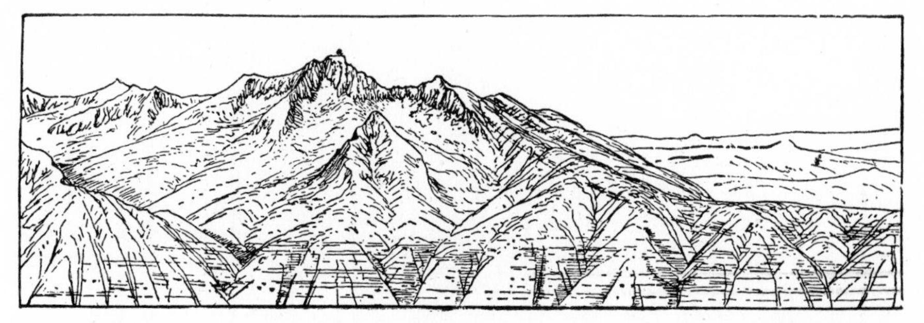

U. S. Geological Survey

WEST ELK MOUNTAINS, COLO., SHOWING UPTURNED STRATA ON FLANK OF LACCOLITE

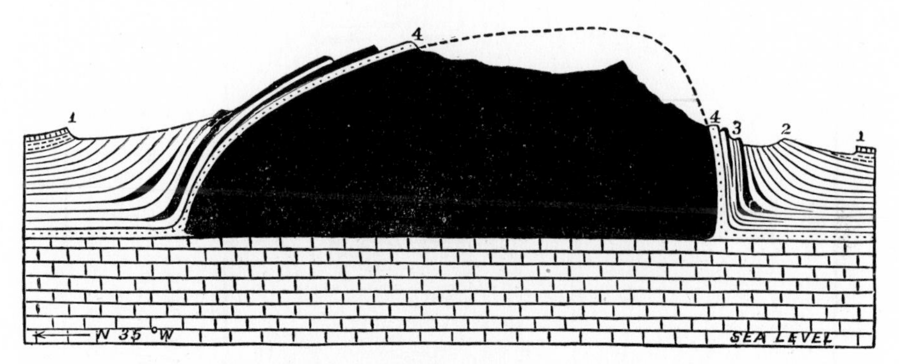


SECTION OF HILLERS LACCOLITE, MOUNT HENRY REGION, UTAH, SHOWING SIMPLE STRUCTURE

Plateau in Colorado, Utah, and Arizona. Among these, the Henry Mountains are the largest. They have been famous since the days they were first described in 1877.

In the Black Hills there is every gradation in a series ranging from an unbroken dome of stratified rock arching over the summit of a concealed mass of plutonic rock, to imposing towers of columnar rhyolite exposed by the removal of the softer strata into which it was intruded. The first in the series is *Little Sun Dance Dome,* a mile in diameter, the outer layers of which have been removed and the inner ones, now forming the crown of the dome, deeply gashed by erosion, but not enough to expose the top of the igneous plug which presumably exists beneath. The other extreme is the *Devil's Tower*. Here the arch of stratified rock which once surmounted the summit of the plutonic plug has been completely removed.

*Holmes, U. S. Geological Survey*

ABAJO MOUNTAINS

A maturely dissected laccolithic dome in the Colorado Plateau, eastern Utah. This small dome has a diameter of about six miles, and the highest summits rise 3,000 to 4,000 feet above the surrounding plateau. The sharply upturned Dakota sandstone beds are shown at *c*, *c* in the middle of the view at the base of Abajo Peak. Elsewhere the Cretaceous shales *d*, *d*, *d* dip away more gently from the mountains.

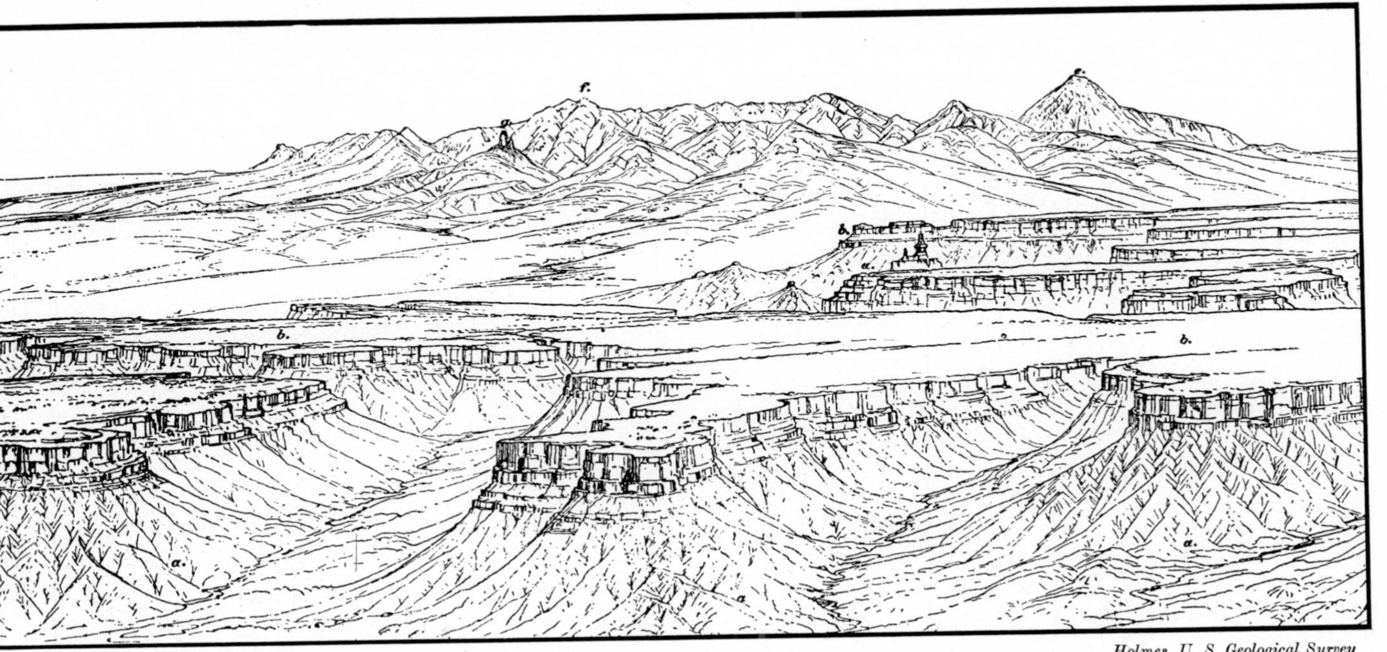

*Holmes, U. S. Geological Survey*

EL LATE MOUNTAINS

A maturely dissected laccolithic dome in the Colorado Plateau, southwestern Colorado. The plateau consists of horizontal rocks which are turned up at the base of the mountains. The higher peaks rise somewhat more than 3,000 feet above the plateau surface. The central core of the mountain mass consists of several varieties of igneous rocks, and there are also layers of fine-grained igneous rocks interbedded between the upturned sandstones and shales. These form minor hogbacks and ridges.

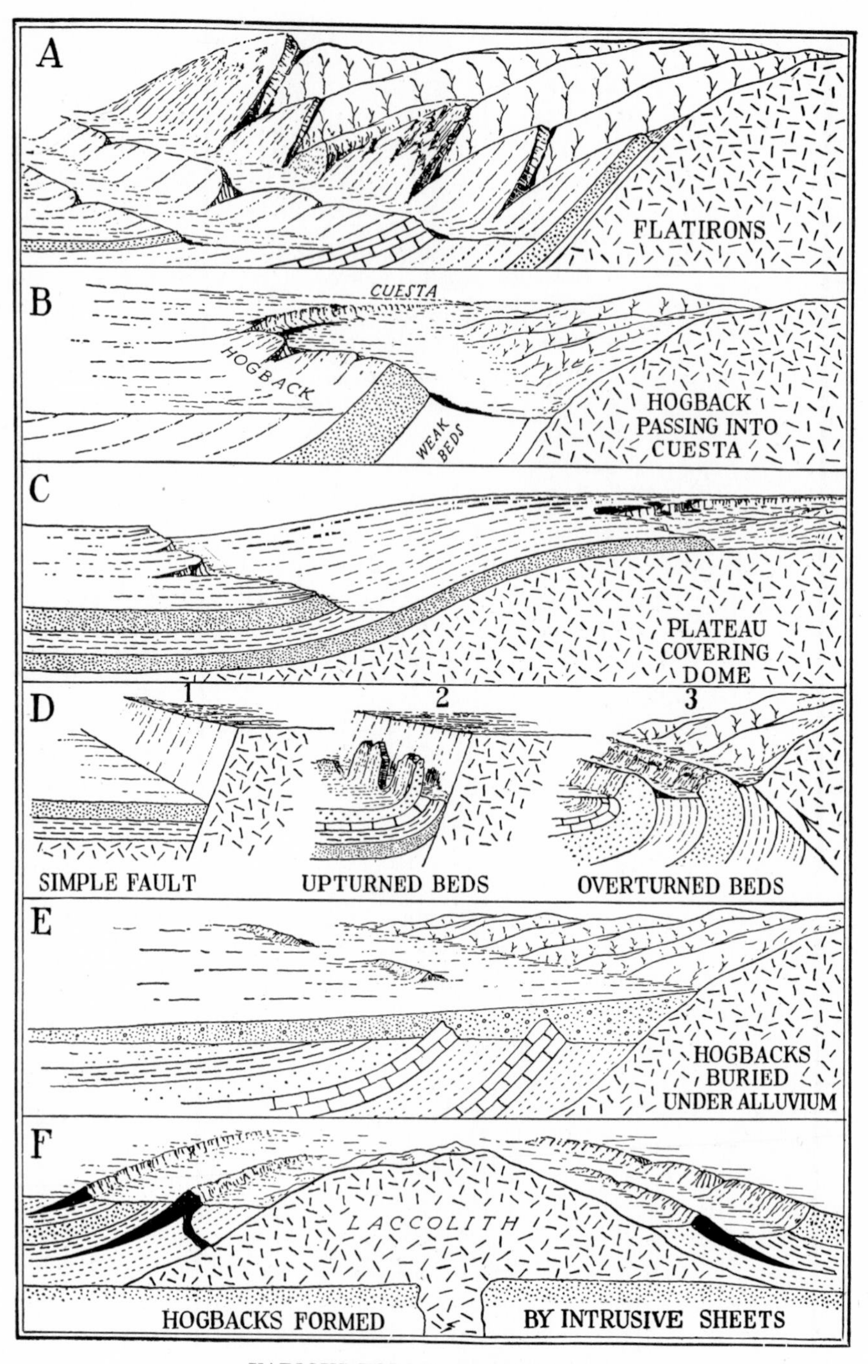


VARIOUS FORMS OF HOGBACKS

## KINDS OF HOGBACKS

FLATIRONS. When the lowest member of the sedimentary layers in a dome mountain is resistant, then it behaves as if it were an integral part of the underlying crystalline rock and the first hogback is not separated from the central core of the dome by a valley. The result is a series of so-called *flatirons*, which appear to be plastered on the ends of the spurs, as shown in Fig. *A*. A flatiron of this type may actually project higher than the top of the main range, as Ed Peak does above the Bighorns.

HOGBACKS AND CUESTAS. When the lowest layer of sedimentary beds is weak, a lowland is eroded between the hogback and the crystalline massive. Figure *B* shows how this hogback may gradually pass into a cuesta where the beds are not so sharply upturned. In that case the lowland at the foot of the cuesta will probably have a greater width than it does where the beds are vertical.

PLATEAU UPLANDS. The lowest members of the sedimentary series may rise over the crest of the dome and be preserved as a plateau or upland, higher in altitude than the crystallines, as shown in *C*.

FAULTING. The sedimentaries may be faulted off in different ways. A simple normal fault, without any drag in the sedimentary beds, will cut them off squarely, so that no hogback can be developed, as in *D*1. Or the beds may be sharply upturned to produce vertical wall-like ridges like the entrance pylons to the Garden of the Gods. Still more, if the crystallines are thrust outward, the sedimentary beds may be strongly overturned. After erosion, the resulting hogbacks present their steep faces away from the uplifted mass contrary to what is usually the case. This is pictured in *D*3 and is represented by the remarkable array of overturned hogbacks along the front of the Rockies in Montana south of the Lewis overthrust.

BURIED HOGBACKS. Where extensive alluvial deposits are built out from the mountains, the hogbacks may be completely buried, as in Fig. *E*. They may become exhumed by later erosion.

IGNEOUS HOGBACKS. Finally, in the case of laccolithic intrusions, there may be many accompanying sheets and lenses of igneous rock. These, under erosion, may prove to be more resistant than the adjacent sedimentary beds and may remain in relief as small hogbacks. This type of feature occurs in the Henry Mountains.

LIMESTONE HOGBACKS. An unusual type of hogback scarp is represented in the Karst district of Istria and is mapped on page 539. The central part of this dome has been stripped of its original limestone cover and now consists of beds of shale and clay supporting surface-flowing streams. The limestone, however, outcrops around the margin of the uplift and there the streams coming away from the dome flow into amphitheater-like pockets and disappear underground along the contact between the limestone and the underlying impervious beds.

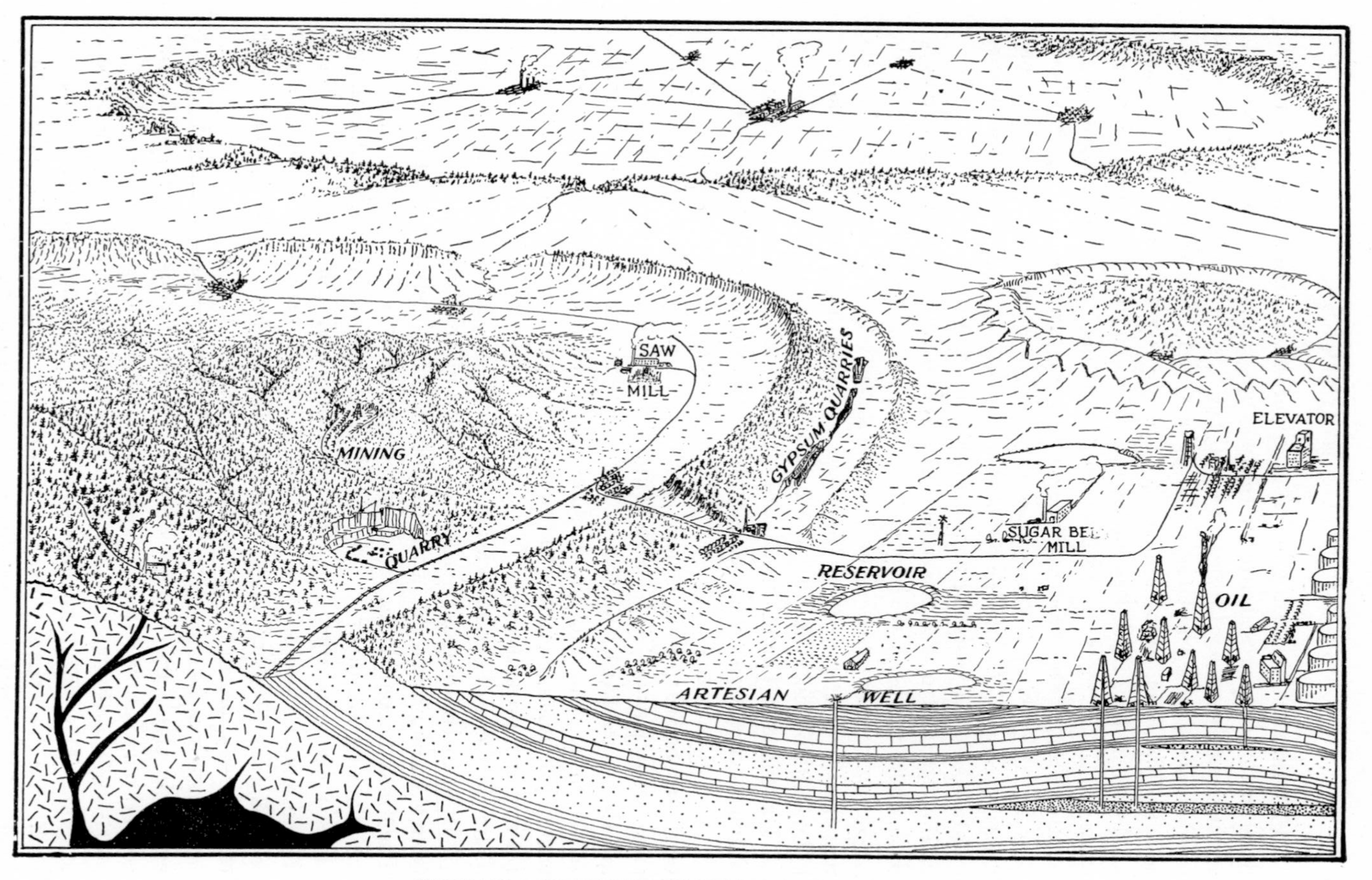


HUMAN ACTIVITIES IN REGIONS OF DOMES

## GEOGRAPHICAL ASPECTS

As in the case of most other types of mountains, the geographic effects of dome structures are due both to topography and to the natural resources of the region.

TOPOGRAPHY. The ringlike pattern of the features in dome mountains is always reflected in the plan of railroads and roads. There is little other choice of location. In the Black Hills the railroads form almost a complete circle about the hills. In the Weald the line from Folkestone to London follows the chief subsequent valley. In the Adirondacks the Ogdensburg division of the New York Central runs along the Black River Valley between the crystalline core of the mountains and the upturned sedimentaries around the western flank.

FORESTS. The central, more rugged portion of most dome mountains is usually well timbered but sparsely settled in spite of its greater natural charm. It usually has a rainfall more copious than the surrounding foothills but the greater degree of relief makes agricultural activities difficult. The name *Black Hills* was suggested by the heavily wooded interior section. The crests of the hogback ridges support an open growth of small pines and other evergreens but the circular valleys are open and parklike.

MINING. In the central portion of dome mountains where crystalline rocks are exposed, mining may be the most significant activity. In the Black Hills several hundred millions of dollars' worth of gold have already been extracted and there are many other minerals, such as silver, lead, copper, iron, tin, and tungsten, as well as many of a nonmetallic character. Smaller domes, due to laccolithic intrusions, are usually not rich in mineral deposits.

OIL AND GAS. Domes customarily provide suitable structures for the occurrence of oil, even though the oil-bearing strata are still deeply buried. This, of course, would not be true in the case of large domes with crystalline cores like the Black Hills, but subsidiary domes around the flanks of larger uplifts frequently have rich yields of oil. The *Teapot Dome* and the *Rock Springs Uplift* in Wyoming contain valuable reservoirs of oil and gas. They are low domes not greatly eroded but are surrounded by high ranges like the Laramie Range and the Wind River Range, from which the sedimentary strata have been stripped away.

Some low domes eroded at the center to form extensive basins may be regions of great agricultural interest. The famous *Blue Grass region* of Kentucky lies in the center of the eroded *Cincinnati Arch* and owes its fertility to the beds of limestone, rich in phosphates, which are there exposed. This unusually productive portion constitutes the *Lexington Plain* which is rimmed by a less fertile tract called the *Outer Blue Grass region*. In the distant portion of the accompanying illustration a low eroded dome is represented with several towns, from each one of which radiate roads in all directions across the rich agricultural lowland.

## MAPS ILLUSTRATING DOME MOUNTAINS

Unusually interesting examples of young salt domes are shown on the *Derouen*, *Belle Isle*, *Jeanerette*, and *Bayou Sale*, *La.*, sheets of the Mississippi River Commission. The circular areas on the *Stuttgart*, *Ark.* (U.S.G.S.), map may be of the same type.

Maturely dissected domes of a size sufficiently small to be represented entirely on one topographic sheet are shown on the following maps of the U.S. Geological Survey: the *Terlingua*, *Texas*, sheet which depicts The Solitario, one of the most perfect circular domes known; the *Watrous*, *N.Mex.*, sheet portraying the Turkey Mountain region with its annular drainage; the *Wingate*, *N Mex.*, sheet which portrays the Zuni Mountains, a particularly fine dome; the *Henry Mountains* and *San Rafael*, *Utah*, sheets, showing the laccoliths of Mount Ellsworth and Mount Ellen with small concentric hogback ridges; the *Medicine Bow*, *Wyo.-Colo.*, sheet, depicting the Sheep Mountain dome with its rimming hogbacks. Several maps portray domes which have been eroded to produce basins in their central parts, as, for example, the *Oregon Basin*, *Meeteetse*, and *Grass Creek Basin*, *Wyo.*, sheets. The *Axial*, *Danforth Hills*, and *Monument Butte*, *Colo.*, maps all display parts of the Axial Basin dome. The famous Kettleman Hills oil domes are shown on the *Avenal Gap* and *Las Viejos Hills*, *Calif.*, sheets. The *Paradox Valley*, *Colo.*, map shows several eroded domes.

Many members of the Rocky Mountain system are great domes, too large to be shown on a single topographic sheet. The topographic maps covering their flanks illustrate hogbacks of several forms. The *Boulder*, *Colo.*, sheet shows hogbacks of the flatiron type; the *Denver*, *Fort Collins*, *Loveland*, and *Livermore*, *Colo.*, sheets all display splendid series of large hogbacks dipping eastward away from the mountainous uplift.

The *Sherman*, *Wyo.*, sheet shows some of the vertical hogbacks along the east side of the Bighorn Range, and the *Heart Butte* and *Saypo*, *Mont.*, sheets show overturned hogbacks dipping to the west because of great thrusting from that direction.

The *Rapid City*, *S. Dak.*, sheet shows hogbacks around the flanks of the Black Hills with a well-defined lowland, part of the Red Valley which separates the hogback belt from the mountainous core. One of the largest of all hogbacks is shown on the *Grand Hogback*, *Colo.*, sheet, the Grand Hogback being transected by several gaps.

Hogbacks are shown less conspicuously on the *Canon City*, *Manitou*, and *Platte Canyon*, *Colo.*, sheets. On the Manitou sheet the hogbacks stand vertically and form the well-known features in the Garden of the Gods.

Dome features in the Folded Appalachians are shown on the *Davis*, *W. Va.-Md.*, sheet in Canaan Valley and on the *Williamsport*, *Pa.*, sheet where Nippenose Valley and Mosquito Cove are eroded domes.

Note also Burke Garden on the *Pocahontas*, *Va.-W. Va.*, sheet.

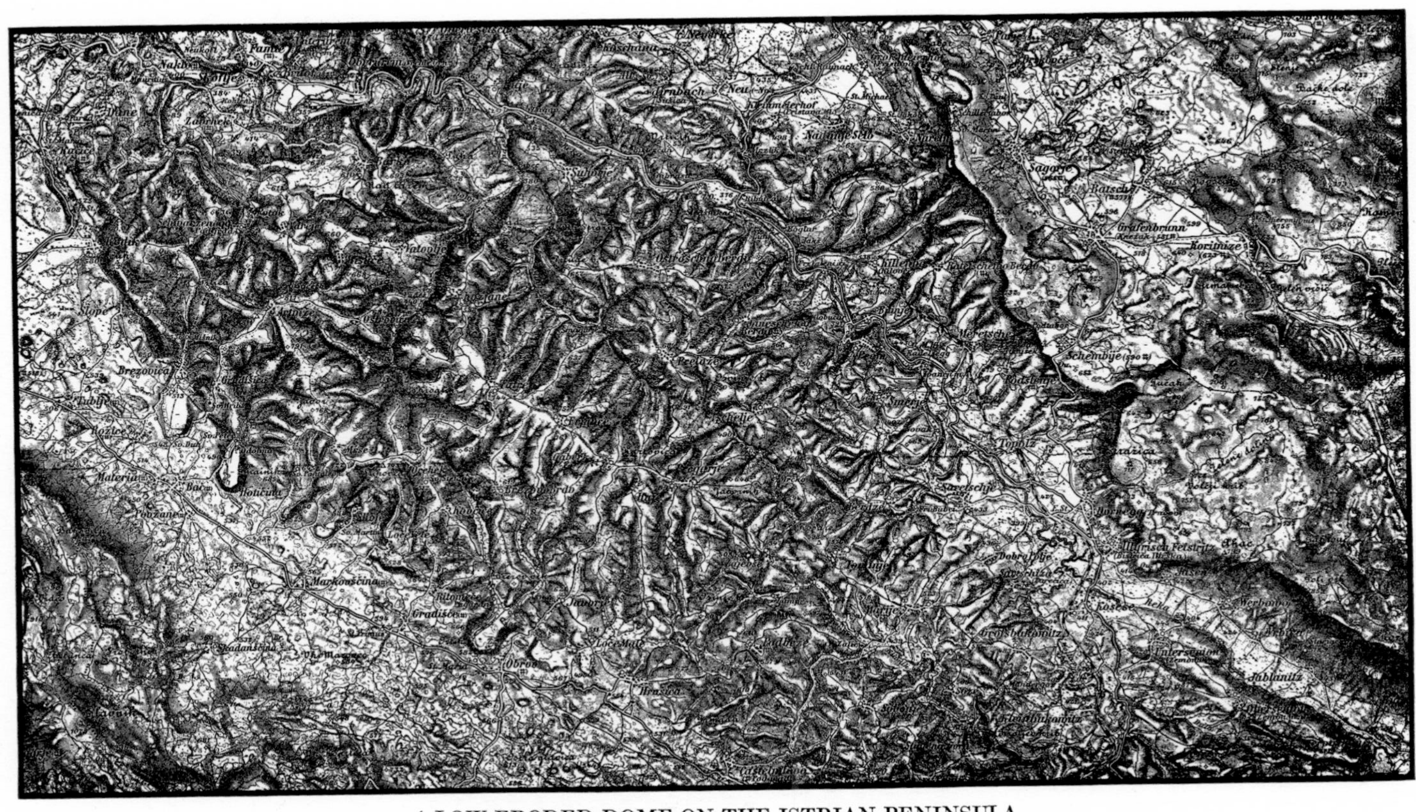


A LOW ERODED DOME ON THE ISTRIAN PENINSULA
Showing rimming limestone scarps and shale hills in the center
Austria; *Sesana and St. Peter* sheet, Zone 23, Col. X (1:75,000), reproduced on one half that scale.

## QUESTIONS

1. Draw cross sections of the following types of dome:
   *a.* An uneroded low dome.
   *b.* A low dome eroded along its axis to produce a topographic basin, floored by sedimentary rocks.
   *c.* A low dome eroded along its axis to produce a double basin, the inner basin being surrounded by a platform or rim. How does this correspond with the Cincinnati Arch?
   *d.* A dome strongly uplifted but not eroded, formed by a large batholith.
   *e.* The same dome eroded to expose high crystalline core. Also hogbacks and lowlands.
   *f.* The same dome with the crystalline core worn down lower than the surrounding rim of sedimentary rocks. Name an example of this.
   *g.* A dome due to laccolithic intrusion, not eroded.
   *h.* The same dome, eroded, with hogbacks formed of both igneous and sedimentary beds.
   *i.* An eroded dome showing flatirons.
   *j.* A compound dome; one having subsidiary domes on its flank.
2. Indicate on a cross section of a low dome the location of a gas well; an oil well; and a well yielding water only. Indicate also how faulting might serve to impound a reservoir of oil.
3. Draw a contour map of a simple ridge-shaped hogback with a wind gap and a water gap.
4. Draw a contour map of a flatiron, or triangular-shaped hogback.
5. Is Stone Mountain, near Atlanta, Ga., a dome mountain? Are the domes in Yosemite Park dome mountains?
6. What and where is Teapot Dome?
7. What is the difference between the Nashville Dome and the Nashville Basin?
8. Is the Paris Basin a dome or a basin?
9. What is the difference between the Michigan Basin and the Nashville Basin?
10. Why do flatirons sometimes form instead of ridgelike hogbacks?
11. Why do hogbacks sometimes dip toward the dome instead of away from it?
12. Draw a map showing the stream system of a maturely dissected dome. Label the various genetic types of streams and indicate where stream capture has probably occurred.
13. What is the difference between a laccolith and a laccolite?
14. What is the scale of the aerial map on page 509? See *Avenal Gap* and *Las Viejos Hills, Calif.*, sheets.
15. Do crystalline rocks outcrop in the high part of the Sheep Mountain dome illustrated on page 505?

## TOPICS FOR INVESTIGATION

1. The mineral resources of the Black Hills.
2. The Blue Grass region of Kentucky. Soils; crops; industries; cities.
3. The Weald; its physiographic development.
4. Dome mountains. Actual structural conditions around the flanks of different domes.
5. Oil domes. Location; geologic character and economic aspects.
6. Laccolithic domes. Their origin and general characteristics.
7. Theories as to oil accumulation. History of the anticlinal theory.

# REFERENCES


## Dome Mountains in General

Davis, W. M. (1898) *Physical geography.* Boston, p. 169–171.

Grabau, A. W. (1920) *Text book of geology.* New York, part 1, p. 199–204, 722–729.

King, P. B. (1932) *An outline of the structural geology of the United States.* 16th Intern. Geol. Cong., Guidebook 28, p. 20–24.

## Salt Domes

Barton, D. C. (1933) *Mechanics of formation of salt domes with special reference to Gulf coast salt domes of Texas and Louisiana.* Am. Assn. Petr. Geol., Bull. 17, p. 1025–1083.

Goldman, M. I. (1933) *Origin of the anhydrite cap rock of American salt domes.* U. S. Geol. Surv., Prof. Paper 175, p. 83–114.

Goldston, W. L., Jr., and Stevens, G. D. (1934) *Esperson dome, Liberty County, Texas.* Am. Assn. Petr. Geol., Bull. 18, p. 1632–1654.

Hanna, M. A. (1934) *Geology of the Gulf coast salt domes.* Problems of petroleum geology (Sidney Powers memorial volume), Am. Assn. Petr. Geol., p. 629–678.

Moody, C. L. (1931) *Chestnut dome, Natchitoches Parish, Louisiana.* Am. Assn. Petr. Geol., Bull. 15, p. 277–278.

Moody, C. L. (1931) *Tertiary history of region of Sabine uplift, Louisiana.* Am. Assn. Petr. Geol., Bull. 15, p. 531–551.

Steinmayer, R. A. (1933) *Salt-dome possibilities.* La. Dept. Conservation, Bull. 22 (General Bull., Handbook, Minerals Div.), p. 17–30.

Torrey, P. D., and Fralich, C. E. (1926) *An experimental study of the origin of salt domes.* Jour. Geol., vol. 34, p. 224–234.

## Laccoliths

Cross, C. W. (1894) *The laccolithic mountain groups of Colorado, Utah, and Arizona.* U. S. Geol. Surv., 14th Ann. Rept., part 2, p. 157–241.

Daly, R. A. (1903–08) *Mechanics of igneous intrusion.* Am. Jour. Sci., 4th ser., vol. 15, p. 269–298; vol. 16, p. 107–126; vol. 26, p. 17–50.

Gilbert, G. K. (1877) *Report on the geology of the Henry Mountains.* U. S. Geog. and Geol. Surv. Rocky Mt. Region (Powell), p. 18–98.

Jaggar, T. A., Jr. (1901) *The laccoliths of the Black Hills.* U. S. Geol. Surv., 21st Ann. Rept., part 3, p. 163–290.

Kelley, V. C., and Soske, J. L. (1936) *Origin of the Salton volcanic domes, Salton Sea, California.* Jour. Geol., vol. 44, p. 496–509.

Knight, G. L., and Landes, K. K. (1932) *Kansas laccoliths.* Jour. Geol., vol. 40, p. 1–15.

MacCarthy, G. R. (1925) *Some facts and theories concerning laccoliths.* Jour. Geol., vol. 33, p. 1–18.

## Dome Mountains and Laccoliths—Specific Regions

Adams, G. I. (1901) *Physiography and geology of the Ozark region.* U. S. Geol. Surv., 22d Ann. Rept., part 2, p. 69–94.

Arkell, W. J. (1936) *Analysis of the Mesozic and Cenozoic folding in England.* 16th Intern. Geol. Cong., C.r., vol. 2, p. 937–952. Structure of Wealdan dome. Many references.

Bevan, A. (1929) *Rocky Mountain front in Montana.* Geol. Soc. Am., Bull. 40, p. 427–456. Overturned hogbacks.

Cross, C. W. (1905) *Description of the Rico quadrangle, Colorado.* U. S. Geol. Surv., Folio 130.

Cross, C. W., and Spencer, A. C. (1899) *Description of the La Plata quadrangle, Colorado.* U. S. Geol. Surv., Folio 60.

Cross, C. W., and Spencer, A. C. (1900) *Geology of the Rico Mountains, Colorado.* U. S. Geol. Surv., 21st Ann. Rept., part 2, p. 7–165.

DARTON, N. H. (1901, 1909) . . . *geology and water resources . . . of the Black Hills . . .* U. S. Geol. Surv., 21st Ann. Rept., part 4, p. 489–599; Prof. Paper 65, 105 p.

DARTON, N. H., and PAIGE, S. (1925) *Central Black Hills.* U. S. Geol. Surv., Folio 219.

DUTTON, C. E. (1885) *Mount Taylor and the Zuñi Plateau.* U. S. Geol. Surv., 6th Ann. Rept., p. 105–198.

HERSHEY, O. H. (1901) *Peneplains of the Ozark highland.* Am. Geol., vol. 27, p. 25–41.

JILLSON, W. R. (1931) *Structural geology of northern central Kentucky.* Pan-Am. Geol., vol. 55, p. 337–341; Ky. Geol. Surv., ser. 6, vol. 43, p. 121–125.

KEMP, J. F. (1906) *The physiography of the Adirondacks.* Pop. Sci. Monthly, vol. 68, p. 195–210.

MATSON, G. C. (1909) *Water resources of the Blue Grass region, Kentucky, . . .* U. S. Geol. Surv., W.-S. Paper 233, p. 1–39.

MICHENER, C. E. (1934) *The northward extension of the Sweetgrass arch.* Jour. Geol., vol. 42, p. 45–61.

NEWTON, H., and JENNEY, W. P. (1880) *Geology of the Black Hills.* U. S. Geog. and Geol. Surv. Rocky Mt. Region (Powell), 566 p.

POWELL, J. W. (1876) *Geology of the eastern portion of the Uinta Mountains.* U. S. Geol. and Geog. Surv. Terr. (Hayden), 218 p.

WEED, W. H. (1901) *Geology of the Little Belt Mountains, Montana.* U. S. Geol. Surv., 20th Ann. Rept., part 3, p. 387–461.

WEED, W. H., and PIRSSON, L. V. (1898) *Geology and mineral resources of the Judith Mountains of Montana.* U. S. Geol. Surv., 18th Ann. Rept., part 3, p. 485–556.

WILSON, C. W., JR. (1935) *The pre-Chattanooga development of the Nashville dome.* Jour. Geol., vol. 43, p. 449–481.

WILSON, C. W., JR., and SPAIN, E. L., JR. (1936) *Upper Paleozoic development of the Nashville dome, Tennessee.* Am. Assoc. Petr. Geol., Bull. 20, p. 1071–1085.

WOOLDRIDGE, S. W., and LINTON, D. L. (1938) *The influence of Pliocene transgression on the geomorphology of southeast England.* Jour. Geomorphology, vol. 1, p. 40–54.

# XVI
# BLOCK MOUNTAINS

*Royal Canadian Air Force*

A LONG FAULT-LINE SCARP, MACKENZIE PROVINCE, NORTHWEST CANADA, NEAR CORONATION GULF
With a general view of the lake-dotted Arctic peneplane.

*Shipler, Salt Lake City*

FRONT OF THE SPANISH WASATCH NEAR PROVO, UTAH

Showing a series of triangular facets at the ends of the spurs. Also, in the middle distance the level of the Provo shoreline, the lower of the two prominent Lake Bonneville beaches.

DEATH VALLEY, CALIF.

A graben surrounded by block mountains which are late mature or old in their stage of dissection.

*Willard*

*A. K. Lobeck*

FRONT OF SONOMA RANGE, NEV., VIEWED FROM ABOVE PLEASANT VALLEY

The white streak at the base of the range is a fault scarp 16 feet high formed during an earthquake in 1915.



THE SONOMA RANGE, NEV., SHOWING FAULT SCARP FORMED IN 1915 AT BASE OF RANGE

The fault appears as a dark sinuous line running up and down over the alluvial fans.

*Eliot Blackwelder*

*C. R. Longwell*

FAULT SCARP 1000 FEET HIGH, WEST OF GRAND CANYON
Upraised Mississippian limestones in the distance. Permian limestones in the foreground. Total throw of fault 5000 feet.



NEAR VIEW OF RECENT FAULT SCARP, BASE OF SONOMA RANGE, NEV.
Showing 16-foot cliff formed in 1915 in the alluvium.

*Paul F. Kerr*

*C. R. Longwell*

FRONT OF SPOTTED RANGE, NEV.

Showing straight base line of block mountain truncating the structure. The strongly dipping beds constitute a fault "splinter" standing in front of the main range.



THE FRONT OF THE WASATCH MOUNTAINS SOUTH OF SALT LAKE CITY

The straight base line of a maturely dissected block mountain. Triangular facets barely discernible.

*U. S. Forest Service*

## BLOCK MOUNTAINS

SYNOPSIS. The outstanding problem for the geomorphologist in the study of block mountains and faulted structures is the recognition of faults from *topographic evidence*. *Geologic evidence* of faulting is to be had only rarely at the surface of the earth.

The following descriptions of block mountains result largely from studies made in the Great Basin of Nevada and Utah.

Block mountains and their rock structures were first studied by Clarence King and interpreted as faulted blocks by G. K. Gilbert between 1870 and 1875. Their complex, internal folded structure was early recognized by Powell as having nothing whatever to do with the present height and form of the ranges. He clearly saw that the region was peneplaned before the present uplift took place. G. D. Louderback made a valuable contribution when he noticed that overlying lava flows were badly broken up and tilted by the faulting which produced the block mountains. In other words, Louderback produced some real geological evidence of faulting when he demonstrated the actual displacement of the lava flows. This fact alone definitely proved that the internal folding of the ranges long antedated the present ranges.

W. M. Davis has analyzed the problem and has accumulated much evidence in favor of the fault theory of origin. In his paper on the Peacock Range he has introduced several terms which are not apt to be adopted by geomorphologists but which serve to emphasize the contributions of the men just mentioned. Thus, one may say that the *King Mountains* of (Mesozoic) compressional deformation were worn down to the *Powell peneplane* (in Tertiary time); then covered with Louderback lava flows, the remnants of which may be called *Louderbacks*, and finally upheaved to form the *Gilbert fault blocks*, which initiated the sculpturing of the Basin Ranges as we know them today.

Opposed to the fault theory is that of J. E. Spurr who regarded the present block mountains as the net result of erosion begun in Mesozoic time, that is, when the first compressional folding took place. He admitted the continued upheaval of the region since that time but emphatically contended that only the more recent faults or folds find direct expression in the topography. He said that in this region, as in most others, deformation lags behind erosion. With this last statement most geomorphologists are in agreement. That is to say, topographic forms, in general, are erosional features rather than deformational features. In this book it is clearly emphasized in the case of dome mountains and folded mountains that the present-day features reflect the resistance of the rocks and not the nature of the original doming and folding. But in the case of block mountains most investigators now agree that deformation has been sufficiently lively to produce the present-day forms, and that erosion accounts for only the canyons which now dissect the range. These are at best only minor features.

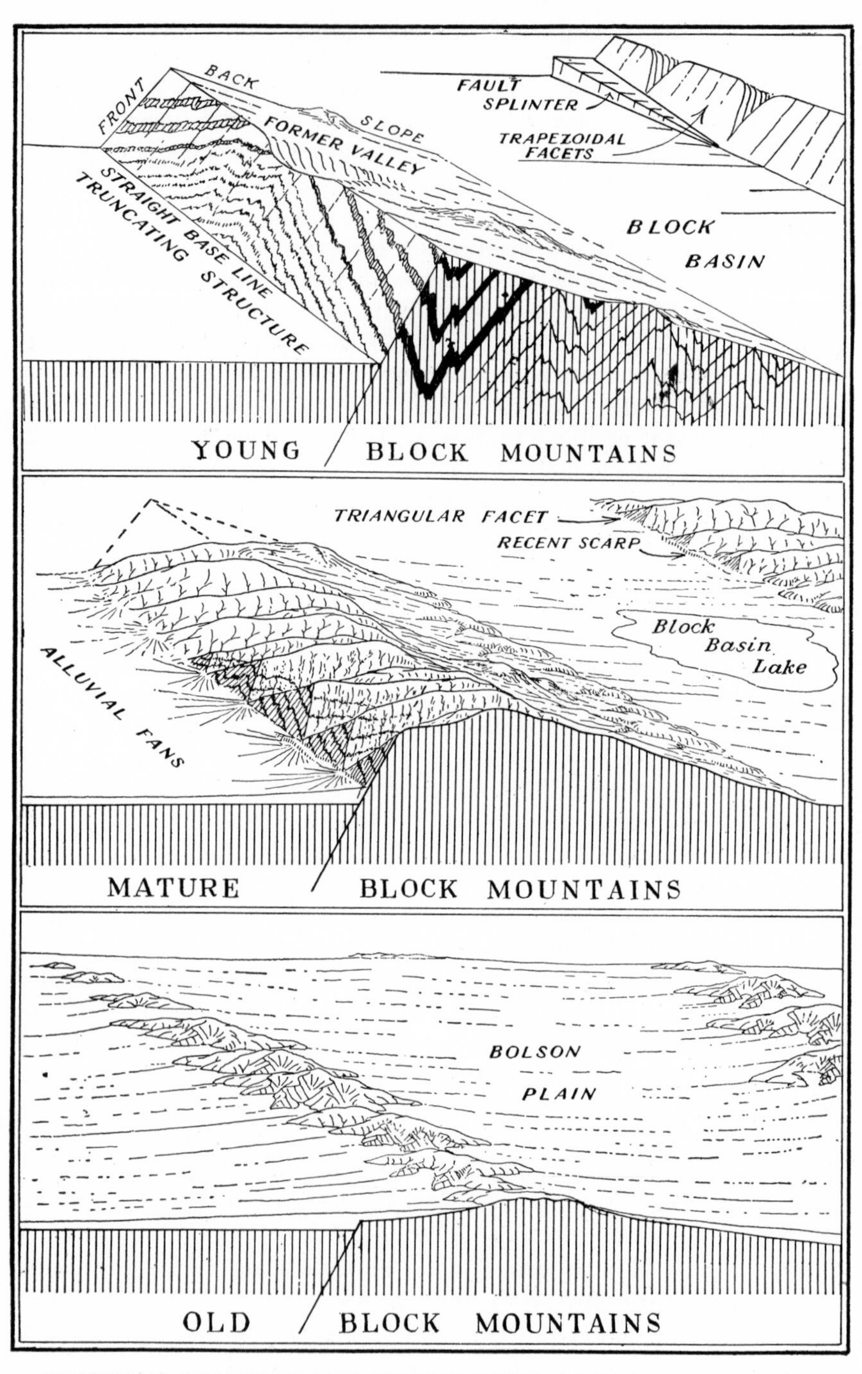


EROSIONAL STAGES IN THE DEVELOPMENT OF BLOCK MOUNTAINS

## THE EROSIONAL DEVELOPMENT OF BLOCK MOUNTAINS

YOUTH. Young block mountains are sharp and angular. They are apt to be unsymmetrical, having a steep front face and a gentle back slope. The face of the block is the fault plane. It may not be a single surface but, because of multiple faults, it may be stepped or it may have *fault splinters* or wedge-shaped blocks standing against it. The back of a young block mountain may slope if the block is strongly tilted, or it may be almost level, like a plateau. Knobs or hills representing monadnocks on a former peneplane may rise above the level of the back slope, and there may be valleys raised to a hanging position but not carrying any drainage.

MATURITY. Maturity brings with it complete dissection of both the front and back slopes. The following features may conveniently be summarized:

*a.* The front of the block is still steeper than the back slope but the divide has been pushed back and there is a tendency toward an equality of the two slopes.

*b.* The base line at the foot of the range tends to remain fairly straight, showing little change from youth.

*c.* The base line and the front of the range, as in youth, truncate the structure. This is evidence that the range is not due to the removal of beds of weaker material, for the range consists of weak and resistant material alike.

*d. Triangular facets* mark the ends of the spurs on the face of the block. These are the remnants of the old fault plane. As the spurs are worn lower and lower, the triangles disappear.

*e.* The canyons, emerging on the face of the block, open abruptly upon the plains; that is, the canyon walls are about as steep near the mouth of the canyon as they are farther upstream.

*f.* Broad sloping plains, made up of coalescing alluvial fans, slope away from the front of the range.

*g.* Indications of recent fault movement may appear in the form of minor terraces and scarps in the alluvial plains at the very base of the range. The streams, too, may show rejuvenation near their mouths.

*h.* Springs—often hot springs—occur along the front of the range, attesting indirectly the presence of a fault.

*i.* Lakes, called *block basin lakes,* may occupy the depressions between upraised or tilted blocks.

OLD AGE. Old block mountains have lost their asymmetrical form. Their front and back slopes are about equal. In fact, the front of the range is now pushed far back from the original fault plane. The spurs have lost their triangular facets and now taper gradually toward the plain. Broad alluvial plains cover the margins of the range, giving it the appearance of being buried up to its ears. In short, the range has lost most of the aspects of a block mountain and is recognized as such only by its isolated position and its general relationship to other ranges whose fault origin is more evident. The block basins between the old block-mountain ranges are filled with alluvium and are known as *bolson* plains, from a Spanish word meaning *purse.*

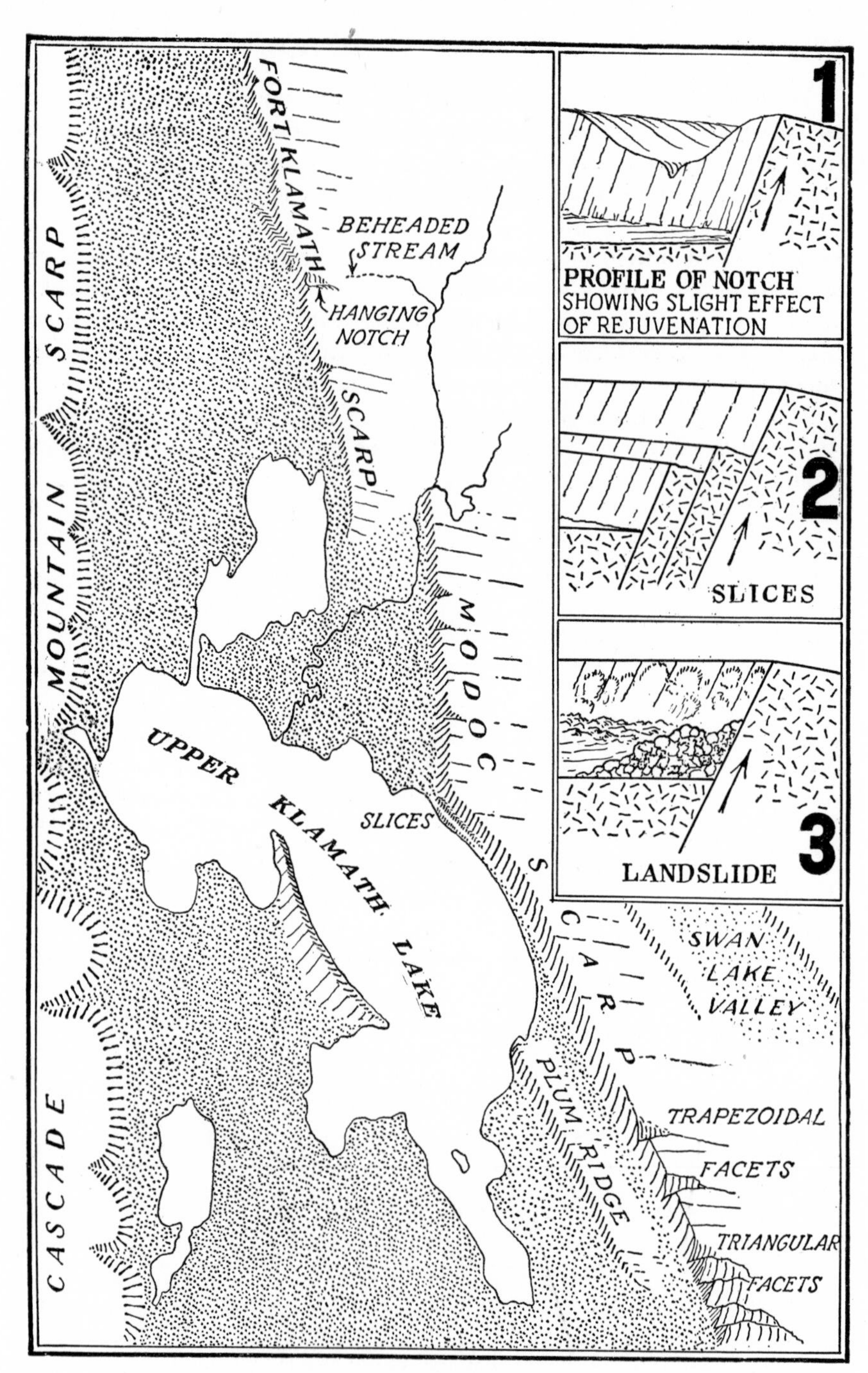


YOUNG BLOCK MOUNTAINS, KLAMATH LAKE REGION, ORE.

In southern Oregon, in the Steens Mountain district and in the region around Warner, Abert, and Klamath Lakes, there are many young block mountains, grabens, and horsts.

THE KLAMATH LAKE REGION, OREGON. This portion of the Columbia Plateau has recently been broken into blocks by north-south faults. Erosion since faulting has been slight and the initial forms of the fault scarps with fault splinters and steps are admirably preserved, notably the scarps bounding the eastern side of the graben occupied by Klamath Lakes. These have been described in detail by Johnson and by Gilbert. They are depicted in the accompanying diagrammatic sketch.

The Fort Klamath scarp is a sharp, clean-cut cliff about 300 feet high, without valleys. The level skyline of the scarp is interrupted by a notch in the form of a broad U (see Insert 1). The sill of the notch is about halfway up the scarp wall, presumably representing the valley of an eastward-flowing stream which was beheaded by the uplift. The bottom of the notch is sharply incised, suggesting that the stream was rejuvenated by the uplift and maintained its course for a short time before being cut off altogether.

STEP FAULTS AND SLICES. Both the Modoc and Plum Ridge scarps face toward the west with slopes of about 50°. In some places they present extremely clean, slickensided surfaces scores of feet long. At the base of the major scarps are small scarps and ridges, called *slices*, due to minor or step faults (Insert 2). The minor fault planes are much softened in outline compared with the larger scarps. This is because they are formed in the weathered material near the surface of the ground, whereas the faulting of the larger scarps exposed unweathered rocks from a greater depth.

The absence of dissecting ravines on the face of the scarp indicates that drainage lines have not even begun to be established. Therefore triangular facets are absent, and alluvial fans are of small extent. The tops of the splinters and minor blocks slope back toward the main scarp so that a trough is produced in which a considerable volume of water may be carried.

Very thin and badly broken blocks and slices sometimes lie between the minor blocks and the main scarp. These, in the field, may closely resemble landslides. Landslides are commonly present at the foot of young fault scarps but rarely in the initial stages. When weathering begins to make headway, landslides become more frequent and, for a short period in the erosional history of a scarp, landslides constitute the most active of the erosional phenomena, more important in the removal of material than streams are. When stream notches are first cut in the scarp face, the facets on the front of the scarp are trapezoidal. They become triangular as they increase in number with the approach of maturity.

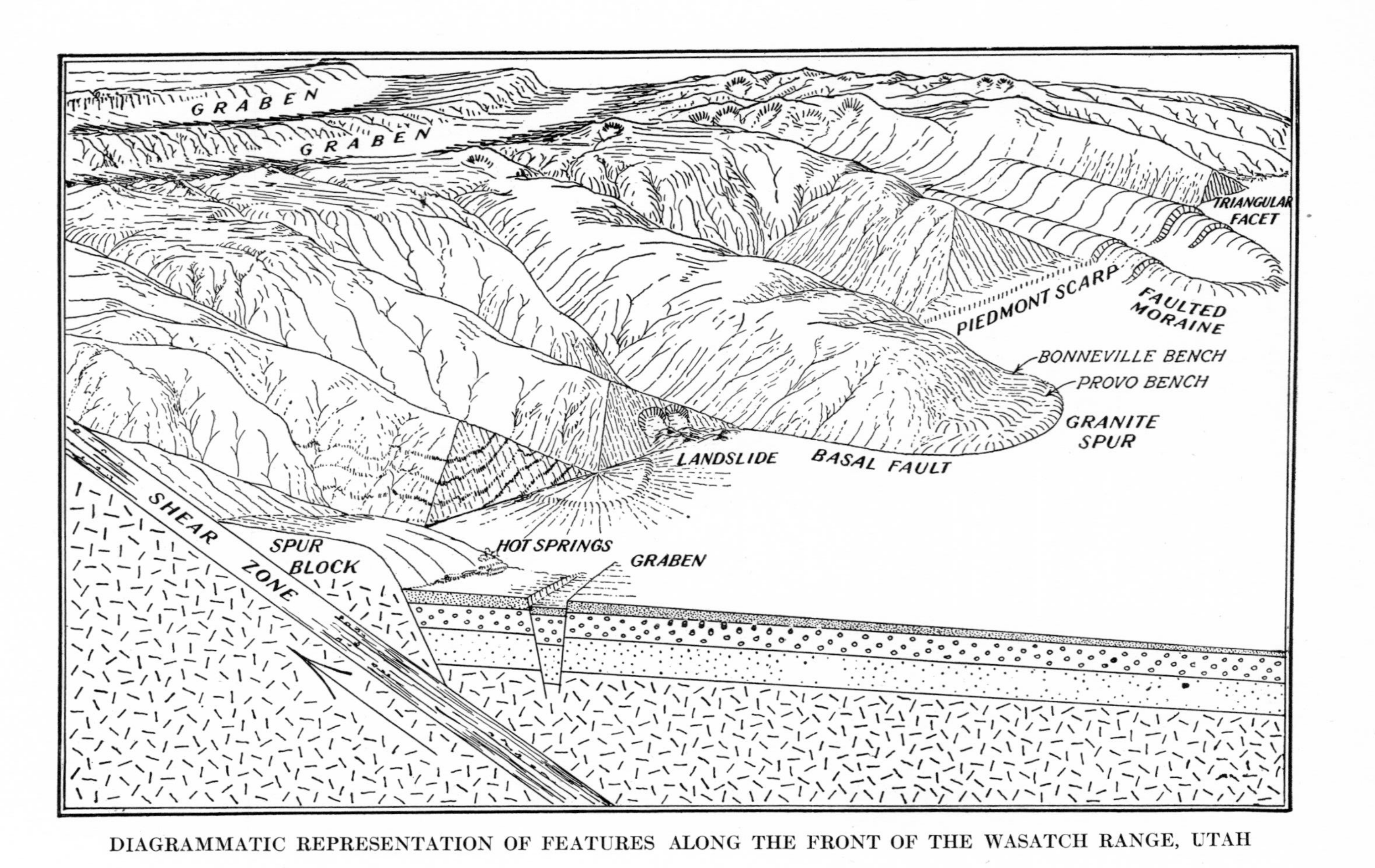


DIAGRAMMATIC REPRESENTATION OF FEATURES ALONG THE FRONT OF THE WASATCH RANGE, UTAH

## MATURE BLOCK MOUNTAINS: AN EXAMPLE, THE WASATCH RANGE

The Wasatch Range in Utah is an unusually large, maturely dissected block mountain. It has a length of 130 miles and a general height above the plains of 4,000 feet. Some peaks rise much higher. The structure is complex. It contains rocks of all ages and of all types. Before uplift these rocks were truncated by a peneplane, remnants of which are preserved on the top and back slope of the range.

THE FRONTAL FAULT. The frontal fault of the Wasatch Range is extremely crooked. Some of the deflections are due to the internal structure of the range. The fault seems to pass around resistant quartzite and granite masses but cuts across sedimentary strata. In some places the fault is actually transverse to the general trend of the range. Cross faults and shear zones are found and occasionally slickensides. The dip of the fault and the slope of the triangular facets are around 35 to 45°.

Projecting beyond the main scarp are many spurs due to granite and quartzite masses and minor blocks, called *spur blocks.* Hot springs commonly occur at the foot of the spur blocks, possibly due to the more open and broken character of the fault around the spurs.

RECENT FAULTING. The front of the Wasatch is embellished by numerous minor details, due to recent faulting. Small scarps, known as *piedmont scarps,* and occasional grabens occur in the alluvial fans and in the moraines. Landslides, occasioned by recent dislocation, are also present. Besides all of the fault-block features, there are the two prominent benches of Lake Bonneville.

The Wasatch Range as a whole is by no means a simple block. It is bounded by faults on both the front and the back, the frontal fault being much the greater. The dislocation resulting from faults on the back slope has produced large grabens. These faults, with the cross faults heretofore mentioned, are partial evidence of the high degree of complexity which a large block mountain may possess.

ANTECEDENT STREAMS. Valleys which completely cross the range are probably courses of antecedent streams. In some cases these streams were forced to abandon their effort to keep pace with the uplift, as witnessed by the cross valleys at high altitudes not at present supporting any stream flow. Other streams appear to have worked headward from the front of the range until they diverted some of the drainage from the back slope.

Many other ranges of the Great Basin, in Utah and Nevada, have been described by Davis, Spurr, Gilbert, Blackwelder, and Louderback in the numerous geological reports of this region.

U. S. Forest Service

FRONT OF SAN BERNARDINO RANGE, CALIF.
A block mountain approaching old age, surrounded by a gravel-covered pediment.

## OLD BLOCK MOUNTAINS: EXAMPLES

In the southern part of the Basin and Range Province, notably in southern California, southern Arizona, and New Mexico, the basin ranges have been almost annihilated by erosion. The presence and position of faults can only be inferred. The ranges are more or less symmetrical in shape and reveal none of the diagnostic features cited as topographic evidences of block faulting.

That this is an arid region doubtless accounts for the actual way these ranges have suffered destruction. It has been accomplished largely by weathering with the simultaneous production of rock pediments by stream erosion. The ranges still rise abruptly from the surrounding plains but the bases are no longer near the original fault.

Rock Pediments. Rock pediments in some cases entirely surround the ranges. They average a mile or two in width and are covered on their outward side by the alluvium of the bajadas or bolson plains. Near the mountains the pediments slope 200 or more feet to the mile and in this zone are occasionally dissected, with valleys averaging 40 feet in depth.

Toward the bajadas the slope of the pediments decreases to about 50 feet ($\frac{1}{2}°$) to the mile. The dissection is much less, the valleys being about 15 feet in depth. The interstream areas are wider and are gravel covered.

The study of this region and the problem of the origin and dissection of the pediments has received attention from Lawson, Kirk Bryan, Johnson, Gilluly, Rich, Ross Field, Keyes, Blackwelder, and W. M. Davis. The essential questions are these: (*a*) Were the pediments formed by the lateral corrasion of the large streams emerging from the mountains or by sheet wash, rill wash, and weathering? (*b*) Does the present dissection of the pediments represent a change in climatic conditions or is the dissection "congenital"; that is, does the dissection of a pediment accompany its development? The present author believes that pediments are formed largely by lateral corrasion, and that the dissection of pediments takes place in the normal cycle of their development as the streams receive less load from the mountains.

Old block mountains, called *inselberge*, island mountains, rise above the almost flat rock pediments, covered by a layer of gravel only one pebble deep, and looking like long sloping alluvial fans.

The bolson plains, with their temporary playa lakes, may be deeply filled with sand and gravel. They serve as reservoirs for supplies of ground water.

Few block mountains of the world occur in humid regions, but under such conditions it is probable that old block mountains would not have pediments surrounding them. They would wear down to subdued forms with low foothills, grading imperceptibly into the surrounding plains.

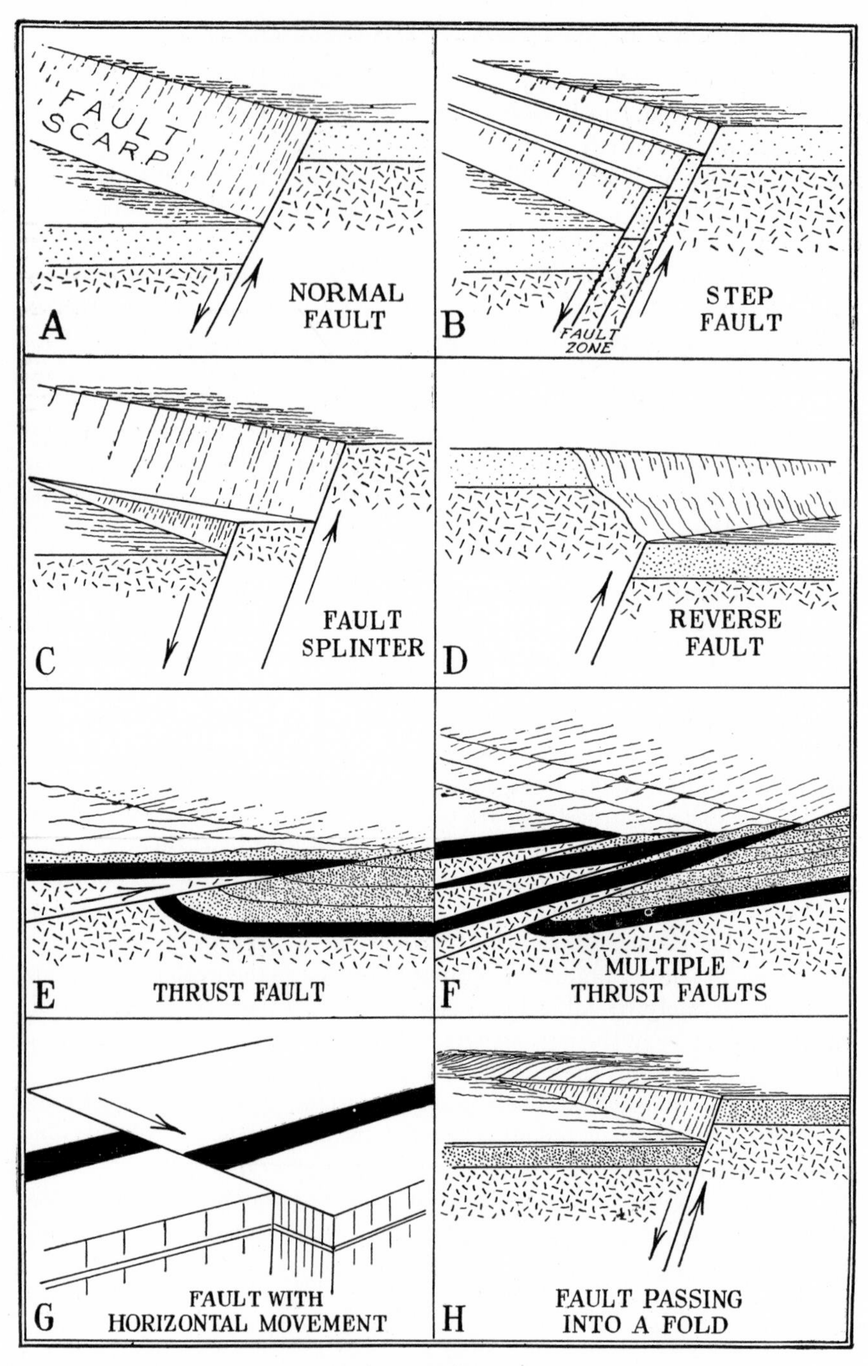


KINDS OF FAULTS

## FAULTED STRUCTURES. KINDS OF FAULTS

The plane of a fault may be vertical or horizontal or at any intermediate angle.

NORMAL FAULTS. If the fault plane is almost vertical and the "hanging wall" or upper side drops down, as in Fig. *A*, a *normal fault* results and the scarp thus formed on the surface of the ground is called a *fault scarp.*

Frequently a number of faults occur together in a belt or *fault zone,* as in *B*, and the resulting scarp is *stepped.* Or the fault may split, being single along part of its course and double elsewhere, as in *C*. This produces a *fault splinter,* which, of course, may be multiple also, if several faults are involved.

REVERSE FAULTS. If the hanging wall or upper side of the block is pushed up, as in *D*, a *reverse fault* results. The topographic cliff thus formed never overhangs the lower block but wears back, and a fault scarp, much like that in Fig. *A*, is developed.

If the fault plane of a reverse fault is gently inclined, as in *E*, it is usually termed a *thrust fault.* Great mountain masses have been moved many miles along low-angle thrust planes of this type. When there is a movement along several thrust planes more or less parallel with each other, an imbricated structure results, as in *F*. This is well exhibited in the southern Appalachians.

Normal faults and thrust faults may amount to many hundreds or thousands of feet. Some of the great thrusts of the world indicate a shift of a score of miles or more.

It may be noted that in low-angle faults (thrust faults) there is practically never any pulling apart of the blocks involved or movement down the slope of the fault plane. The movement always indicates compression rather than tension.

HORIZONTAL MOVEMENT. Vertical or steeply dipping fault planes sometimes exhibit horizontal movement, as in *G*. When this occurs, the total displacement is always a small number of feet, ten or twenty, never hundreds or thousands as with faults having vertical displacement.

The actual amount of vertical movement along a fault plane may vary considerably so that the fault dies out completely at the end; or, what is even more common, it may pass into a fold, as in *H*.

Faults often occur in sets, more or less at right angles with each other, producing rectangular or lozenge-shaped blocks, some uplifted, others down-dropped or tilted. The movement along fault planes, too, may be in any direction, even circular or rotary, as if one block were twisted on the other.

The various degrees of faulting, the direction of movement, the amount and direction of dipping beds affected by the faults, and the amount of erosion which has taken place after the faulting, all profoundly affect the pattern of the outcrops and the plan of the topographic features. Several of these matters will be treated in the following pages.

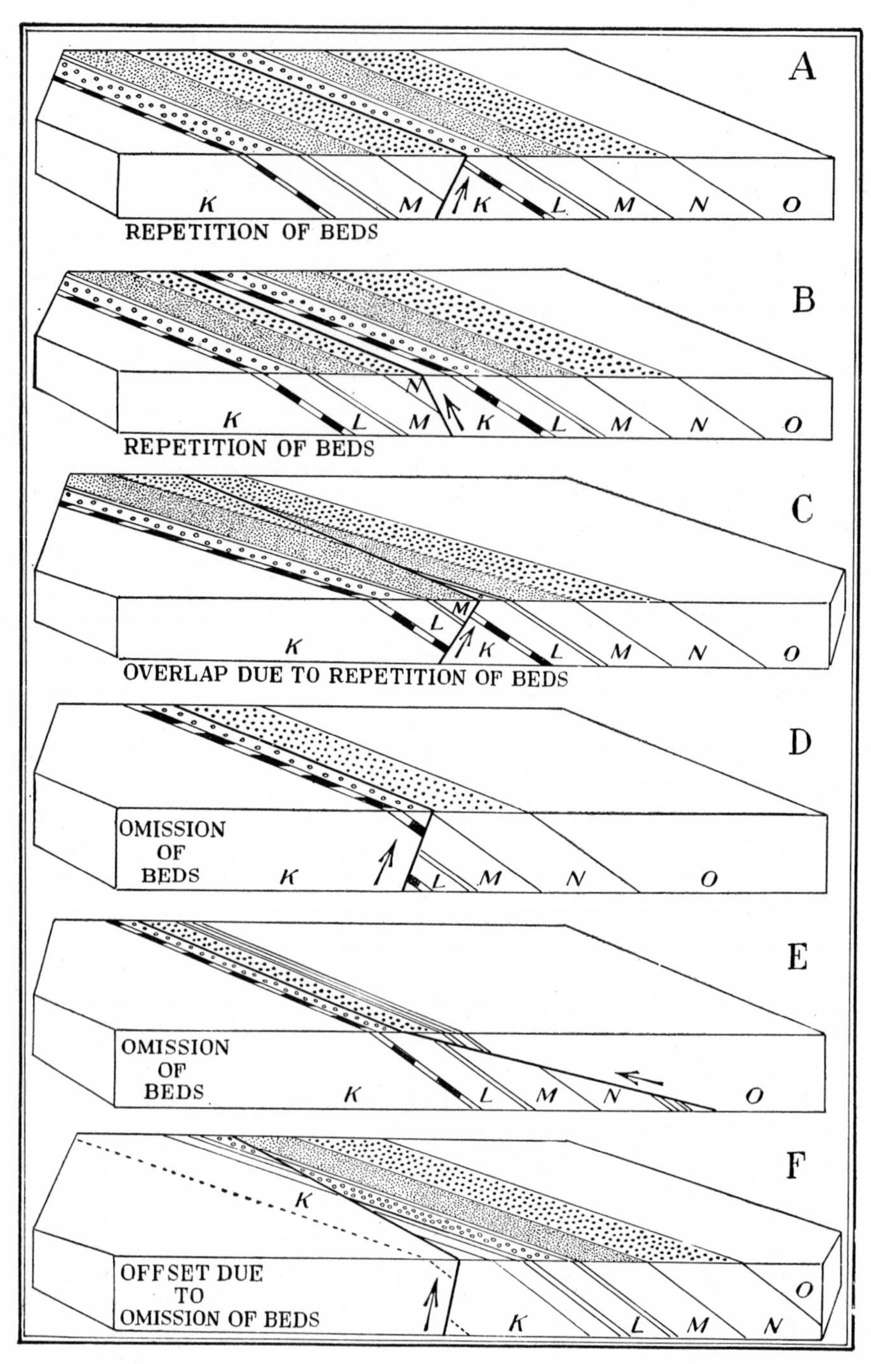


REPETITION AND OMISSION OF BEDS; OVERLAPS AND OFFSETS

## REPETITION AND OMISSION OF BEDS. OVERLAPS AND OFFSETS

REPETITION OF BEDS, AND OVERLAP. If a strike fault, that is, a fault whose strike is the same as the strike of the beds, cuts a series of dipping beds in the manner shown in *A* or *B*, and the block on the down-dip side is moved up the fault plane, whether it be a normal fault, as in *A*, or a thrust fault, as in *B*, and the region then eroded, the outcrop of some of the beds is repeated. The number of beds repeated depends upon the amount of movement along the fault plane, as well as upon the amount of erosion. It is greatest if the movement is extensive and if the erosion is considerable. In order that repetition may occur, the fault plane must dip in a direction opposite to the dip of the beds, as in *A*; or if it dips in the same direction, it must dip more steeply, as in *B*.

In the event that the fault is not strictly a strike fault and intersects the beds diagonally, as in *C*, the beds are not repeated indefinitely along their strike but *overlap* only for a short distance. On the surface the effect appears to be the same as if there had occurred only a horizontal shift along the plane of the fault, without any vertical movement.

OMISSION OF BEDS, AND OFFSET. If, however, in the case of a normal strike fault, as in *D*, intersecting a series of dipping beds at a sharp angle and the block on the up-dip side is raised, and the region then eroded, some of the beds are eliminated. The number of beds eliminated is commensurate with the amount of faulting and the amount of erosion. This is true also in the case of thrust faults dipping in the same direction as the beds but less steeply than the beds do, as shown in Fig. *E*.

Again, in the event that the strike of the fault intersects the strike of the beds diagonally, as in *F*, then there is an *offset*. The same beds are not omitted everywhere along the strike of the fault; just as in *C*, the same beds are not repeated everywhere along the strike of the fault. For instance, in Fig. *C*, the bed *M* is repeated in one place, whereas bed *N* is repeated at another place.

There is another important matter to be noted, namely, that this interruption in the series of beds, which appears on the surface of the ground, may seem to be due simply to horizontal movement. As previously stated, however, horizontal movement along fault planes is very limited. An offset or overlap of topographic features had best be ascribed to vertical displacement of dipping strata until proved otherwise.

The simplest rule to remember in all cases of this kind is that, after erosion, the outcrop of a dipping bed is moved over in the direction of the dip. If in the upthrown block this happens to be away from the fault, there results a repetition of beds. If, however, the dip in the upthrown block happens to be toward the fault, there results an omission of beds.

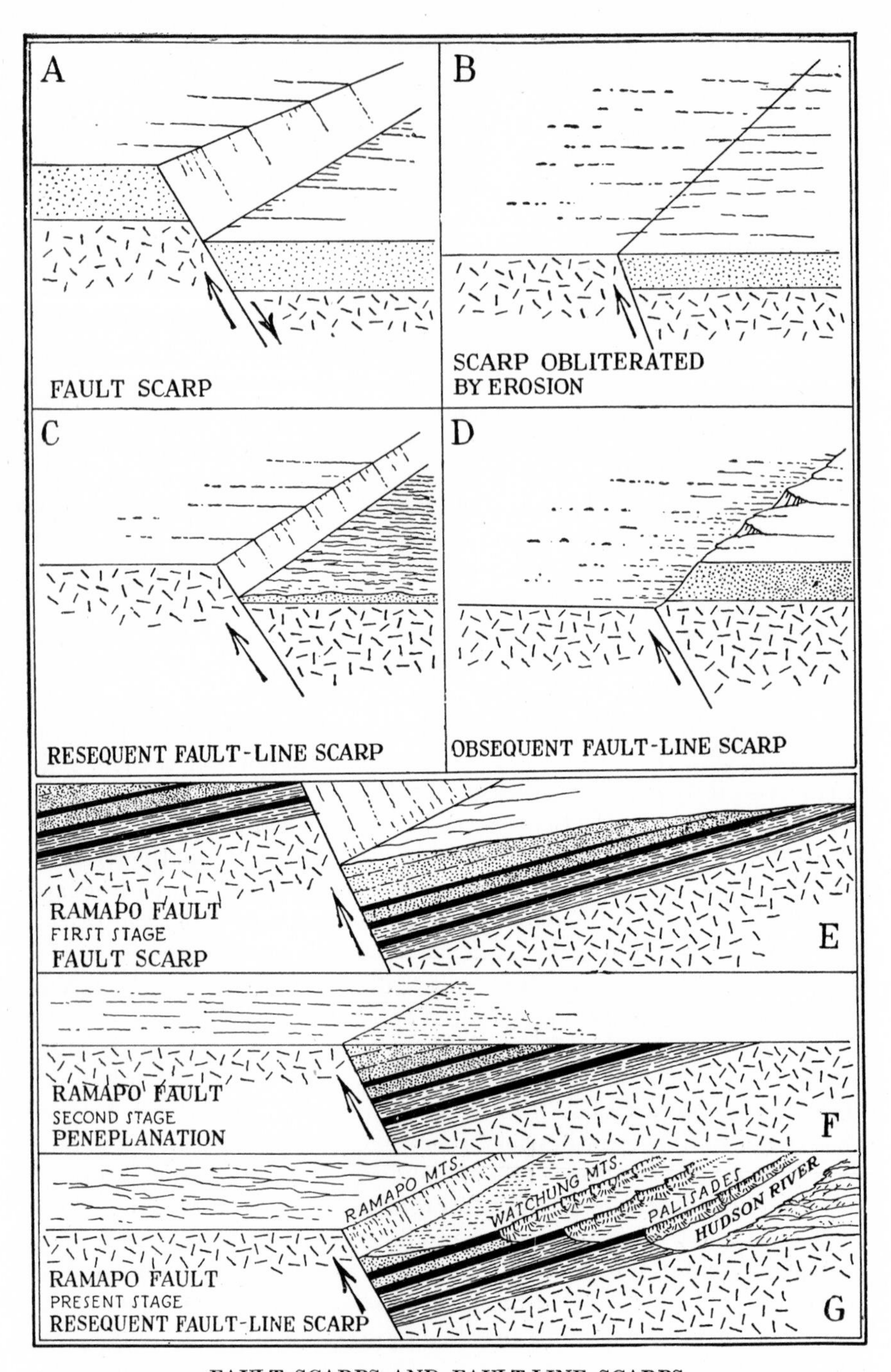


FAULT SCARPS AND FAULT-LINE SCARPS

## FAULT SCARPS AND FAULT-LINE SCARPS

Not all scarps which occur along the line of a fault owe their height to the magnitude of the fault. In fact, it is probably safe to say that the present form of most scarps in such positions—and indeed the actual height of the scarp—is due directly to erosion. The fault merely determines where the scarp shall be.

Figure *A* shows a true fault scarp. The height of the scarp represents the actual amount of faulting.

Figure *B* shows the same region reduced to a peneplane. There is no scarp.

Figure *C* shows the region again after renewed erosion has removed the weaker beds on one side of the fault. This renews the scarp, but it is obvious that the scarp is due directly to erosion. Such a scarp is termed a *fault-line scarp;* and if it faces in the same direction as the original fault scarp did, it is called a *resequent fault-line scarp.*

If, however, as in Fig. *D*, the original uplifted block is eroded to a lower level than the original down-dropped block, because of weaker rock being brought to the surface at the time of peneplanation, then the direction of the scarp is reversed and such a scarp is termed an *obsequent fault-line scarp.*

Fault scarps, as well as fault-line scarps, indicate youth in the erosional history of a region. As an accompaniment of youth, too, landslides are common in both cases. For example, in the state of Washington, parts of the Columbia Plateau, near Ellensburg, have been faulted to produce mountains 2,000 to 3,000 feet high. Along the front of these fault scarps are some remarkable landslides, attesting the comparative recency of this uplift.

Then, again, in the Colorado Plateau there are some mighty scarps along the lines of some of the great faults. These are fault-line scarps. Along their fronts, too, are extensive landslides, which indicate that erosive activities are unusually rapid. The scarps are young scarps; or, what actually seems most likely in those cases, the scarps are being rejuvenated by a recent regional elevation of the whole country and not by renewed faulting.

Other examples of fault-line scarps are along the shores of Lake Superior, the Ramapo scarp (Figs. *E*, *F*, and *G*) bordering the Triassic Lowland in New York and New Jersey, and the scarp marking the eastern side of the Connecticut Lowland in New England. The Swedish scarps along the margin of the lakes in southern Sweden are really fault-line scarps, as are also similar steep shores along some Canadian lakes.

Streams dissecting fault scarps are of the genetic type known as *consequent* streams; those dissecting resequent fault-line scarps are *resequent* streams; and those dissecting obsequent fault-line scarps are *obsequent* streams. Streams eroding basins or lowlands at the base of fault-line scarps, like the Ramapo River at the foot of the Ramapos, are subsequent or in some instances resequent streams.

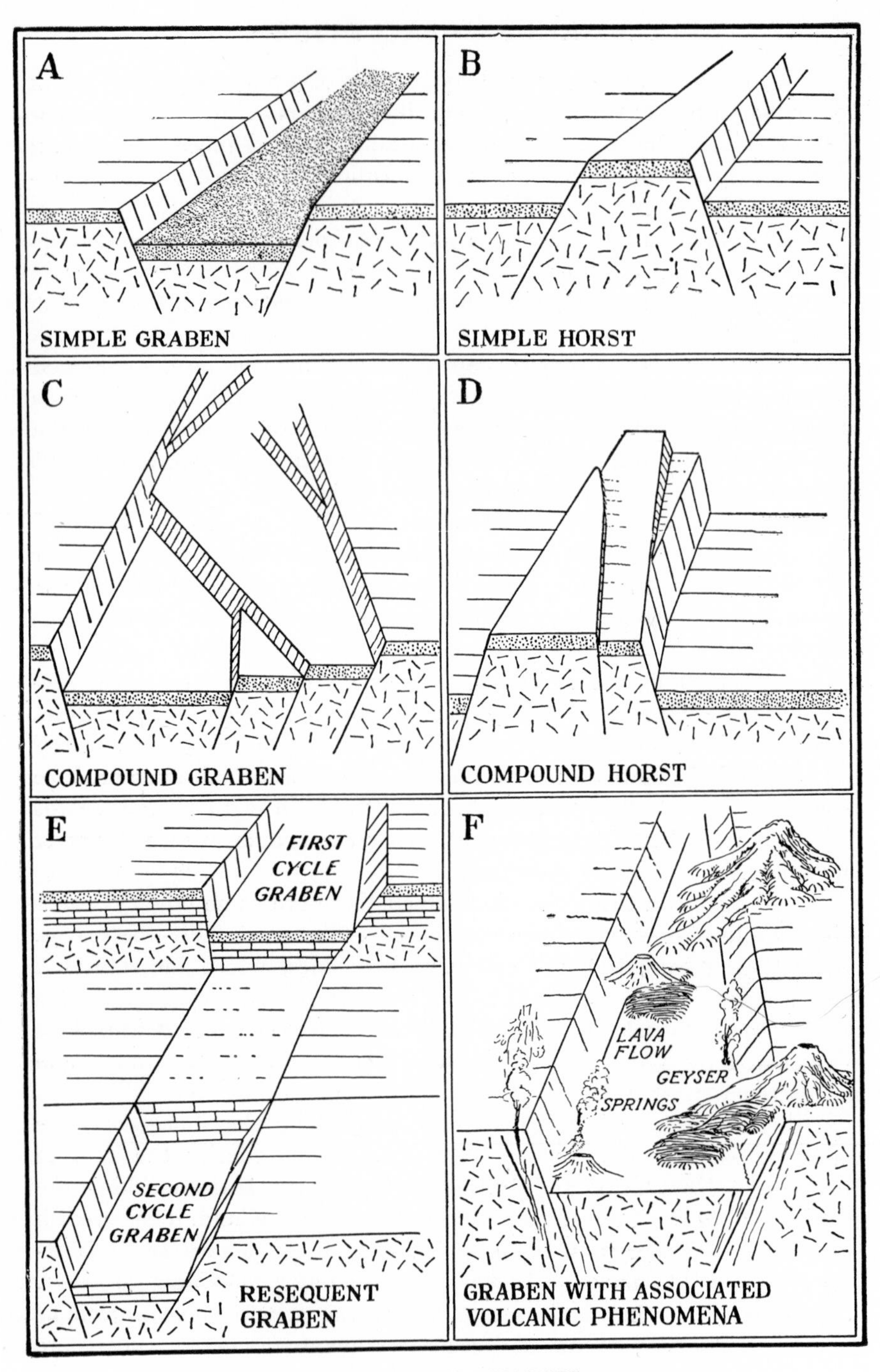


GRABENS AND HORSTS

A *graben* or *fault trough* is a long, narrow depression bounded on both sides by faults (Fig. *A*). A *horst* is a block elevated above adjacent regions along fault lines (Fig. *B*). Grabens and horsts are rarely simple but consist of a number of blocks, slightly dislocated with relation to each other, as in *C* and *D*, like a brick pavement that has settled. The dislocation producing grabens is discussed on the next page.

The term *graben* comes from the German, meaning a ditch or grave, and has as its best known example the Rhine Graben, bounded on the east and west by the Schwarzwald and Vosges, respectively.

EXAMPLES OF GRABENS. Some grabens are very deeply depressed, so that their floors lie below sea level. Death Valley in southern California is apparently of this character. The bottom of the Dead Sea lies 2,600 feet below sea level. Even the surface of the Dead Sea and of Jordan Valley near by is 1,300 feet below sea level. The total depth of this graben is about 1 mile below the plateaus on either side.

Some of the deepest grabens of the world are those which form the great "deeps" or troughs of the ocean, long depressions which usually lie close to land masses or island horsts. The Bartlett Trough, south of Cuba, drops 3 miles below the level of the ocean floor. A similar deep, south of Java, has a depth of 4 miles; and the great Philippine Deep, which forms a long trough close to the Philippine Islands on the east, appears to have the astounding depth of over 5 miles.

It should not be assumed that the present depth of all long depressions bounded by faults indicates the original amount of displacement. The valleys now observed may be the result of later erosion following a period of base-leveling, as shown in *E*. The scarps bounding such a depression are fault-line scarps—resequent fault-line scarps, to be more precise. This is undoubtedly the history of some of the long valleys in the New Jersey Highlands and probably of the Berkshire and Connecticut Lowlands in Massachusetts.

VULCANISM ALONG FAULT LINES. Fault lines, in grabens as well as elsewhere, serve as zones along which molten rock comes to the surface. It is therefore common to find lava flows, cinder cones, and even large volcanic masses arranged along the margin of grabens (Fig. *F*). The Kaiserstuhl is a volcanic neck standing on one of the Rhine Graben faults and the mass of the Vogelsbirge occupies the northern end of the Rhine Graben. The lava flows of the Triassic Lowland and of the Connecticut Valley were undoubtedly controlled by the earlier faulting of those grabens. Attention has also been called to the structure of the Yellowstone Park geyser basins which are structurally grabens, the geysers owing their location to the position of the faults. Still another example of fault-line vulcanism is the series of volcanic cones along the foot of the Balcones fault scarp in southern Texas.

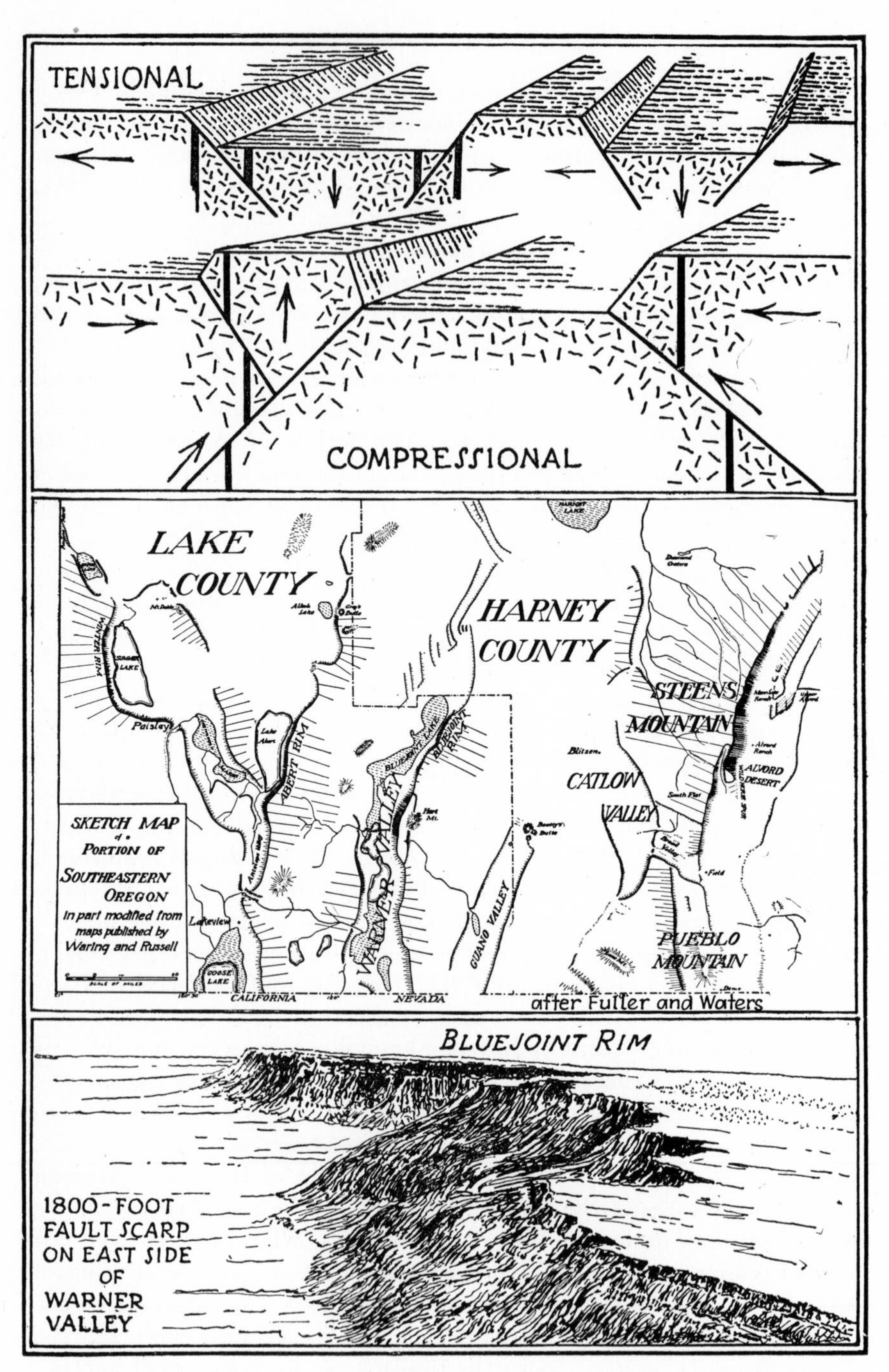


THE TENSIONAL AND COMPRESSIONAL EXPLANATION OF GRABENS
THE GRABENS OF SOUTHERN OREGON

## GRABENS AND HORSTS: EXAMPLES

TENSION VS. COMPRESSION THEORY OF ORIGIN. Two theories for the formation of grabens have been proposed. The *tensional theory*, by which grabens are believed to be down-dropped blocks between normal faults, is the one usually adopted. According to the newer *compressional theory* (held by Bailey Willis and Warren D. Smith), the blocks or horsts on either side of a graben have been pushed up along thrust faults.

The grabens of southern Oregon and the rift valleys of Africa are well-known examples which have been studied by the exponents of the two theories.

SOUTHERN OREGON, EVIDENCES OF TENSION. The Columbia Plateau of southern Oregon participates in the block faulting which is the predominant characteristic of the Great Basin lying to the south. The faulting in Oregon, however, has produced a series of north-south grabens rather than tilted block mountains like those in Nevada. In the belt of country eastward from the Cascades for 250 miles or more, there are seven great north-south tectonic depressions. They have a north-south extent of 100 miles or more, the average width of each depression being about 10 miles.

According to Fuller and Waters the tensional origin of these faults is indicated by definite normal faulting, curving and zigzag faults, step faults, circular fault basins, volcanic activity parallel to the faults, absence of folds, thrust faults or other pressure effects, and by the even distribution of faults rather than their local concentration in zones of initial failure. These investigators think that none of the preceding phenomena could come about by compression. For example, circular fault basins only 3 miles in diameter with rims 1,000 feet high, if due to compression, would require forces working uniformly from all directions, conditions which are more likely to produce a dome than a basin. The even distribution of the faults over a belt hundreds of miles wide favors the idea of tension. Rocks are not able to withstand a tensional pull and they therefore break in many places. For the same reason the shrinking of cooling lava produces many vertical columns evenly spaced instead of a few wide joints in one locality.

As regards the absence of folding in the graben region, it should be noted that in other parts of the Columbia Plateau, notably in central Washington, the lavas are strongly folded and in some places actually overturned. No grabens occur in the region where folding prevails.

Noteworthy, too, is the absence of volcanic vents associated with planes of thrust faulting, but their association with normal faults is known from numerous examples in widely separated parts of the world.

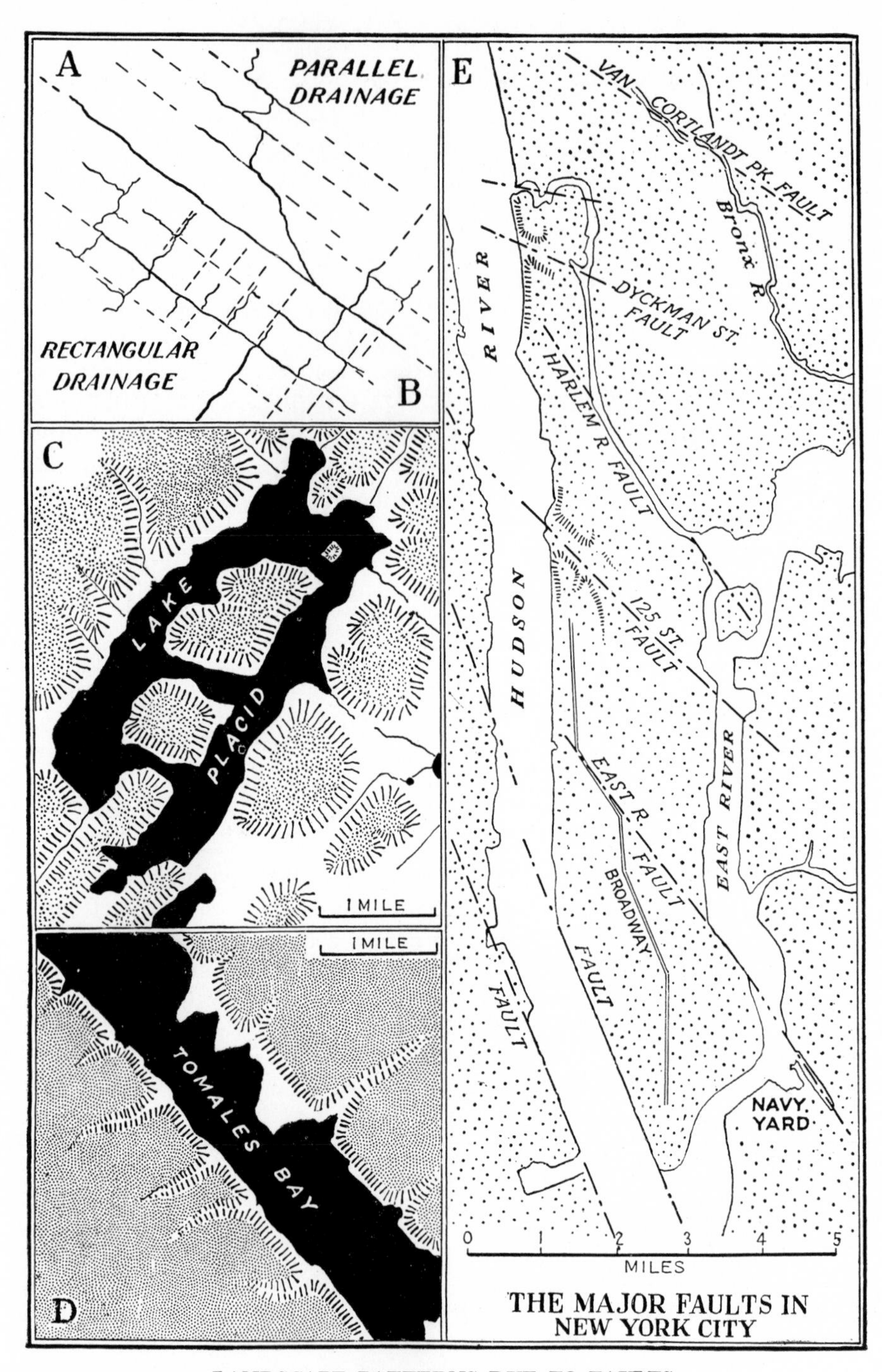


LANDSCAPE PATTERNS DUE TO FAULTS

## FAULTS AS LINES OF WEAKNESS

Faults, as well as joints, are lines of weakness, even where the actual movement is very slight. In the process of stream adjustment, rivers take advantage of these lines of weakness and acquire a pattern which reflects to some extent the pattern of the fault system.

STRAIGHT AND RECTANGULAR DRAINAGE PATTERNS. For example, a stream may develop many straight stretches in its course where it follows for short distances small fault zones or joints, as in Fig. *A*. Or rivers may become strongly rectangular in pattern if there are two sets of faults intersecting at right angles, as in Fig. *B*.

If the streams of a region cut deep valleys along two intersecting sets of faults, the resulting topography may be strongly "checkerboard" in pattern, as in the Adirondacks (Fig. *C*). Lake Placid, for example, shows this effect to a remarkable degree.

Fault zones, or "rifts," as they are sometimes termed, even though they are due to single fault zones and are not down-dropped trenches, may reveal themselves in the topography for long distances by a succession of long straight valleys, lakes, sags, or cols in the hills and by narrow estuaries where they reach the coast. The San Andreas rift in California (Fig. *D*) is a notable example of this.

NEW YORK CITY FAULTS. Within the limits of New York City there are several faults trending parallel with each other in a northwest-southeast direction, as shown in Fig. *E*. Each one of these faults affects the topography. The northernmost is followed by the western end of the Harlem River. The second one determines the Dyckman Street valley. A third one is at One Hundred Twenty-fifth Street, where it causes the Manhattanville depression over which the subway and Riverside Drive are carried on viaducts.

These faults are known, not only by the topography, but by many test drill holes and by tunnels for aqueducts and other purposes.

The One-hundred Twenty-fifth Street fault extends to a great depth and is a very badly weathered zone. Actually, the valley along the fault is 200 feet below sea level but is filled with glacial drift. Where the aqueduct passes across this fault, it is carried along at a depth of 350 feet below sea level in order to avoid the buried valley. Where it actually crosses the crushed and weathered rock of the fault zone, the tunnel is heavily reinforced with steel girders for a distance of 250 feet.

The southern part of Manhattan Island seems to be bounded by a fault on its western side but this is not known through direct evidence. Tunnels under the Hudson River do not go down to bedrock but only through silt. Hence they afford no information as to the rock structure.

Fault Scarps as Distinguished from Lake Terraces. Renewed faulting along old lines of displacement is of common occurrence. Earthquakes in regions of block faulting result from such uplifts. Fault scarps are thus produced, ranging in height from a foot or so to 50 feet or more. When this occurs at the base of block mountains, the observable displacement is not at the very mountain front, but at some distance away in the alluvium which usually buries the actual fault. If the alluvial plain slopes away uniformly, the resulting scarp is horizontal; but if the allu-

SKETCH SHOWING DISTINCTION BETWEEN HORIZONTAL LAKE BEACH AND RECENT FAULT SCARP CURVING OVER ALLUVIAL FAN

vium is in the form of a cone or fan, as illustrated above, the scarp passes up and over the surface of the fan. This arched form of the scarp is even more pronounced if it cuts across a moraine and displaces it in the manner shown on the next page. In the Great Basin, the many horizontal terraces formed by old lake beaches can readily be distinguished from the fault scarps which curve up and down over the topography. The fault scarps, when viewed from above, in the plane of the fault, appear straight. Shore-line terraces, on the other hand, are much more curving when seen from above, but they are perfectly straight when seen on a level with the eye.

There is evidence that the recent fault scarps have been produced not by a single movement but by stages of uplift. For instance, where they cross the flood plains of streams, they are low. This suggests that only the latest displacement is here preserved, the record of all earlier movements having been erased by the stream.

The Sonoma Range Fault in Nevada. In 1915 there occurred in central Nevada an earthquake of great intensity. A fault scarp was produced in the alluvial apron, or bajada, at the base of the Sonoma Range, running for a length of 22 miles and having a height of 5 to 15 feet.

Air travelers passing between Salt Lake City and San Francisco can readily identify it as a white streak at the base of the mountain. The effect of this disturbance was perceptible from Oregon to southern California and from the Pacific Coast to east of Salt Lake City. It was an earthquake as severe as that which rocked California in 1906. Had it occurred in a more densely settled district, it probably would have been numbered among the more destructive earthquakes of history. Many

RECENT FAULTS ALONG BASE OF WASATCH CUTTING ACROSS ALLUVIAL PLAIN AND DISTANT MORAINES

secondary cracks developed in the alluvium, which was so greatly agitated as to cause much of the underground water to reach the surface. Springs and streams increased their flow to seven times their normal amount. The playas became covered with water and many mud craterlets developed. In some places, with each lurch of the ground, the water spurted into the air to a height of several feet. A considerable body of alluvium remained clinging to the uplifted mountain mass.

Numerous other so-called *recent* fault scarps, which traverse alluvial fans or separate them from the hard rock at the foot of the ranges, are to be observed in the Great Basin. Many of them were recorded by the King Survey which explored this region in 1875. Such scarps have been designated *piedmont scarps* or *fan scarps*. Many of them show, in addition to the vertical movement, also a horizontal displacement as great as 3 feet.

Near Provo, Utah, a series of unusually interesting recent scarps may be observed at the base of the Wasatch. Oddly enough, the scarps do not all face away from the mountains, with the result that small grabens are produced in the alluvium, as illustrated on the following pages.

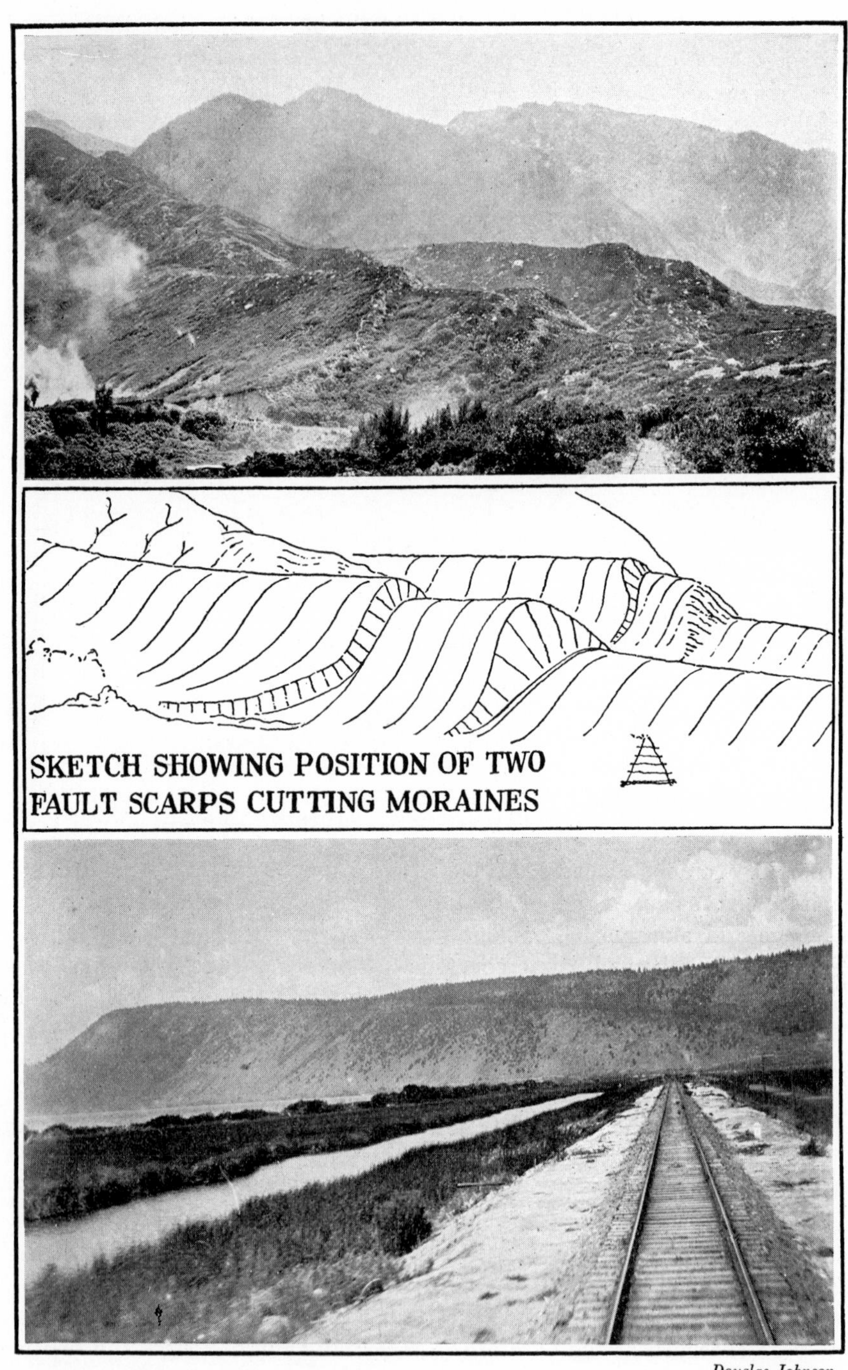


*Douglas Johnson*
*A. K. Lobeck*

UPPER VIEW: LATERAL MORAINES AT FOOT OF WASATCH RANGE, BROKEN BY POST-GLACIAL FAULTING
LOWER VIEW: FACE OF YOUNG BLOCK MOUNTAIN WITH FAULT SPLINTER. KLAMATH LAKE REGION, ORE.

*A. K. Lobeck*

UPPER VIEW: GRABEN AT BASE OF WASATCH RANGE, UTAH, PRODUCED BY RECENT FAULTING
LOWER VIEW: RECENT FAULT SCARP AT BASE OF WASATCH RANGE

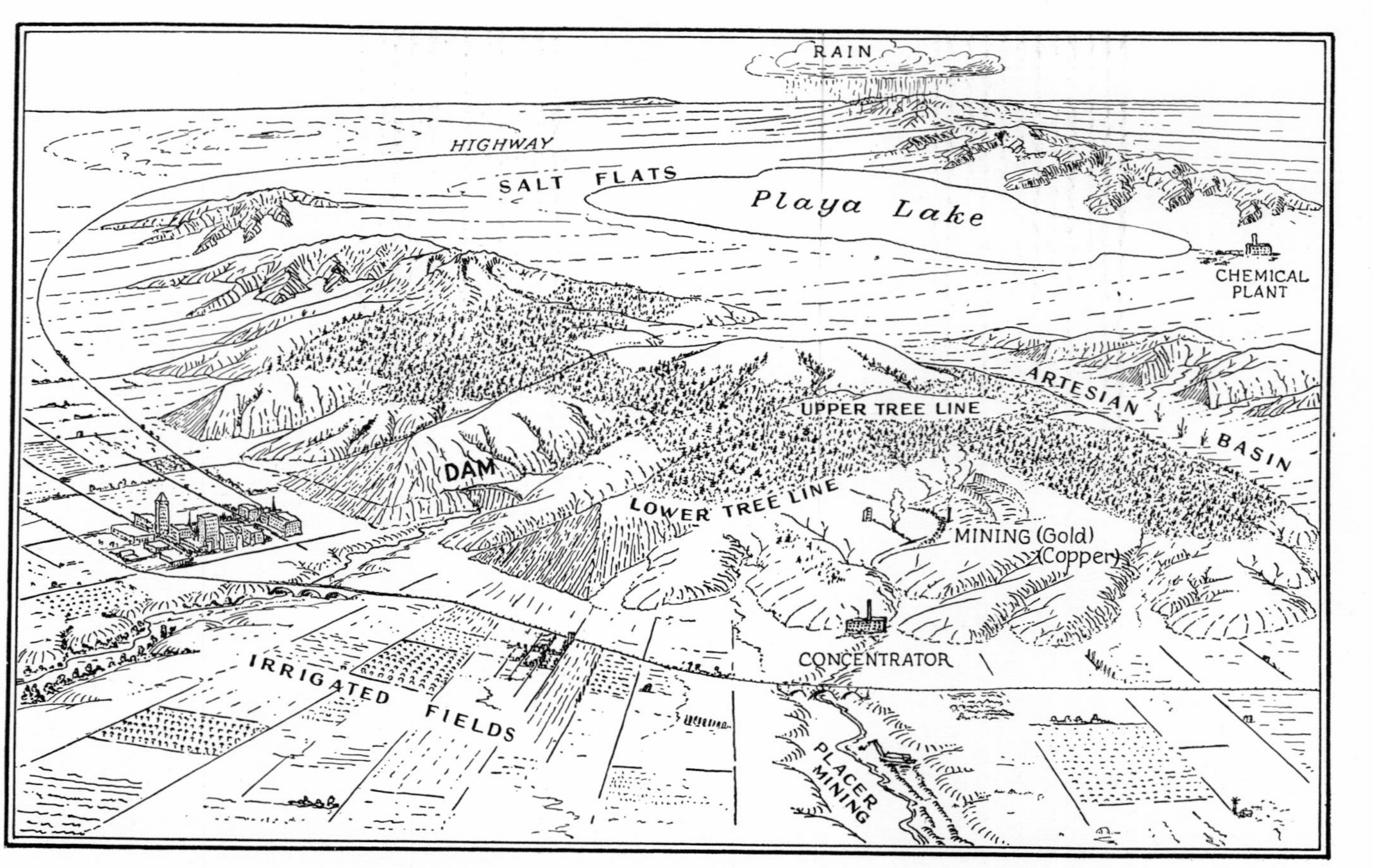


IDEALIZED VIEW OF MATURELY DISSECTED BLOCK MOUNTAINS
Illustrating certain human activities characteristic of such regions.

## SOME GEOGRAPHICAL ASPECTS OF BLOCK MOUNTAINS

Block mountains, rising above surrounding plains and basins, are like islands in the sea. They are places of refuge. The mountains themselves, however, rarely become thickly populated. The people congregate around their bases. There, abundant water may be had from the streams which take their rise in the hills, where the rainfall is more copious than on the plains. Occasionally water is obtained from artesian sources where old lake beds, filling a basin between the ranges, serve to collect the water which falls around its rim. The supply for Ogden, Utah, is obtained in this manner.

At the foot of the mountains the sloping alluvial plains may be rendered immensely fertile by irrigation. A veritable garden is thus produced and agriculture of an intensive type is developed. Fruit growing is important under such conditions and dairying thrives, with alfalfa as the chief supporting crop.

Cities spring up, usually at the mouths of the largest canyons, not so much because of the available water but chiefly because the canyons lead back to the lowest passes through the range to the country beyond.

A strip of country only a few miles long, at the foot of the Wasatch Range in Utah, contains most of the population of that state, in what has been well termed the *Oasis of Utah.* Here are located Logan, Ogden, Salt Lake City, Provo, Manti, and many lesser places.

Block mountains, even in arid locations, usually have enough rainfall to support some forest growth. The larger block mountains may have two timber lines: one, above which trees cannot grow because of the cold and wind; and the lower tree line, below which trees cannot grow because of the lack of moisture. The rains of these regions are often evaporated before they reach the lower slopes.

Some of the most important and productive mining camps of the west are situated in block mountains. For example, the famous Bingham copper mine in the Oquirrh Range near Salt Lake City; Virginia City, Nev.; Eureka, Nev.; and Bisbee, Ariz. In fact in much of the Great Basin, mining is probably as important as all other activities combined.

Block mountains are never serious barriers to transportation routes. Through railroads readily avoid them. Highways, however, commonly cross the smaller ranges. Motorists in Utah, Nevada, and southern California are sometimes impressed with the frequent alternation of grassy parklike rolling topography in the hills, where the vegetation is luscious and where occasional ranches are to be seen, with the long straight stretches on the plains, where all is desert and where only at the rarest intervals one may see a mill or chemical works recovering salts from one of the desert basins.

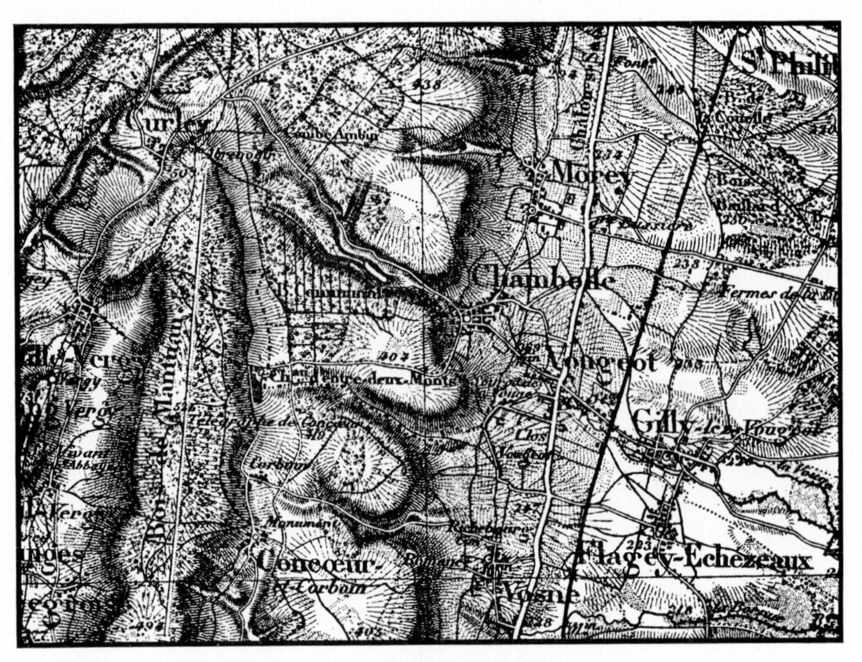


DISSECTED FAULT SCARP OF THE CEVENNES
France; *Beaune* sheet, No. 125 (1:80,000).

## MAPS ILLUSTRATING BLOCK MOUNTAINS AND FAULTING

Young and almost undissected block mountains and fault scarps are shown on the *Klamath Lake, Ore., San Pedro, N. Mex.*, and the *Disaster, Nev.*, sheets. The *Honuapo* and *Kilauea, Hawaii*, sheets show in a striking way several fresh fault scarps. Maturely dissected block mountains with large block basins and occasional playa lakes appear on numerous maps in Nevada and Utah, such as the *Lovelock* and *Roberts Mountains, Nev.*, and the *Hawthorne* and *Silver Peak, Nev.-Calif.*, sheets. The *Granite Range, Nev.*, map shows several parallel block mountains and some very straight-sided lakes. The *Las Vegas, Nev.-Calif.*, sheet portrays many parallel blocks with several clearly defined fault scarps. On the *Logan, Utah*, sheet there appear a few triangular facets south of Logan, and some show also on the *Furnace Creek, Calif.*, sheet, but it is only occasionally that these features may be detected on maps. The *Halleck, Nev.*, sheet shows a magnificent dissected block mountain, maturely glaciated.

Old block mountains, with rimming pediments grading away into alluvial fans, are shown on the *Benson, Ajo, Phoenix*, and *Camelsback, Ariz.*, and on the *Avawatz Mts., Hayes Ranch*, and *Alturas, Calif.*, sheets.

Grabens are represented on the *Mt. Whitney* and *Elsinore, Calif.*, maps; dissected fault scarps are shown on the *Tehama, Calif., Castle Rock* and *Montrose, Colo.*, and *Hamilton, Mont.-Idaho*, sheet. Fault-line scarps and fault-line blocks are represented on the *Meriden, Conn.*, and *Ramapo, N. Y.-N. J.*, sheets, and rift valleys appear on the *Point Reyes, Palmdale, San Mateo*, and *Del Sur, Calif.*, sheets.

## QUESTIONS

1. What does Davis mean by a louderback?
2. Draw a contour map of a triangular facet.
3. How would you distinguish between a terrace formed by waves, such as those of Lake Bonneville, around the base of block mountains and a terrace formed by recent faulting?
4. How would you distinguish between a landslide on the front of a block mountain and minor fault splinters (or slices)?
5. What examples are there of antecedent streams transecting block mountains?
6. How, in the field, would you distinguish between a fault scarp and a fault-line scarp?
7. Do you know of any examples of tilted fault blocks in the eastern United States?
8. How would you distinguish between a fault scarp and an erosional scarp?
9. Why are some faults crooked?
10. What are geyser basins?
11. How would you explain a drag along a fault plane in which the beds were bent in the wrong direction?
12. Draw five cross sections or block diagrams to show the five stages in the development of block mountains outlined in the synopsis to this chapter.
13. How would you account for a geyser on top of the rim of a geyser basin?
14. Name several examples of vulcanism, such as geysers, lava flows, and volcanoes, situated along fault lines, in addition to those cited in the text.
15. A series of eastward dipping weak and resistant beds is cut by a vertical fault striking northeast-southwest, the eastern block being raised. The region is then peneplaned, followed by rejuvenation which permits erosion of the weaker beds. Draw a contour or a hachure map of the new topography. Do the ridges overlap or are they offset?
16. A series of eastward dipping weak and resistant beds is transected by a northwest-southeast vertical fault, the eastern block being raised; the whole area then peneplaned and later eroded along the weaker belts. Do the new ridges overlap or are they offset? Prepare a simple hachure map.
17. A series of beds dipping 30 degrees west is cut by a normal north-south fault (a strike fault), the fault plane dipping 45 degrees west. The western side is downthrown. After peneplanation is there omission or repetition of beds?
18. A series of beds dipping 45 degrees west is cut by a normal north-south fault. The fault plane dips 30 degrees west. The western block is downthrown. After peneplanation is there omission or repetition of beds?
19. What and where is the Great Rift Valley? What lakes does it contain?
20. What significance attaches to the San Andreas rift? The Lewis overthrust? The Ramapo fault?
21. Name some other great faults or faulted regions of the world.
22. Where are the Hanging Hills of Meriden? What is their structure and history?
23. Has any geologically recent faulting occurred in the eastern United States?

## TOPICS FOR INVESTIGATION

1. Fault hypothesis *vs.* erosional hypothesis for origin of block mountains.
2. Graben *vs.* ramp theory for origin of rift valleys.
3. Criteria for the recognition of fault scarps.
4. Examples of recent faulting. Active faults.
5. Earthquakes and their relation to faulting.
6. Fault regions of the world.

# REFERENCES


Block Mountains in General

Davis, W. M. (1903) *Mountain ranges of the Great Basin.* Mus. Comp. Zool., Bull. 42, p. 129–177; Geog. Essays, p. 725–772. A classic treatment of block mountains.

Davis, W. M. (1932) *Basin range types.* Science, new ser., vol. 76, p. 241–245.

Fenneman, N. M. (1932) *Physiographic history of the Great Basin.* Pan-Am. Geol., vol. 57, p. 131–142.

Gilbert, G. K. (1875) *Surveys west of the* 100*th meridian* (Wheeler). U. S. Geol. and Geog. Surv. Terr., part 1, p. 19 *et seq.*

Russell, I. C. (1884) *A geological reconnaissance in southern Oregon.* U. S. Geol. Surv., 4th Ann. Rept., p. 435–464.

Spurr, J. E. (1901) *Origin and structure of the Basin Ranges.* Geol. Soc. Am., Bull. 12, p. 217–270.

Block Mountains—Specific Localities

Davis, W. M. (1905) *The Wasatch, Canyon, and House Ranges, Utah.* Mus. Comp. Zool., Bull. 49, p. 17–56.

Davis, W. M. (1930) *The Peacock Range, Arizona.* Geol. Soc. Am., Bull. 41, p. 293–313. The term *louderbacks* suggested for remnants of lava flows on top of tilted blocks.

Eardley, A. J. (1933) *Strong relief before block faulting in the vicinity of the Wasatch Mountains, Utah.* Jour. Geol., vol. 41, p. 243–267.

Gilbert, G. K. (1928) *Studies in Basin Range structure.* U. S. Geol. Surv., Prof. Paper 153, 92 p.

Johnson, D. W. (1903) *Block mountains in New Mexico.* Am. Geol., vol. 31, p. 135–139.

Johnson, D. W. (1918) *Block faulting in the Klamath Lakes region.* Jour. Geol., vol. 26, p. 229–236.

Louderback, G. D. (1926) *Morphologic features of the Basin Range displacements in the Great Basin.* Univ. Cal., Dept. Geol. Sci., Bull. 16, No. 1, 42 p.

Grabens and Rift Valleys

Barrell, J. *Central Connecticut in the geologic past.* Wyo. Hist. and Geol. Soc., Proc. 12, p. 25–54, 1912; Geol. Surv. Conn., Bull. 23, 44 p., 1915.

Dake, C. L. (1918) *Valley City graben. Utah.* Jour. Geol., vol. 26, p. 569–573.

Davis, W. M. (1889) *Topographic development of the Triassic formations of the Connecticut Valley.* Am. Jour. Sci., 3d ser., vol. 37, p. 423–434.

Fuller, R. E., and Waters, A. C. (1929) *The nature and origin of the horst and graben structure of southern Oregon.* Jour. Geol., vol. 37, p. 204–238.

Gregory, J. W. (1896) *The Great Rift Valley.* London, 422 p.

Johnson, D. W. (1930) *Geomorphologic aspects of rift valleys.* 15th Intern. Geol. Cong., C.r. 2, p. 354–373.

Taber, S. (1922) *The great fault troughs of the Antilles.* Jour. Geol., vol. 30, p. 89–114.

Taber, S. (1927) *Fault troughs.* Jour. Geol., vol. 35, p. 577–606.

Willis, B. (1928) *Dead Sea problem: rift valley or ramp valley?* Geol. Soc Am., Bull. 39, p. 490–542.

Willis, B. (1930) *Living Africa.* New York, 320 p.

Faulting in General

Ambrose, J. W. (1933) *A discussion of the movement of fault blocks.* Am. Jour. Sci., 5th ser., vol. 26, p. 552–563.

Blackwelder, E. (1928) *The recognition of fault scarps.* Jour. Geol., vol. 36, p. 289–311.

Davis, W. M. (1913) *Nomenclature of surface forms on faulted structures.* Geol. Soc. Am., Bull. 24, p. 187–216.

Geikie, J. (1905) *Structural and field geology.* New York, p. 158–182.

LEITH, C. K. (1913) *Structural geology*. New York, p. 31–74.

NEVIN, C. M. (1931) *Structural geology*. New York, p. 78–138.

SMITH, E. A. (1893) *Underthrust folds and faults*. Am. Jour. Sci., 3d ser., vol. 45, p. 305–306.

STOČES, B., and WHITE, C. H. (1935) *Structural geology*. London, p. 177–268.

WILLIS, B., and WILLIS, R. (1929) *Geological structures*. New York, p. 59–124.

WILLIS, R. (1935) *Development of thrust faults*. Geol. Soc. Am., Bull. 46, p. 409–424.

FAULTING—SPECIFIC LOCALITIES

BROWN, J. S. (1922) *Fault features of Salton Basin, California*. Jour. Geol., vol. 30, p. 217–226.

CLARK, B. L. (1930) *Tectonics of the Coast Ranges of middle California*. Geol. Soc. Am., Bull. 41, p. 747–828. Fault patterns described.

DOTT, R. H. (1934) *Overthrusting in Arbuckle Mountains, Oklahoma*. Am. Assoc. Petr. Geol., Bull. 18, p. 567–602.

HODGE, E. T. (1931) *Columbia River fault*. Geol. Soc. Am., Bull. 42, p. 923–984.

HUNTINGTON, E., and GOLDTHWAIT, J. W. (1903) *The Hurricane fault in southwestern Utah*. Jour. Geol., vol. 11, p. 46–63.

HUNTINGTON, E., and GOLDTHWAIT, J. W. (1904) *The Hurricane fault in the Toquerville district, Utah*. Mus. Comp. Zool., Bull. 42, p. 199–259.

KRAMER, W. B. (1934) *En echelon faults in Oklahoma*. Am. Assoc. Petr. Geol., Bull. 18, p. 243–250.

LIVINGSTON, D. C. (1932) *A major overthrust in western Idaho and northeastern Oregon*. Northwest Sci., vol. 6, p. 31–36.

LOUDERBACK, G. D. (1911) *Lake Tahoe, California-Nevada*. Jour. Geol., vol. 9, p. 277–279.

NOBLE, L. F. (1932) *The San Andreas rift in the desert region of southeastern California*. Carnegie Inst. Wash., Year Book 31, p. 355–363.

PACK, F. J. (1926) *New discoveries relating to the Wasatch fault*. Am. Jour. Sci., 5th ser., vol. 11, p. 399–410.

RICH, J. L. (1934) *Mechanics of low-angle overthrust faulting as illustrated by Cumberland thrust block, Virginia, Kentucky, and Tennessee*. Am. Assoc. Petr. Geol., Bull. 18, p. 1584–1596.

SWINNERTON, A. C. (1932) *Structural geology in the vicinity of Ticonderoga, New York*. Jour. Geol., vol. 40, p. 402–416.

WILLIS, R. (1925) *Physiography of the California Coast Ranges*. Geol. Soc. Am., Bull. 36, p. 641–678.

WOODWORTH, J. B. (1907) *Postglacial faults of eastern New York*. N. Y. State Mus., Bull. 107, p. 5–28.

RECENT FAULTING

ANDERSON, A. L. (1934) *A preliminary report on recent block faulting in Idaho*. Northwest Sci., vol. 8, p. 17–28.

GIANELLA, V. P., and CALLAGHAN, E. (1934) *The earthquake of Dec.* 20, 1932, *at Cedar Mt., Nevada, and its bearing on the genesis of Basin Range structure*. Jour. Geol., vol. 42, p 1–22. Numerous references to recent faulting.

JONES, J. C. (1915) *The Pleasant Valley, Nevada, earthquake of Oct.* 2, 1915. Seis. Soc. Am., Bull. 5, p. 190–205.

KERR, P. F., and SCHENCK, H. G. (1925) *Active thrust faults in San Benito County, California*. Geol. Soc. Am., Bull. 36, p. 465–494.

KIRKHAM, V. R. D., and JOHNSON, M. M. (1929) *Active faults near Whitebird, Idaho*. Jour. Geol., vol. 37, p. 700–711.

LAWSON, A. C. (1911) *On some post-glacial faults near Banning, Ontario*. Seis. Soc. Am., Bull. 1, p. 159–166.

PAGE, B. M. (1935) *Basin-range faulting of* 1915 *in Pleasant Valley, Nevada*. Jour. Geol., vol. 43, p. 690–707.

SANDBERG, A. E. (1932) *New fault line at Duluth, Minnesota.* Pan-Am. Geol., vol. 58, p. 271–272.

EARTHQUAKES

DAKE, C. L., and NELSON, L. A. (1933) *Postbolson faulting in New Mexico.* Science, new ser., vol. 78, p. 168–169.

DAVIS, W. M. (1934) *The Long Beach earthquake.* Geog. Rev., vol. 24, p. 1–11.

DAVISON, C. (1936) *Great earthquakes.* London, 286 p.

GIANELLA, V. P., and CALLAGHAN, E. (1933) *The Cedar Mt., Nevada, earthquake of Dec. 20, 1932.* Am. Geoph. Un., 14th Ann. Meeting, Trans., p. 257–260.

GILBERT, G. K. (1884) *A theory of earthquakes of the Great Basin, with a practical application.* Am. Jour. Sci., 3d ser., vol. 27, p. 49–53.

GLENN, L. C. (1933) *The geography and geology of Reelfoot Lake, Tennessee.* Tenn. Acad. Sci., Jour. 8, p. 3–12.

HAYFORD, J. F. 1908) *The earth movements in the California earthquake of* 1906. U. S. Coast Geod. Surv., Rept. for 1907, App. 3, p. 67–104.

HECK, N. H. (1936) *Earthquakes.* London, 226 p.

HEIM, A. (1934) *Earthquake region of Taofu.* Geol. Soc. Am., Bull. 45, p. 1035–1050. Earthquakes and displacements.

HOBBS, W. H. (1910) *The earthquake of* 1872 *in the Owens Valley, California.* Beiträge zur Geophysik, vol. 10, p. 352–385.

LAWSON, A. C., ET AL. (1908) *The California earthquake of April* 18, 1906. Carnegie Inst. Wash., Publ. 87, 2 vols. and atlas; part 1, 254 p.; part 2, p. 255–451. A full account of the great San Andreas fault. Numerous excellent photographs and maps.

REEDS, C. A. (1933) *A 25-year map of major earthquakes.* Nat. Hist., vol. 33, p. 450–451.

REEDS, C. A. (1934) *Earthquakes.* Nat. Hist., vol. 34, p. 733–747.

REID, H. F. (1933) *The mechanics of earthquakes; the elastic rebound theory; regional strain.* Natl. Research Council, Bull. 90, p. 87–103.

SELLARDS, E. H. (1932) *The Wortham-Mexia, Texas, earthquake.* Texas Univ., Bull. 3201, p. 105–112.

SELLARDS, E. H. (1932) *The Valentine, Texas, earthquake.* Texas Univ., Bull. 3201, p. 113–138.

TARR, R. S., and MARTIN, L. (1912) *The Earthquakes at Yakutat Bay, Alaska in September,* 1899. U. S. Geol. Surv., Prof. Paper 69, 135 p.

# XVII
# FOLDED MOUNTAINS

*Courtesy of Ralph L. Miller*

STRIP WORKING ALONG STRIKE OF ANTHRACITE COAL OUTCROP, NEAR COALDALE, PA.
The left foreground shows subsidiary folds of massive sandstone from which the coal has been stripped. Monoclinal ridges in the distance.

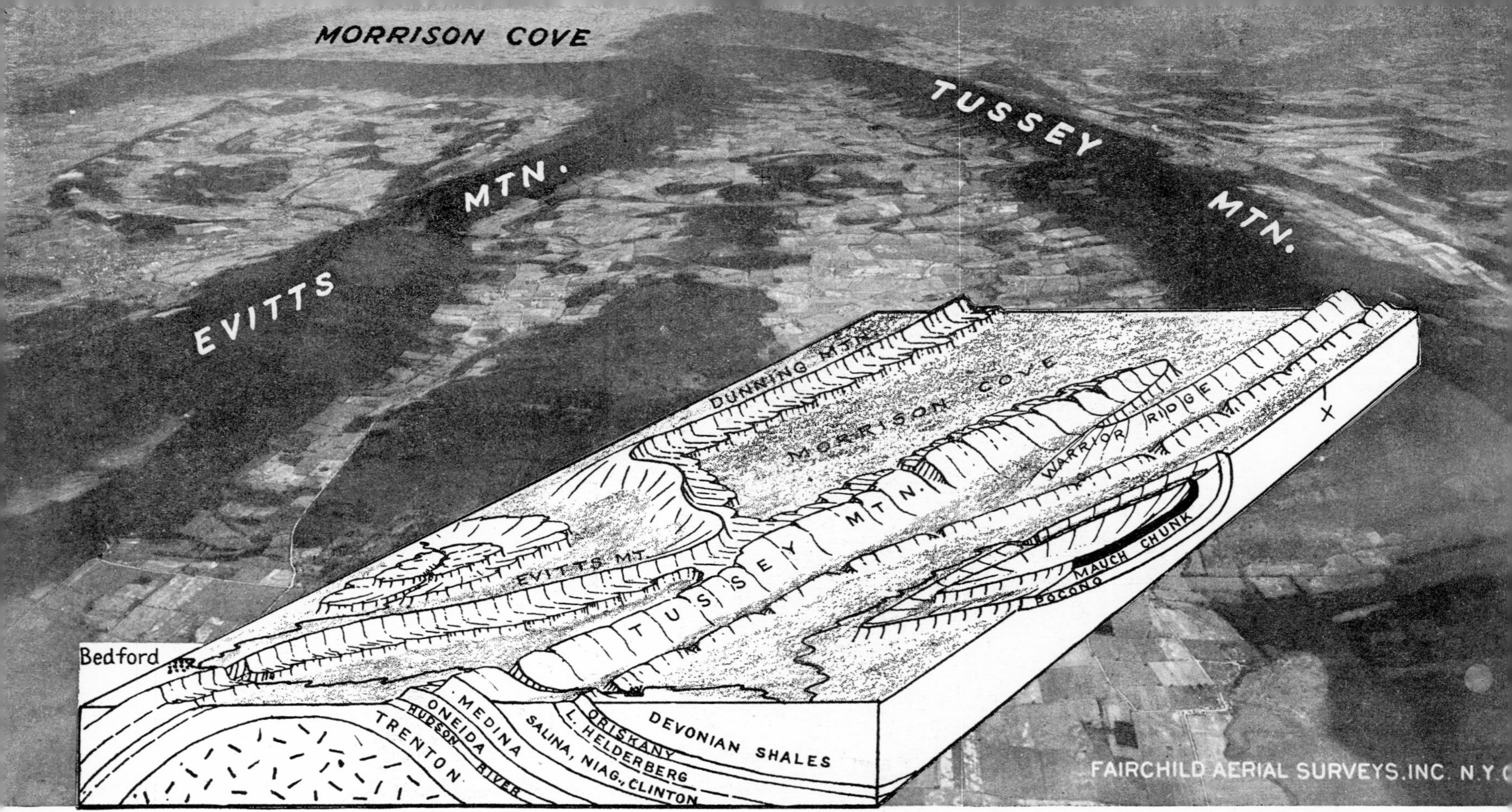

*Fairchild Aerial Surveys*

ZIGZAG RIDGES IN THE FOLDED APPALACHIANS OF PENNSYLVANIA

The diagram shows the topography and structure of the Tussey quadrangle. Note the clearly defined basin structure west of Evitts Mountain on both diagram and picture.

VIEW LOOKING WEST ACROSS TUSSEY QUADRANGLE, PA., SHOWING RIDGES OF THE FOLDED APPALACHIANS

Warrior Ridge is part of a breached anticline pitching south.

*Fairchild Aerial Surveys*

*Swissair-Photo*

WATER-GAP ACROSS PITCHING ANTICLINE; MOUTIER, RÔCHE DE COURT, SUR CELIBERG, AARE, AND BIELERSEE IN THE SWISS JURA



ANTICLINAL RIDGES OF THE FOLDED JURA, SWITZERLAND
THE RAIMEUX FROM THE SOUTHEAST

*Swissair-Photo*

*Stose, U. S. Geological Survey*

GAP OF THE POTOMAC RIVER THROUGH SIDELING HILL, AND OTHER FOLDED MOUNTAINS AT GREAT CACAPON, W. VA.



RIDGES OF THE FOLDED APPALACHIANS IN VIRGINIA

The ridge crests are much less even than those in Pennsylvania.

*U. S. Forest Service*

EVEN-CRESTED RIDGES OF THE FOLDED APPALACHIANS, EASTERN PENNSYLVANIA

Maturely dissected folded mountains once peneplaned. Each ridge is a monoclinal mountain of resistant sandstone or conglomerate.

*A. K. Lobeck*

# FOLDED MOUNTAINS

SYNOPSIS. Folded mountains range in complexity from simple open folds to tightly compressed overturned and faulted structures of large dimensions. Simple folds are characterized by alternating anticlines and synclines. Each anticline, if simple, pitches away at its two ends. Anticlines and synclines, however, may fork and branch.

Young folded mountains, like young dome mountains, are practically unknown. This is probably because folding is a slow process and maturity is reached in the erosion cycle long before folding ceases.

Maturity is characterized by linear monoclinal ridges and monoclinal valleys. At the ends of the folds these ridges and valleys converge, and the characteristic zigzag pattern of folded mountains results. The convergence of the ridges at the nose of an anticline is in the direction of pitch; at the nose of a syncline the convergence is opposite to the direction of pitch.

It is apparent that the number of ridges in a region is due not only to the number of folds but also to the number of resistant beds involved.

An important result of the erosion of folded structures is the reversal of the topography or development of *inverted relief*. The wearing away of the anticlines and excessive denudation along their axes, where weaker strata are exposed at the surface, causes the synclines to occupy the higher levels and thus synclinal ridges are produced. The anticlinal uplifts are replaced by anticlinal valleys.

It appears that most folded mountain regions have been peneplaned and are now mature in one of their later cycles of development. This is borne out by the uniform height of the ridges and by the transverse courses of the larger rivers, which seem to have been superposed either from a recently deposited coastal-plain cover or from alluvial deposits of their own making on a peneplane surface. These are consequent streams. Their larger tributaries, following the weaker belts, are subsequent, and they in turn have obsequent and resequent tributaries. The resequent streams flow either down the dip of the monoclinal ridges or down the pitch along the axes of the synclines. The whole drainage pattern is called *trellis*.

In a study of folded mountains, and especially in the examination of topographic maps of folded structures, numerous minor details are valuable in revealing the structure. Benches occurring along one side of a monoclinal ridge indicate that the beds dip in the opposite direction, not toward the bench; funnel-shaped water gaps across monoclinal ridges converge in the direction of dip; the nose of a pitching anticline is rounded and tapering; the nose of a pitching syncline is steep and abrupt; obsequent streams are usually shorter than resequent streams; a synclinal ridge is flat-topped and often slightly depressed along the axis; the gentle side of monoclinal ridges is on the dip slope.

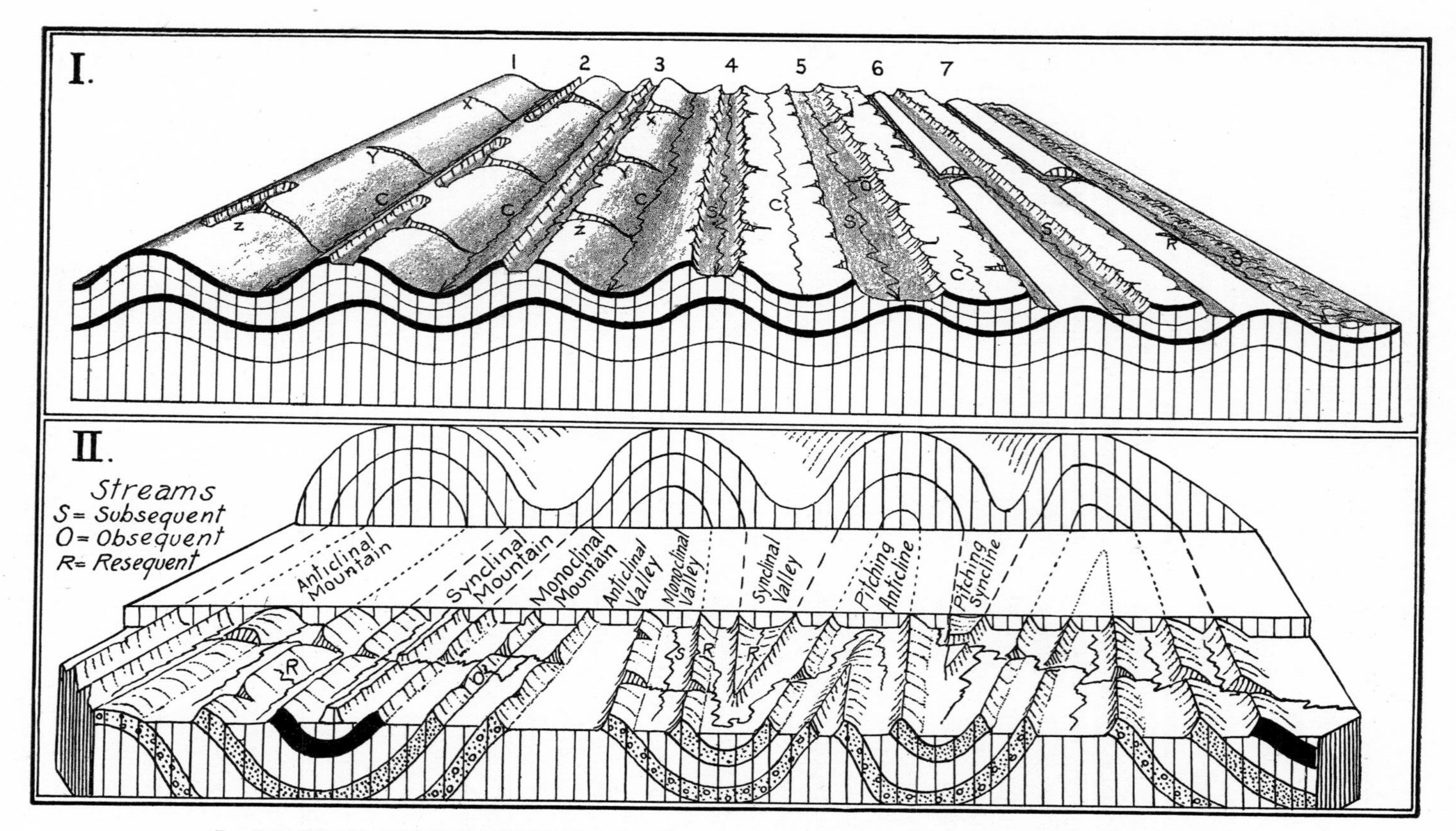


I.—DIAGRAM ILLUSTRATING THE EROSIONAL DEVELOPMENT OF FOLDED MOUNTAINS
II.—FOLDED MOUNTAINS, PENEPLANED AND REJUVENATED, ILLUSTRATING ANTICLINAL, SYNCLINAL, AND MONOCLINAL RIDGES AND VALLEYS

## FOLDED MOUNTAINS. THE CYCLE OF EROSION

The term *folded mountains* is given to those structures which have open and relatively simple folds. There is every gradation between simple open folds and tight folding, in which overturning of the beds and thrust faulting play a prominent part. This latter type, common in the Alps and similar ranges, may be designated as *complexly folded*. There is, in the other direction also, a gradation from the well-defined folded structure of folded mountains to the subdued undulations of a warped plateau.

FOLDED MOUNTAINS. INITIAL STAGE IN THE EROSIONAL CYCLE. In their initial stages, therefore, folded mountains may be pictured as in the left-hand side of Fig. I. A series of sedimentary beds has been regularly bent into arches and troughs, that is, into anticlines and synclines. Several different types of rocks are included in the series, as would be expected where sedimentary layers are involved. For the sake of simplicity, the arches and troughs are assumed to lie almost horizontally with no appreciable pitch.

Two sets of streams take up their courses on these initial folds. One group represented by stream *C* occupies the synclinal depressions, and a second group represented by the streams *X*, *Y*, *Z*, tributary to the first, flows down the flanks of the arches. All of these streams are consequent streams, their direction of flow and their location being determined by the initial slope of the land. The long streams occupying the troughs, such as *C*, are *longitudinal consequents;* the others coming in from the side are *lateral consequents*.

YOUTH. As erosion proceeds, the streams cut down their valleys. This is especially true of the lateral consequents *X*, *Y*, *Z*, because they have steep gradients. In fold 1 of Fig. I, each of these lateral streams *X*, *Y*, and *Z* has developed a gorge on the flank of the fold where the slope is greatest. Stream *Z* has done even more. Its headwaters, having eaten back to the crest of the ridge, have started to work lengthwise. It is, of course, impossible for them to work down on the other side, and their only field for development is along the crest in either direction. Further development, as in fold 3, shows that all of the lateral streams have worked headward to the crest of the ridge and all of them have opened out basins along the axis. This is exactly after the manner of dome-mountain development previously described. More than this, however, stream *Z* is for some reason more vigorous than its fellows and has encroached upon the territory of stream *Y* and diverted its headwaters by capture. This notable achievement is followed by further success of the same sort. In the next fold, No. 4, all of the streams working on the crest of the ridge have been diverted in this manner but stream *Z* has itself been captured by some other stream not appearing in this illustration. In short, a new river, *S*, has come into being, a stream which occupies what was formerly the axis of the fold. This stream *S* is a subsequent stream. It flows upon the weaker, underlying beds of the area, and it follows the strike of the formations. When this stage is reached, the region is mature. That is, with the approach to this condition, folded mountains acquire their greatest degree of relief and perhaps their greatest amount of detail. These are always characteristics of maturity when that term is applied to any land form.

MATURITY. In the next fold, No. 5, stream *S* shows still further its vigorous traits. Flowing, as it does, upon a belt of weaker rock, it is able to cut down more readily than the streams on either side. Its headwaters have reached back into the plateau and have already tapped the drainage region of stream *C* between folds 5 and 6 where stream *C* was originally the master stream. Stream *C* has been completely dismembered by a series of captures and no longer functions as an independent system. Thus a new type of stream has come into being. It is the obsequent stream *O* which has worked back into the narrow plateau. Obsequent streams are usually short streams with steep gradients, often tributary to subsequent streams. Hence they are predatory in their habits. In the next fold, No. 6, stream *S* has stripped away much of the weaker material upon which it was flowing and has exposed an older, underlying, and more resistant bed. In part of its course it actually cuts across this newly exposed ridge. This is a case of *structural superposition* and it is perhaps only one way in which subsequent streams become superimposed. Elsewhere, stream *S* still follows the outcrops of the weaker beds, first on one side and then on the other side of the new ridge.

Some of the tributaries of stream *S* are small streams like *R*, in fold 7, which flow down the flanks of the newly uncovered arch of resistant rock. These are *resequent streams*. They have courses similar in direction to the much earlier consequent streams of the region, but they have been more recently developed; hence *resequent* meaning *recent consequents*.

OLD AGE. In its final stage in the cycle of development a region of folded mountains exhibits little or no relief. A peneplane bevels the structure and the streams flow across it with little regard for the presence of weak or resistant beds. In fact, it is more than likely that in these later stages the streams have spread a thin mantle of alluvium over the area, thus concealing much of the structure.

REJUVENATION. After a peneplane has been formed, a new cycle of development may be introduced by a general uplift of the entire region. This may be accompanied by tilting and as a result the streams enter upon a new regime of cutting. Streams lying transverse to the structure cut down across the harder formations. In general, however, the larger drainage lines adjust themselves to the weaker formations. In this manner the region is etched out. Parallel ridges and valleys alternate with each other. The ridge tops present a rather level profile, the trace of the peneplane surface inherited from the previous cycle. Occasional wind gaps may interrupt the otherwise more or less continuous skyline of the ridges. Such gaps attest the former course of transverse streams long since diverted to other channels. The number of ridges is due not directly to the number of the original folds but rather to the number of resistant formations and to the positions the folds happen to have with relation to the peneplane surface which bevels them.

TYPES OF RIDGES AND VALLEYS. During the erosion of folded mountains three structural types of ridges and three corresponding structural types of valleys are produced. They are known as *anticlinal*, *synclinal*, and *monoclinal ridges*, and *anticlinal*, *synclinal*, and *monoclinal valleys*. In Fig. I on the previous page all of these types are shown.

*Anticlinal ridges* may result from the original upwarping of the strata, as in fold 1, or they may be due to the removal of surrounding weaker material by erosion, as in fold 6. In either case, however, the axis of the anticline follows the axis of the ridge.

*Synclinal ridges* are due entirely to erosion. They have the appearance of narrow, elongated plateaus. The flat area along the crest may, however, be much less extensive than that shown in Fig. I in the ridge between folds 5 and 6. Synclinal ridges drop off abruptly by escarpments to the bordering valleys on either side.

*Monoclinal ridges* are perhaps more common than the other two types. Many of them are shown in Fig. II. If the dip of the formations is less than 45°, usually the dip slope of the ridge is more gentle than the slope on the opposite side of the ridge. But if the dip of the formation is very great, then it is likely that the slopes on the two sides of the ridge will be about the same.

An *anticlinal valley* is shown in fold 4. It follows the axis of an anticline and is drained by a subsequent stream. Anticlinal valleys are always due to erosion. They represent a reversal of the topography, for such a valley now occupies the site of the former ridge. Along the axis of an anticlinal valley there occur the older rocks of the region, just as along the crest of a synclinal mountain are to be found the younger rocks.

*Synclinal valleys*, like that between folds 1 and 2, are due to the initial folds of the region. Like anticlinal mountains, however, which may be due to the stripping off of overlying strata as in fold 7, so also may synclinal valleys be formed by removal of the higher beds from a syncline. To these the term *resequent synclinal valleys* may be applied.

*Monoclinal valleys* are probably more common than the two preceding types of valleys for the same reason that monoclinal ridges are more common than the other types. Monoclinal ridges and monoclinal valleys alternate where there is a series of formations all dipping in the same direction.

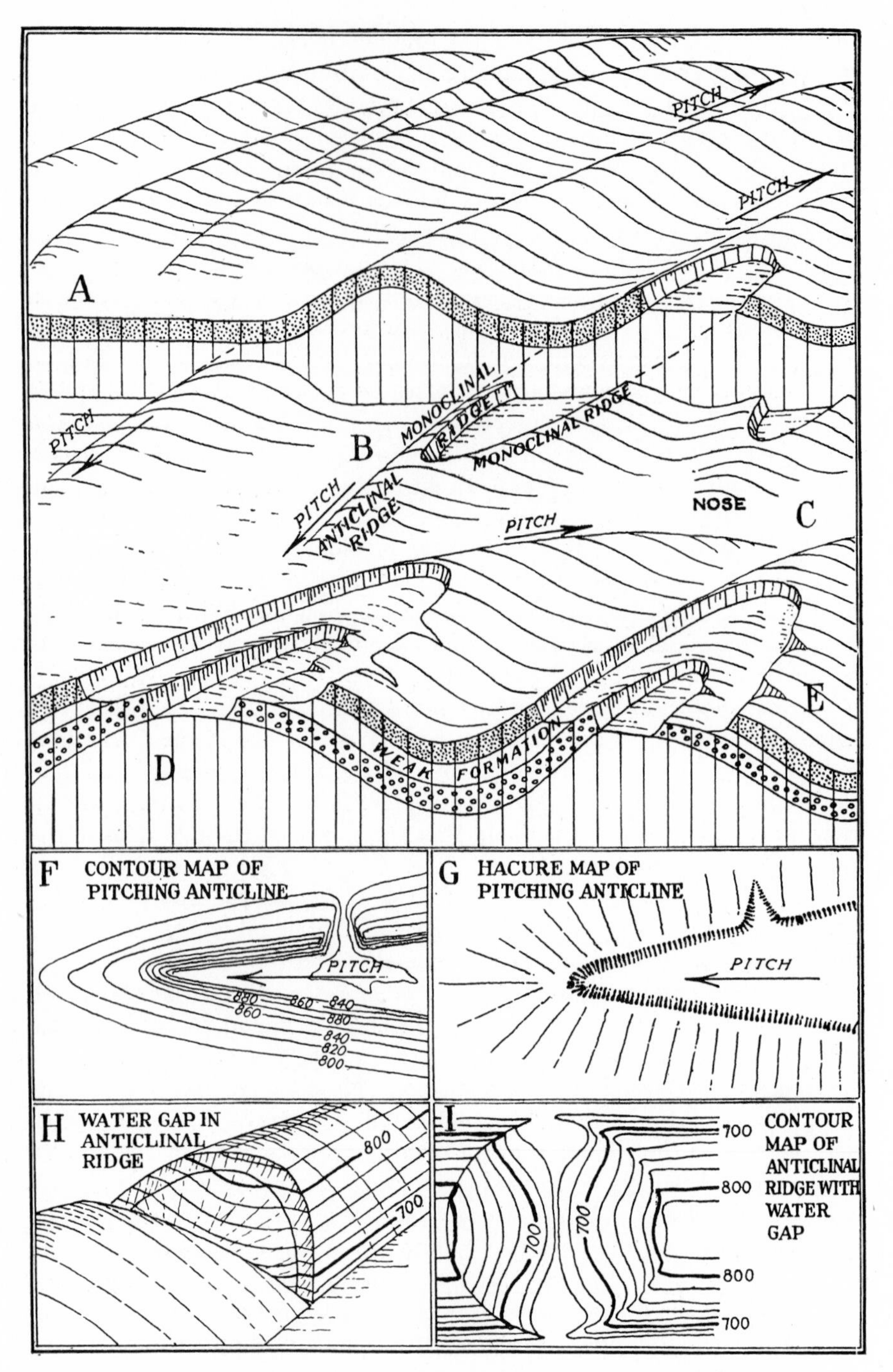


THE EROSIONAL FEATURES OF PITCHING ANTICLINES

THE EROSIONAL FEATURES OF PITCHING ANTICLINES. The initial surface of a folded region, if it is ever developed without much modification by erosion, would have the aspect of Fig. *A* opposite. It would look more like a wrinkled sheet or tablecloth than like a piece of corrugated cardboard. The folds would run not interminably but for short distances and die out, only to be replaced by other folds which appear on either side. These wrinkles or arches in folded-mountain districts have been termed *cigar-shaped mountains*. Each fold plunges at its two ends, or the axis of the arch pitches into the ground at each end. In some cases the wrinkles fork, having branches or prongs. When erosion of such structures takes place, the resulting topographic features are not exactly like those previously described. In Fig. *B* the two monoclinal ridges formed on the two limbs of the anticline do not run strictly parallel with each other but converge and meet, the convergence being in the direction toward which the anticline pitches. If the anticline pitches gently, as at *B*, then the convergence is gradual, the two monoclinal ridges being almost parallel, and the end or nose of the anticline will form an anticlinal mountain which descends gently into the ground. If, however, the anticline pitches steeply, then the convergence of the monoclinal ridges is more rapid and the nose of the anticline plunges underground more abruptly, as in Fig. *C*, the whole structure then resembling a dome. The monoclinal ridges present their steeper slopes toward the axis of the anticline and their gentler slopes away toward the outside. In some cases, on the inside of the anticline, between the two bordering ridges there may be a platform or bench, revealing the presence of another resistant member, as at *D*. This bench may continue indefinitely along the base of the monoclinal ridge. If there are two resistant formations, there is a double set of features, one inside the other, and even more if there are more than two resistant beds.

In asymmetrical or overturned anticlines the monoclinal limbs after erosion, instead of having steeper slopes toward the axis, may have slopes which are more or less the same on both sides of the ridge.

CONTOUR MAPS OF ANTICLINAL STRUCTURES. Figure *F* is a contour map of an eroded pitching anticline. *G* is a hachure map of a similar feature. Figure *H* is a picture of a water gap transecting an anticlinal ridge. Figure *I* shows by contours a similar gap across an anticlinal ridge. On the right-hand side the contours are separated, to show how they follow the face of the cliff a short distance before swinging across the gentle slope in the middle of the gap. The resistant formation of the anticline comes to river level at both ends of the gap but in the middle it is far above the stream. On the left side of the gap the contours are drawn together as they would appear if the cliff were vertical. Such gaps, represented by contours, may be seen on the *Greenland Gap, W. Va.*, topographic map.

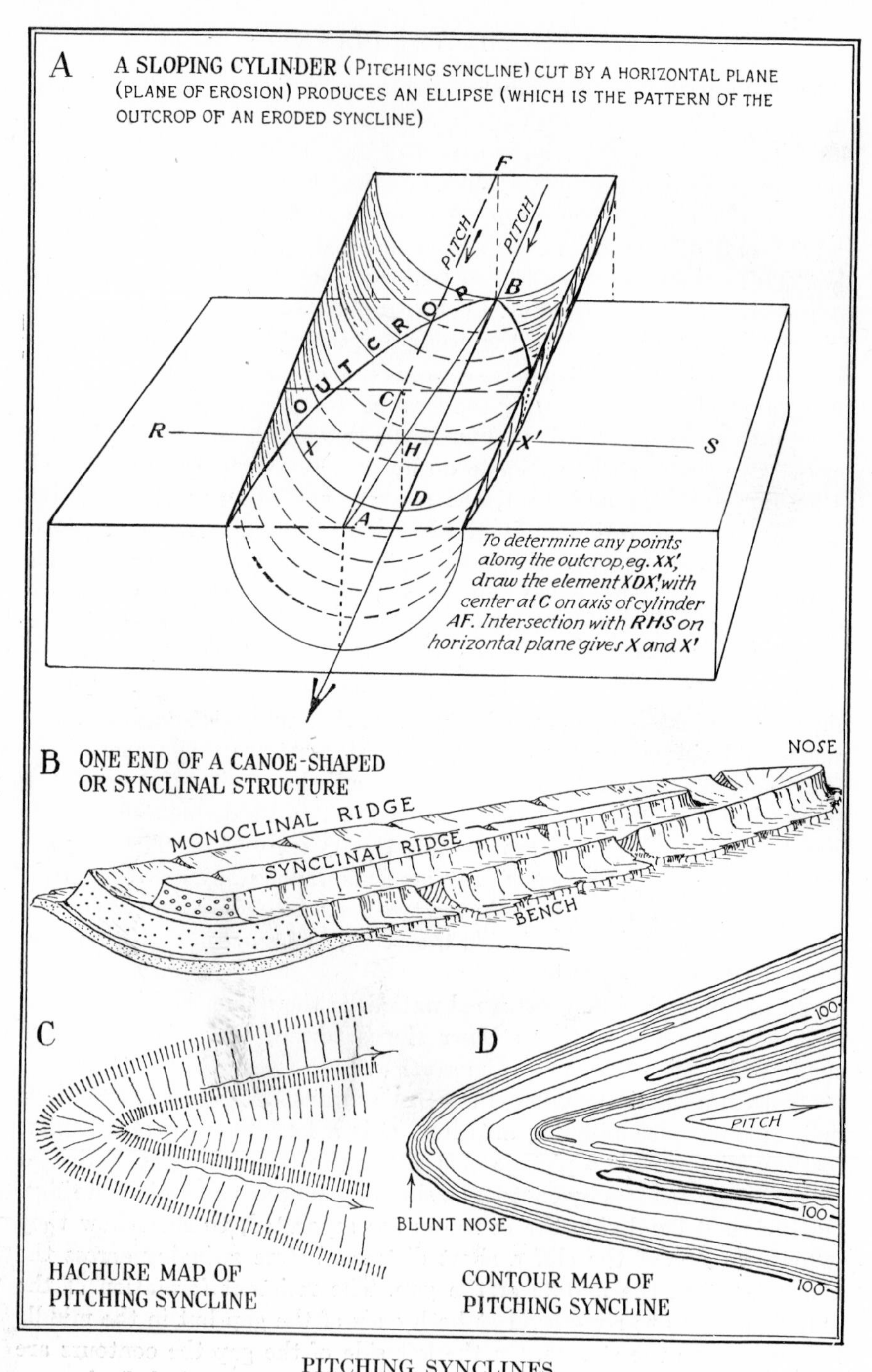


PITCHING SYNCLINES

## PITCHING SYNCLINES

It has been shown on the pages just preceding that the erosion of pitching or plunging folds produces ridges which converge toward an apex on the axis of the fold. In most respects the features developed upon pitching synclines are just the reverse of those developed upon pitching anticlines. Figure *A*, opposite, depicts a pitching syncline which is represented as a perfect cylinder, the pitch being toward the observer. A horizontal plane of erosion cuts this cylinder. The resulting outcrop is shown as a curved line which is an ellipse. If the pitch is very gentle, then the ellipse approaches a parabola in form and the limbs of the fold are almost parallel with each other; but if the pitch is very steep, as in a dome, the pattern of outcrop is almost circular.

THE EROSIONAL FEATURES OF PITCHING SYNCLINES. It may now be observed that the monoclinal ridges which result from the erosion of pitching synclines converge in a direction which is opposite to the pitch of the fold, this being contrary to the facts in regard to pitching anticlines. Moreover, the monoclinal ridges have their steep slopes facing toward the outside, and their gentle slopes directed inward toward the axis of the fold as shown in *B*. The nose of the eroded syncline is steep and blunt and is therefore in contrast with the nose of the pitching anticline which is smooth, rounded, and tapering. Occasionally benches occur at the foot of the steeper slopes just as they do in the case of anticlines, but the benches associated with synclines face away from the structure instead of toward the axis, as in the case of those associated with anticlines.

Synclinal mountains are long narrow ridges, steep on both sides. Occasionally they have flat crests and are then like elongated plateaus, the center of the plateau usually being slightly lower than the rim on either side.

Some synclinal mountains have a width of several miles, as, for example, Broad Top Mountain in south central Pennsylvania and Walden Ridge in Tennessee.

If the syncline is not perfectly symmetrical, then the resulting synclinal mountain may have a sharp ridge on one side, where the structure is strongly upturned, and a simple escarpment on the other, where the formations are flat lying.

In spite of the fact that synclinal mountains represent a reversal of the topography due to the erosion of the original folds, nevertheless synclinal mountains are as common as anticlinal mountains. Practically all large plateau areas are synclinal in structure.

Gaps across synclinal mountains are rarely so characteristic as those across anticlines. Occasionally such gaps show a more flaring opening at the two ends and a somewhat constricted middle portion, this being quite in contrast with the gaps in anticlinal ridges, pictured in *H* and *I* on the preceding page.

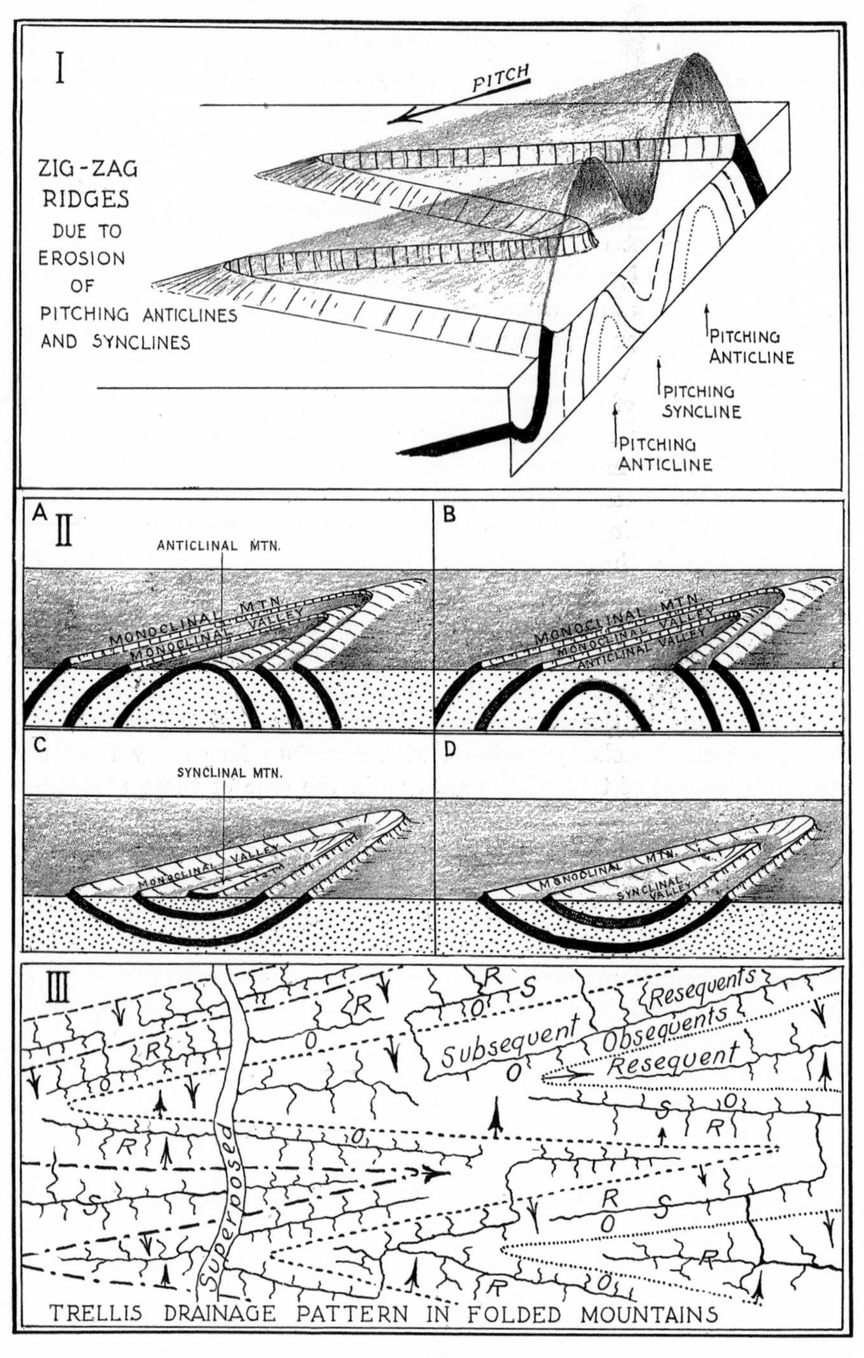


ZIGZAG RIDGES

## ZIGZAG RIDGES

The early settlers who first went into central Pennsylvania adopted the term *Endless Mountains* for the Appalachian ridges. These ridges formed gigantic zigzag patterns on the earth's surface and seemed to be without an end, because at intervals they turned back and forth upon themselves instead of terminating as many mountains do.

THE CAUSE OF ZIGZAG RIDGES. It is the combination of pitching anticlines and synclines which causes the zigzag pattern.

Figure I, opposite, depicts two anticlines and an intervening syncline, all pitching in the same direction toward the left side of the drawing. Erosion of the folds has produced a level-topped zigzag ridge. From this it may correctly be inferred that, whenever a branching or forking fold is eroded, zigzag ridges result. Also the folds must be open folds, that is, not too tightly compressed or overturned or complicated by faulting. Thrust faulting is apt to cause the disappearance of one limb of the fold.

If two or more resistant formations are present, then there may result several parallel sets of zigzags, as in the coal basins of eastern Pennsylvania.

Figure II, on the opposite page, illustrates the variations in the features produced along the axes of pitching folds. In *A* an anticlinal mountain occupies the axis of the anticline but in *B* there is an anticlinal valley, because erosion has not revealed the underlying third resistant formation. In *C* a synclinal mountain appears in the axis, but in *D* there is a synclinal valley, either because a third resistant formation is lacking or because it has been completely removed by erosion.

It appears, therefore, that the presence or absence of an anticlinal or synclinal mountain is fortuitous as it depends upon the shape of the fold, the number of resistant formations, and the degree of erosion, none of which bear any direct relation to each other.

TRELLIS STREAM PATTERNS. The stream pattern in a region of folded structure is typical and is known as *trellis*, as it resembles the branches of a vine on a trellis or arbor. Usually there is a master stream which transects the structure. Such a transverse stream has probably been superposed from a peneplane which beveled the region. It cuts through the ridges in water gaps. Tributary to the master stream are the longitudinal streams. These follow the belts of weaker rocks and are therefore *subsequent* in origin. The subsequent streams in turn have tributaries of two types: One set is short; the other slightly longer. The short tributaries are *obsequent* streams, which flow opposite to the direction of the dip of the beds. The longer are *resequent* streams, which flow down the dip of the beds. A second type of *resequent* stream may also be recognized. It is represented by those streams which flow along the axes of synclines, in the direction of the pitch of the structure. They flow, however, upon a stripped surface, and not in the bottom of an original trough.

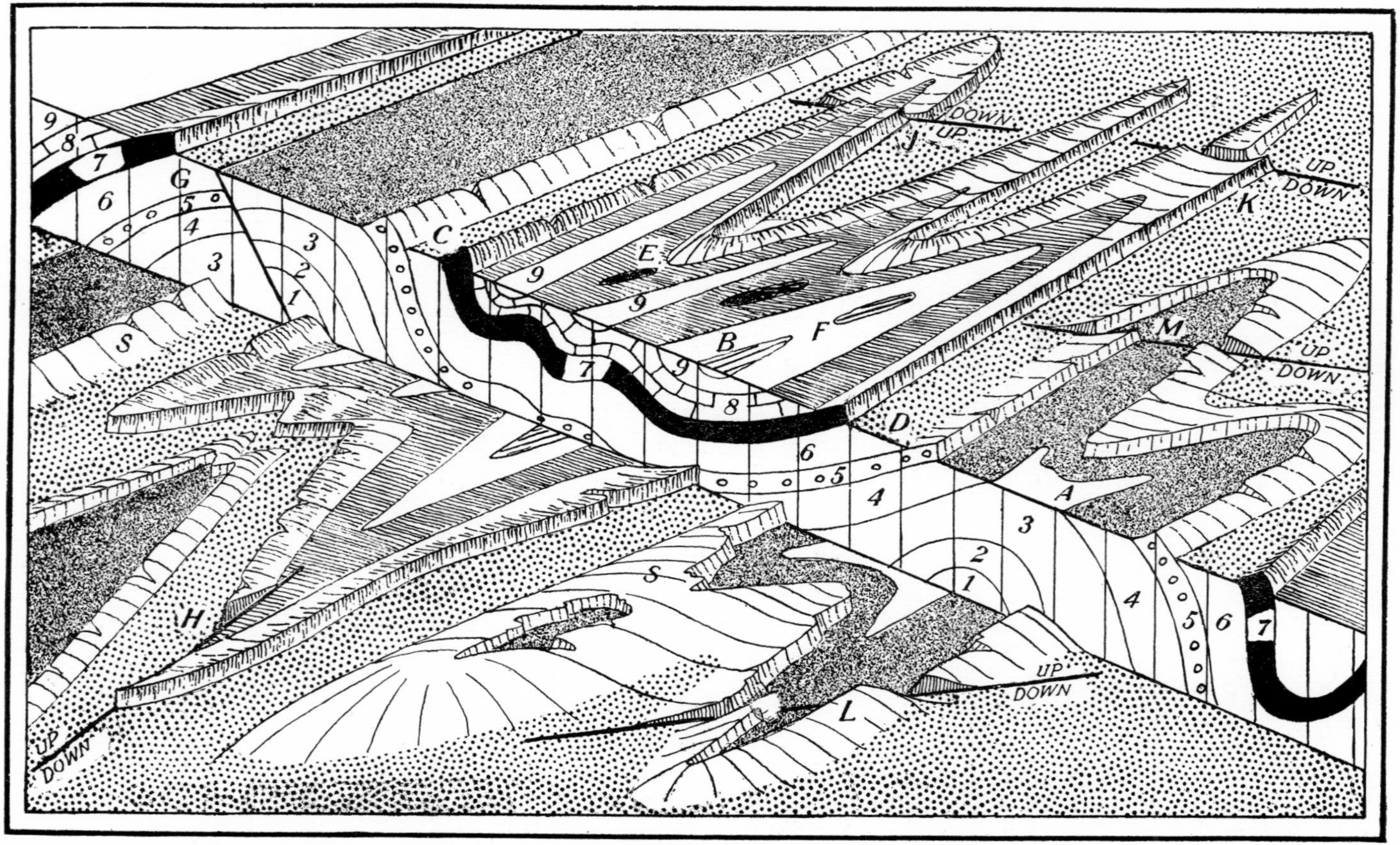

GEOLOGICAL DIAGRAM OF ERODED PITCHING ANTICLINES AND SYNCLINES TO ILLUSTRATE PATTERNS OF OUTCROP AS INFLUENCED BY THE STRUCTURE

## GEOLOGICAL OUTCROPS

The pattern of geological outcrops in regions of folded rocks is not quite so puzzling as the diagram on the opposite page might make it appear.

First of all, it is obvious that in a region of eroded folds the older rocks will appear along the axes of the anticlines, as at *A*, and the younger rocks along the axes of the synclines, as at *B*. Where the beds stand almost vertically, the outcrops will be much narrower, as at *C*, than where the beds are more nearly flat lying, as at *D*.

A little patch of older rocks surrounded by younger rocks, as at *E*, indicates an anticline; while a patch of younger rocks surrounded by older rocks indicates a syncline, as at *F*. Each of these structures may be superimposed upon a larger structure of the same or of a different sort.

On a geological map showing a zigzag pattern of outcrops, if any one of the following four facts be known, the other three may be deduced:

*a*. Pitch of any fold.
*b*. Dip of beds at any point (provided it is known also whether they are overturned).
*c* Relative age of two of the formations at any point.
*d*. Topography at any point.

The student may experiment with this idea on the accompanying diagram and will have little difficulty in discovering how perfectly it works. For instance, if the pitch at any point is given, then the character of the fold is known, whether anticline or syncline. From this the dip is everywhere determinable, which in turn gives the relative ages of the formations and suggests also which side of the ridges is steep and which side is gentle.

Faulting in folded rocks produces interesting dislocations in the ridges. For example, the thrust fault at *G* (which is at the same time a strike fault, as most thrusts in folded rocks are apt to be) eliminates the bed 5 on one side of the valley and produces also a thinning of the outcrop of bed 6.

The normal fault at *H*, cutting sharply across the strike of the beds, causes an offset in the ridge and a displacement of beds so that bed 6 comes in contact with bed 8, at the point *H*.

At *J* and *K* the ends of two synclines are cut by transverse faults. In each case the part which is dropped down has the wider outcrop.

At *L* and *M* the ends of two anticlines are cut by transverse faults. In each case the part which is dropped down has the narrower outcrop.

All other possible faults will be found to be variations of those illustrated. Practically all folded-mountain regions contain numerous examples of faults, on a small scale, serious enough to create important problems in mining, as in the anthracite-coal basins of Pennsylvania.

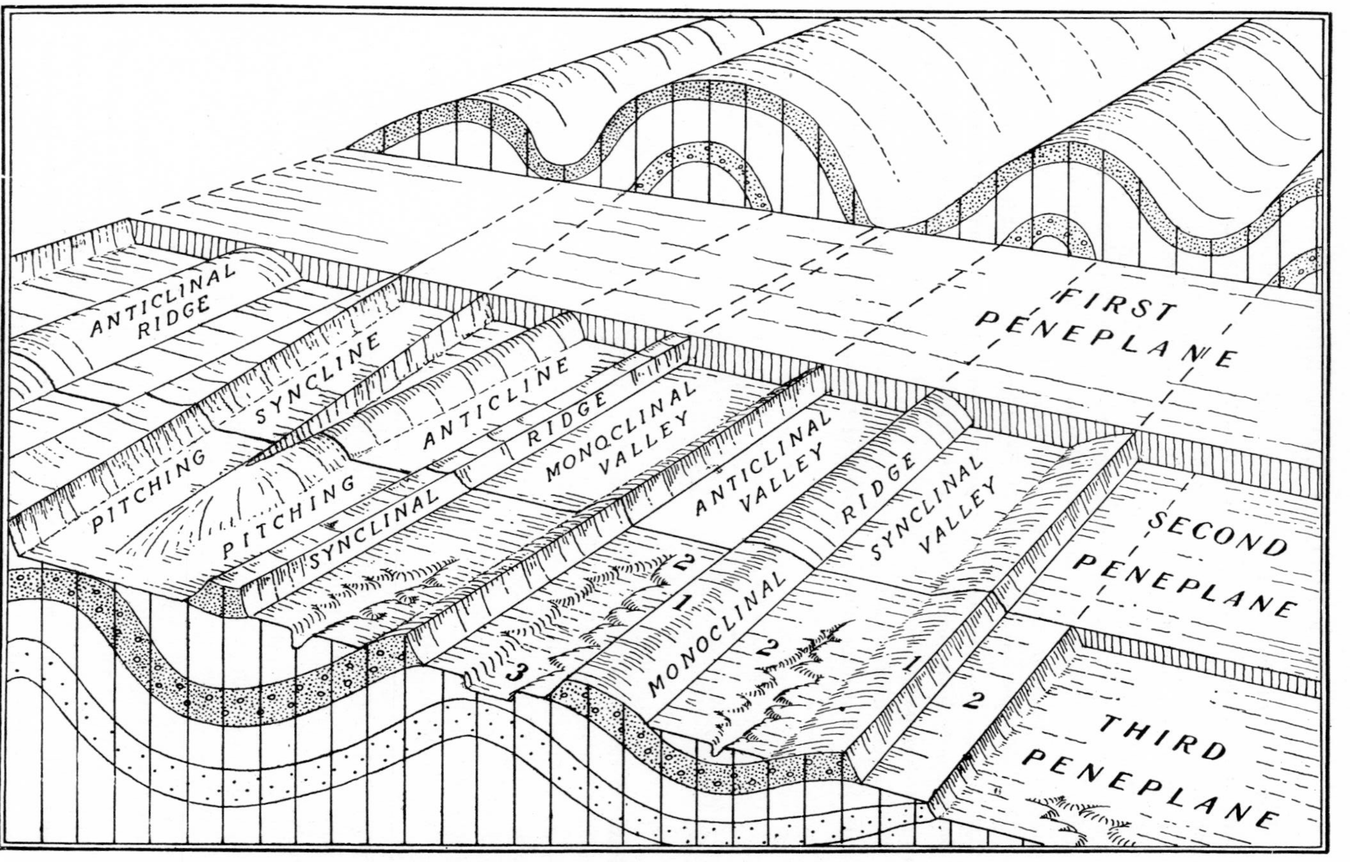


CYCLES OF EROSION IN FOLDED MOUNTAINS

EVIDENCE OF SEVERAL CYCLES OF EROSION. Most existing folded mountain regions are not now young in their first cycle of development. Some folded mountains show evidence of having been reduced to a peneplane and then rejuvenated by uplift so as to cause the weaker areas again to be eroded, perhaps to a second distinct peneplane level.

The accompanying illustration shows in the distance several undissected folds. In the next block the whole region has been reduced to a peneplane, the *first peneplane*. This cuts indiscriminately across weak and resistant beds alike. In the third block from the rear a second and lower peneplane has been formed on the weaker beds. The resistant layers stand up as monadnock ridges, their even crests alone preserving the first peneplane level.

In the foremost block, another period of rejuvenation is recorded by the erosion of valleys below the level of the second peneplane and by the wearing down of the weakest beds to form a third peneplane. Remnants of the second peneplane form the flat valleys between the ridges, although here and there, where the weak limestone comes to the surface, the third peneplane is beginning to appear. There is even a suggestion of some slight dissection of the third peneplane.

Four cycles of erosion are recorded in this region. The first cycle was complete; the second and third were only partially completed before they were interrupted by rejuvenation; and the fourth cycle has just begun.

There is a temptation to ascribe the flat and even crest line of every ridge to a period of peneplanation. This, however, may not necessarily be true. For instance, in the Folded Appalachians, there are smaller ridges with even crest lines below the level of the main ridges. The lower elevation of these small ridges is due to the lesser thickness of the resistant bed which underlies them. Their crests were once at the level of the higher ridges and have been lowered uniformly throughout their length. Their even height has been inherited from higher peneplanes.

FOUR CYCLES OF EROSION IN PENNYSLVANIA. In the folded mountains of Pennsylvania three peneplanes are usually recognized, somewhat like those here illustrated. The highest peneplane is called the *Schooley*, the second is called the *Harrisburg*, and the third is called the *Somerville*. There is also the beginning of a fourth cycle of erosion.

In the study of all regions where the topography is strongly controlled by the structure, it is necessary to guard against the danger of ascribing to peneplanation those effects which are due to differential erosion. Broad horizontal benches and uplands are in some localities taken to be peneplanes when they reflect in no way cycles of erosion. Such benches may all be formed simultaneously by differential erosion upon horizontal structures.

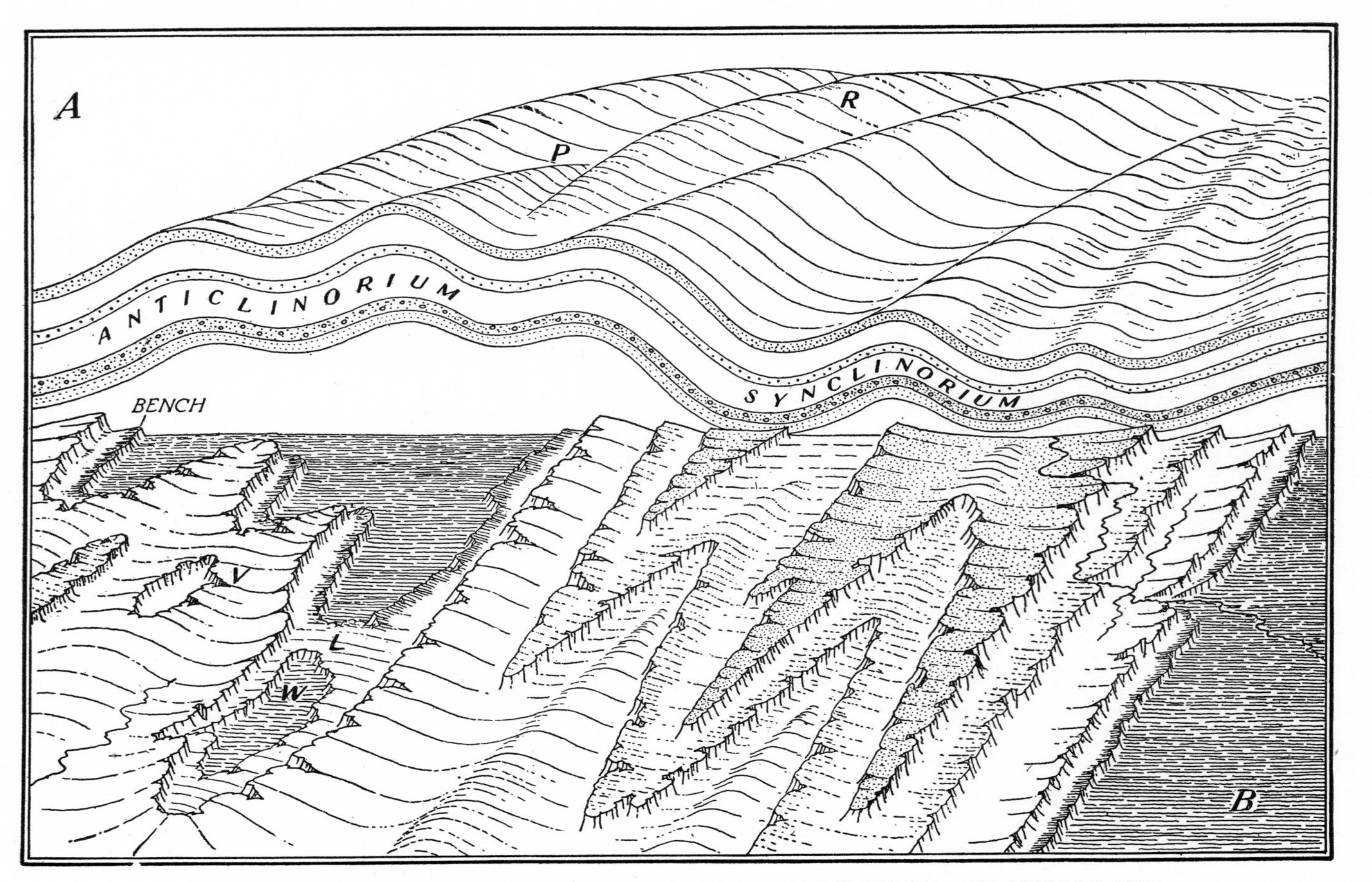


THE FEATURES PRODUCED BY THE EROSION OF ANTICLINORIA AND SYNCLINORIA

## ANTICLINORIA AND SYNCLINORIA

Most folded mountain regions of the world consist of broad zones of upfolding and zones of downfolding upon which are superposed the smaller anticlines and synclines. The large up-warped areas are called *anticlinoria;* the down-warped belts are called *synclinoria.*

The drawing on the opposite page epitomizes the essential features of such structures, without, however, introducing the vast variety of complications due to thrusting and overturning of folds and the heterogeneity due to the presence of folds of all different sizes and kinds.

In Fig. *B*, in the anticlinorium, erosion has carried away most of the higher strata so that only the lowest of the resistant formations is left to produce ridges. This means that an eroded anticlinorium usually exhibits many coalescing anticlinal valleys and relatively few ridges.

The synclinorium preserves fragments of all the resistant formations. This means many ridges. An eroded synclinorium is apt to exhibit several synclinal basins branching from a main axis.

If a synclinorium consists of a resistant member, deeply buried under a thick series of weak deposits, a broad basin entirely devoid of ridges may result. This is true of the so-called *Swiss Plateau* and of the *Mazarn Basin* in the Ouachitas.

If an anticlinorium is not too highly arched and if it consists of but one resistant member overlain by weak beds, erosion may simply strip off the weak cover and leave a series of unbreached anticlinal ridges. This is the aspect of the Jura Mountains in Switzerland.

Plunging anticlines, as shown at *P*, are apt to be replaced laterally by other anticlines plunging in the opposite direction. The crests of anticlines, too, rise and fall as they are followed along their strike, as at *R*. Erosion of such anticlines may produce a succession of topographic basins or windows, as at *W* and *V*, separated from each other by *land bridges*, as at *L*, the bridges occurring where the anticlines are least elevated. The basins may drain outward or the drainage may be underground through sink holes if the underlying formations are porous or soluble.

The entire anticlinorium is apt to be arcuate, the folds at the ends being close together and gradually pinching out entirely, as if folding was prevented by some underlying resistant noncompressible mass.

The illustration depicts many of the details in folded mountains already alluded to. Attention may be called to the benches at the base of the ridges surrounding the anticlinal valleys. Such a bench occurs when two resistant members making up a ridge are separated by a thin weaker layer. The presence of a bench at the foot of a ridge is always an indication that the beds in the ridge are dipping away from the bench.

Note also the several water gaps, which always narrow down in the direction of the dip of the beds, regardless of the direction of flow of the streams.

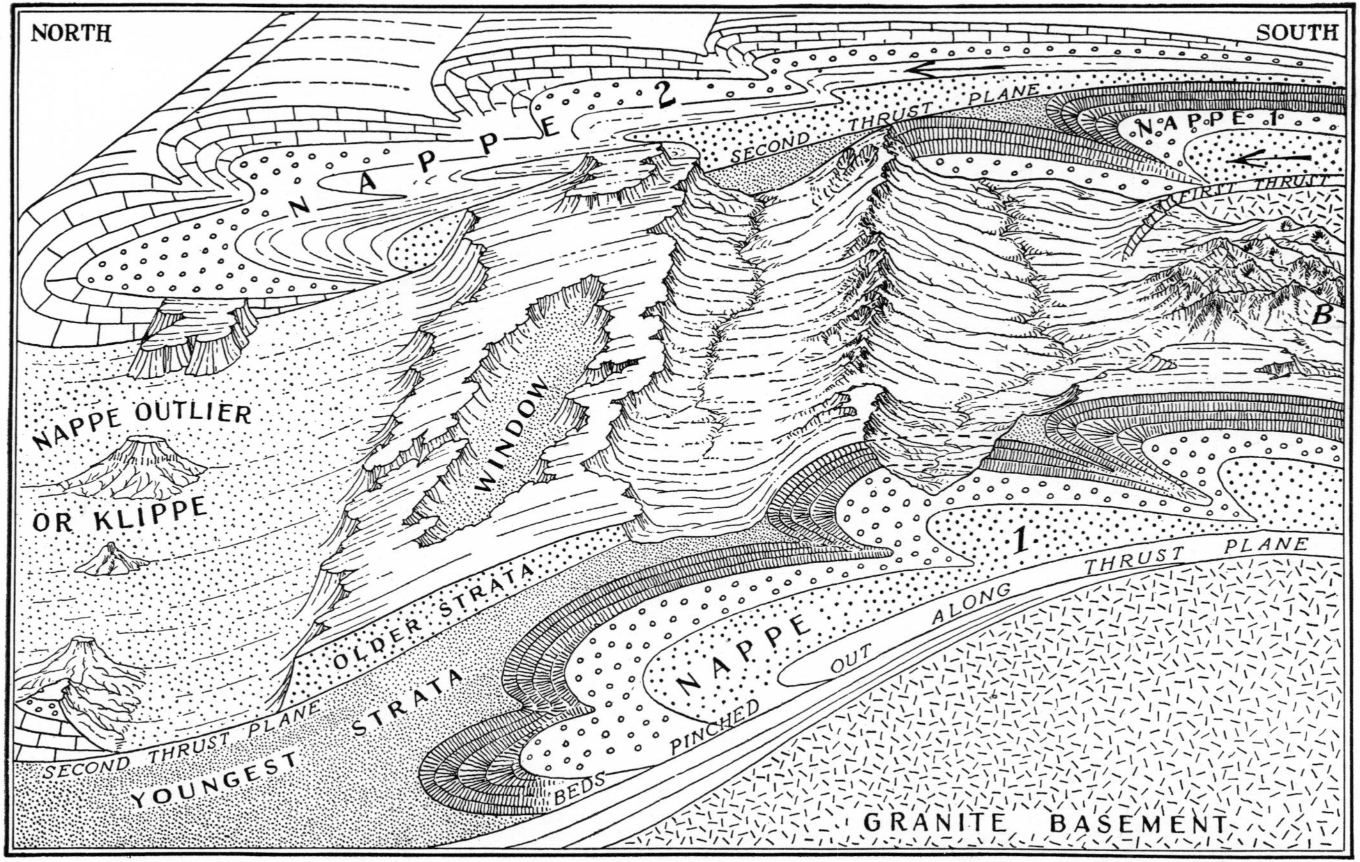

DIAGRAMMATIC REPRESENTATION OF ALPINE FOLDING SHOWING TWO NAPPES, THRUST FROM THE RIGHT, THE SECOND ONE FAR OVERRIDING THE FIRST

NAPPES. The Alps and other mountains of this type, like the Caucasus and the Himalayas, are extremely complicated in detail. In their broader outlines, however, they are easy to understand. Their essential plan consists of a number of great recumbent, that is, strongly overturned, folds thrust from the south, one over the other. These are called *nappes*. Each nappe is many miles long. The back part of the accompanying illustration shows nappe 2 overriding nappe 1, which in turn has been thrust over the core of old crystalline rocks, appearing at the extreme right. On the underside of each nappe, where it has slid along the fault plane, the beds may be very thin, due to stretching, or they may be pinched out entirely.

The large, massive beds, notably the limestone formations, bend into big folds or fracture into large blocks. The softer intermediate shaly beds get squeezed into all manner of small crumplings and in many cases become changed into slate, phyllite, or schist.

Beneath everything is a crystalline base which becomes strongly jointed and broken by minor faults.

The diagram illustrates these points but makes no attempt to show the complexity which results when one fold gets wrapped around another in what is called *involution*.

EROSIONAL FEATURES. The erosion of alpine folds produces an unusually varied and interesting topography.

Where all of the nappes have been eroded away, the cracked-up crystalline base is exposed to form jagged peaks, usually granite, like Mont Blanc, as at *B*. On both sides of the crystalline Alps limestone ranges appear, sometimes in the form of plateaus or tablelands, but more often as ridges and peaks of every conceivable shape. So-called *windows* or *fensters* are produced by the erosion of plateau areas, so that the youngest strata of an underlying nappe are exposed at the bottom of the valley thus formed. The frame of the window is made up of older strata belonging to an upper nappe which was thrust over the lower one.

A *nappe outlier* or *Klippe* is a remnant of a higher nappe spared by erosion. In the field it is recognizable by the fact that older strata cap younger ones.

Windows occur in the higher areas, whereas *Klippen* are more common in the synclinal portions of a nappe. Were it not for the presence of occasional *Klippen*, the reconstruction of the higher nappes would, in many cases, be impossible.

In the Alps and similar ranges, it is necessary to know the true succession of the beds and their fossil content in order to work out the structure of those areas where overturning and faulting occur.

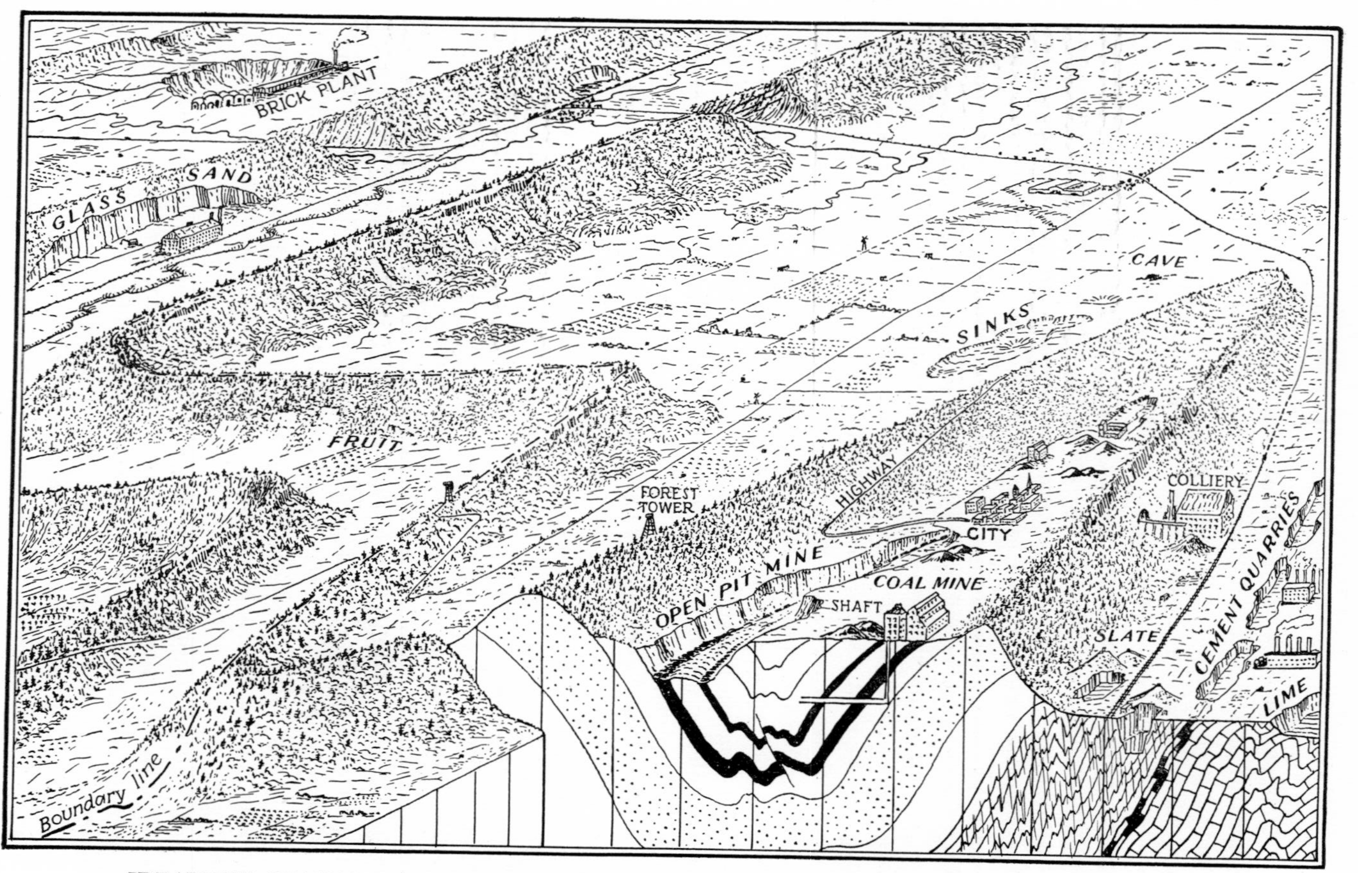


IDEALIZED DIAGRAM SHOWING HUMAN ACTIVITIES IN REGION OF FOLDED MOUNTAINS

## GEOGRAPHICAL ASPECTS

Folded mountains exert an influence upon man's occupation of a region chiefly because of two factors: (*a*) the topography and (*b*) the resources.

INFLUENCE OF TOPOGRAPHY. The topography of folded mountains is so peculiar as to stamp a character all its own upon the pattern of railroads, roads, distribution of towns, and position of political boundaries. Railroads and roads follow the valleys and have a trellis pattern much like that of streams. There are few tunnels of any great extent. The railroads are fairly level but have numerous curves.

Not so with the highways. The Lincoln Highway, for example, crosses the folded ridges of southern Pennsylvania in long ramps. Some of the finest and most picturesque roads cross the ridges without recourse to stream gaps. Occasionally the actual summit traversed may be a wind gap but that is by no means always the case.

In contrast with the transportation routes, which necessarily conform with the valleys, are the political boundaries which follow the ridges. A county map of Pennsylvania indicates how closely the political lines reflect the pattern of the geology and shows in a striking manner the contrast between the folded mountains in the eastern and the plateau structure in the western part of the state. In a similar fashion the international boundary between France and Switzerland jumps from crest to crest of the Jura ridges.

The importance of water gaps in determining the position of towns in folded mountains is evident. Harrisburg and Stroudsburg, Pa., and Cumberland, Md., are good examples. In Galicia, Poland, the fortresses of Cracow, Tarnow, Przemysl, and Lemberg commanded important gaps through the folded foothills of the Carpathians during the war.

The valleys in folded mountains have occasionally served as routes of invasion and military movements. Witness the use of the Great Valley by General Lee before and after the Battle of Gettysburg; likewise the Hudson Valley before the Battle of Saratoga during the Revolution.

INFLUENCE OF NATURAL RESOURCES. The accompanying diagram illustrates some of the activities which depend upon the natural resources of a folded-mountain region. The ridges are heavily wooded; the anticlinal valleys are rich agricultural areas because of the underlying limestone formations; the synclines constitute the coal basins and are the loci of numerous towns. Some coal-mining centers are not on the coal field but in adjacent valleys. The coal is secured by horizontal drifts. Other mineral resources depend upon the presence of limestone for cement; upon the slightly metamorphosed shales for slate; upon the purer sandstone ridges for glass sand and ganister for furnace lining; upon shales which are pulverized for the manufacture of brick. Some resistant sandstones, such as the novaculite of the Ouachitas in Arkansas, serve as abrasives for making grindstones and whetstone.

## MAPS ILLUSTRATING FOLDED MOUNTAINS

There are no areas in the United States which display young, undissected folded mountains in their first cycle of erosion. Numerous maps, however, show folded structures.

Anticlines transected by water gaps are admirably represented on the *Greenland Gap, W. Va.*, sheet, which shows also a monoclinal ridge with several gaps. The *Mount Union, Pa.*, map shows a large anticlinal mountain (Blue Mountain) transected by a large gap. It shows also a splendid synclinal structure in Terrace Mountain. Pitching anticlines are clearly represented on the *Allensville, Loysville, Milton, Hollidaysburg, Everett, Pa.*, and the *Frostburg, Md.-W. Va.-Pa.*, quadrangles. A breached anticline with windows (Nippenose Valley and Mosquito Cove) is shown on the *Williamsport, Pa.*, sheet. Grassy Cove on the *Kingston, Tenn.*, map is a similar feature. The *Davis, W. Va.-Md.*, sheet illustrates a much eroded anticline which has been replaced by an anticlinal valley (Canaan Valley). On the *Everett, Pa.*, sheet Morrison Cove is also an anticlinal valley bordered by monoclinal ridges dipping away to the east and the west. The benches on these ridges are noteworthy. Warrior Ridge on this map is a pitching anticline. The *Pikeville, Pikeville Special*, and *Chattanooga, Tenn.*, maps all showp arts of the remarkable Sequatchie anticlinal valley, as well as the synclinal mountain, Walden Ridge. The *Fort Payne, Ala.-Ga.*, and the *Stevenson, Ala.-Ga.-Tenn.*, maps show a long synclinal mountain.

An especially interesting block of maps consists of the *Hummelstown, Harrisburg, New Bloomfield, Lykens, Millersburg*, and *Pine Grove, Pa.*, sheets, which display together a splendid series of zigzag structures, in which three different resistant formations account for all of the high ridges. Several pitching anticlines and pitching synclines are included. There are also a few water gaps which reveal, by their flaring shape, the direction in which the beds are dipping. A similar gap may be noted on the *Shippensburg, Pa.*, sheet.

Wind gaps are shown on many of the maps but an especially striking one appears on the *Wind Gap, Pa.*, sheet. The maps of the Ouachitas, such as the *Fort Smith* and the *Hot Springs and Vicinity, Ark.*, sheets show many interesting folds but are much less readily interpreted than are those of the Appalachians. For old worn-down folded mountains the *De Queen, Ark.-Okla.*, map is about as satisfactory an example as is available.

One of the most unusual examples of partly submerged folded mountains is shown on the *Belcher Islands* sheet of the Northwest Territory of Canada (scale 1:253,440). Although no topography is represented, the structure can nevertheless be interpreted.

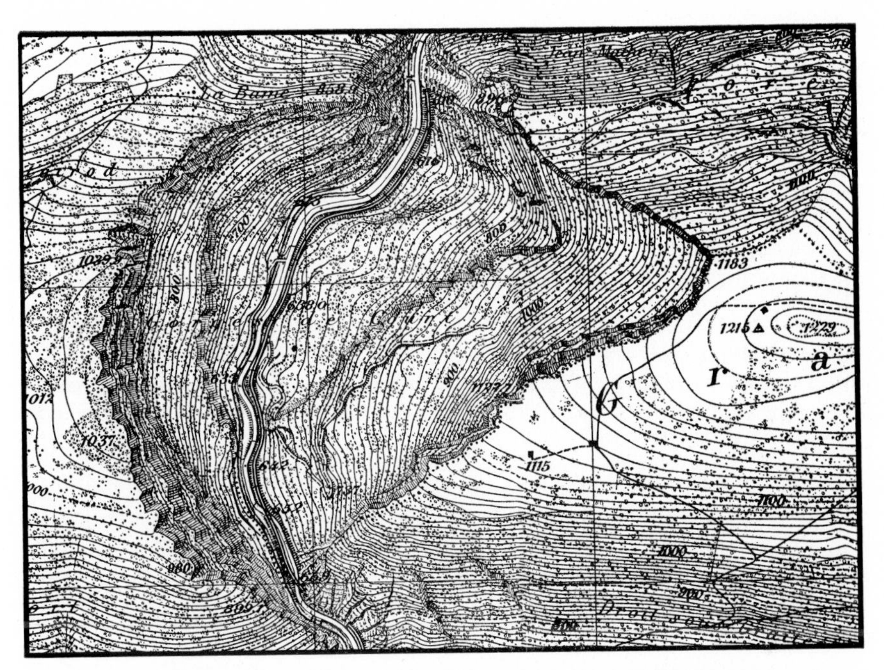


ANTICLINAL GORGE OF LA BIRSE RIVER, SWISS JURA
Switzerland; *Court* and *Gansbrunnen* sheets, Nos. 108 and 109 (1:25,000).

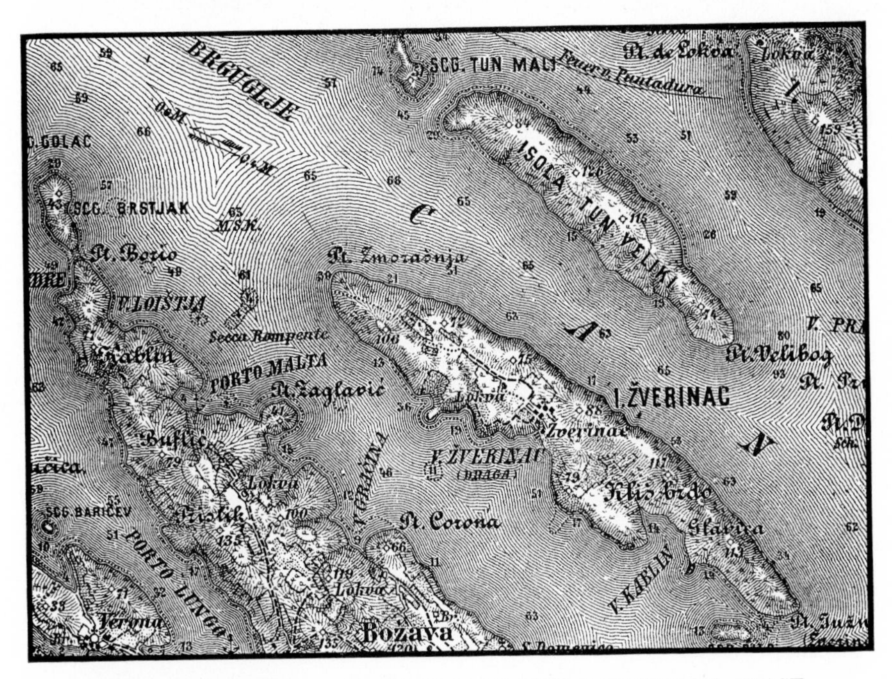


DROWNED FOLDED MOUNTAINS OF THE DALMATIAN COAST
Austria: *Zara* sheet, Zone 29, Col. XII (1:75,000).

## QUESTIONS

1. How would you account for a double crest on a monoclinial ridge?
2. What would cause a monoclinal ridge gradually to disappear when followed along its strike?
3. Anticlinal ridges and synclinal valleys are common in the Jura. Does that necessarily indicate the first cycle of erosion? Explain.
4. How can several cycles or partial cycles of development be represented in a folded-mountain region?
5. Draw a geological cross section of a monoclinal ridge with bench on one side.
6. Draw a contour map of the nose of a pitching anticline; of a pitching syncline.
7. In a region of eroded anticlines and synclines, where would you expect to find the oldest rocks? The youngest?
8. In a region where the folds are overturned so that the rocks all dip in the same direction (homoclinal dips), how could you tell from a topographic map which are the anticlines and which the synclines?
9. Is the direction in which a water gap flares in crossing a monoclinal ridge affected at all by the direction in which the stream flows?
10. Is a stream which follows the axis of a syncline a resequent or a subsequent stream, or may it be either?
11. Why do many folded mountains have such even crest lines?
12. Explain why a symmetrical anticline, an overturned anticline, and a thrust fault may be considered successive stages in the process of folding.
13. Draw a zigzag line representing map of a zigzag ridge. On one side of the ridge mark "older" and on the other side mark "younger" to indicate relative age of the beds. From this information show the axes of the anticlines and synclines; mark the dip of the beds in several places, the pitch of each fold, and show by hachures the steep side of the ridge.
14. Draw the stream pattern only of an eroded folded mountain region. The pattern of the streams should reveal clearly an eroded anticline and an eroded syncline with a number of monoclinal ridges. Label the various genetic types of streams and also mark the anticlinal and synclinal axes.
15. What is meant by *en echelon* folds?
16. What is meant by *similar* folds? By *parallel* folds?
17. What connection is there between folded mountains and isostasy?
18. Select typical quadrangles of the Pennsylvania folded mountains and prepare block diagrams to show topography and structure.
19. What rich limestone valleys occur in the Folded Appalachians? What is the Great Valley?
20. What is the structure of Baraboo Range, Wis.? Shawangunk Mountain, N. Y.? Lookout Mountain, Tenn.? Kittatinny Mountain, N. J.? Hot Springs Mountain, Ark.?

## TOPICS FOR INVESTIGATION

1. Folding. Cause of folding. Factors determining localization of folds.
2. Folded-mountain regions of the United States. The Ouachitas; the Appalachians.
3. Folded-mountain regions of the world. The Jura; the Alps; the Carpathians; the Atlas.
4. The development of drainage systems in folded regions.
5. Contrast between northern and southern Folded Appalachians in regard to structure.

## REFERENCES


### General

Burrard, S. (1922) *Folding of mountain ranges—the argument from isostasy.* Geol. Soc. Am., Bull. 33, p. 333–336.

Claypole, E. W. (1885) *Pennsylvania before and after the elevation of the Appalachian Mountains.* Am. Naturalist, vol. 19, p. 257–268.

Fermor, L. L. (1924) *The pitch of rock folds.* Econ. Geol., vol. 19, p. 559–562.

Ickes, E. L. (1923) *Similar, parallel, and neutral surface types of folding.* Econ. Geol., vol. 18, p. 575–591.

Lawson, A. C. (1927) *Folded mountains and isostasy.* Geol. Soc. Am., Bull. 38, p. 253–273.

Link, T. A. (1928) *En echelon folds and arcuate mountains.* Jour. Geol., vol. 36, p. 526–538.

Lowe, W. F. (1931) *Relation of minor folds to earth deformation.* Int. Petr. Tech., vol. 8, p. 245–248.

Price, P. H. (1931) *The Appalachian structural front.* Jour. Geol., vol. 39, p. 24–44.

Prouty, W. F. (1932) *Origin of folded mountains.* Elisha Mitchell Sci. Soc., Jour. 47, p. 33–46.

Sherrill, R. E. (1934) *Symmetry of northern Appalachian foreland folds.* Jour. Geol., vol. 42, p. 225–247.

Swanson, C. O. (1928) *Isostasy and mountain building.* Jour. Geol., vol. 36, p. 411–433.

Van Hise, C. R. (1895) *Principles of North American pre-Cambrian geology.* U. S. Geol. Surv., 16th Ann. Rept., part 1. *Analysis of folds,* p. 603–633.

Woodworth, J. B. (1932) *Contributions to the study of mountain building.* Am. Jour. Sci., 5th ser., vol. 23, p. 155–171.

Also texts of Structural Geology listed in Chap. II.

### Drainage Development

Ashley, G. H. (1933) *The scenery of Pennsylvania, its origin and development, based on recent studies of physiographic and glacial geology.* Pa. Geol. Surv., 4th ser., Bull. G6, 91 p.

Ashley, G. H. (1935) *Studies in Appalachian mountain sculpture.* Geol. Soc. Am., Bull. 46, p. 1395–1436. Pictures.

Davis, W. M. (1889) *The rivers and valleys of Pennsylvania.* Natl. Geog. Mag., vol. 1, p. 183–253; Geog. Essays, p. 413–484.

Davis, W. M. (1923) *The cycle of erosion and the summit level of the Alps.* Jour. Geol., vol. 31, p. 1–41.

Johnson, D. W. (1934) *How rivers cut gateways through mountains.* Sci. Monthly, vol. 38, p. 129–135.

Meyerhoff, H. A., and Olmsted, E. W. (1934) *Wind gaps and water gaps in Pennsylvania.* Am. Jour. Sci., 5th ser., vol. 27, p. 410–416.

Meyerhoff, H. A., and Olmsted, E. W. (1936) *The origins of Appalachian drainage.* Am. Jour. Sci., 5th ser., vol. 32, p. 21–42.

Ver Steeg, K. (1930) *Wind gaps and water gaps of the northern Appalachians, their characteristics and significance.* N. Y. Acad. Sci., Annals 32, p. 87–220. Many illustrations.

### Folding Experiments

Chamberlin, R. T., and Shepard, F. P. (1923) *Some experiments in folding.* Jour. Geol., vol. 31, p. 490–512.

Willis, B. (1893) *Mechanics of Appalachian structure.* U. S. Geol. Surv., 13th Ann. Rept., part 2, p. 211–281.

### Regions Described

Anderson, R. van Vleck (1936) *Geology in the coastal Atlas of western Algeria.* Geol. Soc. Am., Mem. 4, 450 p.

Bowman, I. (1911) *Forest physiography.* New York, p. 665–684.

Butts, C., Stose, G. W., and Jonas, A. I. (1932) *Southern Appalachian region.* 16th Intern. Geol. Cong., Guidebook 3.

Campbell, M. R. (1903) *Geographic development of northern Pennsylvania and southern New York.* Geol. Soc. Am., Bull. 14, p. 277–296.

Chamberlin, R. T. (1910) *The Appalachian folds of central Pennsylvania.* Jour. Geol., vol. 18, p. 228–251.

Chittenden, A. P. (1897) *Mountain structures of Pennsylvania.* Am. Geog. Soc., Bull. 29, p. 175–180.

Collet, L. W. (1927) *The structure of the Alps.* London, 289 p.

Crawford, R. D., Willson, K. M., and Perini, V. C. (1920) *Some anticlines of Routt County, Colorado.* Colo. Geol. Surv., Bull. 23, 61 p.

Darton, N. H. (1894) *Shawangunk Mountain.* Natl. Geog. Mag., vol. 6, p. 23–34.

Davis, W. M. (1906) *The mountains of southernmost Africa.* Am. Geog. Soc., Bull. 38, p. 593–623.

Eaton, H. N. (1919) *Some subordinate ridges of Pennsylvania.* Jour. Geol., vol. 27, p. 121–127.

Fenneman, N. M. (1938) *Physiography of eastern United States.* New York, p. 195–278.

Geikie, J. (1911) *The architecture and origin of the Alps.* Scot. Geog. Mag., vol. 27, p. 393–417.

Hayes, C. W., and Campbell, M. R. (1894) *Geomorphology of the southern Appalachians.* Natl. Geog. Mag., vol. 6, p. 63–126.

Hayes, C. W. (1899) *Physiography of the Chattanooga district in Tennessee, Georgia, and Alabama.* U. S. Geol. Surv., 19th Ann. Rept., part 2, p. 1–58.

Keith, A. (1923) *Outlines of Appalachian structure.* Geol. Soc. Am., Bull. 34, p. 309–380.

Moody, C. L., and Taliaferro, N. L. (1918) *Anticlines near Sunshine, Park County, Wyoming.* Univ. Calif., Dept. Geol., Bull. 10, p. 445–459.

Rich, J. L. (1933) *Physiography and structure at Cumberland Gap.* Geol. Soc. Am., Bull. 44, p. 1219–1236.

Smith, G-H. (1931) *Physiography of the Baraboo Range of Wisconsin.* Pan-Am. Geol., vol. 56, p. 123–140.

U. S. Geological Survey. Folios no. 2, 12, 25, 26, 27, 33, 35, 59, 61, 75, 78, 118, 151, 170, 175, 179, 215, 225, et al.

Willis, B. (1895) *The northern Appalachians.* Natl. Geog. Soc., Mon. 1, p. 169–202.

### Geographical Aspects

Blanchard, R. (1921) *The natural regions of the French Alps.* Geog. Rev., vol. 11, p. 31–49.

Brigham, A. P. (1924) *The Appalachian Valley.* Scot. Geog. Mag., vol. 40, p. 218–230.

Martonne, E. de (1917) *The Carpathians: physiographic features controlling human geography.* Geog. Rev., vol. 3, p. 417–437.

Peattie, R. (1936) *Mountain geography.* Cambridge, Mass., 257 p.

Tower, W. S. (1906) *Regional and economic geography of Pennsylvania. Central Province.* Geog. Soc. Phila., Bull. 4, p. 113–136, 193–204.

# XVIII
# COMPLEX MOUNTAINS

*Royal Canadian Air Force*

AIR VIEW OF REGION ABOUT 10 BY 15 MILES IN EXTENT, NORTHEAST OF YELLOWKNIFE RIVER, NORTHWEST TERRITORIES, CANADA

A region of complex pre-Cambrian metamorphics intruded by white granite veins and sills.

A. K. Lobeck

THE STEPPES OF SIBERIA NEAR KRASSNOIARSK

An old complex mountain region. Part of the upland belt which extends for several thousand miles in southern Siberia.

THE SAWTOOTH MOUNTAINS, IDAHO

A maturely dissected mountain area of granite rocks formerly peneplaned. Part of the Idaho batholith.

*U. S. Forest Service*

THE OLDER APPALACHIANS, N. C.

Maturely dissected complex mountains, formerly peneplaned, now deeply covered with soil, and richly forested.

THE BLUE RIDGE MOUNTAINS, SHENANDOAH NATIONAL PARK, VA.

Maturely dissected complex mountains of quartzite and crystalline rocks, showing full-bodied forms and flowing profiles.

*Norfolk and Western Railway*

*U. S. Army Air Service*

THE TETON RANGE, WYO.

Maturely dissected and strongly glaciated complex mountains, sharp crested arêtes, matterhorn peaks, and moraines.

LAKE PLACID AND WHITEFACE MOUNTAIN; THE ADIRONDACKS, N. Y.
Maturely dissected, complex mountains, strongly jointed, with rectangular block-like pattern of lakes and hills.

*Fairchild Aerial Surveys*

*A. K. Lobeck*

THE CORDILLERA OF PUERTO RICO

Maturely dissected complex mountains, once peneplaned. Tuffs and other volcanics strongly folded; extremely steep slopes deeply soil-covered.



SCHOOLEY MOUNTAIN, N. J.

Peneplaned complex mountains, moderately dissected; gentle slopes, heavy soil cover, non-glaciated.

*A. K. Lobeck*

# COMPLEX MOUNTAINS

Synopsis. Complex mountains consist of combinations of structures which, if considered in detail, would be called *simple.* When the variety of structures is great and when different kinds of structures are intimately admixed with each other, they defy classification and the ensemble is said to be *complex.*

Or, from a slightly different viewpoint, it may be said that, when the initial forms can be observed and when the structures are undergoing the first cycle of erosion, classification is possible, even if the variety be great. But when a region, even though it may not involve many different kinds of structures, is worn down to its roots so that the foundation rocks are exposed with only fragments of the one-time overlying strata infolded here and there and with the deeper parts of the igneous bodies revealed—perhaps igneous bodies which never reached the surface—then it becomes quite impossible to assign such a region to a simple category.

To the geologist, the organization of a complex mountain mass is not a haphazard mixture of miscellaneous structures. The origin of mountains is ascribed to definite, but not always well-understood, causes. The arrangement of mountains on the earth's surface is believed to follow certain plans which have been repeated again and again. For that reason a comprehension of the existing plan of the continents enables us to have some insight into the structure of complex areas formed in earlier geological time but preserving now only fragmentary bits of the past.

Origin of Continents. Some geologists see in the continents only a haphazard distribution of land and sea, dating from the origin of the earth itself and reflecting original inequalities in the accumulation of planetesimals, or caused by the unequal distribution of gaseous pressure early in the earth's history while still molten. Others see more system in the arrangement of continents and ocean basins. The *tetrahedral theory* is based on the idea that the earth, in shrinking, assumed something of a tetrahedral form, the geometrical form which has the largest surface in proportion to its volume. The continents appear to be placed at the corners and along the edges of the tetrahedron. Other theories see in the continents certain definite structures, geanticlines and geosynclines, apparently obeying some law of arrangement. The theory of *continental drift,* usually called the *Wegener hypothesis,* is an important idea which has many advocates but probably even more opponents.

The geomorphologist, however, is concerned less with the origin and behavior of the continents than he is with the origin of mountain ranges and with their distribution.

Origin of Mountains. Mountains and uplands may be formed in the following ways:

*A*. By epeirogenetic (continent-building) forces, producing change in elevation with relation to sea level.
 1. Isostatic or local.
 2. Eustatic or world wide.

*B*. By orogenetic (mountain-building) forces:
 1. Compression, causing folds and thrusts.
 2. Tension, causing faults.

*The theory of isostasy* is based on the idea that there is a state of balance between large high areas of the earth's crust, such as mountain ranges, uplands, and plateaus, on the one hand, and adjacent lowlands and plains, on the other. This condition of balance is thought to be due to the existence of relatively light rock material in the mountainous areas and relatively heavy rocks beneath the plains. Moreover, the same idea of balance is carried over to explain the presence of continents and ocean basins, the continents being made up largely of granitic and sedimentary rocks, whereas the oceanic areas, with extremely few exceptions, reveal the presence of the heavier, basic igneous rocks beneath the sea.

Two types of evidence supporting this theory may be mentioned. First is what is known as *gravity anomalies*. This means simply that the attractive effect of mountain masses upon delicate pendulums is less than the amount calculated for such a volume of rock if it had the average specific gravity of the earth's crust. This indicates the presence in that mountain area of rocks of lighter weight. On the other hand, the low areas on the earth are found to have an attractive pull greater than that which a similar estimate would expect. At a certain depth, termed the *level of compensation*, it is inferred that the weight of two adjacent areas of highland and lowland must be the same.

In view of this condition of balance it is believed that upland areas, as they are worn down by erosion, will constantly rise; and that basins of accumulation will constantly sink under the added load.

When the continental ice sheet accumulated over North America and Europe, the state of equilibrium was disturbed and the earth's crust was depressed. When the ice melted, the crust of the earth gradually rose, and this uplift is probably still continuing. It is recognized by the tilted beaches in the Great Lakes region and by the uplifted terraces of the Champlain-Hudson Valley. Going northward, the terraces range in elevation from sea level near New York to 150 feet along the Hudson, to 300 feet at Lake Champlain, and about 700 feet in Canada.

Not all apparent changes of level are actually real. The encroachment of the sea upon the land does not necessarily signify a lowering of the land. The gradual compression of marsh lands, or the breaking away of barrier beaches, may permit the sea to wash inland over areas formerly beyond the reach of the waves and thus merely simulate a change of sea level.

*Eustatic or world-wide* changes of sea level are due to several causes, such as changes in the volume of water in the sea brought about by glaciation. These, of course, have nothing to do with crustal disturbance. During the time of ice accumulation the sea level everywhere became lower. When the ice again melted, the sea level everywhere became higher. Conservative estimates indicate that the sea level was lowered more than 200 feet during the last glacial period. Extreme estimates place the figure at several times this amount. It is possible also to imagine the sinking of an ocean basin sufficiently great to withdraw water from all the shorelines of the world, combined with some compensating continental uplift. In that case, the sea level would be lowered everywhere but the amount would vary widely in different places.

*Folding and faulting due to compression* is localized in well-defined belts and produces mountain ranges. A significant fact about the highest mountain ranges of the world is the great thickness of sedimentary strata represented in the mountains, as compared with the much thinner amount of the same formations outside the mountain zone. In other words, it appears that long trenches, troughs, or *geosynclines* were at one time regions of excessive accumulation. The troughs probably subsided as more and more material was deposited in them from the surrounding highlands. Then came a period when the continental area on one side of the trough slid toward the trough, and produced crumpling and thrusting of the strata accumulated in the trough. The Wegener hypothesis of continental drift invokes extreme migration of continents to explain the folding of certain zones, although it does not show why the continents should move. Daly has introduced a modification of the Wegener hypothesis, which may be termed the *landslide explanation*. This simply means that the continents have been raised and the ocean basins lowered due to contraction of the earth's crust from cooling. The elevated position of the continents then caused them to slide toward the oceans and to fold the sedimentary beds along the borders. The essence of the theory lies in its recognition of compression as the chief cause of mountain building and its endeavor to provide some explanation for compression. The Wegener hypothesis and its various modifications, notably the idea as first evolved by Taylor and also by Suess, attempt to explain the localization of the arcuate belts of the mountain ranges of the world, such as the great Alpine belt which extends from the Pyrenees across southern Europe into the heart of Asia; as well as the almost uninterrupted chain of mountain folds which border the Pacific Ocean.

*Tension* does not raise mountains but it does permit the collapse of segments of the earth's crust. This results in faults and other disturbances, with the formation of blocks and folds. Such mountains, usually in the form of horsts and block mountains, are the remnants of larger portions of the earth's crust previously elevated and then later disrupted. The height of blocks and horsts above the sea is due not solely to local uplift but to regional uplift followed by local fracturing.

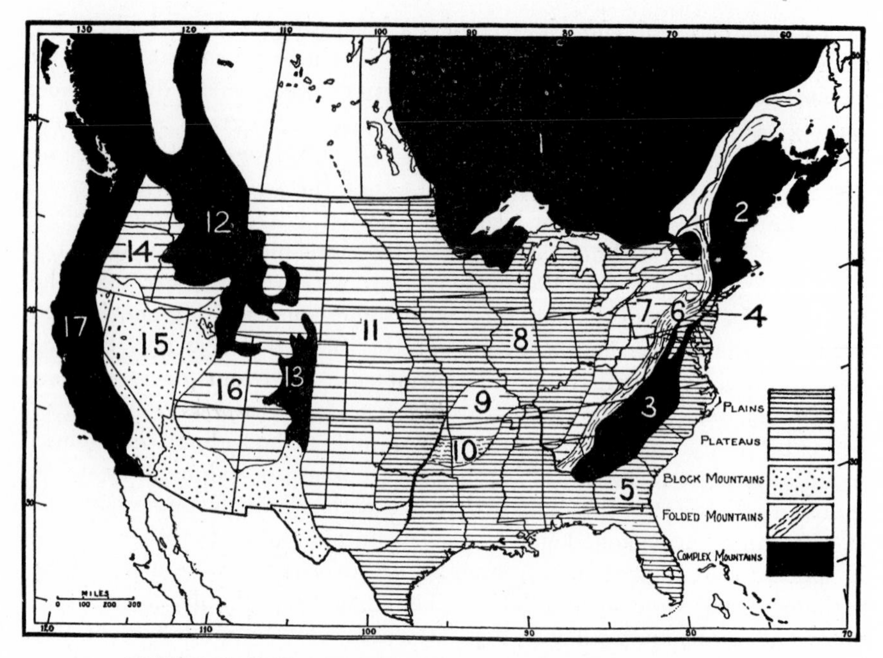


PHYSIOGRAPHIC PROVINCES OF THE UNITED STATES

Those shown in black are complex mountain areas, namely 1. The Laurentian upland; 2. New England; 3. The older Appalachians; 12. The Northern Rockies; 13. The Southern Rockies; 17. The Pacific Ranges.

## COMPLEX MOUNTAINS DEFINED

Complex mountains may be defined as those made up wholly of igneous rocks; of metamorphic rocks; of strongly disturbed sedimentary beds; or, as is often the case, of combinations of these.

Igneous Areas. An example of a complex mountain area of igneous rocks is the Sierra Nevada block in the Yosemite region. A gray or almost white granite forms a large batholithic mass, out of which the present topography has been carved. The chief structural features are the joints which break up this mass into gigantic pieces. Half Dome in the Yosemite owes its flat face to one of these joints. The complex Sierra block contains many quartz veins bearing gold, the mother lodes of the famous placer deposits of 1849.

Another complex mountain area is the Pikes Peak region of the southern Rocky Mountains. Many square miles consist of a pink granite which behaves as a homogeneous, uniform mass. The streams flow indiscriminately in all directions, for there appears to be little structural control.

The Idaho batholith in the northern Rockies, consisting in part of the Salmon River Mountains and the Coeur d'Alene Mountains, is largely granite. Parts of this region, like many parts of the southern Rocky

Mountains, are complicated by intrusions which have introduced valuable ores of silver, copper, lead, and gold.

METAMORPHIC AREAS. An example of a complex mountain area of metamorphic rock is the White Mountain Range in New Hampshire. The rock is an andalusite schist with large veins of milky-white quartz. The structure of the schist apparently exerts little control upon the topographic features. The larger "Notches" may be due to the main joint system but the streams of the region are without any definite pattern or well-ordered system.

Many other parts of New England are regions of metamorphic rocks, such as the *Manhattan Prong* which extends into southern New York. Three types of rock, namely, a gneiss, a crystalline limestone, and a schist, are intimately folded. There is a great contrast in the resistance of these rocks, the limestone being very weak and the others resistant. The limestone outcrops in long bands now eroded to form valleys, and giving therefore a decided trend to the features.

A somewhat more complicated region including both igneous and metamorphic types, is represented by parts of the northern Rockies in Montana. The Anaconda Range, for example, within a distance of 10 to 15 miles, comprises six different kinds of igneous rocks and several metamorphics. The term *crystalline* rock is used to embrace both of these types where closely intermingled.

The illustrations on the following page depict this range viewed from the south. The rocks are granodiorities, strongly sheared. The topography is extremely rugged, the smaller valleys and ravines being controlled by the angular pattern of the many joints and shearing planes. The geological map shows the extent of the different igneous bodies. Their involved structural relations are shown in the cross section. The igneous rocks represented are all of deep-seated types, being diorites, granites, and granodiorites, in places merging toward gneisses.

STRONGLY DISTURBED SEDIMENTARY AREAS. More highly complicated are certain mountain regions largely of sedimentary rocks profoundly disturbed by folding and faulting. This is the character of much of the northern Rockies. Where the complications are not too great, the structure takes on an aspect of folded mountains, with pitching anticlines and synclines. The part illustrated on page 627 is almost entirely lacking in plan. Even the fault systems intersect in a disconcerting manner, as indicated on the geological map. The faults are mainly normal faults and the region is broken into numerous small blocks, some less than a mile across. Few thrust faults appear. The sketch of some of the topography depicts the closely folded character of the sedimentary beds and shows the effect of two of the faults upon the topography. Little or no metamorphism has occurred in this area, the sedimentary beds being mostly limestones and shales. Igneous rocks are infrequent.

*U. S. Geological Survey*

PICTURE, GEOLOGICAL MAP, AND SECTION OF COMPLEX MOUNTAINS IN STRONGLY JOINTED IGNEOUS AND METAMORPHIC ROCKS, ANACONDA RANGE, MONT.

*U. S. Geological Survey*

PICTURE, GEOLOGICAL MAP, AND SECTION OF COMPLEX MOUNTAINS IN STRONGLY CONTORTED SEDIMENTS, ROCKY MOUNTAINS, MONT.

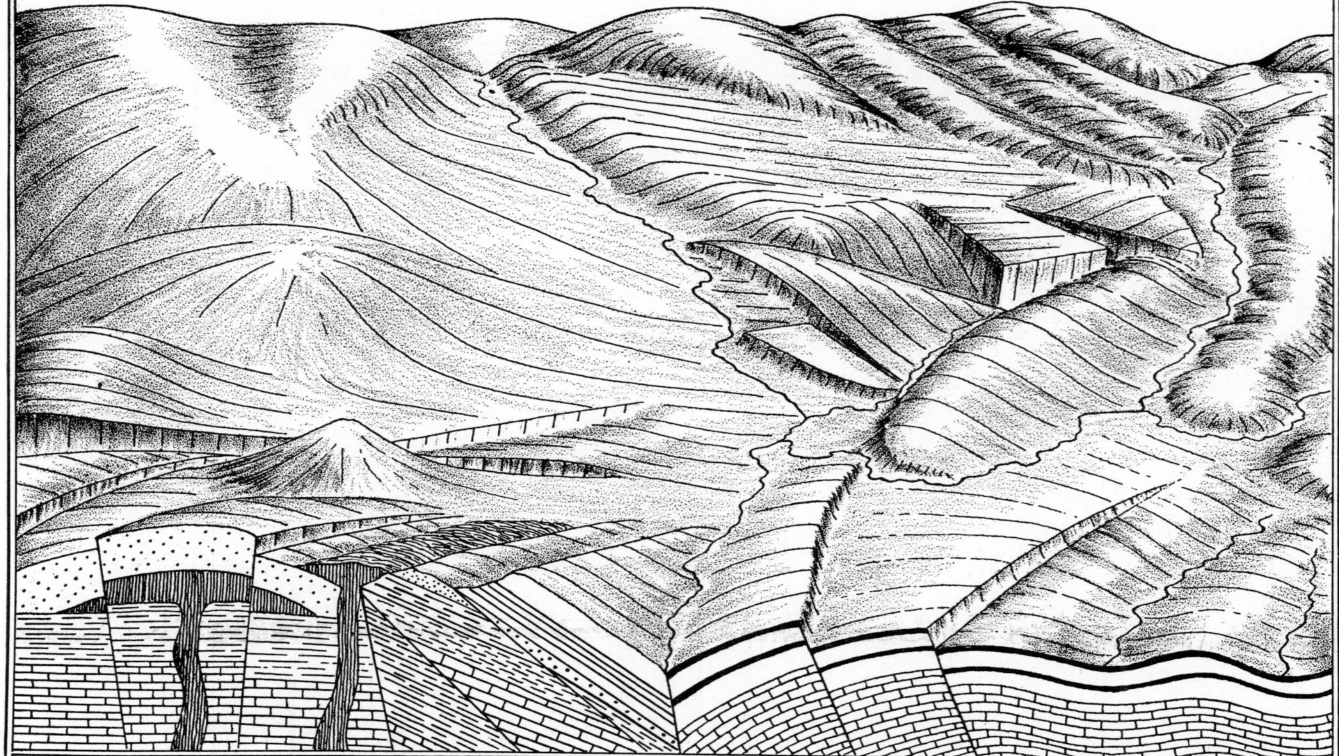

YOUNG COMPLEX MOUNTAINS

DIAGRAMMATIC REPRESENTATION OF YOUNG COMPLEX MOUNTAINS SHOWING SEVERAL TYPES OF STRUCTURES. DEEP-SEATED IGNEOUS BODIES NOT VISIBLE

## YOUNG COMPLEX MOUNTAINS

NO ACTUAL EXAMPLES. First of all, it is probably correct to say that no young complex mountains exist, that is, no complex mountains young in the first cycle of erosion. Nor is it likely that complex mountains have ever existed in the youthful stage. The mere idea of complexity involves also the idea of a long period of growth by constructional forces, such as by folding, faulting, and thrusting, accompanied by vulcanism with intrusions of all sorts. This takes time and during this period of constructional growth the agencies of destruction are constantly at work. It is probable that the growth of complex mountains starts with a little fault here, a little warping there, followed by a long period of rest before the disturbances are resumed. During the periods of rest, erosion wears away the initial form of the block or fold before there is another episode of faulting or folding. And so it goes on, each incident of dislocation followed by a time of wearing away, keeping the mass always in a stage of mature dissection.

AN IMAGINARY EXAMPLE. However, for our purpose, in order better to visualize the structural aspects of a complex region and its drainage problems, let us assume an initial stage like the illustration on the opposite page. The region consists of two parts. At the left is a large domelike mountain mass due to the upwelling of a great batholith. Near the surface the molten magmas have intruded the overlying sediments to form laccolithic bodies. There are some extrusive flows. Deep beneath the ground we may assume all sorts of disturbances with contact and dynamic metamorphism and several periods of intrusions of magmas of varying character.

The right-hand side of the picture represents a region of great overthrusts, folding, and faulting, just the conditions likely to produce profound metamorphism of sedimentary structures at depth, with development of schist and gneiss, accompanied probably by a multitudinous array of minor veins and dikes. These deep-seated conditions we shall know nothing about until erosion exposes them during the mature stage of dissection.

THE INITIAL DRAINAGE SYSTEM. The streams which originate upon the forms illustrated are all consequent streams. Their courses have been determined by the initial slopes of the land. That they flow in all conceivable directions and are tributary to each other does not alter their consequent character. We may give them such designations as *longitudinal* consequents, *lateral* consequents, *radial* consequents, etc., to indicate their positions and patterns, but many of them must remain without definite designations.

COMPLEX MOUNTAIN AREA MATURELY DISSECTED, EXPOSING CRYSTALLINE CORE IN BACKGROUND

## MATURE COMPLEX MOUNTAINS

No type of landscape offers a greater variety of interesting detail than a maturely dissected complex mountain mass. Here we see exposed the core, the very roots of the mountains. The various types of rocks and structures all affect the topography. The streams are in almost perfect adjustment with these rocks and structures. The original consequent drainage has been gradually replaced by streams which are largely subsequent. That is to say, there have been numerous stream captures whereby the subsequent drainage lines have grown at the expense of the earlier consequent pattern.

Maturity is the time of greatest relief. Though much has been worn away from the tops of the great uplifts, the streams have cut deep canyons and gorges. The courses of the master streams are apt to follow the original structural depressions or intermontane basins. Such streams, cutting down and discovering various types of rock from place to place, have alternating stretches showing young and mature characteristics. Gorges occur where the streams have been superposed upon resistant rocks; open valleys occur where the rocks are weak.

The exposure of great granite batholiths, as in the left-hand side of the illustration, results in mountain masses whose summits are rounded and full bodied, provided glaciation has not yet occurred. Erosion is accompanied by weathering, which may produce a deep soil. With favorable climatic conditions such an area may be forested to the summit. On the other hand, some granite areas, because of the strong jointing of the rocks, may display much minor detail in the form of benches, sharp cliffs, or walls, as well as a rectangular arrangement of the larger features.

Regions of schist and gneiss are apt to be less full bodied than granite areas. They show a parallelism of features. There is a "grain" to the country. This is more noticeable as the region passes into old age because the features due to schistosity and bedding are small and are apt to be lost in the larger topographic aspects of the region during the mature stage. Schistose regions often display crags and pinnacles, and sharp divides, even when not glaciated.

It is not possible to state whether granite or schist (or gneiss) areas show greater resistance to erosion. Some granite areas are notably weak. Especially in high mountain summits the granite disintegrates rapidly, much more so than does schist or gneiss, especially if these latter rocks contain abundant quartz. Generally these three crystalline rocks are more resistant than sedimentary rocks and therefore in complex mountain areas the dominant uplands and commanding peaks are usually composed of crystalline rocks. Nevertheless, the present height of existing maturely dissected complex mountain areas is in most cases due to recency of uplift rather than the resistance of the rocks which compose them.

OLD COMPLEX MOUNTAIN AREA. A PENEPLANE WITH MONADNOCKS MAINLY OF CRYSTALLINE ROCKS

## OLD COMPLEX MOUNTAINS

A complex mountain mass reaches old age when it is reduced to a peneplane. There is undisputed evidence from virtually all parts of the world that rocks of the most diverse character and varied structure have been worn down to an almost level surface which, when first formed, was close to the base-level of the streams of the region.

BASE-LEVEL. Actual base-level of a region is the level of the ocean or lake into which the rivers flow. If the region is neither uplifted nor depressed, this is the permanent *base-level* of the region. There may, of course, be a *temporary base-level* of a region if the main river flows over a resistant mass of rock, but this ultimately gives way in favor of the permanent level of the lowest body of water into which the river flows.

A peneplane is not strictly a level plane but must retain some slope if the streams on its surface are to continue to flow. Further discussion of peneplanes continues on the next page. Meanwhile, let us note the general aspect of an old complex mountain area.

It has a remarkably uniform surface beveling all kinds of structures. *Monadnocks,* that is, residual remnants of the former mountain mass not completely eroded away, occur where there is unusually resistant rock. In old complex mountain areas granite bosses often form monadnocks. Stone Mountain on the Piedmont Upland near Atlanta, Ga., is a granite monadnock. So are Dartmoor and Exmoor and other granite masses in the Cornwall peninsula of southern England. Pikes Peak is also a great granite monadnock range. Mount Monadnock, the type example, in southern New Hampshire, is made of schist—andalusite schist, a very firm and resistant rock. Of all rocks, quartzite probably forms the most persistent monadnocks. Many of the hills and ranges in northern Wisconsin rising above the Laurentian peneplane are quartzite knobs or ridges, such as Baraboo Ridge and Rib Hill.

Some monadnocks owe their presence not so much to the resistance of the rocks which compose them as to their position along the divides and at the headwaters of the different river systems. Groups of monadnocks are called *unakas* after the Unaka Mountains in the southern Appalachians. Linear monadnocks, due to projecting resistant beds, are called *catoctins* after Mount Catoctin on the Piedmont in Virginia.

Finally, it should be pointed out that practically all old complex mountains have been either rejuvenated or buried. This is true of most other constructional forms after reaching old age. Peneplanes produced in the present cycle of erosion are practically nonexistent. They have all been raised, and renewed erosion has set in; or they have been buried by submergence. This is, of course, to be expected when it is realized that the surface of the earth is in constant oscillation and that the slightest movements are recorded by lands lying close to sea level, as peneplanes are when first formed.

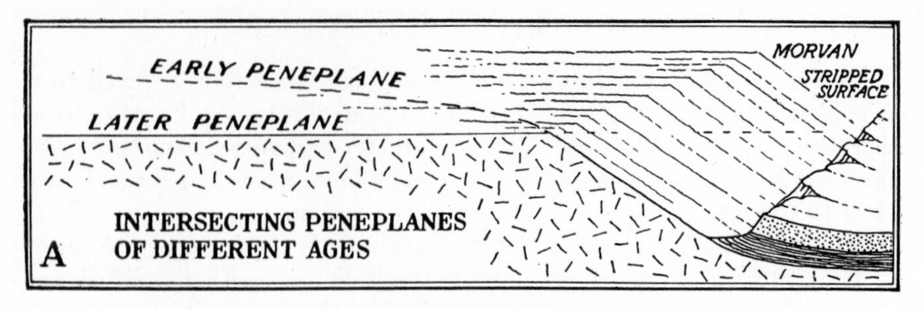


A MORVAN
Two intersecting peneplanes of different ages.

## PENEPLANES

The literature of physiography abounds with references to peneplanes. There is also a considerable divergence of opinion as to their origin. It is admitted by most investigators that peneplanes may be formed subaerially by streams, or by marine planation, or by wind action under arid conditions. Some authorities restrict the term *peneplane* to surfaces developed only by stream action, but in this text it refers to an almost flat surface produced by destructive forces.

THE TERMS BASE-LEVEL, PENEPLANE, AND PENEPLAIN. The term *base-level*, mentioned on preceding pages, is applied to the level surface of the body of water into which the rivers of a region flow. It is the limit to which stream erosion can reduce the region. The term *peneplane* means an "almost-plane" surface and therefore represents varying degrees of reduction toward base-level. Base-level is the end limit of all peneplanes. When first introduced, "peneplane" was spelled "peneplain." It was later pointed out, however, that it was not intended to mean "almost a plain," or region of horizontal structure, but rather "almost a plane," that is, almost a flat surface.

Wind erosion can theoretically develop a peneplane below the base-level of stream erosion. The term *base-level,* therefore, is relevant mainly with regard to streams inasmuch as their erosive activity is controlled by the body of water into which they flow.

Most peneplanes which are now exposed to observation stand well above sea level. Indeed, many great mountain tracts show a decided accordance of summit levels. It has, therefore, occurred to some observers that this accordance may have been brought about while the mountains stood at their present altitude. This has been called *peneplanation without base-leveling.* Weathering of mountain summits above timber line is much more rapid than where the soil is held in place by vegetation. A peneplane surface, beveling high mountain areas, may therefore represent the approximate tree line either now or in the past. Objections to this idea are based largely on the fact that a tree line is rarely of

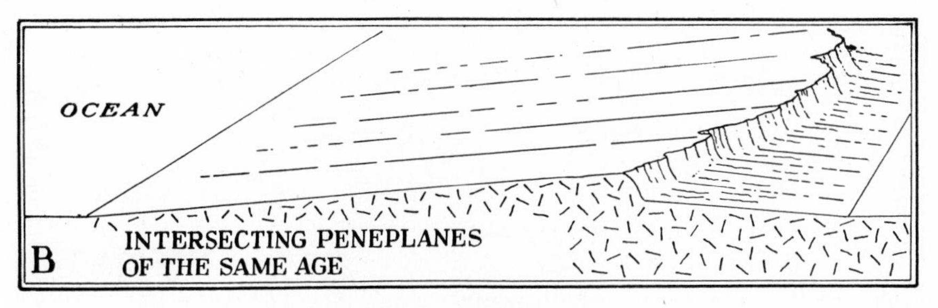


AN ESCARPMENT
Produced by the intersection of two peneplanes of the same age.

uniform elevation over wide areas but may vary several thousand feet in a few miles, depending upon slope and exposure. Another theory for accordance of level involves the notion of erosion down to the almost level roof of an extensive batholith. Still another observer points out the fact that in regions of more or less uniform structure, where the streams are evenly spaced, there is a tendency for the divides to wear down to the same elevation.

The distinctive aspects of peneplanes produced by streams, in contrast with those produced by marine abrasion or by wind erosion, can be deduced. Stream-eroded peneplanes will probably reveal a fairly well-organized drainage pattern adjusted to the rock structure. The monadnocks will be due to the presence of resistant rocks or to location near the divides. Marine planes of erosion may reveal various levels or marine benches, due to stages in the process of uplift. Monadnocks will be steeper on the side exposed to wave action. The location of monadnocks may be related more to protected position than to resistance of rocks. Wind-eroded peneplanes will probably be less extensive than those formed by streams or waves and would show varying levels in their different parts, with no definite relation between the different levels.

Peneplanes may become warped or tilted and beveled by a later peneplane, as in Fig. *A*. Intersecting peneplanes, thus formed, are called *morvans,* after the Morvan region of central France. The intersection between the fall-zone peneplane and the Piedmont peneplane is a morvan. So also is the intersection of the pre-Triassic peneplane and the Schooley peneplane east of the Hudson River north of New York City. Along the Rocky Mountain front in Colorado the Rocky Mountain upland intersects sharply the pre-Paleozoic peneplane which slopes eastward beneath the plains.

Another type of intersecting peneplanes is represented in Fig. *B*. Two peneplanes are shown with equal gradients toward the sea. One is much longer than the other. Hence, at the line where the two peneplanes intersect, there is a sharp break, or scarp. This is one of the explanations for the Blue Ridge escarpment.

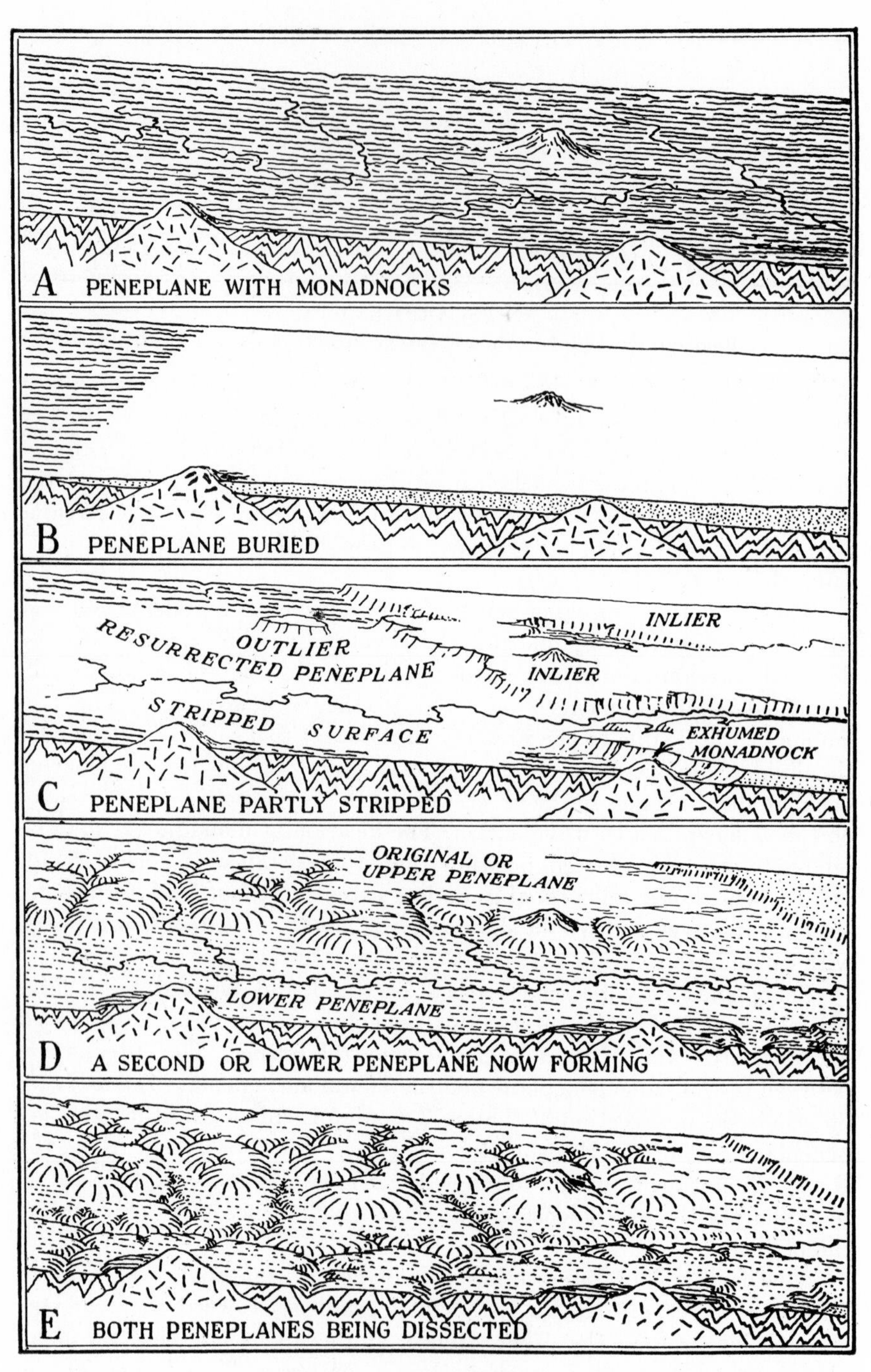


OLD, BURIED, AND RESURRECTED COMPLEX MOUNTAIN AREAS

## OLD, BURIED, AND RESURRECTED, COMPLEX MOUNTAINS

A complex mountain region, after reaching old age, may be depressed beneath the sea and buried, or it may be elevated and rejuvenated, or it may go through both these experiences, being buried first and resurrected later.

The constant, though slight, movements of the earth's crust are recorded most clearly by lands standing near sea level. Many peneplanes have become submerged and covered with marine deposits. Whether the ocean has aided in the peneplanation is difficult to say. But most coastal-plain deposits attest a constant unrest by the many disconformities which interrupt the continuous succession of beds.

Figure *A* represents an old mountain area just before submergence. It is probably not a single continuous plane, as represented, but rather a number of facets intersecting each other at slight angles, each facet or plane representing the work of a more or less distinct river system, all having a somewhat dendritic pattern. In *B* the region has been submerged and buried. Some of the monadnocks project to form islandlike hills, called *mendips*.

In the next stage (*C*), uplift or warping has occurred and erosion has stripped off some of the coastal-plain cover. The peneplane is being resurrected and the monadnocks exhumed. *Outliers* of the coastal plain remain on the *stripped belt* and *inliers* of the oldland show along the stream courses within the coastal plain.

In Fig. *D* the original peneplane has been tilted seaward and thus rejuvenated. The larger streams have widened their valleys to produce a lower peneplane; the smaller streams are still young; the upper peneplane is passing from youth to maturity of dissection. The region is in late youth, or early maturity, in the second cycle of development. The expression $n + 1$ cycle, instead of second cycle, may be used. $n$ stands for "any number" because it is not known how many cycles preceded the present one.

In the final stage (Fig. *E*) the region has again been raised; the previous cycle has in that manner been interrupted. The lower peneplane has been rejuvenated. The region may be described as mature in the third cycle of erosion or, better, as mature in the $n + 2$ cycle.

In maturely dissected complex mountain regions the presence of a rock bench along the valley walls always suggests a partial peneplane, inferring a previous uncompleted cycle of erosion. In plateaus, however, rock benches due to differential erosion on horizontal beds cannot readily be distinguished from partial peneplanation. Therefore in regions of horizontal structure it is necessary to guard against ascribing horizontal benches to partial peneplanation such as may be assumed in regions of complex structures.

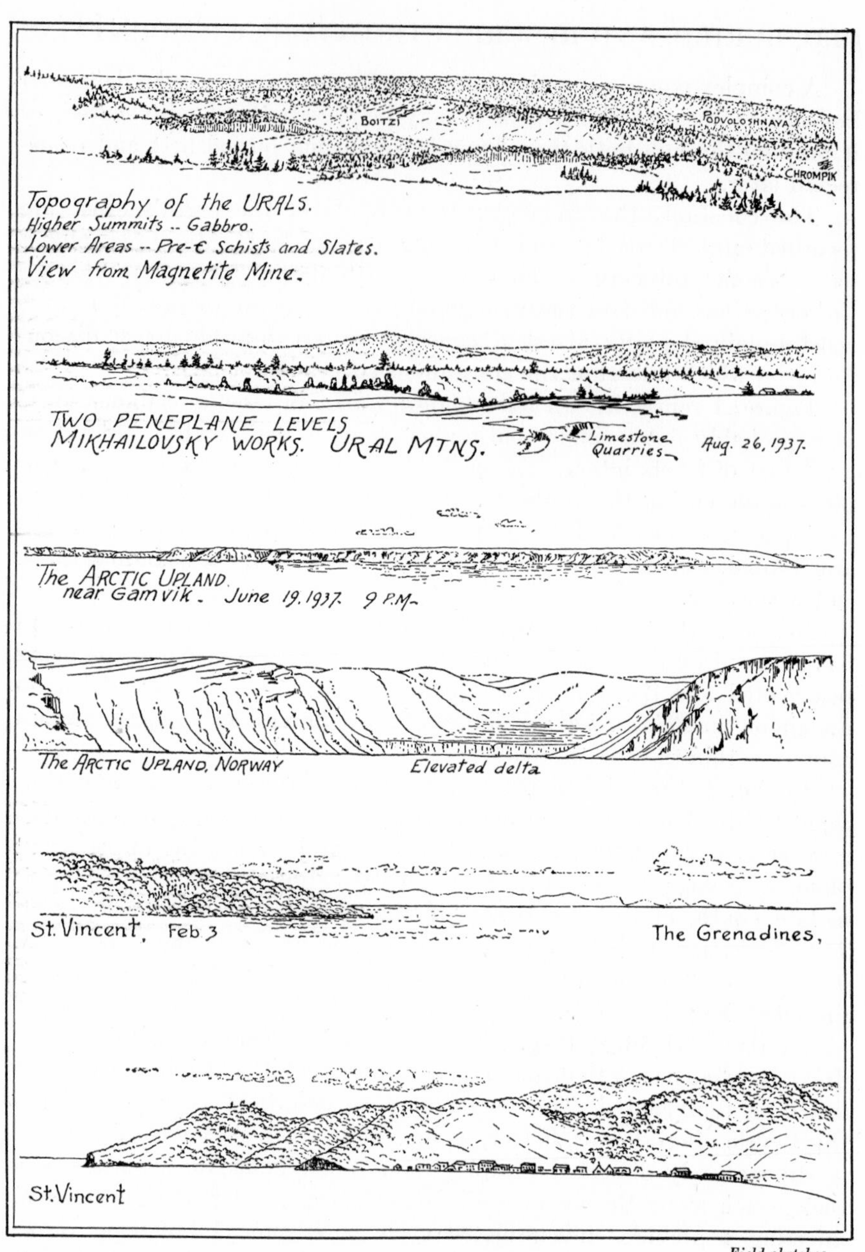


*Field sketches*

PENEPLANES IN COMPLEX MOUNTAIN AREAS IN SIBERIA, SCANDINAVIA, AND THE WEST INDIES

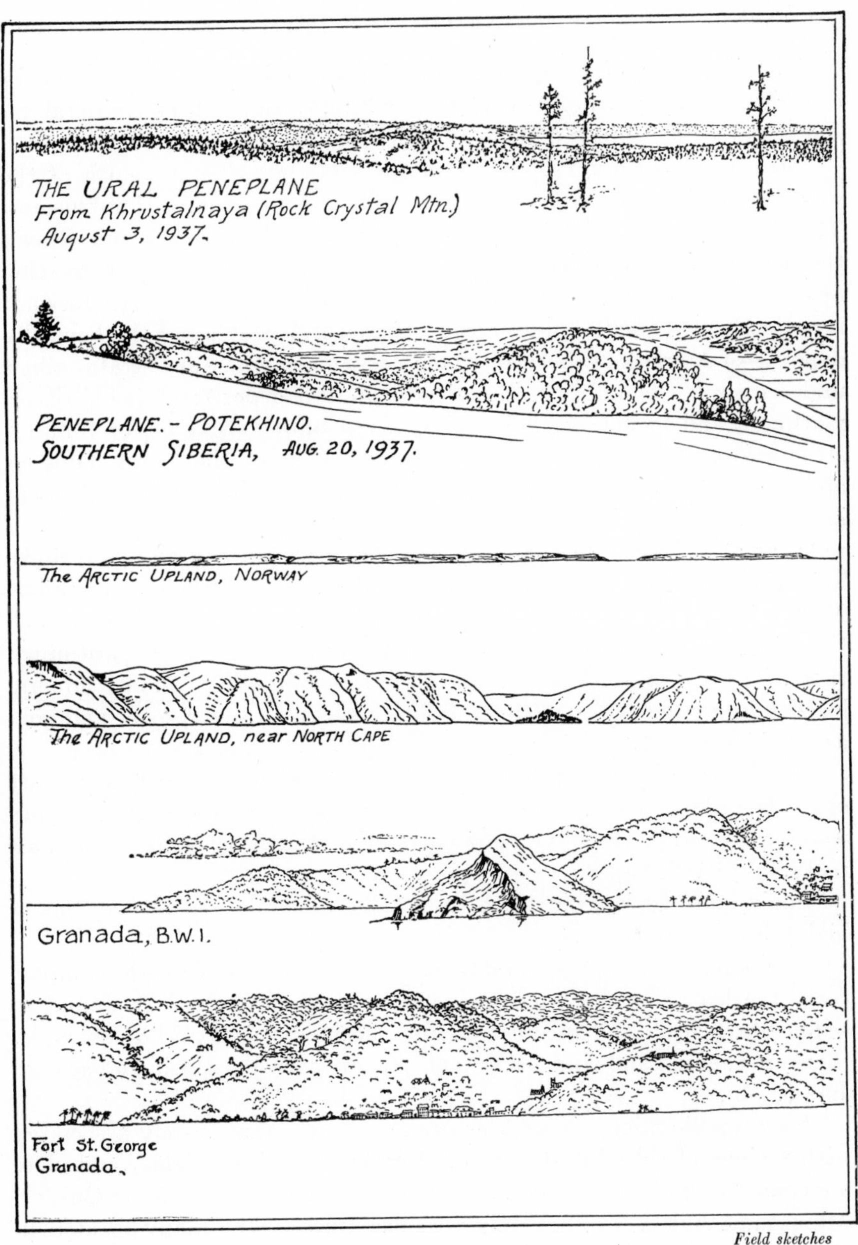


*Field sketches*

PENEPLANES IN SIBERIA, SCANDINAVIA, AND THE WEST INDIES

## GEOGRAPHICAL ASPECTS

The influence of complex mountains upon human affairs may be considered largely from two standpoints: (*a*) the natural mineral resources of such regions; and (*b*) the topographic influences.

MINERAL RESOURCES. Most of the metallic mineral wealth of the world occurs in complex mountains, usually in connection with intrusions. This includes the great iron deposits in northern Sweden and Brazil, the magnetite mines of the Adirondacks, and the historically interesting ores of the highlands of the Hudson River and New Jersey. It does not include the so-called *Clinton ores* of the eastern states nor the hematite deposits of Minnesota, both of which occur in sedimentary strata, albeit under complex conditions. Practically all the deposits of gold, silver, platinum, and nickel, occur in veins in crystalline rocks under complex structural conditions. Witness the central core of the Black Hills as an important mining district, whereas mining has no importance whatever in the surrounding sedimentary areas. The great mineral wealth of Colorado, as at Cripple Creek and also in the Silverton and Ouray regions, is in the complex core of the southern Rockies. Outstanding is Butte, Mont., in one of the most complicated mountain regions with which the miner has ever contended. Copper, gold, and other metals are mined there. The Coeur d'Alene section of Idaho is another famous mineral locality in a region of rugged crystalline rocks with silver, lead, and gold as its chief output.

In Germany the Erzgebirge, or Ore Mountains, constitute a part of the complex massive of Bohemia. In England the important tin mines were in Cornwall, one of the few parts of that country built of complex rocks.

Spain and Mexico, both rich in metallic minerals, are also regions of highly involved structure.

The chief metallic minerals not always associated with complex rocks are lead and zinc, and occasionally iron which may be found in sedimentary beds and often in nonmountainous regions. For example, there are the lead and zinc districts of southwestern Wisconsin and around Joplin, Mo.

Nonmetallic minerals, such as coal, and also oil are rarely associated with regions of complex structure. However, granite, marble, and slate are common products of complex regions. The granite quarries of Quincy, Mass., and Barre, Vt., are only two examples of the many located in the complicated New England area. Similarly the granite quarries of Tennessee are situated in the eastern part of that state in the province of the Older Appalachians.

Marble and slate, both products of dynamic metamorphism, occur in regions where the rocks have been much disturbed, as in Vermont and also in Wales.

It is clear that the vast majority of the mining camps of the world are in complex mountain areas, with the outstanding exception of coal and some other nonmetallic products.

Topographic Considerations. Complex mountains are rarely regions of important agricultural development, especially in maturely dissected regions with a high degree of relief. Exceptions are found in the tropics, as in Puerto Rico, Central America, and Brazil, where coffee plantations extend to the very summits and where the steep mountain sides are used for the cultivation of tobacco. Valleys in complex mountains which have reached maturity and have developed some extent of flat valley floor may attract some measure of population, as in the mountains of western Montana, in the Missoula district, or in New England, as along the Winooski River in northern Vermont. But this is unusual. The river valleys in maturely dissected complex mountain regions are commonly young gorges or canyons, along which even roads and railroads are built with difficulty. Consider the engineering feats required to build highways and railroads across the Colorado Rockies or the Storm King Highway through the Highlands of the Hudson.

The pattern of the river systems in complex mountain areas is usually dendritic. Occasionally in regions of strongly metamorphic rocks, as in Vermont, Westchester County, New York, Brittany, or the Slate Mountains of Germany, the mountains and valleys are arranged in belts and there is a "grain" to the topography. Here the railroads and roads invariably follow the stream valleys; but where no well-defined regularity occurs, the roads are apt to pass over the divides from one valley to another, as in the Rocky Mountain Upland in Colorado.

Old complex mountains, rejuvenated but not too deeply dissected, given proper climatic conditions, may have good agricultural possibilities. This is true of the Piedmont Upland of Virginia and the Carolinas. Here it is the upland peneplane level, rather than the valleys, which is the populous part of the country. The deep soil extends sometimes 50 feet to bedrock on the upland, but in the valleys it has been stripped away by erosion and the unweathered rock is exposed. Some of these facts may suggest the undesirability of trying to utilize the steep slopes of mountain areas for agricultural purposes.

In many parts of Europe the complex mountain masses are so characteristically timbered that the name of the mountain range is identical with that of the forest, as for example, the Schwarzwald, or Black Forest; the Thuringerwald; and the Böhmerwald, or Bohemian Forest. These names all refer to the mountains as topographic features. Indeed, in some cases the term *woods* or *forest* is still used for rugged tracts of country from which the forest has long since been stripped off. Some of the "forests" of Scotland are now only upland heaths and, as mentioned in certain English dictionaries, "there are forests in England without a stick of timber upon them."

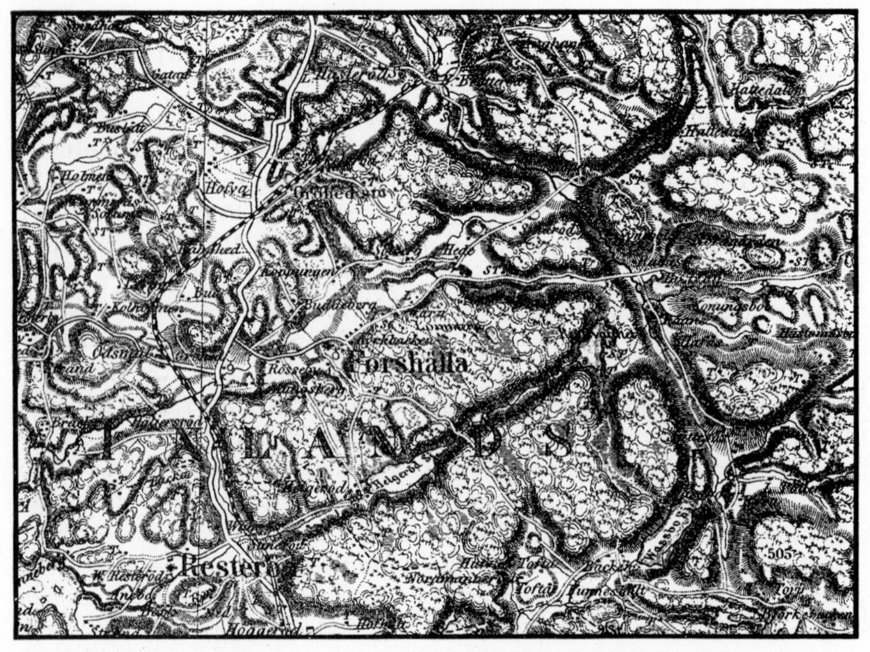


RECTANGULAR DRAINAGE PATTERN DUE TO JOINTING; OLD COMPLEX MOUNTAINS OF SOUTHERN SWEDEN
Sweden; *Uddevalla* sheet, No. 41 (1:100,000).

## MAPS ILLUSTRATING COMPLEX MOUNTAINS

Maturely dissected complex mountains of medium texture are represented on the *Casto, Idaho,* sheet. As this map does not clearly suggest by upland remnants an earlier erosion cycle, the region may be considered mature in its first cycle. The *Cowee, N. Car.-S. Car.,* and *Tujunga, Calif.,* maps both depict a complex mountain area maturely dissected by young valleys which have a distinct dendritic pattern. The *Gorham, N. H.,* area is mature and coarse textured. The *Opelika, Ala.-Ga.,* region is a maturely dissected complex mountain area in its second (or, better, $n + 1$) cycle of development, as it displays a well-defined upland peneplane. A maturely dissected peneplane is also clearly shown on the *Derby, Conn.,* sheet. Maturely dissected peneplanes with monadnocks are represented by the *Atlanta, Ga., Wedowee, Ala.-Ga., Gastonia, N. Car., Kings Mountains, N. Car.-S. Car., Wausau, Wis.,* and *Worcester* and *Dedham, Mass.,* sheets.

The *Hawley, Mass.-Vt.,* region may be taken as an example of a peneplane submaturely dissected by young streams in deep canyons, whereas the *West Chester, Pa.-Del.,* region is a submaturely dissected peneplane of slight relief.

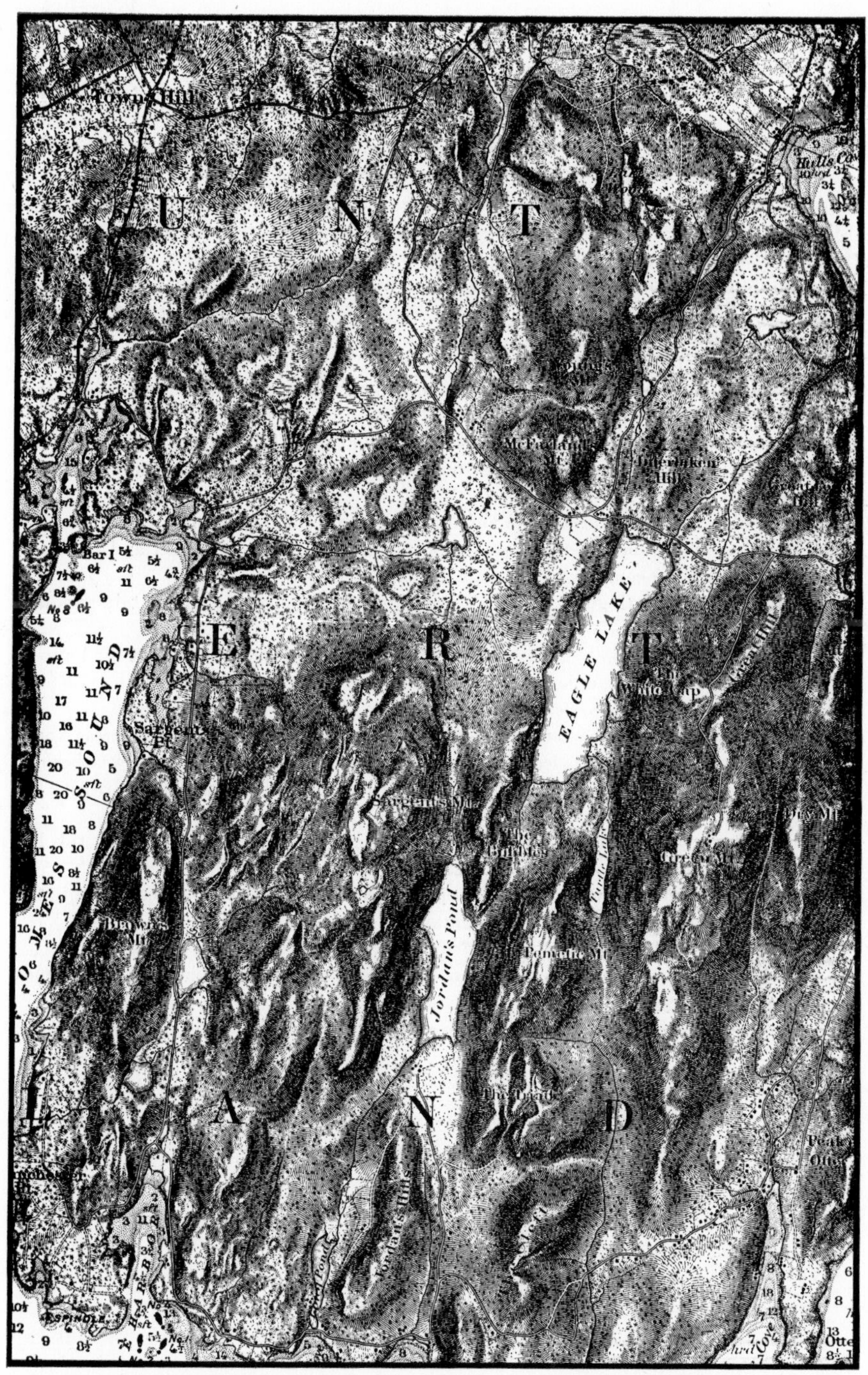


MATURELY DISSECTED COMPLEX MOUNTAIN AREA OF MOUNT DESERT ISLAND, MAINE, SHOWING SOME STRUCTURAL CONTROL
United States; Coast and Geodetic Survey Chart, No. 103 (1:80,000).

## QUESTIONS

1. What stream patterns are characteristic of complex mountains?
2. How would you distinguish, on a topographic map, a granite region from a schist region?
3. How would you account for depression contours in a region of crystalline rocks?
4. How is strong jointing made evident on a topographic map?
5. Do complex mountains usually have a coarse- , medium- , or fine-textured topography?
6. What mineral resources are often found in complex mountains? Does the age of a complex mountain region influence its mineral resources? That is to say, do old complex mountains have mineral resources materially different from young complex mountains?
7. Under what conditions are complex mountains, even in the mature stage, very deeply covered with soil?
8. What is laterite?
9. Are earthquakes more frequent in granite regions than in sedimentary regions, or is there no connection?
10. To what kinds of rocks are monadnocks in complex mountain regions likely to be due?
11. Do the minor structures in complex mountain regions suggest in any way the major structures of the region?
12. Can a block mountain or a dome mountain be also a complex mountain?
13. How do you account for the presence of coal near Worcester, Mass., and in Rhode Island within the complex mountain area of New England?
14. How would you distinguish, on a topographic map an old complex mountain region (a peneplane) from a plain of sedimentary rocks?
15. Why are the many lakes with two outlets in Canada used as evidence that the Laurentian peneplane has undergone recent warping?
16. Name several theories to account for peneplanation without base-leveling.
17. Why do English geologists largely favor the idea of marine peneplanation whereas American students support the idea of subaerial peneplanation by streams and weathering?
18. List all the sketches and pictures in this book which illustrate peneplanes. Criticize the first diagram on page 130. Is this region young in the first cycle?
19. Can peneplanation be accomplished by glaciation? Is there any evidence that this has ever happened?
20. How would you account for the flat-topped benches or alps just above tree line in the Presidential Range of the White Mountains of New Hampshire?
21. Do you think it is misleading to call New York City an old mountain region?
22. Draw an imaginary geological section to show every conceivable kind of complex geological structure.
23. What kind of rocks do you think makes up the region shown on page 643?

## TOPICS FOR INVESTIGATION

1. Theories for the origin of continents and ocean basins.
2. The theory of isostasy (see Wooldridge).
3. The origin of mountain ranges (see Daly).
4. The theory of continental drift (see Daly).
5. The landslide theory of mountain origin (see Daly).
6. The mineral localities of the United States in complex mountains.

## REFERENCES


### Origin

Andrews, E. C. (1923) *Contributions to the hypothesis of mountain formation.* Geol. Soc. Am., Bull. 34, p. 381–399.

Daly, R. A. (1926) *Our mobile earth.* New York, 342 p.

Dana, J. D. (1873) *On the origin of mountains.* Am Jour. Sci., 3d ser., vol. 5, p. 347–350, 423–443, 474–475; vol. 6, p. 6–14, 104–115, 161–172, 304, 381–382.

Davis, W. M. (1912) *Die erklärende Beschreibung der Landformen.* Leipzig, p. 246–315.

Du Toit, A. L. (1937) *Our wandering continents.* Edinburgh, 366 p.

Geikie, J. (1913) *Mountains: their origin, growth, and decay.* Edinburgh, 311 p. Not up to date on all points but has good descriptions, well written.

Hobbs, W. H. (1931) *Earth features and their meaning.* New York. The origin and the forms of mountains, p. 433–445.

Le Conte, J. (1889) *On the origin of normal faults and of the structure of the Basin region.* Am. Jour. Sci., 3d ser., vol. 38, p. 257–263.

Le Conte, J. (1893) *Theories of the origin of mountain ranges.* Jour. Geol., vol. 1, p. 543–573.

Penck, A. (1894) *Morphologie der Erdoberfläche.* Stuttgart. part 2, Die Gebirge, p. 327–438.

Powell, J. W. (1876) *Types of orographic structure.* Am. Jour. Sci., 3d ser., vol. 12, p. 414–428.

Rice, W. N. (1905) *The classification of mountains.* 8th Intern. Geog. Cong., p. 185–189.

Snider, L. C. (1932) *Earth history.* New York. *Theories of mountain making,* p. 192–199. Summarizes the main characteristics and the arguments for and against the following theories of mountain making: (*a*) contraction; (*b*) asthenolith; (*c*) isostasy: (*d*) continental drift; (*e*) sliding continents; (*f*) periodic melting of the subcrust.

Suess, E. (1904–12) *The face of the earth.* Tr. by Sollas, Oxford, vol. 4, p. 498–542. Mountain arcs or festoons. A comprehensive review of the mountain systems of the earth.

Upham, W. (1891) *A classification of mountain ranges according to their structure, origin, and age.* Appalachia, vol. 6, p. 191–207.

Van Hise, C. R. (1904) *A treatise on metamorphism.* U. S. Geol. Surv., Mon. 47. Relations of rock flowage to mountain making, p. 924–931.

Wegener, A. (1924) *Origin of continents and oceans.* London, 205 p.

Willis, B. (1906) *Studies in mountain growth.* Carn. Inst. Wash., Year Book 4, p. 195–203.

Woodworth, J. B. (1932) *Contributions to the study of mountain building.* Am. Jour. Sci., 5th ser., vol. 23, p. 155–171.

### Old Complex Mountains

Daly, R. A. (1905) *Accordance of summit levels among Alpine mountains.* Jour. Geol., vol. 13, p. 105–125.

Davis, W. M. (1896) *Physical geography of southern New England.* Natl. Geog. Soc., Mon. 1, p. 269–304.

Martin, L. (1911) *Physical geography of the Lake Superior region.* U. S. Geol. Surv., Mon. 52, p. 85–117.

Tarr, R. S. (1898) *The peneplain.* Am. Geol., vol. 21, p. 351–370.

Thwaites, F. T. (1931) *Buried pre-Cambrian of Wisconsin.* Geol. Soc. Am., Bull. 42, p. 719–750. Buried old complex mountains.

Weidman, S. (1907) *Geology of north central Wisconsin.* Wisc. Geol. Surv., Bull. 16, 697 p.

Wilson, A. W. G. (1903) *The Laurentian peneplain.* Jour. Geol., vol. 11, p. 615–669.

Wooldridge, S. W., and Morgan, R. S. (1937) *The physical basis of geography. An outline of geomorphology.* London, p. 210–256. Study of peneplanes.

### Regions Described

Bowman, I. (1911) *Forest physiography.* New York. Laurentian Plateau, p. 554–572; New England, p. 636–684; Piedmont Plateau, p. 623–635; etc.

Bowman, I. (1914) *Results of an expedition to the central Andes*. Am. Geog. Soc., Bull. 46, p. 161–183.

Bowman, I. (1914) *The Andes of central Peru*. Am. Geog. Soc., Bull. 46, p. 202.

Davis, W. M. (1911) *Colorado Front Range*. Assn. Am. Geog., Ann. 1, p. 21–83.

Dawson, G. M. (1890) *Later physiographic geology of the Rocky Mountain region in Canada*. Royal Soc. Can., Trans. 8, sec. 4, p. 3–74.

Dawson, G. M. (1901) *Geological record of the Rocky Mountain region in Canada*. Geol. Soc. Am., Bull. 12, p. 57–92.

Emmons, S. F. (1890) *Orographic movements in the Rocky Mountains*. Geol. Soc. Am., Bull. 1, p. 245–286.

Emmons, S. F., et al. (1893) *Geological guide book of the Rocky Mountain excursion*. 5th Intern. Geol. Cong., p. 253–487.

Fenneman, N. M. (1931) *Physiography of western United States*. New York, Chaps. 2, 4, 5, 9.

Fenneman, N. M. (1938) *Physiography of eastern United States*. New York, Chaps. 3, 6, 7, 13.

Hobbs, W. H. (1904) *Tectonic geography of eastern Asia*. Am. Geol., vol. 34, p. 69–80, 141–151, 214–226, 283–291, 371–378.

Joerg, W. (1910) *The tectonic lines of the northern part of the North American Cordillera*. Am. Geog. Soc., Bull. 42, p. 161–179.

Lobeck, A. K. (1917) *Position of the New England peneplain in the White Mountains region*. Geog. Rev., vol. 3, p. 53–60. Explains projected sections.

Pumpelly, R., Wolff, J. E., and Dale, T. N. (1894) *Geology of the Green Mountains in Massachusetts*. U. S. Geol. Surv., Mon. 23, 203 p.

Stark, J. T., and Barnes, F. F. (1932) *The structure of the Sawatch Range*. Am. Jour. Sci., 5th ser., vol. 24, p. 471–480.

Willis, B. (1912) *Geologic structure of the Alps*. Smithsonian Inst., Misc. Coll., No. 2067, vol. 56, 13 p.

Geographical Aspects

Allen, W. E. D. (1929) *The march-lands of Georgia (U.S.S.R.)*. Geog. Jour., vol. 74, p. 135–156.

Atwood, W. W. (1927) *Utilization of the rugged San Juans*. Econ. Geog., vol. 3, p. 193–209.

Huntington, E. (1905) *The mountains of Turkestan*. Geog. Jour., vol. 25, p. 139–158.

Peattie, R. (1936) *Mountain geography. A critique and field study*. Cambridge, Mass., 257 p.

Semple, E. C. (1911) *Influences of geographic environment*. New York, p. 557–606.

# XIX
# VOLCANOES

*Darton, U. S. Geological Survey*

THE DEVIL'S TOWER, NEAR THE BLACK HILLS, WYO.
Remnant of volcanic plug, showing remarkable columnar structure.

C. R. Miller

MT. SHASTA, CALIF., A SUBMATURELY DISSECTED VOLCANO

Shastina, a subsidiary cone on its right flank; cinder cones and lava flows at its base.

CRATER LAKE, ORE.

The caldera of the great volcano Mazama; Wizard Island, a recently formed cinder cone.

*C. R. Miller*

*Publishers' Photo Service*

CRATER OF VESUVIUS, ITALY

Recently-formed young cinder cone built within the present crater; Monte Somma, the rim of the old caldera, in the far distance.

FORMER CRATER OF LASSEN PEAK, CALIF.
Stratified accumulations of volcanic ash in walls of crater, and scattered volcanic ejecta on the slopes.

*P. J. Thompson*

U. S. Geological Survey

MT. EDGECOMB, KRUZOF ISLAND, ALASKA

A young, only slightly dissected volcano with single unbreached crater.

DOUBLE CINDER CONE, NEAR MEXICO CITY, MEXICO
Old lake basins in the craters, now cultivated. Both inner and outer slopes scored by numerous ravines.

*Fairchild Aerial Surveys*

*Barnum Brown, American Museum of Natural History*

SHIPROCK NEAR FARMINGTON, N. M.

A volcanic neck with minor necks and a long radiating dike, the remnants of a former great volcano.

FRONT OF LAVA FLOW; BASE OF CINDER CONE NEAR LASSEN PEAK, CALIF.

The aa type of lava, a jumbled mass of jagged blocks encroaching upon older ash-covered surface.

*Diller, U. S. Geological Survey*

Mendenhall, U. S. Geological Survey

CRATER OF KILAUEA, HAWAII NATIONAL PARK

Showing comparatively smooth surface of pahoehoe or ropy lava, unlike the jagged surface of less fluid lava flows.

LAVA CASCADES, KILAUEA, HAWAII NATIONAL PARK

Ropy or pahoehoe lava, fluid at a comparatively low temperature, congealed at margin of lava flow.

*Mendenhall, U. S. Geological Survey*

*U. S. Department of the Interior*

WIZARD ISLAND, CRATER LAKE, ORE.
Lava flow emerging from western base.



CRATERS OF THE MOON NATIONAL MONUMENT, IDAHO
Spatter cones south of the Big Craters along the Great Rift.

*U. S. Department of the Interior*

*Howard Coombs*

"STEPTOES," McKENZIE PASS, CASCADES, ORE.

White granite hilltops not covered by recent volcanic flows.



LAVA FLOW FROM CINDER CONE, NEAR LASSEN PEAK, CALIF.

At the left an earlier flow, ash covered; Snag Lake in the distance, blocked by recent flow.

*A. K. Lobeck*

*M. Henry Donald, England*

BASIC DIKES NEAR SALAIR, CENTRAL SIBERIA

The resistant dike maintains the higher crest of the limestone hill.

## VOLCANOES

SYNOPSIS. Volcanic forms, like other land forms, pass through a series of erosional stages from youth to old age. The upward growth of a volcano by eruptions, from a small cinder cone to a great mountain, should not be considered as part of the cycle of development from youth to maturity. During all of this time the volcano is young, if it has not been changed by erosion.

Volcanic forms differ from other land forms in that the initial shapes are more varied and are more apt to be unaltered by destructive forces. This is because volcanic forms, cones, craters, calderas, and lava flows develop with such rapidity that erosion does not affect them during the process of growth. Other land forms, like mountains produced by folding, warping, and faulting, are raised so slowly that almost from the outset a mature stage of erosion is attained. The initial stage of volcanic forms is therefore commonly met with, but for some of the other constructional types it is unknown.

Volcanic forms differ also from other constructional types in that the forces of growth are paroxysmal. There are frequent interruptions to the orderly progress of the erosion cycle. A volcano which is mature as a result of stream dissection may again become young because of new volcanic outbursts. Another complication is that the constructive agencies of volcanism are sometimes destructive in their violence, as when the top of a volcano is blown off. The resulting calderas and associated deposits should be treated as constructional forms, due to essentially constructive agencies. It is clear, therefore, that a study of volcanoes demands the recognition and classification of all the various initial forms, on the one hand, and of all the various erosional forms, on the other hand.

The initial forms range from small cinder cones, with minor lava flows exuded from their bases, to large compound volcanoes, with multiple peaks, the whole mass built up of many alternating layers of ash deposits and lava flows. There are, also, complex volcanoes consisting of peaks erected inside calderas which represent the wrecks of earlier mountains. Successive mountains are thus built upon the same site, the remnants of the earlier ones still persisting. The gradations from lava flows to lava domes and volcanic peaks depend upon the degree of violence involved in the eruption.

The erosional forms of volcanoes are best developed in extinct volcanoes of large size. Maturity is attained when streams and glaciers have carved canyons on their flanks and have completely destroyed the initial form.

Further erosion exposes the congealed core. Only a volcanic neck or plug then remains, with perhaps some of the radiating dikes also stripped of their weaker surrounding rocks. This is the stage of old age.

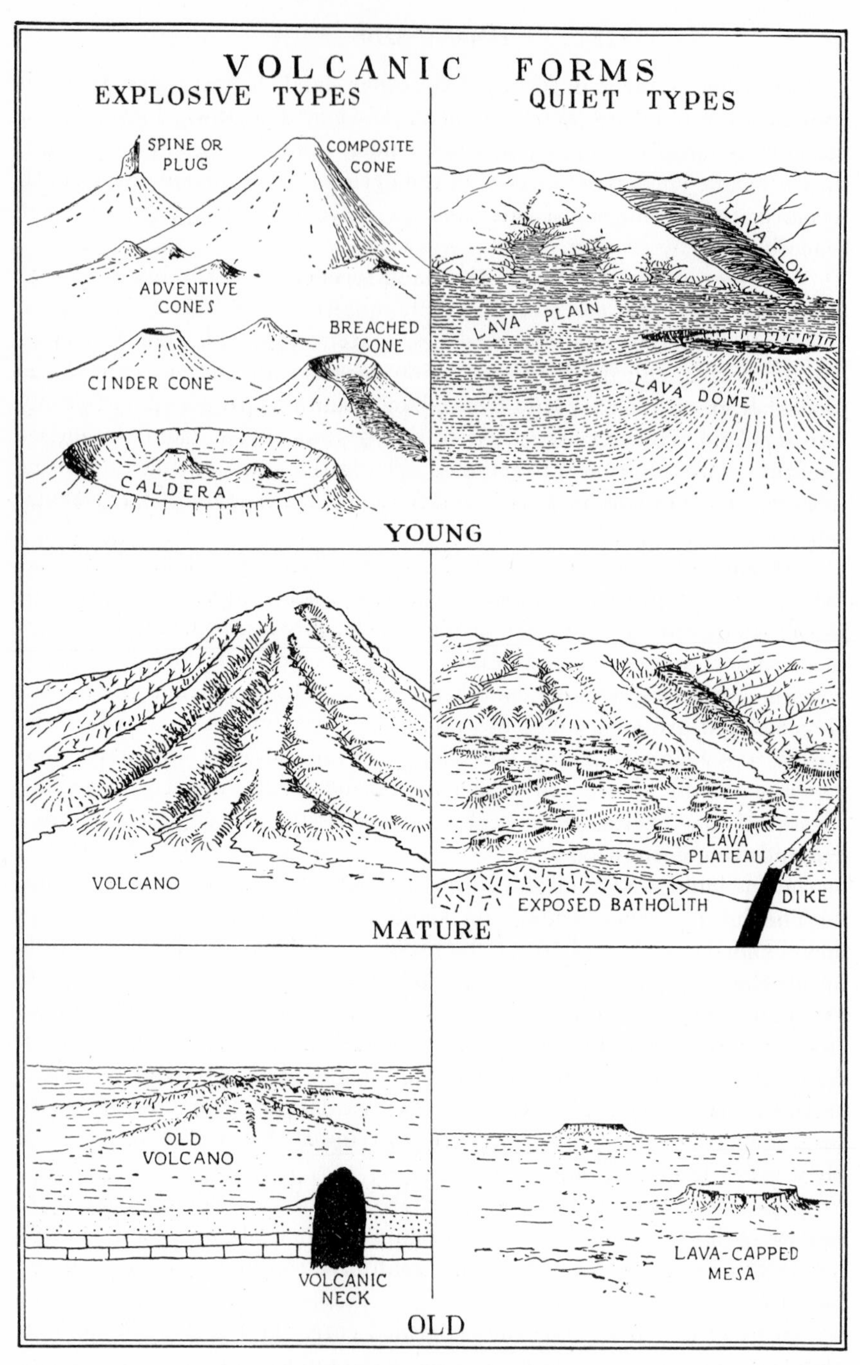


DIAGRAMS SHOWING CLASSIFICATION OF VOLCANIC FORMS ACCORDING TO THEIR ORIGIN AND EROSIONAL STAGE

## CLASSIFICATION OF VOLCANIC FORMS

Land forms resulting from volcanic activity may be listed in two categories: those due mainly to explosive action, and those due to the quiet emission of lava. Each member of these two types may be considered in its youth, its maturity, and its old age as it is changed, through the cycle of development, by erosive agents. As in most fields of natural science, there are transitional or composite forms. These forms embody the characteristics of both the explosive and the quiet type of volcanic activity.

FORMS DUE TO EXPLOSIVE ERUPTIONS. The explosive types are represented by *tuff* or *cinder cones*. *Composite cones* result from the combined effect of explosions and quiet flows and constitute the great volcanoes of the world. There is great variety of size, form, and distribution of cones with relation to each other. *Adventive* or *parasitic cones* may occupy the flanks of large cones. Simple cones usually exhibit a simple *crater* or *explosion crater*, but this may become a *breached crater* when broken through by a lava flow. Explosions at intervals and of diminishing intensity may produce a series of *nested craters*. *Volcanic bombs*, *lava flows*, *mud flows*, and *ash deposits* are commonly associated with young volcanic cones. Unusually violent explosions may completely wreck a volcano, leaving a *caldera*, much larger than a normal crater. New cones may in turn develop within the caldera.

*Maturely dissected cones* are usually extinct cones. Many of the great volcanoes of the world are deeply scarred by streams and glaciers. Some are so greatly worn down as to expose their internal structure, often exhibiting radiating dikes.

*Old volcanoes* are those which have been so completely destroyed as to exhibit no longer their conical form. They have been reduced to merely the core of their former bulk and are termed *volcanic necks*.

FORMS DUE TO QUIET ERUPTIONS. Quiet eruptions, taking place through fissures or dikes, result in lava plateaus or *lava domes* and *lava fields*. In some instances lava flows are confined to valleys and produce *lava tongues*. The foundering or sinking in of part of the lava surface produces open vertical-walled pits, called *lava sinks*, in which molten lava may be visible. These are quite unlike craters in origin although they often bear that name. Numerous details are associated with lava flows, such as *lava bridges or tunnels*, *spatter cones*, *driblet cones*, *lava cascades*, and *tumuli*.

Modification of lava flows by erosion produces such features as *maturely dissected lava plateaus* with the exposure of dikes and other deep-seated igneous bodies like sheets, laccoliths, and batholiths. Old age results when only few remnants are left standing to form *buttes* and *lava-capped mesas*.

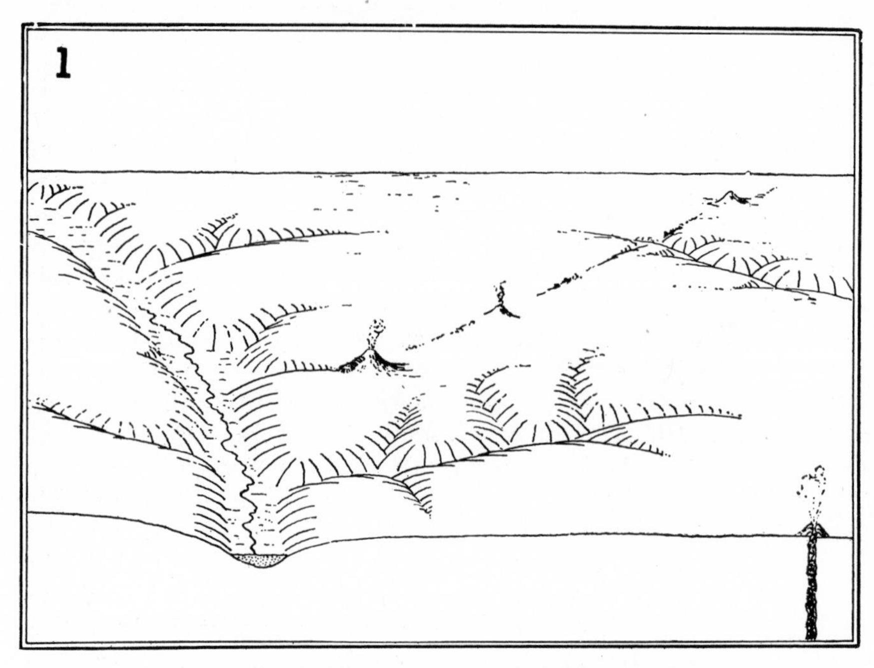


## YOUNG VOLCANOES. CINDER CONES

The initial stages in the growth of cinder cones have frequently been witnessed by man. Activity starts usually with the emission of gas from a fissure on the surface of the ground accompanied occasionally by earthquakes. Sand and particles of rock are thrown out and assume a form something like an ant hill, at first only a few inches in height. The gases may be incandescent and be visible at night. There is no molten rock. The stones and sand thrown up may be simply fragments torn from the walls of the fissure and may consist of limestone or shale or any other type of country rock. This material is termed *cinders* or *ash* even though it may not be of igneous origin.

The rapidity of growth may be such that in the course of a week a cone 400 feet or more in height is built, as occurred in Italy just west of Naples where Monte Nuovo came into being September 29, 1538.

Most cinder cones slope away from the crater at angles of 26 to 30°. As the cones are made up of loose and porous material, erosion does not alter their outline and they remain fresh for centuries. The craters of such cones are usually deep, going down almost to the base of the volcano itself. A cone 400 to 500 feet high would have a base about 2,000 to 2,500 feet in diameter.

It is not unusual for several cones to arise along one fissure. In Mexico on September 28, 1759, a fissure appeared on a level plain, 2,000 feet above sea level in the midst of sugar fields and 35 miles distant from any volcano. Rocks were thrown to a great height and built up six conical hills on the line of the chasm, the smallest being 300 feet high, and the

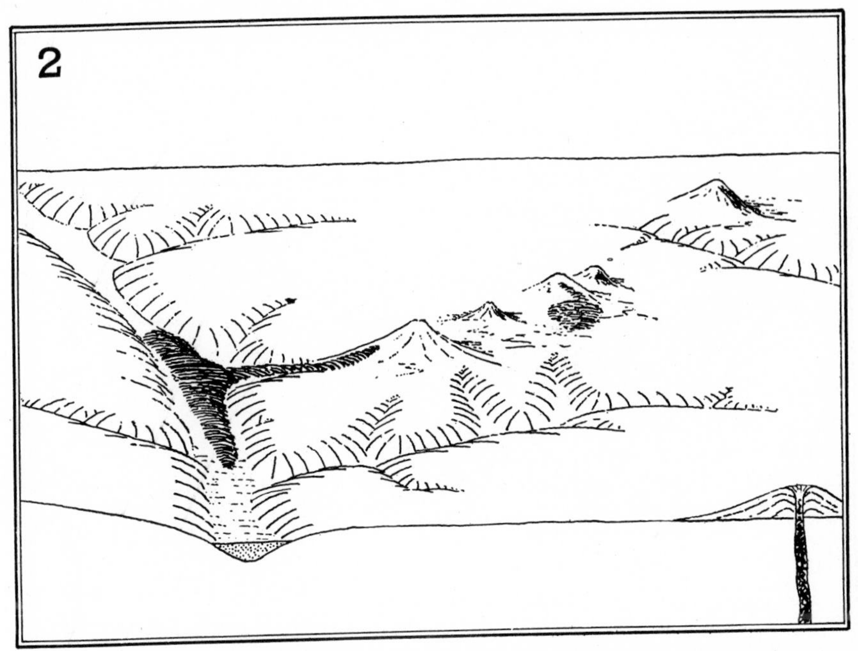


largest, called *Jorullo*, 1,600 feet above the plain. For several preceding months there were earthquakes and subterranean rumbling. Unlike Monte Nuovo, there were also streams of lava poured out from the flanks of Jorullo, continuing for half a year. Twenty years later the lava was still hot enough, below the surface, to light a cigar.

Another example of a cone whose growth was observed from the very beginning is Mount Izalco in San Salvador which started in 1770 as a small orifice on a tract of cultivated ground. It is now 3,000 feet high. In 1871 the volcano of Camiguin started from a fissure in a level plain on one of the small islands north of Luzon in the Philippines. It continued active for four years and attained a height of 1,800 feet.

In cross section a cinder cone is made up of layers of loose fragments, the larger pieces being near the crater. The bedding planes slope away from the crater and also toward the crater along their inner edges. The crater is maintained by the violence of the gaseous outbursts and rarely becomes filled up until lava rises into it from beneath. If lava finds its way up into the cone, it usually breaks through the loose layers that make up the cone and emerges near the base or along the flank of the cone rather than at the summit. Occasionally *breached craters* are formed when the lava breaks through the rim.

Cinder cones have been known to form beneath the sea and in a few instances to rise above sea level, where they are soon destroyed by the waves. Sabrina Island in the Azores and Graham Island near Sicily, as well as occurrences in the Bogoslof Islands off of Alaska and in the South Seas, are the results of submarine activity.

YOUNG VOLCANOES. LAVA FLOWS, COMPOSITE CONES, VOLCANIC PLUGS, ETC.

Lava flows alone do not build up high volcanic peaks. They are only incidental to the development of cinder cones, and many cinder cones do not exhibit this phenomenon.

The lava emitted by cinder cones or even by large volcanoes is rarely enough to cover much country. The great lava fields of the world, like those of the Columbia Plateau and of the Dekkan, have resulted from fissure eruptions. Small tongues of lava break through the sides of cinder cones and volcanoes and find their way down valleys, blocking them to form lakes. River systems are displaced and new streams develop on each side of a lava flow, between it and the valley wall.

A large volcano of the *composite* type is built up by the combined effect of violent gaseous outbursts with the accumulation of cinders and ash, and numerous individual lava flows, most of which occupy only a small part of the side of the cone. The lava beds of most volcanic areas, even at a distance from the volcanoes, are separated from the overlying ones by layers of ash and occasionally by lake deposits. Small cinder cones sometimes hurl forth large projectilelike masses of viscous lava which are known as *volcanic bombs*. They are usually spiral or spindle shaped and sometimes accumulate in large numbers at the foot of the cone. Finer and more molten material may be shot to a greater height and, cooling rapidly, form drops, beads, and filaments, called *Pelée's hair*.

The sequence of diagrams on pages 664–670 follows illustrations by W. M. Davis.

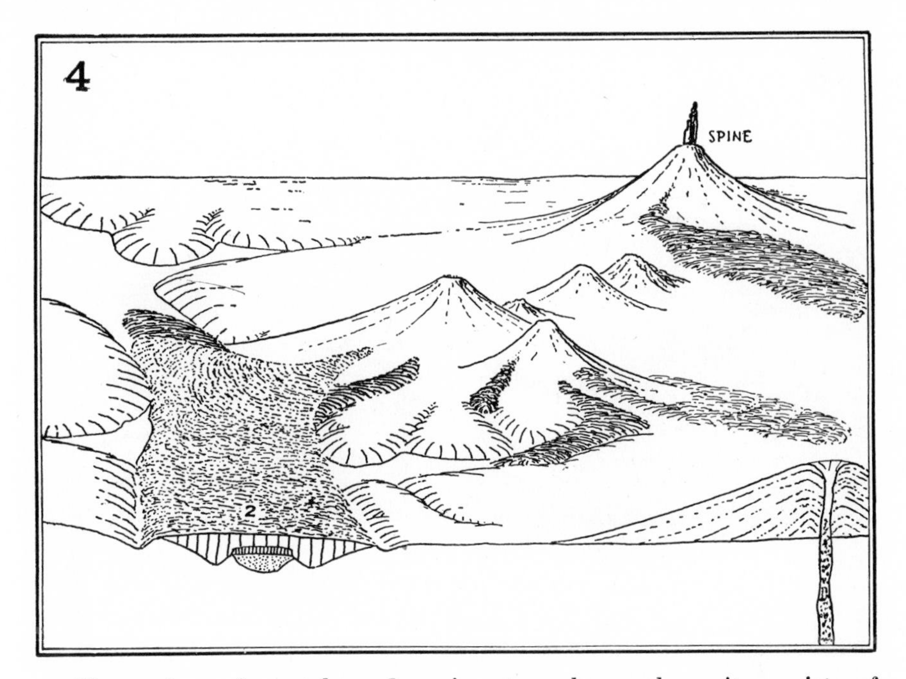


The surface of most lava flows is extremely rough, as it consists of the congealed surface of the flow which has later been broken into blocks and rolled in with the more fluid material beneath. The front of the lava flow is usually steep and jagged, standing about as high as a house, and may be almost unscalable. Where the molten lava runs out from beneath a congealed crust, a natural bridge or *lava tunnel* results. Lava surfaces which have a ropy aspect are called *pahoehoe* and, when such surfaces are caused to bulge up with many fractures and with a bread-crust appearance, the resulting hump or hill is called a *tumulus*.

In rare instances during the growth of a volcano the core or neck becomes a solid or highly viscous mass and is pushed out bodily like a great piston by the tremendous pressure beneath. Such a feature is called a *plug* or *spine*. The growth of a spine was actually observed during the eruption of Mount Pelée on the island of Martinique in the West Indies. It began in October, 1902, and reached its maximum elevation in seven months. It attained a height of 700 to 1,000 feet above the crater and retained the form of the tube in which it had solidified. After that, it slowly crumbled away under the influence of the atmosphere and its own weight and from the explosion of gases beneath it.

It is rare that volcanoes grow so rapidly and so uniformly as to develop a perfect cone without showing any effects of erosion. Cones like Vesuvius, Fujiyama, and Mayon in the Philippines, although several thousand feet in height, are young cones. They show little effect of stream or glacial erosion. Most of the great volcanoes of the world, and especially those which are extinct, are moderately dissected and have reached a submature or even a mature stage of development.

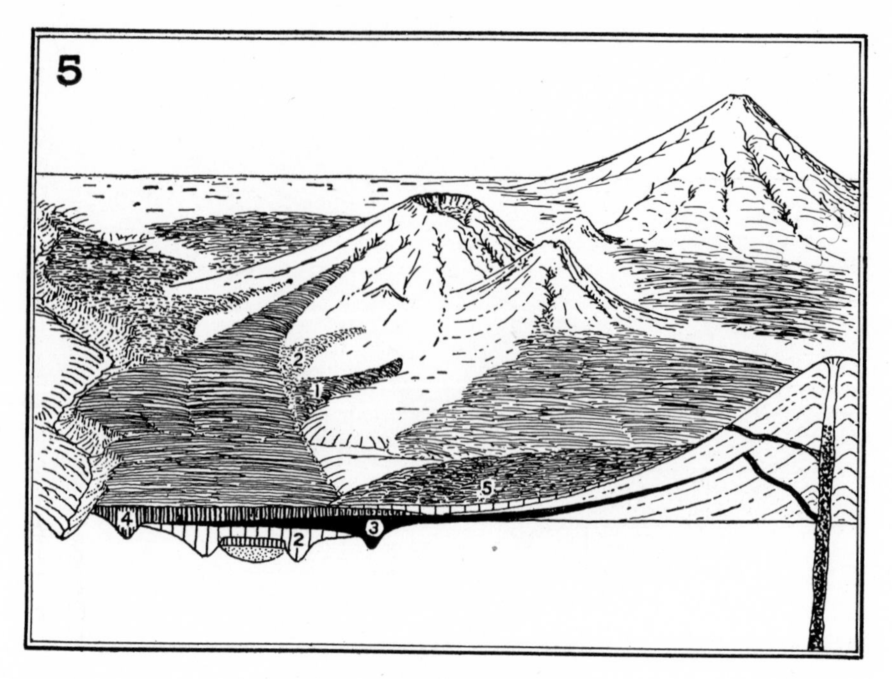


## MATURE VOLCANOES

The great volcanoes of the world, built up slowly through many years by both violent eruptions and the quiet outpouring of numerous lava flows from their flanks, have most of them been much altered by streams and glaciers. They are mature not because they have reached maximum proportions but because they are dissected and have lost their fresh conical form.

Many large volcanoes support, on their lower slopes, smaller cones, called *parasitic*, *lateral*, or *adventive cones*. These may actually develop to so great a size as to vie with the major cone in importance, as Shastina does on the side of Mount Shasta.

During the process of growth, great canyons formed on the slopes of large volcanoes may become filled with later flows. Glacial troughs may form and these in turn be occupied by lava streams. Ash and fragmental material representing periods of violence may be interbedded with the lava layers.

Some large volcanoes, as well as many cinder cones, are not symmetrical but elongated in one direction owing to the prevailing wind movement. Large volcanoes, too, may lie along some great fissure or line of faulting. The volcanoes of central France, the volcanoes of Iceland, and the chain of great peaks in Nicaragua show a linear arrangement.

During the eruptions of a large volcano, tremendous pressures are exerted and these develop radial cracks in the cone which are filled with lava to form dikes. These later appear in relief as the volcano is worn down. In the Spanish Peaks in southern Colorado there have been several

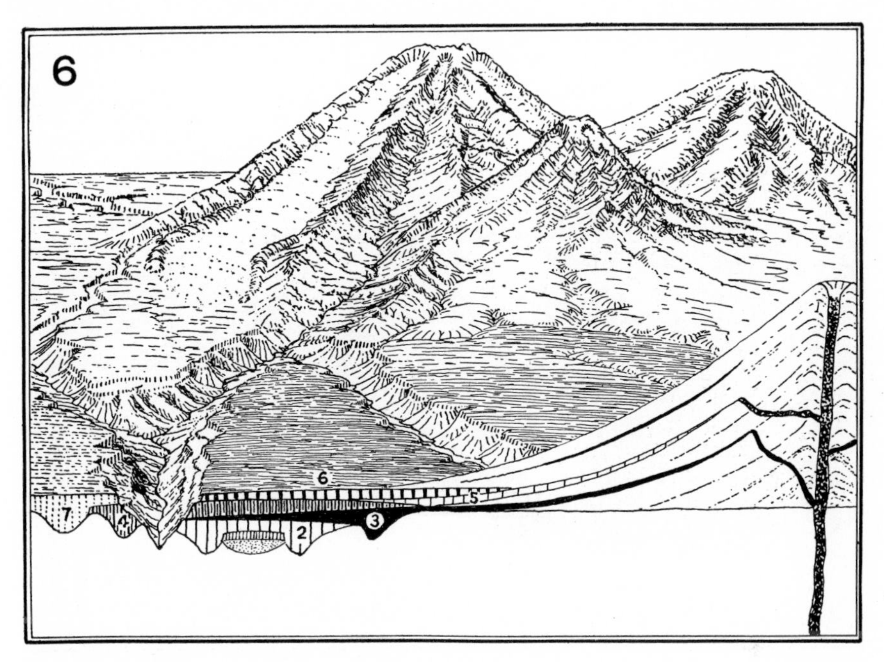


periods of dike formation, different types of lava having been intruded in each case.

Large volcanoes do not grow at a uniform rate throughout their whole career but show cycles of activity. Vesuvius, the most carefully studied volcano in the world, has, since recorded history, exhibited several major outbursts separated by times of quiescence. During the quiet periods there are slight emissions of gas and frequent extrusions of lava. During the great eruptions, as in 1872 and 1906, the crater became much enlarged and at times the entire head of the volcano was blown off and a new cinder cone developed in the crater thus formed. A series of craters, one within the other, called *nested craters*, is sometimes met with among young cinder cones.

Accompanying the great eruptions of volcanoes are copious floods of mud from the condensed vapor and ash. This mud flows rapidly down the sides of the cone because it does not congeal upon cooling and, when it inundates farms and villages, causes much damage. Floods also occur from the sudden melting of extensive fields of snow on the mountain summits when hot gases are emitted and sweep down the slopes. These produce violent erosive effects in a short time.

Some of the largest mountains in the world are of volcanic origin and rise 10,000 to 15,000 feet above their bases. A region of this character, when dissected to a mature stage, is one of extreme topographic complexity. The original volcanic forms may be completely disguised by the numerous canyons which have been formed and only by noting the attitude of the ash deposits and lava beds in the canyon walls can the original location of the various volcanic centers be determined.

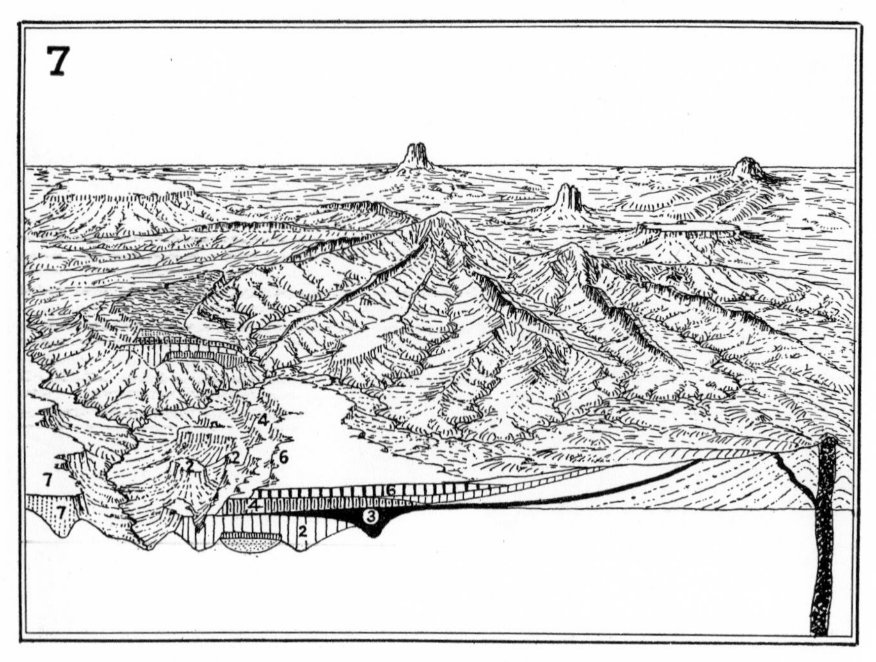


## OLD VOLCANOES. VOLCANIC NECKS. DIKES

The late stages in the dissection of a volcanic area are typified by volcanic necks and by wall-like dikes rising above a generally flat country. Slightly earlier stages reveal the radial plan of the volcanoes' spurs. The earlier radial drainage pattern still persists but in time even this disappears and only the central cores of the volcanoes remain.

VOLCANIC NECKS. A large volcanic neck may rise 2,000 feet or more above the surrounding country and usually appears to be standing on a pedestal of nonvolcanic rocks. It is evident that these formations at the base have been protected from erosion by the resistant mass which surmounts them, for elsewhere they may be completely removed. In any given region, as in the Mount Taylor district of New Mexico, the various volcanic necks rise to different heights and their bases stand at various altitudes, indicating clearly that they have no connection with each other as would be the case if they were once parts of a lava flow.

The diameter of a volcanic neck may be as much as 1,000 to 2,000 feet, the general form of the neck being cylindrical.

Most volcanic necks show vertical columnar structure in the upper part, which at the base curves outward toward the enclosing walls. This is the result of cooling, the joints forming approximately at right angles to the cooling surface. Near the top of the neck the cooling surface was at the crater, hence the vertical joints, but at the base of the neck

the enclosing walls constituted the cooling plane, and there the radial structure developed. The famous Devil's Tower of the Black Hills region shows this phenomenon. If volcanic necks occur in a region of horizontal sedimentary strata, the sedimentary beds show little disturbance due to volcanism because the neck was not pushed up suddenly in its present size but was formed by the gradual enlargement of a small fissure.

DIKES. The presence of wall-like dikes indicates that much of the enclosing rock has been worn away, and that the dikes remain because of their greater resistance. Narrow dikes, no matter how great the durability of the rock, rarely produce any topographic effect, whereas thick dikes are almost certain to.

The Spanish Peaks district of Colorado is famous for its system of dikes which have been revealed by erosion. The dikes are from 2 to 100 feet or more in thickness and dip steeply, first one way and then the other. They rise as smooth wall-like masses 50 to over 100 feet in height and extend from a few hundred yards to 10 or 15 miles across the country. Frequently they intersect each other. The dikes are most abundant at lower levels and it appears that they rarely have penetrated into the upper strata.

There are two distinct sets of dikes in this region, one radiating from West Spanish Peak. The other set radiates from some much more remote point and the dikes in this set are therefore almost parallel with each other. Naturally they intersect those of the first group.

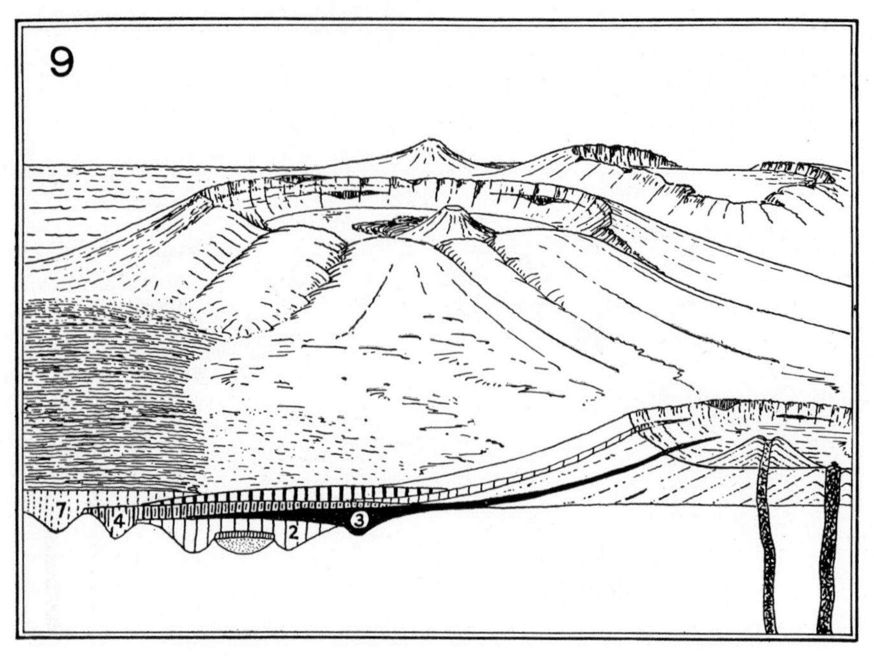


## WRECKED VOLCANOES. CALDERAS AND NESTED CALDERAS

Many volcanoes, after attaining large size by numerous small violent outbursts and quiet lava outpourings, are more or less destroyed by an explosion of unusual vigor. In some instances, only the summit is torn away but in other cases the violence of the eruption blows off most of the volcano. The resulting depression is a *caldera.* It is much larger than a crater for its diameter is several times as great as the neck or core of the volcano. A caldera may be several miles across, but most craters are less than 1,000 feet in diameter and rarely more than 1,000 feet deep. When first formed, craters may be fissurelike but the relatively small amount of explosive gases which come to the surface tend to find the single weakest means of exit. This becomes enlarged to form the neck of the volcano. The craters of cinder cones have the same slope on the inside and outside, about 25 to 30°, this being the angle of rest of tuffaceous material. But calderas, because they are formed by much more violent eruptions, have steeper slopes on the inside, this slope being due not to the settling down of fragmental debris but to the disruption of the beds of lava and tuff which constitute the volcano. A vast amount of fine material is deposited up to a distance of several miles from a volcano which has been blown to pieces.

Resumption of volcanic activity within a caldera results in new cinder cones and in time there arises another large volcano which may

equal or exceed the original one in size. Successive destructive explosions may produce several *nested calderas*, one within the other.

The inside of a caldera provides a cross section of the structure of the volcano and reveals the stages in its growth. The transverse view thus provided shows that lava flows fill valleys which radiated from the center of the volcano and extended down its flanks. Some lava flows now fill glaciated valleys like those on the sides of Mount Mazama which occupied the site of Crater Lake. Other valleys now filled with lava were not glaciated prior to the lava flows.

Dikes also show in the walls of wrecked volcanoes, and there are numerous irregular intrusions and veins which serve to bind the unconsolidated material into a substantial mass.

The calderas produced by the decapitation of large volcanoes are quite different in origin from those which contain the lava lakes in the Hawaiian Islands. The former are due to explosion; and the latter are due to engulfment. It has been suggested that Crater Lake occupies a depression caused by the foundering or engulfment of a large volcano but this belief seems to be quite untenable.

Among the well-known examples of volcanoes which have been destroyed by prodigious explosions in historic time are Krakatoa in the Dutch East Indies in 1883, Mount Katmai in Alaska in 1912, Bandai-San in Japan in 1888, and Vesuvius in A.D. 79.

YOUNG LAVA DOMES, NOT DISSECTED

## YOUNG LAVA DOMES

The quiet type of volcano built up by innumerable lava flows is represented by the lava domes of Hawaii and some other Pacific Islands. Parts of the Columbia Plateau exhibit similar forms. Domes of this type in the Hawaiian Islands attain elevations of 13,000 feet. They extend over great areas, usually elliptical in plan and having diameters of 50 to 70 miles. The slope of the surface of these domes is about 5° near the summit and even less near the periphery. Rising from the ocean floor which is 16,000 feet deep, these domes may have actual elevations approximately 30,000 feet and diameters of 100 miles or more. As the name implies, they are merely domes, and in no respect do they resemble the volcanic cones already described.

Usually at the summit of lava domes there is a large depression or volcanic sink, bounded by steep inward-facing cliffs. This sink is actually formed by the collapse of part of the dome and the engulfment of the rock. It may be 2 to 3 miles across and 1,000 feet deep in its lowest part. Somewhere on the floor of the sink, which is made up of different levels, a boiling cauldron of lava is exposed, if the volcano is still active. This level of the lava rises and falls and sometimes the lava breaks through the rim of the cauldron and flows down the outside of the dome. More often the lava emerges from a fissure on the flank of the dome and flows

YOUNG LAVA DOMES, SLIGHTLY DISSECTED

away in a narrow stream, usually for many miles because of its exceptionally fluid condition.

In area the lava flows may cover 10 to 20 square miles. The whole great dome is built up of a network of lava streams. Two types of lava are recognized. The *pahoehoe*, which is most common, has a satiny or smooth, shining surface, which is also hummocky or billowy. After the surface has congealed, the molten lava beneath may continue to flow out and leave a bridge or tunnel lined with pendants of dripping lava, resembling stalactites.

*Aa* is rough broken lava from a few inches to a thousand feet in size, piled up in a chaotic mass. It is formed by the breaking up and pushing along of the solidified lava by the molten stream beneath.

All of these lavas are basic and contain very little gas. Nevertheless, most lava domes show some evidence of explosive activity in the form of small cinder cones, either within the caldera (or crater, as it is usually called) or on the side of the dome. These sometimes have a linear arrangement as if placed along a fault or fissure.

Where gases sputter out through the side of the dome, a *spatter* or *driblet cone* may be built up 10 to 12 feet above the ground.

Small swellings or domical hills which rise above the floor of the lava sinks, because of pressure beneath, are called *tumuli*.

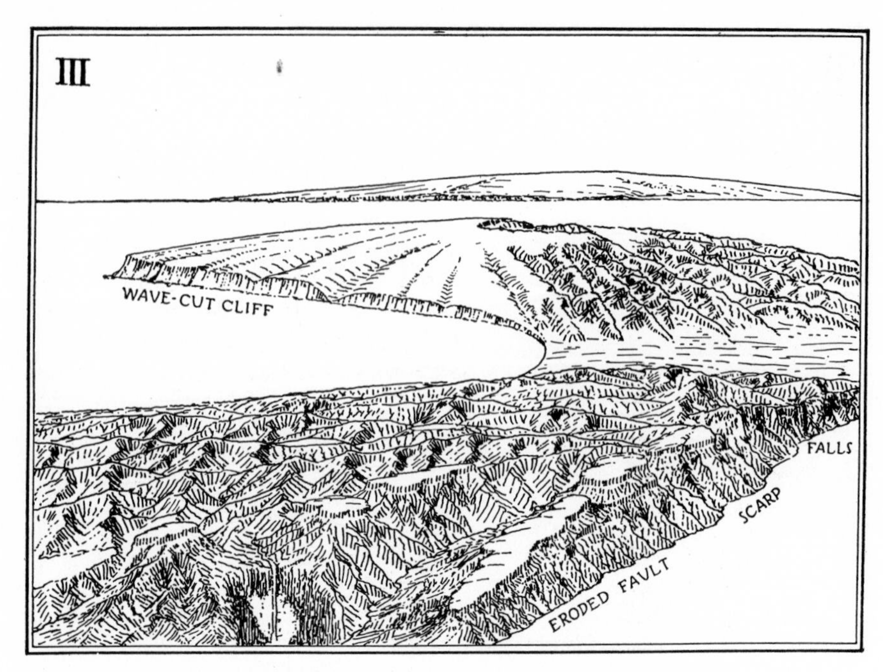


MATURELY DISSECTED LAVA DOMES

## MATURELY DISSECTED AND OLD LAVA DOMES

Young lava domes are drained by numerous small radiating, intermittent, evenly spaced streams. With the cessation of lava eruptions the streams cut deep sharp-walled valleys into the upland. Between the gorges are broad flat interfluves.

MATURE LAVA DOMES. The growth of the stream systems is accomplished by headward erosion. Sapping along the less resistant lava beds causes the unsupported masses above the weak beds to break off in landslides. The more resistant flows and sills form sheer cliffs many hundreds of feet high. The heads of the canyons are huge amphitheaters 1,000 to 3,000 feet high. Hundreds of streams tumbling from the plateau summits deeply furrow the walls of the amphitheaters and the upper portions of the canyons.

In time the interfluves become steep, sharply serrated ridges and the valleys may become wide and flat floored and covered with alluvium.

The stage of erosion may vary greatly on the two sides of a dome if there is a difference in rainfall, as in Hawaii. The rainy windward side may reach an advanced stage of dissection and acquire a very rugged appearance, while the leeward side retains its smooth domelike outline.

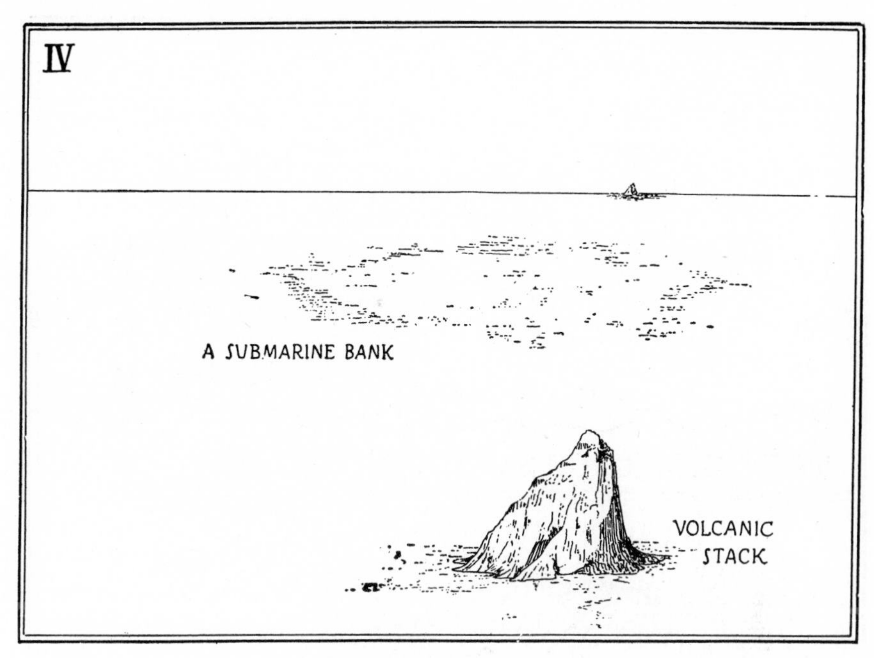


OLD LAVA DOMES

The waves are also effective agents of destruction. The margin of the dome is cut back and a wave-cut platform is formed, bordered by a sea cliff. Tilting and uplift of the entire dome may bring the wave-cut bench above sea level.

In some cases faulting on a large scale occurs not only in the vicinity of the central caldera, but the whole dome may be transected. Half of it may actually founder beneath the sea, leaving a fault scarp of magnificent proportions, in some instances 3,000 to 4,000 feet in height.

Streams reaching the coast may cascade over scarps thus formed, as well as over the scarps produced by wave erosion.

The crater, or caldera, at the summit of the dome becomes obscured by the disintegration of its steep walls, and under suitable conditions it may support a heavy growth of forest.

Old Lava Domes. In old age, lava domes are reduced to volcanic stacks and finally to submarine banks. These latest stages are largely the work of wave action, the streams having dwindled greatly in size with the reduced rainfall that comes from reduced elevation. One of the largest stacks remaining in the Hawaiian Islands is ¾ mile long, ¼ mile wide, and almost 900 feet high. The banks representing the final stage of destruction stand usually about 240 feet below sea level and may extend over 60 miles, depending on the size of the original dome.

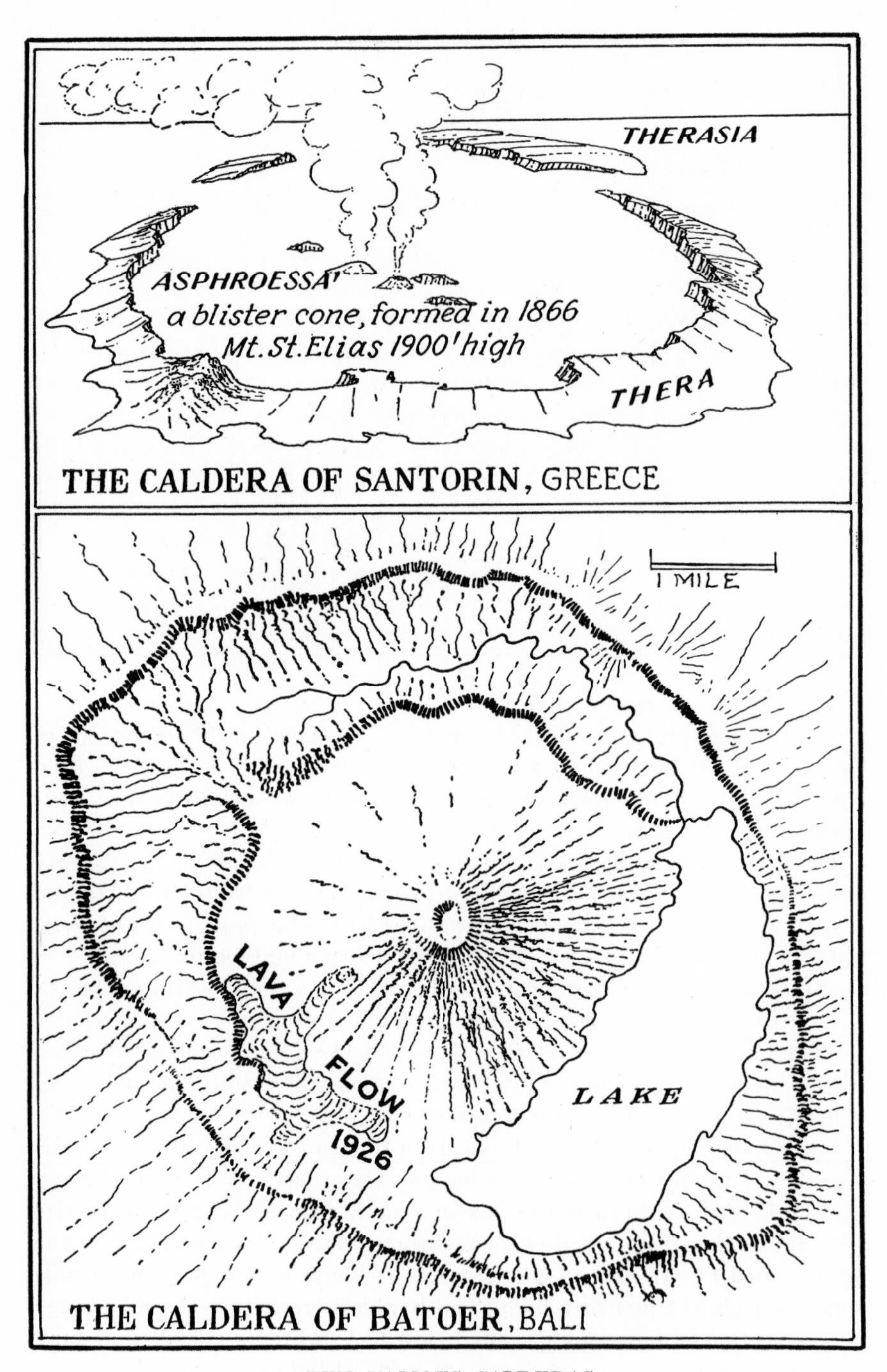


TWO FAMOUS CALDERAS

Calderas, Craters, and Volcanic Sinks. Calderas are depressions similar to craters in form, but of much larger size. The term *caldera* is not used by all authors for precisely the same feature. Some writers apply the term only to those large depressions formed by the subsidence of a volcano or volcanic area. Others call such areas *volcanic sinks,* like the large lava-filled depressions of the Hawaiian volcanoes. The term caldera might best be applied to large depressions due to explosive activity of unusual violence, which has completely destroyed a former volcano and left a depression in its place. Some great calderas of the world, like that of Aso in Japan (described by Robert Anderson) and that of Crater Lake, have been usually accounted for by subsidence or engulfment. This appears highly improbable. Subsidence in the Hawaiian lava fields is of quite a different nature. It does not mean the engulfment of any great amount of material, nor the destruction of any volcanic peak. The Hawaiian volcanic sinks are simply parts of a lava field or lava dome which has collapsed to form a depression or lava lake. Ordinarily when collapse occurs or when the lava in the lake becomes lower, there is also a breaking out of lava at some lower level along the side of the dome and a discharge on to the surface of the plain.

Recently Formed Calderas. Several great explosions have produced giant calderas in recent times. On the island of Java alone, three volcanoes were wrecked during the past century. In one case over 1 cubic mile of finely broken material was ejected and the volcano reduced in height by 4,000 feet. In another case 50 cubic miles of material was blown away by explosions. When Krakatoa was destroyed, 5 cubic miles were blown away and in 1912, when Mount Katmai in Alaska was destroyed, the same amount was thrown into the air. The destruction of volcanoes by explosion is a common occurrence; however, no case is known of the actual engulfment of a volcano in historic time.

It would seem that the wrecking of a volcano by explosion would cause a vast amount of pulverized rock, ash, or pumice to settle over the near-by country and this appears in every instance to have been true. It was true at Vesuvius in the year 79 and at Katmai in 1912. Northeast of Crater Lake, in line with the prevailing winds, are extensive deposits of recent pumice, some fragments as much as an inch in diameter having been hurled as far as 80 miles from the volcano. Near Crater Lake the accumulation is 200 feet deep. Several cubic miles of pumice have thus been accounted for. At Krakatoa the immediate region was not deeply buried in ash at the time of its destruction but millions of tons of fine dust were carried thousands of miles distant from the volcano, and some of it even around the world, causing ruddy sunsets in remote lands for the next year.

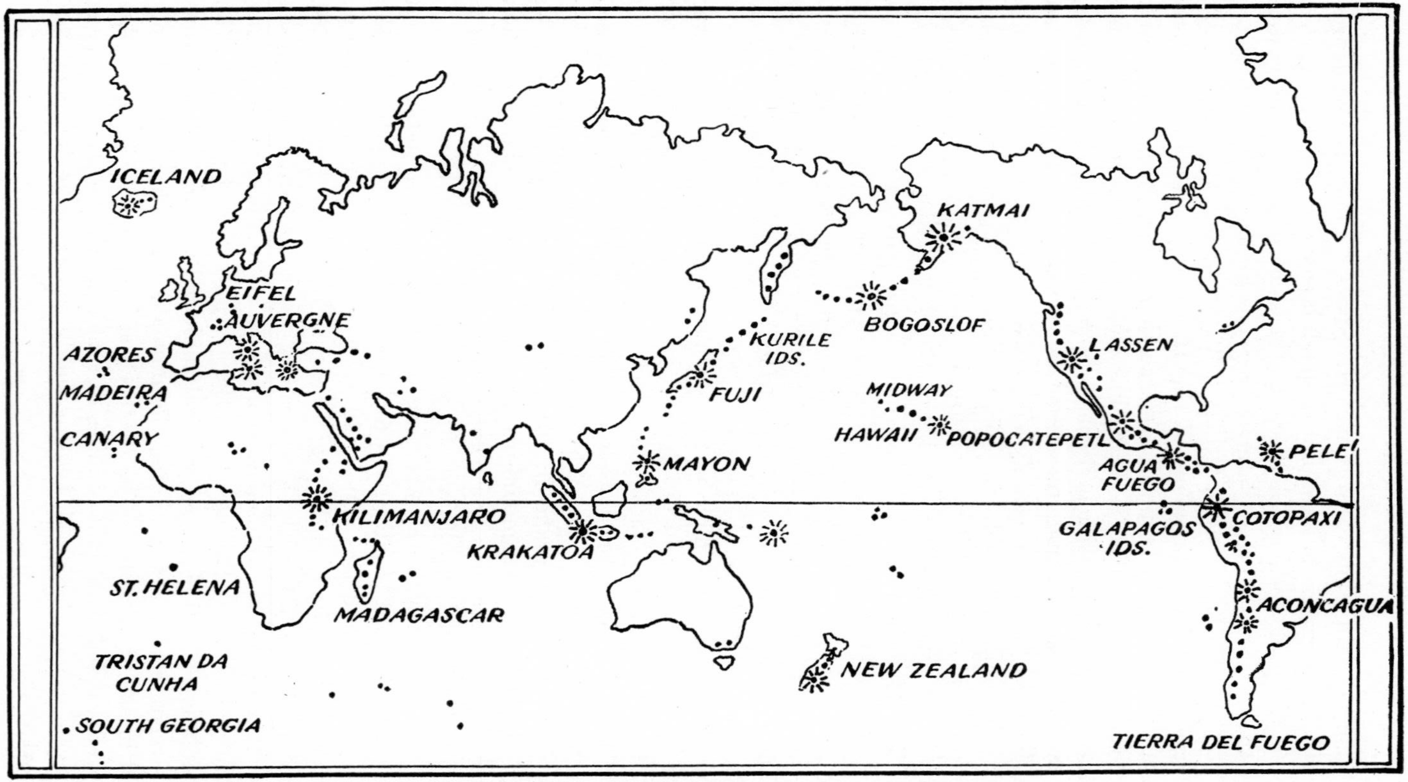

THE VOLCANOES OF THE WORLD

## VOLCANOES OF THE WORLD: DISTRIBUTION

Volcanoes appear to be arranged along well-defined belts or arcs, corresponding undoubtedly with belts of weakness in the outer crust of the earth. From Tierra del Fuego to Alaska there is an almost unbroken chain of volcanic peaks—the *ring of fire*—many of which have been active in historic time. The least active portion is in the United States.

No active cones occur as far south as Cape Horn but the High Andes are dominated by peaks of impressive height, their activity progressively greater as one goes northward, most of those in Chile being dormant. Aconcagua rises to a height of 23,000 feet but its crater has disappeared through erosion. Several Bolivian peaks attain elevations over 20,000 feet and some near Lake Titicaca are active. In Peru and Ecuador the volcanoes are arranged in two roughly parallel lines. Many approach an elevation of 20,000 feet, several being moderately active. In Central America the ranges trend east and west and consist of scores of volcanoes, several being among the most active in the world. Volcan de Agua and Volcan de Fuego, both in Guatemala, have frequently been active in historic time. In 1541 Agua destroyed the former capital of Guatemala, the city of Antigua. In San Salvador the famous cone of Izalco first appeared in 1770 and now rises 3,000 feet above the cattle fields. Popocatepetl, Ixtaccihuatl, and Orizaba cap the Mexican tableland. Many small cones have developed in recent times.

Among the volcanic areas in the western United States are the Mount Taylor region of New Mexico; the San Francisco peaks in Arizona; Mount Lassen, the most active of them all; Shasta; and the spectacular chain dominating the Cascade upland including Mounts Pitt, Jefferson, Hood, Rainier, St. Helens, and Baker. Hundreds of small cones and recent lava flows occur in New Mexico, Arizona, Utah, Nevada, and California, many of them being definitely related to the faults of the region. In Alaska, Katmai on the mainland and Bogoslof among the Aleutian Islands are examples of recent activity.

From Kamchatka there is an almost continuous belt through the Kurile Islands, Japan, the Philippines, Celebes, New Guinea, the Solomon Islands, New Caledonia, and New Zealand.

Scattered throughout the Pacific are other volcanic chains. The Hawaiian Islands are the most extensive and many of the South Sea islands are of volcanic origin. Parallel to the South American coast are the Galapagos and Juan Fernandez Islands, which may be related to each other.

Around the Indian Ocean is another circle of volcanoes, represented by Timor, Flores, Sumbawa, Lombok, Bali, Java, and Sumatra, and by the chain which runs through Arabia, the rift valley of eastern Africa, and Madagascar. Several of the volcanoes in the Dutch islands have had spectacular eruptions, notably that of Zimboro on the island of Sumbawa in 1815 and of Krakatoa in 1883. Most of the peaks of Java and Sumatra are less than 10,000 feet in height.

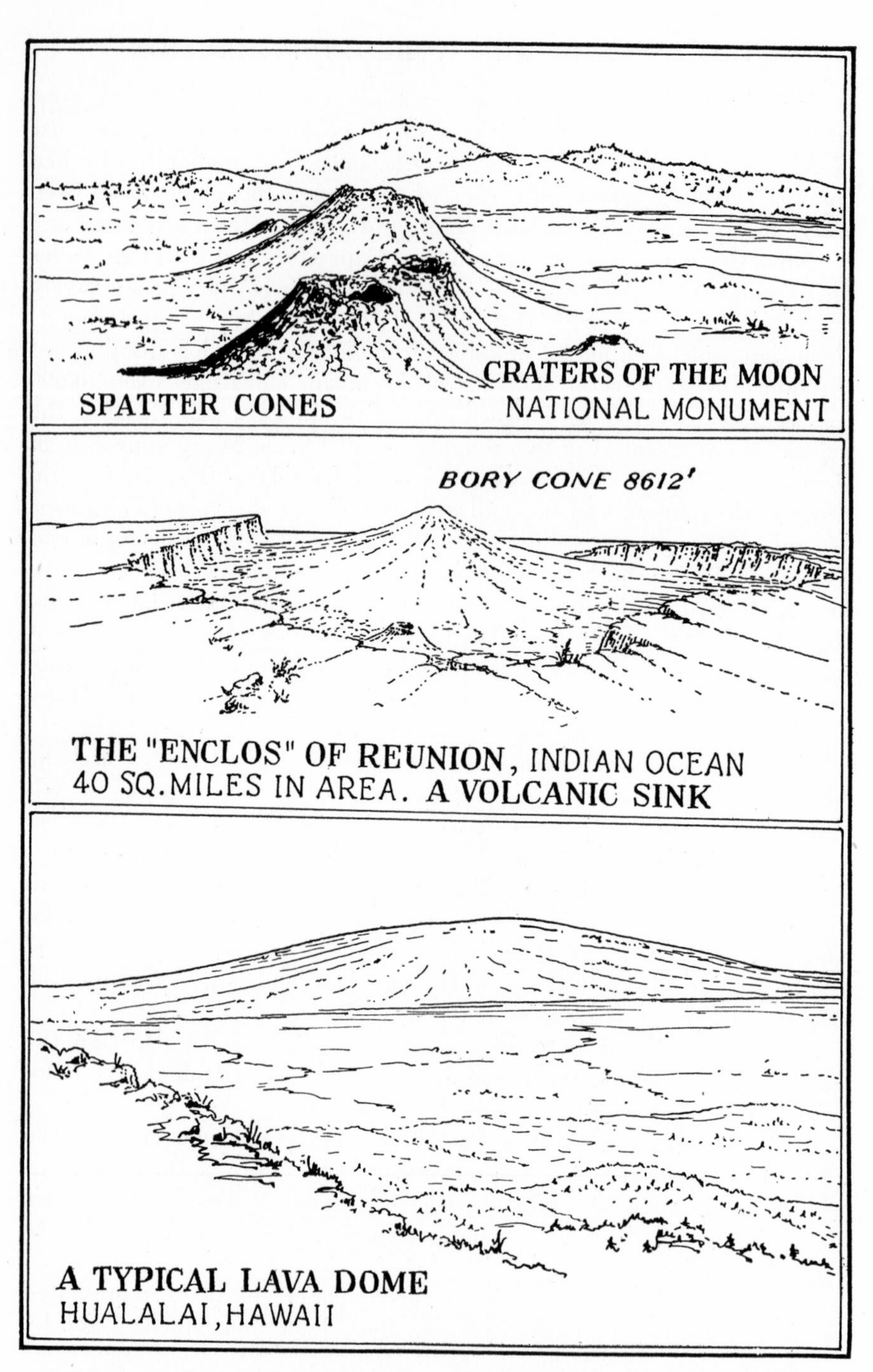


EXAMPLES OF VOLCANIC FORMS, ILLUSTRATING SPATTER CONES, VOLCANIC SINK, AND LAVA DOME

A number of volcanoes occur in Persia and in central Asia.

In the rift valley of Africa Mount Kilimanjaro rises some 14,000 feet high into a region of perpetual snow. It presents some unusual climatic conditions, not exactly arctic in character because each day is made up of sunshine and darkness and there is no long period of great cold. The floor of one of the calderas has become a vast game refuge.

The Mediterranean Sea is bordered by a volcanic belt of unusual activity. Starting with Mount Ararat, whose elevation is 17,000 feet, in Armenia near the Caspian Sea, the chain of peaks extends through Asia Minor and includes several islands of the Greek archipelago, notably Santorin. This is part of an ancient caldera, now largely submerged but containing several smaller islands formed during historic time. The chain then follows the peninsula of Italy into central France with an offshoot to the Lipari Islands, Sicily, and Africa. Vesuvius stands at the junction of these two chains. Mount Etna on Sicily is the largest volcano of Europe and the most destructive. It rises 11,000 feet directly above the sea on a base 90 miles in circumference, its flanks covered with parasitic cones.

The extinct volcanoes of central France are arranged along several belts. The western belt includes the great masses of Mount Cantal and Mount Doré, both completely dissected so as to reveal the roots of the whole volcanic system. North of Mt. Doré is the chain of Puys extending for about 60 miles, the word *puy* meaning a hill or mound. Dozens of these hills show well-defined craters and many of them have lava flows apparently very recently formed. Besides the perfect crater cones there are great intumescences or domelike blisters of viscous acidic lavas.

The Eifel region of western Germany contains a number of low craters called *maars*, often lake filled, but no large hills and no extensive lava flows. Other ancient volcanic groups along the Rhine include the Siebengebirge, or Seven Mountains, between Coblenz and Cologne.

Quite separate from the places just mentioned are the volcanic areas extending from Iceland through the eastern Atlantic and including northern Ireland and the Faroes, the Azores, the Madeira and Canary Islands, the Cape Verde Islands, St. Paul, St. Helena, Tristan da Cunha, and South Georgia. The Iceland volcanoes are the most active of this series and are represented by volcanic peaks and by lava domes, or shield volcanoes. These have a raised rim around the summit crater, due to the spattering of the lava lake which fills the crater. The Kólotta Gyngja, a typical Icelandic shield, rises 1,400 feet above its surroundings with a diameter of about 4 miles. Its surface slope is 7° and its crater is ⅓ mile across. Numerous thin lava flows have broken out from the sides of the dome.

The West Indian volcanoes form a continuous belt from St. Christopher southward to Grenada and include Mount Pelée on Martinique and Soufrière on Guadeloupe, each of which is about 5,000 feet high.

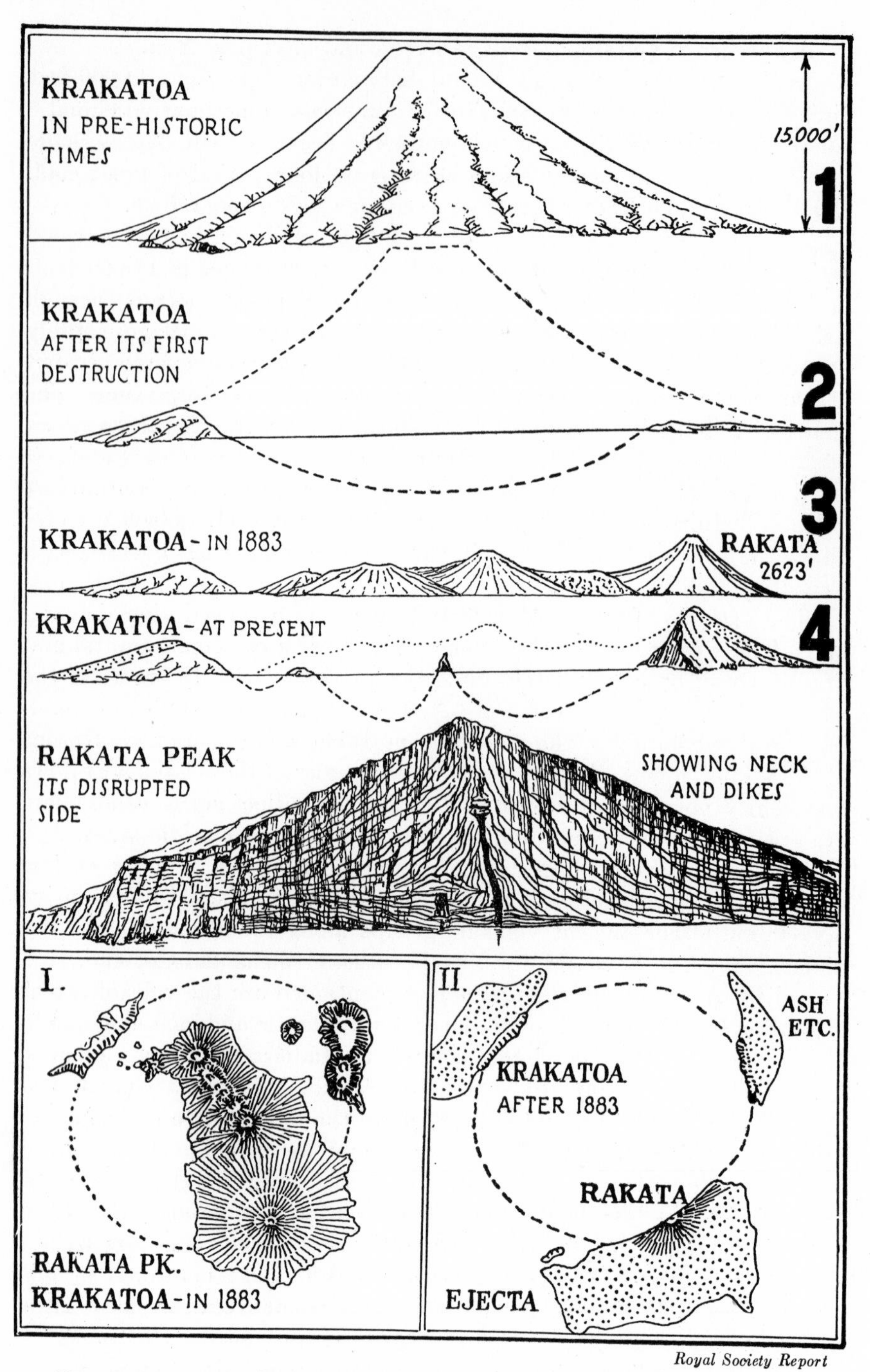


*Royal Society Report*

KRAKATOA, BEFORE AND AFTER ITS GREAT ERUPTION IN 1883

KRAKATOA. In August, 1883, the volcanic island group of Krakatoa, between Sumatra and Java, was destroyed by a series of terrific explosions, some of which were heard more than 3,000 miles away. A mass of material estimated at 18 cubic miles was thrown into the air, some of it to a height of 17 miles, in the form of lapilli, ashes, and dust. Part of this settled and buried what was left of the islands but much remained in suspension for many months and caused brilliant sunsets all over the earth. Dust fell on the decks of vessels 1,600 miles distant three days after the eruption. Great sea waves called *tsunamis* were generated, one of which rose 100 feet and destroyed 36,000 people and over 1,000 villages along the shores of Java and Sumatra. By their force a large ship was carried inland for 1½ miles and left stranded 30 feet above sea level. These waves were several feet high after crossing several thousand miles of the Indian Ocean. About half of the main volcanic peak of Rakata was left and showed in its cross section the main neck and also numerous dikes which bound the mass together.

KATMAI. The eruption of Mount Katmai in Alaska, situated where the Alaskan peninsula joins on to the mainland, occurred on June 6, 1912. Very few people lived anywhere near this volcano, nor were there any detailed maps available to show the actual form of the land before the eruption. The explosion, however, was heard 750 miles away at Juneau and even far north at Dawson and Fairbanks across the Alaska Range. The quantity of dust thrown into the air was appalling and is estimated to have been at least 5 cubic miles. Day was turned into the darkest night. For three days the dust continued to fall in large quantities. At Kodiak, 100 miles away, it made a deposit 10 inches deep. At Katmai, about 12 miles away, it was 3 feet thick.

Had this occurred at New York the city would have been buried under many feet of ash. Philadelphia would have been covered with a foot of ash and would have been in inky darkness for 60 hours. Ash would have been distributed over the landscape as far as Buffalo and Washington. The sounds of the explosions would have been heard at Atlanta and St. Louis, and the fumes would have been noticed in Denver and Miami.

The explosion at Mount Katmai produced a craterlike abyss 2 to 2½ miles in diameter and from 2,000 to 3,700 feet in depth. Half the area of the crater floor was soon occupied by a lake of warm water. At one of the parasitic cones, called *Novarupta,* a volcanic plug 200 feet high was extruded. A valley northwest of Mount Katmai was, just before the main explosion, deeply buried in sand and fragments of all kinds of rocks, obscurely stratified, changing the previous rough topography into a flat plain. This material was forced up along numerous fissures formed by the great pressure beneath. This plain, with its many fumaroles, is now called the *Valley of* 10,000 *Smokes.*

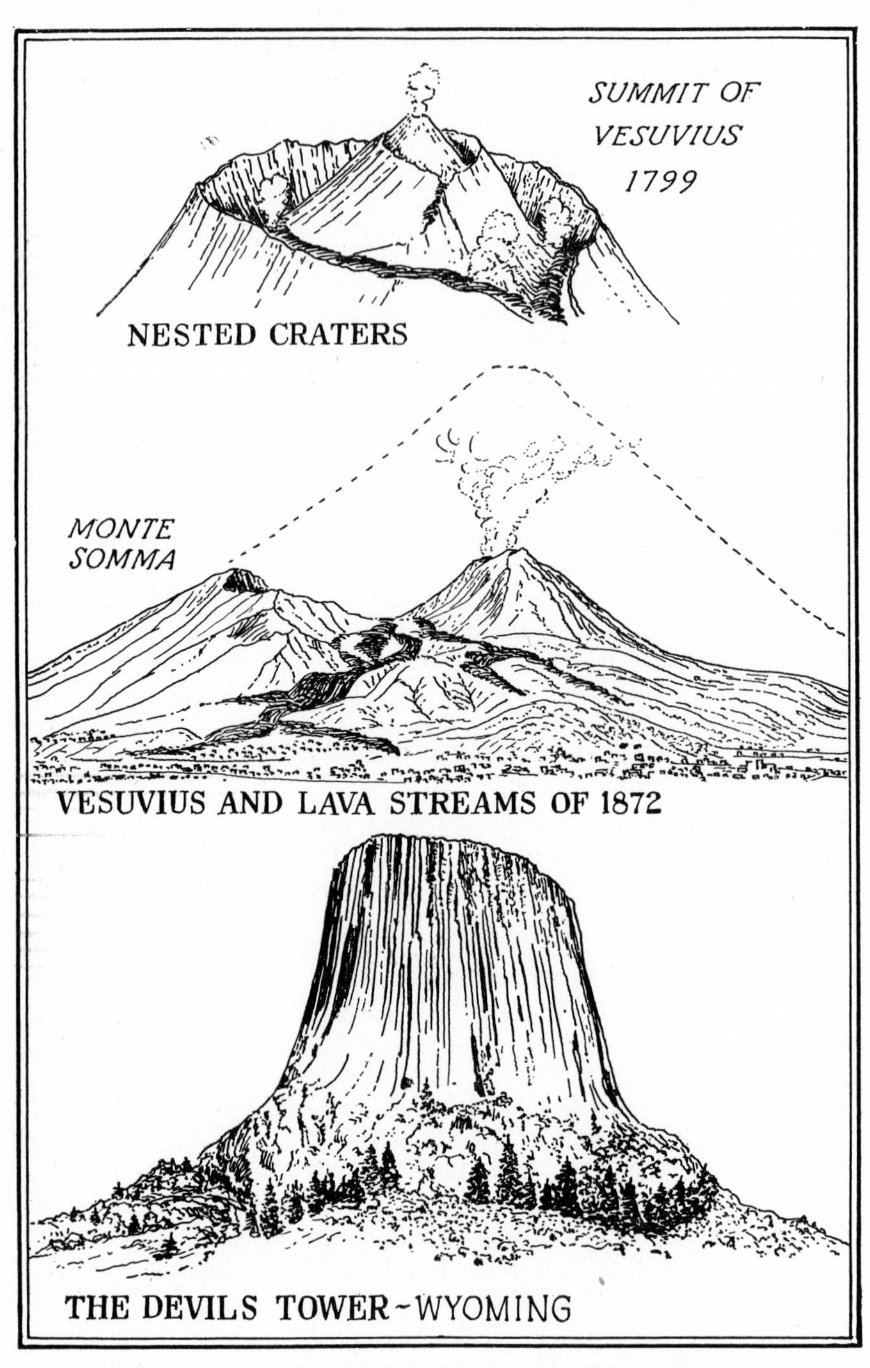


VESUVIUS, AN EXAMPLE OF A RECENT CONE IN AN OLD CALDERA
THE DEVIL'S TOWER. A VOLCANIC NECK

VESUVIUS. Vesuvius is the best known and longest known of active volcanoes. Before the great eruption of A.D. 79 it had been dormant for many years, a towering mountain covered with vineyards and villas. At its foot lay the populous cities of Herculaneum and Pompeii. For several years prior to A.D. 79 there were frequent earthquake shocks but in August of that year the disturbances became more violent and terminated in the great eruption recorded by Pliny the Younger. This shattered the ancient volcano and only parts of it now remain to form Monte Somma, a portion of the rim of a large caldera. The tremendous quantity of pulverized rock which was scattered far and wide buried the surrounding country and fell out at sea. No lava was ejected. The history of Vesuvius after A.D. 79 is characterized by many explosions and two or three long periods of quiescence, the longest of which extended from 1139 to 1631. The greatest recent activity has occurred in 1872 and 1906. On both these occasions large parts of the top were blown off and several lava streams broke out on the side of the cone, destroying some towns. Lava-dammed lakes were also formed. This alternation between periods of great violence, when the summit was completely blown away, and periods of more moderate activity has caused Vesuvius to exhibit at times a series of *nested craters*.

The present Vesuvius, which has a height of about 4,000 feet, has been built largely since the year A.D. 79. The structure of the ancient cone of Monte Somma and of the modern cone has clearly been seen in the calderas at different times. The layers of cinders, ash, and lava dip away from the rim at angles of 26 to 40° and are cut vertically by numerous dikes which bind the entire mass together.

DEVIL'S TOWER, OR MATO TEPEE, NEAR THE BLACK HILLS IN WYOMING. THE PROBLEM OF ITS ORIGIN. This is regarded by some as an ancient volcanic neck. By others, however, it is considered a butte, or erosion remnant of an old lava sheet or laccolith. The peculiar columnar structure, which is almost vertical, favors the belief that it is part of a horizontal bed. On the other hand, the neck character of such a mass would be proved by showing that the igneous rock continues downward through the rocks of the earth's crust, whereas an erosion remnant of a lava sheet or intruded mass would rest upon rocks which are continuous beneath it.

Certain undoubted volcanic necks in the Mount Taylor region have columnar structure similar to that in the Devil's Tower. In both the Devil's Tower and in the Mount Taylor necks the columns are vertical in the upper part of the butte but flare outward in the lower part. Such a structure is compatible with the structure of a volcanic neck, where the cooling surface would at the top be the surface of the ground (usually the floor of the crater) and near the base would be the walls of the neck. The isolated occurrence of the Devil's Tower, with no other volcanic necks near by, is perhaps the reason that its origin as a volcanic neck has not been generally accepted.

EXAMPLES OF RIFT LINE VOLCANOES

## RIFT-LINE VOLCANOES

Volcanoes rarely stand isolated from other volcanoes. Usually several are alined along a fracture or rift in the earth's crust. Small cinder cones, when first formed, in many cases are elongate because they have grown up over a fissure. Chains of craters are common in Iceland. The Chain of Laki, formed in 1783, extends 20 miles and embraces about 100 separate craters or cones ranging up to 500 feet in height. The Mono Craters in California, just east of the Sierra Nevada fault scarp, are similar to those in Iceland and appear to be located on one of the minor fissures associated with that great scarp. The volcanoes of Central America, magnificent peaks rising thousands of feet above the plain, stand on a line parallel with the trend of the adjacent highlands and also with the minor ridges and the coast line. The famous Tarawera Rift in New Zealand was formed in 1886 by an explosion which wrecked a series of volcanic peaks. It is 9 miles long and goes directly across the centers of these cones. The largest part of the rift is an elongated caldera. When it was first formed, there were a number of geysers and fumaroles along its floor but the depression has since been largely filled with water.

An extended row of active volcanoes rising to heights of 13,000 feet exists in the great "Rift Valley" of East Africa, which is a long graben. Similarly, in the Rhine Graben, the Kaiserstuhl stands practically on a fault line and along its northern continuation is the volcanic mass of the Vogelsbirge. The volcanoes of central France are arranged in chains, notably the chain of the Puys or cinder cones north of Mount Cantal. In the Eifel of Germany a similar arrangement may be noted. The volcanoes of Japan, of Sumatra, and of Java are in each case definitely alined along important fissures; and this is probably true also of the volcanic peaks of the Cascades. The Andes, the volcanoes of Mexico, and those of New Zealand are still other examples.

Along the eastern front of the Colorado Rockies the Spanish Peaks constitute but one evidence of volcanism situated on a great fault. In Texas at the foot of the Balcones fault scarp are similar cases.

The volcanic islands of the world are almost without exception arranged along a fairly definite line which is usually arcuate rather than straight. The Aleutian Islands extend through an arc 1,000 miles long, which is continued in the Alaskan peninsula several hundred miles more to Mount Katmai. The Hawaiian Islands range through a belt 1,500 miles long from Midway Island to Hawaii. The Kurile Islands between Kamchatka and Japan stretch for 600 miles and the Lesser Antilles have about the same length.

It cannot be affirmed that all the great faults of the world are loci of volcanic activity. Nevertheless, it is clear that many of them are. On the other hand, undoubtedly it is true that all the great volcanoes of the world are definitely alined along zones of fracturing or faulting.

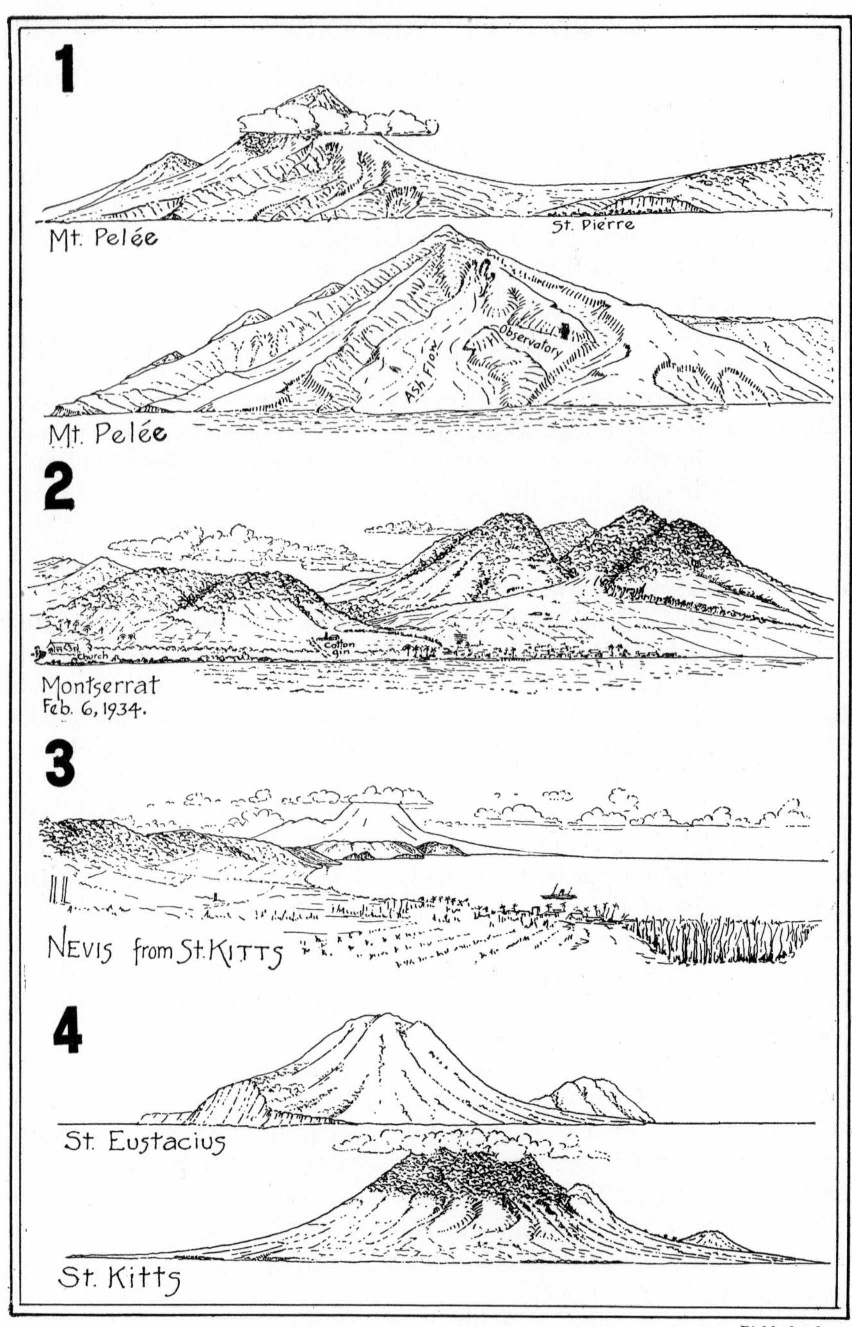


*Field sketches*

VOLCANOES OF THE WEST INDIES

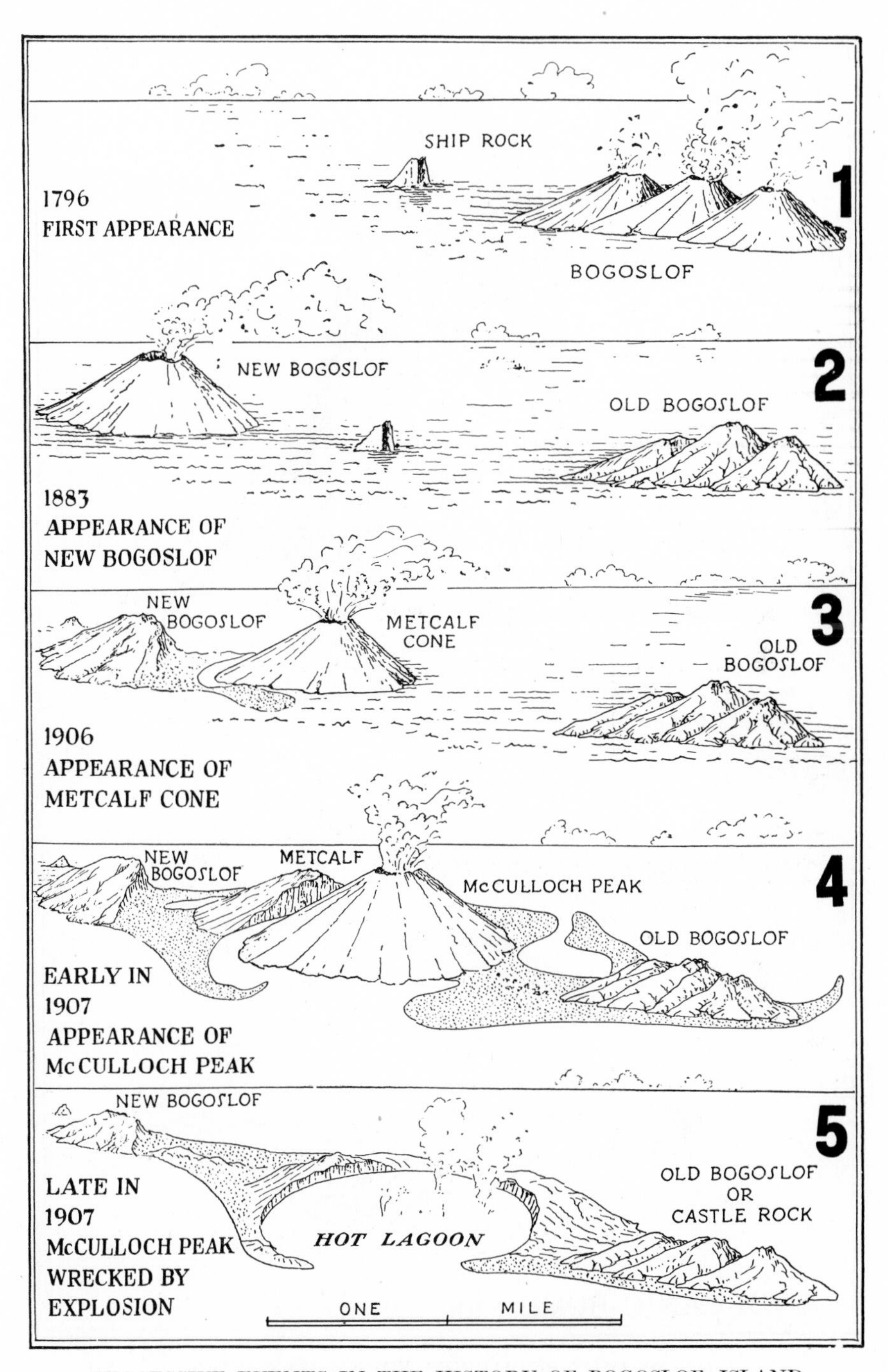


SUCCESSIVE EVENTS IN THE HISTORY OF BOGOSLOF ISLAND

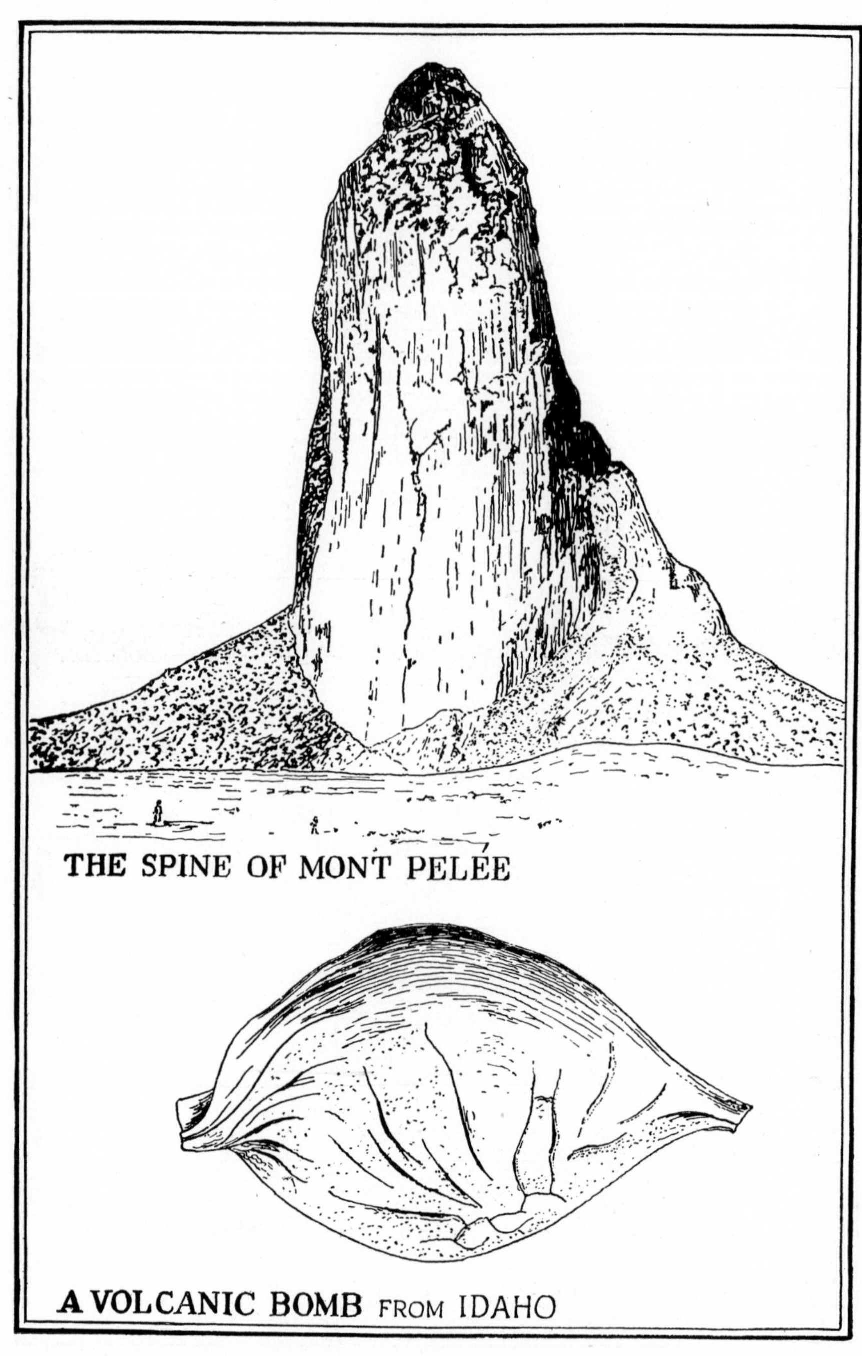


FORMS OF EJECTA FROM YOUNG VOLCANOES

*From photographs by Douglas Johnson*

TWO TYPES OF VOLCANIC NECKS

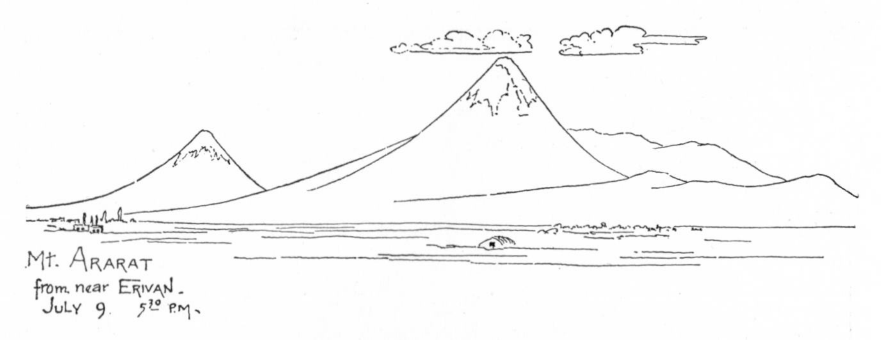


## VOLCANOES AND MAN

For many centuries the culture and thought of the human race was centered around the Mediterranean, where active volcanoes and earthquakes were common. As a result, men assumed that changes on the earth's surface were always accompanied with violence. From the earliest days man was a cataclysmist. Uniformitarianism was first clearly enunciated by men who lived in those parts of the earth where geological activities proceeded in a subtle and quiet manner.

The destructive effect of volcanoes is represented by the three contrasting examples of Vesuvius, Mount Pelée, and Agua in Guatemala.

The great eruption of Vesuvius in A.D. 79 was much like that of Krakatoa in 1883, Mount Katmai in 1912, and Mount Conseguina, Central America, in 1835. In each case the volcano had remained dormant for a long time and then was literally blown away. The vast amount of debris buried the country for miles around.

The destruction caused by Pelée was of quite a different type. A terrific burst of hot gas swept down the mountainside over the city of St. Pierre with its thousands of inhabitants. Every soul but one was killed, the survivor being protected by prison walls. The high temperature of this gas scorched everything. Clothing, fruits, coffee berries, food, furniture, the people themselves were suddenly turned into carbon by the process of destructive distillation. Burning could not take place because of the lack of oxygen but everything volatile disappeared.

In the third case, Mount Agua in Guatemala caused destruction indirectly. It was an ancient volcano with a large caldera filled with a lake, like Crater Lake. In the year 1541, after an earthquake, the wall of the caldera gave way and the water escaped. The deluge of water and mud overwhelmed the old capital of Guatemala (now called *Ciudad Vieja*), which had been established 24 years before.

Lava flows rarely cause loss of life although towns have been engulfed. The movement is always so slow that the inhabitants have ample time to escape. On one occasion, however, a stream of lava issued from Mount Etna and invaded cultivated fields. Many people were near its advancing front, engaged in the rescue of property. Suddenly its extremity

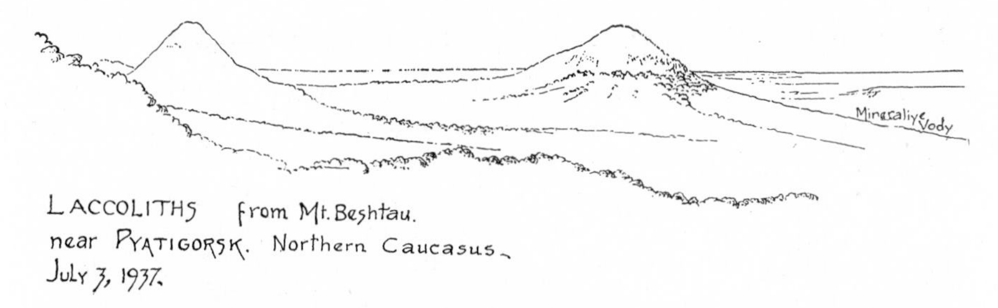


was seen to swell up like an enormous blister and then to burst, discharging a quantity of steam with a volley of fragments, solid and liquid. Sixty-nine persons were killed. The catastrophe was caused by the lava flowing over a subterranean reservoir of ground water which was suddenly converted into steam.

Although volcanoes in many parts of the world have caused much loss of life, nevertheless this has not deterred people from returning time and time again to the region of their former homes because volcanic areas have remarkably fertile soil. One of the most fertile regions of the world is the Auvergne district of central France.

A unique example of life on an isolated volcano is provided by the island of Saba in the West Indies. This is a young volcanic cone, slightly more than 2 miles in diameter, and rising more than ½ mile above sea level. Not having any ports, it is rarely visited. It was in fact the last stronghold of the buccaneers. At the present time it is inhabited by descendants of the first Dutch settlers and has a population of 1,661 people, over 300 to the square mile. This relatively dense population is made possible by several special industries, the most important of which is boat building. The boats are built of native wood in the old crater, where the village of Bottom is situated, and lowered down the cliffs to the harborless shore and sold for interisland traffic.

Among the great volcanoes of the world which have played an important part in the religion and artistic thoughts of the people are Mount Ararat and Fujiyama. Neither one of these peaks has shown much volcanic activity since man has known them. Ararat is one of the highest volcanoes in the world, rising over 17,000 feet, a half mile higher than the highest peak in the United States. Little wonder that its beautifully symmetrical form and snow-splashed summit should have had woven about them stories intimately concerned with the history of the human race.

To Popocatepetl in Mexico the Aztecs offered sacrifices to propitiate the fulminating wrath of the devils who dwelt in the "Smoking Mountain." No mountain in the world has gained so important a place in the artistic life of a country as has Fujiyama. The remarkable grace of this mountain's curve makes it one of the most beautiful in the world.

## MAPS ILLUSTRATING VOLCANOES AND VOLCANIC FORMS

Young undissected cinder cones and small volcanoes are shown on the *Newberry Crater, Ore.*, the *San Francisco Mountain, Ariz.*, and the *Lassen Volcanic National Park, Calif.*, maps. The Lassen Park map shows a large volcano with crater and many small cones with craters as well as recent lava flows and lava-dammed lakes (*e.g.*, Snag Lake). The San Francisco Mountain sheet reveals by its crescent-shaped contours the presence of many cinder cones. An unusual map is the *Modoc Lava Bed, Calif.*, sheet which displays volcanic cones, lava flows, and faulted lava flows. The *Craters of the Moon National Monument, Idaho,* map has a wealth of detail, showing cones, craters, flows, tree molds, etc. The *Mt. Lyell, Calif.*, sheet shows a row of unusually fine cones, called the Mono craters.

Numerous small cones, as well as larger volcanoes, are shown on the *Three Sisters* and *Maiden Peak, Ore.*, sheets.

Maturely dissected volcanoes of large size are represented on the *Mt. Shasta, Calif.*, and the *Mount Hood* and *Mount Jefferson, Ore.*, sheets.

Old volcanic necks appear on the *Mount Taylor, N. Mex.*, and the *Black Butte, Calif.*, sheets, the latter showing also a dike. The *Shiprock, N. Mex.*, sheet shows a remarkably fine volcanic neck with associated dikes, by far the best map available to illustrate such features. Numerous dikes appear also on the *Spanish Peaks, Colo.*, map.

Among the Hawaiian maps which illustrate the peculiar volcanic features of that region are the *Puna, Hawaii* (lava dome, craters, cracks, flows, caves), *Kahoolawe* (complete lava dome), the *Lahaina* (complete dome maturely dissected), *Mauna Loa* (an almost complete dome with crater, cracks, flows, etc.), *Honuapo* (lava flows and faults), *Waipio* (maturely dissected dome), and the *Kaohe* (flows and minor cones) sheets.

The *Ashland, Ore.*, sheet illustrates a caldera, situated along the line of a distinct fault scarp. The caldera with recent cinder cone and lava flows appears in greater detail on the *Crater Lake National Park, Ore.*, map.

A young lava plateau is represented on the *Bisuka, Idaho,* sheet and a mature lava plateau is shown in the *Pullman, Wash.-Idaho,* region. A good example of lava-capped mesa appears on the *Elmoro, Colo.*, map.

Laccoliths occur on the *Sundance* and *Aladdin, S. Dak.-Wyo.*, maps; and trap ridges due to tilted lava beds appear on the *Granby, Meriden,* and *New Haven, Conn.*, the *Holyoke, Mass.-Conn.*, and the *Passaic* and *Raritan, N. J.*, sheets, as well as on the *Gettysburg, Pa.*, map.

Several unusual examples of steam-produced explosion craters are shown on the *Noria, N. Mex.*, quadrangle, the steam having accumulated in sand under the lava sheet until the pressure was sufficient to cause the explosion. Most of the map represents a deeply sand-covered lava plain.

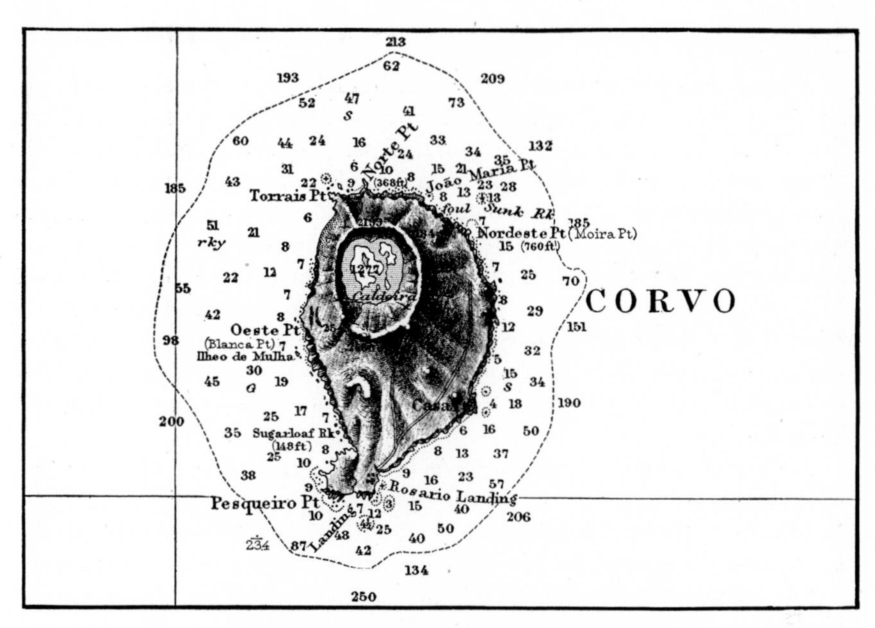


CORVO, A VOLCANO IN THE AZORES
U. S. Hydrographic Chart No. 1373 (1:150,000).

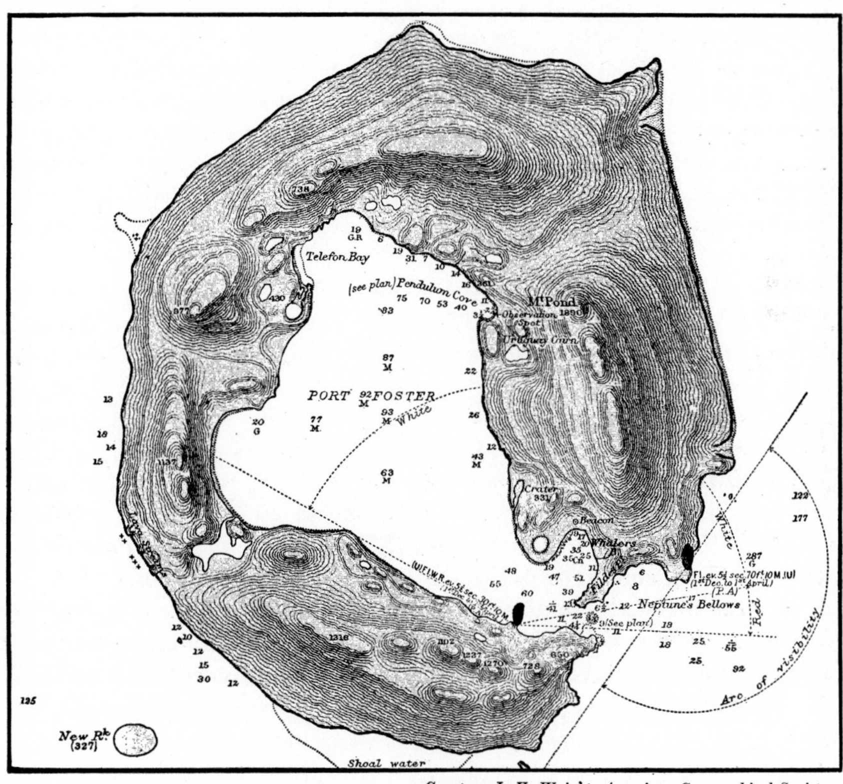


*Courtesy J. K. Wright, American Geographical Society*

DECEPTION ISLAND, A SUBMERGED CALDERA
British Admiralty Chart No. 3205 (1:150,000).

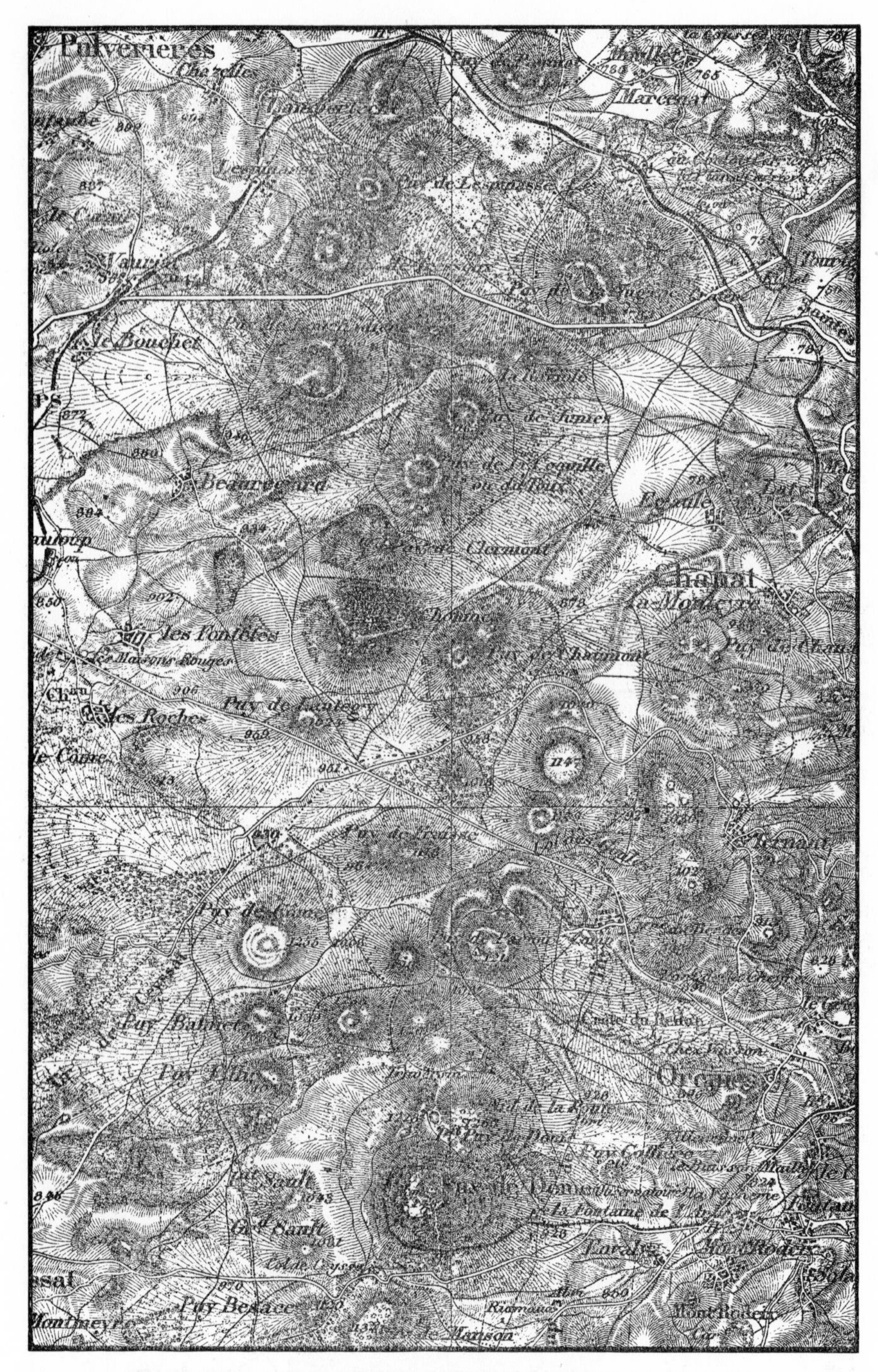


VOLCANOES AND CINDER CONES IN CENTRAL FRANCE
France; *Clermont* sheet (1:80,000).

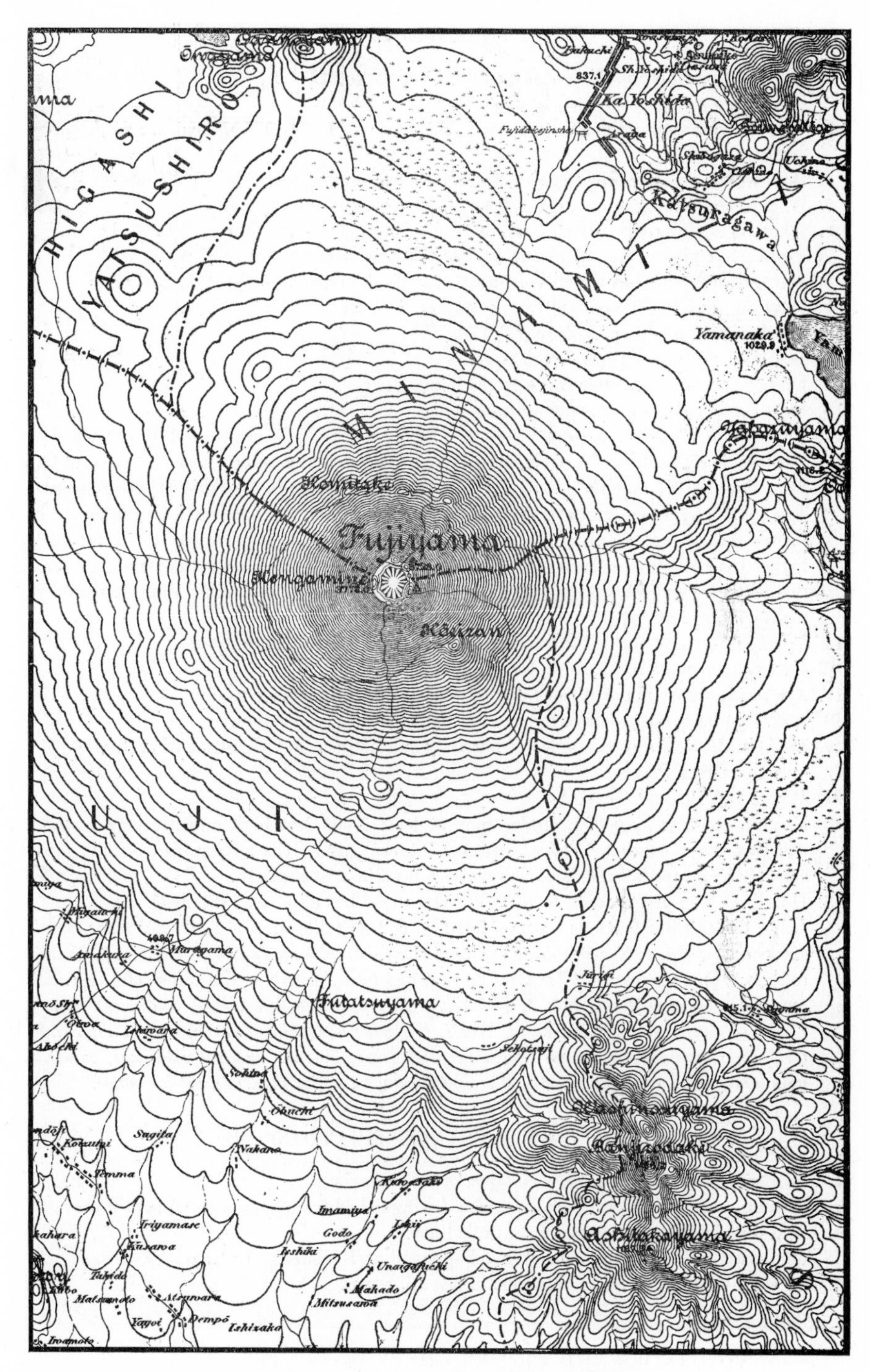


FUJIYAMA AND ADJACENT CONES
Japan; *Fuji* sheet, Zone 9, Col. XI (1:200,000).

## QUESTIONS

1. What is the difference between a caldera and a crater?
2. Is a young volcano necessarily small?
3. Why do you suppose some volcanoes (like Mayon in the Philippines) are so perfectly conical?
4. How would you account for an elliptical-shaped cinder cone?
5. What is the difference between a spine and a volcanic neck?
6. What is the difference between aa and pahoehoe lava, and how is each formed?
7. From what part of a cinder cone are lava flows emitted? From the crater?
8. Name all the ways in which volcanoes may do damage to life and property?
9. Why do dikes often radiate from a center?
10. Is a laccolith a type of volcano?
11. Is a mud volcano a type of volcano?
12. Do active volcanoes ever support glaciers?
13. How rapidly do cinder cones grow?
14. The volume of a circular cone is equal to the area of its base multiplied by one-third of its altitude. (The area of a circle is equal to $\pi r^2$.) How many cubic miles of rock would be torn loose by the complete explosion of a volcano 6 miles in diameter at the base and 3 miles high? How much was torn away when Mount Mazama was destroyed to form Crater Lake, which is 5 miles in diameter? The original peak was at least 2 miles higher than the present rim of Crater Lake. If this material were spread evenly over a distance everywhere within 100 miles of the explosion, how thick a deposit would it make?
15. What is pumice? Scoria? How are they formed? What is bentonite?
16. Can volcanoes be built up underwater?
17. Are violent volcanic eruptions due to steam or to some other gas?
18. Are large volcanoes always found in a region of large lava flows?
19. Do you know of any examples of partially submerged cinder cones and craters?
20. What is a ring dike?
21. Where is Vulcano? Geysir?
22. Draw a contour map of a cinder cone at least 1,000 feet high and a crater at least 200 feet deep, using a 100-foot contour interval. Show the appearance of this same cone if the crater is breached. Show also a lava flow at its base.
23. Do all dikes, when eroded, form walls? Or may dikes appear as ditches?
24. Label the ten illustrations on pages 664 to 673 as follows: No. 1, *Initial stage, small cinder cones;* No. 2, *Young volcanoes and cinder cones with small lava flows;* No. 3, *Young volcanoes, cinder cones with breached craters, and lava flows;* No. 4, *Young volcanoes with many overlapping flows;* No. 5, *Volcanoes with extensive flows, early maturity;* No. 6, *Large volcanoes maturely dissected;* No. 7, *Early old age; volcanic necks and dikes;* No. 8, *Old age, peneplane surmounted by necks and dikes;* No. 9, *Wrecked volcanoes, calderas with young cinder cones;* No. 10, *Wrecked volcanoes, nested calderas, cinder cones.*

## TOPICS FOR INVESTIGATION

1. Did Crater Lake result from a collapse or an explosion?
2. Volcanic necks. How are volcanic necks distinguished from remnants of lava flows?
3. Location of volcanoes. Relation to fault lines.
4. Some great volcanic eruptions. What happened at Krakatoa? Katmai? Vesuvius?
5. Old worn-down volcanoes. What examples are there?

## REFERENCES


### General

ANDERSON, T. (1903) *Volcanic studies in many lands, being reproductions of photographs by the author, with explanatory notes.* London, 200 p.

BONNEY, T. G. (1913) *Volcanoes: their structure and significance.* New York, 3d ed., 379 p.

DAVIS, W. M. (1912) *Die erklärende Beschreibung der Landformen.* Leipzig. Vulkanische Formen, p. 316–351.

DAY, A. L., ET AL. (1931) *Physics of the earth.* I. Volcanology. Natl. Research Council, Bull. 77, 71 p. A stimulating presentation of the present status of our knowledge of active volcanism.

HULL, E. (1892) *Volcanoes past and present.* London, 266 p.

JAGGAR, T. A., JR. (1931) *The mechanism of volcanoes.* Natl. Research Council, Bull. 77, p. 49–71.

SAPPER, K. T. (1927) *Vulkankunde.* Stuttgart, 424 p.

TYRRELL, G. W. (1931) *Volcanoes.* London, 252 p.

VON WOLFF, F. L. (1913–31) *Der Vulkanismus.* Stuttgart, 711 p. Band I: Allgemeiner Teil; Band II: Specieller Teil. Technical, exhaustive, with ample bibliographies.

WASHINGTON, H. S., ET AL. (1928) *Present volcanic activity over the earth.* Wash. Acad. Sci., Jour. 18, p. 509–515.

### Eruptions—Hot Springs

ALEXANDER, W. D. (1933) *Mauna Loa's greatest eruption.* Mid Pacific Mag., vol. 45, p. 302–328.

COULTER, J. W. (1932) *The eruption of Kilauea, Dec. 23, 1931.* Geog. Soc. Phila., Bull. 30, p. 195–199.

CURTISS, G. C. (1903) *Secondary phenomena of the West Indian volcanic eruptions of* 1902. Jour. Geol., vol. 11, p. 199–215.

DALY, R. A. (1911) *The nature of volcanic action.* Am. Acad. Arts and Sci., Proc. 47, p. 47–122.

DAY, A. L., and ALLEN, E. T. (1925) *The volcanic activity and hot springs of Lassen Peak.* Carnegie Inst. Wash., Publ. 360, 190 p.

HEILPRIN, A. (1903) *Mont Pelée and the tragedy of Martinique.* Philadelphia (and London), 325 p.

HILL, R. T. (1902) *Report on the volcanic disturbances in the West Indies.* Natl. Geog. Mag., vol. 13, p. 225–267.

HOBBS, W. H. (1906) *The grand eruption of Vesuvius in* 1906. Jour. Geol., vol. 14, p. 636–655.

JAGGAR, T. A., JR., and FINCH, R. H. (1924) *The explosive eruption of Kilauea in Hawaii,* 1924. Am. Jour. Sci., 5th ser., vol. 8, p. 353–374.

MARTIN, G. C. (1913) *The recent eruption of Katmai Volcano in Alaska.* Natl. Geog. Mag., vol. 24, p. 131–181.

NICHOLS, R. L. (1934) *Pebbles rounded in geyser tubes.* Jour. Geol., vol. 42, p. 430–432.

ROYAL SOCIETY OF LONDON. (1888) *The eruption of Krakatoa and subsequent phenomena.* Report of the Krakatoa Comm., 494 p.

RUSSELL, I. C. (1902) *The recent volcanic eruptions in the West Indies.* Natl. Geog. Mag., vol. 13, p. 267–285.

RUSSELL, I. C. (1902) *Volcanic eruptions on Martinique and St. Vincent.* Natl. Geog. Mag., vol. 13, p. 415–436.

WASHINGTON, H. S. (1926) *Santorini eruption of* 1925. Geol. Soc. Am., Bull. 37, p. 349–384.

### Calderas and Craters

ANDERSON, T. (1905) *On certain recent changes in the crater of Stromboli.* Geog. Jour., vol. 25, p. 123–138. Many illustrations.

BARRINGER, D. M. (1906) *Coon Mountain and its crater.* Acad. Nat. Sci. Phila., Proc. 57, p. 861–886.

Diller, J. S., and Patton, H. B. (1902) *Crater Lake.* U. S. Geol. Surv., Prof. Paper 3, 167 p.

Diller, J. S. (1923) *Did Crater Lake, Oregon, originate by a volcanic subsidence or an explosive eruption?* Jour. Geol., vol. 31, p. 226–227.

Hodge, E. T. (1925) *Mount Multnomah, ancient ancestor of the Three Sisters.* Ore. Univ., Pub. vol. 3, no. 2 (vol. 2, no. 10), 160 p.

Hodge, E. T. (1926) *Geology of Mount Jefferson (Oregon).* Mazama, Ann. Ser., vol. 7, p. 25–58. Good illustrations. Caldera type.

Smith, W. D., and Swartzlow, C. R. (1936) *Mount Mazama: explosion versus collapse.* Geol. Soc. Am., Bull. 47, p. 1809–1830.

Stearns, H. T. (1924) *Craters of the Moon National Monument, Idaho.* Geog. Rev., vol. 14, p. 362–372.

Stearns, H. T. (1928) *Craters of the Moon National Monument, Idaho.* Idaho Bur. Mines and Geol., Bull. 13, 57 p.

Stearns, N. D. (1928) *Exploring the Craters of the Moon, Idaho,* Geog. Soc. Phila., Bull. 26, p. 279–290.

von Engeln, O. D. (1932) *The Ubehebe craters and explosion breccias in Death Valley, California.* Jour. Geol., vol. 40, p. 726–734.

Williams, H. (1935) *Newberry Volcano of central Oregon.* Geol. Soc. Am., Bull. 46, p. 253–304. A great caldera.

Laccoliths and Domes

Gilbert, G. K. (1880) *Report on the geology of the Henry Mountains.* U. S. Geog. and Geol. Surv. Rocky Mt. Region (Powell), 170 p.

Gilbert, G. K. (1896) *Laccolites in southeastern Colorado.* Jour. Geol., vol. 4, p. 816–825.

Jaggar, T. A., Jr., and Howe, E. (1901) *The laccoliths of the Black Hills.* U. S. Geol. Surv., 21st Ann. Rept., part 3, p. 163–303.

Russell, I. C. (1896) *Igneous intrusions in the neighborhood of the Black Hills.* Jour. Geol., vol. 4, p. 23–43.

Williams, H. (1932) *The history and character of volcanic domes.* Calif. Univ., Dept. Geol. Sci., Bull. 21, p. 51–146.

Dikes, etc., and Lava Flows

Davis, W. M. (1898) *The Triassic formation of Connecticut.* U. S. Geol. Surv., 18th Ann. Rept., part 2, 192 p.

Dutton, C. E. (1882) *Lava flows (Grand Canyon district).* U. S. Geol. Surv., Mon. 2, p. 101–112.

Emerson, O. H. (1926) *The formation of aa and pahoehoe.* Am. Jour. Sci., 5th ser., vol. 12, p. 109–114.

Geikie, A. (1892) *Geological sketches.* New York, The lava fields of northwestern Europe, p. 239–249.

Oldham, R. D. (1893) *Lava plateau of the Deccan.* Medlicott and Blanford's *Manual of the geology of India.* Calcutta, p. 255–284.

Peacock, M. A. (1931) *The Modoc lava field, northern California.* Geog. Rev., vol. 21, p. 259–275.

Powers, H. A. (1932) *The lavas of the Modoc Lava Bed quadrangle, California.* Am. Min., vol. 17, p. 253–294.

Ransome, F. L. (1898) *Some lava flows of the western slope of the Sierra Nevada, California.* U. S. Geol. Surv., Bull. 89, 74 p.

Russell, I. C. (1893) *Lava plateau of Columbia and Snake Rivers.* U. S. Geol. Surv., Bull. 108, 108 p.

Shaler, N. S., and Tarr, R. S. (1889) *Dikes of the Cape Ann district, Massachusetts.* U. S. Geol. Surv., 9th Ann. Rept., p. 579–602.

Smith, G. O. (1903) *Geology and physiography of central Washington.* U. S. Geol. Surv., Prof. Paper 19, p. 9–39.

STEARNS, H. T. (1928) *Lava Beds National Monument, California.* Geog. Soc. Phila., Bull. 26, p. 239–253.

TARR, R. A. (1891) *A recent lava flow in New Mexico.* Am. Naturalist, vol. 25, p. 524–527.

WILLIAMS, H. (1936) *Pliocene volcanoes of the Navajo-Hopi country.* Geol. Soc. Am., Bull. 47, p. 111–172. Pictures of necks and dikes.

VOLCANIC NECKS

DUTTON, C. E., and SCHWARTZ, G. M. (1936) *Notes on the jointing of the Devil's Tower, Wyoming.* Jour. Geol., vol. 44, p. 717–728.

DUTTON, C. E. (1885) *Volcanic necks of Zuñi plateaus.* U. S. Geol. Surv., 6th Ann. Rept. p. 105–198.

JOHNSON, D. W. (1907) *Volcanic necks of the Mount Taylor region, New Mexico.* Geol. Soc. Am., Bull. 18, p. 303–324.

AMERICAN VOLCANOES

DILLER, J. S. (1891) *A late eruption in northern California.* U. S. Geol. Surv., Bull. 79, 33 p.

DILLER, J. S. (1895) *Geology of Lassen Peak quadrangle, California.* U. S. Geol. Surv. Folio 15.

GRIGGS, R. F. (1922) *The Valley of Ten Thousand Smokes (Alaska).* Natl. Geog. Soc., 340 p. Well illustrated.

HOLWAY, R. S. (1914) *Recent volcanic activity of Lassen Peak.* Univ. Cal., Publ. in Geog., vol. 1, p. 307–330; Am. Geog. Soc., Bull. 46, p. 740–755.

IDDINGS, J. P., ET AL. (1899) *Yellowstone Park.* U. S. Geol. Surv., Mon. 32, part 2, p. 1–164, 215–440. Petrologic.

JAGGAR, T. A., JR. (1908) *The evolution of Bogoslof Volcano.* Am. Geog. Soc., Bull. 40, p. 385–400.

JILLSON, W. R. (1917) *The volcanic activity of Mount St. Helens and Mount Hood in historical time.* Geog. Rev., vol. 3, p. 481–485.

JILLSON, W. R. (1921) *Physiographic effects of the volcanism of Mt. St. Helens, Washington.* Geog. Rev., vol. 11, p. 398–405.

JOHNSON, D. W. (1907) *A recent volcano in the San Francisco Mountain region, Arizona.* Geog. Soc. Phila., Bull. 5, p. 6–11.

ROBINSON, H. H. (1913) *The San Franciscan volcanic field, Arizona.* U. S. Geol. Surv., Prof. Paper 76, 213 p.

RUSSELL, I. C. (1897) *Volcanoes of North America.* New York, 346 p.

SHIPPEE, R. (1932) *Lost valleys of Peru.* Geog. Rev., vol. 22, p. 562–581.

STONE, J. B., and INGERSON, E. (1934) *Some volcanoes of southern Chile.* Am. Jour. Sci., 5th ser., vol. 28, p. 269–287.

WILLIAMS, H. (1932) *Mount Shasta, a Cascade volcano.* Jour. Geol., vol. 40, p. 417–429.

WEST INDIAN VOLCANOES

GILBERT, G. K. (1904) *The mechanism of the Mont Pelée spine.* Science, new ser., vol. 19, p. 927–928.

HEILPRIN, A. (1904) *The tower of Pelée.* Philadelphia, 62 p.

HOVEY, E. O. (1902) *Observations on the eruptions of 1902 of Le Soufrière, St. Vincent, and Mt. Pelée.* Am. Jour. Sci., 4th ser., vol. 14, p. 319–358; Am. Mus. Nat. Hist., Bull. 16, p. 333–372; Natl. Geog. Mag., vol. 13, p. 444–459.

HOVEY, E. O. (1903) *The new cone of Mt. Pelée and the gorge of the Rivière Blanche.* Am. Jour. Sci., 4th ser., vol. 16, p. 269–281.

HOVEY, E. O. (1904) *The Grande Soufrière of Guadeloupe.* Am. Geog. Soc., Bull. 36, p. 513–530.

HOVEY, E. O. (1908) *Ten days in camp on Mount Pelée, Martinique.* Am. Geog. Soc., Bull. 40, p. 662–679.

HOVEY, E. O. (1909) *Camping on the Soufrière of St. Vincent.* Am. Geog. Soc., Bull. 41, p. 72–83.

European Volcanoes

Eastman, C. R. (1906) *Vesuvius during the Early Middle Ages.* Pop. Sci. Monthly, vol. 69, p. 558–566.

Geikie, A. (1882) *Geological sketches.* New York, Among the volcanoes of central France, p. 74–108.

Geikie, A. (1897) *The ancient volcanoes of Great Britain.* London, 2 vol., 955 p.

Geikie, J. (1907) *Old Scottish volcanoes.* Scot. Geog. Mag., vol. 23, p. 449–463.

Jaggar, T. A., Jr. (1906) *The volcano Vesuvius in* 1906. Tech. Quart., vol. 19, p. 105–115.

Johnston-Lavis, H. J. (1891) *Geological map of Monte Somma and Vesuvius.* London, 22 p., 6 pl.

Johnston-Lavis, H. J. (1891) *The South Italian volcanoes.* Naples, 342 p.

Johnston-Lavis, H. J. (1909) *The eruption of Vesuvius of April,* 1906. Royal Soc. Dublin, Trans., ser. 2, vol. 9, p. 139–200.

Johnston-Lavis, H. J. (1918) *Bibliography of the geology and eruptive phenomena of the . . . volcanoes of southern Italy.* London, 2d ed., 374 p.

Lobley, J. L. (1889) *Mount Vesuvius; a descriptive, historical and geological account of the volcano and its surroundings.* London, 2d ed., 385 p.

Perrett, F. A. (1909) *Vesuvius: characteristics and phenomena of the present repose period.* Am. Jour. Sci., 4th ser., vol. 4., p. 413–430.

Scrope, G. P. (1858) *Geology and extinct volcanoes of central France.* London, 258 p.

Thoroddsen, T. (1906) *Volcanoes of Iceland.* Petermann's Geog. Mitt., Erg'band 32, Erg'heft 152, p. 106–159.

Vom Rath, J. J. G. (1872) *Der Aetna.* Bonn, p. 1–33.

Pacific Ocean Volcanoes

Anderson, R. (1907–08) *The great Japanese volcano Aso.* Pop. Sci. Monthly, vol. 71, p. 29–49; Jour. Geol., vol. 16, p. 499–526.

Bell, J. M. (1906) *The great Tarawera volcanic rift, New Zealand.* Geog. Jour., vol. 27, p. 369–382.

Dana, J. D. (1890) *Characteristics of volcanoes with contributions of facts and principles from the Hawaiian Islands.* New York, 399 p.

Dutton, C. E. (1884) *The Hawaiian volcanoes.* U. S. Geol. Surv., 4th Ann. Rept., p. 75–219.

Hinds, N. E. A. (1931) *The relative ages of the Hawaiian landscapes.* Calif. Univ., Dept. Geol. Sci., Bull. 20, p. 143–260.

Hitchcock, C. H. (1909) *Hawaii and its volcanoes.* Honolulu, 314 p.

Jensen, H. I. (1906) *The geology of Samoa and the eruptions in Savaii.* Linn. Soc. New South Wales, Proc. 31, p. 641–672.

Milne, J., and Burton, W. K. (1894) *The great earthquake in Japan,* 1891. Yokohama, 70 p.

Omori, F. (1911–13) *The Usu-san eruption and the earthquake and elevation phenomena.* Imperial Earthquake Investig. Comm., Bull. 5, p. 1–38, 101–107.

Pratt, W. E., et al. (1911) *The eruption of Taal volcano, Jan.* 30, 1911. Philippine Jour. Sci., vol. 6, p. 63–97.

Stone, J. B. (1926) *The products and structure of Kilauea.* Bishop Mus. (Honolulu), Bull. 33, 59 p.

Symons, G. J., Judd, J. W., et al. (1888) *The eruption of Krakatoa and subsequent phenomena.* Royal Soc. London, 494 p.

Geographical Aspects

Colton, H. S. (1932) *Sunset Crater. The effects of a volcanic eruption on an ancient Pueblo people.* Geog. Rev., vol. 22, p. 582–590.

Gadow, H. (1930) *Jorullo; the history of the volcano of Jorullo and the reclamation of the devastated district by animals and plants.* Cambridge, England, 100 p.

# XX
# METEOR CRATERS

*U. S. War Department*

MAN MADE CRATERS

Formed by shells exploding beneath the surface of the ground. Inert shells ordinarily produce no craters in soft soil though large high velocity low angle shells may form slightly elliptical ones.

*Fairchild Aerial Surveys and U. S. Geological Survey*

THE CAROLINA BAYS NORTH OF SUMTER, S. C.
Airview showing a large number of the elliptical solution depressions, suggesting meteorite scars. Scale: approximately two inches equals one mile.

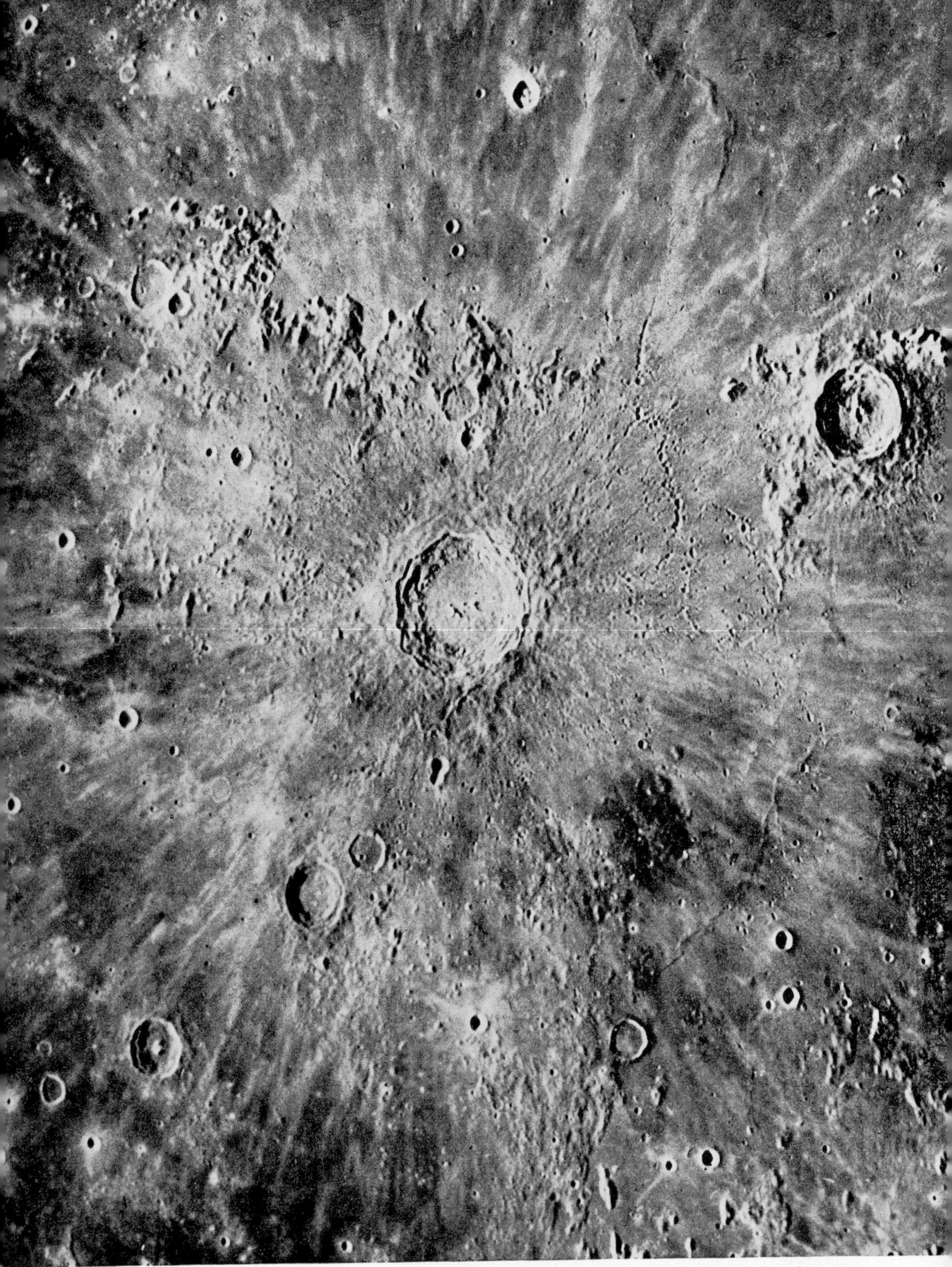

*Moore and Chappell, Lick Observatory*

THE SURFACE OF THE MOON

Showing craters ranging up to 50 or more miles in diameter with rims one to four miles high. Some astronomers believe these craters were formed by volcanic outbursts but others consider them meteorite scars.

*American Museum of Natural History*

METEOR CRATER, ARIZ.

This is the most definitely proved meteor crater in the world. Thousands of meteoritic fragments have been found in the immediate vicinity.

METEOR CRATER, ARIZ.

Showing the sandstone beds slightly upturned around the rim, and the drilling camp at the bottom where unsuccessful efforts have been made to locate the large mass of meteoritic iron supposed to be imbedded there.

*Atchison, Topeka, and Santa Fe Railway*

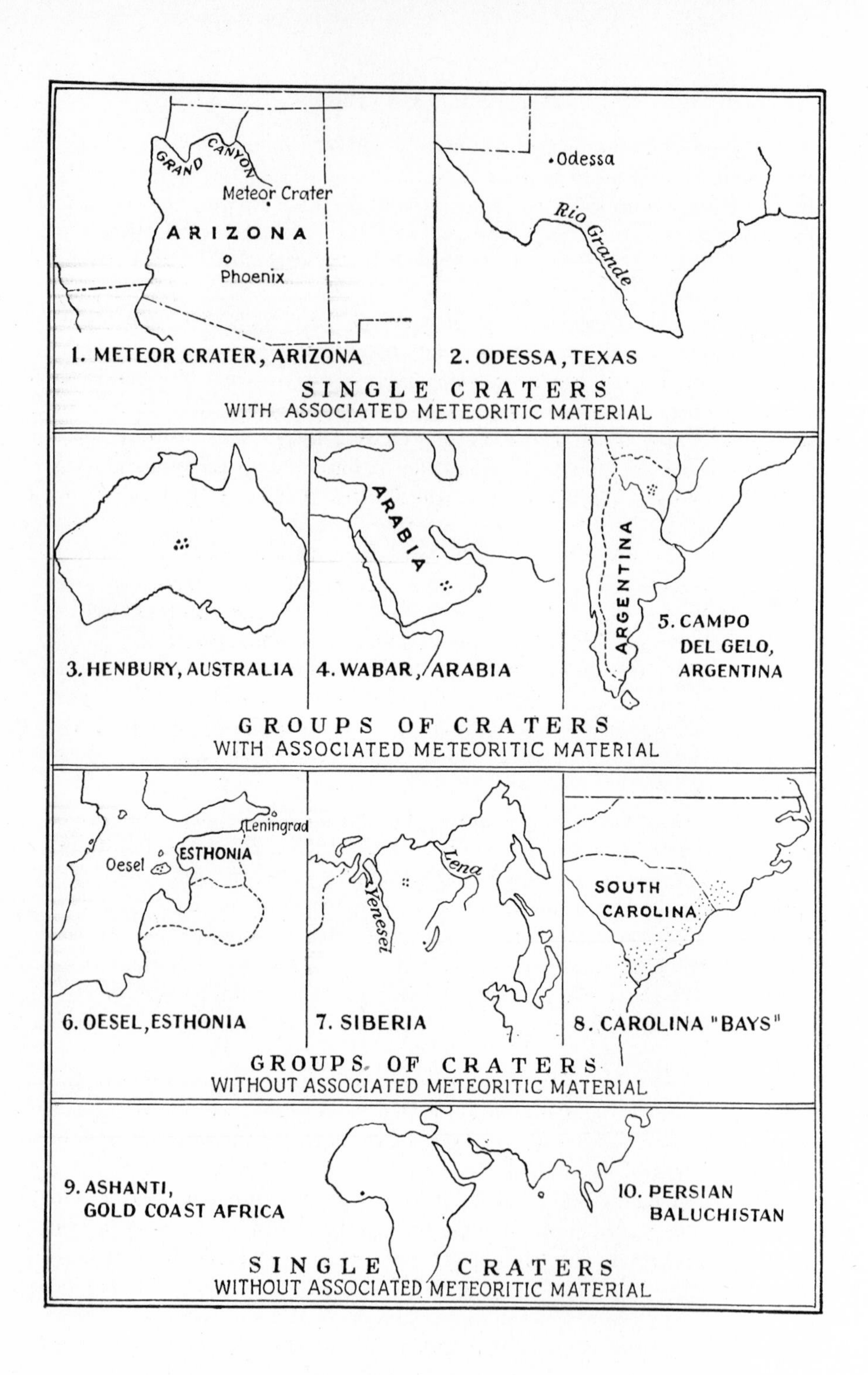
GRAND CANYON
Meteor Crater
ARIZONA
Phoenix
1. METEOR CRATER, ARIZONA
Odessa
Rio Grande
2. ODESSA, TEXAS
SINGLE CRATERS
WITH ASSOCIATED METEORITIC MATERIAL
3. HENBURY, AUSTRALIA
ARABIA
4. WABAR, ARABIA
ARGENTINA
5. CAMPO DEL GELO, ARGENTINA
GROUPS OF CRATERS
WITH ASSOCIATED METEORITIC MATERIAL
Leningrad
Oesel
ESTHONIA
6. OESEL, ESTHONIA
Lena
Yenesei
7. SIBERIA
SOUTH CAROLINA
8. CAROLINA "BAYS"
GROUPS OF CRATERS
WITHOUT ASSOCIATED METEORITIC MATERIAL
9. ASHANTI, GOLD COAST AFRICA
10. PERSIAN BALUCHISTAN
SINGLE CRATERS
WITHOUT ASSOCIATED METEORITIC MATERIAL

## METEOR CRATERS

The supposed meteor craters of the world occur singly and in groups. With some is associated meteoritic material. The location of practically all the known examples in the world is shown on the accompanying map. Only the first five may be considered well-authenticated cases of meteor craters. For all the others there appear to be alternative explanations equally plausible.

In none of the cases has a large mass of meteoritic iron been found within the crater. In two cases small fragments have been found. In all the first five cases, however, many pieces (running into the hundreds or even thousands) of meteoritic iron have been found within a mile or two of the crater, some of the pieces weighing 100 pounds or more.

The craters are believed to have been formed by explosions caused by the immediate volatilization of the masses of iron upon striking the earth.

It has been shown that an iron meteorite with a diameter of 100 feet would be retarded very little in its passage through the earth's atmosphere. Smaller bodies would have their speed reduced. Meteorites the size of hailstones strike the earth at about the same speed that hailstones do. They fall about 230 feet a second. But they are never deeply imbedded in the ground, most not more than 1 or 2 feet in depth. The largest meteorite seen to fall is a stone of 820 pounds which fell in Arkansas in 1930. This penetrated clayey soil to a depth of 8 feet, scattering clods to a distance of 50 feet.

Meteorites are known to enter the earth's atmosphere at speeds between 10 and 40 miles a second. Striking the earth at the minimum speed of 10 miles a second, a meteorite has sufficient kinetic energy, if it were converted into heat, not only to melt the meteorite but vaporize most of it and the surrounding rock with explosive suddenness. In the resulting explosion small unmelted fragments would be blown from the crater over the adjoining country; many others would presumably be buried in the crater. But it is hard to conceive of any large mass remaining unshattered.

The upward force of such an explosion is very much greater than the downward force of percussion. For this reason the rocks around the crater walls are blown up and dip away radially outward instead of toward the center, as might be expected.

In several cases much silica glass is found within the crater and there are other evidences of high temperatures. In most cases a rim largely of shattered rock, from 15 to 100 feet in height, has been thrown up around the crater.

Whether the craters of the moon have been formed by the impact of meteorites, unretarded by any blanket of atmosphere, has been widely discussed by astronomers and geologists, some of whom lean to the meteorite theory.

# METEOR CRATERS OF THE WORLD

*Meteor Crater, Arizona,* formerly known as *Coon Butte,* is 3/4 mile in diameter and almost 600 feet deep, with a rim about 150 feet above the surrounding plain. Fragments and blocks of rock are scattered for 6 miles over the adjacent country. With this material there have been found thousands of pieces of meteoric iron ranging in weight up to 1,000 pounds. Twenty tons of this iron have been collected, but only a few small pieces from within the crater, the bottom of which is flat floored and covered with lake deposits almost 100 feet thick.

Boreholes and shafts put down in the bottom of the crater passed through crushed and metamorphosed sandstone and abundant rock flour and finally reached undisturbed sandstone at a depth of 620 feet. Several hundred thousand dollars have been spent in exploration of this sort, without finding any large body of metal.

It is conceded by virtually all who have investigated Meteor Crater that it was formed by the explosion of a great meteorite. The presence of volcanoes 30 miles away in the San Francisco Mountains and of lava flows within 10 miles, however, suggested the earlier theory that the crater had been formed by a volcanic steam or gas explosion without the extrusion of any lava. But this would not explain the presence of meteoric iron and silica glass.

*The Odessa Crater in Texas* is comparatively small and very shallow. It is roughly circular in outline and has a diameter of 500 feet and a depth of only 18 feet. Its rim is only 2 or 3 feet above the surrounding desert plain where horizontally bedded limestone is exposed. On the steep inner slopes the limestone outcrops and dips in all directions away from the crater at angles of 20 to 30°. The rim is distinctly buff colored as compared with the light gray of the surrounding soil and is richly impregnated with minute particles of iron oxide. Over 1,000 small nickel-iron fragments and several meteorites weighing up to 8 pounds were found in this immediate area by Dr. H. H. Nininger, who in a similar way established the meteoritic origin of the *Haviland Crater* in Kansas.

*The Henbury, Australia, craters* comprise some thirteen depressions within an area of 1/4 square mile, ranging in diameter from 30 to 250 feet and having depths of 3 to 50 feet. Hundreds of pieces of meteoric iron, with much iron shale and black glassy material, have been collected in the immediate vicinity. Like other examples, the rims of these craters have steep inner sides and gentle outer slopes. The rims consist of powdered rock, shattered blocks of sandstone, and quartzite. The craters serve as collecting basins for the slight rainfall of the region and therefore support more vegetation than prevails elsewhere in this arid country.

*The Wabar Craters of Arabia* occur in an area less than 1/2 mile in extent, in the midst of a sandy desert. Only two of them are distinct but there appear to be several others buried in the dunes. The diameters range from 100 to 300 feet, the depth being about 40 feet. The outer slopes are gentle and the inner slopes steep. The rims are built up of silica glass. There are no rock fragments. A few small pieces of meteoric iron have been recovered. The large amount of silica glass is no doubt due to the large masses of iron falling on clean desert sand. Considerable iron and nickel are present in the black and highly glazed surface of the glassy pearls, suggesting that a rain of molten silica was shot out from the craters through an atmosphere of silica, iron, and nickel produced by the vaporization of the desert sand and part of the meteorite.

*The Argentine craters* occur in a region of many depressions and small lakes. The largest is 200 to 300 feet in diameter and has a low rim only 4 feet above the surrounding pampa. Silica glass occurs in the rims and small fragments of rusted meteoric iron. Native iron has been found in this region for many years, one piece weighing over 1,400 pounds. This case, though little known, seems well authenticated.

*The Esthonian craters*, on the island of Oesel in the Baltic, are at best doubtful examples. No meteoric material, iron shale, or silica glass has been found, but this last omission may be due to the lack of quartz in the surrounding dolomite rocks. The main crater has many of the characteristics of the craters just mentioned: steep inner walls, shattered rocks, and a rim with beds dipping away in all directions. Alternative theories include man-made earthworks, weathering of limestone, and solution of salt in salt domes.

*The Siberian craters* lie in a region of peat bogs and permanently frozen ground. Numerous round depressions, in diameter up to 150 feet, occur in a swamp. No bedrock outcrops in the immediate vicinity. However, a great meteorite was known to fall in this vicinity on June 30, 1908, which devastated the forests for miles around. Pine trees were felled radially outward for a distance of 40 miles from the center, the area of devastation covering over 1,000 square miles. A plausible alternative theory is that water froze in cavities between the permanently frozen ground and that frozen only in winter, causing expansion and bursting.

The single *Ashanti Crater* of the Gold Coast is occupied by a large circular lake about 6½ miles in diameter and 250 feet deep, the surface of the lake being 1,000 feet below the crater rim. No good evidence of meteoric origin has been reported.

*The Persian Crater* is about 150 feet in diameter and 50 feet deep and has little to suggest a meteoric origin except the tale of a native who had seen a meteor fall in that locality.

Summary. It is obvious that no meteoric craters have been reported from even moderately rugged country and virtually none from country long inhabited. Erosion and the activities of man have doubtless done much to obscure them. It is highly probable, however, that a familiarity with the characteristics of known meteor craters will help to reveal additional examples.

# THE CAROLINA BAYS, NOT METEORITE SCARS

The Carolina bays are elliptical depressions which occur by the thousands on the coastal plain of North Carolina, South Carolina, and Georgia. Although known for many years, they received no especial attention until F. A. Melton made the ingenious suggestion that they might be meteorite scars. This suggestion arose from the unusual appearance these features have when viewed from the air, and it came directly as a result of the splendid air photographs of the region when they first made their appearance a few years ago. Keen interest in the area followed, not only among geologists but among laymen, who began reading articles in the popular magazines, and whose imagination was stirred by the idea of such a visitation from outer space.

Unfortunately this fascinating hypothesis has to be discarded, for it is now evident that the bays are solution depressions or in some cases merely original irregularities of the coastal-plain surface. Known as *savannas* in the Carolinas, they closely resemble the *prairies* of Florida and similar grassy glades in the coastal-plain forests of Mississippi. The small ones are like the Kentucky sink holes and just as numerous. The meteoritic hypothesis was intended to explain only the elliptical bays surrounded by higher rims but it is impossible to consider these without recognizing them as the more advanced members of a whole sequence of forms ranging from small sinks of varied pattern to the largest bays with regular outlines.

Character of the Bays. Many of the bays are perfect elliptical depressions showing a remarkable degree of parallelism, most of them trending in a northwest-southeast direction. There is every gradation from small, round depressions an acre in area to large elliptical ones a half mile or more in length. The sides slope gently only a few feet to the bottoms of the depressions which are flat floored and usually covered with tall grass. Some are marshy, some contain lakes, and some of the deeper ones bear heavy growths of timber. Usually there is no rim but some of the larger bays have sandy rims on the southeastern sides, clearly due to current and wind action at a time when lakes were present.

Many of the elliptical depressions are pointed or at least narrower at their southeastern ends, and their inward-facing sides in that sector are steeper than elsewhere around the depression. This is due to the gradual enlargement of the depression in that direction by underground solution or by movement of water down the dip of the beds after the manner of sink-hole enlargement.

The depressions are most numerous and largest on the flat inter-stream areas where surface drainage is poor and where most of the precipitation seeps underground to reappear in the well-defined and sharply incised stream courses near by.

Other Solution Depressions. Solution depressions near the inner margin of the coastal plain of South Carolina were studied by Laurence L. Smith before the advent of the meteoritic hypothesis and apparently were not recognized as having any bearing upon the explanation of the elliptical scars. Smith demonstrated clearly that the depressions near Columbia, S. C., which now appear to be identical in shape and in origin with the smaller bays, are a result of solution of iron and aluminum from sandy sediments by ground water. The solution is effected by organic acids resulting from decay of vegetation. Once started, as in some slight initial depression, solution is accelerated by increased dampness and increased growth of vegetation. Many of the depressions contain thin lake deposits with spicules of fresh-water sponges.

On the plains of western Texas, near the Odessa crater, there are many lakes and depressions closely resembling the Carolina bays, some being alined in rows as if following underground water courses. But they all differ from the Odessa crater in the fact that none of them possess elevated rims.

Limonite Deposits: an Explanation for the Magnetic Highs. Of unusual significance are the deposits of limonite which are especially abundant around the southeastern ends of the depressions,—irregular sink holes and symmetrical bays alike—and usually overlying impervious clay beds. Doubtless the flat floor of many of the bays is due to leaching to the level of such impervious layers. The movement of ground water down the dip of the beds to the southeast explains the location of the limonite deposits at that end of the bays. The presence of these limonite deposits (although small amounts of limonite are usually not magnetic) may possibly explain the magnetic highs found by magnetometer surveys and ascribed to the presence of meteoritic iron by the advocates of the meteoritic hypothesis. Magnetic surveys of known limonite areas are therefore desirable if further proof is needed to invalidate the hypothesis of meteoritic origin.

Coastal Lagoons. The studies of coastal estuaries prosecuted by C. Wythe Cooke add much to an understanding of the variety of elliptical bays along the coast where current and wind action are the dominating factors. These features are in many respects different from the bays heretofore discussed, although superficially resembling them.

The author of this book finds it impossible to accept the meteoritic explanation for the Carolina bays and gives full support to the solution theory already advocated by Douglas Johnson. Study of these depressions as well as of the Florida sinks and prairies, the glades of the coastal plain forests, the sink holes of Kentucky, the buffalo wallows of the Great Plains, and the countless lakes and sunken "parks" of western Siberia indicate that they are all due mainly to solution and underground circulation, modified in some instances by lake currents and wind action.

Madigan, C. T. (1937) *The Boxhole crater and Muckitta meteorite (central Australia)*. Royal Soc. South Australia, Trans. 61, p. 187–190; abst., Jour. Geomorphology, vol. 1, p. 173, 1938.

Merrill, G. P. (1922) *Meteoritic iron from Odessa, Ector County, Texas*. Am. Jour. Sci., 5th ser., vol. 3, p. 335–337.

Monnig, O. E. (1935) *The Odessa, Tex., meteorite crater*. Pop. Astronomy, vol. 43, p. 34–37.

Nininger, H. H., and Figgins, J. D. (1933) *The excavation of a meteor crater near Haviland, Kansas*. Colo. Mus. Nat. Hist., Proc. 12, p. 9–15.

Nininger, H. H. (1934) *The Odessa, Texas, meteor crater*. Pop. Astronomy, vol. 42, p. 46–47.

Nininger, H. H. (Feb. 1939) *Odessa Meteorite Crater*. Sky Mag. of Hayden Planetarium, vol. 3, p. 6–7, 23.

Sellards, E. H. (1927) *Unusual structural feature in the plains region of Texas*. Geol. Soc. Am., Bull. 38, p. 149 (abst.).

Whipple, F. J. W. (1930) *The great Siberian meteor and the waves, seismic and aerial, which it produced*. Royal Meteor. Soc. London, Quart. Jour., vol. 46, p. 287–304.

The "Meteorite" Scars of South Carolina

Cooke, C. W. (1933) *Origin of the so-called meteorite scars of South Carolina*. Wash. Acad. Sci., Jour. 23, p. 569–570 (abst.).

Cooke, C. W. (1934) *Discussion of the origin of the supposed meteorite scars of South Carolina*. Jour. Geol., vol. 42, p. 88–96.

Cooke, C. W. (1936) *Geology of the Coastal Plain of South Carolina*. U. S. Geol. Surv., Bull. 867. 196 p.

Glenn, L. C. (1895) *Some Darlington, South Carolina, bays*. Science, new ser., vol. 2, p. 472–475.

Johnson, D. W. (1936) *Origin of the supposed meteorite scars of South Carolina*. Science, new ser., vol. 84, p. 15–18.

Johnson, D. W. (1941) *Mysterious craters of the Carolina coast*. New York.

MacCarthy, G. R. (1937) *The Carolina Bays*. Geol. Soc. Am., Bull. 48, p. 1211–1226.

Melton, F. A., and Schriever, W. (1933) *The Carolina bays—are they meteorite scars?* Jour. Geol., vol. 41, p. 52–66.

Muldrow, Edna. (Dec. 1933) *The comet that struck the Carolinas*. Harper's Mag., vol. 168, p. 83–89.

Prouty, W. F. (1934) *Carolina bays*. (abst.) Elisha Mitchell Sci. Soc., Jour. 50, p. 59–60.

Prouty, W. F. (1935) *Carolina bays and elliptical lake basins*. Jour. Geol., vol. 43, p. 200–207.

Smith, Laurence L. (1931) *Solution depressions in sandy sediments of the coastal plain in South Carolina*. Jour. Geol., vol. 39, p. 641–652.

Wylie, C. C. (1933) *Iron meteorites and Carolina bays*. Pop. Astronomy, vol. 41, p. 410–412.

# INDEX

D

E



F

N



O

P

Q



R



S